Peridynamics and Its Applications Using Ansys

Erdogan Madenci •
Sundaram Vinod Kumar Anicode • Yanan Zhang

Peridynamics and Its Applications Using Ansys

Springer

Erdogan Madenci
Department of Aerospace and Mechanical Engineering
The University of Arizona
Tucson, AZ, USA

Sundaram Vinod Kumar Anicode
Department of Aerospace and Mechanical Engineering
The University of Arizona
Tucson, AZ, USA

Yanan Zhang
Department of Aerospace and Mechanical Engineering
The University of Arizona
Tucson, AZ, USA

ISBN 978-3-031-59895-1 ISBN 978-3-031-59896-8 (eBook)
https://doi.org/10.1007/978-3-031-59896-8

This Springer imprint is published by the registered company Springer Nature Switzerland AG
The registered company address is: Gewerbestrasse 11, 6330 Cham, Switzerland

Preface

Since its inception by Dr. Stewart A. Silling from Sandia National Laboratories in 2000, the nonlocal peridynamic (PD) theory has been extended and applied to many challenging problems from various disciplines. The PD equilibrium and field equations in the form of integro-differential are free of spatial derivatives without any conditions on the field variables. This is a significant departure from the classical form appearing as partial differential equations with smoothness requirement of the field variable. Therefore, PD has become an unrivaled approach for multiphysics analysis with damage prediction capability across varying length scales.

This book presents a unified implementation of bond- and state-based PD in a commercial finite element framework, Ansys, while addressing common issues associated with nonlocal theories. Its goal is to provide students, researchers, and practicing engineers with a theoretical and practical knowledge of the PD theory and the skills required to analyze engineering problems with Ansys. It is introductory and self-contained in order to minimize the need for additional reference material.

This book begins with an overview of the peridynamic theory. It presents the derivation of PD equation of motion and PD field equation for a constant and variable horizon based on the Euler-Lagrange equation. Also, it describes different strategies for imposing the relevant boundary conditions. Chapter 2 presents the peridynamic differential operator (PDDO). It provides the nonlocal representation of the local derivatives of up to second order in 2D and 3D analyses. Chapter 3 introduces an approach to apply various combinations of boundary conditions when solving the strong form of PD equilibrium and field equations. This approach prevents unphysical concentrations near the boundaries without the use of a fictitious layer. Chapter 4 presents the weak form of PD equilibrium and field equations based on the variational approach for direct imposition of essential and natural boundary conditions. Chapter 5 presents a coupling approach for the solution of equilibrium and field equations with uniform and non-uniform discretization in a commercial finite element analysis framework, Ansys. It is based on the variational principles. The PD and FE regions share the same nodes along their interface without any constraint conditions and/or an overlap zone. It is a general framework to perform bond-based (BB), ordinary state-based (OSB), and non-ordinary state-based (NOSB) PD analyses coupled with traditional elements. MATRIX27 elements native to Ansys are used for the PD region and the traditional elements for the remaining region. This coupling approach does not suffer from any unphysical concentrations near the interface in either steady-state or transient analysis. Chapter 6 presents the extension of coupled PD/FE analysis for deformation due to initial strain field. Chapter 7 presents a peridynamic framework to solve different field equations coupled among them or coupled with the deformation field. Chapter 8 presents the Ansys elements to model the PD interactions for structural analysis and coupled field analysis. These elements such as COMBIN14, MASS21, MASS71, LINK31, LINK33, LINK34, and MATRIX27 are native to Ansys. They are invoked in the analysis through the use of Ansys APDL commands. Their selection depends on the analysis type. Chapter 9 provides a description of the pre-processing operations to perform PD analysis by using the Ansys GUI. It describes the customized Ansys submenus that are developed using Ansys UIDL (User Interface Design Language) for PD modeling. The customized submenus are grouped under a menu named PD Preprocessor which is located under the Ansys Preprocessor menu. Chapter 10 presents the implementation of peridynamic (PD) functions in Ansys framework to evaluate local derivatives of up to second order. Applications concern the differentiation of one-, two-, and three- dimensional mathematical expressions with and without a jump discontinuity. Chapter 11 presents applications of PD to construct the solutions to field equations concerning thermal and moisture diffusions. The specific simulations concern the thermal shock in a plate and desorption in an initially fully saturated bar composed of two different materials. Chapter 12 presents the applications of single- and dual-horizon PD through uniform and nonuniform discretization, respectively, by considering an elastic plate with or without a crack under static loading. Chapter 13 presents the applications of coupled PD/FE analysis by considering an elastic plate with or without a crack under static and dynamic loading through uniform and nonuniform discretization. Also, this coupling approach is demonstrated by considering a plate with or without a crack subjected to initial thermal strain through uniform discretization. Chapter 14 presents applications for constructing solutions to the coupled field equations concerning dual-phase-lag thermal-moisture, electrodeposition, galvanic corrosion, and electromigration. Finally, Chapter 15 presents an approach to numerically determine the critical bond stretch value for BB-, OSB-, and NOSB-PD models with nonuniform discretization. Also, it demonstrates the applications of critical stretch criteria for failure prediction in the Ansys framework through implicit bond breakage algorithm. The specific applications concern double cantilever beam and three-point bending specimens modeled through uniform and nonuniform discretization. For computational efficiency, the models are constructed by coupling the PD/FE analysis.

The specific steps in pre-processing, solution, and post-processing phases of the analysis are displayed through the Graphical User Interface (GUI) in a step-by-step fashion for each application in Chaps. 10, 11, 12, and 13. The PD operations of the pre-processing, solution, and post-processing for each application in Chaps. 10, 11,

12, 13, 14, and 15 are provided in the form of Ansys input files. The PD operations are executed through the macro files. The input files containing the encrypted macro files to generate the simulations are provided as supplementary material through a link at the end of relevant chapters. The researchers and graduate students can modify the supplementary material and construct solutions to other specific problems. We hope that this book further accelerates the growth of peridynamics.

The material presented in this book was developed as part of the ongoing research at the MURI Center for Material Failure Prediction through Peridynamics at the University of Arizona (AFOSR Grant No. FA9550-14-1-0073; Program Managers, Drs. D. Stargel, A. Sayir and F. Fahroo). Also, we appreciate the encouragement and support of Dr. Nam Phan of NAVAIR in further development of PD theory and its implementation in a commercial finite element framework.

Tucson, AZ, USA

Erdogan Madenci
Sundaram Vinod Kumar Anicode
Yanan Zhang

Contents

1 Peridynamic Theory 1
1.1 Introduction 1
1.2 Peridynamic Concept 3
1.3 Peridynamic Equations of Motion 5
1.4 Displacement and Traction Boundary Conditions 9
1.5 Peridynamic Scalar Field Equation 12
1.6 Field and Flux Boundary Conditions 16
1.7 Periodic Boundary Conditions 18
1.8 Uniform Discretization 19
1.9 Discontinuities 20
References 21

2 Peridynamic Differential Operator 23
2.1 Introduction 23
2.2 PD Functions 26
2.2.1 Two-Dimensional Analysis 27
2.2.2 Three-Dimensional Analysis 33
2.3 Nonlocal Form of Bond Direction 40
2.3.1 Two-Dimensional Analysis 40
2.3.2 Three-Dimensional Analysis 42
2.4 Nonlocal Form of Divergence of a Second-Order Tensor 44
References 46

3 Unified Peridynamic Approach 47
3.1 Introduction 47
3.2 PD Equation of Motion 49
3.3 PD Internal Force Density Vector 50
3.3.1 Bond-Based (BB) Interactions 50
3.3.2 Ordinary State-Based (OSB) Interactions 55

3.3.3 Nonordinary State-Based (NOSB) Interactions 59
3.3.4 Peridynamic Differential Operator (PDDO) Interactions 62
3.4 PD Equilibrium Equations Under Homogeneous Deformation 64
3.4.1 BB Interactions 65
3.4.2 OSB Interactions 65
3.4.3 NOSB Interactions 66
3.4.4 PDDO Interactions 67
3.5 Unification of PD Equilibrium Equations 67
3.6 PD Field Equation 68
3.7 PD Flow Density of Field 69
3.7.1 BB Interactions 69
3.7.2 PDDO Interactions 71
3.8 PD Field Equations Under Linear Distribution 72
3.8.1 BB Interactions 72
3.8.2 PDDO Interactions 72
3.9 Unification of PD Field Equations 72
3.10 Global Discretization of PD Equilibrium Equations 73
3.11 Global Discretization of PD Field Equations 77
3.12 Local Discretization of PD Equilibrium Equations 81
3.12.1 BB-PD Equilibrium Equation 81
3.12.2 OSB-PD Equilibrium Equation 83
3.12.3 NOSB-PD Equilibrium Equation 88
3.12.4 PDDO Equilibrium Equation 93
3.13 Local Discretization of PD Field Equations 99
3.13.1 BB-PD Field Equation 99
3.13.2 PDDO Field Equation 101
3.13.3 BB-PD Transient Field Variation 105
3.13.4 BB-PD Transient Material Property 106
References 107

4 Weak Form of Peridynamic Equilibrium and Field Equations 109
4.1 Introduction 109
4.2 Weak Form of PD Equilibrium Equations 109
4.3 Weak Form of PD Field Equations 112
4.4 Discrete Form of PD Equilibrium Equations 116
4.5 Discrete Form of PD Field Equations 120
4.6 Discrete Form of PD Internal Force Vector 124
4.6.1 BB-PD Interactions 124
4.6.2 OSB-PD Interactions 126
4.6.3 NOSB-PD Interactions 128
4.6.4 PDDO Interactions 132
4.6.5 PD Internal Traction 134
4.6.6 PD Inertial Force 140

4.7 Discrete Form of PD Internal Net Field Flow . . . 141
4.7.1 BB-PD Interactions . . . 141
4.7.2 PDDO Interactions . . . 142
4.7.3 PD Internal Field Flux . . . 144
4.7.4 PD Transient Field Variation . . . 148
4.7.5 PD Transient Material Property . . . 149
References . . . 149

5 Coupling of Peridynamics with Finite Elements . . . 151
5.1 Introduction . . . 151
5.2 Coupling of PD with FE Regions for Deformation Analysis . . . 152
5.3 Coupling of PD with FE Regions for Field Analysis . . . 156
5.4 Discretization for PD/FEM Coupling . . . 161
5.4.1 Deformation Analysis . . . 161
5.4.2 Field Analysis . . . 167
References . . . 175

6 Coupled Peridynamic and Finite Element Analysis in the Presence of Initial Strain . . . 177
6.1 Introduction . . . 177
6.2 PD Internal Force Vectors in the Presence of Initial Deformation . . . 178
6.3 PD Internal Force Vectors for Small Deformation Kinematics . . . 183
6.3.1 BB Interactions . . . 184
6.3.2 OSB Interactions . . . 185
6.3.3 NOSB Interactions . . . 187
6.3.4 PDDO Interactions . . . 188
6.4 PD Internal Force Vectors for Uniform Discretization . . . 188
6.5 Governing Equations for a Coupled PD and FE Analysis . . . 190
References . . . 194

7 Peridynamics for Multiphysics Analysis . . . 195
7.1 Introduction . . . 195
7.2 Coupling of Different Field and Equilibrium Equations . . . 197
7.3 Boundary Conditions for Coupled Field Variables and Displacement . . . 199
7.4 PDDO Form of Coupled Equations . . . 200
7.5 PDDO Form of Boundary Conditions . . . 201
7.6 Numerical Discretization . . . 201
7.6.1 Discrete Form of Coupled Equations . . . 202
7.6.2 Discrete Form of Boundary Conditions . . . 212
7.6.3 Global Form of Coupled Equations . . . 217
References . . . 220

8 Ansys Elements for Peridynamic Interactions . . . 223
8.1 Introduction . . . 223
8.2 Matrix Form of PD Equilibrium and Field Equations . . . 223
8.3 COMBIN14 . . . 224
8.4 MASS21 . . . 227
8.5 MATRIX27 . . . 229
8.6 LINK33 . . . 233
8.7 MASS71 . . . 234
8.8 LINK31 . . . 236
8.9 LINK34 . . . 238
Reference . . . 239

9 Ansys Menus for PD Analysis . . . 241
9.1 Introduction . . . 241
9.2 Field Selection . . . 241
9.3 Discretization . . . 243
9.4 Region Identification . . . 247
9.4.1 Domain . . . 247
9.4.2 Boundary Interactions in Strong Form . . . 249
9.4.3 Boundary Interactions in Weak Form . . . 251
9.5 Cutout Insertion . . . 252
9.6 Family Generation . . . 255
9.7 Crack Insertion . . . 257
9.8 Element Coefficients . . . 259
9.9 Element Generation . . . 260
9.9.1 Domain Elements . . . 261
9.9.2 Boundary Elements . . . 265
9.10 Imposition of Nonlocal Boundary Conditions . . . 268
9.10.1 Direct Approach . . . 269
9.10.2 Mirror Node Approach . . . 271
9.11 Field Damping . . . 287
9.12 Thermal Dual Phase Lag (DPL) Damping . . . 289
9.13 Dual Phase Lag (DPL) Inertia . . . 292
9.14 Material Thermal Properties for Moisture Concentration . . . 294
9.15 Peridynamic Solution . . . 295
9.15.1 Setup CSAT Table . . . 295
9.15.2 CSAT . . . 296
9.15.3 Time Dep. CSAT . . . 297
References . . . 299

10 Implementation of PDDO in Ansys Framework . . . 301
10.1 Introduction . . . 301
10.2 Evaluation of PD Functions . . . 302
10.3 Derivatives of a Function . . . 306

10.4 Applications 306
10.4.1 One-Dimensional Functions 307
10.4.2 Two-Dimensional Function 311
10.4.3 Three-Dimensional Function 316
References 321

11 Applications of Field Equations 323
11.1 Introduction 323
11.2 Simulation of Thermal Shock in a Plate 323
11.2.1 Modeling in Ansys Framework 324
11.2.2 Strong Form of Thermal Diffusion Equation 325
11.2.3 Weak Form of Thermal Diffusion Equation 333
11.2.4 Coupled PD/FE Form of Thermal Diffusion Equation 335
11.3 Simulation of Moisture Diffusion in a Nonhomogeneous Bar 339
11.3.1 Modeling in Ansys Framework 342
11.3.2 Moisture Diffusion with Time-Varying and Spatially Uniform Temperature 343
11.3.3 Moisture Diffusion with Time-Varying and Spatially Nonuniform Temperature 350
References 352

12 Applications of Peridynamics for Static Structural Analysis 353
12.1 Introduction 353
12.2 Plate Under Tension 353
12.2.1 Modeling in Ansys Framework 354
12.2.2 Strong Form of BB-PD with Uniform Discretization 354
12.2.3 Strong Form of OSB- and NOSB-PD with Uniform Discretization 364
12.2.4 Strong Form of NOSB-PD with Nonuniform Discretization 367
12.3 Plate with a Crack Under Tension 369
12.3.1 Strong Form of OSB-PD with Nonuniform Discretization 371
12.3.2 Weak Form of NOSB-PD with Nonuniform Discretization 381
References 387

13 Coupled Peridynamics and Finite Element Method for Static and Transient Structural Analysis 389
13.1 Introduction 389
13.2 Modeling in Ansys Framework 390
13.3 Static Response of a Plate Under Traction 390
13.3.1 Model Setup 390

13.3.2 Preprocessing 392
13.3.3 Post-processing 401
13.4 Static Response of a Plate Under Displacement Loading 403
13.4.1 Model Setup 403
13.4.2 Preprocessing 404
13.4.3 Post-processing 406
13.5 Static Response of a Plate with a Crack Under Tension 408
13.5.1 Model Setup 408
13.5.2 Preprocessing 409
13.5.3 Post-processing 412
13.6 Dynamic Response of a Plate Under Compression 412
13.6.1 Model Setup 412
13.6.2 Uniform Discretization 413
13.6.3 Nonuniform Discretization 419
13.7 Static Response of a Plate Under Initial Thermal Strain 424
13.7.1 Preprocessing 425
13.7.2 Post-processing 433
13.8 Static Response of a Plate with a Crack Due to Combined Initial Thermal Strain and Mechanical Loading 435
References 441

14 Applications of Coupled Field Equations 443
14.1 Introduction 443
14.2 Coupled Dual-Phase-Lag Thermal-Moisture Fields 444
14.2.1 Ultrashort-Pulse Laser Heating of Nanoscale Metal Film 445
14.2.2 A Bar Under Sudden Pulse of Hygro-Thermal Loading 447
14.3 Electrodeposition Process 448
14.4 Galvanic Corrosion Process 455
14.5 Electromigration Process 460
References 467

15 Peridynamics for Failure Prediction 469
15.1 Introduction 469
15.2 Critical Stretch Criteria for Crack Initiation and Growth 470
15.3 Numerical Evaluation of Critical Stretch Value 471
15.4 Implicit Algorithm for Crack Growth 476
15.5 Numerical Results 477
15.5.1 Double Cantilever Beam Specimen 477
15.5.2 Three-Point Bending Specimen 483
References 488

Index 491

Chapter 1
Peridynamic Theory

1.1 Introduction

In reality, the intrinsic interactions in a physical system or a phenomenon are dependent on time and space with a natural time scale and a length scale. However, the governing equations to describe such phenomena are derived by employing the balance laws and concepts from calculus such as differentiation and integration; thus, the interactions are averaged out. Therefore, the resulting field equations are referred to as "local" without an internal length scale. They are expressed in terms of temporal and spatial derivatives, thus undefined along discontinuities. However, many physical problems have a natural length scale and a discontinuity.

In classical continuum theory, a material point is assumed to interact only with its immediate neighbors; hence, it is a local theory. Therefore, a material point exchanges mass, momentum, and energy with only its closest neighbors. The classical continuum mechanics (CCM) provides a mathematical structure to describe the behavior of materials at macroscopic length scales when subjected to external loads. The macroscopic properties of the material are invariant with respect to length and time scales. The material behavior is evaluated by considering short-range interactions, that is, interactions are restricted to immediate neighbors. As a result, in classical mechanics, the stress state at a point depends on the deformation at that point only. The validity of this assumption becomes questionable across different length scales. In general, this assumption is acceptable at the macroscale. However, theoretical and experimental studies reported the existence of long-range interactions resulting in dispersion of elastic waves, hence dependence of material properties on the length scale.

Classical mechanics based on contact forces are mathematically represented in the form of derivatives; thus, modeling fracture is an inherent limitation because the spatial derivatives are not defined across a discontinuity such as a crack. Despite the development of many important concepts to predict crack initiation and its growth in materials, it is still a major challenge within the framework of classical continuum

E. Madenci et al., *Peridynamics and Its Applications Using Ansys*,
https://doi.org/10.1007/978-3-031-59896-8_1

mechanics. Solutions to the local field equations in complex domains are commonly achieved through computational techniques that require discretization and numerical differentiation. As a result, the field equations are recast in the form of algebraic equations without a unified treatment of discontinuities. In the presence of discontinuities, existing methods are not satisfactory.

The nonlocal peridynamic (PD) theory introduced by Silling (2000) and Silling et al. (2007) restores long-range interactions and presents an internal length scale in the form of an integral equation. The integrand is free of spatial derivatives; thus, it is valid across a crack. In PD theory, internal forces are expressed through nonlocal interactions between the material points within a solid body, and damage is a part of the constitutive model. The damage can propagate when and where it is energetically favorable. This feature allows damage initiation and propagation at multiple sites with arbitrary paths. The nucleation and propagation of a crack are simulated by breaking a PD bond (interaction). Therefore, it facilitates modeling the formation of cracks and examination of failure patterns.

The main difference between PD and the classical continuum mechanics (CCM) is that the former is formulated using integral equations without any smoothness requirement as opposed to the derivatives of displacement components. Unlike the CCM, it introduces an internal length parameter referred to as the horizon of a point. The PD equation of motion converges to its classical counterpart for a smooth deformation field as the horizon approaches zero at a material point with a radially symmetric domain of interaction (Silling & Lehoucq, 2010).

The PD equations of motion can be broadly classified as Bond-Based (BB), Ordinary State-Based (OSB), and Non-Ordinary State-Based (NOSB). The BB-PD introduced by Silling (2000) does not distinguish between distortion and dilatation, thus leading to one independent material constant, referred to as the *micro-modulus*. The Poisson's ratio has a fixed value depending on the analysis dimension. In the BB-PD, the interaction of a pair of material points is independent of the influence of other points in their domain of interaction. In order to remove this restriction and to include general material models, Silling et al. (2007) introduced the state-based PD. The OSB- and NOSB-PD are dependent on the influence of other points within their individual domains of interaction. Silling (2000) and Silling et al. (2007) derived these equations of motion for a uniform horizon. Therefore, the numerical evaluation of integrals appearing in the PD equations of motion can be readily evaluated by employing uniform discretization with a constant (single) horizon size. Uniform discretization with sufficient refinement may require substantial computational time. When suitable, nonuniform discretization is attractive for reducing the computational time. However, nonuniform discretization inherently presents a variable horizon and suffers from the imbalance of linear momentum due to the presence of ghost forces. Therefore, Ren et al. (2017) introduced the concept of dual horizons in PD theory based on a sound theoretical foundation. It is simple and efficient for non-uniform discretization with a variable horizon and free of ghost forces.

This chapter presents an overview of the peridynamic (PD) theory. Considering primary and secondary PD bonds along with the concept of dual horizon, it presents the derivation of PD equation of motion and PD field equation for a constant and variable horizon based on the Euler-Lagrange equation. Also, it describes different strategies for imposing the relevant boundary conditions.

1.2 Peridynamic Concept

Peridynamics introduced by Silling (2000) employs the concept of nonlocal bond interactions in a continuum. It invokes the effect of other points on the point of interest within its range of interaction. Assuming that the continuum can be composed of infinitely many material points, $\mathbf{x}_{(i)}$ for $(i = 1, 2, .., \infty)$ with an incremental volume, $V_{(i)}$, the displacement field $\mathbf{u}_{(k)}$ at point $\mathbf{x}_{(k)}$ is influenced by $\mathbf{u}_{(j)}$ at point $\mathbf{x}_{(j)}$ within its horizon, $H_{(k)}$ (interaction domain), as shown in Fig. 1.1 The interaction domain of point $\mathbf{x}_{(k)}$ is defined as $H_{(k)} = \{j; j \in H_{(k)}\}$. The points, $\mathbf{x}_{(j)}$ located in $H_{(k)}$ are called the family members of $\mathbf{x}_{(k)}$. The bond, $\boldsymbol{\xi}_{(k)(j)} = \mathbf{x}_{(j)} - \mathbf{x}_{(k)}$ between these points is referred to as the "*primary*" bond.

As shown in Fig. 1.2, the bonds, $\boldsymbol{\xi}_{(j)(k)} = \mathbf{x}_{(k)} - \mathbf{x}_{(j)}$ and $\boldsymbol{\xi}_{(p)(k)} = \mathbf{x}_{(k)} - \mathbf{x}_{(p)}$ between points $\mathbf{x}_{(j)}$ and $\mathbf{x}_{(k)}$ and $\mathbf{x}_{(p)}$ and $\mathbf{x}_{(k)}$, respectively, are referred to as the "*secondary*" bonds provided that $\mathbf{x}_{(k)}$ belongs to $H_{(j)}$ and $H_{(p)}$. For a nonuniform discretization, the material point, $\mathbf{x}_{(k)}$ can be within the horizon of other material points, $\mathbf{x}_{(p)}$; however, these points may not be within the horizon of point, $\mathbf{x}_{(k)}$. It is worth noting that $H_{(k)}$ includes material point $\mathbf{x}_{(j)}$ but not $\mathbf{x}_{(p)}$. The interaction domain for the secondary bonds is defined as $\widetilde{H}_{(k)} = \{m; k \in H_{(m)}\}$. It is referred to as the "dual" horizon (Ren et al., 2017).

In PD theory, the bonds between the material points can be broken (removed) based on a particular criterion. The status of bonds is monitored through a status parameter defined as

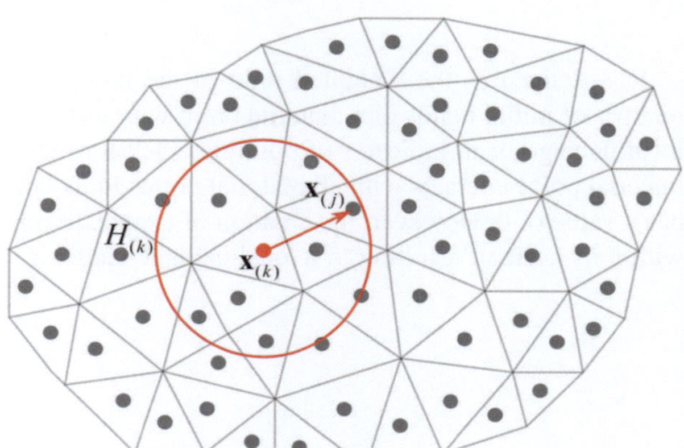

Fig. 1.1 Primary PD bond between material point $\mathbf{x}_{(k)}$ and the family member $\mathbf{x}_{(j)}$ in its horizon, $H_{(k)} = \{j; j \in H_{(k)}\}$

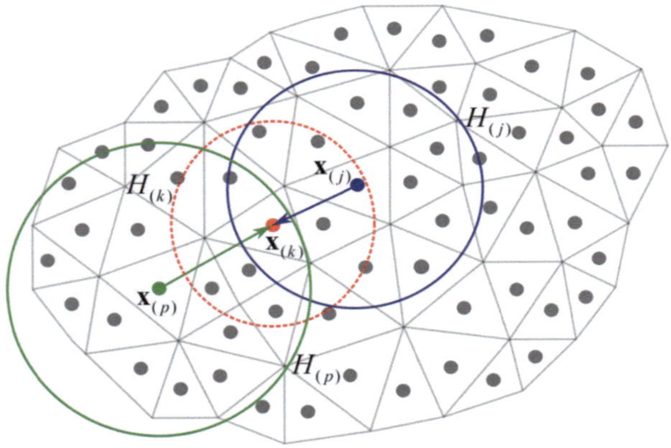

Fig. 1.2 Secondary bonds $\boldsymbol{\xi}_{(j)(k)}$ and $\boldsymbol{\xi}_{(p)(k)}$ associated with point $\mathbf{x}_{(k)}$ belonging to the interaction domain (" *dual* " horizon) $\widetilde{H}_{(k)} = \{m; k \in H_{(m)}\}$ with $m = j$ and p

$$\widehat{\mu}\left(\mathbf{x}_{(j)} - \mathbf{x}_{(k)}\right) = \begin{cases} 1, & \text{intact} \\ 0, & \text{broken} \end{cases} \tag{1.1}$$

where $\widehat{\mu}\left(\mathbf{x}_{(j)} - \mathbf{x}_{(k)}\right)$ represents the status of a bond connecting point $\mathbf{x}_{(k)}$ with $\mathbf{x}_{(j)}$. Based on the interaction status of point $\mathbf{x}_{(k)}$ with all the bonds in its family, the local damage, $\phi(\mathbf{x}_{(k)})$ can be evaluated as (Silling & Askari, 2005)

$$\phi\left(\mathbf{x}_{(k)}\right) = 1 - \frac{\sum\limits_{j \in H_{(k)}} \widehat{\mu}\left(\mathbf{x}_{(j)} - \mathbf{x}_{(k)}\right) V_{(j)} + \sum\limits_{p \in \widetilde{H}_{(k)}} \widehat{\mu}\left(\mathbf{x}_{(k)} - \mathbf{x}_{(p)}\right) V_{(p)}}{\sum\limits_{j \in H_{(k)}} V_{(j)} + \sum\limits_{p \in \widetilde{H}_{(k)}} V_{(p)}} \quad \text{for dual horizon} \tag{1.2a}$$

and

$$\phi\left(\mathbf{x}_{(k)}\right) = 1 - \frac{\sum\limits_{j \in H_{(k)}} \widehat{\mu}\left(\mathbf{x}_{(j)} - \mathbf{x}_{(k)}\right) V_{(j)}}{\sum\limits_{j \in H_{(k)}} V_{(j)}} \quad \text{for single horizon} \tag{1.2b}$$

It is defined as the weighted ratio of the number of broken bonds to the total number of initial bonds of a material point with its family members. The measure of local damage is an indicator of possible crack formation within a body. Initially, a material point interacts with all material points in its horizon; thus, the local damage has a value of 0. However, the creation of a crack terminates half of the interactions within its horizon resulting in a local damage value of ½ for single horizon PD.

1.3 Peridynamic Equations of Motion

The primary bond, $\boldsymbol{\xi}_{(k)(j)}$ between points $\mathbf{x}_{(k)}$ and $\mathbf{x}_{(j)}$ and the secondary bond, $\boldsymbol{\xi}_{(j)(k)}$ between $\mathbf{x}_{(j)}$ and $\mathbf{x}_{(k)}$ are distinct from each other. Their deformation and bond status (broken or not) may be different. The micropotential, $w_{(k)(j)}$ associated with the primary bond, $\boldsymbol{\xi}_{(k)(j)}$ may not necessarily be equal to $w_{(j)(k)}$ associated with the secondary bond, $\boldsymbol{\xi}_{(j)(k)}$, that is, $w_{(j)(k)} \neq w_{(k)(j)}$, because $w_{(j)(k)}$ depends on the state of material points within the family of material point $\mathbf{x}_{(j)}$. Due to the interaction between material points $\mathbf{x}_{(k)}$ and $\mathbf{x}_{(j)}$, a scalar-valued micropotential, $w_{(k)(j)}$, develops; it depends on the material properties as well as the stretch between point $\mathbf{x}_{(k)}$ and all other material points, $\mathbf{x}_{(j)}$ in its family.

Conforming to the Lagrangian description of motion, each material point is identified by its coordinates, $\mathbf{x}_{(i)}$, and is associated with an incremental volume, $V_{(i)}$, and a mass density of $\rho(\mathbf{x}_{(i)})$. As shown in Fig. 1.3, the material point $\mathbf{x}_{(i)}$ experiences displacement, $\mathbf{u}_{(i)}$, and its location is described by the position vector, $\mathbf{y}_{(i)}$ in the deformed state with respect to a Cartesian coordinate system. The displacement and body load vectors at material point $\mathbf{x}_{(i)}$ are represented by $\mathbf{u}_{(i)}(\mathbf{x}_{(i)}, t)$ and $\mathbf{b}_{(i)}(\mathbf{x}_{(i)}, t)$, respectively.

As suggested by Madenci and Oterkus (2014), the PD equations of motion at material point $\mathbf{x}_{(k)}$ can be derived by considering the Euler-Lagrange equation as

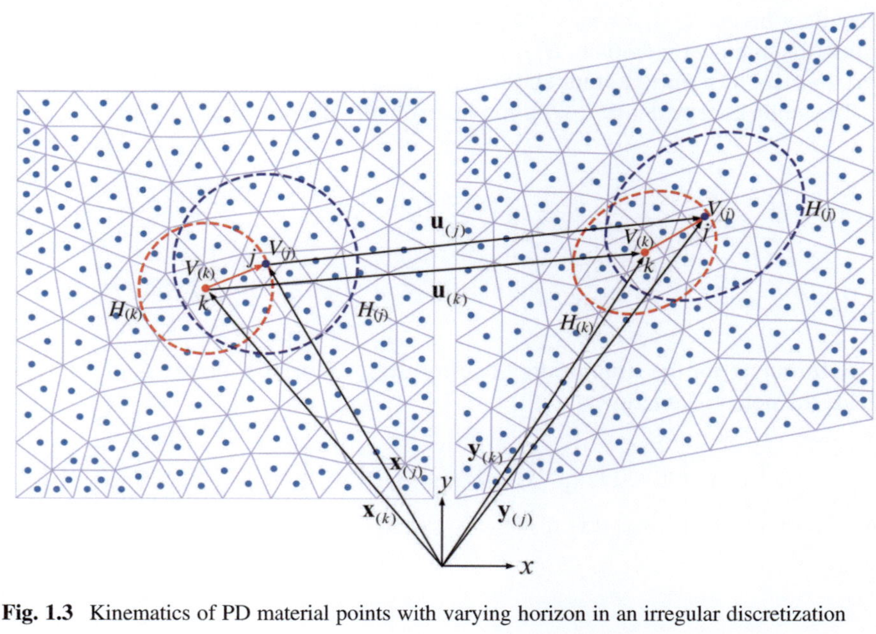

Fig. 1.3 Kinematics of PD material points with varying horizon in an irregular discretization

$$\frac{d}{dt}\left(\frac{\partial L}{\partial \dot{\mathbf{u}}_{(k)}}\right)-\frac{\partial L}{\partial \mathbf{u}_{(k)}}=0 \tag{1.3}$$

in which the Lagrangian, L is defined as

$$L=T-U. \tag{1.4}$$

where T and U represent the total kinetic and potential energies, respectively, in the body. They can be obtained by summation of kinetic and potential energies of all material points as

$$T=\sum_{i=1}^{\infty}\frac{1}{2}\rho_{(i)}\dot{\mathbf{u}}_{(i)}\cdot\dot{\mathbf{u}}_{(i)}V_{(i)} \tag{1.5}$$

and

$$U=\sum_{i=1}^{\infty}W_{(i)}V_{(i)}-\sum_{i=1}^{\infty}\left(\mathbf{b}_{(i)}\cdot\mathbf{u}_{(i)}\right)V_{(i)}. \tag{1.6}$$

The strain energy density, $W_{(i)}$, of material point $\mathbf{x}_{(i)}$ can be expressed as a summation of micropotentials, $w_{(i)(j)}$, due to the deformation of primary bonds, $\boldsymbol{\xi}_{(i)(j)}=\mathbf{x}_{(j)}-\mathbf{x}_{(i)}$ between material points $\mathbf{x}_{(i)}$ and $\mathbf{x}_{(j)}$ located within the interaction domain $H_{(i)}$ and summation of micropotentials, $w_{(j)(i)}$, due to the deformation of secondary bonds, $\boldsymbol{\xi}_{(j)(i)}=\mathbf{x}_{(i)}-\mathbf{x}_{(j)}$ between material points $\mathbf{x}_{(j)}$ and $\mathbf{x}_{(i)}$ located within the interaction domain, $H_{(j)}$. For a nonuniform discretization with a varying horizon size, it is worth noting that $w_{(j)(i)}\neq w_{(i)(j)}$.

The strain energy density, $W_{(i)}$, can be explicitly expressed as

$$\begin{aligned}W_{(i)}=&\frac{1}{2}\sum_{j=1}^{\infty}w_{(i)(j)}\left(\mathbf{y}_{(1^i)}-\mathbf{y}_{(i)},\ \mathbf{y}_{(2^i)}-\mathbf{y}_{(i)},\cdots\right)V_{(j)}\\&+\frac{1}{2}\sum_{j=1}^{\infty}w_{(j)(i)}\left(\mathbf{y}_{(1^j)}-\mathbf{y}_{(j)},\ \mathbf{y}_{(2^j)}-\mathbf{y}_{(j)},\cdots\right)V_{(j)}\end{aligned} \tag{1.7}$$

Substituting for the strain energy density, $W_{(i)}$, of material point $\mathbf{x}_{(i)}$ in Eq. (1.6) yields the potential energy as

$$U=\sum_{i=1}^{\infty}\left\{\frac{1}{2}\sum_{j=1}^{\infty}\begin{bmatrix}w_{(i)(j)}\left(\mathbf{y}_{(1^i)}-\mathbf{y}_{(i)},\ \mathbf{y}_{(2^i)}-\mathbf{y}_{(i)},\cdots\right)\\+w_{(j)(i)}\left(\mathbf{y}_{(1^j)}-\mathbf{y}_{(j)},\ \mathbf{y}_{(2^j)}-\mathbf{y}_{(j)},\cdots\right)\end{bmatrix}V_{(j)}V_{(i)}-\left(\mathbf{b}_{(i)}\cdot\mathbf{u}_{(i)}\right)V_{(i)}\right\} \tag{1.8}$$

The Lagrangian can be written in an expanded form by showing only the terms associated with the material point, $\mathbf{x}_{(k)}$ as

$$
\begin{aligned}
L= & \ldots\ldots+\frac{1}{2}\rho_{(k)}\,\dot{\mathbf{u}}_{(k)}\cdot\dot{\mathbf{u}}_{(k)}\,V_{(k)}+\ldots\ldots \\
& -\frac{1}{2}\sum_{j=1}^{\infty}\left[w_{(k)(j)}\left(\mathbf{y}_{(1^k)}-\mathbf{y}_{(k)},\mathbf{y}_{(2^k)}-\mathbf{y}_{(k)},\cdots\right)\right. \\
& \left.+w_{(j)(k)}\left(\mathbf{y}_{(1^j)}-\mathbf{y}_{(j)},\mathbf{y}_{(2^j)}-\mathbf{y}_{(j)},\cdots\right)\right]V_{(j)}\,V_{(k)}\ldots\ldots \\
& -\frac{1}{2}\sum_{i=1}^{\infty}\left[w_{(i)(k)}\left(\mathbf{y}_{(1^i)}-\mathbf{y}_{(i)},\mathbf{y}_{(2^i)}-\mathbf{y}_{(i)},\cdots\right)\right. \\
& \left.+w_{(k)(i)}\left(\mathbf{y}_{(1^k)}-\mathbf{y}_{(k)},\mathbf{y}_{(2^k)}-\mathbf{y}_{(k)},\cdots\right)\right]V_{(k)}V_{(i)}\ldots\ldots \\
& +\left(\mathbf{b}_{(k)}\cdot\mathbf{u}_{(k)}\right)V_{(k)}\ldots.
\end{aligned}
\tag{1.9}
$$

or

$$
\begin{aligned}
L= & \ldots\ldots+\frac{1}{2}\rho_{(k)}\,\dot{\mathbf{u}}_{(k)}\cdot\dot{\mathbf{u}}_{(k)}\,V_{(k)}+\ldots\ldots \\
& -\sum_{j=1}^{\infty}\left[w_{(k)(j)}\left(\mathbf{y}_{(1^k)}-\mathbf{y}_{(k)},\ \mathbf{y}_{(2^k)}-\mathbf{y}_{(k)},\cdots\right)V_{(j)}\right]\ V_{(k)}\ldots\ldots \\
& -\sum_{i=1}^{\infty}\left[w_{(i)(k)}\left(\mathbf{y}_{(1^i)}-\mathbf{y}_{(i)},\ \mathbf{y}_{(2^i)}-\mathbf{y}_{(i)},\cdots\right)V_{(i)}\right]\ V_{(k)}\ldots\ldots \\
& +\left(\mathbf{b}_{(k)}\cdot\mathbf{u}_{(k)}\right)V_{(k)}\ldots.
\end{aligned}
\tag{1.10}
$$

Substituting from Eq. (1.10) into Eq. (1.3) results in the Euler-Lagrange equation at material point $\mathbf{x}_{(k)}$ as

$$
\begin{aligned}
\rho_{(k)}\ddot{\mathbf{u}}_{(k)}\,V_{(k)}+ & \left(\sum_{j=1}^{\infty}\frac{\partial w_{(k)(j)}}{\partial\left(\mathbf{y}_{(j)}-\mathbf{y}_{(k)}\right)}\frac{\partial\left(\mathbf{y}_{(j)}-\mathbf{y}_{(k)}\right)}{\partial\mathbf{u}_{(k)}}V_{(j)}\right. \\
& \left.+\sum_{i=1}^{\infty}\frac{\partial w_{(i)(k)}}{\partial\left(\mathbf{y}_{(k)}-\mathbf{y}_{(i)}\right)}\frac{\partial\left(\mathbf{y}_{(k)}-\mathbf{y}_{(i)}\right)}{\partial\mathbf{u}_{(k)}}V_{(i)}-\mathbf{b}_{(k)}\right)V_{(k)}=0
\end{aligned}
\tag{1.11}
$$

or

$$
\rho_{(k)}\ddot{\mathbf{u}}_{(k)}=\sum_{j=1}^{\infty}\frac{\partial w_{(k)(j)}}{\partial\left(\mathbf{y}_{(j)}-\mathbf{y}_{(k)}\right)}V_{(j)}-\sum_{i=1}^{\infty}\frac{\partial w_{(i)(k)}}{\partial\left(\mathbf{y}_{(k)}-\mathbf{y}_{(i)}\right)}V_{(i)}+\mathbf{b}_{(k)}
\tag{1.12}
$$

Based on the dimensional analysis of this equation, it is apparent that $\partial w_{(k)(j)}/\partial(\mathbf{y}_{(j)} - \mathbf{y}_{(k)})$ represents the force density that material point $\mathbf{x}_{(j)}$ exerts on material point $\mathbf{x}_{(k)}$ through the primary bond and $\partial w_{(i)(k)}/\partial(\mathbf{y}_{(k)} - \mathbf{y}_{(i)})$ represents the force density that material point $\mathbf{x}_{(k)}$ exerts on material point $\mathbf{x}_{(i)}$ through the secondary bond. With this interpretation, Eq. (1.12) can be rewritten as

$$\begin{aligned}\rho_{(k)}\ddot{\mathbf{u}}_{(k)} = & \sum_{j=1}^{\infty}\mathbf{t}_{(k)(j)}\left(\mathbf{u}_{(j)} - \mathbf{u}_{(k)}, \mathbf{x}_{(j)} - \mathbf{x}_{(k)}, t\right)V_{(j)} \\ & - \sum_{i=1}^{\infty}\mathbf{t}_{(i)(k)}\left(\mathbf{u}_{(k)} - \mathbf{u}_{(i)}, \mathbf{x}_{(k)} - \mathbf{x}_{(i)}, t\right)V_{(i)} + \mathbf{b}_{(k)},\end{aligned} \tag{1.13}$$

where

$$\mathbf{t}_{(k)(j)} = \frac{\partial w_{(k)(j)}}{\partial\left(\mathbf{y}_{(j)} - \mathbf{y}_{(k)}\right)} \tag{1.14a}$$

and

$$\mathbf{t}_{(i)(k)} = \frac{\partial w_{(i)(k)}}{\partial\left(\mathbf{y}_{(k)} - \mathbf{y}_{(i)}\right)} \tag{1.14b}$$

According to the PD theory introduced by Silling (2000), the influence of material points interacting with $\mathbf{x}_{(k)}$ is assumed to vanish beyond its horizon (family members), $H_{(k)}$. Similarly, material point $\mathbf{x}_{(i)}$ interacts only with material points in its own family, $H_{(i)}$. Thus, Eq. (1.13) can be rewritten as

$$\rho_{(k)}\ddot{\mathbf{u}}_{(k)} = \sum_{j\in H_{(k)}}\mathbf{t}_{(k)(j)}V_{(j)} - \sum_{i\in \widetilde{H}_{(k)}}\mathbf{t}_{(i)(k)}V_{(i)} + \mathbf{b}_{(k)} \tag{1.15}$$

in which $H_{(k)} = \{j; j \in H_{(k)}\}$ and $\widetilde{H}_{(k)} = \left\{i; k \in H_{(i)}\right\}$ are the horizon and "dual" horizon of material point, $\mathbf{x}_{(k)}$, as described in Figs. 1.1 and 1.2. Ren et al. (2017) coined Eq. (1.15) as "dual horizon" PD equilibrium equation. It includes all the forces within the horizon of $\mathbf{x}_{(k)}$ and forces arising from any point, $\mathbf{x}_{(i)}$ whose horizon includes point, $\mathbf{x}_{(k)}$.

The volume of each material point $V_{(j)}$ is infinitesimally small, for the limiting case of $V_{(j)} \to 0$, the summation can be replaced with integration. Thus, Eq. (1.15) can be recast as an integral equation with dual horizons, $H_{\mathbf{x}}$ and $\widetilde{H}_{\mathbf{x}}$ as (Ren et al., 2017)

$$\rho(\mathbf{x})\ddot{\mathbf{u}}(\mathbf{x}, t) = \int_{H_{\mathbf{x}}}\mathbf{t}(\mathbf{x}' - \mathbf{x}, t)dV_{\mathbf{x}'} - \int_{\widetilde{H}_{\mathbf{x}}}\mathbf{t}(\mathbf{x} - \mathbf{x}', t)dV_{\mathbf{x}'} + \mathbf{b}(\mathbf{x}, t) \tag{1.16}$$

in which $H_{\mathbf{x}}$ is the horizon of point $\mathbf{x}$ associated with the primary bonds and $\widetilde{H}_{\mathbf{x}} = \{\mathbf{x}'; \mathbf{x} \in H_{\mathbf{x}'}\}$ is the dual horizon of point $\mathbf{x}$ associated with the secondary bonds. The volume of material point $\mathbf{x}'$ is represented by $dV_{\mathbf{x}'}$. The interaction domain $H_{\mathbf{x}}$ of point $\mathbf{x}$ can be different than that of $H_{\mathbf{x}'}$ for point $\mathbf{x}'$.

1.4 Displacement and Traction Boundary Conditions

Unlike the CCM, the derivation of PD equilibrium equation based on the principle of virtual work does not lead to the relevant boundary conditions. Also, the traction vector, $\mathbf{T}^* = \mathbf{Pn}$, with $\mathbf{P}$ being the first Piola-Kirchoff stress tensor and $\mathbf{n}$ being the unit vector normal to the surface, does not appear in PD theory. However, the boundary conditions in the form of displacement constraints on $\mathcal{S}_{\mathbf{u}}$, external tractions on $\mathcal{S}_{\boldsymbol{\sigma}}$, as shown in Fig. 1.4, are unavoidable in practical engineering applications. Also, the construction of solution to the PD equilibrium equation usually requires numerical techniques. Therefore, the boundary conditions are enforced at the discretized level and there exist different strategies.

One approach is to use a fictitious material layer along the boundary and apply displacement boundary conditions on the material points in the fictitious layer. As shown in Fig. 1.5, the width of the fictitious layer for applying displacement constraints is denoted as $d_{\mathbf{u}}$. Macek and Silling (2007) suggested the use of one horizon as the width of the fictitious layer for BB-PD, that is, $d_{\mathbf{u}} = \delta$. For state-based PD, the width of the fictitious material layer is twice the size of the horizon, that is, $d_{\mathbf{u}} = 2\delta$. The displacement constraints are applied by assigning the known values, $\mathbf{U}^*$ to the fictitious material points in region $\mathcal{R}_f$ as

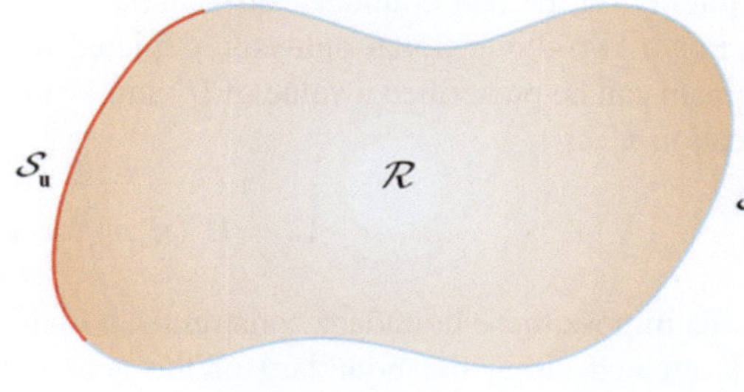

Fig. 1.4 A body, $\mathcal{R}$ subjected to displacement and traction boundary conditions along $\mathcal{S}_{\mathbf{u}}$ and $\mathcal{S}_{\boldsymbol{\sigma}}$, respectively

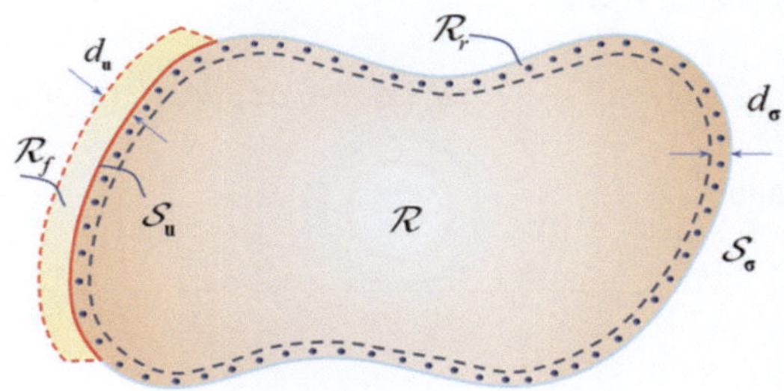

Fig. 1.5 Fictitious and real boundary regions for imposing displacement constraints and traction conditions

$$\mathbf{u}(\mathbf{x},t) = \mathbf{U}^* \text{ for } \mathbf{x} \in \mathscr{R}_f \tag{1.17}$$

In PD theory, the traction-free conditions are satisfied in a natural way. The non-zero traction boundary condition can be applied as a body force in a "real material layer", $\mathscr{R}_r$ along the boundary of a nonzero volume. Based on numerical experiments, the extent of the boundary layer, $\mathscr{R}_r$ should be as close to the boundary as possible. Therefore, a boundary layer for external load application, $\mathscr{R}_r$, with depth $d_{\boldsymbol{\sigma}} = \Delta$, is introduced along the boundary of the material region $\mathscr{R}$, as shown in Fig. 1.5.

In the case of applied traction, $\mathbf{T}^*(\mathbf{x}, t)$ or a point force, $\mathbf{P}(t)$, over the surface, $\mathcal{S}_{\boldsymbol{\sigma}}$ of the boundary layer, $\mathscr{R}_r$, the body force density vector can be expressed as

$$\mathbf{b}(\mathbf{x},t) = \frac{1}{\Delta}\mathbf{T}^*(\mathbf{x},t) \tag{1.18a}$$

or

$$\mathbf{b}(\mathbf{x},t) = \frac{1}{\mathcal{S}_{\boldsymbol{\sigma}}\Delta}\mathbf{P}(t) \tag{1.18b}$$

However, the accuracy of displacements at material points near the boundaries depends on the displacements of the material points in the fictitious layer. Unfortunately, there is no direct way to determine the displacement of material points in the fictitious layer.

Alternatively, Madenci and Oterkus (2016) adopted a finite-difference-based extrapolation function to determine the displacement of fictitious points by using their mirror image values near the boundary in the real domain, $\mathscr{R}$. The extrapolations are performed in such a way that the specified boundary values, $\mathbf{U}^*$ are satisfied explicitly on the real boundary surface, $\mathcal{S}_{\mathbf{u}}$.

For a two-dimensional analysis, depicted in Fig. 1.6, the left boundary of the domain can be prescribed a value of U^* and V^* in horizontal and vertical directions at location $\mathbf{x}^*$ as

$$\mathbf{U}^* = \mathbf{U}^*(\mathbf{x}^*,t) \text{ for } \mathbf{x} \in \mathcal{S}_{\mathbf{u}} \tag{1.19}$$

To impose these boundary constraints, a fictitious boundary layer, $\mathscr{R}_f$ with depth δ is created along the boundary of the actual material region, $\mathscr{R}$. The boundary condition can then be imposed in the fictitious region, $\mathscr{R}_f$ along the boundary of material surface, $\mathcal{S}_{\mathbf{u}}$ as

$$u_f\left(\mathbf{x}_f, t+\Delta t\right) = 2U^*(\mathbf{x}^*, t+\Delta t) - u(\mathbf{x},t) \tag{1.20}$$

and

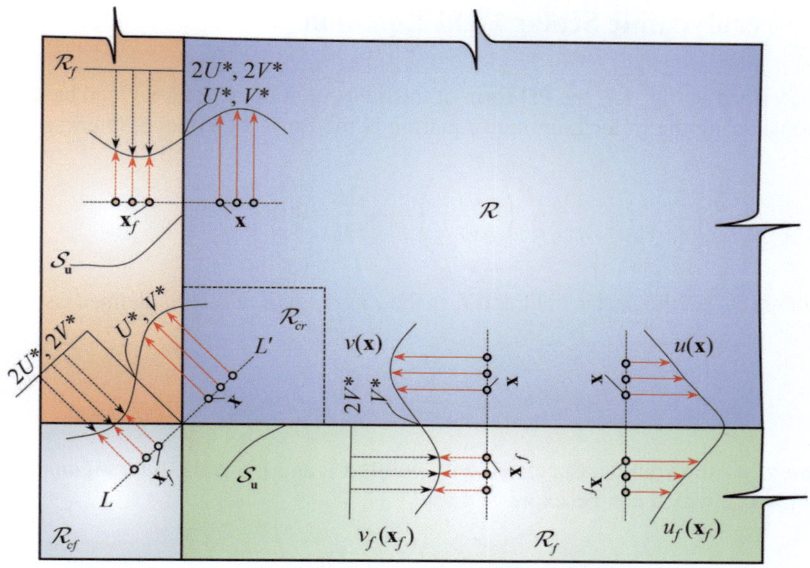

Fig. 1.6 Imposition of displacement boundary conditions with mirror PD points in the fictitious region

$$v_f\left(\mathbf{x}_f, t+\Delta t\right) = 2V^*\left(\mathbf{x}^*, t+\Delta t\right) - v(\mathbf{x}, t) \tag{1.21}$$

in which u_f and v_f represent the horizontal and vertical displacement components in the fictitious layer at location $\mathbf{x}_f$. In the case of prescribed values of $U^* = V^* = 0$, it represents the clamped or fixed boundary conditions.

The displacement in the vertical direction along the bottom boundary in Fig. 1.6 is prescribed with a value of V^*, whereas the displacement in the horizontal direction is unconstrained. Such a scenario constitutes where only one component of displacement is desired to be constrained and the other direction is constraint-free. As a result, the boundary condition can be imposed as

$$u_f\left(\mathbf{x}_f, t+\Delta t\right) = u(\mathbf{x}, t) \tag{1.22}$$

and

$$v_f\left(\mathbf{x}_f, t+\Delta t\right) = 2V^*\left(\mathbf{x}^*, t+\Delta t\right) - v(\mathbf{x}, t) \tag{1.23}$$

It represents the roller support boundary conditions for the prescribed value, $V^* = 0$. In the fictitious corner region, $\mathscr{R}_{cf}$, the displacement values of the material points are obtained through extrapolation of the values of material points in the corner region, $\mathscr{R}_{cr}$, carried out along the line LL' and lines parallel to it.

1.5 Peridynamic Scalar Field Equation

As described in Fig. 1.7, the PD form of the field equation at point $\mathbf{x}_{(k)}$ can be derived by considering the Euler-Lagrange equation in the form (Madenci & Oterkus, 2014)

$$\frac{d}{dt}\left(\frac{\partial L}{\partial \dot{\Theta}_{(k)}}\right) - \frac{\partial L}{\partial \Theta_{(k)}} = 0 \tag{1.24}$$

in which $\Theta_{(k)}$ is the scalar field and L is the Lagrangian, which is defined as

$$L = \sum_{i=1}^{\infty} L_{(i)} V_{(i)} \tag{1.25}$$

where $V_{(i)}$ is the volume associated with points $\mathbf{x}_{(i)}$ and $L_{(i)}$ is the Lagrangian density at point $\mathbf{x}_{(i)}$, which can be expressed as

$$L_{(i)} = Z_{(i)} + s_{(i)}\Theta_{(i)} \tag{1.26}$$

in which $s_{(i)}$ is the field source per unit volume at point $\mathbf{x}_{(i)}$. The scalar field potential $Z_{(i)}$ at point $\mathbf{x}_{(i)}$ depends on interactions of point $\mathbf{x}_{(i)}$ with the other points, $\mathbf{x}_{(j)}$ in its domain of interaction. It is the summation of all interactions associated with point $\mathbf{x}_{(i)}$.

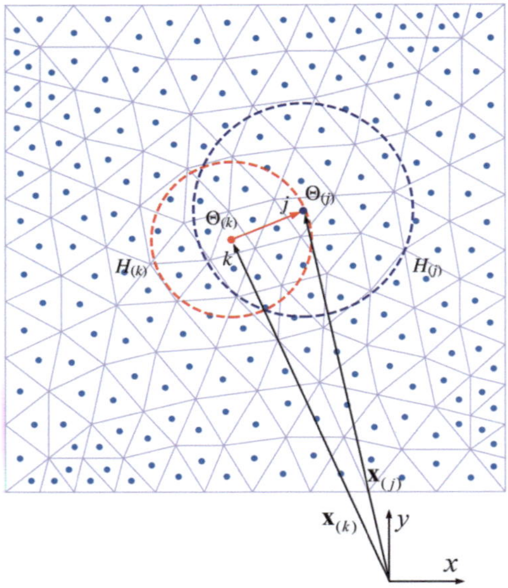

Fig. 1.7 PD field interaction with varying horizon in an irregular discretization

The interaction referred to as the scalar micro-field potential, $z_{(i)(j)}$ depends on the difference between field values of point $\mathbf{x}_{(i)}$ and all other points $\mathbf{x}_{(j)}$ in its family. Note that the scalar micro-field potential $z_{(j)(i)} \neq z_{(i)(j)}$, as $z_{(j)(i)}$ depends on the state of material points that interact with material point $\mathbf{x}_{(j)}$. The scalar micro-field potential is denoted as

$$z_{(i)(j)} = z_{(i)(j)}\left(\Theta_{(1^i)} - \Theta_{(i)},\ \Theta_{(2^i)} - \Theta_{(i)}, \cdots\right) \tag{1.27a}$$

and

$$z_{(j)(i)} = z_{(j)(i)}\left(\Theta_{(1^j)} - \Theta_{(j)},\ \Theta_{(2^j)} - \Theta_{(j)}, \cdots\right) \tag{1.27b}$$

where $\Theta_{(i)}$ is the scalar field value at point $\mathbf{x}_{(i)}$, $\Theta_{(1^i)}$ is the scalar field value of the first point that interacts with point $\mathbf{x}_{(i)}$, and, similarly, $\Theta_{(j)}$ is the scalar field value at point $\mathbf{x}_{(j)}$ while $\Theta_{(1^j)}$ is the scalar field value of the first point that interacts with point $\mathbf{x}_{(j)}$. The scalar field potential $Z_{(i)}$ is defined as

$$\begin{aligned} Z_{(i)} = \frac{1}{2}\sum_{j=1}^{\infty}\Big(& z_{(i)(j)}\left(\Theta_{(1^i)} - \Theta_{(i)},\ \Theta_{(2^i)} - \Theta_{(i)}, \cdots\right) \\ & + z_{(j)(i)}\left(\Theta_{(1^j)} - \Theta_{(j)},\ \Theta_{(2^j)} - \Theta_{(j)}, \cdots\right)\Big) V_{(j)} \end{aligned} \tag{1.28}$$

Consequently, after invoking Eqs. (1.27a, 1.27b) into Eq. (1.28), Eq. (1.26) can be rewritten as

$$L = \sum_{i=1}^{\infty}\left\{\frac{1}{2}\sum_{j=1}^{\infty}\begin{bmatrix} z_{(i)(j)}\left(\Theta_{(1^i)} - \Theta_{(i)}, \Theta_{(2^i)} - \Theta_{(i)}, \cdots\right) \\ + z_{(j)(i)}\left(\Theta_{(1^j)} - \Theta_{(j)}, \Theta_{(2^j)} - \Theta_{(j)}, \cdots\right)\end{bmatrix} V_{(j)} + s_{(i)}\Theta_{(i)}\right\} V_{(i)} \tag{1.29}$$

It can be expressed in an expanded form by considering only the terms associated with point $\mathbf{x}_{(k)}$ as

$$
\begin{aligned}
L = & \cdots \frac{1}{2}\sum_{j=1}^{\infty}\left\{\left[z_{(k)(j)}\left(\Theta_{(1^k)}-\Theta_{(k)},\ \Theta_{(2^k)}-\Theta_{(k)},\cdots\right)\right.\right. \\
& \left.\left.+z_{(j)(k)}\left(\Theta_{(1^j)}-\Theta_{(j)},\ \Theta_{(2^j)}-\Theta_{(j)},\cdots\right)\right]\right\}V_{(j)}V_{(k)}\cdots \\
& \cdots+\frac{1}{2}\sum_{i=1}^{\infty}\left\{\left[z_{(i)(k)}\left(\Theta_{(1^i)}-\Theta_{(i)},\ \Theta_{(2^i)}-\Theta_{(i)},\cdots\right)\right.\right. \\
& \left.\left.+z_{(k)(i)}\left(\Theta_{(1^k)}-\Theta_{(k)},\ \Theta_{(2^k)}-\Theta_{(k)},\cdots\right)\right]\right\}V_{(k)}V_{(i)}\cdots \\
& \cdots+s_{(k)}\Theta_{(k)}V_{(k)}\cdots
\end{aligned}
\tag{1.30}
$$

or

$$
\begin{aligned}
L = & \cdots\sum_{j=1}^{\infty}\left[z_{(k)(j)}\left(\Theta_{(1^k)}-\Theta_{(k)},\ \Theta_{(2^k)}-\Theta_{(k)},\cdots\right)V_{(j)}\right]V_{(k)} \\
& \cdots+\sum_{i=1}^{\infty}\left[z_{(i)(k)}\left(\Theta_{(1^i)}-\Theta_{(i)},\ \Theta_{(2^i)}-\Theta_{(i)},\cdots\right)V_{(i)}\right]V_{(k)} \\
& \cdots+s_{(k)}\Theta_{(k)}V_{(k)}\cdots
\end{aligned}
\tag{1.31}
$$

Substituting from Eq. (1.31) into Eq. (1.24) results in the Euler-Lagrange equation at point $\mathbf{x}_{(k)}$ as

$$
\begin{aligned}
& \sum_{j=1}^{\infty}\frac{\partial z_{(k)(j)}}{\partial\left(\Theta_{(j)}-\Theta_{(k)}\right)}\frac{\partial\left(\Theta_{(j)}-\Theta_{(k)}\right)}{\partial\Theta_{(k)}}V_{(j)}V_{(k)} \\
& \quad+\sum_{i=1}^{\infty}\frac{\partial z_{(i)(k)}}{\partial\left(\Theta_{(k)}-\Theta_{(i)}\right)}\frac{\partial\left(\Theta_{(k)}-\Theta_{(i)}\right)}{\partial\Theta_{(k)}}V_{(i)}V_{(k)}+s_{(k)}V_{(k)}=0
\end{aligned}
\tag{1.32}
$$

or

$$
-\sum_{j=1}^{\infty}\frac{\partial z_{(k)(j)}}{\partial\left(\Theta_{(j)}-\Theta_{(k)}\right)}V_{(j)}+\sum_{i=1}^{\infty}\frac{\partial z_{(i)(k)}}{\partial\left(\Theta_{(k)}-\Theta_{(j)}\right)}V_{(i)}+s_{(k)}=0 \tag{1.33}
$$

Based on the dimensional analysis of this equation, it is apparent that $\partial z_{(k)(j)}/\partial(\Theta_{(j)}-\Theta_{(k)})$ represents the flow density of a field from point $\mathbf{x}_{(j)}$ to point $\mathbf{x}_{(k)}$ through the primary bond and $\partial z_{(i)(k)}/\partial(\Theta_{(k)}-\Theta_{(j)})$ represents the flow density of a field from point $\mathbf{x}_{(i)}$ to $\mathbf{x}_{(k)}$ through the secondary bond. Based on this interpretation, $h_{(k)(j)}$ and $h_{(i)(k)}$ are defined as

$$h_{(k)(j)} = \frac{\partial z_{(k)(j)}}{\partial \left(\Theta_{(j)} - \Theta_{(k)}\right)} \quad \text{and} \quad h_{(i)(k)} = \frac{\partial z_{(i)(k)}}{\partial \left(\Theta_{(k)} - \Theta_{(j)}\right)} \tag{1.34}$$

With these definitions, Eq. (1.33) representing the internal net field flow can be rewritten as

$$\sum_{j=1}^{\infty} \left(h_{(k)(j)} V_{(j)}\right) - \sum_{i=1}^{\infty} \left(h_{(i)(k)} V_{(i)}\right) - s_{(k)} = 0 \tag{1.35}$$

Considering only the influence of material points (family members) in $H_{(k)}$ interacting with $\mathbf{x}_{(k)}$ and point $\mathbf{x}_{(i)}$ interacting with points in its own family, $H_{(i)}$, as shown in Fig. 1.7, Eq. (1.35) can be recast as

$$\sum_{j \in H_{(k)}} \left(h_{(k)(j)} V_{(j)}\right) - \sum_{i \in \widetilde{H}_{(k)}} \left(h_{(i)(k)} V_{(i)}\right) - s_{(k)} = 0 \tag{1.36}$$

in which $H_{(k)} = \{j; j \in H_{(k)}\}$ and $\widetilde{H}_{(k)} = \left\{i; k \in H_{(i)}\right\}$ are the horizon and "dual" horizon of point $\mathbf{x}_{(k)}$, as described in Figs. 1.1 and 1.2. This equation represents the steady state PD form of the field equation at point $\mathbf{x}_{(k)}$. As outlined by Bathe (1996), when the specified flux of the field input changes significantly over a short period of time, the source term $s_{(k)}$ in Eq. (1.36) requires its modification as

$$s_{(k)} = s_e\left(\mathbf{x}_{(k)}, t\right) - s_b\left(\mathbf{x}_{(k)}, t\right) \tag{1.37}$$

where $s_b(\mathbf{x}_{(k)}, t)$ is the generated field source and $s_e(\mathbf{x}_{(k)}, t)$ is defined as

$$s_e\left(\mathbf{x}_{(k)}, t\right) = C\left(\mathbf{x}_{(k)}, t\right)\dot{\Theta}\left(\mathbf{x}_{(k)}, t\right) + \dot{C}\left(\mathbf{x}_{(k)}, t\right)\Theta\left(\mathbf{x}_{(k)}, t\right) \tag{1.38}$$

in which $C(\mathbf{x}_{(k)}, t)$ represents the material property, and that for most field equations it is constant, that is, $\dot{C}\left(\mathbf{x}_{(k)}, t\right) = 0$.

Substituting from Eq. (1.37) along with Eq. (1.38) into Eq. (1.36) results in the transient field equation of the form

$$C\left(\mathbf{x}_{(k)}, t\right)\dot{\Theta}\left(\mathbf{x}_{(k)}, t\right) + \dot{C}\left(\mathbf{x}_{(k)}, t\right)\Theta\left(\mathbf{x}_{(k)}, t\right) = \sum_{j \in H_{(k)}} \left(h_{(k)(j)} V_{(j)}\right) - \sum_{i \in \widetilde{H}_{(k)}} \left(h_{(i)(k)} V_{(i)}\right) + s_b\left(\mathbf{x}_{(k)}, t\right) \tag{1.39}$$

The volume of each material point $V_{(j)}$ or $V_{(i)}$ is infinitesimally small, for the limiting case of $V_{(j)}$, $V_{(i)} \rightarrow 0$, the summation can be replaced with integration. Thus, Eq. (1.39) can be recast as an integral with dual horizons, $H_{\mathbf{x}}$ and $\widetilde{H}_{\mathbf{x}}$ as

$$C(\mathbf{x}_{(k)}, t)\dot{\Theta}(\mathbf{x}, t) + \dot{C}(\mathbf{x}, t)\Theta(\mathbf{x}, t) = \int\limits_{H_{\mathbf{x}}} h(\mathbf{x}' - \mathbf{x}, t)dV_{\mathbf{x}'} - \int\limits_{\widetilde{H}_{\mathbf{x}}} h(\mathbf{x} - \mathbf{x}', t)dV_{\mathbf{x}'} + s_b(\mathbf{x}, t) \quad (1.40)$$

in which $H_{\mathbf{x}}$ is the horizon of point $\mathbf{x}$ associated with the primary bonds and $\widetilde{H}_{\mathbf{x}} = \{\mathbf{x}'; \mathbf{x} \in H_{\mathbf{x}'}\}$ is the dual horizon of point $\mathbf{x}$ associated with the secondary bonds.

1.6 Field and Flux Boundary Conditions

The PD field equation does not contain any spatial derivatives; thus, boundary conditions are, in general, not necessary for the solution of an integro-differential equation. However, the boundary conditions in the form of field variables on $\mathcal{S}_\Theta$ and external field flux on $\mathcal{S}_\mathbf{q}$, as shown in Fig. 1.8, are unavoidable in practical engineering applications. The PD boundary conditions on the field variable can be imposed through a "fictitious layer," $\mathcal{R}_f$ along the boundary, as shown in Fig. 1.9.

For applying the specified field value, the width of fictitious boundary layer d_Θ can be defined as $d_\Theta = \delta$ (Macek & Silling, 2007), as shown in Fig. 1.9. The field flux does not directly appear in the PD field equation. Therefore, the application of field flux is also different from that of the classical field equations. Hence, the field flux cannot be applied as a boundary condition since its volume integration results in a zero value. Therefore, the field flux can be applied as a field source in a "real material layer", $\mathcal{R}_r$ along the "real" boundary of a nonzero volume of size $d_\mathbf{q} = \Delta$.

The field source, $s_b(\mathbf{x}, t)$ can be defined as

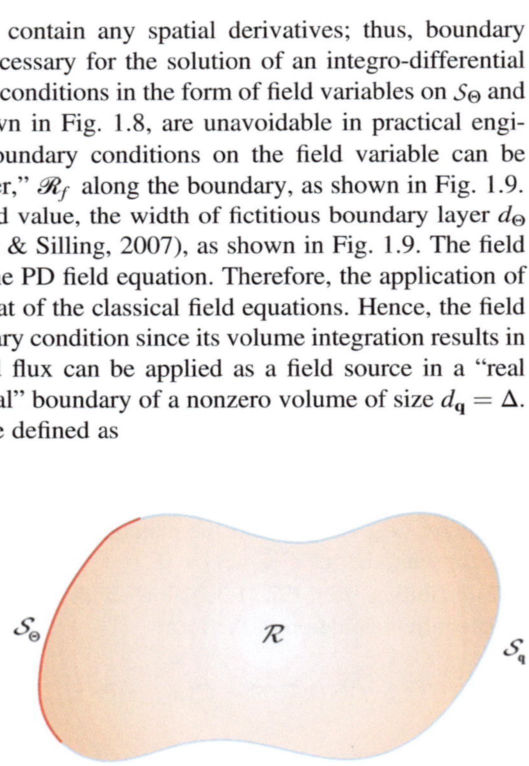

Fig. 1.8 A body subjected to field and flux boundary conditions

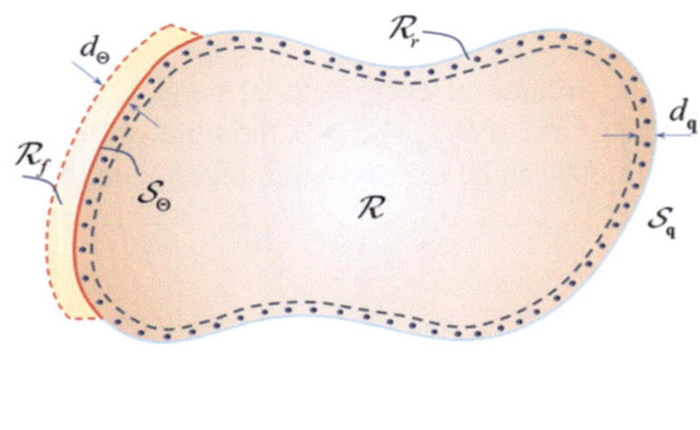

Fig. 1.9 Fictitious and real boundary regions for imposing field and flux conditions

$$s_b(\mathbf{x},t)=\frac{\dot{Q}}{V_{\mathbf{q}}}=\frac{-\int_{\mathcal{S}_{\mathbf{q}}}\mathbf{q}.\mathbf{n}d\mathcal{S}}{\mathcal{S}_{\mathbf{q}}\Delta}=-\frac{\mathbf{q}\cdot\mathbf{n}\mathcal{S}_{\mathbf{q}}}{\mathcal{S}_{\mathbf{q}}\Delta}=-\frac{\mathbf{q}\cdot\mathbf{n}}{\Delta} \tag{1.41}$$

where $\dot{Q}$ is the field flux rate, $V_{\mathbf{q}}$ is the volume of the "real" boundary region, $\mathbf{q}$ is the field flux over the surface $\mathcal{S}_{\mathbf{q}}$ of the boundary layer $\mathcal{R}_r$ with depth, Δ and unit normal vector, $\mathbf{n}$. For a specified field flux, $\mathbf{q}^*(\mathbf{x}, t)$, over the surface $\mathcal{S}_{\mathbf{q}}$, the field source, $s_b(\mathbf{x}, t)$ becomes

$$s_b(\mathbf{x},t)=-\frac{1}{\Delta}\mathbf{q}^*(\mathbf{x},t)\cdot\mathbf{n} \text{ for } \mathbf{x}\in\mathcal{R}_r. \tag{1.42}$$

If there exists no specified flux, $\mathbf{q}^*(\mathbf{x}, t) = 0$, it can be viewed as imposing a zero-valued field source.

If the field equation represents heat conduction in a domain, the imposition of convection and radiation flux boundary conditions may become necessary. Similar to the specified field flux boundary condition, they can be imposed as a field source, $s_b(\mathbf{x}, t)$ in the form of

$$s_b(\mathbf{x},t)=\frac{1}{\Delta}\widehat{h}(\mathbf{x},t)(\Theta(\mathbf{x},t)-\Theta_\infty(\mathbf{x},t)) \text{ for } \mathbf{x}\in\mathcal{R}_r \tag{1.43}$$

and

$$s_b(\mathbf{x},t)=\frac{1}{\Delta}\widehat{\varepsilon}(\mathbf{x},t)\left(\Theta^4(\mathbf{x},t)-\Theta^4_\infty(\mathbf{x},t)\right) \text{ for } \mathbf{x}\in\mathcal{R}_r \tag{1.44}$$

where $\widehat{h}(\mathbf{x},t)$ and $\widehat{\varepsilon}(\mathbf{x},t)$ are the convection and radiation heat transfer coefficients, respectively.

Alternative to this approach, Madenci and Oterkus (2016) introduced a finite-difference-based extrapolation functions to determine the field in the fictitious region, $\mathcal{R}_f$ by using their mirror image values near the boundary in the actual domain, $\mathcal{R}$. As shown in Fig. 1.10, the prescribed boundary field variable, $\Theta_s = \Theta(\mathbf{x}^*, t)$ can be imposed in a layer of fictitious region, $\mathcal{R}_f$ along the boundary of the actual material surface, $\mathcal{S}_\Theta$ as

$$\Theta_f\left(\mathbf{x}_f,t+\Delta t\right)=2\Theta_s(\mathbf{x}^*,t+\Delta t)-\Theta(\mathbf{x},t), \tag{1.45}$$

in which Θ_f is field value in the fictitious layer at location $\mathbf{x}_f$ with $\mathbf{x}^*\in\mathcal{S}_\Theta, \mathbf{x}_f\in\mathcal{R}_f, \mathbf{x}\in\mathcal{R}$. As shown in Fig. 1.10, the left boundary of the domain is prescribed a value of Θ_s at location $\mathbf{x}^*$.

Similarly, the imposition of gradient $\mathbf{q}_g(\mathbf{x}^*)$ on the boundary can be achieved by assigning the mirror image of field values near the boundary from the real domain to the points in the fictitious region, $\mathcal{R}_f$ as

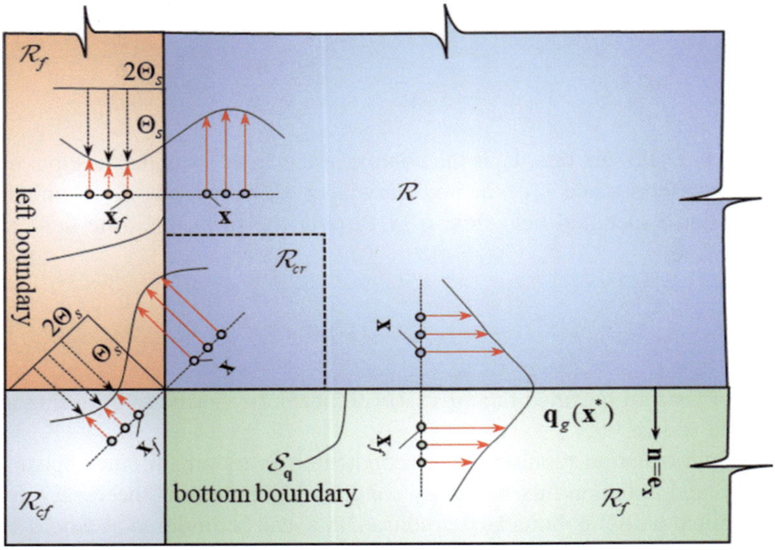

Fig. 1.10 Imposition of Dirichlet and Neumann boundary conditions for PD field equation

$$\Theta_f(\mathbf{x}_f, t + \Delta t) = \mathbf{q}_g(\mathbf{x}^*, t + \Delta t) \cdot (\mathbf{x}_f - \mathbf{x}) + \Theta(\mathbf{x}, t) \tag{1.46}$$

with $\mathbf{x}^* \in \mathcal{S}_\mathbf{q}, \mathbf{x}_f \in \mathcal{R}_f, \mathbf{x} \in \mathcal{R}$.

For zero flux condition, this equation reduces to

$$\Theta_f(\mathbf{x}_f, t + \Delta t) = \Theta(\mathbf{x}, t) \tag{1.47}$$

with $\mathbf{x}^* \in \mathcal{S}_\mathbf{q}, \mathbf{x}_f \in \mathcal{R}_f, \mathbf{x} \in \mathcal{R}$

1.7 Periodic Boundary Conditions

As explained by Madenci et al. (2018, 2020), the periodic boundary conditions in PD models can be imposed in a natural way. The continuity of the displacement and field variable on the surfaces defining the region is achieved by assigning the field values in the fictitious regions from the real domain as shown in Fig. 1.11. For the field variable, this is achieved by the following constraints defined as

$$\Theta_f(x_{fL}, y_{fL}) = \Theta(x_{fL} + L, y_{fL}) \quad \text{for} \quad (x_{fL}, y_{fL}) \in \mathcal{R}_{fL} \tag{1.48a}$$

and

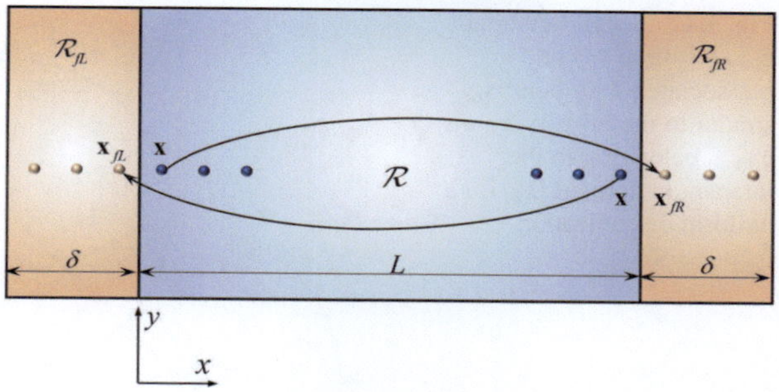

Fig. 1.11 Imposition of periodic boundary condition by assigning values from the real domain to the points in the fictitious regions

$$\Theta_f\left(x_{fR}, y_{fR}\right) = \Theta\left(x_{fR} - L, y_{fR}\right) \quad \text{for} \quad \left(x_{fR}, y_{fR}\right) \in \mathscr{R}_{fR} \tag{1.48b}$$

Similarly, the constraints for the displacement field can be enforced as

$$\mathbf{u}_f\left(x_{fL}, y_{fL}\right) = \mathbf{u}\left(x_{fL} + L, y_{fL}\right) \quad \text{for} \quad \left(x_{fL}, y_{fL}\right) \in \mathscr{R}_{fL} \tag{1.49a}$$

and

$$\mathbf{u}_f\left(x_{fR}, y_{fR}\right) = \mathbf{u}\left(x_{fR} + L, y_{fR}\right) \quad \text{for} \quad \left(x_{fR}, y_{fR}\right) \in \mathscr{R}_{fR} \tag{1.49b}$$

The size of the fictitious regions, that is, $\mathscr{R}_{fR}$ and $\mathscr{R}_{fL}$ corresponding to the right and the left boundaries, is specified as δ.

1.8 Uniform Discretization

For uniform discretization as shown in Fig. 1.12, the PD points have equal volumes. Furthermore, the radius of the interaction domain $H_{(k)}$ of point $\mathbf{x}_{(k)}$ is same as that of its family members, $\mathbf{x}_{(\alpha)}$, that is, $H_{(\alpha)}\ \forall\ \alpha \in H_{(k)}$. As a result, all the family members of point $\mathbf{x}_{(k)}$ interact with point, $\mathbf{x}_{(k)}$.

Hence, Eq. (1.16) and (1.40) can be rewritten as

$$\rho_{(k)}\ddot{\mathbf{u}}_{(k)} = \sum_{j \in H_{(k)}} \mathbf{t}_{(k)(j)} V_{(j)} - \sum_{j \in H_{(k)}} \mathbf{t}_{(j)(k)} V_{(j)} + \mathbf{b}_{(k)} \tag{1.50}$$

and

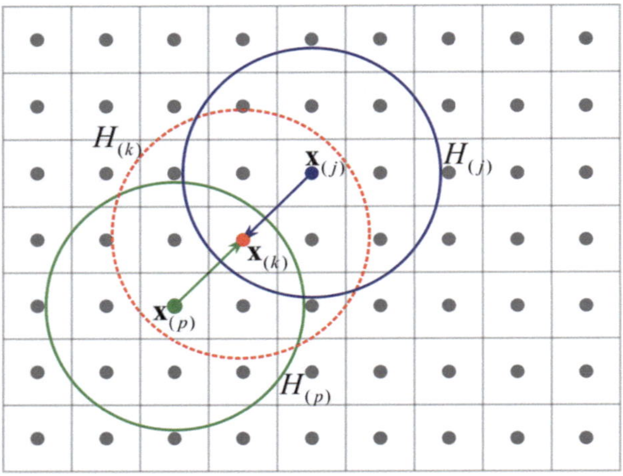

Fig. 1.12 Secondary bonds $\boldsymbol{\xi}_{(m)(k)}$ with $m = j$ and p associated with point $\mathbf{x}_{(k)}$ similar to the primary bonds $\boldsymbol{\xi}_{(k)(m)}$ belonging to the interaction domain $H_{(k)}$ for a uniform discretization

$$C\big(\mathbf{x}_{(k)}, t\big)\dot{\Theta}\big(\mathbf{x}_{(k)}, t\big) + \dot{C}\big(\mathbf{x}_{(k)}, t\big)\Theta\big(\mathbf{x}_{(k)}, t\big) = \sum_{j \in H_{(k)}} \big(h_{(k)(j)} V_{(j)}\big) - \sum_{j \in H_{(k)}} \big(h_{(j)(k)} V_{(j)}\big) + s_b\big(\mathbf{x}_{(k)}, t\big), \tag{1.51}$$

In the limit of $V_{(j)} \to 0$, these equations reduce to

$$\rho(\mathbf{x})\ddot{\mathbf{u}}(\mathbf{x}, t) = \int_{H_\mathbf{x}} (\mathbf{t}(\mathbf{x}' - \mathbf{x}, t) - \mathbf{t}(\mathbf{x} - \mathbf{x}', t)) dV_{\mathbf{x}'} + \mathbf{b}(\mathbf{x}, t) \tag{1.52}$$

and

$$C(\mathbf{x}, t)\dot{\Theta}(\mathbf{x}, t) + \dot{C}(\mathbf{x}, t)\Theta(\mathbf{x}, t) = \int_{H_\mathbf{x}} (h(\mathbf{x}' - \mathbf{x}, t) - h(\mathbf{x} - \mathbf{x}', t)) dV_{\mathbf{x}'} + s_b(\mathbf{x}, t) \tag{1.53}$$

1.9 Discontinuities

In a discretized model, geometric discontinuities such as cracks are simply introduced through the elimination of PD bonds between the material points located across the crack surfaces as shown in Fig. 1.13. The bond between material points is removed if both material points are located directly across the crack surfaces; otherwise, their interactions remain intact. The status parameter, $\widehat{\mu}\big(\mathbf{x}_{(j)} - \mathbf{x}_{(k)}\big)$ defined in Eq. (1.1) facilitates the insertion of a crack.

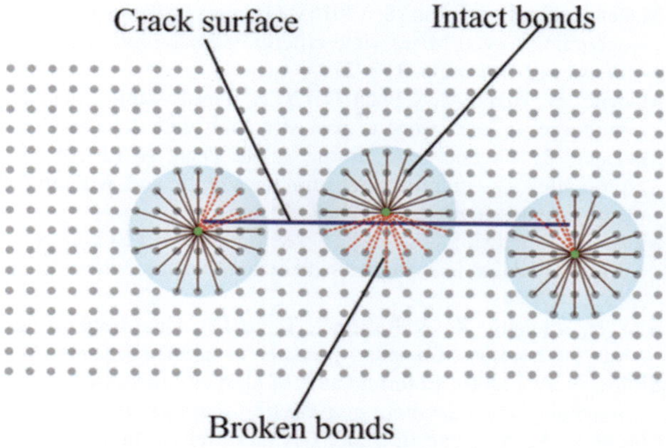

Fig. 1.13 Creation of a crack by removal of bonds across the crack surfaces

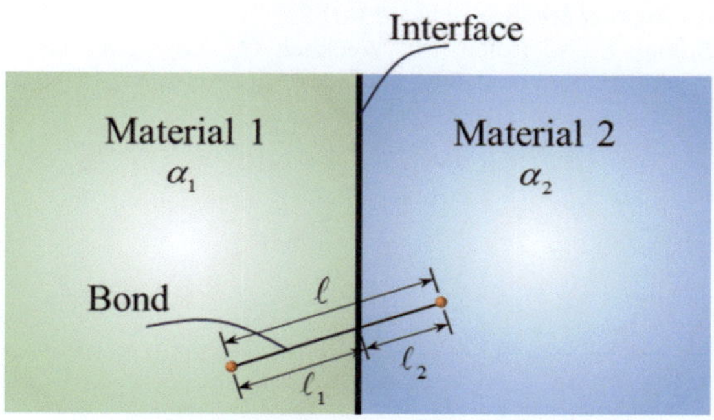

Fig. 1.14 Interface bond between material points in dissimilar materials

In the presence of an interface between dissimilar materials, the average bond property can be approximated as (Madenci & Oterkus, 2014; Mitts et al., 2020)

$$\alpha = \frac{\alpha_1 \ell_1 + \alpha_2 \ell_2}{\ell} \tag{1.54}$$

where ℓ_1 is the length of the bond in material 1, ℓ_2 is the length of the bond in material 2, and $\ell = \ell_1 + \ell_2$ is the total length of the bond as shown in Fig 1.14. The parameters α_1 and α_2 represent the property of materials.

References

Bathe, K. (1996). *Finite element procedures*. Prentice-Hall.

Macek, R. W., & Silling, S. A. (2007). Peridynamics via finite element analysis. *Finite Elements in Analysis and Design, 43*, 1169–1178.

Madenci, E., & Oterkus, E. (2014). *Peridynamic theory and its applications*. Springer.

Madenci, E., & Oterkus, S. (2016). Ordinary state-based peridynamics for plastic deformation according to von Mises yield criteria with isotropic hardening. *Journal of the Mechanics and Physics of Solids, 86*, 192–219.

Madenci, E., Barut, A., & Phan, N. (2018). Peridynamic unit cell homogenization for thermoelastic properties of heterogenous microstructures with defects. *Composite Structures, 188*, 104–115.

Madenci, E., Yaghoobi, A., Barut, A., & Phan, N. (2020). Peridynamic unit cell for effective properties of complex microstructures with and without defects. *Theoretical and Applied Fracture Mechanics, 110*, 102835.

Mitts, C., Naboulsi, S., Przybyla, C., & Madenci, E. (2020). Axisymmetric peridynamic analysis of crack deflection in a single strand ceramic matrix composite. *Engineering Fracture Mechanics, 235*, 107074.

Ren, H., Zhuang, X., & Rabczuk, T. (2017). Dual horizon peridynamics: A stable solution to varying horizons. *Computer Methods in Applied Mechanics and Engineering, 318*, 762–782.

Silling, S. A. (2000). Reformulation of elasticity theory for discontinuities and long-range forces. *Journal of the Mechanics and Physics of Solids, 48*, 175–209.

Silling, S. A., & Askari, E. (2005). A meshfree method based on the peridynamic model of solid mechanics. *Computers & Structures, 83*, 1526–1535.

Silling, S. A., & Lehoucq, R. B. (2010). Peridynamic theory of solid mechanics. *Advances in Applied Mechanics, 44*, 73–168.

Silling, S. A., Epton, M., Weckner, O., Xu, J., & Askari, E. (2007). Peridynamic states and constitutive modeling. *Journal of Elasticity, 88*, 151–184.

Chapter 2
Peridynamic Differential Operator

2.1 Introduction

Solutions to the governing differential equations of many physical fields are commonly constructed by employing standard numerical techniques such as finite-difference, finite-element, finite-volume, and smoothed particle hydrodynamics methods. Such techniques require the approximation of derivatives at discrete points in the domain. These techniques yield satisfactory predictions if the field variables satisfy the smoothness requirement. Otherwise, the predictions suffer from the presence of a singularity arising from the spatial derivative of a discontinuous field.

The finite-difference method approximates the derivatives at the spatial grid points through subtraction and division and transforms the differential equation to its discrete form. It is simple and intuitive for computing the first- and second-order derivatives using forward and central difference. The finite element method divides the domain into smaller subdivisions (elements) with nodal points. The field variable is approximated within each element in terms of nodal unknowns. Assembling the nodal unknowns from each element results in a system of algebraic equations; they represent the weak form of a differential equation. It is extremely suitable to consider domains with complex geometries, and to enforce natural boundary conditions. The finite volume method approximates the spatial derivatives by integrating across a discrete (finite) control volume associated with each grid point while satisfying the integral conservation law for each control volume. This is achieved by converting the volume integrals to surface integrals through the divergence theorem. These integrals are evaluated as fluxes at the surfaces of each finite volume, and the flux entering a specific finite volume is identical to tha tof leaving the adjacent volume. The finite volume method enables the use of unstructured grid; thus, it is suitable for considering domains with complex geometry. These methods face numerical challenges in the presence of singularities emerging from the spatial derivatives of a discontinuous field because differentiation is a local process involving division.

E. Madenci et al., *Peridynamics and Its Applications Using Ansys*,
https://doi.org/10.1007/978-3-031-59896-8_2

Nonlocal mesh-free methods such as Smoothed Particle Hydrodynamics (SPH), Reproducing Kernel Particle Method (RKPM), and Generalized Finite Difference Method (GFDM) remove the inherent limitations of the aforementioned local differentiation methods. They account for the nonlocal interactions among the points (particles) located within a region of finite distance (domain of interaction).

The SPH introduced by Lucy (1977) and Gingold and Monaghan (1977) divides the domain into a finite set of particles. It approximates the function and its derivatives by using a smoothing function through these particles over a domain of interaction defined by a characteristic or a smoothing length. The field variable at each point (particle) is smoothed by a kernel function over the domain of interaction. Therefore, the choice of kernel functions is crucial for accurate approximation. Also, the implementation of boundary conditions may present difficulties due to the necessity of ghost particles near the boundary. In order to remove the drawbacks of the SPH method, Liu et al. (1995) proposed the RKPM by introducing a correction function in the kernel of integral transformation along with the corresponding reproducing conditions. It requires different correction functions and reproducing conditions for different spatial derivatives. The correction functions associated with the higher-order derivatives are influenced only by those of the lower-order derivatives. Both the SPH and RKPM require symmetric kernels and are limited to spatial functions and their derivatives. Therefore, they are not applicable to differentiation with respect to time.

The GFDM by Liszka and Orkisz (1980) employs the Taylor Series Expansion (TSE) together with an irregular net of grid points within the local support region (interaction domain). It provides the approximation of arbitrary-order derivatives by solving a system of algebraic equations resulting from the Least Squares Minimization (LSM) with arbitrary weighting coefficients. It minimizes a residual based on the discrete summation of the error between the functional values of the point of interest and its neighboring points in the interaction domain. The accuracy of GFDM relies on the number points, the selection of star shape (interaction domain), and the weight functions. In these nonlocal methods, the approximation of the function and its derivatives depends on the values of other nearby points.

Alternatively, the nonlocal Peridynamic Differential Operator (PDDO) introduced by Madenci et al. (2016, 2017, 2019) converts the local differentiation to its nonlocal integral form in a unified manner regardless of the presence of jump discontinuities or singularities. It is a paradigm shift in computational analysis. The peridynamic (PD) theory introduced by Silling (2000) and Silling et al. (2007) incorporates the nonlocal interactions of points within a finite domain. The extent of this interaction domain is defined by an internal parameter referred to as the "horizon" of a point.

The PDDO employs the concept of PD interactions and uses a set of orthogonal functions known as PD functions to evaluate derivatives with respect to spatial variables and time. It restores the nonlocal interactions at a point by considering its association with other points in an arbitrary domain of interaction. The PDDO is expressed in terms of only integration (summation) over the domain of interaction. Therefore, it is not prone to singularities arising from the presence of discontinuities.

However, it requires the construction of PD functions at each point. The PD functions are determined directly by making them orthogonal to each term in the Taylor Series Expansion (TSE), except for the derivative of interest. Both the lower- and higher-order derivatives influence each other while determining the PD functions in the presence of a nonsymmetric family. The PDDO permits any order of the derivatives directly from the orthogonality condition of the PD functions without performing any differentiation. It is simply a bridge between differentiation and integration. Therefore, the PDDO enables numerical differentiation through integration. It converts the local form of differentiation to its nonlocal PD form. Also, it recovers the local differentiation as the size of the interaction domain approaches zero.

There exist similarities between the Reproducing Kernel (RK) (Liu et al., 1995; Aluru, 2000; Bessa et al., 2014) and Gradient Reproducing Kernel (G-RK) operators by Chi et al. (2013) and the PDDO. However, there are certainly fundamental differences. For the approximation of zeroth-order derivative (function itself), the PDDO, RK and G-RK are all equivalent. The PDDO and G-RK are also equivalent for the approximation of first-order derivatives. Pertinent to the zeroth- and first-order derivatives, the reproducing conditions and the correction functions of the RK and G-RK are the same as the orthogonality conditions and the PD functions of the PDDO, respectively. Otherwise, there exists no correspondence when approximating the higher-order derivatives.

The RK operator requires different correction functions and reproducing conditions for different order spatial derivatives. Their determination requires the derivatives of the "moment matrix" as derived by Aluru (2000). Also, the correction functions associated with the higher-order derivatives are influenced only by those of the lower-order derivatives. The G-RK operator provides the second-order derivatives by differentiating the first-order derivative approximations.

The PDDO employs neither a kernel function nor reproducibility conditions for different orders of derivatives. It enables accurate determination of any arbitrary order of partial derivatives of the spatial and temporal functions. The PDDO is free of the requirement of symmetric kernels. This feature removes the necessity of ghost points near the boundary. Therefore, it is not a special case of RK or G-RK operators.

As introduced by Liszka (1984), the GFDM employs the TSE in conjunction with LSM to approximate the function and its derivatives based on a set of functional values within the local support region. Although the PDDO appears to be similar to GFDM, it is fundamentally different. The PDDO includes the effect of entity such as the volume of each point in the interaction domain. Therefore, its residual is derived based on the analytical integration over the continuous interaction domain. However, the GFDM employs a residual based on the discrete summation of the error between the functional values (without an entity) of the point of interest and its neighboring points in the interaction domain. Unlike the GFDM, the PDDO leads to analytical expressions in integral form for arbitrary-order derivatives. As a result, these derivatives recover the PD equation of motion derived by Silling (2000) and Silling et al. (2007) at a point with a symmetric position in a spherical or circular interaction domain.

This chapter describes the construction of peridynamic (PD) functions and provides the PD form of local derivatives of up to second order in 2D and 3D analyses. Also, it presents their analytical expressions for gradient, divergence, gradient of divergence, and the Laplacian of a displacement vector at a point with a complete horizon in the form of a sphere or a circle in accordance with the dimension of the analysis. Furthermore, it provides the PD form of the bond direction vector and its dyadic product with itself as well as the PD form of the divergence of a second-order tensor.

2.2 PD Functions

The PD functions for the derivatives are determined directly by making them orthogonal to every term in the Taylor Series Expansion (TSE), except for the term with the desired derivative (Madenci et al., 2016, 2019). As shown in Fig. 2.1, point $\mathbf{x}_{(k)}$ interacts with other points, $\mathbf{x}_{(j)}$ in its interaction domain, $H_{(k)} = \{j; j \in H_{(k)}\}$. The points, $\mathbf{x}_{(j)}$ located in $H_{(k)}$ are called the family members of $\mathbf{x}_{(k)}$. The bond, $\boldsymbol{\xi}_{(k)(j)} = \mathbf{x}_{(j)} - \mathbf{x}_{(k)}$ between these points is referred to as a "*primary*" bond. The degree of interactions between the points in each family can be specified through a nondimensional weight function, $w(|\boldsymbol{\xi}_{(k)(j)}|)$. Although not a limitation, the commonly accepted weight functions are of the form

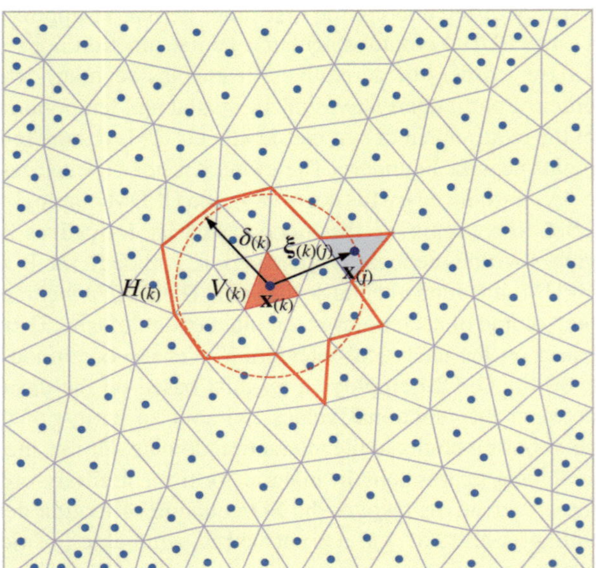

Fig. 2.1 Primary PD bond between material point $\mathbf{x}_{(k)}$ and its family member $\mathbf{x}_{(j)}$ in its horizon, $H_{(k)} = \{j; j \in H_{(k)}\}$

$$w\left(|\boldsymbol{\xi}_{(k)(j)}|\right) = \exp\left(-\left(a|\boldsymbol{\xi}_{(k)(j)}|/\delta_{(k)}\right)^2\right) \tag{2.1}$$

or

$$w_n\left(|\boldsymbol{\xi}_{(k)(j)}|\right) = \frac{\delta_{(k)}^{n+1}}{\left|\boldsymbol{\xi}_{(k)(j)}\right|^{n+1}} \tag{2.2}$$

in which the horizon size, $\delta_{(k)}$ specifies the extent of the interaction domain and a is a specified constant. As shown in Fig. 2.1, the points encompassed by the horizon of point, $\mathbf{x}_{(k)}$ establish the family population. If the centroid of a finite volume is included within the horizon, it is included in the family of $\mathbf{x}_{(k)}$. Alternatively, the family population of point $\mathbf{x}_{(k)}$ can be established by selecting the closest number of $N_{(k)}$ points. The parameter $N_{(k)}$ represents the number of points (population) in the family of $\mathbf{x}_{(k)}$.

2.2.1 Two-Dimensional Analysis

In order to derive the expressions for the first- and second-order derivatives, the TSE of a scalar-valued function, $f(\mathbf{x}_{(j)})$ can be expressed in a 2D space as

$$\begin{aligned} f(\mathbf{x}_{(j)}) - f(\mathbf{x}_{(k)}) &= \xi_{1(k)(j)} \frac{\partial f(\mathbf{x}_{(k)})}{\partial x_1} + \xi_{2(k)(j)} \frac{\partial f(\mathbf{x}_{(k)})}{\partial x_2} + \frac{1}{2!}\xi_{1(k)(j)}^2 \frac{\partial^2 f(\mathbf{x}_{(k)})}{\partial x_1^2} \\ &+ \frac{1}{2!}\xi_{2(k)(j)}^2 \frac{\partial^2 f(\mathbf{x}_{(k)})}{\partial x_2^2} + \xi_{1(k)(j)}\xi_{2(k)(j)} \frac{\partial^2 f(\mathbf{x}_{(k)})}{\partial x_1 \partial x_2} + R_2(\mathbf{x}_{(k)}) \end{aligned} \tag{2.3}$$

where $\mathbf{x}_{(k)}^T = \left(x_{1(k)}, x_{2(k)}\right)$, $\xi_{1(k)(j)}$ and $\xi_{2(k)(j)}$ are the components of the relative position vector, $\boldsymbol{\xi}_{(k)(j)}$, and $R_2(\mathbf{x}_{(k)})$ represents the remainder of the second-order approximation. Although not necessary, function $f(\mathbf{x}_{(k)})$ is treated as known in order to consider the relative change in the function, $f(\mathbf{x}_{(j)}) - f(\mathbf{x}_{(k)})$. Multiplying each term with the PD functions, $g_2^{p_1 p_2}\left(\mathbf{x}_{(k)}, \boldsymbol{\xi}_{(k)(j)}\right)$ while considering $R_2(\mathbf{x}_{(k)})$ to be negligible, and summing over the domain of interaction (family), $H_{(k)}$ result in

$$\begin{aligned}
&\sum_{j\in H_{(k)}}\left(f\left(\mathbf{x}_{(j)}\right)-f\left(\mathbf{x}_{(k)}\right)\right)g_2^{p_1p_2}\left(\mathbf{x}_{(k)},\boldsymbol{\xi}_{(k)(j)}\right)A_{(j)}=\\
&\quad\frac{\partial f\left(\mathbf{x}_{(k)}\right)}{\partial x_1}\sum_{j\in H_{(k)}}\xi_{1(k)(j)}g_2^{p_1p_2}\left(\mathbf{x}_{(k)},\boldsymbol{\xi}_{(k)(j)}\right)A_{(j)}\\
&\quad+\frac{\partial f\left(\mathbf{x}_{(k)}\right)}{\partial x_2}\sum_{j\in H_{(k)}}\xi_{2(k)(j)}g_2^{p_1p_2}\left(\mathbf{x}_{(k)},\boldsymbol{\xi}_{(k)(j)}\right)A_{(j)}\\
&\quad+\frac{1}{2}\frac{\partial f^2\left(\mathbf{x}_{(k)}\right)}{\partial x_1^2}\sum_{j\in H_{(k)}}\xi_{1(k)(j)}^2g_2^{p_1p_2}\left(\mathbf{x}_{(k)},\boldsymbol{\xi}_{(k)(j)}\right)A_{(j)}\\
&\quad+\frac{1}{2}\frac{\partial f^2\left(\mathbf{x}_{(k)}\right)}{\partial x_2^2}\sum_{j\in H_{(k)}}\xi_{2(k)(j)}^2g_2^{p_1p_2}\left(\mathbf{x}_{(k)},\boldsymbol{\xi}_{(k)(j)}\right)A_{(j)}\\
&\quad+\frac{\partial f^2\left(\mathbf{x}_{(k)}\right)}{\partial x_1\partial x_2}\sum_{j\in H_{(k)}}\xi_{1(k)(j)}\xi_{2(k)(j)}g_2^{p_1p_2}\left(\mathbf{x}_{(k)},\boldsymbol{\xi}_{(k)(j)}\right)A_{(j)}
\end{aligned}\tag{2.4}$$

Except for the term with the desired order derivative, the PD functions must be orthogonal to the remaining terms in the TSE as

$$\frac{1}{n_1!n_2!}\sum_{j\in H_{(k)}}\xi_{1(k)(j)}^{n_1}\xi_{2(k)(j)}^{n_2}g_2^{p_1p_2}\left(\mathbf{x}_{(k)},\boldsymbol{\xi}_{(k)(j)}\right)A_{(j)}=\delta_{n_1p_1}\delta_{n_2p_2}\tag{2.5}$$

with (n_1, n_2, p_1, p_2 = 0, 1, 2) and $\delta_{n_ip_i}$ is the Kronecker delta symbol. The PD functions, $g_N^{p_1p_2}(\mathbf{x},\boldsymbol{\xi})$ with $p_1 + p_2 \leq N = 2$ are explained by Madenci et al. (2019). The superscript p_i denotes the order of differentiation with respect to the variable x_i for $i = 1, 2$.

Applying the orthogonality conditions, Eq. (2.5) results in the PD form of the derivatives as

$$\begin{Bmatrix}f_{,1}\left(\mathbf{x}_{(k)}\right)\\ f_{,2}\left(\mathbf{x}_{(k)}\right)\end{Bmatrix}=\sum_{j\in H_{(k)}}\left(f\left(\mathbf{x}_{(j)}\right)-f\left(\mathbf{x}_{(k)}\right)\right)\begin{Bmatrix}g_2^{10}\left(\mathbf{x}_{(k)},\boldsymbol{\xi}_{(k)(j)}\right)\\ g_2^{01}\left(\mathbf{x}_{(k)},\boldsymbol{\xi}_{(k)(j)}\right)\end{Bmatrix}A_{(j)}\tag{2.6}$$

and

$$\begin{Bmatrix}f_{,11}\left(\mathbf{x}_{(k)}\right)\\ f_{,22}\left(\mathbf{x}_{(k)}\right)\\ f_{,12}\left(\mathbf{x}_{(k)}\right)\end{Bmatrix}=\sum_{j\in H_{(k)}}\left(f\left(\mathbf{x}_{(j)}\right)-f\left(\mathbf{x}_{(k)}\right)\right)\begin{Bmatrix}g_2^{20}\left(\mathbf{x}_{(k)},\boldsymbol{\xi}_{(k)(j)}\right)\\ g_2^{02}\left(\mathbf{x}_{(k)},\boldsymbol{\xi}_{(k)(j)}\right)\\ g_2^{11}\left(\mathbf{x}_{(k)},\boldsymbol{\xi}_{(k)(j)}\right)\end{Bmatrix}A_{(j)}\tag{2.7}$$

Thus, the gradient and Laplacian of $f(\mathbf{x})$ can be expressed as

$$\nabla f\left(\mathbf{x}_{(k)}\right)=\sum_{j\in H_{(k)}}\left(f\left(\mathbf{x}_{(j)}\right)-f\left(\mathbf{x}_{(k)}\right)\right)\mathbf{g}_{(k)(j)}A_{(j)} \tag{2.8}$$

and

$$\nabla^2 f\left(\mathbf{x}_{(k)}\right)=\sum_{j\in H_{(k)}}\mathrm{tr}\mathbf{G}_{(k)(j)}\left(f\left(\mathbf{x}_{(j)}\right)-f\left(\mathbf{x}_{(k)}\right)\right)A_{(j)} \tag{2.9}$$

where the vector $\mathbf{g}_{(k)(j)}$ and matrix $\mathbf{G}_{(k)(j)}$ are defined as

$$\mathbf{g}_{(k)(j)}=\left\{\begin{array}{c} g_2^{10}\left(\mathbf{x}_{(k)},\boldsymbol{\xi}_{(k)(j)}\right) \\ g_2^{01}\left(\mathbf{x}_{(k)},\boldsymbol{\xi}_{(k)(j)}\right)\end{array}\right\} \tag{2.10}$$

and

$$\mathbf{G}_{(k)(j)}=\begin{bmatrix} g_2^{20}\left(\mathbf{x}_{(k)},\boldsymbol{\xi}_{(k)(j)}\right) & g_2^{11}\left(\mathbf{x}_{(k)},\boldsymbol{\xi}_{(k)(j)}\right) \\ g_2^{11}\left(\mathbf{x}_{(k)},\boldsymbol{\xi}_{(k)(j)}\right) & g_2^{02}\left(\mathbf{x}_{(k)},\boldsymbol{\xi}_{(k)(j)}\right)\end{bmatrix} \tag{2.11}$$

The PD functions can be constructed as a linear combination of polynomial basis functions

$$\begin{aligned} g_2^{p_1p_2}\left(\mathbf{x}_{(k)},\boldsymbol{\xi}_{(k)(j)}\right)=&\,a_{10}^{p_1p_2}w_{10}\left(\left|\boldsymbol{\xi}_{(k)(j)}\right|\right)\xi_{1(k)(j)}+a_{01}^{p_1p_2}w_{01}\left(\left|\boldsymbol{\xi}_{(k)(j)}\right|\right)\xi_{2(k)(j)} \\ &+a_{20}^{p_1p_2}w_{20}\left(\left|\boldsymbol{\xi}_{(k)(j)}\right|\right)\xi_{1(k)(j)}^2+a_{02}^{p_1p_2}w_{02}\left(\left|\boldsymbol{\xi}_{(k)(j)}\right|\right)\xi_{2(k)(j)}^2 \\ &+a_{11}^{p_1p_2}w_{11}\left(\left|\boldsymbol{\xi}_{(k)(j)}\right|\right)\xi_{1(k)(j)}\xi_{2(k)(j)}\end{aligned} \tag{2.12}$$

where $a_{q_1q_2}^{p_1p_2}$ are the unknown coefficients and $w_{q_1q_2}\left(\left|\boldsymbol{\xi}_{(k)(j)}\right|\right)$ are the specified weight functions. Substituting from Eq. (2.12) into Eq. (2.5) yields

$$\begin{aligned} &\frac{1}{n_1!n_2!}\sum_{j\in H_{(k)}}\left(\begin{array}{l} a_{10}^{p_1p_2}w_{10}\left(\left|\boldsymbol{\xi}_{(k)(j)}\right|\right)\xi_{1(k)(j)}^{n_1+1}\xi_{2(k)(j)}^{n_2}+a_{01}^{p_1p_2}w_{01}\left(\left|\boldsymbol{\xi}_{(k)(j)}\right|\right)\xi_{1(k)(j)}^{n_1}\xi_{2(k)(j)}^{n_2+1} \\ +a_{20}^{p_1p_2}w_{20}\left(\left|\boldsymbol{\xi}_{(k)(j)}\right|\right)\xi_{1(k)(j)}^{n_1+2}\xi_{2(k)(j)}^{n_2}+a_{02}^{p_1p_2}w_{02}\left(\left|\boldsymbol{\xi}_{(k)(j)}\right|\right)\xi_{1(k)(j)}^{n_1}\xi_{2(k)(j)}^{n_2+2} \\ +a_{11}^{p_1p_2}w_{11}\left(\left|\boldsymbol{\xi}_{(k)(j)}\right|\right)\xi_{1(k)(j)}^{n_1+1}\xi_{2(k)(j)}^{n_2+1}\end{array}\right)A_{(j)} \\ &=\delta_{n_1p_1}\delta_{n_2p_2}\end{aligned} \tag{2.13}$$

With a particular choice of weight functions from Eq. (2.2), this equation can be recast as a system of algebraic equations for the determination of the coefficients $a_{q_1q_2}^{p_1p_2}$ in the form

$$\mathbf{A}\left(\mathbf{x}_{(k)}\right)\mathbf{a}_{(k)}^{p_1p_2} = \mathbf{b}^{p_1p_2} \tag{2.14}$$

with

$$\mathbf{A}\left(\mathbf{x}_{(k)}\right) = \begin{bmatrix} \mathbf{A}_{11}\left(\mathbf{x}_{(k)}\right) & \mathbf{A}_{12}\left(\mathbf{x}_{(k)}\right) \\ \mathbf{A}_{21}\left(\mathbf{x}_{(k)}\right) & \mathbf{A}_{22}\left(\mathbf{x}_{(k)}\right) \end{bmatrix}; \ \mathbf{a}_{(k)}^{p_1p_2} = \begin{Bmatrix} \mathbf{a}_{1(k)}^{p_1p_2} \\ \mathbf{a}_{2(k)}^{p_1p_2} \end{Bmatrix}; \ \mathbf{b}^{p_1p_2} = \begin{Bmatrix} \mathbf{b}_1^{p_1p_2} \\ \mathbf{b}_2^{p_1p_2} \end{Bmatrix} \tag{2.15}$$

The known submatrices can be expressed as

$$\mathbf{A}_{11}(\mathbf{x}_{(k)}) = \sum_{j \in H_{(k)}} w_1(|\boldsymbol{\xi}_{(k)(j)}|) \begin{bmatrix} \xi_{1(k)(j)}^2 & \xi_{1(k)(j)}\xi_{2(k)(j)} \\ \xi_{1(k)(j)}\xi_{2(k)(j)} & \xi_{2(k)(j)}^2 \end{bmatrix} A_{(j)} \tag{2.16}$$

$$\mathbf{A}_{12}\left(\mathbf{x}_{(k)}\right) = \mathbf{A}_{21}^T\left(\mathbf{x}_{(k)}\right) = \sum_{j \in H_{(k)}} w_2\left(|\boldsymbol{\xi}_{(k)(j)}|\right) \begin{bmatrix} \xi_{1(k)(j)}^3 & \xi_{1(k)(j)}\xi_{2(k)(j)}^2 & \xi_{1(k)(j)}^2\xi_{2(k)(j)} \\ \xi_{1(k)(j)}^2\xi_{2(k)(j)} & \xi_{2(k)(j)}^3 & \xi_{1(k)(j)}\xi_{2(k)(j)}^2 \end{bmatrix} A_{(j)} \tag{2.17}$$

and

$$\mathbf{A}_{22}\left(\mathbf{x}_{(k)}\right) = \sum_{j \in H_{(k)}} w_2\left(|\boldsymbol{\xi}_{(k)(j)}|\right) \begin{bmatrix} \xi_{1(k)(j)}^4 & \xi_{1(k)(j)}^2\xi_{2(k)(j)}^2 & \xi_{1(k)(j)}^3\xi_{2(k)(j)} \\ \xi_{1(k)(j)}^2\xi_{2(k)(j)}^2 & \xi_{2(k)(j)}^4 & \xi_{1(k)(j)}\xi_{2(k)(j)}^3 \\ \xi_{1(k)(j)}^3\xi_{2(k)(j)} & \xi_{1(k)(j)}\xi_{2(k)(j)}^3 & \xi_{1(k)(j)}^2\xi_{2(k)(j)}^2 \end{bmatrix} A_{(j)} \tag{2.18}$$

The unknown vectors, $\mathbf{a}_{1(k)}^{p_1p_2}$ and $\mathbf{a}_{2(k)}^{p_1p_2}$ are defined as

$$\mathbf{a}_{1(k)}^{p_1p_2} = \begin{Bmatrix} a_{10}^{p_1p_2} \\ a_{01}^{p_1p_2} \end{Bmatrix}_{(k)} \tag{2.19}$$

and

$$\mathbf{a}_{2(k)}^{p_1p_2} = \begin{Bmatrix} a_{20}^{p_1p_2} \\ a_{02}^{p_1p_2} \\ a_{11}^{p_1p_2} \end{Bmatrix}_{(k)} \tag{2.20}$$

The known vectors, $\mathbf{b}_1^{p_1p_2}$ and $\mathbf{b}_2^{p_1p_2}$ are defined as

$$\mathbf{b}_1^{p_1p_2} = \begin{Bmatrix} p_1 \\ p_2 \end{Bmatrix} \text{ and } \mathbf{b}_2^{p_1p_2} = \mathbf{0} \text{ for } p_1 + p_2 = 1 \tag{2.21}$$

and

$$\mathbf{b}_1^{p_1p_2} = \mathbf{0} \text{ and } \mathbf{b}_2^{p_1p_2} = \begin{Bmatrix} p_1 - \delta_{p_1p_2} \\ p_2 - \delta_{p_1p_2} \\ p_1p_2 \end{Bmatrix} \text{ for } p_1 + p_2 = 2 \tag{2.22}$$

After determining the coefficients, $a_{q_1q_2}^{p_1p_2}$ via $\mathbf{a}_{(k)} = \mathbf{A}_{(k)}^{-1}\mathbf{b}$, the PD functions $g_2^{p_1p_2}\left(\mathbf{x}_{(k)}, \boldsymbol{\xi}_{(k)(j)}\right)$ can be constructed by using Eq. (2.12).

Similarly, the gradient, divergence, gradient of divergence, and the Laplacian of a displacement vector, $\mathbf{u}(\mathbf{x}_{(k)})$ can be expressed as

$$\nabla \mathbf{u}\left(\mathbf{x}_{(k)}\right) = \sum_{j \in H_{(k)}} \left(\mathbf{u}\left(\mathbf{x}_{(j)}\right) - \mathbf{u}\left(\mathbf{x}_{(k)}\right)\right) \otimes \mathbf{g}_{(k)(j)} A_{(j)}, \tag{2.23}$$

$$\nabla \cdot \mathbf{u}\left(\mathbf{x}_{(k)}\right) = \sum_{j \in H_{(k)}} \left(\mathbf{u}\left(\mathbf{x}_{(j)}\right) - \mathbf{u}\left(\mathbf{x}_{(k)}\right)\right) \cdot \mathbf{g}_{(k)(j)} A_{(j)}, \tag{2.24}$$

$$\nabla \nabla \cdot \mathbf{u}\left(\mathbf{x}_{(k)}\right) = \sum_{j \in H_{(k)}} \mathbf{G}_{(k)(j)} \left(\mathbf{u}\left(\mathbf{x}_{(j)}\right) - \mathbf{u}\left(\mathbf{x}_{(k)}\right)\right) A_{(j)} \tag{2.25}$$

and

$$\nabla^2 \mathbf{u}\left(\mathbf{x}_{(k)}\right) = \sum_{j \in H_{(k)}} \text{tr}\mathbf{G}_{(k)(j)} \left(\mathbf{u}\left(\mathbf{x}_{(j)}\right) - \mathbf{u}\left(\mathbf{x}_{(k)}\right)\right) A_{(j)} \tag{2.26}$$

The symbol $\otimes$ represents the dyadic product of two vectors. The area of each material point $A_{(j)}$ is infinitesimally small, for the limiting case of $A_{(j)} \to 0$, and the summation can be replaced with integration. These expressions can be rewritten as

$$\nabla \mathbf{u} = \int_{H_\mathbf{x}} (\mathbf{u}(\mathbf{x}') - \mathbf{u}(\mathbf{x})) \otimes \mathbf{g}(\mathbf{x}, \boldsymbol{\xi})\, dA_{\mathbf{x}'} \tag{2.27}$$

$$\nabla \cdot \mathbf{u} = \int_{H_\mathbf{x}} (\mathbf{u}(\mathbf{x}') - \mathbf{u}(\mathbf{x})) \cdot \mathbf{g}(\mathbf{x}, \boldsymbol{\xi})\, dA_{\mathbf{x}'} \tag{2.28}$$

$$\nabla\nabla\cdot\mathbf{u}=\int_{H_{\mathbf{x}}}\mathbf{G}(\mathbf{x},\boldsymbol{\xi})(\mathbf{u}(\mathbf{x}+\boldsymbol{\xi})-\mathbf{u}(\mathbf{x}))dA_{\mathbf{x}'} \tag{2.29}$$

and

$$\nabla^2\mathbf{u}=\int_{H_{\mathbf{x}}}\mathrm{tr}\mathbf{G}(\mathbf{x},\boldsymbol{\xi})(\mathbf{u}(\mathbf{x}+\boldsymbol{\xi})-\mathbf{u}(\mathbf{x}))dA_{\mathbf{x}'} \tag{2.30}$$

in which $H_{\mathbf{x}} = \{\mathbf{x}'; \mathbf{x}' \in H_{\mathbf{x}}\}$ is the horizon of point $\mathbf{x}$ and $\mathbf{x}' = \mathbf{x} + \boldsymbol{\xi}$. The area of point $\mathbf{x}'$ is represented by $dA_{\mathbf{x}'}$. The vector $\mathbf{g}(\mathbf{x}, \boldsymbol{\xi})$ and matrix $\mathbf{G}(\mathbf{x}, \boldsymbol{\xi})$ are defined as

$$\mathbf{g}(\mathbf{x},\boldsymbol{\xi})=\begin{Bmatrix} g_2^{10}(\mathbf{x},\boldsymbol{\xi}) \\ g_2^{01}(\mathbf{x},\boldsymbol{\xi}) \end{Bmatrix} \tag{2.31}$$

and

$$\mathbf{G}(\mathbf{x},\boldsymbol{\xi})=\begin{bmatrix} g_2^{20}(\mathbf{x},\boldsymbol{\xi}) & g_2^{11}(\mathbf{x},\boldsymbol{\xi}) \\ g_2^{11}(\mathbf{x},\boldsymbol{\xi}) & g_2^{02}(\mathbf{x},\boldsymbol{\xi}) \end{bmatrix} \tag{2.32}$$

For point $\mathbf{x}$ with a symmetric position in a circular interaction domain, the unknown coefficients of the matrix, $\mathbf{A}$ can be evaluated analytically. Also, the off-diagonal submatrices, $\mathbf{A}_{12}$ and $\mathbf{A}_{21}$ in matrix $\mathbf{A}$ vanish; thus, it decouples the first- and second-order derivatives in Eq. (2.14).

As derived by Madenci et al. (2022), the gradient, divergence, gradient of divergence, and the Laplacian of a displacement vector, $\mathbf{u}(\mathbf{x})$ at point $\mathbf{x}$ with a complete circular horizon can be evaluated as

$$\nabla\mathbf{u}(\mathbf{x})=\frac{2}{\pi\delta^4}\int_{H_{\mathbf{x}}}w_1(|\boldsymbol{\xi}|)(\mathbf{u}(\mathbf{x}+\boldsymbol{\xi})-\mathbf{u}(\mathbf{x}))\otimes\boldsymbol{\xi}dA_{\mathbf{x}'} \tag{2.33}$$

$$\nabla\cdot\mathbf{u}(\mathbf{x})=\frac{2}{\pi\delta^4}\int_{H_{\mathbf{x}}}w_1(|\boldsymbol{\xi}|)(\mathbf{u}(\mathbf{x}+\boldsymbol{\xi})-\mathbf{u}(\mathbf{x}))\cdot\boldsymbol{\xi}dA_{\mathbf{x}'} \tag{2.34}$$

$$\begin{aligned}\nabla\nabla\cdot\mathbf{u}(\mathbf{x})=&-\frac{3}{\pi\delta^6}\int_{H_{\mathbf{x}}}w_2(|\boldsymbol{\xi}|)(\boldsymbol{\xi}\cdot\boldsymbol{\xi})(\mathbf{u}(\mathbf{x}+\boldsymbol{\xi})-\mathbf{u}(\mathbf{x}))dA_{\mathbf{x}'}\\&+\frac{12}{\pi\delta^6}\int_{H_{\mathbf{x}}}w_2(|\boldsymbol{\xi}|)[\boldsymbol{\xi}\cdot(\mathbf{u}(\mathbf{x}+\boldsymbol{\xi})-\mathbf{u}(\mathbf{x}))]\boldsymbol{\xi}dA_{\mathbf{x}'}\end{aligned} \tag{2.35}$$

and

$$\nabla^2 \mathbf{u}(\mathbf{x}) = \frac{6}{\pi\delta^6} \int_{H_\mathbf{x}} w_2(|\boldsymbol{\xi}|)(\boldsymbol{\xi}\cdot\boldsymbol{\xi})(\mathbf{u}(\mathbf{x}+\boldsymbol{\xi}) - \mathbf{u}(\mathbf{x}))dA_{\mathbf{x}'} \tag{2.36}$$

Also, the PD forms of gradient and Laplacian of $f(\mathbf{x})$ can be evaluated as

$$\nabla f(\mathbf{x}) = \frac{2}{\pi\delta^4} \int_{H_\mathbf{x}} w_1(|\boldsymbol{\xi}|)(f(\mathbf{x}+\boldsymbol{\xi}) - f(\mathbf{x}))\boldsymbol{\xi} dA_{\mathbf{x}'} \tag{2.37}$$

and

$$\nabla^2 f(\mathbf{x}) = \frac{6}{\pi\delta^6} \int_{H_\mathbf{x}} w_2(|\boldsymbol{\xi}|)(\boldsymbol{\xi}\cdot\boldsymbol{\xi})(f(\mathbf{x}+\boldsymbol{\xi}) - f(\mathbf{x}))dA_{\mathbf{x}'} \tag{2.38}$$

2.2.2 Three-Dimensional Analysis

Similar to the steps in the preceding section, the TSE of a function, $f(\mathbf{x}_{(j)})$ can be expressed in a 3D space as

$$\begin{aligned} f(\mathbf{x}_{(j)}) - f(\mathbf{x}_{(k)}) &= \xi_{1(k)(j)} \frac{\partial f(\mathbf{x}_{(k)})}{\partial x_1} + \xi_{2(k)(j)} \frac{\partial f(\mathbf{x}_{(k)})}{\partial x_2} + \xi_{3(k)(j)} \frac{\partial f(\mathbf{x}_{(k)})}{\partial x_3} \\ &+ \frac{1}{2!}\xi_{1(k)(j)}^2 \frac{\partial^2 f(\mathbf{x}_{(k)})}{\partial x_1^2} + \frac{1}{2!}\xi_{2(k)(j)}^2 \frac{\partial^2 f(\mathbf{x}_{(k)})}{\partial x_2^2} + \frac{1}{2!}\xi_{3(k)(j)}^2 \frac{\partial^2 f(\mathbf{x}_{(k)})}{\partial x_3^2} \\ &+ \xi_{1(k)(j)}\xi_{2(k)(j)} \frac{\partial^2 f(\mathbf{x}_{(k)})}{\partial x_1 \partial x_2} + \xi_{1(k)(j)}\xi_{3(k)(j)} \frac{\partial^2 f(\mathbf{x}_{(k)})}{\partial x_1 \partial x_3} + \xi_{2(k)(j)}\xi_{3(k)(j)} \frac{\partial^2 f(\mathbf{x}_{(k)})}{\partial x_2 \partial x_3} + R_2(\mathbf{x}_{(k)}) \end{aligned} \tag{2.39}$$

where $\mathbf{x}_{(k)}^T = (x_{1(k)}, x_{2(k)}, x_{3(k)})$, $\xi_{1(k)(j)}$, $\xi_{2(k)(j)}$, and $\xi_{3(k)(j)}$ are the components of the relative position vector, $\boldsymbol{\xi}_{(k)(j)}$ and $R_2(\mathbf{x}_{(k)})$ represents the remainder of the second-order approximation. Multiplying each term with the PD functions, $g_2^{p_1p_2p_3}(\mathbf{x}_{(k)}, \boldsymbol{\xi}_{(k)(j)})$ while considering $R_2(\mathbf{x}_{(k)})$ to be negligible, and summing over the domain of interaction (family), $H_{(k)}$ result in

$$
\begin{aligned}
&\sum_{j\in H_{(k)}}\left(f\left(\mathbf{x}_{(j)}\right)-f\left(\mathbf{x}_{(k)}\right)\right)g_2^{p_1p_2p_3}\left(\mathbf{x}_{(k)},\boldsymbol{\xi}_{(k)(j)}\right)V_{(j)}= \\
&\quad\frac{\partial f\left(\mathbf{x}_{(k)}\right)}{\partial x_1}\sum_{j\in H_{(k)}}\xi_{1(k)(j)}g_2^{p_1p_2p_3}\left(\mathbf{x}_{(k)},\boldsymbol{\xi}_{(k)(j)}\right)V_{(j)} \\
&\quad+\frac{\partial f\left(\mathbf{x}_{(k)}\right)}{\partial x_2}\sum_{j\in H_{(k)}}\xi_{2(k)(j)}g_2^{p_1p_2p_3}\left(\mathbf{x}_{(k)},\boldsymbol{\xi}_{(k)(j)}\right)V_{(j)} \\
&\quad+\frac{\partial f\left(\mathbf{x}_{(k)}\right)}{\partial x_3}\sum_{j\in H_{(k)}}\xi_{3(k)(j)}g_2^{p_1p_2p_3}\left(\mathbf{x}_{(k)},\boldsymbol{\xi}_{(k)(j)}\right)V_{(j)} \\
&\quad+\frac{1}{2}\frac{\partial f^2\left(\mathbf{x}_{(k)}\right)}{\partial x_1^2}\sum_{j\in H_{(k)}}\xi_{1(k)(j)}^2g_2^{p_1p_2p_3}\left(\mathbf{x}_{(k)},\boldsymbol{\xi}_{(k)(j)}\right)V_{(j)} \\
&\quad+\frac{1}{2}\frac{\partial f^2\left(\mathbf{x}_{(k)}\right)}{\partial x_2^2}\sum_{j\in H_{(k)}}\xi_{2(k)(j)}^2g_2^{p_1p_2p_3}\left(\mathbf{x}_{(k)},\boldsymbol{\xi}_{(k)(j)}\right)V_{(j)} \\
&\quad+\frac{1}{2}\frac{\partial f^2\left(\mathbf{x}_{(k)}\right)}{\partial x_3^2}\sum_{j\in H_{(k)}}\xi_{3(k)(j)}^2g_2^{p_1p_2p_3}\left(\mathbf{x}_{(k)},\boldsymbol{\xi}_{(k)(j)}\right)V_{(j)} \\
&\quad+\frac{\partial f^2\left(\mathbf{x}_{(k)}\right)}{\partial x_1\partial x_2}\sum_{j\in H_{(k)}}\xi_{1(k)(j)}\xi_{2(k)(j)}g_2^{p_1p_2p_3}\left(\mathbf{x}_{(k)},\boldsymbol{\xi}_{(k)(j)}\right)V_{(j)} \\
&\quad+\frac{\partial f^2\left(\mathbf{x}_{(k)}\right)}{\partial x_1\partial x_3}\sum_{j\in H_{(k)}}\xi_{1(k)(j)}\xi_{3(k)(j)}g_2^{p_1p_2p_3}\left(\mathbf{x}_{(k)},\boldsymbol{\xi}_{(k)(j)}\right)V_{(j)} \\
&\quad+\frac{\partial f^2\left(\mathbf{x}_{(k)}\right)}{\partial x_2\partial x_3}\sum_{j\in H_{(k)}}\xi_{2(k)(j)}\xi_{3(k)(j)}g_2^{p_1p_2p_3}\left(\mathbf{x}_{(k)},\boldsymbol{\xi}_{(k)(j)}\right)V_{(j)}
\end{aligned}
\tag{2.40}
$$

The PD functions must be orthogonal to each term in the TSE as

$$
\frac{1}{n_1!n_2!n_3!}\sum_{j\in H_{(k)}}\xi_{1(k)(j)}^{n_1}\xi_{2(k)(j)}^{n_2}\xi_{3(k)(j)}^{n_3}g_2^{p_1p_2p_3}\left(\mathbf{x}_{(k)},\boldsymbol{\xi}_{(k)(j)}\right)V_{(j)}=\delta_{n_1p_1}\delta_{n_2p_2}\delta_{n_3p_3}
\tag{2.41}
$$

with $(n_1, n_2, n_3, p_1, p_2, p_3 = 0, 1, 2)$ and $\delta_{n_ip_i}$ as the Kronecker delta symbol. The PD functions, $g_N^{p_1p_2p_3}(\mathbf{x},\boldsymbol{\xi})$ with $p_1 + p_2 + p_3 \leq N = 2$ are explained by Madenci et al. (2019). The superscript p_i denotes the order of differentiation with respect to the variable x_i for $i = 1, 2, 3$.

Applying the orthogonality conditions, Eq. (2.41) results in the PD form of the derivatives as

$$\left\{\begin{array}{l} f_{,1}\left(\mathbf{x}_{(k)}\right) \\ f_{,2}\left(\mathbf{x}_{(k)}\right) \\ f_{,3}\left(\mathbf{x}_{(k)}\right) \end{array}\right\} = \sum_{j\in H_{(k)}} \left(f\left(\mathbf{x}_{(j)}\right) - f\left(\mathbf{x}_{(k)}\right)\right) \left\{\begin{array}{l} g_2^{100}\left(\mathbf{x}_{(k)}, \boldsymbol{\xi}_{(k)(j)}\right) \\ g_2^{010}\left(\mathbf{x}_{(k)}, \boldsymbol{\xi}_{(k)(j)}\right) \\ g_2^{001}\left(\mathbf{x}_{(k)}, \boldsymbol{\xi}_{(k)(j)}\right) \end{array}\right\} V_{(j)} \tag{2.42}$$

and

$$\left\{\begin{array}{l} f_{,11}\left(\mathbf{x}_{(k)}\right) \\ f_{,22}\left(\mathbf{x}_{(k)}\right) \\ f_{,33}\left(\mathbf{x}_{(k)}\right) \\ f_{,12}\left(\mathbf{x}_{(k)}\right) \\ f_{,13}\left(\mathbf{x}_{(k)}\right) \\ f_{,23}\left(\mathbf{x}_{(k)}\right) \end{array}\right\} = \sum_{j\in H_{(k)}} \left(f\left(\mathbf{x}_{(j)}\right) - f\left(\mathbf{x}_{(k)}\right)\right) \left\{\begin{array}{l} g_2^{200}\left(\mathbf{x}_{(k)}, \boldsymbol{\xi}_{(k)(j)}\right) \\ g_2^{020}\left(\mathbf{x}_{(k)}, \boldsymbol{\xi}_{(k)(j)}\right) \\ g_2^{002}\left(\mathbf{x}_{(k)}, \boldsymbol{\xi}_{(k)(j)}\right) \\ g_2^{110}\left(\mathbf{x}_{(k)}, \boldsymbol{\xi}_{(k)(j)}\right) \\ g_2^{101}\left(\mathbf{x}_{(k)}, \boldsymbol{\xi}_{(k)(j)}\right) \\ g_2^{011}\left(\mathbf{x}_{(k)}, \boldsymbol{\xi}_{(k)(j)}\right) \end{array}\right\} V_{(j)} \tag{2.43}$$

Thus, the gradient and Laplacian of $f(\mathbf{x})$ can be expressed as

$$\nabla f\left(\mathbf{x}_{(k)}\right) = \sum_{j\in H_{(k)}} \left(f\left(\mathbf{x}_{(j)}\right) - f\left(\mathbf{x}_{(k)}\right)\right) \mathbf{g}_{(k)(j)} V_{(j)} \tag{2.44}$$

and

$$\nabla^2 f\left(\mathbf{x}_{(k)}\right) = \sum_{j\in H_{(k)}} \mathrm{tr}\mathbf{G}_{(k)(j)} \left(f\left(\mathbf{x}_{(j)}\right) - f\left(\mathbf{x}_{(k)}\right)\right) V_{(j)} \tag{2.45}$$

where the vector $\mathbf{g}$ and matrix $\mathbf{G}$ are defined as

$$\mathbf{g}_{(k)(j)} = \left\{\begin{array}{l} g_2^{100}\left(\mathbf{x}_{(k)}, \boldsymbol{\xi}_{(k)(j)}\right) \\ g_2^{010}\left(\mathbf{x}_{(k)}, \boldsymbol{\xi}_{(k)(j)}\right) \\ g_2^{001}\left(\mathbf{x}_{(k)}, \boldsymbol{\xi}_{(k)(j)}\right) \end{array}\right\} \tag{2.46}$$

and

$$\mathbf{G}_{(k)(j)}=\begin{bmatrix} g_2^{200}(\mathbf{x}_{(k)},\boldsymbol{\xi}_{(k)(j)}) & g_2^{110}(\mathbf{x}_{(k)},\boldsymbol{\xi}_{(k)(j)}) & g_2^{101}(\mathbf{x}_{(k)},\boldsymbol{\xi}_{(k)(j)}) \\ g_2^{110}(\mathbf{x}_{(k)},\boldsymbol{\xi}_{(k)(j)}) & g_2^{020}(\mathbf{x}_{(k)},\boldsymbol{\xi}_{(k)(j)}) & g_2^{011}(\mathbf{x}_{(k)},\boldsymbol{\xi}_{(k)(j)}) \\ g_2^{101}(\mathbf{x}_{(k)},\boldsymbol{\xi}_{(k)(j)}) & g_2^{011}(\mathbf{x}_{(k)},\boldsymbol{\xi}_{(k)(j)}) & g_2^{002}(\mathbf{x}_{(k)},\boldsymbol{\xi}_{(k)(j)}) \end{bmatrix} \tag{2.47}$$

The PD functions can be constructed as a linear combination of polynomial basis functions as

$$\begin{aligned} &g_2^{p_1p_2p_3}\left(\mathbf{x}_{(k)},\boldsymbol{\xi}_{(k)(j)}\right)= \\ &\quad a_{100}^{p_1p_2p_3}\,w_{100}\left(\left|\boldsymbol{\xi}_{(k)(j)}\right|\right)\xi_{1(k)(j)}+a_{010}^{p_1p_2p_3}\,w_{010}\left(\left|\boldsymbol{\xi}_{(k)(j)}\right|\right)\xi_{2(k)(j)} \\ &\quad +a_{001}^{p_1p_2p_3}\,w_{001}\left(\left|\boldsymbol{\xi}_{(k)(j)}\right|\right)\xi_{3(k)(j)}+a_{200}^{p_1p_2p_3}\,w_{200}\left(\left|\boldsymbol{\xi}_{(k)(j)}\right|\right)\xi_{1(k)(j)}^2 \\ &\quad +a_{020}^{p_1p_2p_3}\,w_{020}\left(\left|\boldsymbol{\xi}_{(k)(j)}\right|\right)\xi_{2(k)(j)}^2+a_{002}^{p_1p_2p_3}\,w_{002}\left(\left|\boldsymbol{\xi}_{(k)(j)}\right|\right)\xi_{3(k)(j)}^2 \\ &\quad +a_{110}^{p_1p_2p_3}\,w_{110}\left(\left|\boldsymbol{\xi}_{(k)(j)}\right|\right)\xi_{1(k)(j)}\xi_{2(k)(j)} \\ &\quad +a_{101}^{p_1p_2p_3}\,w_{101}\left(\left|\boldsymbol{\xi}_{(k)(j)}\right|\right)\xi_{1(k)(j)}\xi_{3(k)(j)} \\ &\quad +a_{011}^{p_1p_2p_3}\,w_{011}\left(\left|\boldsymbol{\xi}_{(k)(j)}\right|\right)\xi_{2(k)(j)}\xi_{3(k)(j)} \end{aligned} \tag{2.48}$$

where $a_{q_1q_2q_3}^{p_1p_2p_3}$ are the unknown coefficients and $w_{q_1q_2q_3}\left(\left|\boldsymbol{\xi}_{(k)(j)}\right|\right)$ are the specified weight functions. Substituting from Eq. (2.48) into Eq. (2.41) yields

$$\begin{aligned} &\frac{1}{n_1!n_2!n_3!}\sum_{j\in H_{(k)}}\left(\begin{array}{l} a_{100}^{p_1p_2p_3}\,w_{100}\xi_{1(k)(j)}^{n_1+1}\xi_{2(k)(j)}^{n_2}\xi_{3(k)(j)}^{n_3}+a_{010}^{p_1p_2p_3}\,w_{010}\xi_{1(k)(j)}^{n_1}\xi_{2(k)(j)}^{n_2+1}\xi_{3(k)(j)}^{n_3} \\ +a_{001}^{p_1p_2p_3}\,w_{001}\xi_{1(k)(j)}^{n_1}\xi_{2(k)(j)}^{n_2}\xi_{3(k)(j)}^{n_3+1}+a_{200}^{p_1p_2p_3}\,w_{200}\xi_{1(k)(j)}^{n_1+2}\xi_{2(k)(j)}^{n_2}\xi_{3(k)(j)}^{n_3} \\ +a_{020}^{p_1p_2p_3}\,w_{020}\xi_{1(k)(j)}^{n_1}\xi_{2(k)(j)}^{n_2+2}\xi_{3(k)(j)}^{n_3}+a_{002}^{p_1p_2p_3}\,w_{002}\xi_{1(k)(j)}^{n_1}\xi_{2(k)(j)}^{n_2}\xi_{3(k)(j)}^{n_3+2} \\ +a_{110}^{p_1p_2p_3}\,w_{110}\xi_{1(k)(j)}^{n_1+1}\xi_{2(k)(j)}^{n_2+1}\xi_{3(k)(j)}^{n_3}+a_{101}^{p_1p_2p_3}\,w_{101}\xi_{1(k)(j)}^{n_1+1}\xi_{2(k)(j)}^{n_2}\xi_{3(k)(j)}^{n_3+1} \\ +a_{011}^{p_1p_2p_3}\,w_{011}\xi_{1(k)(j)}^{n_1}\xi_{2(k)(j)}^{n_2+1}\xi_{3(k)(j)}^{n_3+1} \end{array}\right)V_{(j)} \\ &=\delta_{n_1p_1}\delta_{n_2p_2}\delta_{n_3p_3} \end{aligned} \tag{2.49}$$

With a particular choice of weight functions from Eq. (2.2), this equation can be recast as a system of algebraic equations for the determination of the coefficients $a_{q_1q_2q_3}^{p_1p_2p_3}$ in the form

$$\mathbf{A}\left(\mathbf{x}_{(k)}\right)\mathbf{a}_{(k)}^{p_1p_2p_3}=\mathbf{b}^{p_1p_2p_3} \tag{2.50}$$

with

$$\mathbf{A}\left(\mathbf{x}_{(k)}\right)=\begin{bmatrix}\mathbf{A}_{11}\left(\mathbf{x}_{(k)}\right) & \mathbf{A}_{12}\left(\mathbf{x}_{(k)}\right)\\ \mathbf{A}_{21}\left(\mathbf{x}_{(k)}\right) & \mathbf{A}_{22}\left(\mathbf{x}_{(k)}\right)\end{bmatrix};\ \mathbf{a}_{(k)}^{p_1p_2p_3}=\begin{Bmatrix}\mathbf{a}_{1(k)}^{p_1p_2p_3}\\ \mathbf{a}_{2(k)}^{p_1p_2p_3}\end{Bmatrix};\ \mathbf{b}^{p_1p_2p_3}=\begin{Bmatrix}\mathbf{b}_1^{p_1p_2p_3}\\ \mathbf{b}_2^{p_1p_2p_3}\end{Bmatrix} \tag{2.51}$$

The known submatrices can be expressed as

$$\mathbf{A}_{11}\left(\mathbf{x}_{(k)}\right)=\sum_{j\in H_{(k)}} w_1\left(|\boldsymbol{\xi}_{(k)(j)}|\right)\left(\boldsymbol{\xi}_{(k)(j)}\otimes\boldsymbol{\xi}_{(k)(j)}\right)V_{(j)} \tag{2.52}$$

$$\mathbf{A}_{12}\left(\mathbf{x}_{(k)}\right)=\mathbf{A}_{21}^T\left(\mathbf{x}_{(k)}\right)=\sum_{j\in H_{(k)}} w_2\left(|\boldsymbol{\xi}_{(k)(j)}|\right)\left[\left(\boldsymbol{\xi}_{(k)(j)}\otimes\widehat{\boldsymbol{\xi}}_{(k)(j)}\right)\ \left(\boldsymbol{\xi}_{(k)(j)}\otimes\widetilde{\boldsymbol{\xi}}_{(k)(j)}\right)\right]V_{(j)} \tag{2.53}$$

and

$$\mathbf{A}_{22}\left(\mathbf{x}_{(k)}\right)=\sum_{j\in H_{(k)}} w_2\left(|\boldsymbol{\xi}_{(k)(j)}|\right)\begin{bmatrix}\left(\widehat{\boldsymbol{\xi}}_{(k)(j)}\otimes\widehat{\boldsymbol{\xi}}_{(k)(j)}\right) & \left(\widehat{\boldsymbol{\xi}}_{(k)(j)}\otimes\widetilde{\boldsymbol{\xi}}_{(k)(j)}\right)\\ \left(\widetilde{\boldsymbol{\xi}}_{(k)(j)}\otimes\widehat{\boldsymbol{\xi}}_{(k)(j)}\right) & \left(\widetilde{\boldsymbol{\xi}}_{(k)(j)}\otimes\widetilde{\boldsymbol{\xi}}_{(k)(j)}\right)\end{bmatrix}V_{(j)} \tag{2.54}$$

in which the vectors $\boldsymbol{\xi}$, $\widehat{\boldsymbol{\xi}}$, and $\widetilde{\boldsymbol{\xi}}$ are defined as

$$\boldsymbol{\xi}_{(k)(j)}=\begin{Bmatrix}\xi_{1(k)(j)}\\ \xi_{2(k)(j)}\\ \xi_{3(k)(j)}\end{Bmatrix},\ \widehat{\boldsymbol{\xi}}_{(k)(j)}=\begin{Bmatrix}\xi_{1(k)(j)}^2\\ \xi_{2(k)(j)}^2\\ \xi_{3(k)(j)}^2\end{Bmatrix}\ \text{and}\ \widetilde{\boldsymbol{\xi}}_{(k)(j)}=\begin{Bmatrix}\xi_{1(k)(j)}\xi_{2(k)(j)}\\ \xi_{1(k)(j)}\xi_{3(k)(j)}\\ \xi_{2(k)(j)}\xi_{3(k)(j)}\end{Bmatrix} \tag{2.55}$$

The unknown vectors, $\mathbf{a}_{1(k)}^{p_1p_2p_3}$ and $\mathbf{a}_{2(k)}^{p_1p_2p_3}$ are defined as

$$\mathbf{a}_{1(k)}^{p_1p_2p_3}=\begin{Bmatrix}a_{100}^{p_1p_2p_3}\\ a_{010}^{p_1p_2p_3}\\ a_{001}^{p_1p_2p_3}\end{Bmatrix}_{(k)} \tag{2.56}$$

and

$$\mathbf{a}_{2(k)}^{p_1p_2p_3} = \begin{Bmatrix} a_{200}^{p_1p_2p_3} \\ a_{020}^{p_1p_2p_3} \\ a_{002}^{p_1p_2p_3} \\ a_{110}^{p_1p_2p_3} \\ a_{101}^{p_1p_2p_3} \\ a_{011}^{p_1p_2p_3} \end{Bmatrix}_{(k)} \tag{2.57}$$

The known vectors, $\mathbf{b}_1^{p_1p_2p_3}$ and $\mathbf{b}_2^{p_1p_2p_3}$ are defined as

$$\mathbf{b}_1^{p_1p_2p_3} = \begin{Bmatrix} p_1 \\ p_2 \\ p_3 \end{Bmatrix} \text{ and } \mathbf{b}_2^{p_1p_2p_3} = \mathbf{0} \quad \text{for } p_1 + p_2 + p_3 = 1 \tag{2.58}$$

and

$$\mathbf{b}_1^{p_1p_2p_3} = \mathbf{0} \text{ and } \mathbf{b}_2^{p_1p_2p_3} = \begin{Bmatrix} p_1 - \frac{(p_1+p_2)}{2}\delta_{p_1p_2} - \frac{(p_1+p_3)}{2}\delta_{p_1p_3} \\ p_2 - \frac{(p_1+p_2)}{2}\delta_{p_1p_2} - \frac{(p_2+p_3)}{2}\delta_{p_2p_3} \\ p_3 - \frac{(p_1+p_3)}{2}\delta_{p_1p_3} - \frac{(p_2+p_3)}{2}\delta_{p_2p_3} \\ p_1p_2 \\ p_1p_3 \\ p_2p_3 \end{Bmatrix} \text{ for } p_1 + p_2 + p_3 = 2 \tag{2.59}$$

The PD functions $g_2^{p_1p_2p_3}(\mathbf{x}, \boldsymbol{\xi})$ are readily constructed after the determination of the coefficients $a_{q_1q_2q_3}^{p_1p_2p_3}$. After determining the coefficients, $a_{q_1q_2}^{p_1p_2}$ via $\mathbf{a}_{(k)} = \mathbf{A}_{(k)}^{-1}\mathbf{b}$, the PD functions $g_2^{p_1p_2}\left(\mathbf{x}_{(k)}, \boldsymbol{\xi}_{(k)(j)}\right)$ can be constructed by using Eq. (2.48).

The gradient, divergence, gradient of divergence, and the Laplacian of a displacement vector, $\mathbf{u}(\mathbf{x}_{(k)})$ can be expressed as

$$\nabla \mathbf{u}\left(\mathbf{x}_{(k)}\right) = \sum_{j \in H_{(k)}} \left(\mathbf{u}\left(\mathbf{x}_{(j)}\right) - \mathbf{u}\left(\mathbf{x}_{(k)}\right)\right) \otimes \mathbf{g}_{(k)(j)} V_{(j)} \tag{2.60}$$

$$\nabla \cdot \mathbf{u}\left(\mathbf{x}_{(k)}\right) = \sum_{j \in H_{(k)}} \left(\mathbf{u}\left(\mathbf{x}_{(j)}\right) - \mathbf{u}\left(\mathbf{x}_{(k)}\right)\right) \cdot \mathbf{g}_{(k)(j)} V_{(j)} \tag{2.61}$$

$$\nabla \nabla \cdot \mathbf{u}\left(\mathbf{x}_{(k)}\right) = \sum_{j \in H_{(k)}} \mathbf{G}_{(k)(j)} \left(\mathbf{u}\left(\mathbf{x}_{(j)}\right) - \mathbf{u}\left(\mathbf{x}_{(k)}\right)\right) V_{(j)} \tag{2.62}$$

and

$$\nabla^2 \mathbf{u}\left(\mathbf{x}_{(k)}\right) = \sum_{j \in H_{(k)}} \mathrm{tr}\mathbf{G}_{(k)(j)}\left(\mathbf{u}\left(\mathbf{x}_{(j)}\right) - \mathbf{u}\left(\mathbf{x}_{(k)}\right)\right) V_{(j)} \tag{2.63}$$

The symbol $\otimes$ represents the dyadic product of two vectors. The volume of each point $V_{(j)}$ is infinitesimally small, for the limiting case of $V_{(j)} \to 0$, and the summation can be replaced with integration. These expressions can be rewritten as

$$\nabla \mathbf{u}(\mathbf{x}) = \int_{H_\mathbf{x}} (\mathbf{u}(\mathbf{x}') - \mathbf{u}(\mathbf{x})) \otimes \mathbf{g}(\mathbf{x}, \boldsymbol{\xi})\, dV_{\mathbf{x}'} \tag{2.64}$$

$$\nabla \cdot \mathbf{u}(\mathbf{x}) = \int_{H_\mathbf{x}} (\mathbf{u}(\mathbf{x}') - \mathbf{u}(\mathbf{x})) \cdot \mathbf{g}(\mathbf{x}, \boldsymbol{\xi}) dV_{\mathbf{x}'} \tag{2.65}$$

$$\nabla \nabla \cdot \mathbf{u}(\mathbf{x}) = \int_{H_\mathbf{x}} \mathbf{G}(\mathbf{x}, \boldsymbol{\xi})(\mathbf{u}(\mathbf{x} + \boldsymbol{\xi}) - \mathbf{u}(\mathbf{x})) dV_{\mathbf{x}'} \tag{2.66}$$

and

$$\nabla^2 \mathbf{u}(\mathbf{x}) = \int_{H_\mathbf{x}} \mathrm{tr}\mathbf{G}(\mathbf{x}, \boldsymbol{\xi})(\mathbf{u}(\mathbf{x} + \boldsymbol{\xi}) - \mathbf{u}(\mathbf{x})) dV_{\mathbf{x}'} \tag{2.67}$$

in which $H_\mathbf{x} = \{\mathbf{x}'; \mathbf{x}' \in H_\mathbf{x}\}$ is the horizon of point $\mathbf{x}$ and $\mathbf{x}' = \mathbf{x} + \boldsymbol{\xi}$. The volume of point $\mathbf{x}'$ is represented by $dV_{\mathbf{x}'}$. The vector $\mathbf{g}(\mathbf{x}, \boldsymbol{\xi})$ and matrix $\mathbf{G}(\mathbf{x}, \boldsymbol{\xi})$ are defined as

$$\mathbf{g}(\mathbf{x}, \boldsymbol{\xi}) = \begin{Bmatrix} g_2^{100}(\mathbf{x}, \boldsymbol{\xi}) \\ g_2^{010}(\mathbf{x}, \boldsymbol{\xi}) \\ g_2^{001}(\mathbf{x}, \boldsymbol{\xi}) \end{Bmatrix} \tag{2.68}$$

and

$$\mathbf{G}(\mathbf{x}, \boldsymbol{\xi}) = \begin{bmatrix} g_2^{200}(\mathbf{x}, \boldsymbol{\xi}) & g_2^{110}(\mathbf{x}, \boldsymbol{\xi}) & g_2^{101}(\mathbf{x}, \boldsymbol{\xi}) \\ g_2^{110}(\mathbf{x}, \boldsymbol{\xi}) & g_2^{020}(\mathbf{x}, \boldsymbol{\xi}) & g_2^{011}(\mathbf{x}, \boldsymbol{\xi}) \\ g_2^{101}(\mathbf{x}, \boldsymbol{\xi}) & g_2^{011}(\mathbf{x}, \boldsymbol{\xi}) & g_2^{002}(\mathbf{x}, \boldsymbol{\xi}) \end{bmatrix}. \tag{2.69}$$

As derived by Madenci et al. (2022), the gradient, divergence, gradient of divergence, and the Laplacian of a displacement vector, $\mathbf{u}(\mathbf{x})$ at point $\mathbf{x}$ with a complete spherical horizon can be expressed as

$$\nabla \mathbf{u}(\mathbf{x}) = \frac{9}{4\pi\delta^5} \int_{H_{\mathbf{x}}} w_1(|\boldsymbol{\xi}|)(\mathbf{u}(\mathbf{x}+\boldsymbol{\xi}) - \mathbf{u}(\mathbf{x})) \otimes \boldsymbol{\xi} dV_{\mathbf{x}'} \tag{2.70}$$

$$\nabla \cdot \mathbf{u}(\mathbf{x}) = \frac{9}{4\pi\delta^5} \int_{H_{\mathbf{x}}} w_1(|\boldsymbol{\xi}|)(\mathbf{u}(\mathbf{x}+\boldsymbol{\xi}) - \mathbf{u}(\mathbf{x})) \cdot \boldsymbol{\xi} dV_{\mathbf{x}'} \tag{2.71}$$

$$\begin{aligned}\nabla\nabla \cdot \mathbf{u}(\mathbf{x}) = & -\frac{3}{\pi\delta^7} \int_{H_{\mathbf{x}}} w_2(|\boldsymbol{\xi}|)(\boldsymbol{\xi}\cdot\boldsymbol{\xi})(\mathbf{u}(\mathbf{x}+\boldsymbol{\xi}) - \mathbf{u}(\mathbf{x})) dV_{\mathbf{x}'} \\ & + \frac{15}{\pi\delta^7} \int_{H_{\mathbf{x}}} w_2(|\boldsymbol{\xi}|)[\boldsymbol{\xi}\cdot(\mathbf{u}(\mathbf{x}+\boldsymbol{\xi}) - \mathbf{u}(\mathbf{x}))]\boldsymbol{\xi} dV_{\mathbf{x}'}\end{aligned} \tag{2.72}$$

and

$$\nabla^2 \mathbf{u}(\mathbf{x}) = \frac{6}{\pi\delta^7} \int_{H_{\mathbf{x}}} w_2(|\boldsymbol{\xi}|)(\boldsymbol{\xi}\cdot\boldsymbol{\xi})(\mathbf{u}(\mathbf{x}+\boldsymbol{\xi}) - \mathbf{u}(\mathbf{x})) dV_{\mathbf{x}'} \tag{2.73}$$

Also, the PD forms of gradient and Laplacian of $f(\mathbf{x})$ can be expressed as

$$\nabla f(\mathbf{x}) = \frac{9}{4\pi\delta^5} \int_{H_{\mathbf{x}}} w_1(|\boldsymbol{\xi}|)(f(\mathbf{x}+\boldsymbol{\xi}) - f(\mathbf{x}))\boldsymbol{\xi} dV_{\mathbf{x}'} \tag{2.74}$$

and

$$\nabla^2 f(\mathbf{x}) = \frac{6}{\pi\delta^7} \int_{H_{\mathbf{x}}} w_2(|\boldsymbol{\xi}|)(\boldsymbol{\xi}\cdot\boldsymbol{\xi})(f(\mathbf{x}+\boldsymbol{\xi}) - f(\mathbf{x})) dV_{\mathbf{x}'} \tag{2.75}$$

2.3 Nonlocal Form of Bond Direction

2.3.1 Two-Dimensional Analysis

For first- and second-order derivatives, the PD functions at point $\mathbf{x}$ with a symmetric and complete circular interaction domain can be expressed as (Madenci et al., 2022)

$$\mathbf{g}(\mathbf{x}, \boldsymbol{\xi}) = \begin{Bmatrix} g_2^{10}(\mathbf{x}, \boldsymbol{\xi}) \\ g_2^{01}(\mathbf{x}, \boldsymbol{\xi}) \end{Bmatrix} = \frac{2w_1(|\boldsymbol{\xi}|)}{\pi\delta^4} \begin{Bmatrix} \xi_1 \\ \xi_2 \end{Bmatrix} \tag{2.76}$$

and

$$\begin{Bmatrix} g_2^{20}(\mathbf{x},\boldsymbol{\xi}) \\ g_2^{02}(\mathbf{x},\boldsymbol{\xi}) \\ g_2^{11}(\mathbf{x},\boldsymbol{\xi}) \end{Bmatrix} = \frac{3w_2(|\boldsymbol{\xi}|)}{\pi\delta^6} \begin{Bmatrix} 3\xi_1^2 - \xi_2^2 \\ -\xi_1^2 + 3\xi_2^2 \\ 4\xi_1\xi_2 \end{Bmatrix} \tag{2.77}$$

It is observed that $g_2^{20}(\mathbf{x},\boldsymbol{\xi}) + g_2^{02}(\mathbf{x},\boldsymbol{\xi})$ can be evaluated as

$$g_2^{20}(\mathbf{x},\boldsymbol{\xi}) + g_2^{02}(\mathbf{x},\boldsymbol{\xi}) = \frac{6w_2(|\boldsymbol{\xi}|)}{\pi\delta^6}(\boldsymbol{\xi}\cdot\boldsymbol{\xi}) \tag{2.78}$$

After substituting for the weight function $w_1(|\boldsymbol{\xi}|) = \delta^2/|\boldsymbol{\xi}|^2$, the vector $\mathbf{g}(\mathbf{x},\boldsymbol{\xi})$ in Eq. (2.76) can be expressed as

$$\mathbf{g}(\mathbf{x},\boldsymbol{\xi}) = \frac{2}{\pi\delta^2|\boldsymbol{\xi}|^2}\begin{Bmatrix} \xi_1 \\ \xi_2 \end{Bmatrix} = \frac{2}{\pi\delta^2\,|\,\boldsymbol{\xi}\,|}\begin{Bmatrix} \dfrac{\xi_1}{|\,\boldsymbol{\xi}\,|} \\ \dfrac{\xi_2}{|\,\boldsymbol{\xi}\,|} \end{Bmatrix} = \frac{2}{\pi\delta^2\,|\,\boldsymbol{\xi}\,|}\mathbf{n}(\mathbf{x},\boldsymbol{\xi}) \tag{2.79}$$

Thus, the nonlocal form of bond direction $\mathbf{n}(\mathbf{x},\boldsymbol{\xi})$ can be expressed as

$$\mathbf{n}(\mathbf{x},\boldsymbol{\xi}) = \frac{\boldsymbol{\xi}}{|\,\boldsymbol{\xi}\,|} = \frac{\pi\delta^2}{2}\mathbf{g}(\mathbf{x},\boldsymbol{\xi})\,|\,\boldsymbol{\xi}\,| \tag{2.80}$$

Similarly, after substituting for the weight function $w_2(|\boldsymbol{\xi}|) = \delta^3/|\boldsymbol{\xi}|^3$, the second-order derivatives in Eq. (2.77) can be expressed as

$$\begin{Bmatrix} g_2^{20}(\mathbf{x},\boldsymbol{\xi}) \\ g_2^{02}(\mathbf{x},\boldsymbol{\xi}) \\ g_2^{11}(\mathbf{x},\boldsymbol{\xi}) \end{Bmatrix} = \frac{3}{\pi\delta^3}\frac{1}{|\,\boldsymbol{\xi}\,|}\begin{Bmatrix} 3n_1^2 - n_2^2 \\ -n_1^2 + 3n_2^2 \\ 4n_1n_2 \end{Bmatrix} = \frac{3}{\pi\delta^3}\frac{1}{|\,\boldsymbol{\xi}\,|}\begin{Bmatrix} 4n_1^2 - \mathbf{n}(\mathbf{x},\boldsymbol{\xi})\cdot\mathbf{n}(\mathbf{x},\boldsymbol{\xi}) \\ 4n_2^2 - \mathbf{n}(\mathbf{x},\boldsymbol{\xi})\cdot\mathbf{n}(\mathbf{x},\boldsymbol{\xi}) \\ 4n_1n_2 \end{Bmatrix} \tag{2.81}$$

This expression leads to the determination of the components of $\mathbf{n}(\mathbf{x},\boldsymbol{\xi})\otimes\mathbf{n}(\mathbf{x},\boldsymbol{\xi})$ as

$$\begin{Bmatrix} n_1n_1 \\ n_2n_2 \\ n_1n_2 \end{Bmatrix} = \frac{\pi\delta^3\,|\,\boldsymbol{\xi}\,|}{12}\begin{Bmatrix} g_2^{20}(\mathbf{x},\boldsymbol{\xi}) + \dfrac{3}{\pi\delta^3}\dfrac{1}{|\,\boldsymbol{\xi}\,|}\mathbf{n}(\mathbf{x},\boldsymbol{\xi})\cdot\mathbf{n}(\mathbf{x},\boldsymbol{\xi}) \\ g_2^{02}(\mathbf{x},\boldsymbol{\xi}) + \dfrac{3}{\pi\delta^3}\dfrac{1}{|\,\boldsymbol{\xi}\,|}\mathbf{n}(\mathbf{x},\boldsymbol{\xi})\cdot\mathbf{n}(\mathbf{x},\boldsymbol{\xi}) \\ g_2^{11}(\mathbf{x},\boldsymbol{\xi}) \end{Bmatrix} \tag{2.82}$$

Invoking the weight function $w_2(|\boldsymbol{\xi}|) = \delta^3/|\boldsymbol{\xi}|^3$ into Eq. (2.78) results in

$$\mathbf{n}(\mathbf{x},\boldsymbol{\xi}) \cdot \mathbf{n}(\mathbf{x},\boldsymbol{\xi}) = \frac{\pi\delta^3}{6} \mid \boldsymbol{\xi} \mid \left(g_2^{20}(\mathbf{x},\boldsymbol{\xi}) + g_2^{02}(\mathbf{x},\boldsymbol{\xi})\right) \tag{2.83}$$

Substituting from Eq. (2.83) into (2.82) along with Eq. (2.32) leads to an explicit form of $\mathbf{n}(\mathbf{x},\boldsymbol{\xi}) \otimes \mathbf{n}(\mathbf{x},\boldsymbol{\xi})$ as

$$\mathbf{n}(\mathbf{x},\boldsymbol{\xi}) \otimes \mathbf{n}(\mathbf{x},\boldsymbol{\xi}) = \frac{\pi\delta^3 \mid \boldsymbol{\xi} \mid}{12} \left(\mathbf{G}(\mathbf{x},\boldsymbol{\xi}) + \frac{1}{2}\mathrm{tr}\mathbf{G}(\mathbf{x},\boldsymbol{\xi})\mathbf{I}\right) \tag{2.84}$$

Hence, the nonlocal form of the dyadic product of bond direction vectors $(\mathbf{n}(\mathbf{x},\boldsymbol{\xi}) \otimes \mathbf{n}(\mathbf{x},\boldsymbol{\xi}))$ and $(\mathbf{n}(\mathbf{x},\boldsymbol{\xi}) \cdot \mathbf{n}(\mathbf{x},\boldsymbol{\xi}))$ can be expressed as

$$\mathbf{n}(\mathbf{x},\boldsymbol{\xi}) \otimes \mathbf{n}(\mathbf{x},\boldsymbol{\xi}) = \frac{\pi\delta^3}{24}(2\mathbf{G}(\mathbf{x},\boldsymbol{\xi}) + \mathrm{tr}\mathbf{G}(\mathbf{x},\boldsymbol{\xi})\mathbf{I}) \mid \boldsymbol{\xi} \mid \tag{2.85}$$

and

$$\mathbf{n}(\mathbf{x},\boldsymbol{\xi}) \cdot \mathbf{n}(\mathbf{x},\boldsymbol{\xi}) = \frac{\pi\delta^3 \mid \boldsymbol{\xi} \mid}{6} \mathrm{tr}\mathbf{G}(\mathbf{x},\boldsymbol{\xi}) \tag{2.86}$$

2.3.2 *Three-Dimensional Analysis*

Similar to two-dimensional analysis, the PD functions at point, $\mathbf{x}$ with a symmetric and complete spherical interaction domain can be expressed as (Madenci et al., 2022)

$$\mathbf{g}(\mathbf{x},\boldsymbol{\xi}) = \left\{ \begin{array}{c} g_2^{100}(\mathbf{x},\boldsymbol{\xi}) \\ g_2^{010}(\mathbf{x},\boldsymbol{\xi}) \\ g_2^{001}(\mathbf{x},\boldsymbol{\xi}) \end{array} \right\} = \frac{9w_1(|\boldsymbol{\xi}|)}{4\pi\delta^5} \left\{ \begin{array}{c} \xi_1 \\ \xi_2 \\ \xi_3 \end{array} \right\} \tag{2.87}$$

and

$$\left\{ \begin{array}{c} g_2^{200}(\mathbf{x},\boldsymbol{\xi}) \\ g_2^{020}(\mathbf{x},\boldsymbol{\xi}) \\ g_2^{002}(\mathbf{x},\boldsymbol{\xi}) \\ g_2^{110}(\mathbf{x},\boldsymbol{\xi}) \\ g_2^{101}(\mathbf{x},\boldsymbol{\xi}) \\ g_2^{011}(\mathbf{x},\boldsymbol{\xi}) \end{array} \right\} = \frac{3w_2(|\boldsymbol{\xi}|)}{\pi\delta^7} \left\{ \begin{array}{l} 4\xi_1^2 - \xi_2^2 - \xi_3^2 \\ -\xi_1^2 + 4\xi_2^2 - \xi_3^2 \\ -\xi_1^2 - \xi_2^2 + 4\xi_3^2 \\ 5\xi_1\xi_2 \\ 5\xi_1\xi_3 \\ 5\xi_2\xi_3 \end{array} \right\} \tag{2.88}$$

It is observed that $g_2^{200}(\mathbf{x},\boldsymbol{\xi}) + g_2^{020}(\mathbf{x},\boldsymbol{\xi}) + g_2^{002}(\mathbf{x},\boldsymbol{\xi})$ can be evaluated as

$$g_2^{200}(\mathbf{x},\boldsymbol{\xi}) + g_2^{020}(\mathbf{x},\boldsymbol{\xi}) + g_2^{002}(\mathbf{x},\boldsymbol{\xi}) = \frac{6w_2(|\boldsymbol{\xi}|)}{\pi\delta^7}(\boldsymbol{\xi}\cdot\boldsymbol{\xi}) \tag{2.89}$$

After substituting for the weight function $w_1(|\boldsymbol{\xi}|) = \delta^2/|\boldsymbol{\xi}|^2$, the vector $\mathbf{g}(\mathbf{x},\boldsymbol{\xi})$ in Eq. (2.87) can be expressed as

$$\mathbf{g}(\mathbf{x},\boldsymbol{\xi}) = \frac{9}{4\pi\delta^3|\boldsymbol{\xi}|^2}\begin{Bmatrix}\xi_1\\ \xi_2\\ \xi_3\end{Bmatrix} = \frac{9}{4\pi\delta^3\mid\boldsymbol{\xi}\mid}\begin{Bmatrix}\dfrac{\xi_1}{\mid\boldsymbol{\xi}\mid}\\ \dfrac{\xi_2}{\mid\boldsymbol{\xi}\mid}\\ \dfrac{\xi_2}{\mid\boldsymbol{\xi}\mid}\end{Bmatrix} = \frac{9}{4\pi\delta^3\mid\boldsymbol{\xi}\mid}\mathbf{n}(\mathbf{x},\boldsymbol{\xi}) \tag{2.90}$$

Thus, the nonlocal form of the bond direction $\mathbf{n}(\mathbf{x},\boldsymbol{\xi})$ can be expressed as

$$\mathbf{n}(\mathbf{x},\boldsymbol{\xi}) = \frac{4\pi\delta^3}{9}\mathbf{g}(\mathbf{x},\boldsymbol{\xi})\mid\boldsymbol{\xi}\mid \tag{2.91}$$

Substituting for the weight function $w_2(|\boldsymbol{\xi}|) = \delta^3/|\boldsymbol{\xi}|^3$, the second-order derivatives in Eq. (2.88) can be expressed as

$$\begin{Bmatrix}n_1^2\\ n_2^2\\ n_3^2\\ n_1n_2\\ n_1n_3\\ n_2n_3\end{Bmatrix} = \frac{\pi\delta^4}{15}\mid\boldsymbol{\xi}\mid\begin{Bmatrix}g_2^{200}(\mathbf{x},\boldsymbol{\xi}) + \dfrac{3}{\pi\delta^4\mid\boldsymbol{\xi}\mid}\mathbf{n}(\mathbf{x},\boldsymbol{\xi})\cdot\mathbf{n}(\mathbf{x},\boldsymbol{\xi})\\ g_2^{020}(\mathbf{x},\boldsymbol{\xi}) + \dfrac{3}{\pi\delta^4\mid\boldsymbol{\xi}\mid}\mathbf{n}(\mathbf{x},\boldsymbol{\xi})\cdot\mathbf{n}(\mathbf{x},\boldsymbol{\xi})\\ g_2^{002}(\mathbf{x},\boldsymbol{\xi}) + \dfrac{3}{\pi\delta^4\mid\boldsymbol{\xi}\mid}\mathbf{n}(\mathbf{x},\boldsymbol{\xi})\cdot\mathbf{n}(\mathbf{x},\boldsymbol{\xi})\\ g_2^{110}(\mathbf{x},\boldsymbol{\xi})\\ g_2^{101}(\mathbf{x},\boldsymbol{\xi})\\ g_2^{011}(\mathbf{x},\boldsymbol{\xi})\end{Bmatrix} \tag{2.92}$$

Invoking the weight function $w_2(|\boldsymbol{\xi}|) = \delta^3/|\boldsymbol{\xi}|^3$ into Eq. (2.89) results in

$$\mathbf{n}(\mathbf{x},\boldsymbol{\xi})\cdot\mathbf{n}(\mathbf{x},\boldsymbol{\xi}) = \frac{\pi\delta^4}{6}\mid\boldsymbol{\xi}\mid\left(g_2^{200}(\mathbf{x},\boldsymbol{\xi}) + g_2^{020}(\mathbf{x},\boldsymbol{\xi}) + g_2^{002}(\mathbf{x},\boldsymbol{\xi})\right) \tag{2.93}$$

Substituting from Eq. (2.93) into (2.92) along with Eq. (2.69) leads to an explicit form of $(\mathbf{n}(\mathbf{x}, \boldsymbol{\xi}) \otimes \mathbf{n}(\mathbf{x}, \boldsymbol{\xi}))$ as

$$\mathbf{n}(\mathbf{x}, \boldsymbol{\xi}) \otimes \mathbf{n}(\mathbf{x}, \boldsymbol{\xi}) = \frac{\pi\delta^4 \mid \boldsymbol{\xi} \mid}{15}\left(\mathbf{G}(\mathbf{x}, \boldsymbol{\xi}) + \frac{1}{2}\mathrm{tr}\mathbf{G}(\mathbf{x}, \boldsymbol{\xi})\mathbf{I}\right) \tag{2.94}$$

Hence, the nonlocal form of the dyadic product of bond direction vectors $(\mathbf{n}(\mathbf{x}, \boldsymbol{\xi}) \otimes \mathbf{n}(\mathbf{x}, \boldsymbol{\xi}))$ and $(\mathbf{n}(\mathbf{x}, \boldsymbol{\xi}) \cdot \mathbf{n}(\mathbf{x}, \boldsymbol{\xi}))$ can be expressed as

$$\mathbf{n}(\mathbf{x}, \boldsymbol{\xi}) \otimes \mathbf{n}(\mathbf{x}, \boldsymbol{\xi}) = \frac{\pi\delta^4}{30}(2\mathbf{G}(\mathbf{x}, \boldsymbol{\xi}) + \mathrm{tr}\mathbf{G}(\mathbf{x}, \boldsymbol{\xi})\mathbf{I}) \mid \boldsymbol{\xi} \mid \tag{2.95}$$

and

$$\mathbf{n}(\mathbf{x}, \boldsymbol{\xi}) \cdot \mathbf{n}(\mathbf{x}, \boldsymbol{\xi}) = \frac{\pi\delta^4 \mid \boldsymbol{\xi} \mid}{6}\mathrm{tr}\mathbf{G}(\mathbf{x}, \boldsymbol{\xi}) \tag{2.96}$$

2.4 Nonlocal Form of Divergence of a Second-Order Tensor

The components of divergence of a second-order tensor, $\mathbf{P}(\mathbf{x})$ can be written as

$$\nabla \cdot \mathbf{P}(\mathbf{x}) \sim L_i(\mathbf{x}) = \frac{\partial P_{i1}}{\partial x_1} + \frac{\partial P_{i2}}{\partial x_2} + \frac{\partial P_{i3}}{\partial x_3} \quad \text{for } i = 1, 2, 3 \tag{2.97}$$

in which the PDDO form of the spatial derivatives in accordance with Eq. (2.44) can be expressed as

$$\frac{\partial P_{i1}(\mathbf{x})}{\partial x_1} = \int_{H_\mathbf{x}} (P_{i1}(\mathbf{x}') - P_{i1}(\mathbf{x}))g_2^{100}(\mathbf{x}, \boldsymbol{\xi})dV_{\mathbf{x}'} \tag{2.98}$$

$$\frac{\partial P_{i2}(\mathbf{x})}{\partial x_2} = \int_{H_\mathbf{x}} (P_{i2}(\mathbf{x}') - P_{i2}(\mathbf{x}))g_2^{010}(\mathbf{x}, \boldsymbol{\xi})dV_{\mathbf{x}'} \tag{2.99}$$

and

$$\frac{\partial P_{i3}(\mathbf{x})}{\partial x_3} = \int_{H_\mathbf{x}} (P_{i3}(\mathbf{x}') - P_{i3}(\mathbf{x}))g_2^{001}(\mathbf{x}, \boldsymbol{\xi})dV_{\mathbf{x}'} \tag{2.100}$$

Substituting from Eqs. (2.98) to (2.100) into Eq. (2.97) yields

$$\nabla \cdot \mathbf{P}(\mathbf{x}) \sim L_i(\mathbf{x}) = \int\limits_{H_{\mathbf{x}}} (P_{i1}(\mathbf{x}') - P_{i1}(\mathbf{x})) g_2^{100}(\mathbf{x}, \boldsymbol{\xi}) dV_{\mathbf{x}'} + \int\limits_{H_{\mathbf{x}}} (P_{i2}(\mathbf{x}') - P_{i2}(\mathbf{x})) g_2^{010}(\mathbf{x}, \boldsymbol{\xi}) dV_{\mathbf{x}'} + \int\limits_{H_{\mathbf{x}}} (P_{i3}(\mathbf{x}') - P_{i3}(\mathbf{x})) g_2^{001}(\mathbf{x}, \boldsymbol{\xi}) dV_{\mathbf{x}'} \quad \text{for } i = 1, 2, 3 \tag{2.101}$$

or

$$L_i(\mathbf{x}) = \int\limits_{H_{\mathbf{x}}} \{(P_{i1}(\mathbf{x}') - P_{i1}(\mathbf{x})), (P_{i2}(\mathbf{x}') - P_{i2}(\mathbf{x})), (P_{i3}(\mathbf{x}') - P_{i3}(\mathbf{x}))\} \times \begin{Bmatrix} g_2^{100}(\mathbf{x}, \boldsymbol{\xi}) \\ g_2^{010}(\mathbf{x}, \boldsymbol{\xi}) \\ g_2^{001}(\mathbf{x}, \boldsymbol{\xi}) \end{Bmatrix} dV_{\mathbf{x}'} \text{ for } i = 1, 2, 3 \tag{2.102}$$

In matrix form, these expressions can be rewritten as

$$\begin{Bmatrix} L_1(\mathbf{x}) \\ L_2(\mathbf{x}) \\ L_3(\mathbf{x}) \end{Bmatrix} = \int\limits_{H_{\mathbf{x}}} \begin{bmatrix} (P_{11}(\mathbf{x}') - P_{11}(\mathbf{x})) & (P_{12}(\mathbf{x}') - P_{12}(\mathbf{x})) & (P_{13}(\mathbf{x}') - P_{13}(\mathbf{x})) \\ (P_{21}(\mathbf{x}') - P_{21}(\mathbf{x})) & (P_{22}(\mathbf{x}') - P_{22}(\mathbf{x})) & (P_{23}(\mathbf{x}') - P_{23}(\mathbf{x})) \\ (P_{31}(\mathbf{x}') - P_{31}(\mathbf{x})) & (P_{32}(\mathbf{x}') - P_{32}(\mathbf{x})) & (P_{33}(\mathbf{x}') - P_{33}(\mathbf{x})) \end{bmatrix} \begin{Bmatrix} g_2^{100}(\mathbf{x}, \boldsymbol{\xi}) \\ g_2^{010}(\mathbf{x}, \boldsymbol{\xi}) \\ g_2^{001}(\mathbf{x}, \boldsymbol{\xi}) \end{Bmatrix} dV_{\mathbf{x}'} \tag{2.103}$$

or

$$\nabla \cdot \mathbf{P}(\mathbf{x}) = \mathbf{L}(\mathbf{x}) = \int\limits_{H_{\mathbf{x}}} (\mathbf{P}(\mathbf{x}') - \mathbf{P}(\mathbf{x})) \mathbf{g}(\mathbf{x}, \boldsymbol{\xi}) dV_{\mathbf{x}'} \tag{2.104}$$

In discrete form, it can be expressed as

$$\nabla \cdot \mathbf{P}\left(\mathbf{x}_{(k)}\right) = \mathbf{L}\left(\mathbf{x}_{(k)}\right) = \sum_{j \in H_{(k)}} \left(\mathbf{P}\left(\mathbf{x}_{(j)}\right) - \mathbf{P}\left(\mathbf{x}_{(k)}\right)\right) \mathbf{g}_{(k)(j)} V_{(j)} \tag{2.105}$$

References

Aluru, N. R. (2000). A point collocation method based on reproducing kernel approximations. *International Journal for Numerical Methods in Engineering, 47*, 1083–1121.

Bessa, M., Foster, J. T., Belytschko, T., & Liu, W. K. (2014). A meshfree unification: Reproducing kernel peridynamics. *Computational Mechanics, 53*, 1251–1264.

Chi, S. W., Chen, J. S., Hu, H. Y., & Yang, J. P. (2013). A gradient reproducing kernel collocation method for boundary value problems. *International Journal for Numerical Methods in Engineering, 93*(13), 1381–1402.

Gingold, R. A., & Monaghan, J. J. (1977). Smoothed particle hydrodynamics: Theory and application to non-spherical stars. *Monthly Notices of the Royal Astronomical Society, 181*, 375–389.

Liszka, T. (1984). An interpolation method for an irregular net of nodes. *International Journal for Numerical Methods in Engineering, 20*, 1599–1612.

Liszka, T., & Orkisz, J. (1980). The finite difference method at arbitrary irregular grids and its application in applied mechanics. *Computers and Structures, 11*, 83–95.

Liu, W. K., Jun, S., & Zhang, Y. F. (1995). Reproducing kernel particle methods. *International Journal for Numerical Methods in Fluids, 20*, 1081–1106.

Lucy, L. B. (1977). A numerical approach to the testing of the fission hypothesis. *Astronomical Journal, 82*, 1013–1024.

Madenci, E., Barut, A., & Futch, M. (2016). Peridynamic differential operator and its applications. *Computer Methods in Applied Mechanics and Engineering, 304*, 408–451.

Madenci, E., Dorduncu, M., Barut, A., & Futch, M. (2017). Numerical solution of linear and nonlinear partial differential equations using the peridynamic differential operator. *Numerical Methods for Partial Differential Equations, 33*, 1726–1753.

Madenci, E., Barut, A., & Dorduncu, M. (2019). *Peridynamic differential operator for numerical analysis*. Springer.

Madenci, E., Roy, P., & Behera, D. (2022). *Advances in peridynamics*. Springer.

Silling, S. A. (2000). Reformulation of elasticity theory for discontinuities and long-range forces. *Journal of the Mechanics and Physics of Solids, 48*, 175–209.

Silling, S. A., Epton, M., Weckner, O., Xu, J., & Askari, A. (2007). Peridynamics states and constitutive modeling. *Journal of Elasticity, 88*, 151–184.

Chapter 3
Unified Peridynamic Approach

3.1 Introduction

As explained in Chap. 1, the PD equation of motion and PD field equation are derived at a point with a complete horizon in the "interior" of the domain away from the boundary. The essential (forced) boundary conditions can be applied on the fictitious layer along the boundary. Similarly, the natural boundary conditions can be applied in the form of a body force or a field source on the real layer along the boundary. The accuracy of displacement and field variable at points near the boundaries depends on their values in these layers. It may lead to unphysical concentrations near the boundaries. This chapter presents a unified approach to enforce different combinations of boundary conditions when solving for the strong form of PD equilibrium and field equations without a fictitious layer.

The domain shown in Fig. 3.1 is divided into three regions, that is, inner, $\mathcal{D}$, outer, $\mathcal{R}$, and "real" boundary layer region, $\mathcal{B}$. The inner region is governed by the PD bond-wise interactions, and the outer region and the boundary layer region are governed by the PDDO point-wise interactions. The inner and the outer regions are also referred to as PD and PDDO regions, respectively.

As shown in Fig. 3.2, the definition of internal force density vectors, $\mathbf{t}(\mathbf{x}' - \mathbf{x})$ and $\mathbf{t}(\mathbf{x}' - \mathbf{x})$ at points $\mathbf{x}$ and $\mathbf{x}'$ classifies the formulation as the bond-based peridynamic (BB-PD), the ordinary state-based peridynamic (OSB-PD) and the nonordinary state-based peridynamic (NOSB-PD). The classification "ordinary" indicates that the force density vector is aligned with the line of interaction.

Conceptually rather simple, the BB-PD introduced by Silling (2000) does not distinguish between dilatation and distortion. Therefore, it presents only one independent elastic constant; the second elastic constant has a fixed value depending on the analysis dimension. The independent PD material parameter is referred to as the *micro-modulus*. This micro-modulus can be determined through calibration against the classical strain energy density (SED) by considering a simple loading condition such as isotropic expansion.

E. Madenci et al., *Peridynamics and Its Applications Using Ansys*,
https://doi.org/10.1007/978-3-031-59896-8_3

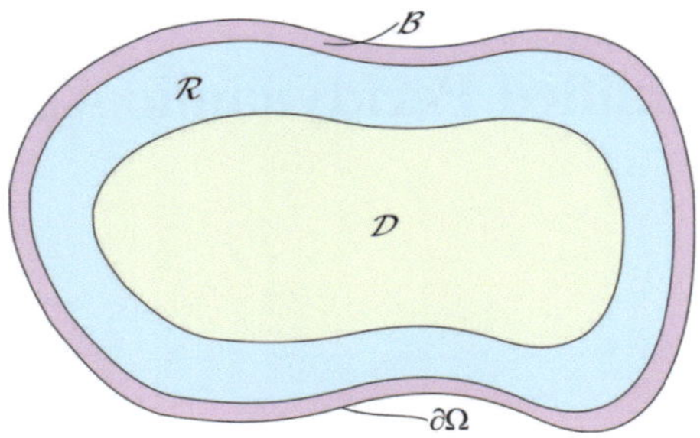

Fig. 3.1 The PD region, $\mathcal{D}$ and PDDO region, $\mathcal{R}$ of the domain with a "real" boundary layer region, $\mathcal{B}$

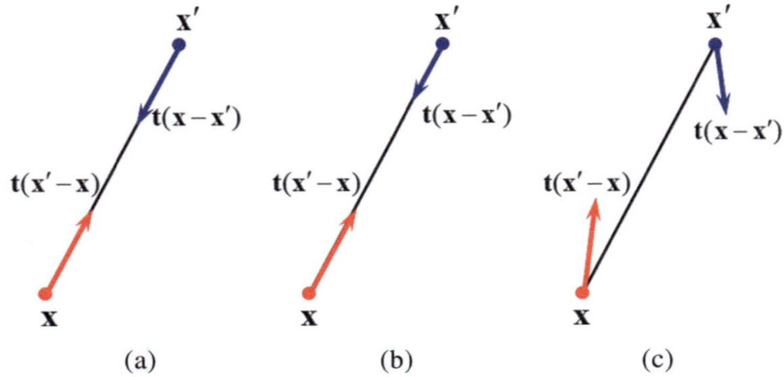

Fig. 3.2 The PD force density vectors: (**a**) BB (single horizon), (**b**) OSB, and (**c**) NOSB

The state-based PD introduced by Silling et al. (2007) retains the two independent elastic constants such as Young's modulus and Poisson's ratio. However, the determination of the force density-stretch relation requires an explicit PD form of the SED function. Silling et al. (2007) suggested a SED function for three-dimensional analysis based on the deformation state and force state for a material referred to as a *"linear peridynamic solid."* This material model is also calibrated so that it corresponds to the classical isotropic linear elastic material response for all deformations with no dilatation. In order to avoid the calibration procedure, Madenci (2017) presented the PD form of the strain invariants through the introduction of PD integrals. Therefore, the PD form of the existing classical SED functions can readily be constructed for linearly elastic and hyperelastic isotropic materials. It leads to the force density vector in the PD equation of motion without any calibration.

The NOSB-PD models utilizing the existing classical continuum constitutive relations can also be referred to as "the PD correspondence material models." Such NOSB-PD models provide a simple way to convert the classical continuum constitutive models into the PD framework with the PD representation of the deformation gradient tensor and stress tensor, which involves the classical constitutive models. Specifically, the NOSB-PD defines a nonlocal integral representation of deformation gradient tensor, which approximates the local one to obtain the stress tensor. Subsequently, the stress tensor enables the determination of the force density vector that appears in the PD equation of motion.

3.2 PD Equation of Motion

Considering the bond-based (BB), ordinary state-based (OSB), and nonordinary state-based (NOSB) PD interactions for varying horizon due to nonuniform discretization, the dual horizon PD equilibrium equations, Eq. (1.16) can be expressed as (Anicode & Madenci, 2022, 2023)

$$\rho\ddot{\mathbf{u}}(\mathbf{x},t)=\mathbf{L}^{\alpha}(\mathbf{x},t)+\mathbf{b}(\mathbf{x},t) \text{ for } \mathbf{x}\in(\mathcal{D}+\mathcal{R}+\mathcal{B})\ \alpha=BB,OSB,NOSB \tag{3.1}$$

with

$$\mathbf{L}^{\alpha}(\mathbf{x},t)=\int_{H_{\mathbf{x}}}\mathbf{t}^{\alpha}(\mathbf{x}'-\mathbf{x},t)dV_{\mathbf{x}'}-\int_{\widetilde{H}_{\mathbf{x}}}\mathbf{t}^{\alpha}(\mathbf{x}-\mathbf{x}',t)dV_{\mathbf{x}'} \tag{3.2}$$

The internal force vector, $\mathbf{L}^{\alpha}$ replaces the divergence of the stress tensor, $\mathbf{P}(\mathbf{x},t)$ in Classical Continuum Mechanics (CCM) as (Silling & Lehoucq, 2010)

$$\mathbf{L}(\mathbf{x},t)=\nabla\cdot\mathbf{P}(\mathbf{x},t) \text{ for } \mathbf{x}\in\Omega \tag{3.3}$$

with

$$\mathbf{L}(\mathbf{x},t)=\lim_{\delta\to 0}\mathbf{L}^{\alpha}(\mathbf{x},t) \tag{3.4}$$

where $\mathbf{P}(\mathbf{x},t)$ is the first Piola-Kirchhoff stress tensor at point $\mathbf{x}$. Alternative to Eq. (3.2), Madenci et al. (2022) derived the nonlocal form of Eq. (3.1) and the expression for the traction vector, $\mathbf{T}^{PD}$ by using the PDDO as

$$\mathbf{L}^{PDDO}(\mathbf{x},t)=\int_{H_{\mathbf{x}}}\widehat{\mu}(\mathbf{x}'-\mathbf{x})\left(\mathbf{P}^{PD}(\mathbf{x}',t)-\mathbf{P}^{PD}(\mathbf{x},t)\right)\mathbf{g}(\mathbf{x},\boldsymbol{\xi})dV_{\mathbf{x}'} \text{ for } \mathbf{x}\in(\mathcal{D}+\mathcal{R}) \tag{3.5}$$

and

$$\mathbf{T}^{PD}(\mathbf{x},t)=\mathbf{P}^{PD}(\mathbf{x},t)\mathbf{n} \text{ for } \mathbf{x}\in\mathcal{B} \tag{3.6}$$

where $\mathbf{P}^{PD}(\mathbf{x},t)$ is the PD form of the first Piola-Kirchhoff stress tensor, $\mathbf{P}(\mathbf{x},t)$. As shown in Fig. 3.1, $\mathcal{B}$ represents a “real boundary layer” along the boundary of nonzero volume. It is worth noting that Eq. (3.5) has a single horizon even for irregular discretization without suffering from ghost (residual) forces. However, it is based on point-wise PD interactions rather than bond-wise.

3.3 PD Internal Force Density Vector

The primary and secondary PD bonds shown in Fig. 3.3 are employed to compute the PD internal force density vectors, $\mathbf{t}(\mathbf{x}'-\mathbf{x},t)$ and $\mathbf{t}(\mathbf{x}'-\mathbf{x},t)$ at points $\mathbf{x}$ and $\mathbf{x}'$, respectively. For a nonuniform discretization, material point $\mathbf{x}$ can be within the horizon of other material points, $\mathbf{x}'$ and $\mathbf{x}''$; however, these points may not be within the horizon of point, $\mathbf{x}$. It is worth noting that $H_{\mathbf{x}}$ includes the material point $\mathbf{x}'$ but not $\mathbf{x}''$.

The points $\mathbf{x}'$ located in $H_{\mathbf{x}}$ constitute the family members of $\mathbf{x}$. The bond, $\boldsymbol{\xi}=\mathbf{x}'-\mathbf{x}$ between these points is referred to as a "*primary*" bond. The bonds, $\boldsymbol{\xi}'=\mathbf{x}-\mathbf{x}'=-\boldsymbol{\xi}$ and $\boldsymbol{\xi}''=\mathbf{x}-\mathbf{x}''$ between points $\mathbf{x}'$ and $\mathbf{x}$, and $\mathbf{x}''$ and $\mathbf{x}$, respectively, are referred to as "*secondary*" bonds, provided that $\mathbf{x}$ belongs to $H_{\mathbf{x}'}$ and $H_{\mathbf{x}''}$. The interaction domain for secondary bonds defined as $\widetilde{H}_{\mathbf{x}}$ is referred to as the "dual" horizon (Ren et al., 2017).

3.3.1 *Bond-Based (BB) Interactions*

As shown in Fig. 3.4, the BB-PD force density vectors, $\mathbf{t}^{BB}(\mathbf{x}'-\mathbf{x},t)$ and $\mathbf{t}^{BB}(\mathbf{x}-\mathbf{x}',t)$ for primary and secondary bonds, respectively, can be expressed as

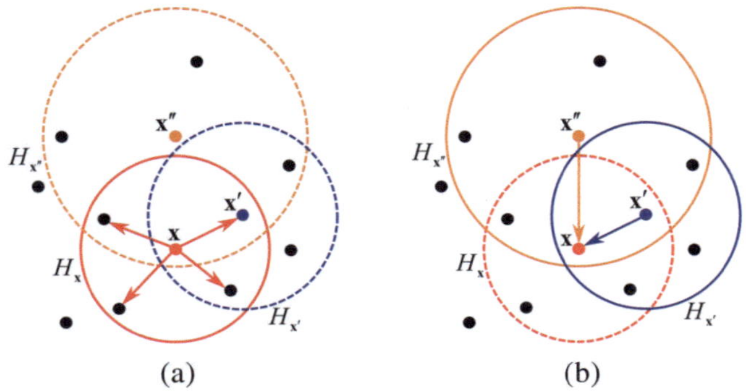

Fig. 3.3 PD interactions due to: (**a**) primary bonds in $H_{\mathbf{x}}$, and (**b**) secondary bonds in $H_{\mathbf{x}'}$ and $H_{\mathbf{x}''}$ or $\widetilde{H}_{\mathbf{x}}$

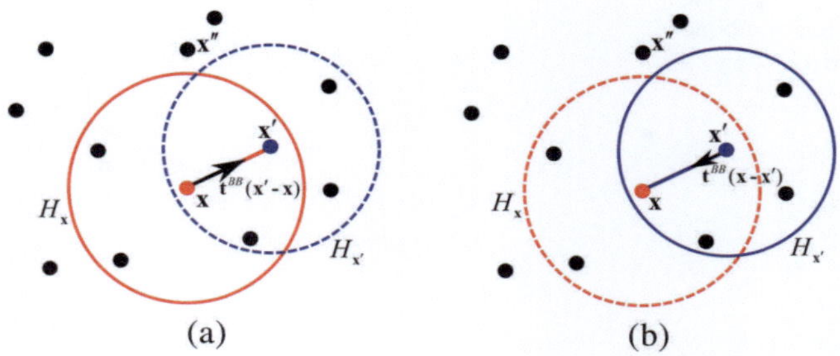

Fig. 3.4 The BB PD internal force density vectors arising from: (**a**) primary bonds (red) in $H_{\mathbf{x}}$ and (**b**) secondary bonds (blue) in $\widetilde{H}_{\mathbf{x}}$

$$\mathbf{t}^{BB}(\mathbf{x}' - \mathbf{x}, t) = \widehat{\mu}(\mathbf{x}' - \mathbf{x}) \frac{c(\delta_{\mathbf{x}})}{2} s \frac{\mathbf{y}' - \mathbf{y}}{|\mathbf{y}' - \mathbf{y}|} \quad \text{for } \mathbf{x}' \in H_{\mathbf{x}} \tag{3.7}$$

and

$$\mathbf{t}^{BB}(\mathbf{x} - \mathbf{x}', t) = \widehat{\mu}(\mathbf{x} - \mathbf{x}') \frac{c(\delta_{\mathbf{x}'})}{2} s \frac{\mathbf{y} - \mathbf{y}'}{|\mathbf{y} - \mathbf{y}'|} \quad \text{for } \mathbf{x}' \in \widetilde{H}_{\mathbf{x}} \tag{3.8}$$

where $\widehat{\mu}(\mathbf{x}' - \mathbf{x})$ and $\widehat{\mu}(\mathbf{x} - \mathbf{x}')$ are the status parameters and $c(\delta_{\mathbf{x}})$ and $c(\delta_{\mathbf{x}'})$ represent the varying primary and secondary bond constants, respectively. In accordance with Eq. (1.1), the status parameters control the interactions between material points $\mathbf{x}$ and $\mathbf{x}'$. For the intact bond, their values are equal to unity; otherwise, equal to zero.

The bond stretch s is defined as (Silling, 2000)

$$s = \frac{|\mathbf{y}' - \mathbf{y}| - |\mathbf{x}' - \mathbf{x}|}{|\mathbf{x}' - \mathbf{x}|} = \frac{|\mathbf{y}' - \mathbf{y}| - |\boldsymbol{\xi}|}{|\boldsymbol{\xi}|} \tag{3.9}$$

The bond constant depends on the horizon size and specific value of the Poisson's ratio depending on the dimension of the analysis. For a point with a complete horizon size of δ, its explicit relation to the engineering constants can be derived as (Silling, 2000; Madenci & Oterkus, 2014)

$$c(\delta) = \begin{cases} \dfrac{6}{\pi\delta^4} \dfrac{E}{(1 - 2\nu)} = \dfrac{30}{\pi\delta^4}\mu & \text{3D with } \nu = 1/4 \\ \dfrac{1}{\pi h\delta^3} \dfrac{6E}{(1 + \nu)(1 - 2\nu)} = \dfrac{24}{\pi h\delta^3}\mu & \text{plane strain with } \nu = 1/4 \\ \dfrac{6}{\pi h\delta^3} \dfrac{E}{(1 - \nu)} = \dfrac{24}{\pi h\delta^3}\mu & \text{plane stress with } \nu = 1/3 \end{cases} \tag{3.10}$$

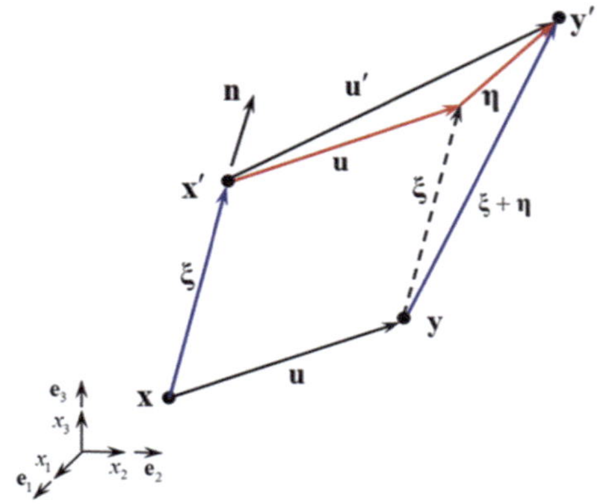

Fig. 3.5 Stretch of a bond between $\mathbf{x}$ and $\mathbf{x}'$

Thus, the PD internal force vector for BB-PD can be evaluated as

$$\mathbf{L}^{BB}(\mathbf{x},t)=\int_{H_\mathbf{x}}\mathbf{t}^{BB}(\mathbf{x}'-\mathbf{x},t)dV_{\mathbf{x}'}-\int_{\widetilde{H}_\mathbf{x}}\mathbf{t}^{BB}(\mathbf{x}-\mathbf{x}',t)dV_{\mathbf{x}'} \tag{3.11}$$

As shown in Fig. 3.5, the material points $\mathbf{x}$ and $\mathbf{x}'$ experience displacements, $\mathbf{u}=\mathbf{u}(\mathbf{x})$ and $\mathbf{u}'=\mathbf{u}(\mathbf{x}')$, respectively. Their positions become $\mathbf{y}=\mathbf{x}+\mathbf{u}$ and $\mathbf{y}'=\mathbf{x}'+\mathbf{u}'$ in the deformed state. For "small deformation", the relative position vector, $\mathbf{y}'-\mathbf{y}$ can be approximated as

$$\mathbf{y}'-\mathbf{y}\approx\boldsymbol{\xi}+(\boldsymbol{\eta}\cdot\mathbf{n})\mathbf{n}=\boldsymbol{\xi}+\left((\mathbf{u}(\mathbf{x}')-\mathbf{u}(\mathbf{x}))\cdot\frac{\boldsymbol{\xi}}{|\boldsymbol{\xi}|}\right)\frac{\boldsymbol{\xi}}{|\boldsymbol{\xi}|} \tag{3.12}$$

in which $\boldsymbol{\eta}=\mathbf{u}'-\mathbf{u}=(\mathbf{y}'-\mathbf{y})-\boldsymbol{\xi}$ is the relative displacement vector and $\boldsymbol{\xi}=\mathbf{x}'-\mathbf{x}$ is the relative position vector of points $\mathbf{x}$ and $\mathbf{x}'$.

The unit vector, $\mathbf{n}(\mathbf{x},\boldsymbol{\xi})$ is defined as $\mathbf{n}(\mathbf{x},\boldsymbol{\xi})=\boldsymbol{\xi}/|\boldsymbol{\xi}|$. Also, the assumption of "small deformation" permits

$$\frac{(\mathbf{y}'-\mathbf{y})}{|\mathbf{y}'-\mathbf{y}|}\cdot\frac{\boldsymbol{\xi}}{|\boldsymbol{\xi}|}=\frac{(\mathbf{y}'-\mathbf{y})}{|\mathbf{y}'-\mathbf{y}|}\cdot\mathbf{n}(\mathbf{x},\boldsymbol{\xi})\approx 1 \tag{3.13}$$

or

$$(\mathbf{y}'-\mathbf{y})\cdot\boldsymbol{\xi}\approx|\mathbf{y}'-\mathbf{y}||\boldsymbol{\xi}| \tag{3.14}$$

This approximation leads to the evaluation of $(\mathbf{u}(\mathbf{x}')-\mathbf{u}(\mathbf{x}))\cdot\boldsymbol{\xi}$ as

$$(\mathbf{u}(\mathbf{x}')-\mathbf{u}(\mathbf{x}))\cdot\boldsymbol{\xi}=((\mathbf{y}'-\mathbf{y})-\boldsymbol{\xi})\cdot\boldsymbol{\xi}\approx|\mathbf{y}'-\mathbf{y}||\boldsymbol{\xi}|-|\boldsymbol{\xi}||\boldsymbol{\xi}| \tag{3.15}$$

Using Eqs. (3.9) and (3.15) yields the approximation of bond stretch for small deformation as

$$s = \frac{|\mathbf{y}' - \mathbf{y}| - |\boldsymbol{\xi}|}{|\boldsymbol{\xi}|} \approx \frac{(\mathbf{u}(\mathbf{x}') - \mathbf{u}(\mathbf{x})) \cdot \mathbf{n}(\mathbf{x}, \boldsymbol{\xi})}{|\boldsymbol{\xi}|} \tag{3.16}$$

Based on the identity $(\mathbf{a} \otimes \mathbf{b})\mathbf{c} = (\mathbf{b} \cdot \mathbf{c})\mathbf{a}$, the expression for the bond stretch vector can be rewritten as

$$s\mathbf{n}(\mathbf{x}, \boldsymbol{\xi}) = \frac{(\mathbf{u}(\mathbf{x}') - \mathbf{u}(\mathbf{x})) \cdot \mathbf{n}(\mathbf{x}, \boldsymbol{\xi})}{|\boldsymbol{\xi}|}\mathbf{n}(\mathbf{x}, \boldsymbol{\xi}) = \frac{1}{|\boldsymbol{\xi}|}(\mathbf{n}(\mathbf{x}, \boldsymbol{\xi}) \otimes \mathbf{n}(\mathbf{x}, \boldsymbol{\xi}))(\mathbf{u}(\mathbf{x}') - \mathbf{u}(\mathbf{x})) \tag{3.17}$$

Thus, the force density vectors can be recast as

$$\mathbf{t}^{BB}(\mathbf{x}' - \mathbf{x}, t) = \widehat{\mu}(\mathbf{x}' - \mathbf{x}) \frac{c(\delta_{\mathbf{x}})}{2|\boldsymbol{\xi}|}(\mathbf{n}(\mathbf{x}, \boldsymbol{\xi}) \otimes \mathbf{n}(\mathbf{x}, \boldsymbol{\xi}))(\mathbf{u}(\mathbf{x}') - \mathbf{u}(\mathbf{x})) \;\; \text{for } \mathbf{x}' \in H_{\mathbf{x}} \tag{3.18}$$

and

$$\mathbf{t}^{BB}(\mathbf{x} - \mathbf{x}', t) = \widehat{\mu}(\mathbf{x} - \mathbf{x}') \frac{c(\delta_{\mathbf{x}'})}{2|\boldsymbol{\xi}|}(\mathbf{n}(\mathbf{x}', \boldsymbol{\xi}') \otimes \mathbf{n}(\mathbf{x}', \boldsymbol{\xi}'))(\mathbf{u}(\mathbf{x}) - \mathbf{u}(\mathbf{x}')) \;\; \text{for } \mathbf{x}' \in \widetilde{H}_{\mathbf{x}} \tag{3.19}$$

This bond force is derived for a material point with a complete horizon and away from the boundary. Otherwise, the bond constants, $c(\delta_{\mathbf{x}})$ or $c(\delta_{\mathbf{x}'})$ require correction when the material point is located near the boundary as explained by Madenci and Oterkus (2014). However, this requirement can be removed by substituting for the nonlocal representation of $\mathbf{n}(\mathbf{x}, \boldsymbol{\xi}) \otimes \mathbf{n}(\mathbf{x}, \boldsymbol{\xi})$ in terms of the PD functions introduced by Madenci et al. (2021). In nonlocal form, $\mathbf{n}(\mathbf{x}, \boldsymbol{\xi}) \otimes \mathbf{n}(\mathbf{x}, \boldsymbol{\xi})$ in Eqs. (2.85) and (2.95) can be expressed as

$$\mathbf{n}(\mathbf{x}, \boldsymbol{\xi}) \otimes \mathbf{n}(\mathbf{x}, \boldsymbol{\xi}) = \frac{\pi \delta_{\mathbf{x}}^4}{30} |\boldsymbol{\xi}| \mathbf{S}(\mathbf{x}, \boldsymbol{\xi}) \qquad \text{3D analysis} \tag{3.20}$$

and

$$\mathbf{n}(\mathbf{x}, \boldsymbol{\xi}) \otimes \mathbf{n}(\mathbf{x}, \boldsymbol{\xi}) = \frac{\pi h \delta_{\mathbf{x}}^3}{24} |\boldsymbol{\xi}| \mathbf{S}(\mathbf{x}, \boldsymbol{\xi}) \qquad \text{2D analysis} \tag{3.21}$$

where the matrix, $\mathbf{S}(\mathbf{x}, \boldsymbol{\xi})$ is defined as

$$\mathbf{S}(\mathbf{x}, \boldsymbol{\xi}) = \left(\mathrm{tr}\widehat{\mathbf{G}}(\mathbf{x}, \boldsymbol{\xi})\mathbf{I} + 2\widehat{\mathbf{G}}(\mathbf{x}, \boldsymbol{\xi})\right) \tag{3.22}$$

Similarly, the PD form of the unit vector, $\mathbf{n}(\mathbf{x}, \boldsymbol{\xi})$ in Eqs. (2.80) and (2.91) can be expressed as

$$\mathbf{n}(\mathbf{x}, \boldsymbol{\xi}) = \frac{4\pi\delta_{\mathbf{x}}^3}{9} |\boldsymbol{\xi}| \widehat{\mathbf{g}}(\mathbf{x}, \boldsymbol{\xi}) \qquad \text{3D analysis} \tag{3.23}$$

and

$$\mathbf{n}(\mathbf{x}, \boldsymbol{\xi}) = \frac{\pi h \delta_{\mathbf{x}}^2}{2} |\boldsymbol{\xi}| \widehat{\mathbf{g}}(\mathbf{x}, \boldsymbol{\xi}) \qquad \text{2D analysis} \tag{3.24}$$

The matrix $\widehat{\mathbf{G}}(\mathbf{x}, \boldsymbol{\xi})$ and vector $\widehat{\mathbf{g}}(\mathbf{x}, \boldsymbol{\xi})$ are expressed in terms of the PD functions as derived for 3D and 2D analyses in Chap. 2. However, it is worth noting that the off-diagonal submatrices $\mathbf{A}_{12} = \mathbf{A}_{21} = \mathbf{0}$ of the shape matrix, $\mathbf{A}$ in Eqs. (2.15) and (2.51) are specified as zero while generating the PD functions for matrix $\widehat{\mathbf{G}}(\mathbf{x}, \boldsymbol{\xi})$ and vector $\widehat{\mathbf{g}}(\mathbf{x}, \boldsymbol{\xi})$ in order to mitigate the imbalance of forces near the boundaries.

After substituting from Eq. (3.10) along with the nonlocal representation of $\mathbf{n}(\mathbf{x}, \boldsymbol{\xi}) \otimes \mathbf{n}(\mathbf{x}, \boldsymbol{\xi})$, the bond forces can be rewritten as

$$\mathbf{t}^{BB}(\mathbf{x}' - \mathbf{x}, t) = \widehat{\mu}(\mathbf{x}' - \mathbf{x}) \frac{\mu}{2} \mathbf{S}(\mathbf{x}, \boldsymbol{\xi}) (\mathbf{u}(\mathbf{x}') - \mathbf{u}(\mathbf{x})) \ \text{ for } \ \mathbf{x}' \in H_{\mathbf{x}}, \tag{3.25}$$

and

$$\mathbf{t}^{BB}(\mathbf{x} - \mathbf{x}', t) = \widehat{\mu}(\mathbf{x} - \mathbf{x}') \frac{\mu}{2} \mathbf{S}(\mathbf{x}', \boldsymbol{\xi}') (\mathbf{u}(\mathbf{x}) - \mathbf{u}(\mathbf{x}')) \ \text{ for } \ \mathbf{x}' \in \widetilde{H}_{\mathbf{x}}, \tag{3.26}$$

These bond forces do not require any surface correction. It is worth noting that $\mathbf{S}(\mathbf{x}, \boldsymbol{\xi}) = \mathbf{S}(\mathbf{x}', \boldsymbol{\xi}')$ for material points $\mathbf{x}$ and $\mathbf{x}'$ with identical horizons.

Thus, the internal force vector, $\mathbf{L}^{BB}(\mathbf{x}, t)$ in Eq. (3.2) can be expressed as

$$\begin{aligned} \mathbf{L}^{BB}(\mathbf{x}, t) = & \int\limits_{H_{\mathbf{x}}} \widehat{\mu}(\mathbf{x}' - \mathbf{x}) \frac{\mu}{2} \mathbf{S}(\mathbf{x}, \boldsymbol{\xi}) (\mathbf{u}(\mathbf{x}') - \mathbf{u}(\mathbf{x})) dV_{\mathbf{x}'} \\ & - \int\limits_{\widetilde{H}_{\mathbf{x}}} \widehat{\mu}(\mathbf{x} - \mathbf{x}') \frac{\mu}{2} \mathbf{S}(\mathbf{x}', \boldsymbol{\xi}') (\mathbf{u}(\mathbf{x}) - \mathbf{u}(\mathbf{x}')) dV_{\mathbf{x}'} \end{aligned} \tag{3.27}$$

For uniform discretization, Eq. (3.27) reduces to

$$\mathbf{L}^{BB}(\mathbf{x}, t) = \int\limits_{H_{\mathbf{x}}} \widehat{\mu}(\mathbf{x}' - \mathbf{x}) \mu \overline{\mathbf{S}}(\mathbf{x}, \mathbf{x}') (\mathbf{u}(\mathbf{x}') - \mathbf{u}(\mathbf{x})) dV_{\mathbf{x}'} \tag{3.28}$$

where

$$\overline{\mathbf{S}}(\mathbf{x}, \mathbf{x}') = \frac{\mathbf{S}(\mathbf{x}, \boldsymbol{\xi}) + \mathbf{S}(\mathbf{x}', \boldsymbol{\xi}')}{2} \tag{3.29}$$

3.3.2 Ordinary State-Based (OSB) Interactions

As shown in Fig. 3.6, the OSB-PD force density vectors, $\mathbf{t}^{OSB}(\mathbf{x}' - \mathbf{x}, t)$ and $\mathbf{t}^{OSB}(\mathbf{x} - \mathbf{x}', t)$ for primary and secondary bonds, respectively, can be derived as (Silling et al., 2007; Madenci, 2017)

$$\mathbf{t}^{OSB}(\mathbf{x}' - \mathbf{x}, t) = \widehat{\mu}(\mathbf{x}' - \mathbf{x})\left\{ad(\delta_{\mathbf{x}})\frac{1}{|\boldsymbol{\xi}|}\theta(\mathbf{x}) + b(\delta_{\mathbf{x}})\frac{1}{|\boldsymbol{\xi}|}s\right\}\frac{\mathbf{y}' - \mathbf{y}}{|\mathbf{y}' - \mathbf{y}|} \tag{3.30}$$

and

$$\mathbf{t}^{OSB}(\mathbf{x} - \mathbf{x}', t) = \widehat{\mu}(\mathbf{x} - \mathbf{x}')\left\{ad(\delta_{\mathbf{x}'})\frac{1}{|\boldsymbol{\xi}|}\theta(\mathbf{x}') + b(\delta_{\mathbf{x}'})\frac{1}{|\boldsymbol{\xi}|}s\right\}\frac{\mathbf{y} - \mathbf{y}'}{|\mathbf{y} - \mathbf{y}'|} \tag{3.31}$$

in which the volume change (dilatation) is expressed as

$$\theta(\mathbf{x}, t) = d(\delta_{\mathbf{x}})\int_{H_{\mathbf{x}}}\widehat{\mu}(\mathbf{x}'' - \mathbf{x})s dV_{\mathbf{x}''} \tag{3.32}$$

and

$$\theta(\mathbf{x}', t) = d(\delta_{\mathbf{x}'})\int_{H_{\mathbf{x}'}}\widehat{\mu}(\mathbf{x}''' - \mathbf{x}')s dV_{\mathbf{x}'''} \tag{3.33}$$

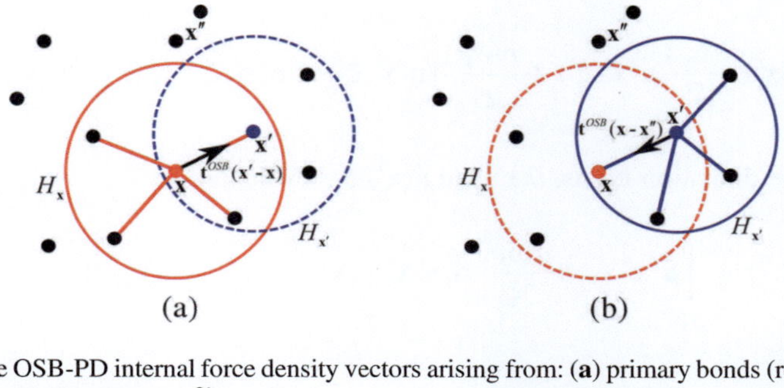

Fig. 3.6 The OSB-PD internal force density vectors arising from: (**a**) primary bonds (red) in $H_{\mathbf{x}}$ and (**b**) secondary bonds (blue) in $\widetilde{H}_{\mathbf{x}}$

The PD material parameters, a, $b(\delta)$ and $d(\delta)$ can be derived as (Madenci, 2017)

$$a = \begin{cases} \mu\left(\dfrac{4\nu - 1}{1 - 2\nu}\right) & \text{3D and plane strain} \\ \mu\left(\dfrac{3\nu - 1}{1 - \nu}\right) & \text{plane stress} \end{cases} \tag{3.34}$$

$$b(\delta) = \begin{cases} \dfrac{45\mu}{4\pi\delta^3} & \text{for 3D} \\ \dfrac{8\mu}{h\pi\delta^2} & \text{for 2D} \end{cases} \tag{3.35}$$

and

$$d(\delta) = \begin{cases} \dfrac{9}{4\pi\delta^3} & \text{for 3D} \\ \dfrac{2}{h\pi\delta^2} & \text{for 2D} \end{cases} \tag{3.36}$$

Thus, the PD internal force vector for OSB-PD in Eq. (3.2) can be evaluated as

$$\mathbf{L}^{OSB}(\mathbf{x}, t) = \int_{H_\mathbf{x}} \mathbf{t}^{OSB}(\mathbf{x}' - \mathbf{x}, t) dV_{\mathbf{x}'} - \int_{\widetilde{H}_\mathbf{x}} \mathbf{t}^{OSB}(\mathbf{x} - \mathbf{x}', t) dV_{\mathbf{x}'} \tag{3.37}$$

For small deformation, the force density vectors can be rewritten as

$$\begin{aligned} \mathbf{t}^{OSB}(\mathbf{x}' - \mathbf{x}, t) = {} & \widehat{\mu}(\mathbf{x}' - \mathbf{x}) \\ & \times \left(a\theta(\mathbf{x}) \frac{d(\delta_\mathbf{x})}{|\boldsymbol{\xi}|} \mathbf{n}(\mathbf{x}, \boldsymbol{\xi}) + \frac{b(\delta_\mathbf{x})}{|\boldsymbol{\xi}|^2} (\mathbf{n}(\mathbf{x}, \boldsymbol{\xi}) \otimes \mathbf{n}(\mathbf{x}, \boldsymbol{\xi}))(\mathbf{u}(\mathbf{x}') - \mathbf{u}(\mathbf{x})) \right) \end{aligned} \tag{3.38}$$

and

$$\begin{aligned} \mathbf{t}^{OSB}(\mathbf{x} - \mathbf{x}') = {} & \widehat{\mu}(\mathbf{x} - \mathbf{x}') \\ & \times \left(a\theta(\mathbf{x}') \frac{d(\delta_{\mathbf{x}'})}{|\boldsymbol{\xi}|} \mathbf{n}(\mathbf{x}', \boldsymbol{\xi}') + \frac{b(\delta_{\mathbf{x}'})}{|\boldsymbol{\xi}|^2} (\mathbf{n}(\mathbf{x}', \boldsymbol{\xi}') \otimes \mathbf{n}(\mathbf{x}', \boldsymbol{\xi}'))(\mathbf{u}(\mathbf{x}) - \mathbf{u}(\mathbf{x}')) \right) \end{aligned} \tag{3.39}$$

in which the dilatation terms, $\theta(\mathbf{x})$ and $\theta(\mathbf{x}')$ are evaluated as

$$\theta(\mathbf{x}, t) = \int_{H_\mathbf{x}} \widehat{\mu}(\mathbf{x}'' - \mathbf{x}) \frac{d(\delta_{\mathbf{x}'})}{|\boldsymbol{\xi}|} \mathbf{n}(\mathbf{x}, \mathbf{x}'' - \mathbf{x}) \cdot (\mathbf{u}(\mathbf{x}'') - \mathbf{u}(\mathbf{x})) dV_{\mathbf{x}''} \tag{3.40}$$

and

$$\theta(\mathbf{x}', t) = \int_{H_{\mathbf{x}}} \widehat{\mu}(\mathbf{x}''' - \mathbf{x}') \frac{d(\delta_{\mathbf{x}'})}{|\boldsymbol{\xi}|} \mathbf{n}(\mathbf{x}', \mathbf{x}''' - \mathbf{x}') \cdot (\mathbf{u}(\mathbf{x}''') - \mathbf{u}(\mathbf{x}')) dV_{\mathbf{x}'''} \tag{3.41}$$

Substituting for the nonlocal form of $\mathbf{n}(\mathbf{x}, \boldsymbol{\xi}) \otimes \mathbf{n}(\mathbf{x}, \boldsymbol{\xi})$ and $\mathbf{n}(\mathbf{x}, \boldsymbol{\xi})$ at points $\mathbf{x}$ and $\mathbf{x}'$ as given in Eqs. (3.20, 3.21, 3.23, and 3.24), the force density vectors can be rewritten as

$$\mathbf{t}^{OSB}(\mathbf{x}' - \mathbf{x}, t) = \widehat{\mu}(\mathbf{x}' - \mathbf{x}) \left\{ a\theta(\mathbf{x}) \widehat{\mathbf{g}}(\mathbf{x}, \boldsymbol{\xi}) + \frac{n}{2(n+1)} \mu \frac{\delta_{\mathbf{x}}}{|\boldsymbol{\xi}|} \mathbf{S}(\mathbf{x}, \boldsymbol{\xi}) (\mathbf{u}(\mathbf{x}') - \mathbf{u}(\mathbf{x})) \right\} \tag{3.42}$$

and

$$\mathbf{t}^{OSB}(\mathbf{x} - \mathbf{x}', t) = \widehat{\mu}(\mathbf{x} - \mathbf{x}') \left\{ a\theta(\mathbf{x}') \widehat{\mathbf{g}}(\mathbf{x}', \boldsymbol{\xi}') + \frac{n}{2(n+1)} \mu \frac{\delta_{\mathbf{x}'}}{|\boldsymbol{\xi}|} \mathbf{S}(\mathbf{x}', \boldsymbol{\xi}') (\mathbf{u}(\mathbf{x}) - \mathbf{u}(\mathbf{x}')) \right\} \tag{3.43}$$

with $n = 2$ and $n = 3$ for 2D and 3D analysis, respectively. Similarly, the expressions for dilatation at points $\mathbf{x}$ and $\mathbf{x}'$ can be rewritten as

$$\theta(\mathbf{x}, t) = \int_{H_{\mathbf{x}}} \widehat{\mu}(\mathbf{x}'' - \mathbf{x}) \mathbf{g}(\mathbf{x}, \mathbf{x}'' - \mathbf{x}) \cdot (\mathbf{u}(\mathbf{x}'') - \mathbf{u}(\mathbf{x})) dV_{\mathbf{x}''} \tag{3.44}$$

and

$$\theta(\mathbf{x}', t) = \int_{H_{\mathbf{x}'''}} \widehat{\mu}(\mathbf{x}''' - \mathbf{x}') \mathbf{g}(\mathbf{x}', \mathbf{x}''' - \mathbf{x}') \cdot (\mathbf{u}(\mathbf{x}''') - \mathbf{u}(\mathbf{x}')) dV_{\mathbf{x}'''} \tag{3.45}$$

After substituting Eqs. (3.44) and (3.45) into Eqs. (3.42) and (3.43), the force density vectors of the primary and secondary bonds can be recast as

$$\mathbf{t}^{OSB}(\mathbf{x}' - \mathbf{x}, t) = \widehat{\mu}(\mathbf{x}' - \mathbf{x}) \times \left\{ \begin{array}{l} a \int_{H_{\mathbf{x}}} \widehat{\mu}(\mathbf{x}'' - \mathbf{x}) (\widehat{\mathbf{g}}(\mathbf{x}, \boldsymbol{\xi}) \otimes \mathbf{g}(\mathbf{x}, \mathbf{x}'' - \mathbf{x})) (\mathbf{u}(\mathbf{x}'') - \mathbf{u}(\mathbf{x})) dV_{\mathbf{x}''} \\ + \frac{n}{2(n+1)} \mu \frac{\delta_{\mathbf{x}}}{|\boldsymbol{\xi}|} \mathbf{S}(\mathbf{x}, \boldsymbol{\xi}) (\mathbf{u}(\mathbf{x}') - \mathbf{u}(\mathbf{x})) \end{array} \right\} \tag{3.46}$$

and

$$\mathbf{t}^{OSB}(\mathbf{x}-\mathbf{x}',t)=\widehat{\mu}(\mathbf{x}-\mathbf{x}') \times \left\{ \begin{array}{l} a \int\limits_{H_{\mathbf{x}'}} \widehat{\mu}(\mathbf{x}'''-\mathbf{x}')(\widehat{\mathbf{g}}(\mathbf{x}',\boldsymbol{\xi}') \otimes \mathbf{g}(\mathbf{x}',\mathbf{x}'''-\mathbf{x}'))(\mathbf{u}(\mathbf{x}''')-\mathbf{u}(\mathbf{x}'))dV_{\mathbf{x}'''} \\ + \dfrac{n}{2(n+1)}\mu \dfrac{\delta_{\mathbf{x}'}}{|\boldsymbol{\xi}'|}\mathbf{S}(\mathbf{x}',\boldsymbol{\xi}')(\mathbf{u}(\mathbf{x})-\mathbf{u}(\mathbf{x}')) \end{array} \right\} \tag{3.47}$$

Substituting for the force density vectors in the internal force vector, $\mathbf{L}^{OSB}(\mathbf{x},t)$ in Eq. (3.2) results in

$$\begin{aligned} \mathbf{L}^{OSB}(\mathbf{x},t) = & \int\limits_{H_{\mathbf{x}}} \widehat{\mu}(\mathbf{x}'-\mathbf{x}) \left\{ \begin{array}{l} a \int\limits_{H_{\mathbf{x}}} \widehat{\mu}(\mathbf{x}''-\mathbf{x})(\widehat{\mathbf{g}}(\mathbf{x},\boldsymbol{\xi}) \otimes \mathbf{g}(\mathbf{x},\mathbf{x}''-\mathbf{x}))(\mathbf{u}(\mathbf{x}'')-\mathbf{u}(\mathbf{x}))dV_{\mathbf{x}''} \\ + \dfrac{n}{2(n+1)}\mu \dfrac{\delta_{\mathbf{x}}}{|\boldsymbol{\xi}|}\mathbf{S}(\mathbf{x},\boldsymbol{\xi})(\mathbf{u}(\mathbf{x}')-\mathbf{u}(\mathbf{x})) \end{array} \right\} dV_{\mathbf{x}'} \\ & - \int\limits_{\widetilde{H}_{\mathbf{x}}} \widehat{\mu}(\mathbf{x}-\mathbf{x}') \left\{ \begin{array}{l} a \int\limits_{H_{\mathbf{x}'}} \widehat{\mu}(\mathbf{x}'''-\mathbf{x}')(\widehat{\mathbf{g}}(\mathbf{x}',\boldsymbol{\xi}') \otimes \mathbf{g}(\mathbf{x}',\mathbf{x}'''-\mathbf{x}'))(\mathbf{u}(\mathbf{x}''')-\mathbf{u}(\mathbf{x}'))dV_{\mathbf{x}'''} \\ + \dfrac{n}{2(n+1)}\mu \dfrac{\delta_{\mathbf{x}'}}{|\boldsymbol{\xi}'|}\mathbf{S}(\mathbf{x}',\boldsymbol{\xi}')(\mathbf{u}(\mathbf{x})-\mathbf{u}(\mathbf{x}')) \end{array} \right\} dV_{\mathbf{x}'} \end{aligned} \tag{3.48}$$

The material parameters do not require any surface correction. For a constant horizon, δ, Eq. (3.48) reduces to

$$\mathbf{L}^{OSB}(\mathbf{x},t) = \int\limits_{H_{\mathbf{x}}} \widehat{\mu}(\mathbf{x}'-\mathbf{x}) \times \left\{ \begin{array}{l} a \int\limits_{H_{\mathbf{x}}} \widehat{\mu}(\mathbf{x}''-\mathbf{x})(\widehat{\mathbf{g}}(\mathbf{x},\boldsymbol{\xi}) \otimes \mathbf{g}(\mathbf{x},\mathbf{x}''-\mathbf{x}))(\mathbf{u}(\mathbf{x}'')-\mathbf{u}(\mathbf{x}))dV_{\mathbf{x}''} \\ - a \int\limits_{H_{\mathbf{x}'}} \widehat{\mu}(\mathbf{x}'''-\mathbf{x}')(\widehat{\mathbf{g}}(\mathbf{x}',\boldsymbol{\xi}') \otimes \mathbf{g}(\mathbf{x}',\mathbf{x}'''-\mathbf{x}'))(\mathbf{u}(\mathbf{x}''')-\mathbf{u}(\mathbf{x}'))dV_{\mathbf{x}'''} \\ + \dfrac{n}{(n+1)}\mu \dfrac{\delta}{|\boldsymbol{\xi}|}\overline{\mathbf{S}}(\mathbf{x},\boldsymbol{\xi})(\mathbf{u}(\mathbf{x}')-\mathbf{u}(\mathbf{x})) \end{array} \right\} dV_{\mathbf{x}'} \tag{3.49}$$

3.3.3 Nonordinary State-Based (NOSB) Interactions

For NOSB-PD interactions, Silling et al. (2007) and Madenci and Oterkus (2014) derived the explicit form of the internal force density vector for any material model by comparing the PD virtual work expression to that of the classical continuum mechanics. As shown in Fig. 3.7, the NOSB-PD internal force density vectors, $\mathbf{t}^{NOSB}(\mathbf{x}' - \mathbf{x}, t)$ and $\mathbf{t}^{NOSB}(\mathbf{x} - \mathbf{x}', t)$ for the primary and secondary bonds, respectively, can be expressed as (Madenci et al., 2022)

$$\mathbf{t}^{NOSB}(\mathbf{x}' - \mathbf{x}, t) = \widehat{\mu}(\mathbf{x}' - \mathbf{x})\mathbf{P}^{PD}_{\boldsymbol{\xi}}(\mathbf{x}, t)\mathbf{g}(\mathbf{x}, \boldsymbol{\xi}) \tag{3.50}$$

and

$$\mathbf{t}^{NOSB}(\mathbf{x} - \mathbf{x}', t) = \widehat{\mu}(\mathbf{x} - \mathbf{x}')\mathbf{P}^{PD}_{\boldsymbol{\xi}'}(\mathbf{x}', t)\mathbf{g}(\mathbf{x}', \boldsymbol{\xi}') \tag{3.51}$$

where the vector, $\mathbf{g}(\mathbf{x}, \boldsymbol{\xi})$ is defined as

$$\mathbf{g}(\mathbf{x}, \boldsymbol{\xi}) = w(|\boldsymbol{\xi}|)\mathbf{K}^{-1}(\mathbf{x})\boldsymbol{\xi} \tag{3.52}$$

in which $w(|\boldsymbol{\xi}|)$ is the weight function and the shape tensor, $\mathbf{K}$ is defined as

$$\mathbf{K}(\mathbf{x}) = \int_{H_{\mathbf{x}}} w(|\boldsymbol{\xi}|)(\boldsymbol{\xi} \otimes \boldsymbol{\xi})dV_{\mathbf{x}'} \tag{3.53}$$

where the symbol $\otimes$ denotes the dyadic product of two vectors.

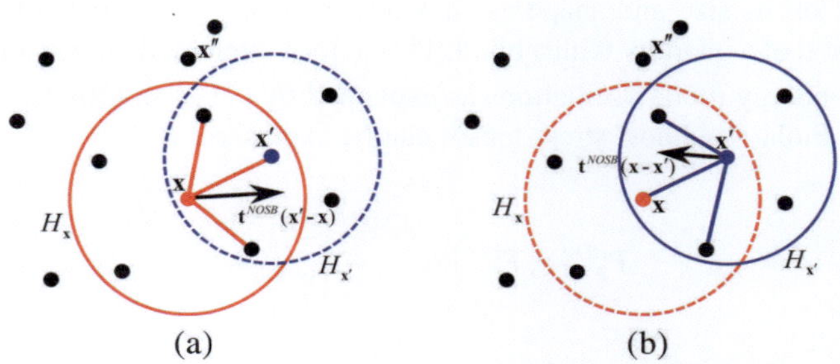

Fig. 3.7 The NOSB-PD internal force density vectors arising from: (**a**) primary bonds (red) in $H_{\mathbf{x}}$ and (**b**) secondary bonds (blue) in $\widetilde{H}_{\mathbf{x}}$

The first Piola-Kirchhoff stress tensor, $\mathbf{P}_{\boldsymbol{\xi}}^{PD}(\mathbf{x},t)$ is evaluated by using the bond-associated deformation gradient tensor, $\mathbf{F}_{\boldsymbol{\xi}}^{PD}(\mathbf{x})$, expressed as (Chen & Spencer, 2019; Madenci et al. 2019a, b; Roy et al., 2020)

$$\mathbf{F}_{\boldsymbol{\xi}}^{PD}(\mathbf{x}) = \int\limits_{H_{\mathbf{x}} \cap H_{\mathbf{x}'}} \widehat{\mu}(\mathbf{x}'' - \mathbf{x})(\mathbf{y}'' - \mathbf{y}) \otimes \mathbf{g}_{\boldsymbol{\xi}}(\mathbf{x}, \mathbf{x}'' - \mathbf{x}) dV_{\mathbf{x}''}. \tag{3.54}$$

or

$$\mathbf{F}_{\boldsymbol{\xi}}^{PD}(\mathbf{x}) = \int\limits_{H_{\mathbf{x}} \cap H_{\mathbf{x}'}} \widehat{\mu}(\mathbf{x}'' - \mathbf{x})(\mathbf{u}'' - \mathbf{u}) \otimes \mathbf{g}_{\boldsymbol{\xi}}(\mathbf{x}, \mathbf{x}'' - \mathbf{x}) dV_{\mathbf{x}''} + \mathbf{I} \tag{3.55}$$

in conjunction with the result of the orthogonality property of PD functions

$$\mathbf{I} = \int\limits_{H_{\mathbf{x}} \cap H_{\mathbf{x}'}} (\mathbf{x}'' - \mathbf{x}) \otimes \mathbf{g}_{\boldsymbol{\xi}}(\mathbf{x}, \mathbf{x}'' - \mathbf{x}) dV_{\mathbf{x}''} \tag{3.56}$$

in which $\boldsymbol{\xi} = \mathbf{x}'' - \mathbf{x}$ and the vector $\mathbf{g}_{\boldsymbol{\xi}}$ is composed of PD functions as (Madenci et al., 2017, 2022)

$$\mathbf{g}_{\boldsymbol{\xi}}(\mathbf{x}, \boldsymbol{\xi}) = \begin{Bmatrix} g_N^{100}(\mathbf{x}, \boldsymbol{\xi}) \\ g_N^{010}(\mathbf{x}, \boldsymbol{\xi}) \\ g_N^{001}(\mathbf{x}, \boldsymbol{\xi}) \end{Bmatrix} \tag{3.57}$$

where superscripts (100), (010), and (001) denote the order of differentiation with respect to the variable, x_1, x_2 and x_3, respectively, and the subscript, $N = 1$ represents the order of TSE in the construction of the PD functions. The PD functions are constructed by considering the intersection region, $H_{\boldsymbol{\xi}} = H_{\mathbf{x}} \cap H_{\mathbf{x}'}$ without any requirement on its size and shape as shown in Fig. 3.8. The subscript $\boldsymbol{\xi}$ indicates the association of a quantity with a bond. This deformation gradient tensor, $\mathbf{F}_{\boldsymbol{\xi}}^{PD}(\mathbf{x})$ is free of zero-energy mode oscillations as explained by Gu et al. (2019).

The first Piola-Kirchhoff stress tensor can be expressed as

$$\mathbf{P}_{\boldsymbol{\xi}}^{PD}\left(\mathbf{x}, \mathbf{F}_{\boldsymbol{\xi}}^{PD}\right) = \frac{\partial W\left(\mathbf{x}, \mathbf{F}_{\boldsymbol{\xi}}^{PD}\right)}{\partial \mathbf{F}_{\boldsymbol{\xi}}^{PD}} \tag{3.58}$$

where $W\left(\mathbf{x}, \mathbf{F}_{\boldsymbol{\xi}}^{PD}\right)$ is the strain energy density function describing the material behavior. The PD internal force vector for NOSB-PD in Eq. (3.2) can be evaluated as

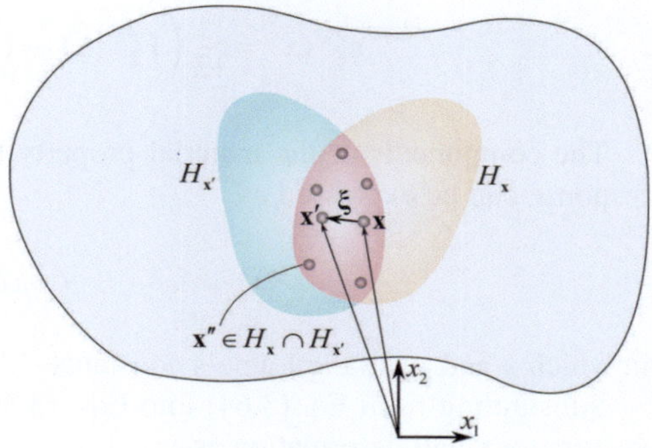

Fig. 3.8 The intersection of interaction domains, $(H_{\mathbf{x}} \cap H_{\mathbf{x}'})$ of material points $\mathbf{x}$ and $\mathbf{x}'$

$$\mathbf{L}^{NOSB}(\mathbf{x}, t) = \int_{H_{\mathbf{x}}} \mathbf{t}^{NOSB}(\mathbf{x}' - \mathbf{x}, t) dV_{\mathbf{x}'} - \int_{\widetilde{H}_{\mathbf{x}}} \mathbf{t}^{NOSB}(\mathbf{x} - \mathbf{x}', t) dV_{\mathbf{x}'} \tag{3.59}$$

For elastic deformation of isotropic materials, the first Piola-Kirchhoff stress tensor can be explicitly expressed as

$$\mathbf{P}_{\xi}^{PD}\left(\mathbf{x}, \mathbf{F}_{\xi}^{PD}\right) = \mathbf{F}_{\xi}^{PD}\mathbf{S}_{\xi}^{PD} = \mathbf{F}_{\xi}^{PD}\left(\mathbf{C} : \mathbf{E}_{\xi}^{PD}(\mathbf{x})\right) \tag{3.60}$$

in which $\mathbf{S}_{\xi}^{PD}$ is the second Piola-Kirchhoff stress tensor, $\mathbf{C}$ is the fourth-order material property tensor, and $\mathbf{E}_{\xi}^{PD}(\mathbf{x})$ represents the bond-associated (BA) Green's strain tensor which can be evaluated as

$$\mathbf{E}_{\xi}^{PD}(\mathbf{x}) = \frac{1}{2}\left(\left(\mathbf{F}_{\xi}^{PD}(\mathbf{x})\right)^{\mathrm{T}}\mathbf{F}_{\xi}^{PD}(\mathbf{x}) - \mathbf{I}\right) \tag{3.61}$$

The force density vectors can be rewritten as

$$\mathbf{t}^{NOSB}(\mathbf{x}' - \mathbf{x}, t) = \widehat{\mu}(\mathbf{x}' - \mathbf{x})\mathbf{F}_{\xi}^{PD}(\mathbf{x}, t)\mathbf{S}_{\xi}^{PD}(\mathbf{x}, t)\mathbf{g}(\mathbf{x}, \boldsymbol{\xi}) \tag{3.62}$$

and

$$\mathbf{t}^{NOSB}(\mathbf{x} - \mathbf{x}', t) = \widehat{\mu}(\mathbf{x} - \mathbf{x}')\mathbf{F}_{\xi'}^{PD}(\mathbf{x}', t)\mathbf{S}_{\xi'}^{PD}(\mathbf{x}', t)\mathbf{g}(\mathbf{x}', \boldsymbol{\xi}') \tag{3.63}$$

Small deformation assumptions lead to

$$\mathbf{P}_{\xi}^{PD}(\mathbf{x}) = \boldsymbol{\sigma}_{\xi}^{PD}(\mathbf{x}) = \frac{\partial W\left(\mathbf{x}, \boldsymbol{\varepsilon}_{\xi}^{PD}\right)}{\partial \boldsymbol{\varepsilon}_{\xi}^{PD}} = \mathbf{C} : \boldsymbol{\varepsilon}_{\xi}^{PD}(\mathbf{x}) \tag{3.64}$$

in which $\boldsymbol{\sigma}_{\xi}^{PD}$ is Cauchy's stress tensor and $\boldsymbol{\varepsilon}_{\xi}^{PD}$ is the linear strain tensor, which can be evaluated as

$$\boldsymbol{\varepsilon}_{\xi}^{PD}(\mathbf{x}) = \left[\frac{1}{2}\left(\mathbf{F}_{\xi}^{PD}(\mathbf{x}) + \left(\mathbf{F}_{\xi}^{PD}(\mathbf{x})\right)^{\mathrm{T}}\right) - \mathbf{I}\right] \tag{3.65}$$

The components of the material property tensor $\mathbf{C}$ for linearly elastic isotropic response can be expressed as

$$\mathbf{C} \sim \lambda\delta_{ij}\delta_{k\ell} + \mu\left(\delta_{ik}\delta_{j\ell} + \delta_{i\ell}\delta_{jk}\right) \tag{3.66}$$

in which λ and μ are the Lame's constants.

Substituting from Eq. (3.64) into Eqs. (3.50) and (3.51) yields the force density vectors for small deformation as

$$\mathbf{t}^{NOSB}(\mathbf{x}' - \mathbf{x}, t) = \widehat{\mu}(\mathbf{x}' - \mathbf{x})\boldsymbol{\sigma}_{\xi}^{PD}(\mathbf{x}, t)\mathbf{g}(\mathbf{x}, \boldsymbol{\xi}) \tag{3.67}$$

and

$$\mathbf{t}^{NOSB}(\mathbf{x} - \mathbf{x}', t) = \widehat{\mu}(\mathbf{x} - \mathbf{x}')\boldsymbol{\sigma}_{\xi'}^{PD}(\mathbf{x}', t)\mathbf{g}(\mathbf{x}', \boldsymbol{\xi}') \tag{3.68}$$

The PD internal force vector for NOSB-PD in Eq. (3.59) can be expressed as

$$\mathbf{L}^{NOSB}(\mathbf{x}, t) = \int_{H_{\mathbf{x}}} \widehat{\mu}(\mathbf{x}' - \mathbf{x})\boldsymbol{\sigma}_{\xi}^{PD}(\mathbf{x}, t)\mathbf{g}(\mathbf{x}, \boldsymbol{\xi})dV_{\mathbf{x}'} - \int_{\widetilde{H}_{\mathbf{x}}} \widehat{\mu}(\mathbf{x} - \mathbf{x}')\boldsymbol{\sigma}_{\xi'}^{PD}(\mathbf{x}', t)\mathbf{g}(\mathbf{x}', \boldsymbol{\xi}')dV_{\mathbf{x}'} \tag{3.69}$$

For a constant horizon, δ, Eq. (3.69) reduces to

$$\mathbf{L}^{NOSB}(\mathbf{x}, t) = \int_{H_{\mathbf{x}}} \widehat{\mu}(\mathbf{x}' - \mathbf{x})\left(\boldsymbol{\sigma}_{\xi}^{PD}(\mathbf{x}, t)\mathbf{g}(\mathbf{x}, \boldsymbol{\xi}) - \boldsymbol{\sigma}_{\xi'}^{PD}(\mathbf{x}', t)\mathbf{g}(\mathbf{x}', \boldsymbol{\xi}')\right)dV_{\mathbf{x}'} \tag{3.70}$$

3.3.4 Peridynamic Differential Operator (PDDO) Interactions

As shown in Fig. 3.9, the explicit form of the PDDO internal force density vectors for any material model can be derived from Eq. (3.5) as

$$\mathbf{t}^{PDDO}(\mathbf{x}, \mathbf{x}' - \mathbf{x}, t) = \widehat{\mu}(\mathbf{x}' - \mathbf{x})\mathbf{P}^{PD}(\mathbf{x}, t)\mathbf{g}(\mathbf{x}, \boldsymbol{\xi}) \tag{3.71}$$

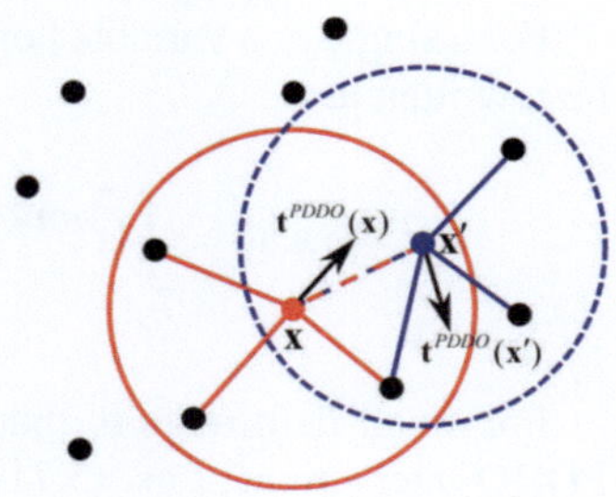

Fig. 3.9 The PDDO internal force density vectors at points $\mathbf{x}$ and $\mathbf{x}'$ due to interactions with only their family members

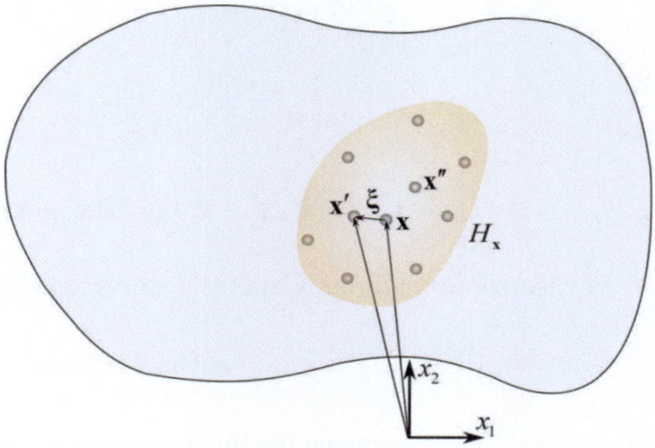

Fig. 3.10 The interaction domain $H_{\mathbf{x}}$ of material point $\mathbf{x}$ for a PDDO interaction

and

$$\mathbf{t}^{PDDO}(\mathbf{x}', \mathbf{x}' - \mathbf{x}, t) = \widehat{\mu}(\mathbf{x}' - \mathbf{x})\mathbf{P}^{PD}(\mathbf{x}', t)\mathbf{g}(\mathbf{x}, \boldsymbol{\xi}) \tag{3.72}$$

The first Piola-Kirchhoff stress tensor can be explicitly expressed as

$$\mathbf{P}^{PD}\left(\mathbf{x}, \mathbf{F}^{PD}\right) = \frac{\partial W\left(\mathbf{x}, \mathbf{F}^{PD}(\mathbf{x})\right)}{\partial \mathbf{F}^{PD}} \tag{3.73}$$

where $W(\mathbf{x}, \mathbf{F}^{PD}(\mathbf{x}))$ is the strain energy density function describing the material behavior and $\mathbf{F}^{PD}(\mathbf{x})$ is the point-wise deformation gradient, as shown in Fig. 3.10.

For elastic deformation of isotropic materials, the first Piola-Kirchhoff stress tensor can be explicitly expressed as

$$\mathbf{P}^{PD}\left(\mathbf{x}, \mathbf{F}^{PD}\right) = \mathbf{F}^{PD}\mathbf{S}^{PD} = \mathbf{F}^{PD}\left(\mathbf{C} : \mathbf{E}^{PD}(\mathbf{x})\right) \tag{3.74}$$

in which $\mathbf{S}^{PD}$ is the second Piola-Kirchhoff stress tensor and $\mathbf{E}^{PD}(\mathbf{x})$ represents Green's strain tensor, which can be evaluated as

$$\mathbf{E}^{PD}(\mathbf{x}) = \frac{1}{2}\left(\left(\mathbf{F}^{PD}(\mathbf{x})\right)^{\mathrm{T}}\mathbf{F}^{PD}(\mathbf{x}) - \mathbf{I}\right) \tag{3.75}$$

For a single or a variable horizon, the PDDO internal force vector in Eq. (3.5) can be rewritten as

$$\mathbf{L}^{PDDO}(\mathbf{x},t) = \int_{H_{\mathbf{x}}} \left(\mathbf{t}^{PDDO}(\mathbf{x}',\mathbf{x}'-\mathbf{x},t) - \mathbf{t}^{PDDO}(\mathbf{x},\mathbf{x}'-\mathbf{x},t)\right) dV_{\mathbf{x}'} \tag{3.76}$$

For small deformation, that is, $\mathbf{P}^{PD}(\mathbf{x}) = \boldsymbol{\sigma}^{PD}(\mathbf{x})$, the force density vectors for PDDO interactions, Eqs. (3.71) and (3.72), can be rewritten as

$$\mathbf{t}^{PDDO}(\mathbf{x},\mathbf{x}'-\mathbf{x},t) = \widehat{\mu}(\mathbf{x}'-\mathbf{x})\boldsymbol{\sigma}^{PD}(\mathbf{x},t)\mathbf{g}(\mathbf{x},\boldsymbol{\xi}) \tag{3.77}$$

and

$$\mathbf{t}^{PDDO}(\mathbf{x}',\mathbf{x}'-\mathbf{x},t) = \widehat{\mu}(\mathbf{x}'-\mathbf{x})\boldsymbol{\sigma}^{PD}(\mathbf{x}',t)\mathbf{g}(\mathbf{x},\boldsymbol{\xi}) \tag{3.78}$$

For isotropic material, Cauchy's stress tensor, $\boldsymbol{\sigma}^{PD}$ can be evaluated as

$$\boldsymbol{\sigma}^{PD}(\mathbf{x}) = \mathbf{C} : \boldsymbol{\varepsilon}^{PD}(\mathbf{x}) \tag{3.79}$$

in which $\boldsymbol{\varepsilon}^{PD}(\mathbf{x})$ represents the linear strain tensor and $\mathbf{C}$ is a material property tensor. The linear strain tensor, $\boldsymbol{\varepsilon}^{PD}(\mathbf{x})$ is evaluated as

$$\boldsymbol{\varepsilon}^{PD}(\mathbf{x}) = \left[\frac{1}{2}\left(\mathbf{F}^{PD}(\mathbf{x}) + \left(\mathbf{F}^{PD}(\mathbf{x})\right)^{\mathrm{T}}\right) - \mathbf{I}\right] \tag{3.80}$$

with $\mathbf{F}^{PD}(\mathbf{x})$ representing the point-wise deformation gradient tensor. Substituting for the force density vectors from Eqs. (3.77) and (3.78) in Eq. (3.76) yields the PDDO internal force vector as

$$\mathbf{L}^{PDDO}(\mathbf{x},t) = \int_{H_{\mathbf{x}}} \widehat{\mu}(\mathbf{x}'-\mathbf{x})\left(\boldsymbol{\sigma}^{PD}(\mathbf{x}',t) - \boldsymbol{\sigma}^{PD}(\mathbf{x},t)\right)\mathbf{g}(\mathbf{x},\boldsymbol{\xi}) dV_{\mathbf{x}'} \tag{3.81}$$

3.4 PD Equilibrium Equations Under Homogeneous Deformation

For a homogeneous isotropic expansion specified as $\mathbf{y} = \mathbf{m}\mathbf{x} + \mathbf{x}$ with $\mathbf{m} = s_0\mathbf{I}$, the deformation gradient tensor can be evaluated as

$$\mathbf{F}_0 = (s_0 + 1)\mathbf{I} \tag{3.82}$$

and the corresponding stretch, s becomes

$$s = \frac{|\mathbf{y}' - \mathbf{y}| - |\boldsymbol{\xi}|}{|\boldsymbol{\xi}|} = s_0 \tag{3.83}$$

3.4.1 BB Interactions

For a single and symmetric horizon, the internal force vector for BB-PD can be evaluated as

$$\begin{aligned} \mathbf{L}^{BB}(\mathbf{x}, t) &= \int\limits_{H_\mathbf{x}} c s_0 \frac{\mathbf{m}(\mathbf{x}' - \mathbf{x}) + \mathbf{x}' - \mathbf{x}}{|\mathbf{m}(\mathbf{x}' - \mathbf{x}) + \mathbf{x}' - \mathbf{x}|} dV_{\mathbf{x}'} \\ &= \int\limits_{H_\mathbf{x}} c s_0 \frac{(s_0 + 1)(\mathbf{x}' - \mathbf{x})}{(s_0 + 1)|(\mathbf{x}' - \mathbf{x})|} dV_{\mathbf{x}'} = c s_0 \int\limits_{H_\mathbf{x}} \frac{\boldsymbol{\xi}}{|\boldsymbol{\xi}|} dV_{\mathbf{x}'} = \mathbf{0} \end{aligned} \tag{3.84}$$

with $\mathbf{L}^{\mathrm{BB}}(\mathbf{x}, t) \neq \mathbf{0}$ for a non-symmetric horizon. Thus, residual forces develop at material points located in region, $\mathscr{R}$ whose extent is δ from the boundary, as shown in Fig. 3.1.

3.4.2 OSB Interactions

The internal force vector for OSB-PD can be evaluated as

$$\mathbf{L}^{OSB}(\mathbf{x}, t) = \int\limits_{H_\mathbf{x}} \left\{ \frac{9}{4\pi\delta^3} \mu \left(\frac{4\nu - 1}{1 - 2\nu} \right) \frac{1}{|\boldsymbol{\xi}|} (\theta(\mathbf{x}) + \theta(\mathbf{x}')) + \frac{45\mu}{2\pi\delta^3} \frac{1}{|\boldsymbol{\xi}|} s \right\} \frac{\mathbf{y}' - \mathbf{y}}{|\mathbf{y}' - \mathbf{y}|} dV_{\mathbf{x}'} \text{ for 3D} \tag{3.85}$$

in which the volume change (dilatation) becomes

$$\theta(\mathbf{x}, t) = 3 s_0 \text{ for 3D.} \tag{3.86}$$

For a single and symmetric horizon, it can be reduced to

$$\mathbf{L}^{OSB}(\mathbf{x}, t) = \frac{9}{2\pi\delta^3} \mu \left\{ 3 \left(\frac{4\nu - 1}{1 - 2\nu} \right) + 5 \right\} s_0 \int\limits_{H_\mathbf{x}} \frac{1}{|\boldsymbol{\xi}|} \frac{\boldsymbol{\xi}}{|\boldsymbol{\xi}|} dV_{\mathbf{x}'} = \mathbf{0} \tag{3.87}$$

However, $\mathbf{L}^{OSB}(\mathbf{x}, t) \neq \mathbf{0}$ for a non-symmetric horizon. Thus, residual forces develop at material points located in region, $\mathscr{R}$ whose extent is 2δ from the boundary, as shown in Fig. 3.1.

3.4.3 NOSB Interactions

For the specified homogeneous deformation, Eq. (3.82), the PD form of the bond-associated deformation gradient tensor, $\mathbf{F}_{\boldsymbol{\xi}}^{PD}$ from Eq. (3.55) along with Eq. (3.57) and (3.58) can be evaluated as

$$\mathbf{F}_{\boldsymbol{\xi}}^{PD}(\mathbf{x}) = \mathbf{F}_0\left(\int_{H_{\mathbf{x}} \cap H_{\mathbf{x}'}} \boldsymbol{\xi} \otimes \mathbf{g}_{\boldsymbol{\xi}}(\mathbf{x}'' - \mathbf{x}) dV_{\mathbf{x}'}\right) \tag{3.88}$$

or

$$\mathbf{F}_{\boldsymbol{\xi}}^{PD}(\mathbf{x}) = \mathbf{F}_0 \mathbf{I} = \mathbf{F}_0 \tag{3.89}$$

Corresponding to this homogeneous deformation gradient, the first Piola-Kirchhoff stress tensor can be evaluated as

$$\mathbf{P}_{\boldsymbol{\xi}}^{PD}(\mathbf{x}, t) = \mathbf{P}_{\boldsymbol{\xi}}^{PD}\left(\mathbf{F}_{\boldsymbol{\xi}}^{PD}\right) = \mathbf{P}_0 \tag{3.90}$$

After substituting from Eq. (3.90) into Eqs. (3.50) and (3.51), the internal force vector $\mathbf{L}^{NOSB}(\mathbf{x}, t)$ for the NOSB-PD, can be evaluated as

$$\begin{aligned} \mathbf{L}^{NOSB}(\mathbf{x}, t) &= \int\limits_{H_{\mathbf{x}}} (\mathbf{t}(\mathbf{x}' - \mathbf{x}, t) - \mathbf{t}(\mathbf{x} - \mathbf{x}', t)) dV_{\mathbf{x}'} \\ &= \mathbf{P}_0 \int\limits_{H_{\mathbf{x}}} \left(w(|\boldsymbol{\xi}|)\mathbf{K}^{-1}(\mathbf{x})\boldsymbol{\xi} - w(|\boldsymbol{\xi}|)\mathbf{K}^{-1}(\mathbf{x}')(-\boldsymbol{\xi})\right) dV_{\mathbf{x}'} \end{aligned} \tag{3.91}$$

or

$$\mathbf{L}^{NOSB}(\mathbf{x}, t) = \mathbf{P}_0 \int\limits_{H_{\mathbf{x}}} w(|\boldsymbol{\xi}|)\left(\mathbf{K}^{-1}(\mathbf{x}) + \mathbf{K}^{-1}(\mathbf{x}')\right)\boldsymbol{\xi} dV_{\mathbf{x}'} \tag{3.92}$$

For a single and symmetric horizon with $\mathbf{K}(\mathbf{x}) = \mathbf{K}(\mathbf{x}') = \mathbf{K}_0$, Eq. (3.92) becomes

$$\mathbf{L}^{NOSB}(\mathbf{x}, t) = 2\mathbf{P}_0 \mathbf{K}_0{}^{-1} \int\limits_{H_{\mathbf{x}}} w(|\boldsymbol{\xi}|)\boldsymbol{\xi} dV_{\mathbf{x}'} \tag{3.93}$$

in which

$$\int_{H_{\mathbf{x}}} w(|\boldsymbol{\xi}|)\boldsymbol{\xi} dV_{\mathbf{x}'} = \mathbf{0} \tag{3.94}$$

Finally, the internal force vector from Eq. (3.93) is evaluated as

$$\mathbf{L}^{NOSB}(\mathbf{x}, t) = \mathbf{0} \tag{3.95}$$

This result is consistent with the evaluation of $\nabla \cdot \mathbf{P}(\mathbf{x}, t) = \mathbf{0}$, Eq. (3.3) corresponding to a homogeneous deformation. However, for a nonsymmetric horizon with $\mathbf{K}(\mathbf{x}) \neq \mathbf{K}(\mathbf{x}')$, Eq. (3.92) becomes

$$\mathbf{L}^{NOSB}(\mathbf{x}, t) = \mathbf{P}_0 \int_{H_{\mathbf{x}}} w(|\boldsymbol{\xi}|)\left(\mathbf{K}^{-1}(\mathbf{x}) + \mathbf{K}^{-1}(\mathbf{x}')\right)\boldsymbol{\xi} dV_{\mathbf{x}'} = \mathbf{P}_0\mathbf{c} \neq \mathbf{0} \tag{3.96}$$

with $\mathbf{c}$ being a nonvanishing vector. It deviates from the zero result of the internal force vector of CCM from Eq. (3.3). Thus, residual forces develop at material points located in region $\mathscr{R}$, whose extent is 2δ from the boundary, as shown in Fig. 3.1.

3.4.4 PDDO Interactions

Corresponding to the homogeneous deformation gradient, the PDDO form of the internal force vector, Eq. (3.5) results in

$$\mathbf{L}^{PDDO}(\mathbf{x}, t) = \int_{H_{\mathbf{x}}} \left(\mathbf{P}^{PD}(\mathbf{x}', t) - \mathbf{P}^{PD}(\mathbf{x}, t)\right)\mathbf{g}(\mathbf{x}, \boldsymbol{\xi}) dV_{\mathbf{x}'} = \mathbf{0} \tag{3.97}$$

with $\mathbf{P}^{PD}(\mathbf{x}', t) = \mathbf{P}^{PD}(\mathbf{x}, t) = \mathbf{P}_0$ for homogeneous deformation; thus, eliminating the residual forces.

3.5 Unification of PD Equilibrium Equations

In order to eliminate the residual forces and impose traction (natural) and displacement (essential) boundary conditions, the domain, Ω can be split into three regions of $\mathscr{D}$, $\mathscr{R}$, and $\mathscr{B}$, as shown in Fig. 3.11.

The extent of the domain, Ω is composed of $\Omega = \mathscr{D} \cup \mathscr{R} \cup \mathscr{B}$ with an interior region, $\mathscr{D}$, outer region, $\mathscr{R}$, and a "real boundary layer" region, $\mathscr{B}$ with nonzero volume. The "real boundary layer" region $\mathscr{B}$ includes the natural boundary region $\mathscr{B}_{\boldsymbol{\sigma}}$ and essential boundary region $\mathscr{B}_{\mathbf{u}}$. The extent of region, $\mathscr{R} \cup \mathscr{B}$ is defined by δ

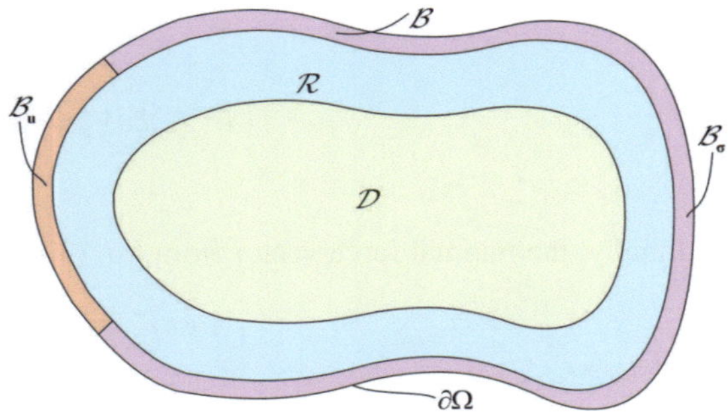

Fig. 3.11 Specific regions of the domain for unified PD equations of motion ($\mathscr{B} = \mathscr{B}_{\mathbf{u}} + \mathscr{B}_{\boldsymbol{\sigma}}$)

for BB- and 2δ for OSB- and NOSB-PD. The PD equations of motion can be recast in respective domains as

$$\rho\ddot{\mathbf{u}}(\mathbf{x},t) = \mathbf{L}^{\alpha}(\mathbf{x},t) + \mathbf{b}(\mathbf{x},t) \text{ for } \mathbf{x} \in \mathcal{D} \text{ and } \alpha = BB, OSB, NOSB \tag{3.98}$$

$$\rho\ddot{\mathbf{u}}(\mathbf{x},t) = \mathbf{L}^{PDDO}(\mathbf{x},t) + \mathbf{b}(\mathbf{x},t) \text{ for } \mathbf{x} \in \mathscr{R} \tag{3.99}$$

$$\mathbf{P}^{PD}(\mathbf{x},t)\mathbf{n} = \mathbf{T}^{*}(\mathbf{x},t) \text{ for } \mathbf{x} \in \mathscr{B}_{\boldsymbol{\sigma}} \tag{3.100}$$

and

$$\mathbf{u}(\mathbf{x},t) = \mathbf{U}^{*}(\mathbf{x},t) \text{ for } \mathbf{x} \in \mathscr{B}_{\mathbf{u}} \tag{3.101}$$

where $\mathbf{T}^{*}$ and $\mathbf{U}^{*}$ represent the applied zero and non-zero tractions on $\mathscr{B}_{\boldsymbol{\sigma}}$ and displacement constraints on $\mathscr{B}_{\mathbf{u}}$, respectively, in the "real boundary layer" region of the domain.

3.6 PD Field Equation

As derived in Chap. 1, corresponding to the BB-PD interactions, the dual horizon PD field equation can be rewritten as

$$C(\mathbf{x},t)\dot{\Theta}(\mathbf{x},t) + \dot{C}(\mathbf{x},t)\Theta(\mathbf{x},t) = L^{BB}(\mathbf{x},t) + s_b(\mathbf{x},t) \text{ for } \mathbf{x} \in \Omega \tag{3.102}$$

with

$$L^{BB}(\mathbf{x},t) = \int\limits_{H_{\mathbf{x}}} h^{BB}(\mathbf{x}' - \mathbf{x},t)dV_{\mathbf{x}'} - \int\limits_{\widetilde{H}_{\mathbf{x}}} h^{BB}(\mathbf{x} - \mathbf{x}',t)dV_{\mathbf{x}'} + s_b(\mathbf{x},t) \tag{3.103}$$

in which $L^{BB}(\mathbf{x},t)$ replaces the divergence of the field flux vector, $\mathbf{q}$ as

$$L(\mathbf{x},t) = \nabla \cdot \mathbf{q}(\mathbf{x},t) \text{ for } \mathbf{x} \in \Omega \tag{3.104}$$

with

$$L(\mathbf{x}, t) = \lim_{\delta \to 0} L^{BB}(\mathbf{x}, t) \tag{3.105}$$

where $\mathbf{q}(\mathbf{x}, t)$ is the field flux at point $\mathbf{x}$.

Alternative to Eq. (3.103), the PDDO form of divergence of field flux vector can be expressed as

$$L^{PDDO}(\mathbf{x}, t) = \int_{H_\mathbf{x}} \widehat{\mu}(\mathbf{x}' - \mathbf{x}) \left(\mathbf{q}^{PD}(\mathbf{x}', t) - \mathbf{q}^{PD}(\mathbf{x}, t) \right) \cdot \mathbf{g}(\mathbf{x}, \boldsymbol{\xi}) dV_{\mathbf{x}'} \text{ for } \mathbf{x} \in (\Omega - \mathscr{B}) \tag{3.106}$$

and

$$\mathbf{q}^{PD}(\mathbf{x}, t) = \mathbf{q}^{*}(\mathbf{x}, t) \text{ for } \mathbf{x} \in \mathscr{B} \tag{3.107}$$

in which $\mathbf{q}^{PD}$ is the PDDO form of the field flux vector and $\mathbf{q}^{*}(\mathbf{x}, t)$ is the specified field flux on the "real" boundary layer region $\mathscr{B}$ of nonzero volume as shown in Fig. 3.11. The vector $\mathbf{g}(\mathbf{x}, \boldsymbol{\xi})$ composed of PD functions is defined as

$$\mathbf{g}(\mathbf{x}, \boldsymbol{\xi}) = \begin{Bmatrix} g_2^{100}(\mathbf{x}, \boldsymbol{\xi}) \\ g_2^{010}(\mathbf{x}, \boldsymbol{\xi}) \\ g_2^{001}(\mathbf{x}, \boldsymbol{\xi}) \end{Bmatrix} \tag{3.108}$$

It is worth noting that Eq. (3.106) has a single horizon even for irregular discretization. However, it is based on point-wise PD interactions rather than bond-wise.

3.7 PD Flow Density of Field

3.7.1 BB Interactions

For BB-PD interactions as shown in Fig 3.12, the flow density of the field in Eq. (3.103) depends only on the field variable, Θ at points $\mathbf{x}$ and $\mathbf{x}'$. Therefore, the flow density of the field for primary, $h(\mathbf{x}' - \mathbf{x}, t)$ and secondary, $h(\mathbf{x} - \mathbf{x}', t)$ bonds can be expressed as (Madenci & Oterkus, 2014)

$$h^{BB}(\mathbf{x}' - \mathbf{x}, t) = \widehat{\mu}(\mathbf{x}' - \mathbf{x}) \frac{\kappa(\delta_\mathbf{x})}{2} \frac{\Theta' - \Theta}{|\mathbf{x}' - \mathbf{x}|} \text{ for } \mathbf{x}' \in H_\mathbf{x} \tag{3.109}$$

and

$$h^{BB}(\mathbf{x} - \mathbf{x}', t) = \widehat{\mu}(\mathbf{x} - \mathbf{x}') \frac{\kappa(\delta_{\mathbf{x}'})}{2} \frac{\Theta - \Theta'}{|\mathbf{x}' - \mathbf{x}|} \text{ for } \mathbf{x}' \in \widetilde{H}_\mathbf{x} \tag{3.110}$$

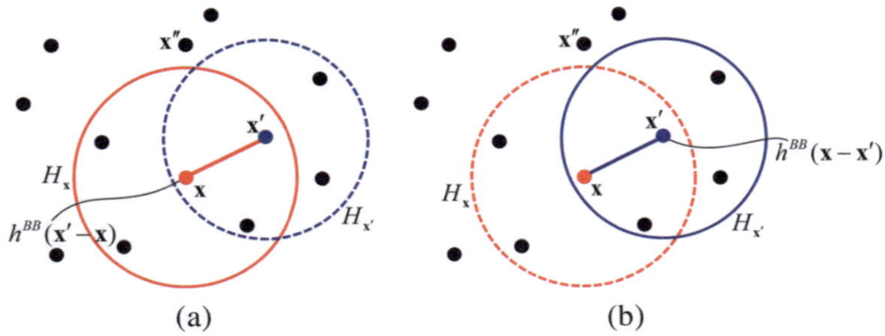

Fig. 3.12 BB PD interactions for flow density of the field through: (**a**) primary bonds (red) in $H_{\mathbf{x}}$ and (**b**) secondary bonds (blue) in $\widetilde{H}_{\mathbf{x}}$

where $\widehat{\mu}(\mathbf{x}' - \mathbf{x})$ and $\widehat{\mu}(\mathbf{x} - \mathbf{x}')$ are the status parameters, $\kappa(\delta_{\mathbf{x}})$ and $\kappa(\delta_{\mathbf{x}'})$ represent the primary and secondary bond constants, respectively.

The bond constant depends on the horizon size, bulk material property, and dimension of the analysis. Its explicit relation to the bulk material property, k can be derived as (Madenci & Oterkus, 2014)

$$\kappa(\delta) = \begin{cases} \dfrac{6k}{\pi\delta^4} & 3\text{D} \\ \dfrac{6k}{\pi h\delta^3} & 2\text{D} \\ \dfrac{2k}{A\delta^2} & 1\text{D} \end{cases} \tag{3.111}$$

Thus, the BB form of the internal net field flow for dual horizon PD can be evaluated as

$$L^{BB}(\mathbf{x}, t) = \int\limits_{H_{\mathbf{x}}} h^{BB}(\mathbf{x}' - \mathbf{x}, t) dV_{\mathbf{x}'} - \int\limits_{\widetilde{H}_{\mathbf{x}}} h^{BB}(\mathbf{x} - \mathbf{x}', t) dV_{\mathbf{x}'} \tag{3.112}$$

For a uniform and single horizon, $h^{BB}(\mathbf{x}' - \mathbf{x}, t) = -\, h^{BB}(\mathbf{x} - \mathbf{x}', t)$, Eq. (3.112) reduces to

$$L^{BB}(\mathbf{x}, t) = \int\limits_{H_{\mathbf{x}}} \left(2h^{BB}(\mathbf{x}' - \mathbf{x}, t)\right) dV_{\mathbf{x}'} = \int\limits_{H_{\mathbf{x}}} f^{BB}(\mathbf{x}' - \mathbf{x}, t) dV_{\mathbf{x}'} = \int\limits_{H_{\mathbf{x}}} \widehat{\mu}(\mathbf{x}' - \mathbf{x}) \kappa \frac{\Theta' - \Theta}{|\mathbf{x}' - \mathbf{x}|} \, dV_{\mathbf{x}'} \tag{3.113}$$

This equation recovers the original form of the field equation by Madenci and Oterkus (2014).

3.7.2 PDDO Interactions

For PDDO interactions as shown in Fig. 3.13, the explicit form of the flow density of the field for any field flux can be written as

$$h^{PDDO}(\mathbf{x}',t) = \widehat{\mu}(\mathbf{x}'-\mathbf{x})\mathbf{q}^{PD}(\mathbf{x}',t)\cdot\mathbf{g}(\mathbf{x},\boldsymbol{\xi}) \tag{3.114}$$

and

$$h^{PDDO}(\mathbf{x},t) = \widehat{\mu}(\mathbf{x}'-\mathbf{x})\mathbf{q}^{PD}(\mathbf{x},t)\cdot\mathbf{g}(\mathbf{x},\boldsymbol{\xi}) \tag{3.115}$$

The field flux vector can be explicitly expressed as

$$\mathbf{q}^{PD}(\mathbf{x},t) = k\nabla\Theta(\mathbf{x}) \tag{3.116}$$

where $\Theta(\mathbf{x})$ is the field variable. The PD form of its gradient can be expressed as

$$\nabla\Theta(\mathbf{x}) = \int\limits_{H_{\mathbf{x}}} (\Theta(\mathbf{x}+\boldsymbol{\xi}) - \Theta(\mathbf{x}))\mathbf{g}(\mathbf{x},\boldsymbol{\xi})dV_{\mathbf{x}'} \tag{3.117}$$

For a single or a variable horizon, the PDDO form of the divergence of field flux vector in Eq. (3.106) can be rewritten as

$$L^{PDDO}(\mathbf{x},t) = \int\limits_{H_{\mathbf{x}}} \left(h^{PDDO}(\mathbf{x}',t) - h^{PDDO}(\mathbf{x},t)\right)\, dV_{\mathbf{x}'} \text{ for } \mathbf{x}\in(\Omega-\mathcal{B}) \tag{3.118}$$

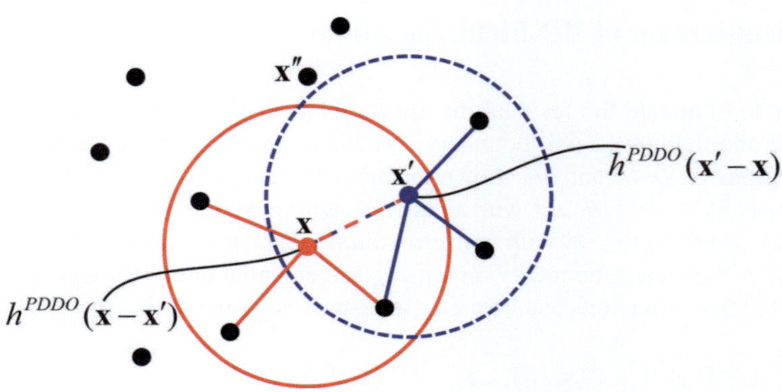

Fig. 3.13 The flow density of the field at points $\mathbf{x}$ and $\mathbf{x}'$ arising from PDDO interactions

3.8 PD Field Equations Under Linear Distribution

3.8.1 BB Interactions

Subjected to a linear field distribution specified as $\Theta(\mathbf{x}) = \mathbf{b} \cdot \mathbf{x} + c$, the BB-PD form of the internal net field flow from Eq. (3.113) at point $\mathbf{x}$ with a constant (single) and complete (symmetric) horizon can be evaluated as

$$L^{BB}(\mathbf{x}, t) = \int\limits_{H_{\mathbf{x}}} \kappa \frac{\mathbf{b} \cdot (\mathbf{x}' - \mathbf{x})}{|\mathbf{x}' - \mathbf{x}|} dV_{\mathbf{x}'} = \int\limits_{H_{\mathbf{x}}} \kappa \frac{\mathbf{b} \cdot \boldsymbol{\xi}}{|\boldsymbol{\xi}|} dV_{\mathbf{x}'} = \kappa \mathbf{b} \cdot \int\limits_{H_{\mathbf{x}}} \frac{\boldsymbol{\xi}}{|\boldsymbol{\xi}|} dV_{\mathbf{x}'} = 0 \tag{3.119}$$

with $L^{BB}(\mathbf{x}, t) \neq 0$ for a non-symmetric horizon resulting in residual internal net field flow. Thus, a residual net field flow develops at material points located in region, $\mathscr{R}$ whose extent is δ from the boundary, as shown in Fig. 3.1.

3.8.2 PDDO Interactions

Corresponding to the linear field distribution, the PDDO form of the divergence of field flux vector in Eq. (3.118) results in

$$L^{PDDO}(\mathbf{x}, t) = \int\limits_{H_{\mathbf{x}}} \widehat{\mu}(\mathbf{x}' - \mathbf{x}) \left(\mathbf{q}^{PD}(\mathbf{x}', t) - \mathbf{q}^{PD}(\mathbf{x}, t) \right) \cdot \mathbf{g}(\mathbf{x}, \boldsymbol{\xi}) dV_{\mathbf{x}'} = 0 \tag{3.120}$$

with $\mathbf{q}^{PD}(\mathbf{x}') = \mathbf{q}^{PD}(\mathbf{x}) = \mathbf{q}_0$ for linear field distribution; thus, eliminating the residual divergence of field flux vector.

3.9 Unification of PD Field Equations

In order to eliminate the residual divergence of field flux vector and impose flux (natural) and field (essential) boundary conditions, the domain, Ω can be split into three regions of $\mathcal{D}$, $\mathscr{R}$ and $\mathscr{B}$, shown in Fig. 3.14. The extent of the domain, Ω is composed of $\Omega = \mathcal{D} \cup \mathscr{R} \cup \mathscr{B}$ with an interior region, $\mathcal{D}$, outer region, $\mathscr{R}$ and a "real boundary layer" region, $\mathscr{B}$ with nonzero volume. The "real boundary layer" region $\mathscr{B}$ includes the natural boundary region $\mathscr{B}_{\mathbf{q}}$ and essential boundary region $\mathscr{B}_{\Theta}$.

The PD field equations can be recast in respective domains as

$$C(\mathbf{x}, t)\dot{\Theta}(\mathbf{x}, t) + \dot{C}(\mathbf{x}, t)\Theta(\mathbf{x}, t) = L^{BB}(\mathbf{x}, t) + s_b(\mathbf{x}, t) \text{ for } \mathbf{x} \in \mathcal{D} \tag{3.121a}$$

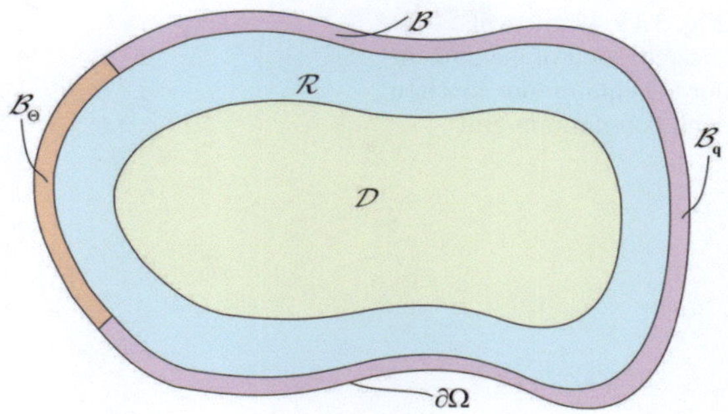

Fig. 3.14 Specific region of the domain for unified PD field equation ($\mathscr{B} = \mathscr{B}_\Theta + \mathscr{B}_\mathbf{q}$)

and

$$C(\mathbf{x},t)\dot{\Theta}(\mathbf{x},t) + \dot{C}(\mathbf{x},t)\Theta(\mathbf{x},t) = L^{PDDO}(\mathbf{x},t) + s_b(\mathbf{x},t) \text{ for } \mathbf{x} \in \mathscr{R} \tag{3.121b}$$

where

$$L^{PDDO}(\mathbf{x},t) = \int_{H_\mathbf{x}} \widehat{\mu}(\mathbf{x}' - \mathbf{x})\left(\mathbf{q}^{PD}(\mathbf{x}',t) - \mathbf{q}^{PD}(\mathbf{x},t)\right) \cdot \mathbf{g}(\mathbf{x},\boldsymbol{\xi}) dV_{\mathbf{x}'} \text{ for } \mathbf{x} \in \mathscr{R} \tag{3.122}$$

$$\mathbf{q}^{PD}(\mathbf{x},t) = \mathbf{q}^*(\mathbf{x},t) \text{ for } \mathbf{x} \in \mathscr{B}_\mathbf{q} \tag{3.123}$$

and

$$\Theta(\mathbf{x},t) = \Theta^*(\mathbf{x},t) \text{ for } \mathbf{x} \in \mathscr{B}_\Theta \tag{3.124}$$

where $\mathbf{q}^*$ and Θ^* represent the applied zero or nonzero field flux on $\mathscr{B}_\mathbf{q}$ and field constraints on $\mathscr{B}_\Theta$, respectively, in the "real boundary layer" region of the domain.

3.10 Global Discretization of PD Equilibrium Equations

Based on the unified PD equilibrium equations, the PD domain $\Omega = \mathcal{D} \cup \mathscr{R} \cup \mathscr{B}$ is discretized with $N = N_\mathcal{D} + N_\mathscr{R} + N_\mathscr{B}$ PD points as shown in Fig. 3.15. The parameters, $N_\mathcal{D}$ and $N_\mathscr{R}$ represent the number of points associated with regions $\mathcal{D}$ and $\mathscr{R}$, respectively. The discretization in these regions can be nonuniform. The parameter $N_\mathscr{B} = N_{\mathscr{B}_\mathbf{u}} + N_{\mathscr{B}_\sigma}$ represents the number of points associated with region $\mathscr{B}$. The parameters $N_{\mathscr{B}_\mathbf{u}}$ and $N_{\mathscr{B}_\sigma}$ represent the number of points associated with regions $\mathscr{B}_\mathbf{u}$ and $\mathscr{B}_\sigma$, respectively. The position of the PD material point k is denoted by $\mathbf{x}_{(k)}$ and its corresponding volume is denoted by $V_{(k)}$.

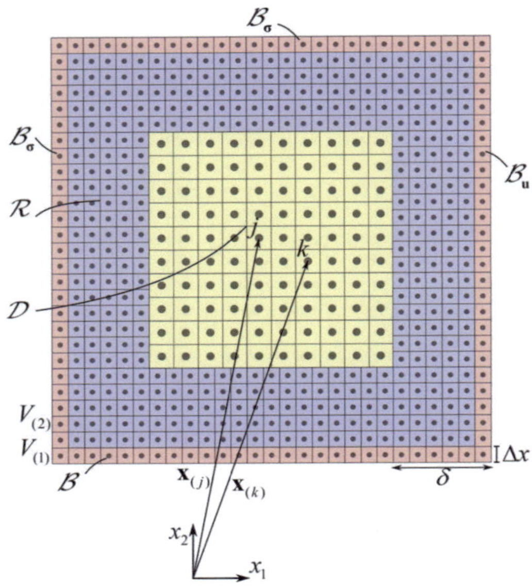

Fig. 3.15 Numerical discretization of the domain for PD equilibrium equation with non-uniform grid

The discrete form of PD internal force vector in the interior region $\mathcal{D}$ and PDDO region $\mathcal{R}$ can be expressed as

$$\mathbf{L}_{(k)} = \begin{cases} \sum_{j=1}^{N_{(k)}} \mathbf{t}^{\alpha}_{(k)(j)} V_{(j)} - \sum_{p=1}^{\widetilde{N}_{(k)}} \mathbf{t}^{\alpha}_{(p)(k)} V_{(p)} & \text{for } \mathbf{x}_{(k)} \in \mathcal{D} \text{ and } \alpha = BB, OSB, NOSB \\ \sum_{j=1}^{N_{(k)}} \left(\mathbf{t}^{PDDO}_{(j)(k)} - \mathbf{t}^{PDDO}_{(k)(j)}\right) V_{(j)} & \text{for } \mathbf{x}_{(k)} \in \mathcal{R} \end{cases} \tag{3.125}$$

where $N_{(k)}$ and $\widetilde{N}_{(k)}$ are the number of primary bonds and secondary bonds, respectively. In the absence of inertia loads, the total virtual work in the domain $\mathcal{D} \cup \mathcal{R}$ can be computed as

$$\sum_{k=1}^{N_{\mathcal{D}}} \delta\mathbf{u}^{T}_{(k)} \left(\sum_{j=1}^{N_{(k)}} \mathbf{t}^{\alpha}_{(k)(j)} V_{(j)} - \sum_{p=1}^{\widetilde{N}_{(k)}} \mathbf{t}^{\alpha}_{(p)(k)} V_{(p)} \right) V_{(k)} + \sum_{k=1}^{N_{\mathcal{R}}} \delta\mathbf{u}^{T}_{(k)} \left(\sum_{j=1}^{N_{(k)}} \left(\mathbf{t}^{PDDO}_{(j)(k)} - \mathbf{t}^{PDDO}_{(k)(j)} \right) V_{(j)} \right) V_{(k)} + \sum_{k=1}^{N_{\mathcal{D}}+N_{\mathcal{R}}} \delta\mathbf{u}^{T}_{(k)} \mathbf{b}_{(k)} V_{(k)} = 0 \tag{3.126}$$

where $\delta \mathbf{u}_{(k)}$ is the virtual displacement vector at point $\mathbf{x}_{(k)}$. This equation can be recast as a global system in matrix form as

$$\delta \mathbf{v}^T \left(\mathbf{F} - \mathbf{F}_{\mathbf{b}}^* \right) = 0 \tag{3.127}$$

The vector of virtual displacements, $\delta \mathbf{v}$, internal forces, $\mathbf{F}$ and the applied body loads, $\mathbf{F}_{\mathbf{b}}^*$ are defined as

$$\delta \mathbf{v} = \{\delta \mathbf{v}_{\mathcal{R}}, \delta \mathbf{v}_{\mathcal{D}}, \delta \mathbf{v}_{\mathcal{B}_{\boldsymbol{\sigma}}}, \delta \mathbf{v}_{\mathcal{B}_{\mathbf{u}}}\}^T \tag{3.128}$$

$$\mathbf{F} = \{ -\mathbf{F}_{\mathcal{R}}, -\mathbf{F}_{\mathcal{D}}, \mathbf{0}, \mathbf{0}\}^T \tag{3.129}$$

and

$$\mathbf{F}_{\mathbf{b}}^* = \left\{ \mathbf{F}_{\mathbf{b}_{\mathcal{R}}}^*, \mathbf{F}_{\mathbf{b}_{\mathcal{D}}}^*, \mathbf{0}, \mathbf{0} \right\}^T \tag{3.130}$$

where the sub-vectors $\delta \mathbf{v}_{\mathcal{R}}$, $\delta \mathbf{v}_{\mathcal{D}}$, $\delta \mathbf{v}_{\mathcal{B}_{\boldsymbol{\sigma}}}$, and $\delta \mathbf{v}_{\mathcal{B}_{\mathbf{u}}}$ contain the unknown displacement components of the nodal points in the PDDO region $\mathcal{R}$, interior region $\mathcal{D}$, and along the PD boundary regions $\mathcal{B}_{\boldsymbol{\sigma}}$ and $\mathcal{B}_{\mathbf{u}}$, respectively. The subvectors $\mathbf{F}_{\mathcal{R}}$, $\mathbf{F}_{\mathcal{D}}$ contain the unknown internal force components of the nodal points in the PDDO region $\mathcal{R}$ and interior region $\mathcal{D}$. The subvectors $\mathbf{F}_{\mathbf{b}_{\mathcal{R}}}^*$,$\mathbf{F}_{\mathbf{b}_{\mathcal{D}}}^*$ are the applied body loads of the nodal points in the PDDO region $\mathcal{R}$ and interior region $\mathcal{D}$. They are defined as

$$\mathbf{F}_{\mathbf{b}_{\mathcal{R}}}^* = \left\{ \mathbf{b}_{(1)} V_{(1)}, \ \mathbf{b}_{(2)} V_{(2)}, \ \mathbf{b}_{(3)} V_{(3)}, \ \cdots \ \mathbf{b}_{(N_{\mathcal{R}})} V_{(N_{\mathcal{R}})} \right\}^T \tag{3.131}$$

and

$$\mathbf{F}_{\mathbf{b}_{\mathcal{D}}}^* = \left\{ \mathbf{b}_{(1)} V_{(1)}, \ \mathbf{b}_{(2)} V_{(2)}, \ \mathbf{b}_{(3)} V_{(3)}, \ \cdots \ \mathbf{b}_{(N_{\mathcal{D}})} V_{(N_{\mathcal{D}})} \right\}^T \tag{3.132}$$

The internal force vector, $\mathbf{F}$ can be written in terms of the global stiffness matrix, $\mathbf{K}$ and the unknown displacement vector, $\mathbf{V}$ as

$$\mathbf{F} = \mathbf{K} \mathbf{V} \tag{3.133}$$

where

$$\mathbf{F} = \sum_{k=1}^{N_{\mathcal{D}} + N_{\mathcal{R}}} \mathbf{L}_{(k)} V_{(k)} \tag{3.134}$$

$$\mathbf{K} = \sum_{k=1}^{N_{\mathcal{D}}} \left(\sum_{j=1}^{N_{(k)}} \mathbf{k}_{(k)(j)}^{\alpha} - \sum_{p=1}^{\widetilde{N}_{(k)}} \mathbf{k}_{(p)(k)}^{\alpha} \right) + \sum_{k=1}^{N_{\mathcal{R}}} \sum_{j=1}^{N_{(k)}} \mathbf{k}_{(k)(j)}^{PDDO} \tag{3.135}$$

and

$$\mathbf{V} = \{\mathbf{V}_{\mathscr{R}}, \mathbf{V}_{\mathcal{D}}, \mathbf{V}_{\mathscr{B}_\sigma}, \mathbf{V}_{\mathscr{B}_\mathrm{u}}\}^T \tag{3.136}$$

The virtual work arising from the applied tractions along the PD boundary region $\Gamma_{\mathscr{B}_\sigma}$ can be expressed as

$$\sum_{k=1}^{N_{\mathscr{B}_\sigma}} \delta\mathbf{u}_{(k)}^T \left(\mathbf{T}_{(k)}^{PD} - \mathbf{T}_{(k)}^*\right) S_{(k)} = 0 \tag{3.137}$$

where $S_{(k)} = h\Delta x$ with $h = 1$ is the boundary area of the PD material point k. It can be recast as

$$\delta\mathbf{v}_{\mathscr{B}_\sigma}^T \left(\sum_{k=1}^{N_{\mathscr{B}_\sigma}} \mathbf{K}_{\sigma(k)}\mathbf{V} - \mathbf{F}_{\mathscr{B}_\sigma}^*\right) = \delta\mathbf{v}_{\mathscr{B}_\sigma}^T \left(\mathbf{F}_{\mathscr{B}_\sigma} - \mathbf{F}_{\mathscr{B}_\sigma}^*\right)$$
$$= \delta\mathbf{v}_{\mathscr{B}_\sigma}^T \left(\mathbf{K}_\sigma\mathbf{V} - \mathbf{F}_{\mathscr{B}_\sigma}^*\right) = 0 \tag{3.138}$$

in which $\mathbf{K}_\sigma$ is the global stiffness matrix due to the applied tractions. The vector, $\mathbf{F}_{\mathscr{B}_\sigma}^*$ arising from the applied traction is defined as

$$\mathbf{F}_{\mathscr{B}_\sigma}^* = \left\{\mathbf{T}_{(1)}^* S_{(1)},, \quad \mathbf{T}_{(2)}^* S_{(2)}, \quad \cdots \quad \mathbf{T}_{(q)}^* S_{(q)} \quad \cdots \quad \mathbf{T}_{(N_{\mathscr{B}_\sigma})}^* S_{(N_{\mathscr{B}_\sigma})}\right\}^T \tag{3.139}$$

Combining Eqs. (3.127) and (3.133) along with Eq. (3.138) results in the final form of the governing equations as

$$\begin{bmatrix} \mathbf{K}_{\mathscr{R}\mathscr{R}} & \mathbf{K}_{\mathscr{R}\mathcal{D}} & \mathbf{K}_{\mathscr{R}\mathscr{B}_\sigma} & \mathbf{K}_{\mathscr{R}\mathscr{B}_\mathrm{u}} \\ \mathbf{K}_{\mathcal{D}\mathscr{R}} & \mathbf{K}_{\mathcal{D}\mathcal{D}} & \mathbf{0} & \mathbf{0} \\ \mathbf{K}_{\mathscr{B}_\sigma\mathscr{R}} & \mathbf{0} & \mathbf{K}_{\mathscr{B}_\sigma\mathscr{B}_\sigma} & \mathbf{K}_{\mathscr{B}_\sigma\mathscr{B}_\mathrm{u}} \\ \mathbf{K}_{\mathscr{B}_\mathrm{u}\mathscr{R}} & \mathbf{0} & \mathbf{K}_{\mathscr{B}_\mathrm{u}\mathscr{B}_\sigma} & \mathbf{K}_{\mathscr{B}_\mathrm{u}\mathscr{B}_\mathrm{u}} \end{bmatrix} \begin{Bmatrix} \mathbf{V}_{\mathscr{R}} \\ \mathbf{V}_{\mathcal{D}} \\ \mathbf{V}_{\mathscr{B}_\sigma} \\ \mathbf{V}_{\mathscr{B}_\mathrm{u}} \end{Bmatrix} = \begin{Bmatrix} \mathbf{F}_{\mathbf{b}_{\mathscr{R}}}^* \\ \mathbf{F}_{\mathbf{b}_{\mathcal{D}}}^* \\ \mathbf{F}_{\mathscr{B}_\sigma}^* \\ \mathbf{0} \end{Bmatrix} \tag{3.140}$$

where the submatrices $\mathbf{K}_{\mathscr{R}\mathscr{R}}$, $\mathbf{K}_{\mathscr{R}\mathcal{D}}$, $\mathbf{K}_{\mathscr{R}\mathscr{B}_\sigma}$, and $\mathbf{K}_{\mathscr{R}\mathscr{B}_\mathrm{u}}$ with the dimensions $2N_{\mathscr{R}} \times 2N_{\mathscr{R}}$, $2N_{\mathscr{R}} \times 2N_{\mathcal{D}}$, $2N_{\mathscr{R}} \times 2N_{\mathscr{B}_\sigma}$, and $2N_{\mathscr{R}} \times 2N_{\mathscr{B}_\mathrm{u}}$, respectively, represent the self-stiffness and coupling stiffness matrices between the unknown vectors of $\mathbf{V}_{\mathscr{R}}$,$\mathbf{V}_{\mathcal{D}}$, $\mathbf{V}_{\mathscr{B}_\sigma}$, and $\mathbf{V}_{\mathscr{B}_\mathrm{u}}$. The submatrices $\mathbf{K}_{\mathcal{D}\mathcal{D}}$ and $\mathbf{K}_{\mathcal{D}\mathscr{R}}$ with the dimensions $2N_{\mathcal{D}} \times 2N_{\mathcal{D}}$ and $2N_{\mathcal{D}} \times 2N_{\mathscr{R}}$, respectively, represent the self-stiffness and coupling stiffness matrices between the unknown vectors of $\mathbf{V}_{\mathcal{D}}$ and $\mathbf{V}_{\mathscr{R}}$. The submatrices $\mathbf{K}_{\mathscr{B}_\sigma\mathscr{B}_\sigma}$, $\mathbf{K}_{\mathscr{B}_\sigma\mathscr{R}}$ and $\mathbf{K}_{\mathscr{B}_\sigma\mathscr{B}_\mathrm{u}}$ with the dimensions $2N_{\mathscr{B}_\sigma} \times 2N_{\mathscr{B}_\sigma}$, $2N_{\mathscr{B}_\sigma} \times 2N_{\mathscr{R}}$ and $2N_{\mathscr{B}_\sigma} \times 2N_{\mathscr{B}_\mathrm{u}}$, respectively, represent the self-stiffness and coupling stiffness matrices between the unknown vectors of $\mathbf{V}_{\mathscr{B}_\sigma}$, $\mathbf{V}_{\mathscr{R}}$ and $\mathbf{V}_{\mathscr{B}_\mathrm{u}}$. The submatrices $\mathbf{K}_{\mathscr{B}_\mathrm{u}\mathscr{B}_\mathrm{u}}$, $\mathbf{K}_{\mathscr{B}_\mathrm{u}\mathscr{R}}$ and $\mathbf{K}_{\mathscr{B}_\mathrm{u}\mathscr{B}_\sigma}$ with the dimensions $2N_{\mathscr{B}_\mathrm{u}} \times 2N_{\mathscr{B}_\mathrm{u}}$, $2N_{\mathscr{B}_\mathrm{u}} \times 2N_{\mathscr{R}}$ and $2N_{\mathscr{B}_\mathrm{u}} \times 2N_{\mathscr{B}_\sigma}$, respectively, represent the self-stiffness and coupling stiffness matrices between the unknown vectors of $\mathbf{V}_{\mathscr{B}_\mathrm{u}}$, $\mathbf{V}_{\mathscr{R}}$, and $\mathbf{V}_{\mathscr{B}_\sigma}$.

3.11 Global Discretization of PD Field Equations

Similar to the PD equilibrium equations, the domain $\Omega = \mathcal{D} \cup \mathcal{R} \cup \mathcal{B}$ is discretized with $N = N_{\mathcal{D}} + N_{\mathcal{R}} + N_{\mathcal{B}}$ PD points as shown in Fig. 3.16 for the solution of PD field equations. The parameters, $N_{\mathcal{D}}$ and $N_{\mathcal{R}}$ represent the number of points associated with regions $\mathcal{D}$ and $\mathcal{R}$, respectively. The discretization in these regions can be nonuniform. The parameter $N_{\mathcal{B}} = N_{\mathcal{B}_\Theta} + N_{\mathcal{B}_\mathbf{q}}$ represents the number of points associated with the real boundary region, $\mathcal{B}$. The parameters $N_{\mathcal{B}_\Theta}$ and $N_{\mathcal{B}_\mathbf{q}}$ represent the number of points associated with regions $\mathcal{B}_\Theta$ and $\mathcal{B}_\mathbf{q}$, respectively. The position of the PD point k is denoted by $\mathbf{x}_{(k)}$ and its corresponding volume is denoted by $V_{(k)}$.

The discrete form of PD internal net field flow in the interior region $\mathcal{D}$ and PDDO region $\mathcal{R}$ can be expressed as

$$L_{(k)} = \begin{cases} \sum\limits_{j=1}^{N_{(k)}} h^{BB}_{(k)(j)} V_{(j)} - \sum\limits_{p=1}^{\widetilde{N}_{(k)}} h^{BB}_{(p)(k)} V_{(p)} & \text{with } \mathbf{x}_{(k)} \in \mathcal{D} \\ \sum\limits_{j=1}^{N_{(k)}} \left(h^{PDDO}_{(j)(k)} - h^{PDDO}_{(k)(j)} \right) V_{(j)} & \text{with } \mathbf{x}_{(k)} \in \mathcal{R} \end{cases} \tag{3.141}$$

Considering the transient field fluctuations and the effect of time rate of change of material property on the flow field, the total virtual field variation in the domain $\mathcal{D} \cup \mathcal{R}$ in Eq. (3.121) can be evaluated as

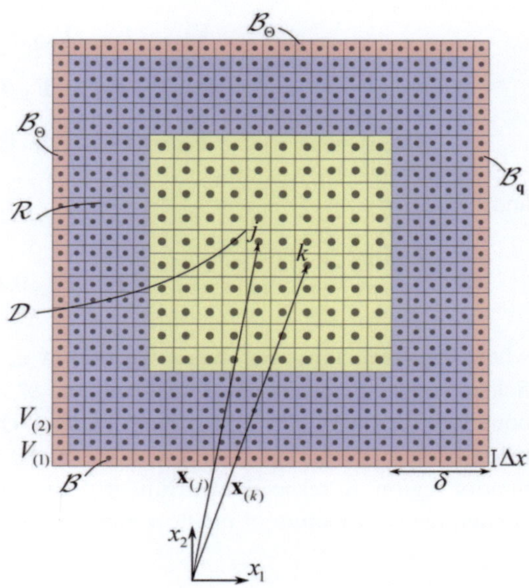

Fig. 3.16 Numerical discretization of the domain for PD field equation with nonuniform grid

$$
\begin{aligned}
&\sum_{k=1}^{N_{\mathcal{D}}+N_{\mathcal{R}}} \left(C_{(k)}\dot{\Theta}_{(k)} + \dot{C}_{(k)}\Theta_{(k)}\right) V_{(k)} \delta\Theta_{(k)} = \\
&\quad \sum_{k=1}^{N_{\mathcal{D}}} \delta\Theta_{(k)} \left(\sum_{j=1}^{N_{(k)}} h^{BB}_{(k)(j)} V_{(j)} - \sum_{p=1}^{\widetilde{N}_{(k)}} h^{BB}_{(p)(k)} V_{(p)} \right) V_{(k)} \\
&\quad + \sum_{k=1}^{N_{\mathcal{R}}} \delta\Theta_{(k)} \left(\sum_{j=1}^{N_{(k)}} \left(h^{PDDO}_{(j)(k)} - h^{PDDO}_{(k)(j)} \right) V_{(j)} \right) V_{(k)} \\
&\quad + \sum_{k=1}^{N_{\mathcal{D}}+N_{\mathcal{R}}} \delta\Theta_{(k)} s_{b(k)} V_{(k)}
\end{aligned}
\tag{3.142}
$$

where $\Theta_{(k)}$ is the unknown field variable, $\delta\Theta_{(k)}$ is the virtual field variable, $C_{(k)}$ and $\dot{C}_{(k)}$ are the field material property and its time rate of change, respectively, at point $\mathbf{x}_{(k)}$. The parameters $N_{(k)}$ and $\widetilde{N}_{(k)}$ are the number of primary and secondary bonds associated with point $\mathbf{x}_{(k)}$, respectively. This equation can be recast in matrix form as

$$
\delta\mathbf{v}^T \left(\mathbf{F}_{\dot{\Theta}} + \mathbf{F} + \mathbf{F}_{\Theta} - \mathbf{F}_s^*\right) = 0 \tag{3.143}
$$

The vectors of $\delta\mathbf{v}$ containing the virtual field, $\delta\Theta_{(k)}$, $\mathbf{F}_{\dot{\Theta}}$ containing the transient field term, $\mathbf{F}$ containing the net field flow, $\mathbf{F}_{\Theta}$ containing the time rate of change of material property and $\mathbf{F}_s^*$ containing the external field source, are defined as

$$
\delta\mathbf{v} = \left\{\delta\mathbf{v}_{\mathcal{R}}, \delta\mathbf{v}_{\mathcal{D}}, \delta\mathbf{v}_{\mathcal{B}_{\mathbf{q}}}, \delta\mathbf{v}_{\mathcal{B}_{\Theta}}\right\}^T \tag{3.144}
$$

$$
\mathbf{F}_{\dot{\Theta}} = \left\{\mathbf{F}_{\dot{\Theta}\mathcal{R}}, \mathbf{F}_{\dot{\Theta}\mathcal{D}}, \mathbf{0}, \mathbf{0}\right\}^T \tag{3.145}
$$

$$
\mathbf{F} = \left\{-\mathbf{F}_{\mathcal{R}}, -\mathbf{F}_{\mathcal{D}}, \mathbf{0}, \mathbf{0}\right\}^T \tag{3.146}
$$

$$
\mathbf{F}_{\Theta} = \left\{\mathbf{F}_{\Theta\mathcal{R}}, \mathbf{F}_{\Theta\mathcal{D}}, \mathbf{0}, \mathbf{0}\right\}^T \tag{3.147}
$$

and

$$
\mathbf{F}_s^* = \left\{\mathbf{F}_{s_{\mathcal{R}}}^*, \mathbf{F}_{s_{\mathcal{D}}}^*, \mathbf{0}, \mathbf{0}\right\}^T \tag{3.148}
$$

where the subvectors $\delta\mathbf{v}_{\mathcal{R}}$, $\delta\mathbf{v}_{\mathcal{D}}$, $\delta\mathbf{v}_{\mathcal{B}_{\mathbf{q}}}$, and $\delta\mathbf{v}_{\mathcal{B}_{\Theta}}$ contain the virtual variation of the field at nodal points in the PDDO region $\mathcal{R}$, interior region $\mathcal{D}$, and along PD boundary region $\Gamma_{\mathcal{B}_{\mathbf{q}}}$ and $\Gamma_{\mathcal{B}_{\Theta}}$, respectively. The subvectors $\mathbf{F}_{\mathcal{R}}$ and $\mathbf{F}_{\mathcal{D}}$ contain the unknown internal net field flow at the nodal points in the PDDO region $\mathcal{R}$ and interior region $\mathcal{D}$, respectively. Similarly, the subvectors $\mathbf{F}_{\Theta\mathcal{R}}, \mathbf{F}_{\dot{\Theta}\mathcal{R}}$ and $\mathbf{F}_{\Theta\mathcal{D}}, \mathbf{F}_{\dot{\Theta}\mathcal{D}}$ contain the contribution of the flow due to transient field effect and the time rate of change of material property at the nodal points in the PDDO region $\mathscr{R}$ and interior region $\mathcal{D}$, respectively.

The subvectors, $\mathbf{F}^*_{s_{\mathscr{R}}}$ and $\mathbf{F}^*_{s_{\mathcal{D}}}$ arising from the external field source applied at the nodal points in the PDDO region $\mathscr{R}$ and interior region $\mathcal{D}$ are defined as

$$\mathbf{F}^*_{s_{\mathscr{R}}} = \left\{ s_{b(1)} V_{(1)}, s_{b(1)} V_{(2)}, \cdots, s_{b(N_{\mathscr{R}})} V_{(N_{\mathscr{R}})} \right\}^T \tag{3.149}$$

and

$$\mathbf{F}^*_{s_{\mathcal{D}}} = \left\{ s_{b(1)} V_{(1)}, s_{b(1)} V_{(2)}, \cdots, s_{b(N_{\mathcal{D}})} V_{(N_{\mathcal{D}})} \right\}^T \tag{3.150}$$

The vector of internal net field flow, $\mathbf{F}$ can be written in terms of the global field coefficient matrix, $\mathbf{K}$ and the unknown field vector, $\mathbf{V}$ as

$$\mathbf{F} = \mathbf{K}\mathbf{V} \tag{3.151}$$

where

$$\mathbf{F} = \sum_{k=1}^{N_{\mathscr{R}}+N_{\mathcal{D}}} L_{(k)} V_{(k)} \tag{3.152}$$

$$\mathbf{K} = \sum_{k=1}^{N_{\mathcal{D}}} \left(\sum_{j=1}^{N_{(k)}} \mathbf{k}^{\alpha}_{(k)(j)} - \sum_{p=1}^{\widetilde{N}_{(k)}} \mathbf{k}^{\alpha}_{(p)(k)} \right) + \sum_{k=1}^{N_{\mathscr{R}}} \sum_{j=1}^{N_{(k)}} \mathbf{k}^{PDDO}_{(k)(j)} \tag{3.153}$$

and

$$\mathbf{V} = \left\{ \mathbf{V}_{\mathscr{R}}, \mathbf{V}_{\mathcal{D}}, \mathbf{V}_{\mathscr{B}_{\mathbf{q}}}, \mathbf{V}_{\mathscr{B}_{\Theta}} \right\}^T \tag{3.154}$$

The contribution to the total virtual field variation from the external applied field flux along the PD boundary region $\Gamma_{\mathscr{B}_{\mathbf{q}}}$ can be expressed as

$$\sum_{k=1}^{N_{\mathscr{B}_{\mathbf{q}}}} \delta\Theta^T_{(k)} \left(\mathbf{q}^{PD}_{(k)} \cdot \mathbf{n}_{(k)} - \mathbf{q}^*_{(k)} \cdot \mathbf{n}_{(k)} \right) A_{(k)} = 0 \tag{3.155}$$

where $S_{(k)} = h\Delta x$ with $h = 1$ is the boundary area of PD point, $\mathbf{x}_{(k)}$. It can be recast as

$$\delta \mathbf{v}^T_{\mathscr{B}_{\mathbf{q}}} \left(\sum_{k=1}^{N_{\mathscr{B}_{\mathbf{q}}}} \mathbf{K}_{\mathbf{q}(k)} \mathbf{V} - \mathbf{F}^*_{\mathscr{B}_{\mathbf{q}}} \right) = \delta \mathbf{v}^T_{\mathscr{B}_{\mathbf{q}}} \left(\mathbf{K}_{\mathbf{q}} \mathbf{V} - \mathbf{F}^*_{\mathscr{B}_{\mathbf{q}}} \right) = 0 \tag{3.156}$$

in which $\mathbf{K_q}$ is the global coefficient matrix due to the field flux. The vector, $\mathbf{F}^*_{\mathscr{B}_\mathbf{q}}$ arising from the applied flux is defined as

$$\mathbf{F}^*_{\mathscr{B}_\mathbf{q}} = \left\{ \mathbf{q}^*_{(1)} \cdot \mathbf{n}_{(1)} S_{(1)}, \mathbf{q}^*_{(2)} \cdot \mathbf{n}_{(2)} S_{(2)}, \cdots, \mathbf{q}^*_{\left(N_{\mathscr{B}_\mathbf{q}}\right)} \cdot \mathbf{n}_{\left(N_{\mathscr{B}_\mathbf{q}}\right)} S_{\left(N_{\mathscr{B}_\mathbf{q}}\right)} \right\}^T \tag{3.157}$$

In region, $\mathcal{D} + \mathscr{R}$, the vector of field flow due to transient field variation, $\mathbf{F}_{\dot{\Theta}}$ and the time rate of change of material property $\mathbf{F}_\Theta$ can be written in terms of the unknown field vectors $\mathbf{V}$ and $\dot{\mathbf{V}}$ as

$$\sum_{k=1}^{N_\mathcal{D}+N_\mathscr{R}} \left(C_{(k)} \dot{\Theta}_{(k)} V_{(k)} + \dot{C}_{(k)} \Theta_{(k)} V_{(k)} \right) \delta\Theta^T_{(k)}$$
$$= \begin{Bmatrix} \delta\mathbf{V}_\mathcal{D} \\ \delta\mathbf{V}_\mathscr{R} \end{Bmatrix}^T \begin{bmatrix} \mathbf{C}_{\mathcal{D}\mathcal{D}} & \mathbf{0} & \mathbf{0} & \mathbf{0} \\ \mathbf{0} & \mathbf{C}_{\mathscr{R}\mathscr{R}} & \mathbf{0} & \mathbf{0} \end{bmatrix} \begin{Bmatrix} \dot{\mathbf{V}}_\mathcal{D} \\ \dot{\mathbf{V}}_\mathscr{R} \\ \dot{\mathbf{V}}_{\mathscr{B}_\mathbf{q}} \\ \dot{\mathbf{V}}_{\mathscr{B}_\Theta} \end{Bmatrix} + \begin{Bmatrix} \delta\mathbf{V}_\mathcal{D} \\ \delta\mathbf{V}_\mathscr{R} \end{Bmatrix}^T \begin{bmatrix} \dot{\mathbf{C}}_{\mathcal{D}\mathcal{D}} & \mathbf{0} & \mathbf{0} & \mathbf{0} \\ \mathbf{0} & \dot{\mathbf{C}}_{\mathscr{R}\mathscr{R}} & \mathbf{0} & \mathbf{0} \end{bmatrix} \begin{Bmatrix} \mathbf{V}_\mathcal{D} \\ \mathbf{V}_\mathscr{R} \\ \mathbf{V}_{\mathscr{B}_\mathbf{q}} \\ \mathbf{V}_{\mathscr{B}_\Theta} \end{Bmatrix} \tag{3.158}$$

where the submatrices $\mathbf{C}_{\mathcal{D}\mathcal{D}}$ and $\mathbf{C}_{\mathscr{R}\mathscr{R}}$ represent diagonal field damping, and $\dot{\mathbf{C}}_{\mathcal{D}\mathcal{D}}$ and $\dot{\mathbf{C}}_{\mathscr{R}\mathscr{R}}$ represent the diagonal transient material property matrices for regions $\mathcal{D}$ and $\mathscr{R}$ with dimensions $N_\mathcal{D} \times N_\mathcal{D}$ and $N_\mathscr{R} \times N_\mathscr{R}$, respectively.

Combining Eqs. (3.153), (3.156), and (3.157) along with Eq. (3.158) results in the final form of the governing equations as

$$\begin{bmatrix} \mathbf{C}_{\mathcal{D}\mathcal{D}} & \mathbf{0} & \mathbf{0} & \mathbf{0} \\ \mathbf{0} & \mathbf{C}_{\mathscr{R}\mathscr{R}} & \mathbf{0} & \mathbf{0} \\ \mathbf{0} & \mathbf{0} & \mathbf{0} & \mathbf{0} \\ \mathbf{0} & \mathbf{0} & \mathbf{0} & \mathbf{0} \end{bmatrix} \begin{Bmatrix} \dot{\mathbf{V}}_\mathcal{D} \\ \dot{\mathbf{V}}_\mathscr{R} \\ \dot{\mathbf{V}}_{\mathscr{B}_\mathbf{q}} \\ \dot{\mathbf{V}}_{\mathscr{B}_\Theta} \end{Bmatrix} + \begin{bmatrix} \mathbf{K}_{\mathscr{R}\mathscr{R}} + \mathbf{C}_{\mathscr{R}\mathscr{R}} & \mathbf{K}_{\mathscr{R}\mathcal{D}} & \mathbf{K}_{\mathscr{R}\mathscr{B}_\mathbf{q}} & \mathbf{K}_{\mathscr{R}\mathscr{B}_\Theta} \\ \mathbf{K}_{\mathcal{D}\mathscr{R}} & \mathbf{K}_{\mathcal{D}\mathcal{D}} + \mathbf{C}_{\mathcal{D}\mathcal{D}} & \mathbf{0} & \mathbf{0} \\ \mathbf{K}_{\mathscr{B}_\mathbf{q}\mathscr{R}} & \mathbf{0} & \mathbf{K}_{\mathscr{B}_\mathbf{q}\mathscr{B}_\mathbf{q}} & \mathbf{K}_{\mathscr{B}_\mathbf{q}\mathscr{B}_\Theta} \\ \mathbf{K}_{\mathscr{B}_\Theta\mathscr{R}} & \mathbf{0} & \mathbf{K}_{\mathscr{B}_\Theta\mathscr{B}_\mathbf{q}} & \mathbf{K}_{\mathscr{B}_\Theta\mathscr{B}_\Theta} \end{bmatrix} \begin{Bmatrix} \mathbf{V}_\mathscr{R} \\ \mathbf{V}_\mathcal{D} \\ \mathbf{V}_{\mathscr{B}_\mathbf{q}} \\ \mathbf{V}_{\mathscr{B}_\Theta} \end{Bmatrix}$$
$$= \begin{Bmatrix} \mathbf{F}^*_{s_\mathscr{R}} \\ \mathbf{F}^*_{s_\mathcal{D}} \\ \mathbf{F}^*_{\mathscr{B}_\mathbf{q}} \\ \mathbf{0} \end{Bmatrix} \tag{3.159}$$

where the submatrices $\mathbf{K}_{\mathscr{R}\mathscr{R}}$, $\mathbf{K}_{\mathscr{R}\mathcal{D}}$, $\mathbf{K}_{\mathscr{R}\mathscr{B}_\mathbf{q}}$, and $\mathbf{K}_{\mathscr{R}\mathscr{B}_\Theta}$ with dimensions $N_\mathscr{R} \times N_\mathscr{R}$, $N_\mathscr{R} \times N_\mathcal{D}$, $N_\mathscr{R} \times N_{\mathscr{B}_\mathbf{q}}$, and $N_\mathscr{R} \times N_{\mathscr{B}_\Theta}$, respectively, represent the self-stiffness and coupling stiffness matrices between the unknown vectors of $\mathbf{V}_\mathscr{R}$, $\mathbf{V}_\mathcal{D}$, $\mathbf{V}_{\mathscr{B}_\mathbf{q}}$, and $\mathbf{V}_{\mathscr{B}_\Theta}$. The submatrices $\mathbf{K}_{\mathcal{D}\mathcal{D}}$ and $\mathbf{K}_{\mathcal{D}\mathscr{R}}$ with the dimensions $N_\mathcal{D} \times N_\mathcal{D}$ and $N_\mathcal{D} \times N_\mathscr{R}$, respectively, represent the self-stiffness and coupling stiffness matrices between the unknown vectors of $\mathbf{V}_\mathcal{D}$ and $\mathbf{V}_\mathscr{R}$. The submatrices $\mathbf{K}_{\mathscr{B}_\mathbf{q}\mathscr{B}_\mathbf{q}}$, $\mathbf{K}_{\mathscr{B}_\mathbf{q}\mathscr{R}}$, and $\mathbf{K}_{\mathscr{B}_\mathbf{q}\mathscr{B}_\Theta}$ with the dimensions $N_{\mathscr{B}_\mathbf{q}} \times N_{\mathscr{B}_\mathbf{q}}$, $N_{\mathscr{B}_\mathbf{q}} \times N_\mathscr{R}$, and $N_{\mathscr{B}_\mathbf{q}} \times N_{\mathscr{B}_\Theta}$, respectively, represent the self-stiffness and coupling stiffness matrices between the unknown vectors of

$\mathbf{V}_{\mathscr{B}_\mathbf{q}}$, $\mathbf{V}_{\mathscr{R}}$ and $\mathbf{V}_{\mathscr{B}_\Theta}$. The submatrices $\mathbf{K}_{\mathscr{B}_\Theta\mathscr{B}_\Theta}$, $\mathbf{K}_{\mathscr{B}_\Theta\mathscr{R}}$, and $\mathbf{K}_{\mathscr{B}_\Theta\mathscr{B}_\mathbf{q}}$ with the dimensions are $N_{\mathscr{B}_\Theta} \times N_{\mathscr{B}_\Theta}$, $N_{\mathscr{B}_\Theta} \times N_{\mathscr{R}}$, and $N_{\mathscr{B}_\Theta} \times N_{\mathscr{B}_\mathbf{q}}$, respectively, represent the self-stiffness and coupling stiffness matrices between the unknown vectors of $\mathbf{V}_{\mathscr{B}_\Theta}$, $\mathbf{V}_{\mathscr{R}}$, and $\mathbf{V}_{\mathscr{B}_\mathbf{q}}$.

3.12 Local Discretization of PD Equilibrium Equations

3.12.1 BB-PD Equilibrium Equation

Figure 3.17a, b show the element connectivity for dual horizon BB-PD interaction between nodes (k and j) and (k and p) denoting the primary bond (red) and the secondary bond (blue), respectively. Using Eq. (3.27), the discrete form of the internal force vector, $\mathbf{L}_{(k)}^{BB}$ in the interior region $\mathcal{D}$ can be expressed as

$$\begin{aligned}\mathbf{L}_{(k)}^{BB} = &\sum_{j \in H_{(k)}} \widehat{\mu}_{(k)(j)} \frac{\mu}{2} \mathbf{S}\left(\mathbf{x}_{(k)}, \boldsymbol{\xi}_{(k)(j)}\right) \left(\mathbf{u}_{(j)} - \mathbf{u}_{(k)}\right) V_{(j)} \\ &+ \sum_{p \in \widetilde{H}_{(k)}} \widehat{\mu}_{(p)(k)} \frac{\mu}{2} \mathbf{S}\left(\mathbf{x}_{(p)}, \boldsymbol{\xi}_{(p)(k)}\right) \left(\mathbf{u}_{(p)} - \mathbf{u}_{(k)}\right) V_{(p)} \end{aligned} \quad \text{for } \mathbf{x}_{(k)} \in \mathcal{D} \quad (3.160)$$

After multiplying with volume $V_{(k)}$, its discrete form can be written as

$$\mathbf{F}_{(k)}^{BB} = \mathbf{L}_{(k)}^{BB} V_{(k)} = \sum_{j \in H_{(k)}} \boldsymbol{\Phi}_{(k)(j)}^{BB} \left(\mathbf{u}_{(j)} - \mathbf{u}_{(k)}\right) + \sum_{p \in \widetilde{H}_{(k)}} \widetilde{\boldsymbol{\Phi}}_{(p)(k)}^{BB} \left(\mathbf{u}_{(p)} - \mathbf{u}_{(k)}\right) \quad (3.161)$$

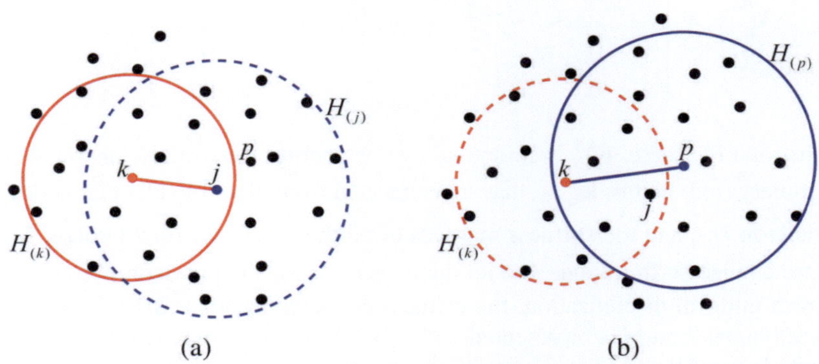

Fig. 3.17 Element connectivity for dual horizon BB-PD interactions: (**a**) primary bond between nodes k and j and (**b**) secondary bond between nodes k and p

with

$$\boldsymbol{\Phi}_{(k)(j)}^{BB} = \widehat{\mu}_{(k)(j)} \frac{\mu}{2} \mathbf{S}\left(\mathbf{x}_{(k)}, \boldsymbol{\xi}_{(k)(j)}\right) V_{(j)} V_{(k)} \tag{3.162}$$

and

$$\widetilde{\boldsymbol{\Phi}}_{(p)(k)}^{BB} = \widehat{\mu}_{(p)(k)} \frac{\mu}{2} \mathbf{S}\left(\mathbf{x}_{(p)}, \boldsymbol{\xi}_{(p)(k)}\right) V_{(p)} V_{(k)} \tag{3.163}$$

The PD force vector, $\mathbf{F}_{(k)}^{BB}$ in component form can be written as

$$\begin{Bmatrix} F_{1(k)}^{BB} \\ F_{2(k)}^{BB} \\ 0 \\ 0 \end{Bmatrix} = \sum_{j \in H_{(k)}} \mathbf{k}_{(k)(j)}^{BB} \begin{Bmatrix} u_{1(k)} \\ u_{2(k)} \\ u_{1(j)} \\ u_{2(j)} \end{Bmatrix} + \sum_{p \in \widetilde{H}_{(k)}} \widetilde{\mathbf{k}}_{(p)(k)}^{BB} \begin{Bmatrix} u_{1(k)} \\ u_{2(k)} \\ u_{1(p)} \\ u_{2(p)} \end{Bmatrix} \tag{3.164}$$

where the stiffness matrices, $\mathbf{k}_{(k)(j)}^{BB}$ and $\widetilde{\mathbf{k}}_{(p)(k)}^{BB}$ are constructed as

$$\mathbf{k}_{(k)(j)}^{BB} = \begin{bmatrix} -\Phi_{11(k)(j)}^{BB} & -\Phi_{12(k)(j)}^{BB} & \Phi_{11(k)(j)}^{BB} & \Phi_{12(k)(j)}^{BB} \\ -\Phi_{21(k)(j)}^{BB} & -\Phi_{22(k)(j)}^{BB} & \Phi_{21(k)(j)}^{BB} & \Phi_{22(k)(j)}^{BB} \\ 0 & 0 & 0 & 0 \\ 0 & 0 & 0 & 0 \end{bmatrix} \tag{3.165}$$

and

$$\widetilde{\mathbf{k}}_{(p)(k)}^{BB} = \begin{bmatrix} -\widetilde{\Phi}_{11(p)(k)}^{BB} & -\widetilde{\Phi}_{12(p)(k)}^{BB} & \widetilde{\Phi}_{11(p)(k)}^{BB} & \widetilde{\Phi}_{12(p)(k)}^{BB} \\ -\widetilde{\Phi}_{21(p)(k)}^{BB} & -\widetilde{\Phi}_{22(p)(k)}^{BB(22)} & \widetilde{\Phi}_{21(p)(k)}^{BB(21)} & \widetilde{\Phi}_{22(p)(k)}^{BB(22)} \\ 0 & 0 & 0 & 0 \\ 0 & 0 & 0 & 0 \end{bmatrix} \tag{3.166}$$

The total PD force, $\mathbf{F}_{(k)}^{BB}$ is computed by considering the stiffness matrices of all the primary (red) bonds, $\mathbf{k}_{(k)(j)}^{BB}$, that are connected from node k to all other nodes j in the horizon $H_{(k)}$ and the stiffness matrices of all the secondary (blue) bonds, $\widetilde{\mathbf{k}}_{(p)(k)}^{BB}$ that are connected from node k to all the nodes p in the dual horizon $\widetilde{H}_{(k)}$.

For a uniform discretization, the element connectivity for a BB-PD interaction between nodes k and j is represented with a bond (blue), as shown in Fig 3.18. The discrete form of the PD force vector $\mathbf{F}_{(k)}^{BB}$ at node k can be written as

$$\mathbf{F}_{(k)}^{BB} = \mathbf{L}_{(k)}^{BB} V_{(k)} = \sum_{j \in H_{(k)}} \left(\boldsymbol{\Phi}_{(k)(j)}^{BB} + \widetilde{\boldsymbol{\Phi}}_{(j)(k)}^{BB}\right)\left(\mathbf{u}_{(j)} - \mathbf{u}_{(k)}\right) \tag{3.167}$$

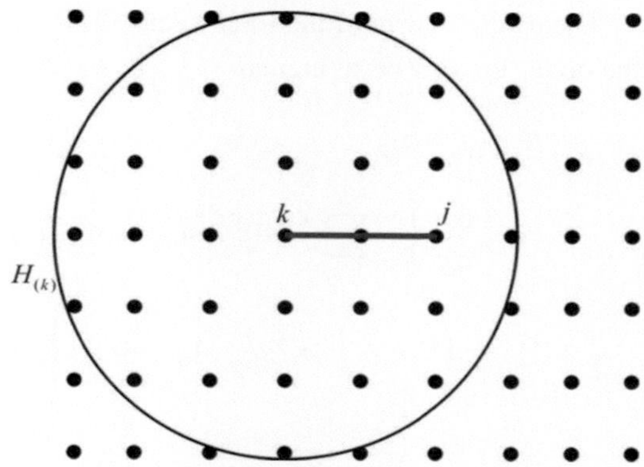

Fig. 3.18 Element connectivity for constant horizon BB-PD interaction between nodes k and j in $H_{(k)}$

or

$$\begin{Bmatrix} F^{BB}_{1(k)} \\ F^{BB}_{2(k)} \\ 0 \\ 0 \end{Bmatrix} = \sum_{j \in H_{(k)}} \left(\mathbf{k}^{BB}_{(k)(j)} + \widetilde{\mathbf{k}}^{BB}_{(j)(k)} \right) \begin{Bmatrix} u_{1(k)} \\ u_{2(k)} \\ u_{1(j)} \\ u_{2(j)} \end{Bmatrix} \tag{3.168}$$

The total PD force vector, $\mathbf{F}^{BB}_{(k)}$ is computed by considering the contribution of all the blue bonds that are connected from node k to all other nodes j in $H_{(k)}$ with their corresponding stiffness matrices $\left(\mathbf{k}^{BB}_{(k)(j)} + \widetilde{\mathbf{k}}^{BB}_{(j)(k)} \right)$.

3.12.2 OSB-PD Equilibrium Equation

Using Eq. (3.48), the discrete form of the internal force vector $\mathbf{L}^{OSB}_{(k)}$ at node k in the interior region $\mathcal{D}$ can be expressed as

$$\begin{aligned} \mathbf{L}^{OSB}_{(k)} = \sum_{j \in H_{(k)}} \widehat{\mu}_{(k)(j)} & \left\{ \begin{aligned} & a_{(k)} \sum_{r=1}^{N_{(k)}} \widehat{\mu}_{(k)(k_r)} \left(\widehat{\mathbf{g}}\left(\mathbf{x}_{(k)}, \boldsymbol{\xi}_{(k)(j)}\right) \otimes \mathbf{g}\left(\mathbf{x}_{(k)}, \boldsymbol{\xi}_{(k)(k_r)}\right) \right) \left(\mathbf{u}_{(k_r)} - \mathbf{u}_{(k)}\right) V_{(k_r)} \\ & + \frac{n}{2(n+1)} \frac{\mu \delta_{(k)}}{\left|\boldsymbol{\xi}_{(k)(j)}\right|} \mathbf{S}\left(\mathbf{x}_{(k)}, \boldsymbol{\xi}_{(k)(j)}\right) \left(\mathbf{u}_{(j)} - \mathbf{u}_{(k)}\right) \end{aligned} \right\} V_{(j)} \\ + \sum_{p \in \widetilde{H}_{(k)}} \widehat{\mu}_{(p)(k)} & \left\{ \begin{aligned} & a_{(p)} \sum_{s=1}^{N_{(p)}} \widehat{\mu}_{(p)(p_s)} \left(\widehat{\mathbf{g}}\left(\mathbf{x}_{(p)}, \boldsymbol{\xi}_{(p)(k)}\right) \otimes \mathbf{g}\left(\mathbf{x}_{(p)}, \boldsymbol{\xi}_{(p)(p_s)}\right) \right) \left(\mathbf{u}_{(p)} - \mathbf{u}_{(p_s)}\right) V_{(p_s)} \\ & + \frac{n}{2(n+1)} \frac{\mu \delta_{(p)}}{\left|\boldsymbol{\xi}_{(p)(k)}\right|} \mathbf{S}\left(\mathbf{x}_{(p)}, \boldsymbol{\xi}_{(p)(k)}\right) \left(\mathbf{u}_{(p)} - \mathbf{u}_{(k)}\right) \end{aligned} \right\} V_{(p)} \end{aligned} \tag{3.169}$$

The discrete form of the force vector, $\mathbf{L}_{(k)}^{OSB}$ after multiplying with the volume of the node, $V_{(k)}$ can be written as

$$\begin{aligned}\mathbf{F}_{(k)}^{OSB} &= \mathbf{L}_{(k)}^{OSB} V_{(k)} \\ &= \sum_{j \in H_{(k)}} \left\{ \sum_{r=1, k_r \neq j}^{N_{(k)}} \mathbf{G}_{(k)(j)}^{(k)(k_r)} \mathbf{u}_{(k_r)} - \boldsymbol{\Phi}_{(k)} \mathbf{u}_{(k)} \right. \\ &\left. + \left(\mathbf{B}_{(k)(j)} + \mathbf{G}_{(k)(j)}^{(k)(j)} \right) \left(\mathbf{u}_{(j)} - \mathbf{u}_{(k)} \right) \right\} + \sum_{p \in \widetilde{H}_{(k)}} \left\{ \widetilde{\boldsymbol{\Phi}}_{(p)} \mathbf{u}_{(p)} \right. \\ &\left. - \sum_{s=1, p_s \neq k}^{N_{(p)}} \widetilde{\mathbf{G}}_{(p)(k)}^{(p)(p_s)} \mathbf{u}_{(p_s)} + \left(\widetilde{\mathbf{B}}_{(p)(k)} + \widetilde{\mathbf{G}}_{(p)(k)}^{(p)(k)} \right) \left(\mathbf{u}_{(p)} - \mathbf{u}_{(k)} \right) \right\}\end{aligned} \tag{3.170}$$

with

$$\mathbf{G}_{(k)(j)}^{(k)(k_r)} = \widehat{\mu}_{(k)(j)} \widehat{\mu}_{(k)(k_r)} a_{(k)} \left(\widehat{\mathbf{g}}\left(\mathbf{x}_{(k)}, \boldsymbol{\xi}_{(k)(j)} \right) \otimes \mathbf{g}\left(\mathbf{x}_{(k)}, \boldsymbol{\xi}_{(k)(k_r)} \right) \right) V_{(k_r)} V_{(j)} V_{(k)} \tag{3.171}$$

$$\mathbf{B}_{(k)(j)} = \widehat{\mu}_{(k)(j)} \frac{n}{2(n+1)} \frac{\mu \delta_{(k)}}{\left| \boldsymbol{\xi}_{(k)(j)} \right|} \mathbf{S}\left(\mathbf{x}_{(k)}, \boldsymbol{\xi}_{(k)(j)} \right) V_{(j)} V_{(k)} \tag{3.172}$$

$$\boldsymbol{\Phi}_{(k)} = \sum_{r=1, k_r \neq j}^{N_{(k)}} \mathbf{G}_{(k)(j)}^{(k)(k_r)} \tag{3.173}$$

$$\widetilde{\mathbf{G}}_{(p)(k)}^{(p)(p_s)} = \widehat{\mu}_{(p)(k)} \widehat{\mu}_{(p)(p_s)} a_{(p)} \left(\widehat{\mathbf{g}}\left(\mathbf{x}_{(p)}, \boldsymbol{\xi}_{(p)(k)} \right) \otimes \mathbf{g}\left(\mathbf{x}_{(p)}, \boldsymbol{\xi}_{(p)(p_s)} \right) \right) V_{(p_s)} V_{(p)} V_{(k)} \tag{3.174}$$

$$\widetilde{\mathbf{B}}_{(p)(k)} = \widehat{\mu}_{(p)(k)} \frac{n}{2(n+1)} \frac{\mu \delta_{(p)}}{\left| \boldsymbol{\xi}_{(p)(k)} \right|} \mathbf{S}\left(\mathbf{x}_{(p)}, \boldsymbol{\xi}_{(p)(k)} \right) V_{(p)} V_{(k)} \tag{3.175}$$

and

$$\widetilde{\boldsymbol{\Phi}}_{(p)} = \sum_{s=1, p_s \neq k}^{N_{(p)}} \widetilde{\mathbf{G}}_{(p)(k)}^{(p)(p_s)} \tag{3.176}$$

Equation (3.170) can be written in the component form as

$$
\begin{Bmatrix} F^{OSB}_{1(k)} \\ F^{OSB}_{2(k)} \\ 0 \\ 0 \end{Bmatrix} = \sum_{j \in H_{(k)}} \left(\mathbf{k}^{OSB}_{(k)(j)} \begin{Bmatrix} u_{1(k)} \\ u_{2(k)} \\ u_{1(j)} \\ u_{2(j)} \end{Bmatrix} + \sum_{r=1, k_r \neq j}^{N_{(k)}} \mathbf{k}^{OSB(k)(k_r)}_{(k)(j)} \begin{Bmatrix} u_{1(k)} \\ u_{2(k)} \\ u_{1(k_r)} \\ u_{2(k_r)} \end{Bmatrix} \right)
$$
$$
+ \sum_{p \in \widetilde{H}_{(k)}} \left(\widetilde{\mathbf{k}}^{OSB}_{(p)(k)} \begin{Bmatrix} u_{1(k)} \\ u_{2(k)} \\ u_{1(p)} \\ u_{2(p)} \end{Bmatrix} + \sum_{s=1, p_s \neq k}^{N_{(p)}} \widetilde{\mathbf{k}}^{OSB(p)(p_s)}_{(p)(k)} \begin{Bmatrix} u_{1(k)} \\ u_{2(k)} \\ u_{1(p_s)} \\ u_{2(p_s)} \end{Bmatrix} \right) \tag{3.177}
$$

in which the stiffness matrices $\mathbf{k}^{OSB}_{(k)(j)}$, $\mathbf{k}^{OSB(k)(k_r)}_{(k)(j)}$, $\widetilde{\mathbf{k}}^{OSB}_{(p)(k)}$, and $\widetilde{\mathbf{k}}^{OSB(p)(p_s)}_{(p)(k)}$ are defined as

$$
\mathbf{k}^{OSB}_{(k)(j)} = \begin{bmatrix} -\Phi_{11(k)} - B_{11(k)(j)} - G^{(k)(j)}_{11(k)(j)} & -\Phi_{12(k)} - B_{12(k)(j)} - G^{(k)(j)}_{12(k)(j)} & B_{11(k)(j)} + G^{(k)(j)}_{11(k)(j)} & B_{12(k)(j)} + G^{(k)(j)}_{12(k)(j)} \\ -\Phi_{21(k)} - B_{21(k)(j)} - G^{(k)(j)}_{21(k)(j)} & -\Phi_{22(k)} - B_{22(k)(j)} - G^{(k)(j)}_{22(k)(j)} & B_{21(k)(j)} + G^{(k)(j)}_{21(k)(j)} & B_{22(k)(j)} + G^{(k)(j)}_{22(k)(j)} \\ 0 & 0 & 0 & 0 \\ 0 & 0 & 0 & 0 \end{bmatrix} \tag{3.178}
$$

$$
\mathbf{k}^{OSB(k)(k_r)}_{(k)(j)} = \begin{bmatrix} 0 & 0 & G^{(k)(k_r)}_{11(k)(j)} & G^{(k)(k_r)}_{12(k)(j)} \\ 0 & 0 & G^{(k)(k_r)}_{21(k)(j)} & G^{(k)(k_r)}_{22(k)(j)} \\ 0 & 0 & 0 & 0 \\ 0 & 0 & 0 & 0 \end{bmatrix} \tag{3.179}
$$

$$
\widetilde{\mathbf{k}}^{OSB}_{(p)(k)} = \begin{bmatrix} -\widetilde{B}_{11(p)(k)} - \widetilde{G}^{(p)(k)}_{11(p)(k)} & -\widetilde{B}_{12(p)(k)} - \widetilde{G}^{(p)(k)}_{12(p)(k)} & \widetilde{\Phi}_{11(p)} + \widetilde{B}_{11(p)(k)} + \widetilde{G}^{(p)(k)}_{11(p)(k)} & \widetilde{\Phi}_{12(p)} + \widetilde{B}_{12(p)(k)} + \widetilde{G}^{(p)(k)}_{12(p)(k)} \\ -\widetilde{B}_{21(p)(k)} - \widetilde{G}^{(p)(k)}_{21(p)(k)} & -\widetilde{B}_{22(p)(k)} - \widetilde{G}^{(p)(k)}_{22(p)(k)} & \widetilde{\Phi}_{21(p)} + \widetilde{B}_{21(p)(k)} + \widetilde{G}^{(p)(k)}_{21(p)(k)} & \widetilde{\Phi}_{22(p)} + \widetilde{B}_{22(p)(k)} + \widetilde{G}^{(p)(k)}_{22(p)(k)} \\ 0 & 0 & 0 & 0 \\ 0 & 0 & 0 & 0 \end{bmatrix} \tag{3.180}
$$

and

$$
\widetilde{\mathbf{k}}^{OSB(p)(p_s)}_{(p)(k)} = \begin{bmatrix} 0 & 0 & -\widetilde{G}^{(p)(p_s)}_{11(p)(k)} & -\widetilde{G}^{(p)(p_s)}_{12(p)(k)} \\ 0 & 0 & -\widetilde{G}^{(p)(p_s)}_{21(p)(k)} & -\widetilde{G}^{(p)(p_s)}_{22(p)(k)} \\ 0 & 0 & 0 & 0 \\ 0 & 0 & 0 & 0 \end{bmatrix} \tag{3.181}
$$

Figure 3.19a, b show the element connectivity for dual horizon OSB-PD interaction between nodes k and j and k and p denoting the primary bond (red) and

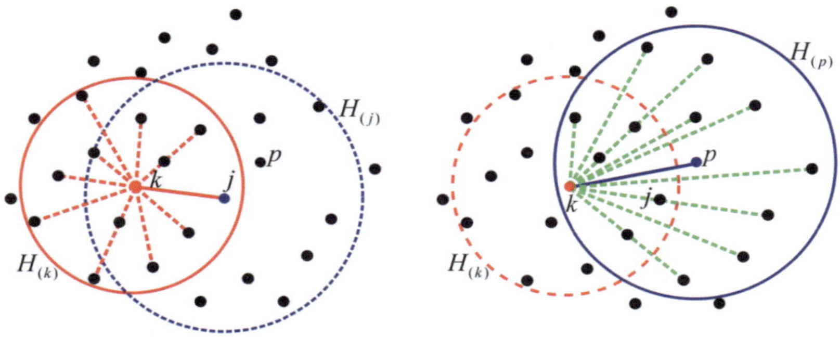

Fig. 3.19 Element connectivity for dual horizon OSB-PD interactions: (**a**) primary bond between nodes k and j and (**b**) secondary bond between nodes k and p

secondary bond (blue), respectively. The interactions can be modeled by considering: (1) an element (shown in red) with the stiffness matrix, $\mathbf{k}_{(k)(j)}^{OSB}$ corresponding to the primary bond and another element (shown in blue) with stiffness matrix, $\widetilde{\mathbf{k}}_{(p)(k)}^{OSB}$ corresponding to the secondary bond, (2) a network of elements around node k (shown in dashed red) with stiffness matrices $\mathbf{k}_{(k)(j)}^{OSB(k)(k_r)}$ due to the direct interaction of node k with its family members, and (3) a network of elements around node k (shown in dashed green) with stiffness matrices, $\widetilde{\mathbf{k}}_{(p)(k)}^{OSB(p)(p_s)}$ due to the indirect interactions of node k with the family members of node p.

The total internal force $\mathbf{F}_{(k)}^{OSB}$ is computed by considering: (1) the stiffness matrices of all the red bonds, $\mathbf{k}_{(k)(j)}^{OSB}$, which are connected from node k to all other nodes j in the horizon $H_{(k)}$, along with their respective direct interactions (dashed red) with stiffness matrices, $\mathbf{k}_{(k)(j)}^{OSB(k)(k_r)}$, and (2) the stiffness matrices, $\widetilde{\mathbf{k}}_{(p)(k)}^{OSB}$ of all the blue bonds that are connected from node k to all the nodes p in its dual horizon $\widetilde{H}_{(k)}$, along with their respective indirect interactions (dashed green) with stiffness matrices, $\widetilde{\mathbf{k}}_{(p)(k)}^{OSB(p)(p_s)}$.

For a uniform discretization, the discrete form of the internal force vector $\mathbf{F}_{(k)}^{OSB}$ at node k can be written as

$$
\begin{aligned}
\mathbf{F}_{(k)}^{OSB} &= \mathbf{L}_{(k)}^{OSB} V_{(k)} \\
&= \sum_{j \in H_{(k)}} \left\{ \begin{array}{l} \displaystyle\sum_{r=1,\, k_r \neq j}^{N_{(k)}} \mathbf{G}_{(k)(j)}^{(k)(k_r)} \mathbf{u}_{(k_r)} - \mathbf{\Phi}_{(k)} \mathbf{u}_{(k)} + \left(\mathbf{B}_{(k)(j)} + \mathbf{G}_{(k)(j)}^{(k)(j)} \right) \left(\mathbf{u}_{(j)} - \mathbf{u}_{(k)} \right) \\ \displaystyle + \widetilde{\mathbf{\Phi}}_{(j)} \mathbf{u}_{(j)} - \sum_{s=1,\, j_s \neq k}^{N_{(j)}} \widetilde{\mathbf{G}}_{(j)(k)}^{(j)(j_s)} \mathbf{u}_{(j_s)} + \left(\widetilde{\mathbf{B}}_{(j)(k)} + \widetilde{\mathbf{G}}_{(j)(k)}^{(j)(k)} \right) \left(\mathbf{u}_{(j)} - \mathbf{u}_{(k)} \right) \end{array} \right\}
\end{aligned}
\tag{3.182}
$$

or

$$\begin{Bmatrix} F_{1(k)}^{OSB} \\ F_{2(k)}^{OSB} \\ 0 \\ 0 \end{Bmatrix} = \sum_{j \in H_{(k)}} \left(\left(\mathbf{k}_{(k)(j)}^{OSB} + \widetilde{\mathbf{k}}_{(j)(k)}^{OSB} \right) \begin{Bmatrix} u_{1(k)} \\ u_{2(k)} \\ u_{1(j)} \\ u_{2(j)} \end{Bmatrix} + \sum_{r=1, k_r \neq j}^{N_{(k)}} \mathbf{k}_{(k)(j)}^{OSB(k)(k_r)} \begin{Bmatrix} u_{1(k)} \\ u_{2(k)} \\ u_{1(k_r)} \\ u_{2(k_r)} \end{Bmatrix} \right.$$
$$\left. + \sum_{s=1, j_s \neq k}^{N_{(j)}} \widetilde{\mathbf{k}}_{(j)(k)}^{OSB(j)(j_s)} \begin{Bmatrix} u_{1(k)} \\ u_{2(k)} \\ u_{1(j_s)} \\ u_{2(j_s)} \end{Bmatrix} \right) \tag{3.183}$$

As shown in Fig. 3.20, the element connectivity for a uniform OSB-PD interaction between nodes k and j can be modeled by considering: (1) an element (shown in blue) with stiffness matrix, $\mathbf{k}_{(k)(j)}^{OSB} + \widetilde{\mathbf{k}}_{(j)(k)}^{OSB}$ for a bond, (2) a network of elements around node k (shown in red) with stiffness matrices, $\mathbf{k}_{(k)(j)}^{OSB(k)(k_r)}$ due to the direct interaction of node k with its family members, and (3) a network of elements around node k (shown in green) with stiffness matrices, $\widetilde{\mathbf{k}}_{(j)(k)}^{OSB(j)(j_s)}$ due to the indirect interactions of node, k with the family members of node j.

The total internal force, $\mathbf{F}_{(k)}^{OSB}$ is computed by considering: (1) the stiffness matrices, $\mathbf{k}_{(k)(j)}^{OSB}$ of all the bonds (blue) that are connected from node k to all other nodes j in the horizon $H_{(k)}$ along with their respective direct interactions (red) with stiffness matrices, $\mathbf{k}_{(k)(j)}^{OSB(k)(k_r)}$ and indirect interactions (green) with stiffness matrices, $\widetilde{\mathbf{k}}_{(j)(k)}^{OSB(j)(j_s)}$.

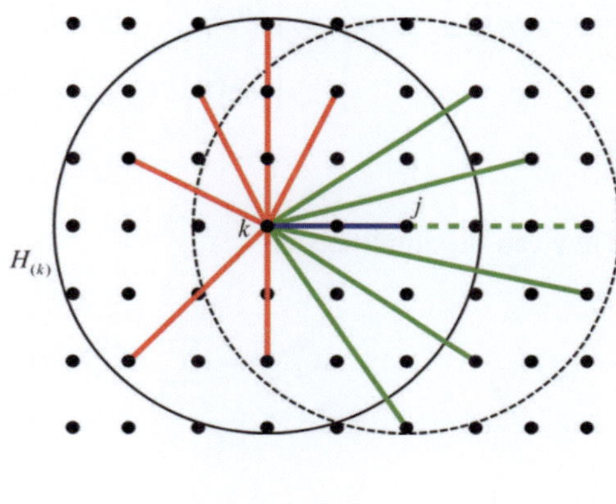

Fig. 3.20 Element connectivity for constant horizon OSB interactions between nodes k and j in $H_{(k)} \cup H_{(j)}$

3.12.3 NOSB-PD Equilibrium Equation

The discrete form of Cauchy's stress tensor for a bond $\boldsymbol{\xi}_{(k)(j)}$ between nodes k and j in Eq. (3.64) can be written as

$$\boldsymbol{\sigma}_{\boldsymbol{\xi}(k)} = \sum_{k_r \in N_{(k\cap j)}} \widehat{\mu}_{(k)(k_r)} \left\{ \begin{array}{l} \lambda \mathrm{tr}\left[\left(\mathbf{u}\left(\mathbf{x}_{(k_r)}\right) - \mathbf{u}\left(\mathbf{x}_{(k)}\right)\right) \otimes \mathbf{g}\left(\mathbf{x}_{(k)}, \boldsymbol{\xi}_{(k)(k_r)}\right)\right]\mathbf{I} \\ +\mu \left[\begin{array}{l} \left(\mathbf{u}\left(\mathbf{x}_{(k_r)}\right) - \mathbf{u}\left(\mathbf{x}_{(k)}\right)\right) \otimes \mathbf{g}\left(\mathbf{x}_{(k)}, \boldsymbol{\xi}_{(k)(k_r)}\right) \\ +\mathbf{g}\left(\mathbf{x}_{(k)}, \boldsymbol{\xi}_{(k)(k_r)}\right) \otimes \left(\mathbf{u}\left(\mathbf{x}_{(k_r)}\right) - \mathbf{u}\left(\mathbf{x}_{(k)}\right)\right) \end{array} \right] \end{array} \right\} V_{(k_r)} \tag{3.184}$$

After substituting from Eq. (3.184) into Eqs. (3.67) and (3.68), the force density vectors for the NOSB-PD corresponding to primary and secondary bonds, respectively, can be expressed as

$$\mathbf{t}^{NOSB}_{(k)(j)} = \widehat{\mu}_{(k)(j)} \sum_{r=1}^{N_{(k\cap j)}} \widehat{\mu}_{(k)(k_r)} \left\{ \begin{array}{l} \lambda \mathrm{tr}\left[\left(\mathbf{u}\left(\mathbf{x}_{(k_r)}\right) - \mathbf{u}\left(\mathbf{x}_{(k)}\right)\right) \otimes \mathbf{g}\left(\mathbf{x}_{(k)}, \boldsymbol{\xi}_{(k)(k_r)}\right)\right]\mathbf{I} \\ +\mu \left[\begin{array}{l} \left(\mathbf{u}\left(\mathbf{x}_{(k_r)}\right) - \mathbf{u}\left(\mathbf{x}_{(k)}\right)\right) \otimes \mathbf{g}\left(\mathbf{x}_{(k)}, \boldsymbol{\xi}_{(k)(k_r)}\right) \\ +\mathbf{g}\left(\mathbf{x}_{(k)}, \boldsymbol{\xi}_{(k)(k_r)}\right) \otimes \left(\mathbf{u}\left(\mathbf{x}_{(k_r)}\right) - \mathbf{u}\left(\mathbf{x}_{(k)}\right)\right) \end{array} \right] \end{array} \right\} V_{(k_r)} \mathbf{g}\left(\mathbf{x}_{(k)}, \boldsymbol{\xi}_{(k)(j)}\right) \tag{3.185}$$

and

$$\mathbf{t}^{NOSB}_{(p)(k)} = \widehat{\mu}_{(p)(k)} \sum_{s=1}^{N_{(p\cap k)}} \widehat{\mu}_{(p)(p_s)} \left\{ \begin{array}{l} \lambda \mathrm{tr}\left[\left(\mathbf{u}\left(\mathbf{x}_{(p_s)}\right) - \mathbf{u}\left(\mathbf{x}_{(p)}\right)\right) \otimes \mathbf{g}\left(\mathbf{x}_{(p)}, \boldsymbol{\xi}_{(p)(p_s)}\right)\right]\mathbf{I} \\ +\mu \left[\begin{array}{l} \left(\mathbf{u}\left(\mathbf{x}_{(p_s)}\right) - \mathbf{u}\left(\mathbf{x}_{(p)}\right)\right) \otimes \mathbf{g}\left(\mathbf{x}_{(p)}, \boldsymbol{\xi}_{(p)(p_s)}\right) \\ +\mathbf{g}\left(\mathbf{x}_{(p)}, \boldsymbol{\xi}_{(p)(p_s)}\right) \otimes \left(\mathbf{u}\left(\mathbf{x}_{(p_r)}\right) - \mathbf{u}\left(\mathbf{x}_{(p)}\right)\right) \end{array} \right] \end{array} \right\} V_{(p_s)} \mathbf{g}\left(\mathbf{x}_{(p)}, \boldsymbol{\xi}_{(p)(k)}\right) \tag{3.186}$$

They can be simplified as

$$\mathbf{t}^{NOSB}_{(k)(j)} = \sum_{r=1}^{N_{(k\cap j)}} \mathbf{H}^{(k)(k_r)}_{\boldsymbol{\xi}(k)(j)} \left(\mathbf{u}_{(k_r)} - \mathbf{u}_{(k)}\right) \tag{3.187}$$

and

$$\mathbf{t}_{(p)(k)}^{NOSB} = \sum_{s=1}^{N_{(p\cap k)}} \mathbf{H}_{\xi(p)(k)}^{(p)(p_s)} \left(\mathbf{u}_{(p_s)} - \mathbf{u}_{(p)} \right) \tag{3.188}$$

in which the matrices $\mathbf{H}_{\xi(k)(j)}^{(k)(k_r)}$ and $\mathbf{H}_{\xi(p)(k)}^{(p)(p_s)}$ are defined as

$$\mathbf{H}_{\xi(k)(j)}^{(k)(k_r)} = \widehat{\mu}_{(k)(j)} \widehat{\mu}_{(k)(k_r)} \begin{bmatrix} \left((\lambda + 2\mu) g_{(k)(k_r)}^{10} g_{(k)(j)}^{10} + \mu g_{(k)(k_r)}^{01} g_{(k)(j)}^{01} \right) & \left(\lambda g_{(k)(k_r)}^{01} g_{(k)(j)}^{10} + \mu g_{(k)(k_r)}^{10} g_{(k)(j)}^{01} \right) \\ \left(\lambda g_{(k)(k_r)}^{10} g_{(k)(j)}^{01} + \mu g_{(k)(k_r)}^{01} g_{(k)(j)}^{10} \right) & \left(\mu g_{(k)(k_r)}^{10} g_{(k)(j)}^{10} + (\lambda + 2\mu) g_{(k)(k_r)}^{01} g_{(k)(j)}^{01} \right) \end{bmatrix} V_{(k_r)} \tag{3.189}$$

and

$$\mathbf{H}_{\xi(p)(k)}^{(p)(p_s)} = \widehat{\mu}_{(p)(k)} \widehat{\mu}_{(p)(p_s)} \begin{bmatrix} \left((\lambda + 2\mu) g_{(p)(p_s)}^{10} g_{(p)(k)}^{10} + \mu g_{(p)(p_s)}^{01} g_{(p)(k)}^{01} \right) & \left(\lambda g_{(p)(p_s)}^{01} g_{(p)(k)}^{10} + \mu g_{(p)(p_s)}^{10} g_{(p)(k)}^{01} \right) \\ \left(\lambda g_{(p)(p_s)}^{10} g_{(p)(k)}^{01} + \mu g_{(p)(p_s)}^{01} g_{(p)(k)}^{10} \right) & \left(\mu g_{(p)(p_s)}^{10} g_{(p)(k)}^{10} + (\lambda + 2\mu) g_{(p)(p_s)}^{01} g_{(p)(k)}^{01} \right) \end{bmatrix} V_{(p_s)} \tag{3.190}$$

Using Eq. (3.69), the discrete form of the internal force vector $\mathbf{L}_{(k)}^{NOSB}$ in the interior region $\mathcal{D}$ can be expressed as

$$\begin{aligned} \mathbf{L}_{(k)}^{NOSB} = & \sum_{j \in H_{(k)}} \left(\sum_{r=1}^{N_{(k\cap j)}} \mathbf{H}_{\xi(k)(j)}^{(k)(k_r)} \left(\mathbf{u}_{(k_r)} - \mathbf{u}_{(k)} \right) V_{(j)} \right) \\ & + \sum_{p \in \widetilde{H}_{(k)}} \left(\sum_{s=1}^{N_{(p\cap k)}} \mathbf{H}_{\xi(p)(k)}^{(p)(p_s)} \left(\mathbf{u}_{(p)} - \mathbf{u}_{(p_s)} \right) V_{(p)} \right) \end{aligned} \tag{3.191}$$

The discrete form of the force vector, $\mathbf{L}_{(k)}^{NOSB}$ after multiplying with the volume, $V_{(k)}$ of node k can be written as

$$
\begin{aligned}
\mathbf{F}_{(k)}^{NOSB} &= \left(\sum_{j \in H_{(k)}} \left(\sum_{r=1}^{N_{(k \cap j)}} \mathbf{H}_{\boldsymbol{\xi}(k)(j)}^{(k)(k_r)} \left(\mathbf{u}_{(k_r)} - \mathbf{u}_{(k)} \right) V_{(j)} \right) \right. \\
&\quad \left. + \sum_{p \in \widetilde{H}_{(k)}} \left(\sum_{s=1}^{N_{(p \cap k)}} \mathbf{H}_{\boldsymbol{\xi}(p)(k)}^{(p)(p_s)} \left(\mathbf{u}_{(p)} - \mathbf{u}_{(p_s)} \right) V_{(p)} \right) \right) V_{(k)} \\
&= \sum_{j \in H_{(k)}} \left(\sum_{r=1,\, k_r \neq j}^{N_{(k \cap j)}} \mathbf{G}_{(k)(j)}^{(k)(k_r)} \mathbf{u}_{(k_r)} - \boldsymbol{\Phi}_{(k)} \mathbf{u}_{(k)} + \mathbf{G}_{(k)(j)}^{(k)(j)} \left(\mathbf{u}_{(j)} - \mathbf{u}_{(k)} \right) \right) \\
&\quad + \sum_{p \in \widetilde{H}_{(k)}} \left(\widetilde{\boldsymbol{\Phi}}_{(p)} \mathbf{u}_{(p)} - \sum_{s=1,\, p_s \neq k}^{N_{(p \cap k)}} \widetilde{\mathbf{G}}_{(p)(k)}^{(p)(p_s)} \mathbf{u}_{(p_s)} + \widetilde{\mathbf{G}}_{(p)(k)}^{(p)(k)} \left(\mathbf{u}_{(p)} - \mathbf{u}_{(k)} \right) \right)
\end{aligned}
\tag{3.192}
$$

with

$$
\mathbf{G}_{(k)(j)}^{(k)(k_r)} = \mathbf{H}_{\boldsymbol{\xi}(k)(j)}^{(k)(k_r)} V_{(j)} V_{(k)} \tag{3.193}
$$

$$
\boldsymbol{\Phi}_{(k)} = \sum_{r=1,\, k_r \neq j}^{N_{(k \cap j)}} \mathbf{G}_{(k)(j)}^{(k)(k_r)} \tag{3.194}
$$

$$
\widetilde{\mathbf{G}}_{(p)(k)}^{(p)(p_s)} = \mathbf{H}_{\boldsymbol{\xi}(p)(k)}^{(p)(p_s)} V_{(p)} V_{(k)} \tag{3.195}
$$

and

$$
\widetilde{\boldsymbol{\Phi}}_{(p)} = \sum_{s=1,\, p_s \neq k}^{N_{(p \cap k)}} \widetilde{\mathbf{G}}_{(p)(k)}^{(p)(p_s)} \tag{3.196}
$$

The internal force vector, $\mathbf{F}_{(k)}^{NOSB}$ in component form can be written as

$$
\begin{aligned}
\begin{Bmatrix} F_{1(k)}^{NOSB} \\ F_{2(k)}^{NOSB} \\ 0 \\ 0 \end{Bmatrix} &= \sum_{j \in H_{(k)}} \left(\mathbf{k}_{(k)(j)}^{NOSB} \begin{Bmatrix} u_{1(k)} \\ u_{2(k)} \\ u_{1(j)} \\ u_{2(j)} \end{Bmatrix} + \sum_{r=1,\, k_r \neq j}^{N_{(k \cap j)}} \mathbf{k}_{(k)(j)}^{NOSB(k)(k_r)} \begin{Bmatrix} u_{1(k)} \\ u_{2(k)} \\ u_{1(k_r)} \\ u_{2(k_r)} \end{Bmatrix} \right) \\
&\quad + \sum_{p \in \widetilde{H}_{(k)}} \left(\left(\widetilde{\mathbf{k}}_{(p)(k)}^{NOSB} \right) \begin{Bmatrix} u_{1(k)} \\ u_{2(k)} \\ u_{1(p)} \\ u_{2(p)} \end{Bmatrix} + \sum_{s=1,\, p_s \neq k}^{N_{(p \cap k)}} \widetilde{\mathbf{k}}_{(p)(k)}^{NOSB(p)(p_s)} \begin{Bmatrix} u_{1(k)} \\ u_{2(k)} \\ u_{1(p_s)} \\ u_{2(p_s)} \end{Bmatrix} \right)
\end{aligned}
\tag{3.197}
$$

with

$$\mathbf{k}_{(k)(j)}^{NOSB} = \begin{bmatrix} -\Phi_{11(k)} - G_{11(k)(j)}^{(k)(j)} & -\Phi_{12(k)} - G_{12(k)(j)}^{(k)(j)} & G_{11(k)(j)}^{(k)(j)} & G_{12(k)(j)}^{(k)(j)} \\ -\Phi_{21(k)} - G_{21(k)(j)}^{(k)(j)} & -\Phi_{22(k)} - G_{22(k)(j)}^{(22)} & G_{21(k)(j)}^{(k)(j)} & G_{22(k)(j)}^{(k)(j)} \\ 0 & 0 & 0 & 0 \\ 0 & 0 & 0 & 0 \end{bmatrix} \tag{3.198}$$

$$\mathbf{k}_{(k)(j)}^{NOSB(k)(k_r)} = \begin{bmatrix} 0 & 0 & G_{11(k)(j)}^{(k)(k_r)} & G_{12(k)(j)}^{(k)(k_r)} \\ 0 & 0 & G_{21(k)(j)}^{(k)(k_r)} & G_{22(k)(j)}^{(k)(k_r)} \\ 0 & 0 & 0 & 0 \\ 0 & 0 & 0 & 0 \end{bmatrix} \tag{3.199}$$

$$\widetilde{\mathbf{k}}_{(p)(k)}^{NOSB} = \begin{bmatrix} -\widetilde{G}_{11(p)(k)}^{(p)(k)} & -\widetilde{G}_{12(p)(k)}^{(p)(k)} & \widetilde{\Phi}_{11(p)} + \widetilde{G}_{11(p)(k)}^{(p)(k)} & \widetilde{\Phi}_{12(p)} + \widetilde{G}_{12(p)(k)}^{(p)(k)} \\ -\widetilde{G}_{(p)(k)}^{(21)} & -\widetilde{G}_{(p)(k)}^{(22)} & \widetilde{\Phi}_{21(p)} + \widetilde{G}_{21(p)(k)}^{(p)(k)} & \widetilde{\Phi}_{22(p)} + \widetilde{G}_{22(p)(k)}^{(p)(k)} \\ 0 & 0 & 0 & 0 \\ 0 & 0 & 0 & 0 \end{bmatrix} \tag{3.200}$$

and

$$\widetilde{\mathbf{k}}_{(p)(k)}^{NOSB(p)(p_s)} = \begin{bmatrix} 0 & 0 & -\widetilde{G}_{11(p)(k)}^{(p)(p_s)} & -\widetilde{G}_{12(p)(k)}^{(p)(p_s)} \\ 0 & 0 & -\widetilde{G}_{21(p)(k)}^{(p)(p_s)} & -\widetilde{G}_{22(p)(k)}^{(p)(p_s)} \\ 0 & 0 & 0 & 0 \\ 0 & 0 & 0 & 0 \end{bmatrix} \tag{3.201}$$

Figure 3.21a, b show the element connectivity for dual horizon NOSB-PD interaction between nodes k and j and k and p denoting the primary (red) and secondary (blue) bonds, respectively. The interactions can be modeled by considering: (1) an element (shown in red) with stiffness matrix $\mathbf{k}_{(k)(j)}^{NOSB}$ corresponding to a primary bond and another element (shown in blue) with stiffness matrix $\widetilde{\mathbf{k}}_{(p)(k)}^{NOSB}$ corresponding to a secondary bond, (2) a network of elements around node k (shown in dashed red) with the stiffness matrices, $\mathbf{k}_{(k)(j)}^{NOSB(k)(k_r)}$ due to the direct interaction of node k with the intersected family members of k and j, and 3) a network of elements around node k (shown in dashed green) with stiffness matrices, $\widetilde{\mathbf{k}}_{(p)(k)}^{NOSB(p)(p_s)}$ due to the indirect interactions of nodes with the intersected family members of k and p.

The total internal force, $\mathbf{F}_{(k)}^{NOSB}$ is computed by considering: (1) the stiffness matrices of all the primary (red) bonds, $\mathbf{k}_{(k)(j)}^{NOSB}$, that are connected from node k to all other nodes j in the domain $H_{(k)} \cap H_{(j)}$ for $j \in H_{(k)}$, along with their respective direct interactions (dashed red) in the intersection domain with stiffness matrices,

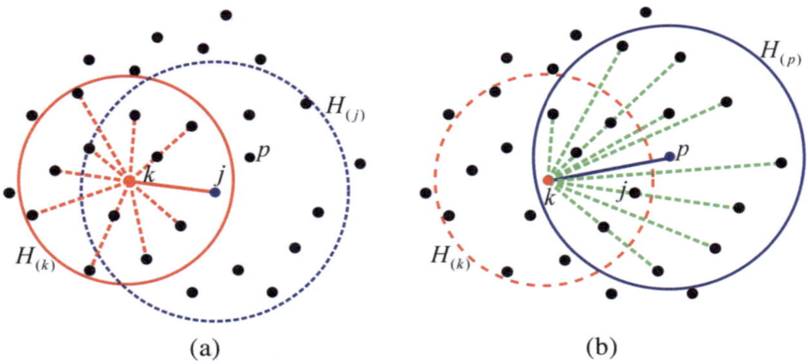

Fig. 3.21 Element connectivity for dual horizon NOSB-PD interactions: (**a**) primary bond between nodes k and j and (**b**) secondary bond between nodes k and p

$\mathbf{k}_{(k)(j)}^{NOSB(k)(k_r)}$, and (2) the stiffness matrices, $\widetilde{\mathbf{k}}_{(p)(k)}^{NOSB}$ of all the secondary (blue) bonds that are connected from node k to all other nodes p in the intersected domain $H_{(p)} \cap H_{(k)}$ for $p \in \widetilde{H}_{(k)}$, along with their respective indirect interactions (dashed green) in the intersected domain with stiffness matrices, $\widetilde{\mathbf{k}}_{(p)(k)}^{NOSB(p)(p_s)}$.

For a uniform discretization, the discrete form of the internal force vector, $\mathbf{F}_{(k)}^{NOSB}$ at node k can be written as

$$\mathbf{F}_{(k)}^{NOSB} = \mathbf{L}_{(k)}^{NOSB} V_{(k)} = \sum_{j \in H_{(k)}} \left\{ \begin{array}{l} \sum\limits_{r=1,\, k_r \neq j}^{N_{(k \cap j)}} \mathbf{G}_{(k)(j)}^{(k)(k_r)} \mathbf{u}_{(k_r)} - \boldsymbol{\Phi}_{(k)} \mathbf{u}_{(k)} + \mathbf{G}_{(k)(j)}^{(k)(j)} \left(\mathbf{u}_{(j)} - \mathbf{u}_{(k)} \right) \\ + \widetilde{\boldsymbol{\Phi}}_{(j)} \mathbf{u}_{(j)} - \sum\limits_{s=1,\, p_s \neq k}^{N_{(j \cap k)}} \widetilde{\mathbf{G}}_{(j)(k)}^{(j)(j_s)} \mathbf{u}_{(j_s)} + \widetilde{\mathbf{G}}_{(j)(k)}^{(j)(k)} \left(\mathbf{u}_{(j)} - \mathbf{u}_{(k)} \right) \end{array} \right\} \tag{3.202}$$

In component form, it can be written as

$$\left\{ \begin{array}{c} F_{1(k)}^{NOSB} \\ F_{2(k)}^{NOSB} \\ 0 \\ 0 \end{array} \right\} = \sum_{j \in H_{(k)}} \left(\begin{array}{l} \left(\mathbf{k}_{(k)(j)}^{NOSB} + \widetilde{\mathbf{k}}_{(j)(k)}^{NOSB} \right) \left\{ \begin{array}{c} u_{1(k)} \\ u_{2(k)} \\ u_{1(j)} \\ u_{2(j)} \end{array} \right\} + \sum\limits_{r=1,\, k_r \neq j}^{N_{(k \cap j)}} \mathbf{k}_{(k)(j)}^{NOSB(k)(k_r)} \left\{ \begin{array}{c} u_{1(k)} \\ u_{2(k)} \\ u_{1(k_r)} \\ u_{2(k_r)} \end{array} \right\} \\ + \sum\limits_{s=1,\, j_s \neq k}^{N_{(j \cap k)}} \widetilde{\mathbf{k}}_{(j)(k)}^{NOSB(j)(j_s)} \left\{ \begin{array}{c} u_{1(k)} \\ u_{2(k)} \\ u_{1(j_s)} \\ u_{2(j_s)} \end{array} \right\} \end{array} \right) \tag{3.203}$$

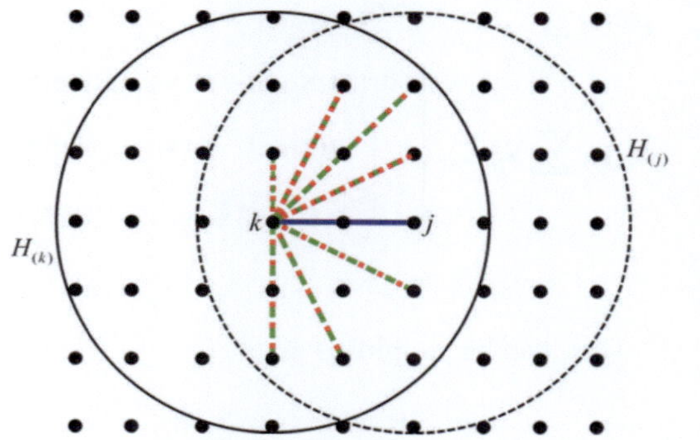

Fig. 3.22 Element connectivity for constant horizon NOSB-PD interactions between nodes k and j in $H_{(k)} \cap H_{(j)}$

As shown in Fig. 3.22, the element connectivity for a NOSB-PD interaction between nodes k and j can be modeled by considering: (1) an element (shown in blue) with stiffness matrix, $\mathbf{k}_{(k)(j)}^{NOSB} + \widetilde{\mathbf{k}}_{(j)(k)}^{NOSB}$ for a bond, (2) a network of elements around node k (shown in red) with the stiffness matrices $\mathbf{k}_{(k)(j)}^{NOSB(k)(k_r)}$ due to the direct interaction of node k, and (3) a network of elements around node k (shown in green) with stiffness matrices, $\widetilde{\mathbf{k}}_{(j)(k)}^{NOSB(j)(j_s)}$ due to the indirect interactions of node, k with the family members of node j in the intersected domain.

3.12.4 PDDO Equilibrium Equation

The discrete form of Cauchy's stress tensor for node k in Eq. (3.79) can be written as

$$\boldsymbol{\sigma}_{(k)} = \sum_{j \in N_{(k)}} \widehat{\mu}_{(k)(j)} \left\{ \begin{array}{l} \lambda \mathrm{tr}\left[\left(\mathbf{u}\left(\mathbf{x}_{(j)}\right) - \mathbf{u}\left(\mathbf{x}_{(k)}\right)\right) \otimes \mathbf{g}\left(\mathbf{x}_{(k)}, \boldsymbol{\xi}_{(k)(j)}\right)\right] \mathbf{I} \\ +\mu \left[\begin{array}{l} \left(\mathbf{u}\left(\mathbf{x}_{(j)}\right) - \mathbf{u}\left(\mathbf{x}_{(k)}\right)\right) \otimes \mathbf{g}\left(\mathbf{x}_{(k)}, \boldsymbol{\xi}_{(k)(j)}\right) \\ +\mathbf{g}\left(\mathbf{x}_{(k)}, \boldsymbol{\xi}_{(k)(j)}\right) \otimes \left(\mathbf{u}\left(\mathbf{x}_{(k)}\right) - \mathbf{u}\left(\mathbf{x}_{(k)}\right)\right) \end{array} \right] \end{array} \right\} V_{(j)} \tag{3.204}$$

After substituting from Eq. (3.204) into Eqs. (3.77) and (3.78), the force density vectors $\mathbf{t}_{(j)(k)}^{PDDO}$ and $\mathbf{t}_{(k)(j)}^{PDDO}$ for a PDDO interaction in the region $\mathscr{R}$ can be expressed as

$$\mathbf{t}_{(j)(k)}^{PDDO} = \widehat{\mu}_{(k)(j)} \sum_{s=1}^{N_{(j)}} \widehat{\mu}_{(j)(j_s)} \left\{ \begin{array}{l} \lambda \left[\mathrm{tr}\left(\mathbf{u}\left(\mathbf{x}_{(j_s)}\right) - \mathbf{u}\left(\mathbf{x}_{(j)}\right)\right) \otimes \mathbf{g}\left(\mathbf{x}_{(j)}, \boldsymbol{\xi}_{(j)(j_s)}\right)\right] \mathbf{I} + \\ \mu \left[\begin{array}{l} \left(\mathbf{u}\left(\mathbf{x}_{(j_s)}\right) - \mathbf{u}\left(\mathbf{x}_{(j)}\right)\right) \otimes \mathbf{g}\left(\mathbf{x}_{(j)}, \boldsymbol{\xi}_{(j)(j_s)}\right) \\ +\mathbf{g}\left(\mathbf{x}_{(j)}, \boldsymbol{\xi}_{(j)(j_s)}\right) \otimes \left(\mathbf{u}\left(\mathbf{x}_{(j_s)}\right) - \mathbf{u}\left(\mathbf{x}_{(j)}\right)\right) \end{array} \right] \end{array} \right\} V_{(j_s)} \mathbf{g}\left(\mathbf{x}_{(k)}, \boldsymbol{\xi}_{(k)(j)}\right) \tag{3.205}$$

and

$$\mathbf{t}_{(k)(j)}^{PDDO} = \widehat{\mu}_{(k)(j)} \sum_{r=1}^{N_{(k)}} \widehat{\mu}_{(k)(k_r)} \left\{ \begin{array}{l} \lambda \mathrm{tr}[(\mathbf{u}(\mathbf{x}_{(k_r)}) - \mathbf{u}(\mathbf{x}_{(k)})) \otimes \mathbf{g}(\mathbf{x}_{(k)}, \xi_{(k)(k_r)})]\mathbf{I} + \\ \mu \left[\begin{array}{l} (\mathbf{u}(\mathbf{x}_{(k_r)}) - \mathbf{u}(\mathbf{x}_{(k)})) \otimes \mathbf{g}(\mathbf{x}_{(k)}, \xi_{(k)(k_r)}) \\ + \mathbf{g}(\mathbf{x}_{(k)}, \xi_{(k)(k_r)}) \otimes (\mathbf{u}(\mathbf{x}_{(k_r)}) - \mathbf{u}(\mathbf{x}_{(k)})) \end{array} \right] \end{array} \right\} V_{(k_r)} \mathbf{g}(\mathbf{x}_{(k)}, \xi_{(k)(j)}) \tag{3.206}$$

They can be simplified as

$$\mathbf{t}_{(j)(k)}^{PDDO} = \sum_{s=1}^{N_{(j)}} \mathbf{H}_{(k)(j)}^{(j)(j_s)} \left(\mathbf{u}_{(j_s)} - \mathbf{u}_{(j)} \right) \tag{3.207}$$

and

$$\mathbf{t}_{(k)(j)}^{PDDO} = \sum_{r=1}^{N_{(k)}} \mathbf{H}_{(k)(j)}^{(k)(k_r)} \left(\mathbf{u}_{(k_r)} - \mathbf{u}_{(k)} \right) \tag{3.208}$$

in which the matrices $\mathbf{H}_{(k)(j)}^{(j)(j_s)}$ and $\mathbf{H}_{(k)(j)}^{(k)(k_r)}$ are defined as

$$\mathbf{H}_{(k)(j)}^{(j)(j_s)} = \widehat{\mu}_{(k)(j)} \widehat{\mu}_{(j)(j_s)} \begin{bmatrix} \left((\lambda + 2\mu) g_{(j)(j_s)}^{10} g_{(k)(j)}^{10} + \mu g_{(j)(j_s)}^{01} g_{(k)(j)}^{01} \right) & \left(\lambda g_{(j)(j_s)}^{01} g_{(k)(j)}^{10} + \mu g_{(j)(j_s)}^{10} g_{(k)(j)}^{01} \right) \\ \left(\lambda g_{(j)(j_s)}^{10} g_{(k)(j)}^{01} + \mu g_{(j)(j_s)}^{01} g_{(k)(j)}^{10} \right) & \left(\mu g_{(j)(j_s)}^{10} g_{(k)(j)}^{10} + (\lambda + 2\mu) g_{(j)(j_s)}^{01} g_{(k)(j)}^{01} \right) \end{bmatrix} V_{(j_s)} \tag{3.209}$$

and

$$\mathbf{H}_{(k)(j)}^{(k)(k_r)} = \widehat{\mu}_{(k)(j)} \widehat{\mu}_{(k)(k_r)} \begin{bmatrix} \left((\lambda + 2\mu) g_{(k)(k_r)}^{10} g_{(k)(j)}^{10} + \mu g_{(k)(k_r)}^{01} g_{(k)(j)}^{01} \right) & \left(\lambda g_{(k)(k_r)}^{01} g_{(k)(j)}^{10} + \mu g_{(k)(k_r)}^{10} g_{(k)(j)}^{01} \right) \\ \left(\lambda g_{(k)(k_r)}^{10} g_{(k)(j)}^{01} + \mu g_{(k)(k_r)}^{01} g_{(k)(j)}^{10} \right) & \left(\mu g_{(k)(k_r)}^{10} g_{(k)(j)}^{10} + (\lambda + 2\mu) g_{(k)(k_r)}^{01} g_{(k)(j)}^{01} \right) \end{bmatrix} V_{(k_r)} \tag{3.210}$$

The discrete form of the internal force vector $\mathbf{L}_{(k)}^{PDDO}$ in the PDDO region $\mathcal{D}$ can be expressed as

$$\mathbf{L}_{(k)}^{PDDO} = \sum_{j \in H_{(k)}} \left(\sum_{s=1}^{N_{(j)}} \mathbf{H}_{(k)(j)}^{(j)(j_s)} \left(\mathbf{u}_{(j_s)} - \mathbf{u}_{(j)} \right) - \sum_{r=1}^{N_{(k)}} \mathbf{H}_{(k)(j)}^{(k)(k_r)} \left(\mathbf{u}_{(k_r)} - \mathbf{u}_{(k)} \right) \right) V_{(j)} \tag{3.211}$$

The discrete form of the force vector $\mathbf{L}_{(k)}^{PDDO}$ after multiplying with the volume of the node, $V_{(k)}$ can be written as

$$\begin{aligned}
\mathbf{F}_{(k)}^{PDDO} &= \mathbf{L}_{(k)}^{PDDO} V_{(k)} \\
&= \sum_{j \in H_{(k)}} \left(\sum_{s=1}^{N_{(j)}} \mathbf{H}_{(k)(j)}^{(j)(j_s)} \left(\mathbf{u}_{(j_s)} - \mathbf{u}_{(j)} \right) + \sum_{r=1}^{N_{(k)}} \mathbf{H}_{(k)(j)}^{(k)(k_r)} \left(\mathbf{u}_{(k)} - \mathbf{u}_{(k_r)} \right) \right) V_{(j)} V_{(k)} \\
&= \sum_{j \in H_{(k)}} \left(\begin{array}{l} -\widetilde{\boldsymbol{\Phi}}_{(j)} \mathbf{u}_{(j)} + \sum\limits_{s=1, j_s \neq k}^{N_{(j)}} \widetilde{\mathbf{G}}_{(k)(j)}^{(j)(j_s)} \mathbf{u}_{(j_s)} + \widetilde{\mathbf{G}}_{(k)(j)}^{(j)(k)} \left(\mathbf{u}_{(k)} - \mathbf{u}_{(j)} \right) \\ - \sum\limits_{r=1, k_r \neq j}^{N_{(k)}} \mathbf{G}_{(k)(j)}^{(k)(k_r)} \mathbf{u}_{(k_r)} + \boldsymbol{\Phi}_{(k)} \mathbf{u}_{(k)} + \mathbf{G}_{(k)(j)}^{(k)(j)} \left(\mathbf{u}_{(k)} - \mathbf{u}_{(j)} \right) \end{array} \right)
\end{aligned} \tag{3.212}$$

with

$$\mathbf{G}_{(k)(j)}^{(k)(k_r)} = \mathbf{H}_{(k)(j)}^{(k)(k_r)} V_{(j)} V_{(k)} \tag{3.213}$$

$$\boldsymbol{\Phi}_{(k)} = \sum_{r=1, k_r \neq j}^{N_{(k)}} \mathbf{G}_{(k)(j)}^{(k)(k_r)} \tag{3.214}$$

and

$$\widetilde{\mathbf{G}}_{(k)(j)}^{(j)(j_s)} = \mathbf{H}_{(k)(j)}^{(j)(j_s)} V_{(j)} V_{(k)} \tag{3.215}$$

$$\widetilde{\boldsymbol{\Phi}}_{(j)} = \sum_{s=1, j_s \neq k}^{N_{(j)}} \widetilde{\mathbf{G}}_{(j)(k)}^{(j)(j_s)} \tag{3.216}$$

The force vector $\mathbf{F}_{(k)}^{PDDO}$ in component form can be written as

$$\begin{Bmatrix} F_{1(k)}^{PDDO} \\ F_{2(k)}^{PDDO} \\ 0 \\ 0 \end{Bmatrix} = \sum_{j \in H_{(k)}} \left(\mathbf{k}_{(k)(j)}^{PDDO} \begin{Bmatrix} u_{1(k)} \\ u_{2(k)} \\ u_{1(j)} \\ u_{2(j)} \end{Bmatrix} + \sum_{r=1, k_r \neq j}^{N_{(k)}} \mathbf{k}_{(k)(j)}^{PDDO(k)(k_r)} \begin{Bmatrix} u_{1(k)} \\ u_{2(k)} \\ u_{1(k_r)} \\ u_{2(k_r)} \end{Bmatrix} \right.$$
$$\left. + \sum_{s=1, j_s \neq k}^{N_{(j)}} \mathbf{k}_{(k)(j)}^{PDDO(j)(j_s)} \begin{Bmatrix} u_{1(k)} \\ u_{2(k)} \\ u_{1(j_s)} \\ u_{2(j_s)} \end{Bmatrix} \right) \tag{3.217}$$

with

$$\mathbf{k}_{(k)(j)}^{PDDO} = \begin{bmatrix} \Phi_{11(k)} + G_{11(k)(j)}^{(k)(j)} + \widetilde{G}_{11(k)(j)}^{(j)(k)} & \Phi_{12(k)} + G_{12(k)(j)}^{(k)(j)} + \widetilde{G}_{12(k)(j)}^{(j)(k)} & -G_{11(k)(j)}^{(k)(j)} - \widetilde{G}_{11(k)(j)}^{(j)(k)} - \widetilde{\Phi}_{12(j)} & -G_{12(k)(j)}^{(k)(j)} - \widetilde{G}_{12(k)(j)}^{(j)(k)} - \widetilde{\Phi}_{12(j)} \\ \Phi_{21(k)} + G_{21(k)(j)}^{(k)(j)} + \widetilde{G}_{21(k)(j)}^{(j)(k)} & \Phi_{22(k)} + G_{22(k)(j)}^{(k)(j)} + \widetilde{G}_{22(k)(j)}^{(j)(k)} & -G_{21(k)(j)}^{(21)} - \widetilde{G}_{21(k)(j)}^{(j)(k)} - \widetilde{\Phi}_{21(j)} & -G_{22(k)(j)}^{(k)(j)} - \widetilde{G}_{22(k)(j)}^{(j)(k)} - \widetilde{\Phi}_{22(j)} \\ 0 & 0 & 0 & 0 \\ 0 & 0 & 0 & 0 \end{bmatrix} \tag{3.218}$$

$$\mathbf{k}_{(k)(j)}^{PDDO(k)(k_r)} = \begin{bmatrix} 0 & 0 & -G_{11(k)(j)}^{(k)(k_r)} & -G_{12(k)(j)}^{(k)(k_r)} \\ 0 & 0 & -G_{21(k)(j)}^{(k)(k_r)} & -G_{22(k)(j)}^{(k)(k_r)} \\ 0 & 0 & 0 & 0 \\ 0 & 0 & 0 & 0 \end{bmatrix} \tag{3.219}$$

and

$$\widetilde{\mathbf{k}}_{(k)(j)}^{PDDO(j)(j_s)} = \begin{bmatrix} 0 & 0 & \widetilde{G}_{11(k)(j)}^{(j)(j_s)} & \widetilde{G}_{12(k)(j)}^{(j)(j_s)} \\ 0 & 0 & \widetilde{G}_{21(k)(j)}^{(j)(j_s)} & \widetilde{G}_{22(k)(j)}^{(j)(j_s)} \\ 0 & 0 & 0 & 0 \\ 0 & 0 & 0 & 0 \end{bmatrix} \tag{3.220}$$

Figure 3.23a, b show the element connectivity for the PDDO interaction between nodes k and j. The interactions can be modeled by considering: (1) an element

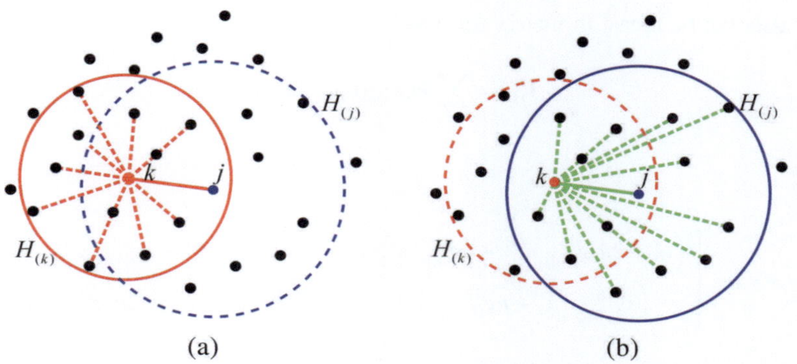

Fig. 3.23 Element connectivity for PDDO interactions between nodes k and j: (**a**) direct interaction and (**b**) indirect interaction

(shown in red) with stiffness matrix $\mathbf{k}_{(k)(j)}^{PDDO}$ corresponding to the bond, (2) a network of elements around node k (shown in dashed red) with stiffness matrices $\mathbf{k}_{(k)(j)}^{PDDO(k)(k_r)}$ due to the direct interaction of node k with its family members, and (3) a network of elements around node j (shown in dashed green) with stiffness matrices, $\mathbf{k}_{(k)(j)}^{PDDO(j)(j_s)}$ due to the indirect interactions of nodes with the family members of node j.

The total internal force vector, $\mathbf{F}_{(k)}^{PDDO}$ is computed by considering: (1) the stiffness matrices, $\mathbf{k}_{(k)(j)}^{PDDO}$ of all the red bonds that are connected from node k to all other nodes j in the domain $H_{(k)}$ along with their respective direct interactions (dashed red) that are connected from node k to other nodes k_r in the domain $H_{(k)}$ with stiffness matrices, $\mathbf{k}_{(k)(j)}^{PDDO(k)(k_r)}$, and (2) the indirect interactions (dashed green) that are connected from node k to all other nodes j_s of the domain $H_{(j)}$ with stiffness matrices, $\mathbf{k}_{(k)(j)}^{PDDO(j)(j_s)}$.

For the PD boundary region $\mathscr{B}_{\boldsymbol{\sigma}}$, the traction vector, $\mathbf{T}_{(k)}^{PD} = \boldsymbol{\sigma}_{(k)}\mathbf{n}_{(k)}$ at node k can be expressed using Eq. (3.204) as

$$\mathbf{T}_{(k)}^{PD} = \left(\sum_{j \in H_{(k)}} \widehat{\mu}_{(k)(j)} \left\{ \begin{array}{l} \lambda \mathrm{tr}\left[\left(\mathbf{u}(\mathbf{x}_{(j)}) - \mathbf{u}(\mathbf{x}_{(k)}) \right) \otimes \mathbf{g}(\mathbf{x}_{(k)}, \xi_{(k)(j)}) \right] \mathbf{I} + \\ \mu \left[\begin{array}{l} \left(\mathbf{u}(\mathbf{x}_{(j)}) - \mathbf{u}(\mathbf{x}_{(k)}) \right) \otimes \mathbf{g}(\mathbf{x}_{(k)}, \xi_{(k)(j)}) \\ + \mathbf{g}(\mathbf{x}_{(k)}, \xi_{(k)(j)}) \otimes \left(\mathbf{u}(\mathbf{x}_{(j)}) - \mathbf{u}(\mathbf{x}_{(k)}) \right) \end{array} \right] \end{array} \right\} V_{(j)} \right) \mathbf{n}(\mathbf{x}_{(k)}) \tag{3.221}$$

It also can be recast in matrix form as

$$\mathbf{T}_{(k)}^{PD} = \sum_{j \in H_{(k)}} \mathbf{H}_{(k)(j)} \left(\mathbf{u}_{(j)} - \mathbf{u}_{(k)} \right) \tag{3.222}$$

with

$$\mathbf{H}_{(k)(j)} = \widehat{\mu}_{(k)(j)} \begin{bmatrix} \left((\lambda + 2\mu) g_{(k)(j)}^{10} n_{(1)}^{(k)} + \mu g_{(k)(j)}^{01} n_{(2)}^{(k)} \right) & \left(\lambda g_{(k)(j)}^{01} n_{(1)}^{(k)} + \mu g_{(k)(j)}^{10} n_{(2)}^{(k)} \right) \\ \left(\lambda g_{(k)(j)}^{10} n_{(2)}^{(k)} + \mu g_{(k)(j)}^{01} n_{(1)}^{(k)} \right) & \left(\mu g_{(k)(j)}^{10} n_{(1)}^{(k)} + (\lambda + 2\mu) g_{(k)(j)}^{01} n_{(2)}^{(k)} \right) \end{bmatrix} V_{(j)} \tag{3.223}$$

The discrete form of force vector, $\mathbf{F}_{(k)}^{\boldsymbol{\sigma}}$ arising from the traction vector can be written as

$$\mathbf{F}_{(k)}^{\boldsymbol{\sigma}} = \mathbf{T}_{(k)}^{PD} A_{(k)} = \sum_{j \in H_{(k)}} \mathbf{H}_{(k)(j)} \left(\mathbf{u}_{(j)} - \mathbf{u}_{(k)} \right) A_{(k)} \tag{3.224}$$

As shown in Fig. 3.24, the force acting at node k due to the nonlocal traction is modeled using a network of elements around node k (shown in red) with the stiffness matrix, $\mathbf{k}_{(k)(j)}^{\boldsymbol{\sigma}}$ corresponding to the direct interactions.

The force vector $\mathbf{F}_{(k)}^{\boldsymbol{\sigma}}$ in component form can be written as

$$\begin{Bmatrix} F_{1(k)}^{\boldsymbol{\sigma}} \\ F_{2(k)}^{\boldsymbol{\sigma}} \\ 0 \\ 0 \end{Bmatrix} = \sum_{j \in H_{(k)}} \mathbf{k}_{(k)(j)}^{\boldsymbol{\sigma}} \begin{Bmatrix} u_{1(k)} \\ u_{2(k)} \\ u_{1(j)} \\ u_{2(j)} \end{Bmatrix} \tag{3.225}$$

with

$$\mathbf{k}_{(k)(j)}^{\boldsymbol{\sigma}} = \begin{bmatrix} -H_{11(k)(j)} A_{(k)} & -H_{12(k)(j)} A_{(k)} & H_{11(k)(j)} A_{(k)} & H_{12(k)(j)} A_{(k)} \\ -H_{21(k)(j)} A_{(k)} & -H_{22(k)(j)} A_{(k)} & H_{21(k)(j)} A_{(k)} & H_{22(k)(j)} A_{(k)} \end{bmatrix} \tag{3.226}$$

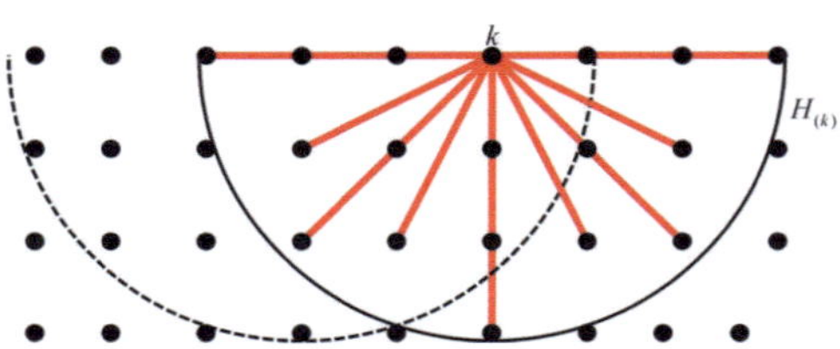

Fig. 3.24 Element connectivity between node k and other nodes in $H_{(k)}$ for PD boundary traction

3.13 Local Discretization of PD Field Equations

3.13.1 BB-PD Field Equation

Figures 3.25a, b show the element connectivity for dual horizon BB-PD field interactions between nodes k and j and k and p denoting the primary (red) and secondary (blue) bonds, respectively.

Using the discrete form of the flow density of field in Eqs. (3.109) and (3.110), the discrete form of the internal net field flow $L^{BB}(\mathbf{x})$ in Eq. (3.112) in the interior region $\mathcal{D}$ can be expressed as

$$L_{(k)}^{BB} = \sum_{j=1}^{N_{(k)}} \widehat{\mu}_{(k)(j)} \frac{\kappa_{(k)}}{2\left|\boldsymbol{\xi}_{(k)(j)}\right|} \left(\Theta_{(j)} - \Theta_{(k)}\right) V_{(j)} - \sum_{p=1}^{\widetilde{N}_{(k)}} \widehat{\mu}_{(p)(k)} \frac{\kappa_{(p)}}{2\left|\boldsymbol{\xi}_{(p)(k)}\right|} \left(\Theta_{(k)} - \Theta_{(p)}\right) V_{(p)} \tag{3.227}$$

Thus, the discrete form of the PD internal net field flow $L_{(k)}^{BB}$ at node k after multiplying with volume $V_{(k)}$ can be written as

$$F_{(k)}^{BB} = L_{(k)}^{BB} V_{(k)} = \sum_{j \in H_{(k)}} \Phi_{(k)(j)}^{BB} \left(\Theta_{(j)} - \Theta_{(k)}\right) + \sum_{p \in \widetilde{H}_{(k)}} \widetilde{\Phi}_{(p)(k)}^{BB} \left(\Theta_{(p)} - \Theta_{(k)}\right) \tag{3.228}$$

with

$$\Phi_{(k)(j)}^{BB} = \widehat{\mu}_{(k)(j)} \frac{\kappa_{(k)}}{2\left|\boldsymbol{\xi}_{(k)(j)}\right|} V_{(j)} V_{(k)} \tag{3.229}$$

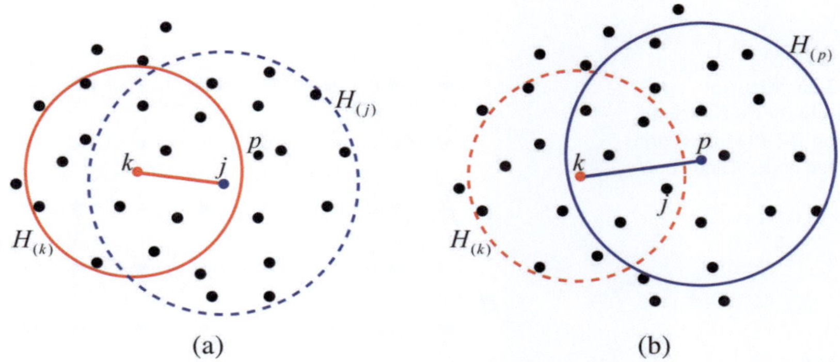

Fig. 3.25 Element connectivity for dual horizon BB-PD interactions: (**a**) primary bond between nodes k and j and (**b**) secondary bond between nodes k and p

and

$$\widetilde{\Phi}_{(p)(k)}^{BB} = \widehat{\mu}_{(p)(k)} \frac{\kappa_{(p)}}{2\left|\boldsymbol{\xi}_{(p)(k)}\right|} V_{(p)} V_{(k)} \tag{3.230}$$

Equation (3.228) can be written in component form as

$$\left\{ \begin{array}{c} F_{(k)}^{BB} \\ 0 \end{array} \right\} = \sum_{j=1}^{N_{(k)}} \mathbf{k}_{(k)(j)}^{BB} \left\{ \begin{array}{c} \Theta_{(k)} \\ \Theta_{(j)} \end{array} \right\} + \sum_{p=1}^{\widetilde{N}_{(k)}} \widetilde{\mathbf{k}}_{(p)(k)}^{BB} \left\{ \begin{array}{c} \Theta_{(k)} \\ \Theta_{(p)} \end{array} \right\} \tag{3.231}$$

where the stiffness matrices $\mathbf{k}_{(k)(j)}^{BB}$ and $\widetilde{\mathbf{k}}_{(p)(k)}^{BB}$ are defined as

$$\mathbf{k}_{(k)(j)}^{BB} = \begin{bmatrix} -\Phi_{(k)(j)}^{BB)} & \Phi_{(k)(j)}^{BB} \\ 0 & 0 \end{bmatrix} \tag{3.232}$$

and

$$\widetilde{\mathbf{k}}_{(p)(k)}^{BB} = \begin{bmatrix} -\widetilde{\Phi}_{(p)(k)}^{BB} & \widetilde{\Phi}_{(p)(k)}^{BB} \\ 0 & 0 \end{bmatrix} \tag{3.233}$$

The total net field flow at node k, $F_{(k)}^{BB}$ is computed by considering the stiffness matrices, $\mathbf{k}_{(k)(j)}^{BB}$ of all the primary (red) bonds that are connected from node k to all other nodes j in the horizon $H_{(k)}$ and the stiffness matrices, $\widetilde{\mathbf{k}}_{(p)(k)}^{BB}$ of all the secondary (blue) bonds that are connected from node k to all nodes p in the dual horizon $\widetilde{H}_{(k)}$.

For a uniform discretization, the element connectivity for a BB-PD interaction between nodes k and j is associated with a bond (blue) is shown in Fig. 3.26. The

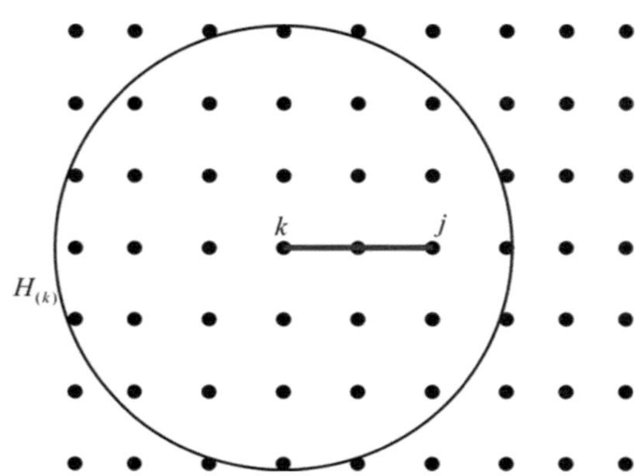

Fig. 3.26 Element connectivity for constant horizon BB PD interactions between nodes k and j in $H_{(k)}$

discrete form of the PD net field flow at node k, $F_{(k)}^{BB}$ after multiplying with volume $V_{(k)}$ can be written as

$$F_{(k)}^{BB} = L_{(k)}^{BB} V_{(k)} = \sum_{j \in H_{(k)}} \left(\Phi_{(k)(j)}^{BB} + \widetilde{\Phi}_{(j)(k)}^{BB} \right) \left(\Theta_{(j)} - \Theta_{(k)} \right) \tag{3.234}$$

or

$$\left\{ \begin{array}{c} F_{(k)}^{BB} \\ 0 \end{array} \right\} = \sum_{j \in H_{(k)}} \left(\mathbf{k}_{(k)(j)}^{BB} + \widetilde{\mathbf{k}}_{(j)(k)}^{BB} \right) \left\{ \begin{array}{c} \Theta_{(k)} \\ \Theta_{(j)} \end{array} \right\} \tag{3.235}$$

The total PD net field flow, $F_{(k)}^{BB}$ is computed by considering the stiffness matrices, $\left(\mathbf{k}_{(k)(j)}^{BB} + \widetilde{\mathbf{k}}_{(j)(k)}^{BB} \right)$ of all the blue bonds that are connected from node k to all other nodes j in $H_{(k)}$.

3.13.2 PDDO Field Equation

After substituting the PDDO form of gradient of the field variable $\Theta(\mathbf{x}, t)$ from Eq. (3.117) into Eqs. (3.114) and (3.115), the discrete form of field flow density at nodes k and j in the PDDO region, $\mathscr{R}$ can be expressed as

$$h_{(j)(k)}^{PDDO} = \widehat{\mu}_{(k)(j)} k_{(j)} \sum_{s=1}^{N_{(j)}} \widehat{\mu}_{(j)(j_s)} \left(\Theta_{(j_s)} - \Theta_{(j)} \right) \mathbf{g}\left(\mathbf{x}_{(j)}, \boldsymbol{\xi}_{(j)(j_s)} \right) V_{(j_s)} \cdot \mathbf{g}\left(\mathbf{x}_{(k)}, \boldsymbol{\xi}_{(k)(j)} \right) \tag{3.236}$$

and

$$h_{(k)(j)}^{PDDO} = \widehat{\mu}_{(k)(j)} k_{(k)} \sum_{r=1}^{N_{(k)}} \widehat{\mu}_{(k)(k_r)} \left(\Theta_{(k_r)} - \Theta_{(k)} \right) \mathbf{g}\left(\mathbf{x}_{(k)}, \boldsymbol{\xi}_{(k)(k_r)} \right) V_{(k_r)} \cdot \mathbf{g}\left(\mathbf{x}_{(k)}, \boldsymbol{\xi}_{(k)(j)} \right) \tag{3.237}$$

They can be recast as

$$h_{(j)(k)}^{PDDO} = \sum_{s=1}^{N_{(j)}} H_{(k)(j)}^{(j)(j_s)} \left(\Theta_{(j_s)} - \Theta_{(j)} \right) \tag{3.238}$$

and

$$h_{(k)(j)}^{PDDO} = \sum_{r=1}^{N_{(k)}} H_{(k)(j)}^{(k)(k_r)} \left(\Theta_{(k_r)} - \Theta_{(k)} \right) \tag{3.239}$$

in which the coefficients $H_{(k)(j)}^{(j)(j_s)}$ and $H_{(k)(j)}^{(k)(k_r)}$ are defined as

$$H_{(k)(j)}^{(j)(j_s)} = \widehat{\mu}_{(k)(j)}\widehat{\mu}_{(j)(j_s)}k_{(j)}\left(g_{(j)(j_s)}^{10}g_{(k)(j)}^{10} + g_{(j)(j_s)}^{01}g_{(k)(j)}^{01}\right)V_{(j_s)} \tag{3.240}$$

and

$$H_{(k)(j)}^{(k)(k_r)} = \widehat{\mu}_{(k)(j)}\widehat{\mu}_{(k)(k_r)}k_{(k)}\left(g_{(k)(k_r)}^{10}g_{(k)(j)}^{10} + g_{(k)(k_r)}^{01}g_{(k)(j)}^{01}\right)V_{(k_r)} \tag{3.241}$$

The discrete form of the PDDO internal net field flow $L_{(k)}^{PDDO}$ at node k in the PDDO region $\mathscr{R}$ can be expressed as

$$L_{(k)}^{PDDO} = \sum_{j\in H_{(k)}}\left(\sum_{s=1}^{N_{(j)}} H_{(k)(j)}^{(j)(j_s)}\left(\Theta_{(j_s)} - \Theta_{(j)}\right) - \sum_{r=1}^{N_{(k)}} H_{(k)(j)}^{(k)(k_r)}\left(\Theta_{(k_r)} - \Theta_{(k)}\right)\right)V_{(j)} \tag{3.242}$$

Thus, the discrete form of the PDDO internal net field flow $L_{(k)}^{PDDO}$ at node k after multiplying with volume $V_{(k)}$ can be written as

$$\begin{aligned}
F_{(k)}^{PDDO} &= L_{(k)}^{PDDO}V_{(k)} \\
&= \sum_{j\in H_{(k)}}\left(\sum_{s=1}^{N_{(j)}} H_{(k)(j)}^{(j)(j_s)}\left(\Theta_{(j_s)} - \Theta_{(j)}\right) - \sum_{r=1}^{N_{(k)}} H_{(k)(j)}^{(k)(k_r)}\left(\Theta_{(k_r)} - \Theta_{(k)}\right)\right)V_{(j)}V_{(k)} \\
&= \sum_{j\in H_{(k)}}\begin{pmatrix} -\widetilde{\Phi}_{(j)}\Theta_{(j)} + \sum\limits_{s=1, j_s\neq k}^{N_{(j)}} \widetilde{G}_{(k)(j)}^{(j)(j_s)}\Theta_{(j_s)} + \widetilde{G}_{(k)(j)}^{(j)(k)}\left(\Theta_{(k)} - \Theta_{(j)}\right) \\ - \sum\limits_{r=1, k_r\neq j}^{N_{(k)}} G_{(k)(j)}^{(k)(k_r)}\Theta_{(k_r)} + \Phi_{(k)}\Theta_{(k)} + G_{(k)(j)}^{(k)(j)}\left(\Theta_{(k)} - \Theta_{(j)}\right)\end{pmatrix}
\end{aligned} \tag{3.243}$$

with

$$G_{(k)(j)}^{(k)(k_r)} = H_{(k)(j)}^{(k)(k_r)}V_{(j)}V_{(k)} \tag{3.244}$$

$$\Phi_{(k)} = \sum_{r=1, k_r\neq j}^{N_{(k)}} G_{(k)(j)}^{(k)(k_r)} \tag{3.245}$$

and

$$\widetilde{G}_{(k)(j)}^{(j)(j_s)} = H_{(k)(j)}^{(j)(j_s)}V_{(j)}V_{(k)} \tag{3.246}$$

$$\widetilde{\Phi}_{(j)} = \sum_{s=1, j_s\neq k}^{N_{(j)}} \widetilde{G}_{(j)(k)}^{(j)(j_s)} \tag{3.247}$$

Equation (3.243) can be written in the component form as

$$\left\{\begin{array}{c} F_{(k)}^{PDDO} \\ 0 \end{array}\right\} = \sum_{j \in H_{(k)}} \left(\mathbf{k}_{(k)(j)}^{PDDO} \left\{\begin{array}{c} \Theta_{(k)} \\ \Theta_{(j)} \end{array}\right\} + \sum_{r=1,\, k_r \neq j}^{N_{(k)}} \mathbf{k}_{(k)(j)}^{PDDO(k)(k_r)} \left\{\begin{array}{c} \Theta_{(k)} \\ \Theta_{(k_r)} \end{array}\right\} + \sum_{s=1, j_s \neq k}^{N_{(j)}} \mathbf{k}_{(k)(j)}^{PDDO(j)(j_s)} \left\{\begin{array}{c} \Theta_{(k)} \\ \Theta_{(j_s)} \end{array}\right\} \right) \tag{3.248}$$

with

$$\mathbf{k}_{(k)(j)}^{PDDO} = \begin{bmatrix} \left(\Phi_{(k)} + G_{(k)(j)}^{(k)(j)} + \widetilde{G}_{(k)(j)}^{(j)(k)}\right) & \left(\Phi_{(k)} + G_{(k)(j)}^{(k)(j)} + \widetilde{G}_{(k)(j)}^{(j)(k)}\right) & \left(-G_{(k)(j)}^{(k)(j)} - \widetilde{G}_{11(k)(j)}^{(j)(k)} - \widetilde{\Phi}_{(j)}\right) & \left(-G_{(k)(j)}^{(k)(j)} - \widetilde{G}_{(k)(j)}^{(j)(k)} - \widetilde{\Phi}_{(j)}\right) \\ 0 & 0 & 0 & 0 \end{bmatrix} \tag{3.249}$$

$$\mathbf{k}_{(k)(j)}^{PDDO(k)(k_r)} = \begin{bmatrix} 0 & 0 & -G_{(k)(j)}^{(k)(k_r)} & -G_{(k)(j)}^{(k)(k_r)} \\ 0 & 0 & 0 & 0 \end{bmatrix} \tag{3.250}$$

And

$$\widetilde{\mathbf{k}}_{(k)(j)}^{PDDO(j)(j_s)} = \begin{bmatrix} 0 & 0 & \widetilde{G}_{(k)(j)}^{(j)(j_s)} & \widetilde{G}_{(k)(j)}^{(j)(j_s)} \\ 0 & 0 & 0 & 0 \end{bmatrix} \tag{3.251}$$

Figures 3.27a, b show the element connectivity for PDDO interaction between nodes k and j. The interactions can be modeled by considering: (1) an element (shown in red) with the stiffness matrix $\mathbf{k}_{(k)(j)}^{PDDO}$ corresponding to the bond, (2) a network of elements around node k (shown in dashed red) with the stiffness matrices, $\mathbf{k}_{(k)(j)}^{PDDO(k)(k_r)}$ due to the direct interaction of node k with its family members, and (3) a network of elements around node j (shown in dashed green) with the stiffness matrices, $\mathbf{k}_{(k)(j)}^{PDDO(j)(j_s)}$ due to the indirect interactions of a node with the family members of node j.

The total PDDO net field flow at node k, $F_{(k)}^{PDDO}$ is computed by considering: (1) the stiffness matrices, $\mathbf{k}_{(k)(j)}^{PDDO}$ of all the blue elements that are connected from node k to all other nodes j in the domain $H_{(k)}$ along with their respective direct interactions (dashed red) in the domain with stiffness matrices, $\mathbf{k}_{(k)(j)}^{PDDO(k)(k_r)}$, and (2) stiffness matrices, $\mathbf{k}_{(k)(j)}^{PDDO(j)(j_s)}$ of the indirect interactions (dashed green) that are connected from node k to all other nodes j_s of the domain $H_{(j)}$.

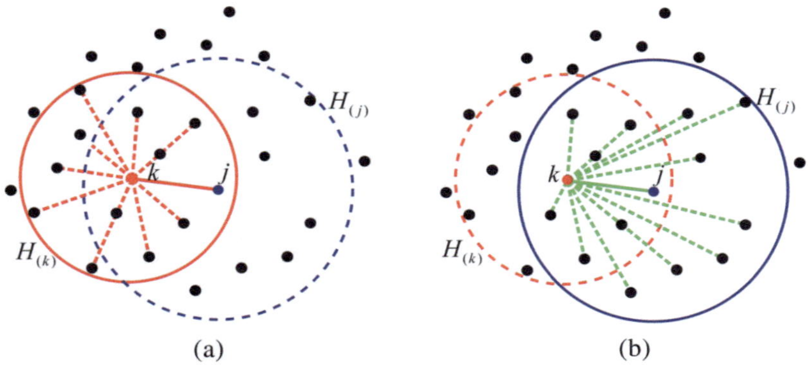

Fig. 3.27 Element connectivity for PDDO interaction between nodes k and j: (**a**) direct interaction and (**b**) indirect interaction

For the PD boundary region $\mathscr{B}_{\mathbf{q}}$, the field flow, $T^{\mathbf{q}}_{(k)} = \mathbf{q}_{(k)} \cdot \mathbf{n}_{(k)}$ at node k can be expressed using the PDDO form of gradient of the field variable $\Theta(\mathbf{x}, t)$ from Eq. (3.117) as

$$T^{\mathbf{q}}_{(k)} = -\mathbf{q}\big(\mathbf{x}_{(k)}\big) \cdot \mathbf{n}\big(\mathbf{x}_{(k)}\big) = \left(-k_{(k)} \sum_{j \in H_{(k)}} \widehat{\mu}_{(k)(j)} \big(\Theta_{(j)} - \Theta_{(k)}\big) \mathbf{g}\Big(\boldsymbol{\xi}_{(k)(j)}\Big) V_{(j)} \right) \cdot \mathbf{n}\big(\mathbf{x}_{(k)}\big) \tag{3.252}$$

It can be recast as

$$T^{\mathbf{q}}_{(k)} = \sum_{j \in H_{(k)}} H_{(k)(j)} \big(\Theta_{(j)} - \Theta_{(k)}\big) \tag{3.253}$$

where the coefficient $H_{(k)(j)}$ is defined as

$$H_{(k)(j)} = \widehat{\mu}_{(k)(j)} k \Big(g^{10}_{(k)(j)} n^{(k)}_{(1)} + g^{01}_{(k)(j)} n^{(k)}_{(2)} \Big) V_{(j)} \tag{3.254}$$

The PDDO net field flow at node k, $F^{\mathbf{q}}_{(k)}$ arising from field flow can be written as

$$F^{\mathbf{q}}_{(k)} = T^{\mathbf{q}}_{(k)} A_{(k)} = \sum_{j \in H_{(k)}} H_{(k)(j)} \big(\Theta_{(j)} - \Theta_{(k)}\big) A_{(k)} \tag{3.255}$$

As shown in Fig. 3.28, the PDDO net field flow at node k is modeled using a network of elements around node k (shown in red) with the field stiffness matrix

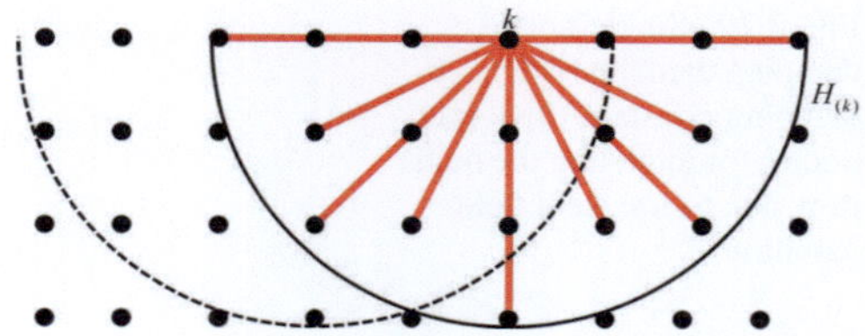

Fig. 3.28 Element connectivity between node k and its members in $H_{(k)}$ for boundary field flow

corresponding to $\mathbf{k}^{\mathbf{q}}_{(k)(j)}$ due to direct interaction. The PDDO net field flow $F^{\mathbf{q}}_{(k)}$ in component form can be written as

$$\left\{ \begin{array}{c} F^{\mathbf{q}}_{(k)} \\ 0 \end{array} \right\} = \sum_{j \in H_{(k)}} \mathbf{k}^{\mathbf{q}}_{(k)(j)} \left\{ \begin{array}{c} \Theta_{(k)} \\ \Theta_{(j)} \end{array} \right\} \tag{3.256}$$

with

$$\mathbf{k}^{\mathbf{q}}_{(k)(j)} = \begin{bmatrix} -H_{(k)(j)}A_{(k)} & H_{(k)(j)}A_{(k)} \\ 0 & 0 \end{bmatrix} \tag{3.257}$$

3.13.3 BB-PD Transient Field Variation

The discrete form of the field flow $F_{\dot{\Theta}(k)}$ due to the transient field variation at node k in the PDDO region $\mathcal{R}$ or interior region $\mathcal{D}$ can be expressed as

$$F_{\dot{\Theta}(k)} = C_{(k)} V_{(k)} \dot{\Theta}_{(k)} \tag{3.258}$$

As shown in Fig. 3.29, the field flow at node k due to transient field variation, $F_{\dot{\Theta}(k)}$ can be modeled using a single-noded damping element at node k (shown in red) with the field stiffness constant corresponding to $C_{(k)}V_{(k)}$.

Alternatively, it can also be modeled using a two-noded damper element with node k and a dummy node k_d (created at the same location as that of node k) as shown in Fig. 3.30 with the stiffness constant corresponding to $C_{(k)}V_{(k)}$. Equation (3.258) can be rewritten in component form as

$$\left\{ \begin{array}{c} F_{\dot{\Theta}(k)} \\ F_{\dot{\Theta}(k_d)} \end{array} \right\} = C_{(k)} V_{(k)} \begin{bmatrix} 1 & -1 \\ -1 & 1 \end{bmatrix} \left\{ \begin{array}{c} \dot{\Theta}_{(k)} \\ \dot{\Theta}_{(k_d)} \end{array} \right\} \tag{3.259}$$

The dummy nodes can be constrained as $\Theta_{(k_d)} = 0$ during the solution process to recover Eq. (3.258).

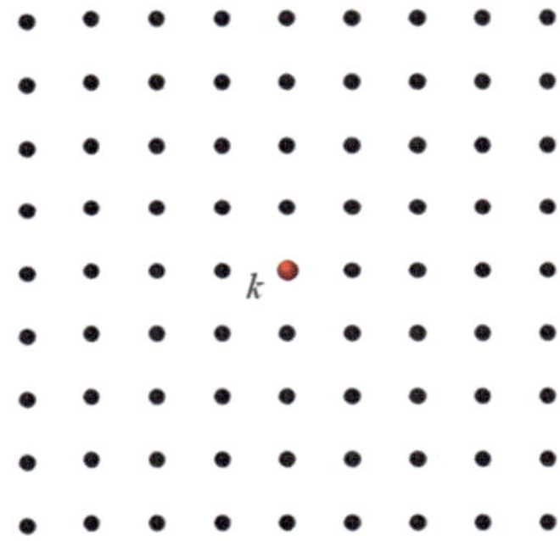

Fig. 3.29 Single-noded damping element with damping constant $C_{(k)}V_{(k)}$ at node k for modeling the field flow due to transient field variation

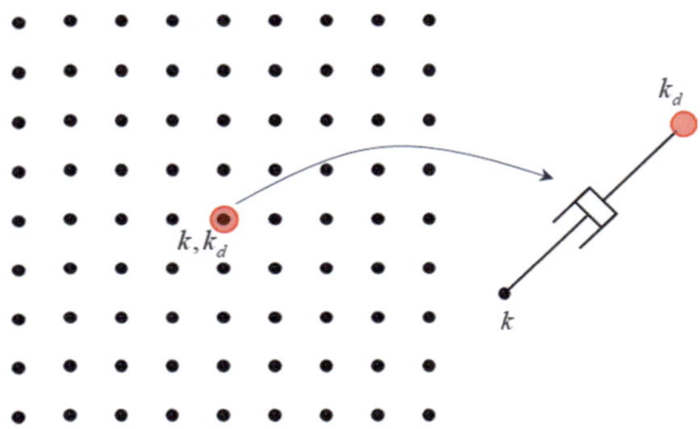

Fig. 3.30 Damper element with damping constant $C_{(k)}V_{(k)}$ between node k and dummy node k_d for modeling the field flow due to transient field variation

3.13.4 BB-PD Transient Material Property

The discrete form of the field flow $F_{\Theta(k)}$ due to the time rate of change of material property at node k in the PDDO region $\mathscr{R}$ or interior region $\mathcal{D}$ can be expressed as

$$F_{\Theta(k)} = \dot{C}_{(k)} V_{(k)} \Theta_{(k)} \tag{3.260}$$

Using a two-noded spring element, Eq. (3.260) can be written in component form as

$$\begin{Bmatrix} F_{\Theta(k)} \\ F_{\Theta(k_d)} \end{Bmatrix} = \dot{C}_{(k)} V_{(k)} \begin{bmatrix} 1 & -1 \\ -1 & 1 \end{bmatrix} \begin{Bmatrix} \Theta_{(k)} \\ \Theta_{(k_d)} \end{Bmatrix} \tag{3.261}$$

As shown in Fig. 3.31, the field flow at node k due to the time rate of change of material property, $F_{\Theta(k)}$ is modeled using a two-noded spring element with node as k and a dummy node k_d (created at the same location as that of node k) with the field stiffness constant corresponding to $\dot{C}_{(k)} V_{(k)}$. During the solution process, the dummy nodes in Eq. (3.261) can be constrained as $\Theta_{(k_d)} = 0$ to recover Eq. (3.260).

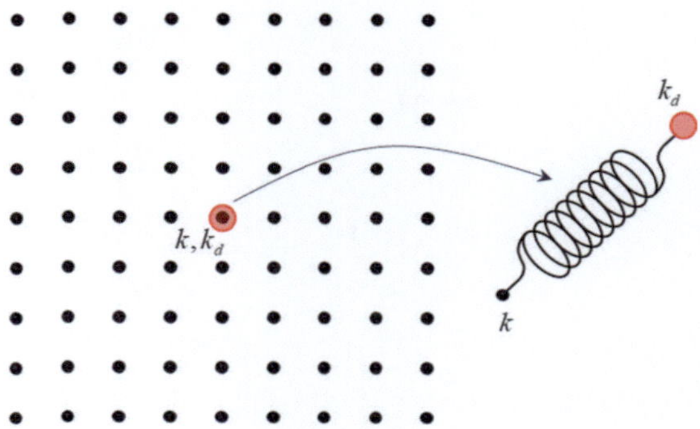

Fig. 3.31 Spring element with stiffness constant $\dot{C}_{(k)}V_{(k)}$ between node k and dummy node k_d for modeling the field flow due to the time rate of change of material property

References

Anicode, S. V. K., & Madenci, E. (2022). Bond-and state-based peridynamic analysis in a commercial finite element framework with native elements. *Computer Methods in Applied Mechanics and Engineering, 398*, 115208.

Anicode, S. V. K., & Madenci, E. (2023). Direct coupling of dual-horizon peridynamics with finite elements for irregular discretization without an overlap zone. *Engineering with Computers, 40*, 1–31.

Chen, H., & Spencer, B. W. (2019). Peridynamic bond-associated correspondence model: Stability and convergence properties. *International Journal for Numerical Methods in Engineering, 117*, 713–727.

Gu, X., Zhang, Q., Madenci, E., & Xia, X. (2019). Possible causes of numerical oscillations in non-ordinary state-based peridynamics and a bond-associated higher-order stabilized model. *Computer Methods in Applied Mechanics and Engineering, 357*, 112592.

Madenci, E. (2017). Peridynamic integrals for strain invariants of homogeneous deformation. *ZAMM Journal of Applied Mathematics and Mechanics/Zeitschrift für Angewandte Mathematik und Mechanik, 97*, 1236–1251.

Madenci, E., & Oterkus, E. (2014). *Peridynamic theory and its applications*. Springer.

Madenci, E., Barut, A., & Futch, M. (2016). Peridynamic differential operator and its applications. *Computer Methods in Applied Mechanics and Engineering, 304*, 408–451.

Madenci, E., Barut, A., & Dorduncu, M. (2019a). *Peridynamic differential operator for numerical analysis*. Springer.

Madenci, E., Dorduncu, M., Phan, N., & Gu, X. (2019b). Weak form of bond-associated non-ordinary state-based peridynamics free of zero energy modes with uniform or non-uniform discretization. *Engineering Fracture Mechanics, 218*, 106613.

Madenci, E., Barut, A., & Phan, N. (2021). Bond-based peridynamics with stretch and rotation kinematics for opening and shearing modes of fracture. *Journal of Peridynamics and Nonlocal Modeling, 3*, 1–44.

Madenci, E., Roy, P., & Behera, D. (2022). *Advances in peridynamics*. Springer.

Ren, H., Zhuang, X., & Rabczuk, T. (2017). Dual-horizon peridynamics: A stable solution to varying horizons. *Computer Methods in Applied Mechanics and Engineering, 318*, 762–782.

Roy, P., Behera, D., & Madenci, E. (2020). Peridynamic simulation of finite elastic deformation and rupture in polymers. *Engineering Fracture Mechanics, 236*, 107226.

Silling, S. A. (2000). Reformulation of elasticity theory for discontinuities and long-range forces. *Journal of the Mechanics and Physics of Solids, 48*, 175–209.

Silling, S. A., & Lehoucq, R. B. (2010). Peridynamic theory of solid mechanics. *Advances in applied mechanics, 44*, 73–168.

Silling, S. A., Epton, M., Weckner, O., Xu, J., & Askari, E. (2007). Peridynamic states and constitutive modeling. *Journal of Elasticity, 88*, 151–184.

Chapter 4
Weak Form of Peridynamic Equilibrium and Field Equations

4.1 Introduction

Unlike the partial differential equations, the solution to the PD form of equations does not require the imposition of boundary conditions. However, the enforcement of essential and natural boundary conditions is unavoidable in real-life applications. The conditions on the dependent field variable are referred to as essential or forced boundary conditions, and the ones involving the derivative of the dependent field variable are referred to as natural boundary conditions. The PD forms of the equilibrium and field equations are expressed in terms of only the dependent field variable as unknown. Therefore, it is suitable to enforce constraints on the field variable directly through a layer of volume along the boundary. The applied natural boundary conditions can be enforced in the form of a body force or a field source distribution through a real boundary layer. Alternatively, natural conditions can be enforced as constraints along the boundary layer by introducing Lagrange multipliers. However, this approach increases the number of unknowns in the governing equations. In order to eliminate the use of Lagrange multipliers, the concept of virtual work can be employed to derive the weak form of PD equilibrium and field equations which permits the enforcement of natural boundary conditions (Madenci et al., 2018, 2022; Anicode & Madenci, 2022, 2023).

4.2 Weak Form of PD Equilibrium Equations

The principle of virtual work can be expressed as:

$$\delta\mathcal{K} + \delta\mathcal{U} + \delta\mathcal{V} = 0 \tag{4.1}$$

where $\delta\mathcal{K}$, $\delta\mathcal{U}$, and $\delta\mathcal{V}$ represent the virtual work due to inertial forces, internal forces, and external forces (traction and body loads), respectively. By considering Eq. (3.1) along with Eq. (3.3), this expression can be rewritten as:

E. Madenci et al., *Peridynamics and Its Applications Using Ansys*,
https://doi.org/10.1007/978-3-031-59896-8_4

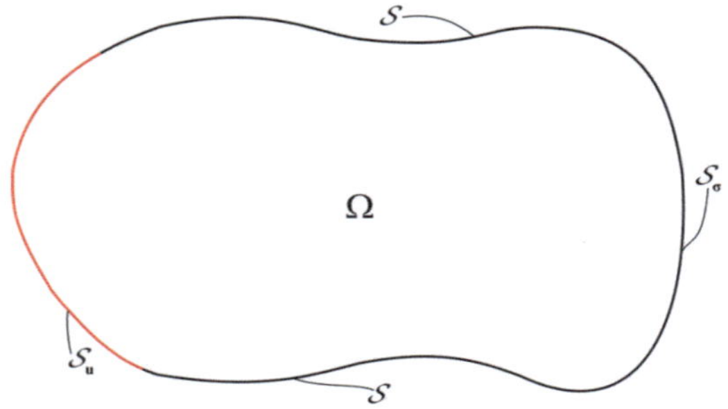

Fig. 4.1 A solid body subjected to displacement constraint on $S_{\mathbf{u}}$ and external traction on $S_{\boldsymbol{\sigma}}$

$$\int_{\Omega} \rho(\mathbf{x},t)\ddot{\mathbf{u}}(\mathbf{x},t)\cdot\delta\mathbf{u}(\mathbf{x})dV + \int_{\Omega} \mathbf{P}^{PD}(\mathbf{x},t) : \delta\nabla\mathbf{u}(\mathbf{x})dV - \int_{S_\sigma} \mathbf{T}^{*PD}(\mathbf{x},t)\cdot\delta\mathbf{u}(\mathbf{x})dS - \int_{\Omega} \mathbf{b}(\mathbf{x},t)\cdot\delta\mathbf{u}(\mathbf{x})dV = 0 \tag{4.2}$$

in which $\rho(\mathbf{x}, t)$ is the density of the material point in the domain, $\boldsymbol{\Omega}$, $\mathbf{P}^{PD}(\mathbf{x}, t)$ is the PD form of the first Piola-Kirchhoff stress tensor, $\mathbf{u}(\mathbf{x}, t)$ is the displacement vector, $\ddot{\mathbf{u}}(\mathbf{x}, t)$ is the acceleration vector, and $\mathbf{b}(\mathbf{x}, t)$ is the body force vector.

As shown in Fig. 4.1, the PD form of vector of external tractions, $\mathbf{T}^{*PD}$, is applied over a segment of the boundary, $S_{\boldsymbol{\sigma}}$, and the segment, $S_{\mathbf{u}}$, is subjected to displacement constraints. The virtual value of the quantity is denoted by $\delta(\ )$. Prior to applying the divergence theorem, Eq. (4.2) can be recast as:

$$\int_{\Omega} \rho(\mathbf{x},t)\ddot{\mathbf{u}}(\mathbf{x},t)\cdot\delta\mathbf{u}(\mathbf{x})dV + \int_{\Omega} \left[\nabla\cdot\left(\mathbf{P}^{PD}(\mathbf{x},t)\delta\mathbf{u}(\mathbf{x})\right) - \left(\nabla\cdot\mathbf{P}^{PD}(\mathbf{x},t)\right)\cdot\delta\mathbf{u}(\mathbf{x})\right]dV - \int_{S_\sigma} \mathbf{T}^{*PD}(\mathbf{x},t)\cdot\delta\mathbf{u}(\mathbf{x})dS - \int_{\Omega} \mathbf{b}(\mathbf{x},t)\cdot\delta\mathbf{u}(\mathbf{x})dV = 0 \tag{4.3}$$

Invoking the divergence theorem along with $(\mathbf{P}^{PD}(\mathbf{x}, t)\delta\mathbf{u}(\mathbf{x}))\cdot\mathbf{n}(\mathbf{x}) = (\mathbf{P}^{PD\mathrm{T}}(\mathbf{x}, t)\mathbf{n}(\mathbf{x}))\cdot\delta\mathbf{u}(\mathbf{x})$ into Eq. (4.3) results in:

$$\int_{\Omega} \rho(\mathbf{x},t)\ddot{\mathbf{u}}(\mathbf{x},t)\cdot\delta\mathbf{u}(\mathbf{x})dV + \oint_{S} \left(\mathbf{P}^{PD\mathrm{T}}(\mathbf{x},t)\mathbf{n}(\mathbf{x})\right)\cdot\delta\mathbf{u}(\mathbf{x})dS - \int_{\Omega} \left(\nabla\cdot\mathbf{P}^{PD}(\mathbf{x},t)\right)\cdot\delta\mathbf{u}(\mathbf{x})dV - \int_{S_\sigma} \mathbf{T}^{*PD}(\mathbf{x},t)\cdot\delta\mathbf{u}(\mathbf{x})dS - \int_{\Omega} \mathbf{b}(\mathbf{x},t)\cdot\delta\mathbf{u}(\mathbf{x})dV = 0 \tag{4.4}$$

in which $\delta\mathbf{u}(\mathbf{x})$ is the virtual displacement vector.

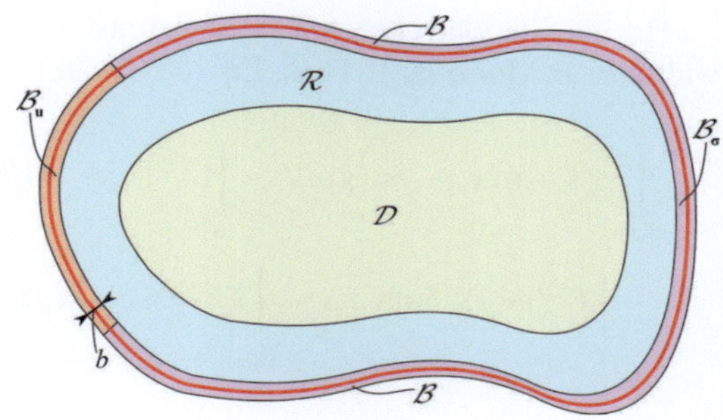

Fig. 4.2 PD computational domain with interior ($\mathcal{D}$) region, outer ($\mathcal{R}$) region, and boundary layer ($\mathcal{B}$)

As shown in Fig. 4.2, the volume of the domain, Ω, is split into inner, $\mathcal{D}$, outer $\mathcal{R}$ and the boundary layer region $\mathcal{B}$. Considering the PD representation of the internal force vector, $\mathbf{L}^{PD}$, and PD form of the Piola-Kirchhoff stress tensor, $\mathbf{P}^{PD}$, Eq. (4.4) can be rewritten as:

$$\begin{aligned}
&\int_{\mathcal{D}\cup\mathcal{R}} \rho(\mathbf{x},t)\ddot{\mathbf{u}}(\mathbf{x},t)\cdot\delta\mathbf{u}(\mathbf{x})dV + \int_{\mathcal{B}} \rho(\mathbf{x},t)\ddot{\mathbf{u}}(\mathbf{x},t)\cdot\delta\mathbf{u}(\mathbf{x})dV + \oint_S \mathbf{T}^{PD}(\mathbf{x},t)\cdot\delta\mathbf{u}(\mathbf{x})dS \\
&\quad - \int_{\mathcal{D}} \mathbf{L}^{PD}(\mathbf{x},t)\cdot\delta\mathbf{u}(\mathbf{x})dV - \int_{\mathcal{R}} \mathbf{L}^{PDDO}(\mathbf{x},t)\cdot\delta\mathbf{u}(\mathbf{x})dV - \int_{\mathcal{B}} \mathbf{L}^{PDDO}(\mathbf{x},t)\cdot\delta\mathbf{u}(\mathbf{x})dV \\
&\quad - \int_{S_\sigma} \mathbf{T}^{*PD}(\mathbf{x},t)\cdot\delta\mathbf{u}(\mathbf{x})dS - \int_{\mathcal{D}\cup\mathcal{R}} \mathbf{b}(\mathbf{x},t)\cdot\delta\mathbf{u}(\mathbf{x})dV - \int_{\mathcal{B}} \mathbf{b}(\mathbf{x},t)\cdot\delta\mathbf{u}(\mathbf{x})dV = 0
\end{aligned} \tag{4.5a}$$

or

$$\begin{aligned}
&\int_{\mathcal{D}\cup\mathcal{R}} \rho(\mathbf{x},t)\ddot{\mathbf{u}}(\mathbf{x},t)\cdot\delta\mathbf{u}(\mathbf{x})dV + b\int_{S} \rho(\mathbf{x},t)\ddot{\mathbf{u}}(\mathbf{x},t)\cdot\delta\mathbf{u}(\mathbf{x})dS + \oint_S \mathbf{T}^{PD}(\mathbf{x},t)\cdot\delta\mathbf{u}(\mathbf{x})dS \\
&\quad - \int_{\mathcal{D}} \mathbf{L}^{PD}(\mathbf{x},t)\cdot\delta\mathbf{u}(\mathbf{x})dV - \int_{\mathcal{R}} \mathbf{L}^{PDDO}(\mathbf{x},t)\cdot\delta\mathbf{u}(\mathbf{x})dV - b\int_{S} \mathbf{L}^{PDDO}(\mathbf{x},t)\cdot\delta\mathbf{u}(\mathbf{x})dS \\
&\quad - \int_{S_\sigma} \mathbf{T}^{*PD}(\mathbf{x},t)\cdot\delta\mathbf{u}(\mathbf{x})dS - \int_{\mathcal{D}\cup\mathcal{R}} \mathbf{b}(\mathbf{x},t)\cdot\delta\mathbf{u}(\mathbf{x})dV - b\int_{S} \mathbf{b}(\mathbf{x},t)\cdot\delta\mathbf{u}(\mathbf{x})dS = 0
\end{aligned} \tag{4.5b}$$

in which $\mathbf{T}^{PD}(\mathbf{x},t) = \mathbf{P}^{PD\mathrm{T}}(\mathbf{x},t)\mathbf{n}(\mathbf{x})$ is the PD representation of the traction vector and $\mathbf{n}(\mathbf{x})$ being the unit vector normal to the surface.

The boundary layer region, $\mathcal{B}$, is defined by the surface of the domain, S, with its width, b, and for $b \ll 1$. Thus, Eq. (4.5b) can be reduced to:

$$\int_{\mathcal{D} \cup \mathcal{R}} \rho(\mathbf{x}, t)\ddot{\mathbf{u}}(\mathbf{x}, t) \cdot \delta\mathbf{u}(\mathbf{x})dV - \int_{\mathcal{D}} \mathbf{L}^{PD}(\mathbf{x}, t) \cdot \delta\mathbf{u}(\mathbf{x})dV - \int_{\mathcal{R}} \mathbf{L}^{PDDO}(\mathbf{x}, t) \cdot \delta\mathbf{u}(\mathbf{x})dV$$
$$+ \oint_{S} \mathbf{T}^{PD}(\mathbf{x}, t) \cdot \delta\mathbf{u}(\mathbf{x})dS - \int_{S_\sigma} \mathbf{T}^{*PD}(\mathbf{x}, t) \cdot \delta\mathbf{u}(\mathbf{x})dS - \int_{\mathcal{D} \cup \mathcal{R}} \mathbf{b}(\mathbf{x}, t) \cdot \delta\mathbf{u}(\mathbf{x})dV = 0 \tag{4.6}$$

For a two-dimensional domain, Eq. (4.6) concerns only area and line integrations and simplifies to:

$$\int_{\mathcal{D} \cup \mathcal{R}} \rho(\mathbf{x}, t)\ddot{\mathbf{u}}(\mathbf{x}, t) \cdot \delta\mathbf{u}(\mathbf{x})dA - \int_{\mathcal{D}} \mathbf{L}^{PD}(\mathbf{x}, t) \cdot \delta\mathbf{u}(\mathbf{x})dA - \int_{\mathcal{R}} \mathbf{L}^{PDDO}(\mathbf{x}, t) \cdot \delta\mathbf{u}(\mathbf{x})dA$$
$$+ \oint_{\Gamma} \mathbf{T}^{PD}(\mathbf{x}, t) \cdot \delta\mathbf{u}(\mathbf{x})d\Gamma - \int_{\Gamma_\sigma} \mathbf{T}^{*PD}(\mathbf{x}, t) \cdot \delta\mathbf{u}(\mathbf{x})d\Gamma - \int_{\mathcal{D} \cup \mathcal{R}} \mathbf{b}(\mathbf{x}, t) \cdot \delta\mathbf{u}(\mathbf{x})dA = 0 \tag{4.7}$$

in which $\mathcal{D}$ and $\mathcal{R}$ represent the area of the interior and outer regions, Γ represents the entire boundary of the domain, Γ_σ and Γ_u represent the segments of the boundary subjected to the external tractions and displacement constraints, respectively. The region $\mathcal{B}$ in Fig. 4.2 with thickness of b is referred to as the boundary layer region. As the width approaches zero, i.e., $b \rightarrow 0$, Γ coincides with the real boundary as shown in Fig. 4.1.

4.3 Weak Form of PD Field Equations

Based on the variational approach, the PD field equation can be expressed in the form

$$\delta\mathcal{U} + \delta\mathcal{V} = 0 \tag{4.8}$$

where $\delta\mathcal{U}$ represents the virtual value of the field potential due to internal field flux and transient field fluctuation, and $\delta\mathcal{V}$ represents the virtual value of the field potential arising from external field interactions due to the known field flux, $\mathbf{q}^{*PD}(\mathbf{x}, t)$, and generated field source, $s_b^{PD}(\mathbf{x}, t)$. This expression can be expanded as:

$$\int_{\Omega} \mathbf{q}^{PD}(\mathbf{x},t)\cdot\delta\nabla\Theta(\mathbf{x})dV + \int_{\Omega}\big(C(\mathbf{x},t)\dot{\Theta}(\mathbf{x},t) + \dot{C}(\mathbf{x},t)\Theta(\mathbf{x},t)\big)\delta\Theta(\mathbf{x})dV \\ - \int_{S_{\mathbf{q}}} \mathbf{q}^{*PD}(\mathbf{x},t)\cdot\mathbf{n}(\mathbf{x})\delta\Theta(\mathbf{x})dS - \int_{\Omega} s_b(\mathbf{x},t)\delta\Theta(\mathbf{x})dV = 0 \tag{4.9}$$

in which $C(\mathbf{x},t)$ and $\dot{C}(\mathbf{x},t)$ represent the physical material property and its time rate of change, respectively, $\mathbf{q}^{PD}(\mathbf{x},t)$ is the PD form of the internal field flux vector, and $\Theta(\mathbf{x},t)$ is the field variable. As shown in Fig. 4.3, the PD form of the vector of external field flux, $\mathbf{q}^{*PD}(\mathbf{x},t)$, is applied over a segment of the boundary, $S_{\mathbf{q}}$, with $\mathbf{n}(\mathbf{x})$ being the unit vector normal to the surface, and the segment, S_{Θ}, is subjected to field constraints. The virtual value of the quantity is denoted by $\delta(\ )$. Prior to applying the divergence theorem, Eq. (4.9) can be recast as

$$\int_{\Omega}\big[\nabla\cdot(\mathbf{q}^{PD}(\mathbf{x},t)\delta\Theta(\mathbf{x})) - (\nabla\cdot\mathbf{q}^{PD}(\mathbf{x},t))\delta\Theta(\mathbf{x})\big]dV - \int_{S_{\mathbf{q}}} \mathbf{q}^{*PD}(\mathbf{x},t)\cdot\mathbf{n}(\mathbf{x})\delta\Theta(\mathbf{x})dS \\ + \int_{\Omega}\big(C(\mathbf{x},t)\dot{\Theta}(\mathbf{x},t) + \dot{C}(\mathbf{x},t)\Theta(\mathbf{x},t)\big)\delta\Theta(\mathbf{x})dV - \int_{\Omega} s_b^{PD}(\mathbf{x},t)\delta\Theta(\mathbf{x})dV = 0 \, . \tag{4.10}$$

Invoking the divergence theorem along with $(\mathbf{q}^{PD}(\mathbf{x},t)\delta\Theta(\mathbf{x}))\cdot\mathbf{n}(\mathbf{x}) = (\mathbf{q}^{PD}(\mathbf{x},t)\cdot\mathbf{n}(\mathbf{x}))\delta\Theta(\mathbf{x})$ into Eq. (4.10) results in

$$\int_{\Omega}\big(C(\mathbf{x},t)\dot{\Theta}(\mathbf{x},t) + \dot{C}(\mathbf{x},t)\Theta(\mathbf{x},t)\big)\delta\Theta(\mathbf{x})dV + \oint_{S}(\mathbf{q}^{PD}(\mathbf{x},t)\cdot\mathbf{n}(\mathbf{x}))\delta\Theta(\mathbf{x})dS \\ - \int_{\Omega}\big(\nabla\cdot\mathbf{q}^{PD}(\mathbf{x},t)\big)\delta\Theta(\mathbf{x})dV - \int_{S_{\mathbf{q}}} \mathbf{q}^{*PD}(\mathbf{x},t)\cdot\mathbf{n}(\mathbf{x})\delta\Theta(\mathbf{x})dS - \int_{\Omega} s_b(\mathbf{x},t)\delta\Theta(\mathbf{x})dV = 0 \, . \tag{4.11}$$

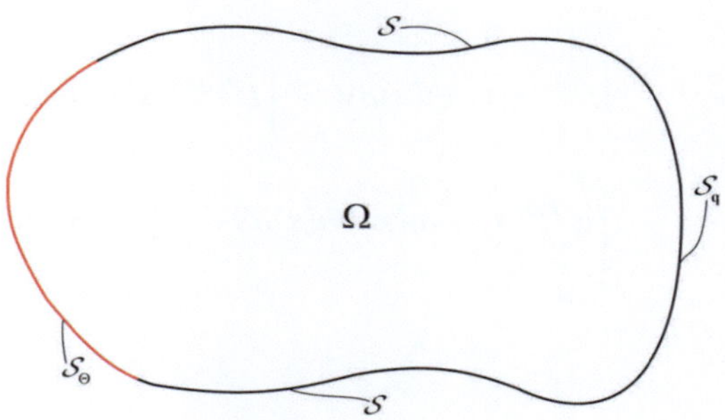

Fig. 4.3 A domain subjected to field constraints on S_{Θ} and external field flux on $S_{\mathbf{q}}$

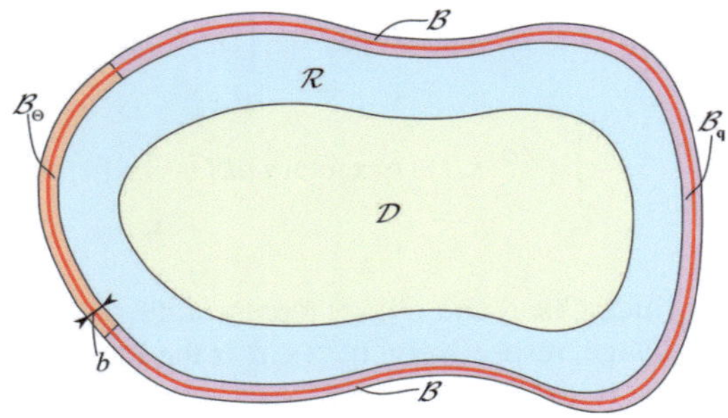

Fig. 4.4 PD computational domain with the interior $\mathcal{D}$, outer $\mathscr{R}$, and boundary layer $\mathscr{B}$

in which $\delta\Theta(\mathbf{x})$ is the virtual field variable. The volume of the domain shown in Fig. 4.4 is split into inner $\mathcal{D}$, outer $\mathscr{R}$, and the boundary layer region $\mathscr{B}$. This equation can be rewritten as

$$\begin{aligned}&\int\limits_{\mathcal{D}\cup\mathscr{R}}\big(C(\mathbf{x},t)\dot{\Theta}(\mathbf{x},t)+\dot{C}(\mathbf{x},t)\Theta(\mathbf{x},t)\big)\delta\Theta(\mathbf{x})dV\\&+\int\limits_{\mathscr{B}}\big(C(\mathbf{x},t)\dot{\Theta}(\mathbf{x},t)+\dot{C}(\mathbf{x},t)\Theta(\mathbf{x},t)\big)\delta\Theta(\mathbf{x})dV+\oint\limits_{S}\mathbf{q}^{PD}(\mathbf{x},t)\cdot\mathbf{n}(\mathbf{x})\delta\Theta(\mathbf{x})dS\\&-\int\limits_{\mathcal{D}}L^{PD}(\mathbf{x},t)\delta\Theta(\mathbf{x})dV-\int\limits_{\mathscr{R}}L^{PDDO}(\mathbf{x},t)\delta\Theta(\mathbf{x})dV-\int\limits_{\mathscr{B}}L^{PDDO}(\mathbf{x},t)\delta\Theta(\mathbf{x})dV\\&-\int\limits_{S_\mathbf{q}}\mathbf{q}^{*PD}(\mathbf{x},t)\cdot\mathbf{n}(\mathbf{x})\delta\Theta(\mathbf{x})dS-\int\limits_{\mathcal{D}\cup\mathscr{R}}s_b(\mathbf{x},t)\delta\Theta(\mathbf{x})dV-\int\limits_{\mathscr{B}}s_b(\mathbf{x},t)\delta\Theta(\mathbf{x})dV=0\end{aligned}\tag{4.12a}$$

or

$$\begin{aligned}&\int\limits_{\mathcal{D}\cup\mathscr{R}}\big(C(\mathbf{x},t)\dot{\Theta}(\mathbf{x},t)+\dot{C}(\mathbf{x},t)\Theta(\mathbf{x},t)\big)\delta\Theta(\mathbf{x})dV\\&+b\int\limits_{\mathscr{B}}\big(C(\mathbf{x},t)\dot{\Theta}(\mathbf{x},t)+\dot{C}(\mathbf{x},t)\Theta(\mathbf{x},t)\big)\delta\Theta(\mathbf{x})dS+\oint\limits_{S}\mathbf{q}^{PD}(\mathbf{x},t)\cdot\mathbf{n}(\mathbf{x})\delta\Theta(\mathbf{x})dS\\&-\int\limits_{\mathcal{D}}L^{PD}(\mathbf{x},t)\delta\Theta(\mathbf{x})dV-\int\limits_{\mathscr{R}}L^{PDDO}(\mathbf{x},t)\delta\Theta(\mathbf{x})dV-b\int\limits_{\mathscr{B}}L^{PDDO}(\mathbf{x},t)\delta\Theta(\mathbf{x})dS\\&-\int\limits_{S_\mathbf{q}}\mathbf{q}^{*PD}(\mathbf{x},t)\cdot\mathbf{n}(\mathbf{x})\delta\Theta(\mathbf{x})dS-\int\limits_{\mathcal{D}\cup\mathscr{R}}s_b(\mathbf{x},t)\delta\Theta(\mathbf{x})dV-b\int\limits_{\mathscr{B}}s_b(\mathbf{x},t)\delta\Theta(\mathbf{x})dS=0\end{aligned}\tag{4.12b}$$

in which

$$L^{PD} = \int_{\mathcal{D}} \mathbf{q}^{PD}(\mathbf{x}, t) \cdot \mathbf{n}(\mathbf{x}) \delta\Theta(\mathbf{x}) dV \tag{4.13}$$

and

$$L^{PDDO} = \int_{\mathcal{R}} \mathbf{q}^{PDDO}(\mathbf{x}, t) \cdot \mathbf{n}(\mathbf{x}) \delta\Theta(\mathbf{x}) dV \tag{4.14}$$

where $\mathbf{q}^{PD}(\mathbf{x}, t)$ is the PD representation of the field flux vector.

The boundary layer region, $\mathcal{B}$, is defined by the surface of the domain, S, with its width, b, and for $b \ll 1$. Thus, Eq (4.12b) can be reduced to

$$\begin{aligned} &\int_{\mathcal{D} \cup \mathcal{R}} \left(C(\mathbf{x}, t)\dot{\Theta}(\mathbf{x}, t) + \dot{C}(\mathbf{x}, t)\Theta(\mathbf{x}, t) \right) \delta\Theta(\mathbf{x}) dV - \int_{\mathcal{D}} L^{PD}(\mathbf{x}, t) \delta\Theta(\mathbf{x}) dV \\ &- \int_{\mathcal{R}} L^{PDDO}(\mathbf{x}, t) \delta\Theta(\mathbf{x}) dV + \oint_{S} \mathbf{q}^{PD}(\mathbf{x}, t) \cdot \mathbf{n}(\mathbf{x}) \delta\Theta(\mathbf{x}) dS \\ &- \int_{S_{\mathbf{q}}} \mathbf{q}^{*PD}(\mathbf{x}, t) \cdot \mathbf{n}(\mathbf{x}) \delta\Theta(\mathbf{x}) dS - \int_{\mathcal{D} \cup \mathcal{R}} s_b(\mathbf{x}, t) \delta\Theta(\mathbf{x}) dV = 0 \end{aligned} \tag{4.15}$$

The PD field equation is expressed only in terms of field variables as unknowns. It is natural to enforce field constraints directly through a material layer of volume along the boundary. For a two-dimensional domain, Eq. (4.15) concerns only area and line integrations and simplifies to

$$\begin{aligned} &\int_{\mathcal{D} \cup \mathcal{R}} \left(C(\mathbf{x}, t)\dot{\Theta}(\mathbf{x}, t) + \dot{C}(\mathbf{x}, t)\Theta(\mathbf{x}, t) \right) \delta\Theta(\mathbf{x}) dA - \int_{\mathcal{D}} L^{PD}(\mathbf{x}, t) \delta\Theta(\mathbf{x}) dA \\ &- \int_{\mathcal{R}} L^{PDDO}(\mathbf{x}, t) \delta\Theta(\mathbf{x}) dA + \oint_{\Gamma} \mathbf{q}^{PD}(\mathbf{x}, t) \cdot \mathbf{n}(\mathbf{x}) \delta\Theta(\mathbf{x}) d\Gamma \\ &- \int_{\Gamma_{\mathbf{q}}} \mathbf{q}^{*PD}(\mathbf{x}, t) \cdot \mathbf{n}(\mathbf{x}) \delta\Theta(\mathbf{x}) d\Gamma - \int_{\mathcal{D} \cup \mathcal{R}} s_b(\mathbf{x}, t) \delta\Theta(\mathbf{x}) dS = 0 \end{aligned} \tag{4.16}$$

in which h is the thickness of the domain, $\mathcal{D}$ and $\mathcal{R}$ represent the area of the interior and outer regions, Γ represents the entire boundary of the domain, and $\Gamma_{\mathbf{q}}$ and Γ_{Θ} represent the segment of the boundary subjected to the external field flux and field constraints, respectively. The region $\mathcal{B}$ in Fig. 4.4 having a width of b is referred to as the boundary layer region. As the width approaches to zero, i.e., $b \rightarrow 0$, Γ coincides with the real boundary as shown in Fig. 4.3.

4.4 Discrete Form of PD Equilibrium Equations

The discretization of the domain for the weak form of PD equilibrium equations is the same as that of the strong form of PD equilibrium equations. As shown in Fig. 4.5, the PD domain $\Omega = \mathcal{D} \cup \mathcal{R} \cup \mathcal{B}$ is discretized with $N = N_{\mathcal{D}} + N_{\mathcal{R}} + N_{\mathcal{B}}$ PD points. The parameters $N_{\mathcal{D}}$ and $N_{\mathcal{R}}$ represent the number of points associated with regions, $\mathcal{D}$ and $\mathcal{R}$, respectively. The discretization in these regions can be nonuniform. The parameter, $N_{\mathcal{B}} = N_{\mathcal{B}_{\mathbf{u}}} + N_{\mathcal{B}_{\boldsymbol{\sigma}}}$, represents the number of points associated with region $\mathcal{B}$. The parameters, $N_{\mathcal{B}_{\mathbf{u}}}$ and $N_{\mathcal{B}_{\boldsymbol{\sigma}}}$, represent the number of points associated with regions $\mathcal{B}_{\mathbf{u}}$ and $\mathcal{B}_{\boldsymbol{\sigma}}$, respectively. The position of the PD material point k is denoted by $\mathbf{x}_{(k)}$. For two-dimensional analysis, its incremental volume, boundary area, and length are denoted by $V_{(k)} = hA_{(k)}$, $S_{(k)} = h\Gamma_{(k)}$, and $\Gamma_{(k)}$, respectively, where $h = 1$ is the thickness of the domain.

Under small deformation assumptions, the discrete form of virtual work in Eq. (4.7) can be expressed as:

$$\begin{aligned}
&\sum_{k=1}^{N_{\mathcal{D}}+N_{\mathcal{R}}} \rho_{(k)} \delta\mathbf{u}_{(k)}^T \ddot{\mathbf{u}}_{(k)} V_{(k)} - \sum_{k=1}^{N_{\mathcal{D}}} \delta\mathbf{u}_{(k)}^T \left(\sum_{j=1}^{N_{(k)}} \mathbf{t}_{(k)(j)}^{\alpha} V_{(j)} - \sum_{p=1}^{\widetilde{N}_{(k)}} \mathbf{t}_{(p)(k)}^{\alpha} V_{(p)} \right) V_{(k)} \\
&- \sum_{k=1}^{N_{\mathcal{R}}} \delta\mathbf{u}_{(k)}^T \left(\sum_{j=1}^{N_{(k)}} \left(\mathbf{t}_{(j)(k)}^{PDDO} - \mathbf{t}_{(k)(j)}^{PDDO} \right) V_{(j)} \right) V_{(k)} + \sum_{k=1}^{N_{\mathcal{B}}} \delta\mathbf{u}_{(k)}^T \left(\boldsymbol{\sigma}_{(k)} \mathbf{n}_{(k)} \right) S_{(k)} \\
&- \sum_{k=1}^{N_{\mathcal{B}_{\boldsymbol{\sigma}}}} \delta\mathbf{u}_{(k)}^T \mathbf{T}_{(k)}^* S_{(k)} - \sum_{k=1}^{N_{\mathcal{D}}+N_{\mathcal{R}}} \delta\mathbf{u}_{(k)}^T \mathbf{b}_{(k)} V_{(k)} = 0
\end{aligned} \tag{4.17}$$

The virtual work due to the PD internal force vector after multiplying with volume $V_{(k)}$ in the interior region $\mathcal{D}$ can be evaluated as:

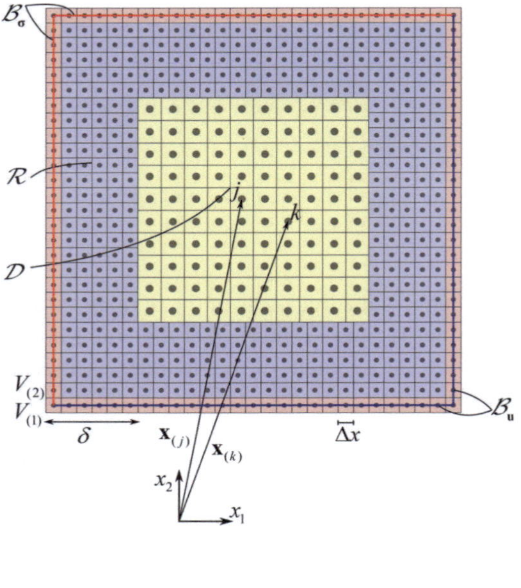

Fig. 4.5 Discretization of the domain for the weak form of PD equilibrium equation with non-uniform grid

$$\sum_{k=1}^{N_{\mathcal{D}}} \delta\mathbf{u}_{(k)}^{T}\left(\sum_{j=1}^{N_{(k)}} \mathbf{t}_{(k)(j)}^{\alpha} V_{(j)} - \sum_{p=1}^{\widetilde{N}_{(k)}} \mathbf{t}_{(p)(k)}^{\alpha} V_{(p)}\right)$$
$$V_{(k)} = \delta\mathbf{V}_{\mathcal{D}}^{T} \sum_{k=1}^{N_{\mathcal{D}}} \left(\sum_{j=1}^{N_{(k)}} \mathbf{k}_{(k)(j)}^{\alpha} - \sum_{p=1}^{\widetilde{N}_{(k)}} \mathbf{k}_{(p)(k)}^{\alpha}\right) \mathbf{V} = \delta\mathbf{V}_{\mathcal{D}}^{T}\mathbf{K}^{\mathcal{D}}\mathbf{V} \tag{4.18}$$

in which $\mathbf{V}$ is the unknown displacement vector. It can be partitioned as:

$$\mathbf{V} = \{\mathbf{V}_{\mathcal{D}} \;\; \mathbf{V}_{\mathcal{R}} \;\; \mathbf{V}_{\mathcal{B}_{\sigma}} \;\; \mathbf{V}_{\mathcal{B}_{\mathbf{u}}}\}^{T} \tag{4.19}$$

in which the vectors, $\mathbf{V}_{\mathcal{D}}$, $\mathbf{V}_{\mathcal{R}}$, $\mathbf{V}_{\mathcal{B}_{\sigma}}$, and $\mathbf{V}_{\mathcal{B}_{\mathbf{u}}}$ include the displacement components at points located in regions $\mathcal{D}$, $\mathcal{R}$, $\mathcal{B}_{\sigma}$, and $\mathcal{B}_{\mathbf{u}}$, respectively. Subsequently, the virtual work due to PD internal force vector, $\mathbf{F}_{\mathcal{D}}$, in the interior region, $\mathcal{D}$, can be recast as:

$$\sum_{k=1}^{N_{\mathcal{D}}} \delta\mathbf{u}_{(k)}^{T}\left(\sum_{j=1}^{N_{(k)}} \mathbf{t}_{(k)(j)}^{\alpha} V_{(j)} - \sum_{p=1}^{\widetilde{N}_{(k)}} \mathbf{t}_{(p)(k)}^{\alpha} V_{(p)}\right) V_{(k)} = \{\delta\mathbf{V}_{\mathcal{D}}^{T}\}\{\mathbf{F}_{\mathcal{D}}\}$$
$$= \{\delta\mathbf{V}_{\mathcal{D}}^{T}\}[\mathbf{K}_{\mathcal{D}\mathcal{D}} \;\; \mathbf{K}_{\mathcal{D}\mathcal{R}} \;\; \mathbf{0} \;\; \mathbf{0}] \begin{Bmatrix} \mathbf{V}_{\mathcal{D}} \\ \mathbf{V}_{\mathcal{R}} \\ \mathbf{V}_{\mathcal{B}_{\sigma}} \\ \mathbf{V}_{\mathcal{B}_{\mathbf{u}}} \end{Bmatrix} \tag{4.20}$$

where the submatrices $\mathbf{K}_{\mathcal{D}\mathcal{D}}$ and $\mathbf{K}_{\mathcal{D}\mathcal{R}}$ with dimensions $2N_{\mathcal{D}} \times 2N_{\mathcal{D}}$ and $2N_{\mathcal{D}} \times 2N_{\mathcal{R}}$, respectively, represent the self-stiffness and coupling stiffness matrices between the unknown vectors of $\mathbf{V}_{\mathcal{D}}$ and $\mathbf{V}_{\mathcal{R}}$. The virtual work due to the internal force vector, $\mathbf{F}_{\mathcal{R}}$, in the exterior region $\mathcal{R}$ and the virtual work due to internal traction, $\mathbf{F}_{\mathcal{B}_{\sigma}}$, around the boundary $\mathcal{B}$ can be evaluated as:

$$\sum_{k=1}^{N_{\mathcal{R}}} \sum_{j=1}^{N_{(k)}} \delta\mathbf{u}_{(k)}^{T}\left(\mathbf{t}_{(j)(k)}^{PDDO} - \mathbf{t}_{(k)(j)}^{PDDO}\right) V_{(j)} V_{(k)} = \{\delta\mathbf{V}_{\mathcal{R}}^{T}\}\{\mathbf{F}_{\mathcal{R}}\}$$
$$= \{\delta\mathbf{V}_{\mathcal{R}}^{T}\}[\mathbf{K}_{\mathcal{R}\mathcal{D}} \;\; \mathbf{K}_{\mathcal{R}\mathcal{R}} \;\; \mathbf{K}_{\mathcal{R}\mathcal{B}_{\sigma}} \;\; \mathbf{K}_{\mathcal{R}\mathcal{B}_{\mathbf{u}}}] \begin{Bmatrix} \mathbf{V}_{\mathcal{D}} \\ \mathbf{V}_{\mathcal{R}} \\ \mathbf{V}_{\mathcal{B}_{\sigma}} \\ \mathbf{V}_{\mathcal{B}_{\mathbf{u}}} \end{Bmatrix} \tag{4.21}$$

and

$$\sum_{k=1}^{N_{\mathscr{B}}} \delta\mathbf{u}_{(k)}^T \left(\boldsymbol{\sigma}_{(k)}\mathbf{n}_{(k)}\right)\Gamma_{(k)} = \left\{ \delta\mathbf{V}_{\mathscr{B}_\sigma}^T \quad \delta\mathbf{V}_{\mathscr{B}_u}^T \right\} \begin{bmatrix} \mathbf{0} & \mathbf{K}_{\mathscr{B}_\sigma\mathscr{R}} & \mathbf{K}_{\mathscr{B}_\sigma\mathscr{B}_\sigma} & \mathbf{K}_{\mathscr{B}_\sigma\mathscr{B}_u} \\ \mathbf{0} & \mathbf{K}_{\mathscr{B}_u\mathscr{R}} & \mathbf{K}_{\mathscr{B}_u\mathscr{B}_\sigma} & \mathbf{K}_{\mathscr{B}_u\mathscr{B}_u} \end{bmatrix} \begin{Bmatrix} \mathbf{V}_{\mathcal{D}} \\ \mathbf{V}_{\mathscr{R}} \\ \mathbf{V}_{\mathscr{B}_\sigma} \\ \mathbf{V}_{\mathscr{B}_u} \end{Bmatrix}$$
$$= \left\{ \delta\mathbf{V}_{\mathscr{B}_\sigma}^T \right\}\left\{ \mathbf{F}_{\mathscr{B}_\sigma} \right\} \tag{4.22}$$

where the submatrices, $\mathbf{K}_{\mathscr{R}\mathscr{R}}$, $\mathbf{K}_{\mathscr{R}\mathcal{D}}$, $\mathbf{K}_{\mathscr{R}\mathscr{B}_\sigma}$ and $\mathbf{K}_{\mathscr{R}\mathscr{B}_u}$ represent the self-stiffness and coupling stiffness matrices between the unknown vectors of $\mathbf{V}_{\mathcal{D}}$, $\mathbf{V}_{\mathscr{R}}$, $\mathbf{V}_{\mathscr{B}_\sigma}$, and $\mathbf{V}_{\mathscr{B}_u}$. Their dimensions are $2N_{\mathscr{R}} \times 2N_{\mathscr{R}}$, $2N_{\mathscr{R}} \times 2N_{\mathcal{D}}$, $2N_{\mathscr{R}} \times 2N_{\mathscr{B}_\sigma}$, and $2N_{\mathscr{R}} \times 2N_{\mathscr{B}_u}$, respectively. Similarly, the submatrices $\mathbf{K}_{\mathscr{B}_\sigma\mathscr{B}_\sigma}$, $\mathbf{K}_{\mathscr{B}_u\mathscr{B}_u}$ and $\mathbf{K}_{\mathscr{B}_\sigma\mathscr{R}}$, $\mathbf{K}_{\mathscr{B}_\sigma\mathscr{B}_u}$, $\mathbf{K}_{\mathscr{B}_u\mathscr{R}}$, $\mathbf{K}_{\mathscr{B}_u\mathscr{B}_\sigma}$ are the corresponding self-stiffness and coupling stiffness matrices between the unknown vectors of $\mathbf{V}_{\mathscr{R}}$, $\mathbf{V}_{\mathscr{B}_\sigma}$, and $\mathbf{V}_{\mathscr{B}_u}$. The dimensions of the submatrices are $2N_{\mathscr{B}_\sigma} \times 2N_{\mathscr{B}_\sigma}$, $2N_{\mathscr{B}_u} \times 2N_{\mathscr{B}_u}$, and $2N_{\mathscr{B}_\sigma} \times 2N_{\mathscr{R}}$, $2N_{\mathscr{B}_\sigma} \times 2N_{\mathscr{B}_u}$, $2N_{\mathscr{B}_u} \times 2N_{\mathscr{R}}$, $2N_{\mathscr{B}_u} \times 2N_{\mathscr{B}_\sigma}$, respectively.

The virtual work due to the inertial force vectors, $\mathbf{F}_{\ddot{\mathbf{u}}\mathcal{D}}$ and $\mathbf{F}_{\ddot{\mathbf{u}}\mathscr{R}}$, in the domain of $\mathcal{D} \cup \mathscr{R}$ can be evaluated as:

$$\sum_{k=1}^{N_{\mathcal{D}}+N_{\mathscr{R}}} \rho_{(k)}\delta\mathbf{u}_{(k)}^T \ddot{\mathbf{u}}_{(k)} V_{(k)} = \begin{Bmatrix} \delta\mathbf{V}_{\mathcal{D}} \\ \delta\mathbf{V}_{\mathscr{R}} \end{Bmatrix}^T \begin{Bmatrix} \mathbf{F}_{\ddot{\mathbf{u}}\mathcal{D}} \\ \mathbf{F}_{\ddot{\mathbf{u}}\mathscr{R}} \end{Bmatrix}$$
$$= \begin{Bmatrix} \delta\mathbf{V}_{\mathcal{D}} \\ \delta\mathbf{V}_{\mathscr{R}} \end{Bmatrix}^T \begin{bmatrix} \mathbf{M}_{\mathcal{D}\mathcal{D}} & \mathbf{0} & \mathbf{0} & \mathbf{0} \\ \mathbf{0} & \mathbf{M}_{\mathscr{R}\mathscr{R}} & \mathbf{0} & \mathbf{0} \end{bmatrix} \begin{Bmatrix} \ddot{\mathbf{V}}_{\mathcal{D}} \\ \ddot{\mathbf{V}}_{\mathscr{R}} \\ \ddot{\mathbf{V}}_{\mathscr{B}_\sigma} \\ \ddot{\mathbf{V}}_{\mathscr{B}_u} \end{Bmatrix} \tag{4.23}$$

in which the submatrix $\mathbf{M}_{\mathcal{D}\mathcal{D}}$ and $\mathbf{M}_{\mathscr{R}\mathscr{R}}$ are diagonal mass matrices in regions $\mathcal{D}$ and $\mathscr{R}$ with dimensions $2N_{\mathcal{D}} \times 2N_{\mathcal{D}}$ and $2N_{\mathscr{R}} \times 2N_{\mathscr{R}}$, respectively.

Due to the applied traction vector, $\mathbf{T}^*$, the virtual work can be evaluated as:

$$\sum_{k=1}^{N_{\mathscr{B}_\sigma}} \delta\mathbf{u}_{(k)}^T \mathbf{T}_{(k)}^* S_{(k)} = \left\{ \delta\mathbf{V}_{\mathscr{B}_\sigma}^T \right\}\left\{ \mathbf{F}_{\mathscr{B}_\sigma}^* \right\} \tag{4.24}$$

where $S_{(k)}$ is the boundary area of the PD material point k, and the force vector $\mathbf{F}_{\mathscr{B}_\sigma}^*$ with dimensions $2N_{\mathscr{B}_\sigma} \times 1$ is defined as:

$$\mathbf{F}_{\mathscr{B}_\sigma}^* = \left\{ \mathbf{T}_{(1)}^* S_{(1)}, \ldots, \mathbf{T}_{(p)}^* S_{(p)}, \ldots, \mathbf{T}_{(N_{\mathscr{B}_\sigma})}^* S_{(N_{\mathscr{B}_\sigma})} \right\}^T \tag{4.25}$$

Due to the applied body force vector, $\mathbf{b}(\mathbf{x}, t)$, the virtual work in the domain of $\mathcal{D} \cup \mathcal{R}$ can be evaluated as:

$$\sum_{k=1}^{N_{\mathcal{D}}+N_{\mathcal{R}}} \delta\mathbf{u}_{(k)}^T \mathbf{b}_{(k)} V_{(k)} = \left\{ \delta\mathbf{V}_{\mathcal{D}}^T \quad \delta\mathbf{V}_{\mathcal{R}}^T \right\} \left\{ \begin{array}{c} \mathbf{F}_{\mathbf{b}_{\mathcal{D}}}^* \\ \mathbf{F}_{\mathbf{b}_{\mathcal{R}}}^* \end{array} \right\} \tag{4.26}$$

in which the force vectors $\mathbf{F}_{\mathbf{b}\mathcal{D}}^*$ and $\mathbf{F}_{\mathbf{b}\mathcal{R}}^*$ are defined as:

$$\mathbf{F}_{\mathbf{b}_{\mathcal{D}}}^* = \left\{ \mathbf{b}_{(1)}^* V_{(1)}, \ldots, \mathbf{b}_{(p)}^* V_{(p)}, \ldots, \mathbf{b}_{(N_{\mathcal{D}})}^* V_{(N_{\mathcal{D}})} \right\}^T \tag{4.27}$$

and

$$\mathbf{F}_{\mathbf{b}_{\mathcal{R}}}^* = \left\{ \mathbf{b}_{(1)}^* V_{(1)}, \ldots, \mathbf{b}_{(p)}^* V_{(p)}, \ldots, \mathbf{b}_{(N_{\mathcal{R}})}^* V_{(N_{\mathcal{R}})} \right\}^T \tag{4.28}$$

After combining Eqs.(4.20) to (4.24) and (4.26), the virtual work expression, Eq. (4.17), can be cast in matrix form as:

$$\delta\mathbf{V}^T (\mathbf{F}_{\ddot{\mathbf{u}}} + \mathbf{F} - \mathbf{F}^*) = \delta\mathbf{V}^T \left(\mathbf{M}\ddot{\mathbf{V}} + \mathbf{K}\mathbf{V} - \mathbf{F}^* \right) = 0 \tag{4.29}$$

in which mass and stiffness matrices and the force vectors are expressed as:

$$\mathbf{M} = \begin{bmatrix} \mathbf{M}_{\mathcal{D}\mathcal{D}} & \mathbf{0} & \mathbf{0} & \mathbf{0} \\ \mathbf{0} & \mathbf{M}_{\mathcal{R}\mathcal{R}} & \mathbf{0} & \mathbf{0} \\ \mathbf{0} & \mathbf{0} & \mathbf{0} & \mathbf{0} \\ \mathbf{0} & \mathbf{0} & \mathbf{0} & \mathbf{0} \end{bmatrix} \tag{4.30}$$

$$\mathbf{K} = \begin{bmatrix} \mathbf{K}_{\mathcal{D}\mathcal{D}} & \mathbf{K}_{\mathcal{D}\mathcal{R}} & \mathbf{0} & \mathbf{0} \\ \mathbf{K}_{\mathcal{R}\mathcal{D}} & \mathbf{K}_{\mathcal{R}\mathcal{R}} & \mathbf{K}_{\mathcal{R}\mathcal{B}_\sigma} & \mathbf{K}_{\mathcal{R}\mathcal{B}_\mathbf{u}} \\ \mathbf{0} & \mathbf{K}_{\mathcal{B}_\sigma\mathcal{R}} & \mathbf{K}_{\mathcal{B}_\sigma\mathcal{B}_\sigma} & \mathbf{K}_{\mathcal{B}_\sigma\mathcal{B}_\mathbf{u}} \\ \mathbf{0} & \mathbf{K}_{\mathcal{B}_\mathbf{u}\mathcal{R}} & \mathbf{K}_{\mathcal{B}_\mathbf{u}\mathcal{B}_\sigma} & \mathbf{K}_{\mathcal{B}_\mathbf{u}\mathcal{B}_\mathbf{u}} \end{bmatrix} \tag{4.31}$$

$$\mathbf{F}_{\ddot{\mathbf{u}}} = \{-\mathbf{F}_{\ddot{\mathbf{u}}\mathcal{R}}, \ -\mathbf{F}_{\ddot{\mathbf{u}}\mathcal{D}}, \mathbf{0}, \mathbf{0}\}^T \tag{4.32a}$$

$$\mathbf{F} = \{-\mathbf{F}_{\mathcal{R}}, \ -\mathbf{F}_{\mathcal{D}}, \mathbf{0}, \mathbf{0}\}^T \tag{4.32b}$$

and

$$\mathbf{F}^* = \left\{ \mathbf{F}_{\mathbf{b}_{\mathcal{D}}}^*, \mathbf{F}_{\mathbf{b}_{\mathcal{R}}}^*, \mathbf{F}_{\mathcal{B}_\sigma}^*, \mathbf{0} \right\}^T \tag{4.32c}$$

4.5 Discrete Form of PD Field Equations

As shown in Fig. 4.6, the PD domain $\Omega = \mathcal{D} \cup \mathcal{R} \cup \mathcal{B}$ is discretized with $N = N_{\mathcal{D}} + N_{\mathcal{R}} + N_{\mathcal{B}}$ PD points. The parameters $N_{\mathcal{D}}$ and $N_{\mathcal{R}}$ represent the number of points associated with regions, $\mathcal{D}$ and $\mathcal{R}$, respectively. The discretization in these regions can be nonuniform. The parameter $N_{\mathcal{B}} = N_{\mathcal{B}_\Theta} + N_{\mathcal{B}_\mathbf{q}}$ represents the number of points associated with region, $\mathcal{B}$. The parameters $N_{\mathcal{B}_\Theta}$ and $N_{\mathcal{B}_\mathbf{q}}$ represent the number of points associated with regions $\mathcal{B}_\Theta$ and $\mathcal{B}_\mathbf{q}$, respectively. The position of the PD point k is denoted by $\mathbf{x}_{(k)}$. For two-dimensional analysis, its incremental volume, boundary area, and length are denoted by $V_{(k)} = hA_{(k)}$, $S_{(k)} = h\Gamma_{(k)}$, and $\Gamma_{(k)}$, respectively, where $h = 1$ is the thickness of the domain.

The discrete form of Eq. (4.15) can be expressed as

$$\begin{aligned}
&\sum_{k=1}^{N_{\mathcal{D}}+N_{\mathcal{R}}} \left(C_{(k)}\dot{\Theta}_{(k)} + \dot{C}_{(k)}\Theta_{(k)}\right)V_{(k)}\delta\Theta_{(k)}^T - \sum_{k=1}^{N_{\mathcal{D}}} \delta\Theta_{(k)} \left(\sum_{j=1}^{N_{(k)}} h_{(k)(j)}^{BB} V_{(j)} - \sum_{p=1}^{\widetilde{N}_{(k)}} h_{(p)(k)}^{BB} V_{(p)} \right) V_{(k)} \\
&- \sum_{k=1}^{N_{\mathcal{R}}} \left(\delta\Theta_{(k)} \sum_{j=1}^{N_{(k)}} \left(h_{(j)(k)}^{PDDO} - h_{(k)(j)}^{PDDO} \right) V_{(j)} \right) V_{(k)} + \sum_{k=1}^{N_{\mathcal{B}_\mathbf{q}}+N_{\mathcal{B}_\Theta}} \delta\Theta_{(k)} \left(\mathbf{q}_{(k)}^{PD} \cdot \mathbf{n}_{(k)} \right) S_{(k)} \\
&- \sum_{k=1}^{N_{\mathcal{B}_\mathbf{q}}} \delta\Theta_{(k)} \left(\mathbf{q}_{(k)}^{*} \cdot \mathbf{n}_{(k)} \right) S_{(k)} - \sum_{k=1}^{N_{\mathcal{D}}+N_{\mathcal{R}}} \delta\Theta_{(k)} s_{b(k)} V_{(k)} = 0
\end{aligned} \tag{4.33}$$

In region $\mathcal{D}$, the total virtual field variation due to the internal net field flow can be evaluated as

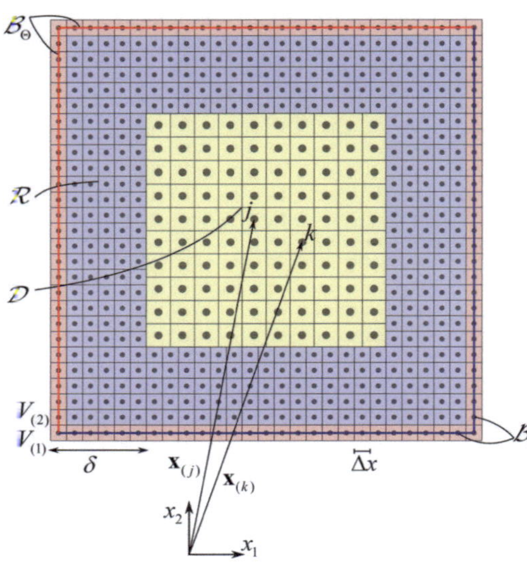

Fig. 4.6 Discretization of the domain for the weak form of PD field equation with non-uniform grid

$$\sum_{k=1}^{N_{\mathcal{D}}} \delta\Theta_{(k)} \left(\sum_{j=1}^{N_{(k)}} h_{(k)(j)}^{BB} V_{(j)} - \sum_{p=1}^{\widetilde{N}_{(k)}} h_{(p)(k)}^{BB} V_{(p)} \right) V_{(k)} \\ = \delta \mathbf{V}^T \sum_{k=1}^{N_{\mathcal{D}}} \left(\sum_{j=1}^{N_{(k)}} \mathbf{k}_{(k)(j)}^{BB} - \sum_{p=1}^{\widetilde{N}_{(k)}} \mathbf{k}_{(p)(k)}^{BB} \right) \mathbf{V} = \delta \mathbf{V}^{\mathrm{T}} \mathbf{K}^{\mathcal{D}} \mathbf{V} \tag{4.34}$$

in which the unknown vector,$\mathbf{V}$, consists of the field variable. It can be partitioned as

$$\mathbf{V} = \left\{ \mathbf{V}_{\mathcal{D}} \quad \mathbf{V}_{\mathcal{R}} \quad \mathbf{V}_{\mathcal{B}_{\mathbf{q}}} \quad \mathbf{V}_{\mathcal{B}_{\Theta}} \right\}^T \tag{4.35}$$

in which the vectors $\mathbf{V}_{\mathcal{D}}$, $\mathbf{V}_{\mathcal{R}}$, $\mathbf{V}_{\mathcal{B}_{\sigma}}$, and $\mathbf{V}_{\mathcal{B}_{\mathbf{u}}}$ include the field values at points located in regions $\mathcal{D}$, $\mathcal{R}$, $\mathcal{B}_{\mathbf{q}}$, and $\mathcal{B}_{\Theta}$, respectively. The total virtual field variation in Eq. (4.34) due to the internal field flow, $\mathbf{F}_{\mathcal{D}}$, in the region $\mathcal{D}$ can be recast as:

$$\sum_{k=1}^{N_{\mathcal{D}}} \delta\Theta_{(k)} \left(\sum_{j=1}^{N_{(k)}} h_{(k)(j)}^{BB} V_{(j)} - \sum_{p=1}^{\widetilde{N}_{(k)}} h_{(p)(k)}^{BB} V_{(p)} \right) V_{(k)} = \left\{ \delta \mathbf{V}_{\mathcal{D}}^T \right\} \{ \mathbf{F}_{\mathcal{D}} \} \\ = \left\{ \delta \mathbf{V}_{\mathcal{D}}^T \right\} [\mathbf{K}_{\mathcal{D}\mathcal{D}} \quad \mathbf{K}_{\mathcal{D}\mathcal{R}} \quad \mathbf{0} \quad \mathbf{0}] \begin{Bmatrix} \mathbf{V}_{\mathcal{D}} \\ \mathbf{V}_{\mathcal{R}} \\ \mathbf{V}_{\mathcal{B}_{\mathbf{q}}} \\ \mathbf{V}_{\mathcal{B}_{\Theta}} \end{Bmatrix} \tag{4.36}$$

where the matrices $\mathbf{K}_{\mathcal{D}\mathcal{D}}$ and $\mathbf{K}_{\mathcal{D}\mathcal{R}}$ with dimensions $N_{\mathcal{D}} \times N_{\mathcal{D}}$ and $N_{\mathcal{D}} \times N_{\mathcal{R}}$, respectively, represent the self-stiffness and coupling stiffness matrices between the unknown vectors of $\mathbf{V}_{\mathcal{D}}$ and $\mathbf{V}_{\mathcal{R}}$.

In the exterior region, $\mathcal{R}$, the total virtual field variation due to the internal field flow, $\mathbf{F}_{\mathcal{R}}$, can be evaluated as:

$$\sum_{k=1}^{N_{\mathcal{R}}} \left(\delta\Theta_{(k)} \sum_{j=1}^{N_{(k)}} \left(h_{(j)(k)}^{PDDO} - h_{(k)(j)}^{PDDO} \right) V_{(j)} \right) V_{(k)} = \left\{ \delta \mathbf{V}_{\mathcal{R}}^T \right\} \{ \mathbf{F}_{\mathcal{R}} \} \\ = \left\{ \delta \mathbf{V}_{\mathcal{R}}^T \right\} \left[\mathbf{K}_{\mathcal{R}\mathcal{D}} \quad \mathbf{K}_{\mathcal{R}\mathcal{R}} \quad \mathbf{K}_{\mathcal{R}\mathcal{B}_{\mathbf{q}}} \quad \mathbf{K}_{\mathcal{R}\mathcal{B}_{\Theta}} \right] \begin{Bmatrix} \mathbf{V}_{\mathcal{D}} \\ \mathbf{V}_{\mathcal{R}} \\ \mathbf{V}_{\mathcal{B}_{\mathbf{q}}} \\ \mathbf{V}_{\mathcal{B}_{\Theta}} \end{Bmatrix} \tag{4.37}$$

In region, $\mathscr{B}$, the total virtual field variation due to internal field flux, $\mathbf{F}_{\mathscr{B}}$, can be evaluated as:

$$
\begin{aligned}
&\sum_{k=1}^{N_{\mathscr{B}_\mathbf{q}}+N_{\mathscr{B}_\Theta}} \delta\Theta_{(k)}\left(\mathbf{q}_{(k)}^{PD}\cdot\mathbf{n}_{(k)}\right)\Gamma_{(k)} \\
&=\left\{\delta\mathbf{V}_{\mathscr{B}_\mathbf{q}}^T \quad \delta\mathbf{V}_{\mathscr{B}_\Theta}^T\right\}\begin{bmatrix}\mathbf{0} & \mathbf{K}_{\mathscr{B}_\mathbf{q}\mathscr{R}} & \mathbf{K}_{\mathscr{B}_\mathbf{q}\mathscr{B}_\mathbf{q}} & \mathbf{K}_{\mathscr{B}_\mathbf{q}\mathscr{B}_\Theta} \\ \mathbf{0} & \mathbf{K}_{\mathscr{B}_\Theta\mathscr{R}} & \mathbf{K}_{\mathscr{B}_\Theta\mathscr{B}_\mathbf{q}} & \mathbf{K}_{\mathscr{B}_\Theta\mathscr{B}_\Theta}\end{bmatrix}\begin{Bmatrix}\mathbf{V}_{\mathscr{D}} \\ \mathbf{V}_{\mathscr{R}} \\ \mathbf{V}_{\mathscr{B}_\mathbf{q}} \\ \mathbf{V}_{\mathscr{B}_\Theta}\end{Bmatrix} \\
&=\left\{\delta\mathbf{V}_{\mathscr{B}}^T\right\}\left\{\mathbf{F}_{\mathscr{B}}\right\}
\end{aligned}
\tag{4.38}
$$

where the matrices, $\mathbf{K}_{\mathscr{R}\mathscr{R}}$, $\mathbf{K}_{\mathscr{R}\mathscr{D}}$, $\mathbf{K}_{\mathscr{R}\mathscr{B}_\mathbf{q}}$, and $\mathbf{K}_{\mathscr{R}\mathscr{B}_\Theta}$ represent the self-stiffness and coupling stiffness matrices between the unknown vectors of $\mathbf{V}_{\mathscr{D}}$, $\mathbf{V}_{\mathscr{R}}$, $\mathbf{V}_{\mathscr{B}_\mathbf{q}}$, and $\mathbf{V}_{\mathscr{B}_\Theta}$. Their dimensions are $N_{\mathscr{R}}\times N_{\mathscr{R}}$, $N_{\mathscr{R}}\times N_{\mathscr{D}}$, $N_{\mathscr{R}}\times N_{\mathscr{B}_\sigma}$, and $N_{\mathscr{R}}\times N_{\mathscr{B}_\mathbf{u}}$, respectively. Similarly, the submatrices $\mathbf{K}_{\mathscr{B}_\mathbf{q}\mathscr{B}_\mathbf{q}}$, $\mathbf{K}_{\mathscr{B}_\Theta\mathscr{B}_\Theta}$ and $\mathbf{K}_{\mathscr{B}_\mathbf{q}\mathscr{R}}$, $\mathbf{K}_{\mathscr{B}_\mathbf{q}\mathscr{B}_\Theta}$, $\mathbf{K}_{\mathscr{B}_\Theta\mathscr{R}}$, $\mathbf{K}_{\mathscr{B}_\Theta\mathscr{B}_\mathbf{q}}$ are the corresponding self-stiffness and coupling stiffness matrices between the unknown vectors of $\mathbf{V}_{\mathscr{R}}$, $\mathbf{V}_{\mathscr{B}_\mathbf{q}}$, and $\mathbf{V}_{\mathscr{B}_\Theta}$. The dimensions of the submatrices are $N_{\mathscr{B}_\mathbf{q}}\times N_{\mathscr{B}_\mathbf{q}}$, $N_{\mathscr{B}_\Theta}\times N_{\mathscr{B}_\Theta}$, $N_{\mathscr{B}_\mathbf{q}}\times N_{\mathscr{R}}$, $N_{\mathscr{B}_\mathbf{q}}\times N_{\mathscr{B}_\Theta}$, $N_{\mathscr{B}_\Theta}\times N_{\mathscr{R}}$, and $N_{\mathscr{B}_\Theta}\times N_{\mathscr{B}_\mathbf{q}}$, respectively.

In region, $\mathscr{D}\cup\mathscr{R}$, the virtual field variation due to the transient field and time rate of change material property in Eq. (4.33) in terms of vectors $\mathbf{F}_{\Theta\mathscr{R}}$, $\mathbf{F}_{\dot{\Theta}\mathscr{R}}$ and $\mathbf{F}_{\Theta\mathscr{D}}$, $\mathbf{F}_{\dot{\Theta}\mathscr{D}}$ can be evaluated as:

$$
\begin{aligned}
&\sum_{k=1}^{N_{\mathscr{D}}+N_{\mathscr{R}}}\left(C_{(k)}\dot{\Theta}_{(k)}V_{(k)}+\dot{C}_{(k)}\Theta_{(k)}V_{(k)}\right)\delta\Theta_{(k)}^T=\begin{Bmatrix}\delta\mathbf{V}_{\mathscr{D}} \\ \delta\mathbf{V}_{\mathscr{R}}\end{Bmatrix}^T\begin{Bmatrix}\mathbf{F}_{\dot{\Theta}\mathscr{D}}+\mathbf{F}_{\Theta\mathscr{D}} \\ \mathbf{F}_{\dot{\Theta}\mathscr{R}}+\mathbf{F}_{\Theta\mathscr{R}}\end{Bmatrix} \\
&=\begin{Bmatrix}\delta\mathbf{V}_{\mathscr{D}} \\ \delta\mathbf{V}_{\mathscr{R}}\end{Bmatrix}^T\begin{bmatrix}\mathbf{C}_{\mathscr{D}\mathscr{D}} & \mathbf{0} & \mathbf{0} & \mathbf{0} \\ \mathbf{0} & \mathbf{C}_{\mathscr{R}\mathscr{R}} & \mathbf{0} & \mathbf{0}\end{bmatrix}\begin{Bmatrix}\dot{\mathbf{V}}_{\mathscr{D}} \\ \dot{\mathbf{V}}_{\mathscr{R}} \\ \dot{\mathbf{V}}_{\mathscr{B}_\mathbf{q}} \\ \dot{\mathbf{V}}_{\mathscr{B}_\Theta}\end{Bmatrix} \\
&\quad+\begin{Bmatrix}\delta\mathbf{V}_{\mathscr{D}} \\ \delta\mathbf{V}_{\mathscr{R}}\end{Bmatrix}^T\begin{bmatrix}\dot{\mathbf{C}}_{\mathscr{D}\mathscr{D}} & \mathbf{0} & \mathbf{0} & \mathbf{0} \\ \mathbf{0} & \dot{\mathbf{C}}_{\mathscr{R}\mathscr{R}} & \mathbf{0} & \mathbf{0}\end{bmatrix}\begin{Bmatrix}\mathbf{V}_{\mathscr{D}} \\ \mathbf{V}_{\mathscr{R}} \\ \mathbf{V}_{\mathscr{B}_\mathbf{q}} \\ \mathbf{V}_{\mathscr{B}_\Theta}\end{Bmatrix}
\end{aligned}
\tag{4.39}
$$

in which submatrices $\mathbf{C}_{\mathcal{DD}}$, $\mathbf{C}_{\mathcal{RR}}$ represent diagonal field damping and $\dot{\mathbf{C}}_{\mathcal{DD}}$, $\dot{\mathbf{C}}_{\mathcal{RR}}$ represent the diagonal transient material property matrices for regions $\mathcal{D}$ and $\mathcal{R}$ with dimensions $N_{\mathcal{D}} \times N_{\mathcal{D}}$ and $N_{\mathcal{R}} \times N_{\mathcal{R}}$, respectively.

For the applied field flux, the virtual field variation in Eq. (4.33) can be evaluated as:

$$\sum_{k=1}^{N_{\mathcal{B}_\mathbf{q}}} \delta\Theta_{(k)} \left(\mathbf{q}^*_{(k)} \cdot \mathbf{n}_{(k)}\right) S_{(k)} = \left\{ \delta\mathbf{V}^T_{\mathcal{B}_\mathbf{q}} \right\} \left\{ \mathbf{F}^*_{\mathcal{B}_\mathbf{q}} \right\} \tag{4.40}$$

in which the vector $\mathbf{F}^*_{\mathcal{B}_\mathbf{q}}$ is defined as

$$\mathbf{F}^*_{\mathcal{B}_\mathbf{q}} = \begin{aligned} &\left\{ \left(\mathbf{q}^*_{(1)} \cdot \mathbf{n}_{(1)}\right) S_{(1)}, \left(\mathbf{q}^*_{(2)} \cdot \mathbf{n}_{(2)}\right) S_{(2)}, \left(\mathbf{q}^*_{(3)} \cdot \mathbf{n}_{(3)}\right) \Gamma_{(3)}, \ldots, \right. \\ &\left. \left(\mathbf{q}^*_{(N_{\mathcal{B}_\mathbf{q}})} \cdot \mathbf{n}_{(N_{\mathcal{B}_\mathbf{q}})}\right) S_{(N_{\mathcal{B}_\mathbf{q}})} \right\}^T \end{aligned} \tag{4.41}$$

In region, $\mathcal{D} \cup \mathcal{R}$, the virtual field variation due to the external field source can be recast as:

$$\sum_{k=1}^{N_{\mathcal{D}}+N_{\mathcal{R}}} \delta\Theta_{(k)} s_{b(k)} V_{(k)} = \left\{ \delta\mathbf{V}^T_{\mathcal{D}} \quad \delta\mathbf{V}^T_{\mathcal{R}} \right\} \begin{Bmatrix} \mathbf{F}^*_{s\mathcal{D}} \\ \mathbf{F}^*_{s\mathcal{R}} \end{Bmatrix} \tag{4.42}$$

in which the vectors $\mathbf{F}^*_{s_{\mathcal{R}}}$ and $\mathbf{F}^*_{s_{\mathcal{D}}}$ respectively, are defined as:

$$\mathbf{F}^*_{s\mathcal{D}} = \left\{ s_{b(1)} V_{(1)}, , \ldots, , s_{b(p)} V_{(p)}, , , \ldots, , s_{b(N_{\mathcal{D}})} V_{(N_{\mathcal{D}})} \right\}^T \tag{4.43}$$

and

$$\mathbf{F}^*_{s\mathcal{R}} = \left\{ s_{b(q)} V_{(q)}, \ldots, s_{b(r)} V_{(r)}, \ldots, s_{b(N_{\mathcal{R}})} V_{(N_{\mathcal{R}})} \right\}^T \tag{4.44}$$

Combining Eqs. (4.36) to (4.40) and (4.42) results in the governing field equation in matrix form as:

$$\delta\mathbf{V}^T \left(\mathbf{F}_{\dot{\Theta}} + \mathbf{F} + \mathbf{F}_{\Theta} - \mathbf{F}^*\right) = \delta\mathbf{V}^T \left(\mathbf{C}\dot{\mathbf{V}} + \mathbf{K}\mathbf{V} - \mathbf{F}^*\right) = 0 \tag{4.45}$$

with

$$\mathbf{C} = \begin{bmatrix} \mathbf{C}_{\mathcal{DD}} & \mathbf{0} & \mathbf{0} & \mathbf{0} \\ \mathbf{0} & \mathbf{C}_{\mathcal{RR}} & \mathbf{0} & \mathbf{0} \\ \mathbf{0} & \mathbf{0} & \mathbf{0} & \mathbf{0} \\ \mathbf{0} & \mathbf{0} & \mathbf{0} & \mathbf{0} \end{bmatrix} \tag{4.46}$$

$$\mathbf{K} = \begin{bmatrix} \mathbf{K}_{\mathcal{D}\mathcal{D}} + \dot{\mathbf{C}}_{\mathcal{D}\mathcal{D}} & \mathbf{K}_{\mathcal{D}\mathcal{R}} & \mathbf{0} & \mathbf{0} \\ \mathbf{K}_{\mathcal{R}\mathcal{D}} & \mathbf{K}_{\mathcal{R}\mathcal{R}} + \dot{\mathbf{C}}_{\mathcal{R}\mathcal{R}} & \mathbf{K}_{\mathcal{R}\mathcal{B}_\mathbf{q}} & \mathbf{K}_{\mathcal{R}\mathcal{B}_\Theta} \\ \mathbf{0} & \mathbf{K}_{\mathcal{B}_\mathbf{q}\mathcal{R}} & \mathbf{K}_{\mathcal{B}_\mathbf{q}\mathcal{B}_\mathbf{q}} & \mathbf{K}_{\mathcal{B}_\mathbf{q}\mathcal{B}_\Theta} \\ \mathbf{0} & \mathbf{K}_{\mathcal{B}_\Theta\mathcal{R}} & \mathbf{K}_{\mathcal{B}_\Theta\mathcal{B}_\mathbf{q}} & \mathbf{K}_{\mathcal{B}_\Theta\mathcal{B}_\Theta} \end{bmatrix} \tag{4.47}$$

$$\mathbf{F}_{\dot{\Theta}} = \left\{\mathbf{F}_{\dot{\Theta}\mathcal{D}}, \mathbf{F}_{\dot{\Theta}\mathcal{R}}, \mathbf{0}, \mathbf{0}\right\}^T \tag{4.48a}$$

$$\mathbf{F} = \{-\mathbf{F}_{\mathcal{D}}, -\mathbf{F}_{\mathcal{R}}, \mathbf{0}, \mathbf{0}\}^T \tag{4.48b}$$

$$\mathbf{F}_{\Theta} = \{\mathbf{F}_{\Theta\mathcal{D}}, \mathbf{F}_{\Theta\mathcal{R}}, \mathbf{0}, \mathbf{0}\}^T \tag{4.48c}$$

and

$$\mathbf{F}^* = \left\{\mathbf{F}^*_{s_\mathcal{D}}, \mathbf{F}^*_{s_\mathcal{R}}, \mathbf{F}^*_{\mathcal{B}_\mathbf{q}}, \mathbf{0}\right\}^T \tag{4.48d}$$

4.6 Discrete Form of PD Internal Force Vector

The discrete form of virtual work due to the internal force in the interior region $\mathcal{D}$ and outer region $\mathcal{R}$ can be expressed as

$$\sum_{k=1}^{N_\mathcal{D}+N_\mathcal{R}} \delta\mathbf{u}^T_{(k)}\left(\nabla\cdot\mathbf{P}^\alpha_{(k)}\right)V_{(k)} = \sum_{k=1}^{N_\mathcal{D}+N_\mathcal{R}} \delta\mathbf{u}^T_{(k)}\mathbf{L}^\alpha_{(k)}V_{(k)} = \sum_{k=1}^{N_\mathcal{D}+N_\mathcal{R}} \delta\mathbf{u}^T_{(k)}\mathbf{F}^\alpha_{(k)} \tag{4.49}$$

in which the $\alpha = BB$, *OSB*, *NOSB*, and *PDDO*.

4.6.1 *BB-PD Interactions*

As derived in Sect. 3.12.1 of Chap. 3, the discrete form of force vector $\mathbf{F}^{BB}_{(k)}$ at node k can be obtained as:

$$\mathbf{F}^{BB}_{(k)} = \mathbf{L}^{BB}_{(k)}V_{(k)} = \sum_{j\in H_{(k)}} \boldsymbol{\Phi}^{BB}_{(k)(j)}\left(\mathbf{u}_{(j)} - \mathbf{u}_{(k)}\right) + \sum_{p\in \widetilde{H}_{(k)}} \widetilde{\boldsymbol{\Phi}}^{BB}_{(p)(k)}\left(\mathbf{u}_{(p)} - \mathbf{u}_{(k)}\right) \tag{4.50}$$

with

$$\boldsymbol{\Phi}^{BB}_{(k)(j)} = \widehat{\mu}_{(k)(j)}\frac{\mu}{2}\mathbf{S}\left(\mathbf{x}_{(k)}, \boldsymbol{\xi}_{(k)(j)}\right)V_{(j)}V_{(k)} \tag{4.51}$$

and

$$\widetilde{\boldsymbol{\Phi}}^{BB}_{(p)(k)} = \widehat{\mu}_{(p)(k)} \frac{\mu}{2} \mathbf{S}\left(\mathbf{x}_{(p)}, \boldsymbol{\xi}_{(p)(k)}\right) V_{(p)} V_{(k)} \tag{4.52}$$

The force vector $\mathbf{F}^{BB}_{(k)}$ in component form can be written as:

$$\begin{Bmatrix} F^{BB}_{1(k)} \\ F^{BB}_{2(k)} \\ 0 \\ 0 \end{Bmatrix} = \sum_{j \in H_{(k)}} \mathbf{k}^{BB}_{(k)(j)} \begin{Bmatrix} u_{1(k)} \\ u_{2(k)} \\ u_{1(j)} \\ u_{2(j)} \end{Bmatrix} + \sum_{p \in \widetilde{H}_{(k)}} \widetilde{\mathbf{k}}^{BB}_{(p)(k)} \begin{Bmatrix} u_{1(k)} \\ u_{2(k)} \\ u_{1(p)} \\ u_{2(p)} \end{Bmatrix} \tag{4.53}$$

where the corresponding stiffness matrices, $\mathbf{k}^{BB}_{(k)(j)}$ and $\widetilde{\mathbf{k}}^{BB}_{(p)(k)}$ are constructed as:

$$\mathbf{k}^{BB}_{(k)(j)} = \begin{bmatrix} -\boldsymbol{\Phi}^{BB}_{11(k)(j)} & -\boldsymbol{\Phi}^{BB}_{12(k)(j)} & \boldsymbol{\Phi}^{BB}_{11(k)(j)} & \boldsymbol{\Phi}^{BB}_{12(k)(j)} \\ -\boldsymbol{\Phi}^{BB}_{21(k)(j)} & -\boldsymbol{\Phi}^{BB}_{22(k)(j)} & \boldsymbol{\Phi}^{BB}_{21(k)(j)} & \boldsymbol{\Phi}^{BB}_{22(k)(j)} \\ 0 & 0 & 0 & 0 \\ 0 & 0 & 0 & 0 \end{bmatrix} \tag{4.54}$$

and

$$\widetilde{\mathbf{k}}^{BB}_{(p)(k)} = \begin{bmatrix} -\widetilde{\boldsymbol{\Phi}}^{BB}_{11(p)(k)} & -\widetilde{\boldsymbol{\Phi}}^{BB}_{12(p)(k)} & \widetilde{\boldsymbol{\Phi}}^{BB}_{11(p)(k)} & \widetilde{\boldsymbol{\Phi}}^{BB}_{12(p)(k)} \\ -\widetilde{\boldsymbol{\Phi}}^{BB}_{21(p)(k)} & -\widetilde{\boldsymbol{\Phi}}^{BB}_{22(p)(k)} & \widetilde{\boldsymbol{\Phi}}^{BB}_{21(p)(k)} & \widetilde{\boldsymbol{\Phi}}^{BB}_{22(p)(k)} \\ 0 & 0 & 0 & 0 \\ 0 & 0 & 0 & 0 \end{bmatrix} \tag{4.55}$$

For uniform discretization, the force vector $\mathbf{F}^{BB}_{(k)}$ can be written as:

$$\mathbf{F}^{BB}_{(k)} = \mathbf{L}^{BB}_{(k)} V_{(k)} = \sum_{j \in H_{(k)}} \left(\boldsymbol{\Phi}^{BB}_{(k)(j)} + \widetilde{\boldsymbol{\Phi}}^{BB}_{(j)(k)} \right) \left(\mathbf{u}_{(j)} - \mathbf{u}_{(k)} \right) \tag{4.56}$$

or

$$\begin{Bmatrix} F^{BB}_{1(k)} \\ F^{BB}_{2(k)} \\ 0 \\ 0 \end{Bmatrix} = \sum_{j \in H_{(k)}} \mathbf{k}^{BB}_{(k)(j)} \begin{Bmatrix} u_{1(k)} \\ u_{2(k)} \\ u_{1(j)} \\ u_{2(j)} \end{Bmatrix} \tag{4.57}$$

4.6.2 OSB-PD Interactions

For the OSB-PD interaction as derived in Sect. 3.12.2 of Chap. 3, the discrete from of force vector $\mathbf{F}_{(k)}^{OSB}$ at point k can be written as:

$$\begin{aligned}\mathbf{F}_{(k)}^{OSB} &= \mathbf{L}_{(k)}^{OSB} V_{(k)} \\ &= \sum_{j\in H_{(k)}} \left\{ \sum_{r=1,\, k_r\neq j}^{N_{(k)}} \mathbf{G}_{(k)(j)}^{(k)(k_r)} \mathbf{u}_{(k_r)} - \boldsymbol{\Phi}_{(k)} \mathbf{u}_{(k)} \right. \\ &\left. + \left(\mathbf{B}_{(k)(j)} + \mathbf{G}_{(k)(j)}^{(k)(j)}\right)\left(\mathbf{u}_{(j)} - \mathbf{u}_{(k)}\right) \right\} + \sum_{p\in \widetilde{H}_{(k)}} \left\{ \widetilde{\boldsymbol{\Phi}}_{(p)} \mathbf{u}_{(p)} \right. \\ &\left. - \sum_{s=1,\, p_s\neq k}^{N_{(p)}} \widetilde{\mathbf{G}}_{(p)(k)}^{(p)(p_s)} \mathbf{u}_{(p_s)} + \left(\widetilde{\mathbf{B}}_{(p)(k)} + \widetilde{\mathbf{G}}_{(p)(k)}^{(p)(k)}\right)\left(\mathbf{u}_{(p)} - \mathbf{u}_{(k)}\right) \right\}\end{aligned} \tag{4.58}$$

with

$$\mathbf{G}_{(k)(j)}^{(k)(k_r)} = \widehat{\mu}_{(k)(j)} \widehat{\mu}_{(k)(k_r)} a_{(k)} \left(\widehat{\mathbf{g}}(\mathbf{x}_{(k)}, \xi_{(k)(j)}) \otimes \mathbf{g}(\mathbf{x}_{(k)}, \xi_{(k)(k_r)})\right) V_{(k_r)} V_{(j)} V_{(k)} \tag{4.59}$$

$$\mathbf{B}_{(k)(j)} = \widehat{\mu}_{(k)(j)} \frac{n}{2(n+1)} \frac{\mu \delta_{(k)}}{\left|\boldsymbol{\xi}_{(k)(j)}\right|} \mathbf{S}\left(\mathbf{x}_{(k)}, \boldsymbol{\xi}_{(k)(j)}\right) V_{(j)} V_{(k)} \tag{4.60}$$

$$\boldsymbol{\Phi}_{(k)} = \sum_{r=1,\, k_r\neq j}^{N_{(k)}} \mathbf{G}_{(k)(j)}^{(k)(k_r)} \tag{4.61}$$

$$\widetilde{\mathbf{G}}_{(p)(k)}^{(p)(p_s)} = \widehat{\mu}_{(p)(k)} \widehat{\mu}_{(p)(p_s)} a_{(p)} \left(\widehat{\mathbf{g}}(\mathbf{x}_{(p)}, \xi_{(p)(k)}) \otimes \mathbf{g}(\mathbf{x}_{(p)}, \xi_{(p)(p_s)})\right) V_{(p_s)} V_{(p)} V_{(k)} \tag{4.62}$$

$$\widetilde{\mathbf{B}}_{(p)(k)} = \widehat{\mu}_{(p)(k)} \frac{n}{2(n+1)} \frac{\mu \delta_{(p)}}{\left|\boldsymbol{\xi}_{(p)(k)}\right|} \mathbf{S}\left(\mathbf{x}_{(p)}, \boldsymbol{\xi}_{(p)(k)}\right) V_{(p)} V_{(k)} \tag{4.63}$$

and

$$\widetilde{\boldsymbol{\Phi}}_{(p)} = \sum_{s=1,\, p_s\neq k}^{N_{(p)}} \widetilde{\mathbf{G}}_{(p)(k)}^{(p)(p_s)} \tag{4.64}$$

The force vector $\mathbf{F}_{(k)}^{OSB}$ in component form can be written as:

$$\begin{Bmatrix} F_{1(k)}^{OSB} \\ F_{2(k)}^{OSB} \\ 0 \\ 0 \end{Bmatrix} = \sum_{j \in H_{(k)}} \left(\mathbf{k}_{(k)(j)}^{OSB} \begin{Bmatrix} u_{1(k)} \\ u_{2(k)} \\ u_{1(j)} \\ u_{2(j)} \end{Bmatrix} + \sum_{r=1, k_r \neq j}^{N_{(k)}} \mathbf{k}_{(k)(j)}^{OSB(k)(k_r)} \begin{Bmatrix} u_{1(k)} \\ u_{2(k)} \\ u_{1(k_r)} \\ u_{2(k_r)} \end{Bmatrix} \right) + \sum_{p \in \widetilde{H}_{(k)}} \left(\widetilde{\mathbf{k}}_{(p)(k)}^{OSB} \begin{Bmatrix} u_{1(k)} \\ u_{2(k)} \\ u_{1(p)} \\ u_{2(p)} \end{Bmatrix} + \sum_{s=1, p_s \neq k}^{N_{(p)}} \widetilde{\mathbf{k}}_{(p)(k)}^{OSB(p)(p_s)} \begin{Bmatrix} u_{1(k)} \\ u_{2(k)} \\ u_{1(p_s)} \\ u_{2(p_s)} \end{Bmatrix} \right) \tag{4.65}$$

in which the stiffness matrices $\mathbf{k}_{(k)(j)}^{OSB}$, $\mathbf{k}_{(k)(j)}^{OSB(k)(k_r)}$, $\widetilde{\mathbf{k}}_{(p)(k)}^{OSB}$, and $\widetilde{\mathbf{k}}_{(p)(k)}^{OSB(p)(p_s)}$ are constructed as

$$\mathbf{k}_{(k)(j)}^{OSB} = \begin{bmatrix} -\Phi_{11(k)} - B_{11(k)(j)} - G_{11(k)(j)}^{(k)(j)} & -\Phi_{12(k)} - B_{12(k)(j)} - G_{12(k)(j)}^{(k)(j)} & B_{11(k)(j)} + G_{11(k)(j)}^{(k)(j)} & B_{12(k)(j)} + G_{12(k)(j)}^{(k)(j)} \\ -\Phi_{21(k)} - B_{21(k)(j)} - G_{21(k)(j)}^{(k)(j)} & -\Phi_{22(k)} - B_{22(k)(j)} - G_{22(k)(j)}^{(k)(j)} & B_{21(k)(j)} + G_{21(k)(j)}^{(k)(j)} & B_{22(k)(j)} + G_{22(k)(j)}^{(k)(j)} \\ 0 & 0 & 0 & 0 \\ 0 & 0 & 0 & 0 \end{bmatrix} \tag{4.66}$$

$$\mathbf{k}_{(k)(j)}^{OSB(k)(k_r)} = \begin{bmatrix} 0 & 0 & G_{11(k)(j)}^{(k)(k_r)} & G_{12(k)(j)}^{(k)(k_r)} \\ 0 & 0 & G_{21(k)(j)}^{(k)(k_r)} & G_{22(k)(j)}^{(k)(k_r)} \\ 0 & 0 & 0 & 0 \\ 0 & 0 & 0 & 0 \end{bmatrix} \tag{4.67}$$

$$\widetilde{\mathbf{k}}_{(p)(k)}^{OSB} = \begin{bmatrix} -\widetilde{B}_{11(p)(k)} - \widetilde{G}_{11(p)(k)}^{(p)(k)} & -\widetilde{B}_{12(p)(k)} - \widetilde{G}_{12(p)(k)}^{(p)(k)} & \widetilde{\Phi}_{11(p)} + \widetilde{B}_{11(p)(k)} + \widetilde{G}_{11(p)(k)}^{(p)(k)} & \widetilde{\Phi}_{12(p)} + \widetilde{B}_{12(p)(k)} + \widetilde{G}_{12(p)(k)}^{(p)(k)} \\ -\widetilde{B}_{21(p)(k)} - \widetilde{G}_{21(p)(k)}^{(p)(k)} & -\widetilde{B}_{22(p)(k)} - \widetilde{G}_{22(p)(k)}^{(p)(k)} & \widetilde{\Phi}_{21(p)} + \widetilde{B}_{21(p)(k)} + \widetilde{G}_{21(p)(k)}^{(p)(k)} & \widetilde{\Phi}_{22(p)} + \widetilde{B}_{22(p)(k)} + \widetilde{G}_{22(p)(k)}^{(p)(k)} \\ 0 & 0 & 0 & 0 \\ 0 & 0 & 0 & 0 \end{bmatrix} \tag{4.68}$$

$$\widetilde{\mathbf{k}}_{(p)(k)}^{OSB(p)(p_s)} = \begin{bmatrix} 0 & 0 & -\widetilde{G}_{11(p)(k)}^{(p)(p_s)} & -\widetilde{G}_{12(p)(k)}^{(p)(p_s)} \\ 0 & 0 & -\widetilde{G}_{21(p)(k)}^{(p)(p_s)} & -\widetilde{G}_{22(p)(k)}^{(p)(p_s)} \\ 0 & 0 & 0 & 0 \\ 0 & 0 & 0 & 0 \end{bmatrix} \tag{4.69}$$

For uniform discretization, the force vector $\mathbf{F}_{(k)}^{OSB}$ can be written as:

$$\mathbf{F}_{(k)}^{OSB} = \mathbf{L}_{(k)}^{OSB} V_{(k)}$$

$$= \sum_{j \in H_{(k)}} \left\{ \begin{array}{l} \sum\limits_{r=1,k_r \neq j}^{N_{(k)}} \mathbf{G}_{(k)(j)}^{(k)(k_r)} \mathbf{u}_{(k_r)} - \boldsymbol{\Phi}_{(k)} \mathbf{u}_{(k)} + \left(\mathbf{B}_{(k)(j)} + \mathbf{G}_{(k)(j)}^{(k)(j)} \right) \left(\mathbf{u}_{(j)} - \mathbf{u}_{(k)} \right) \\ + \widetilde{\boldsymbol{\Phi}}_{(j)} \mathbf{u}_{(j)} - \sum\limits_{s=1,j_s \neq k}^{N_{(j)}} \widetilde{\mathbf{G}}_{(j)(k)}^{(j)(j_s)} \mathbf{u}_{(j_s)} + \left(\widetilde{\mathbf{B}}_{(j)(k)} + \widetilde{\mathbf{G}}_{(j)(k)}^{(j)(k)} \right) \left(\mathbf{u}_{(j)} - \mathbf{u}_{(k)} \right) \end{array} \right\} \tag{4.70}$$

or

$$\left\{ \begin{array}{c} F_{1(k)}^{OSB} \\ F_{2(k)}^{OSB} \\ 0 \\ 0 \end{array} \right\} = \sum_{j \in H_{(k)}} \left(\left(\mathbf{k}_{(k)(j)}^{OSB} + \widetilde{\mathbf{k}}_{(j)(k)}^{OSB} \right) \left\{ \begin{array}{c} u_{1(k)} \\ u_{2(k)} \\ u_{1(j)} \\ u_{2(j)} \end{array} \right\} + \sum_{r=1,k_r \neq j}^{N_{(k)}} \mathbf{k}_{(k)(j)}^{OSB(k)(k_r)} \left\{ \begin{array}{c} u_{1(k)} \\ u_{2(k)} \\ u_{1(k_r)} \\ u_{2(k_r)} \end{array} \right\} \right.$$
$$\left. + \sum_{s=1,j_s \neq k}^{N_{(j)}} \widetilde{\mathbf{k}}_{(j)(k)}^{OSB(j)(j_s)} \left\{ \begin{array}{c} u_{1(k)} \\ u_{2(k)} \\ u_{1(j_s)} \\ u_{2(j_s)} \end{array} \right\} \right) \tag{4.71}$$

4.6.3 NOSB-PD Interactions

For the NOSB-PD interaction as derived in Sect. 3.12.3 of Chap. 3, the discrete from of force vector $\mathbf{F}_{(k)}^{NOSB}$ at point k can be written as:

$$\begin{aligned}\mathbf{F}_{(k)}^{NOSB} = \mathbf{L}_{(k)}^{NOSB} V_{(k)} &= \sum_{j \in H_{(k)}} \left(\sum_{r=1}^{N_{(k \cap j)}} \mathbf{H}_{\boldsymbol{\xi}(k)(j)}^{(k)(k_r)} \left(\mathbf{u}_{(k_r)} - \mathbf{u}_{(k)} \right) V_{(j)} \right) V_{(k)} \\ &+ \sum_{p \in \widetilde{H}_{(k)}} \left(\sum_{s=1}^{N_{(p \cap k)}} \mathbf{H}_{\boldsymbol{\xi}(p)(k)}^{(p)(p_s)} \left(\mathbf{u}_{(p)} - \mathbf{u}_{(p_s)} \right) V_{(p)} \right) V_{(k)} \\ &= \sum_{j \in H_{(k)}} \left(\sum_{r=1,\, k_r \neq j}^{N_{(k \cap j)}} \mathbf{G}_{(k)(j)}^{(k)(k_r)} \mathbf{u}_{(k_r)} - \boldsymbol{\Phi}_{(k)} \mathbf{u}_{(k)} + \mathbf{G}_{(k)(j)}^{(k)(j)} \left(\mathbf{u}_{(j)} - \mathbf{u}_{(k)} \right) \right) \\ &+ \sum_{p \in \widetilde{H}_{(k)}} \left(\widetilde{\boldsymbol{\Phi}}_{(p)} \mathbf{u}_{(p)} - \sum_{s=1,\, p_s \neq k}^{N_{(p \cap k)}} \widetilde{\mathbf{G}}_{(p)(k)}^{(p)(p_s)} \mathbf{u}_{(p_s)} + \widetilde{\mathbf{G}}_{(p)(k)}^{(p)(k)} \left(\mathbf{u}_{(p)} - \mathbf{u}_{(k)} \right) \right)\end{aligned} \tag{4.72}$$

with

$$\mathbf{G}_{(k)(j)}^{(k)(k_r)} = \mathbf{H}_{\boldsymbol{\xi}(k)(j)}^{(k)(k_r)} V_{(j)} V_{(k)} \tag{4.73}$$

$$\boldsymbol{\Phi}_{(k)} = \sum_{r=1,\, k_r \neq j}^{N_{(k \cap j)}} \mathbf{G}_{(k)(j)}^{(k)(k_r)} \tag{4.74}$$

$$\widetilde{\mathbf{G}}_{(p)(k)}^{(p)(p_s)} = \mathbf{H}_{\boldsymbol{\xi}(p)(k)}^{(p)(p_s)} V_{(p)} V_{(k)} \tag{4.75}$$

and

$$\widetilde{\boldsymbol{\Phi}}_{(p)} = \sum_{s=1,\, p_s \neq k}^{N_{(p \cap k)}} \widetilde{\mathbf{G}}_{(p)(k)}^{(p)(p_s)} \tag{4.76}$$

in which the matrices $\mathbf{H}_{\boldsymbol{\xi}(k)(j)}^{(k)(k_r)}$ and $\mathbf{H}_{\boldsymbol{\xi}(p)(k)}^{(p)(p_s)}$ are defined as:

$$\mathbf{H}_{\boldsymbol{\xi}(k)(j)}^{(k)(k_r)} = \widehat{\mu}_{(k)(j)} \widehat{\mu}_{(k)(k_r)} \begin{bmatrix} \left((\lambda + 2\mu) g_{(k)(k_r)}^{10} g_{(k)(j)}^{10} + \mu g_{(k)(k_r)}^{01} g_{(k)(j)}^{01} \right) & \left(\lambda g_{(k)(k_r)}^{01} g_{(k)(j)}^{10} + \mu g_{(k)(k_r)}^{10} g_{(k)(j)}^{01} \right) \\ \left(\lambda g_{(k)(k_r)}^{10} g_{(k)(j)}^{01} + \mu g_{(k)(k_r)}^{01} g_{(k)(j)}^{10} \right) & \left(\mu g_{(k)(k_r)}^{10} g_{(k)(j)}^{10} + (\lambda + 2\mu) g_{(k)(k_r)}^{01} g_{(k)(j)}^{01} \right) \end{bmatrix} V_{(k_r)} \tag{4.77}$$

and

$$\mathbf{H}_{\xi(p)(k)}^{(p)(p_s)} = \widehat{\mu}_{(p)(k)}\widehat{\mu}_{(p)(p_s)} \begin{bmatrix} \left((\lambda+2\mu)g_{(p)(p_s)}^{10}g_{(p)(k)}^{10} + \mu g_{(p)(p_s)}^{01}g_{(p)(k)}^{01}\right) & \left(\lambda g_{(p)(p_s)}^{01}g_{(p)(k)}^{10} + \mu g_{(p)(p_s)}^{10}g_{(p)(k)}^{01}\right) \\ \left(\lambda g_{(p)(p_s)}^{10}g_{(p)(k)}^{01} + \mu g_{(p)(p_s)}^{01}g_{(p)(k)}^{10}\right) & \left(\mu g_{(p)(p_s)}^{10}g_{(p)(k)}^{10} + (\lambda+2\mu)g_{(p)(p_s)}^{01}g_{(p)(k)}^{01}\right) \end{bmatrix} V_{(p_s)} \tag{4.78}$$

The force vector $\mathbf{F}_{(k)}^{NOSB}$ in component form can be written as:

$$\begin{Bmatrix} F_{1(k)}^{NOSB} \\ F_{2(k)}^{NOSB} \\ 0 \\ 0 \end{Bmatrix} = \sum_{j\in H_{(k)}} \left(\mathbf{k}_{(k)(j)}^{NOSB} \begin{Bmatrix} u_{1(k)} \\ u_{2(k)} \\ u_{1(j)} \\ u_{2(j)} \end{Bmatrix} + \sum_{r=1,\,k_r\neq j}^{N_{(k\cap j)}} \mathbf{k}_{(k)(j)}^{NOSB(k)(k_r)} \begin{Bmatrix} u_{1(k)} \\ u_{2(k)} \\ u_{1(k_r)} \\ u_{2(k_r)} \end{Bmatrix} \right) + \sum_{p\in\widetilde{H}_{(k)}} \left(\left(\widetilde{\mathbf{k}}_{(p)(k)}^{NOSB}\right) \begin{Bmatrix} u_{1(k)} \\ u_{2(k)} \\ u_{1(p)} \\ u_{2(p)} \end{Bmatrix} + \sum_{s=1,\,p_s\neq k}^{N_{(p\cap k)}} \widetilde{\mathbf{k}}_{(p)(k)}^{NOSB(p)(p_s)} \begin{Bmatrix} u_{1(k)} \\ u_{2(k)} \\ u_{1(p_s)} \\ u_{2(p_s)} \end{Bmatrix} \right) \tag{4.79}$$

with

$$\mathbf{k}_{(k)(j)}^{NOSB} = \begin{bmatrix} -\Phi_{11(k)} - G_{11(k)(j)}^{(k)(j)} & -\Phi_{12(k)} - G_{12(k)(j)}^{(k)(j)} & G_{11(k)(j)}^{(k)(j)} & G_{12(k)(j)}^{(k)(j)} \\ -\Phi_{21(k)} - G_{21(k)(j)}^{(k)(j)} & -\Phi_{22(k)} - G_{22(k)(j)}^{(k)(j)} & G_{21(k)(j)}^{(k)(j)} & G_{22(k)(j)}^{(k)(j)} \\ 0 & 0 & 0 & 0 \\ 0 & 0 & 0 & 0 \end{bmatrix} \tag{4.80}$$

$$\mathbf{k}_{(k)(j)}^{NOSB(k)(k_r)} = \begin{bmatrix} 0 & 0 & G_{11(k)(j)}^{(k)(k_r)} & G_{12(k)(j)}^{(k)(k_r)} \\ 0 & 0 & G_{21(k)(j)}^{(k)(k_r)} & G_{22(k)(j)}^{(k)(k_r)} \\ 0 & 0 & 0 & 0 \\ 0 & 0 & 0 & 0 \end{bmatrix} \tag{4.81}$$

$$\tilde{\mathbf{k}}_{(p)(k)}^{NOSB} = \begin{bmatrix} -\tilde{G}_{11(p)(k)}^{(p)(k)} & -\tilde{G}_{12(p)(k)}^{(p)(k)} & \tilde{\Phi}_{11(p)} + \tilde{G}_{11(p)(k)}^{(p)(k)} & \tilde{\Phi}_{12(p)} + \tilde{G}_{12(p)(k)}^{(p)(k)} \\ -\tilde{G}_{21(p)(k)}^{(p)(k)} & -\tilde{G}_{22(p)(k)}^{(p)(k)} & \tilde{\Phi}_{21(p)} + \tilde{G}_{21(p)(k)}^{(p)(k)} & \tilde{\Phi}_{22(p)} + \tilde{G}_{22(p)(k)}^{(p)(k)} \\ 0 & 0 & 0 & 0 \\ 0 & 0 & 0 & 0 \end{bmatrix} \tag{4.82}$$

and

$$\tilde{\mathbf{k}}_{(p)(k)}^{NOSB(p)(p_s)} = \begin{bmatrix} 0 & 0 & -\tilde{G}_{11(p)(k)}^{(p)(p_s)} & -\tilde{G}_{12(p)(k)}^{(p)(p_s)} \\ 0 & 0 & -\tilde{G}_{21(p)(k)}^{(p)(p_s)} & -\tilde{G}_{22(p)(k)}^{(p)(p_s)} \\ 0 & 0 & 0 & 0 \\ 0 & 0 & 0 & 0 \end{bmatrix} \tag{4.83}$$

For uniform discretization, the force vector $\mathbf{F}_{(k)}^{NOSB}$ can be written as:

$$\mathbf{F}_{(k)}^{NOSB} = \mathbf{L}_{(k)}^{NOSB} V_{(k)} = \sum_{j \in H_{(k)}} \left\{ \begin{array}{l} \sum\limits_{r=1,\, k_r \neq j}^{N_{(k \cap j)}} \mathbf{G}_{(k)(j)}^{(k)(k_r)} \mathbf{u}_{(k_r)} - \mathbf{\Phi}_{(k)} \mathbf{u}_{(k)} + \mathbf{G}_{(k)(j)}^{(k)(j)} \left(\mathbf{u}_{(j)} - \mathbf{u}_{(k)} \right) \\ + \tilde{\mathbf{\Phi}}_{(j)} \mathbf{u}_{(j)} - \sum\limits_{s=1,\, p_s \neq k}^{N_{(j \cap k)}} \tilde{\mathbf{G}}_{(j)(k)}^{(j)(j_s)} \mathbf{u}_{(j_s)} + \tilde{\mathbf{G}}_{(j)(k)}^{(j)(k)} \left(\mathbf{u}_{(j)} - \mathbf{u}_{(k)} \right) \end{array} \right\} \tag{4.84}$$

or

$$\begin{Bmatrix} F_{1(k)}^{NOSB} \\ F_{2(k)}^{NOSB} \\ 0 \\ 0 \end{Bmatrix} = \sum_{j \in H_{(k)}} \left(\begin{array}{l} \left(\mathbf{k}_{(k)(j)}^{NOSB} + \tilde{\mathbf{k}}_{(j)(k)}^{NOSB} \right) \begin{Bmatrix} u_{1(k)} \\ u_{2(k)} \\ u_{1(j)} \\ u_{2(j)} \end{Bmatrix} + \sum\limits_{r=1,\, k_r \neq j}^{N_{(k \cap j)}} \mathbf{k}_{(k)(j)}^{NOSB(k)(k_r)} \begin{Bmatrix} u_{1(k)} \\ u_{2(k)} \\ u_{1(k_r)} \\ u_{2(k_r)} \end{Bmatrix} \\ + \sum\limits_{s=1,\, j_s \neq k}^{N_{(j \cap k)}} \tilde{\mathbf{k}}_{(j)(k)}^{NOSB(j)(j_s)} \begin{Bmatrix} u_{1(k)} \\ u_{2(k)} \\ u_{1(j_s)} \\ u_{2(j_s)} \end{Bmatrix} \end{array} \right) \tag{4.85}$$

4.6.4 PDDO Interactions

For the PDDO interaction as derived in Sect. 3.12.4 of Chap. 3, the discrete form of force vector $\mathbf{F}_{(k)}^{PDDO}$ at the point k can be written as:

$$\begin{aligned}\mathbf{F}_{(k)}^{PDDO} &= \mathbf{L}_{(k)}^{PDDO} V_{(k)} \\ &= \sum_{j\in H_{(k)}} \left(\sum_{s=1}^{N_{(j)}} \mathbf{H}_{(k)(j)}^{(j)(j_s)} \left(\mathbf{u}_{(j_s)} - \mathbf{u}_{(j)}\right) + \sum_{r=1}^{N_{(k)}} \mathbf{H}_{(k)(j)}^{(k)(k_r)} \left(\mathbf{u}_{(k)} - \mathbf{u}_{(k_r)}\right) \right) V_{(j)} V_{(k)} \\ &= \sum_{j\in H_{(k)}} \left(\begin{array}{c} -\widetilde{\mathbf{\Phi}}_{(j)}\mathbf{u}_{(j)} + \sum\limits_{s=1, j_s \neq k}^{N_{(j)}} \widetilde{\mathbf{G}}_{(k)(j)}^{(j)(j_s)} \mathbf{u}_{(j_s)} + \widetilde{\mathbf{G}}_{(k)(j)}^{(j)(k)} \left(\mathbf{u}_{(k)} - \mathbf{u}_{(j)}\right) \\ -\sum\limits_{r=1, k_r \neq j}^{N_{(k)}} \mathbf{G}_{(k)(j)}^{(k)(k_r)} \mathbf{u}_{(k_r)} + \mathbf{\Phi}_{(k)}\mathbf{u}_{(k)} + \mathbf{G}_{(k)(j)}^{(k)(j)} \left(\mathbf{u}_{(k)} - \mathbf{u}_{(j)}\right) \end{array} \right)\end{aligned} \tag{4.86}$$

with

$$\mathbf{G}_{(k)(j)}^{(k)(k_r)} = \mathbf{H}_{(k)(j)}^{(k)(k_r)} V_{(j)} V_{(k)} \tag{4.87}$$

$$\mathbf{\Phi}_{(k)} = \sum_{r=1, k_r \neq j}^{N_{(k)}} \mathbf{G}_{(k)(j)}^{(k)(k_r)} \tag{4.88}$$

and

$$\widetilde{\mathbf{G}}_{(k)(j)}^{(j)(j_s)} = \mathbf{H}_{(k)(j)}^{(j)(j_s)} V_{(j)} V_{(k)} \tag{4.89}$$

$$\widetilde{\mathbf{\Phi}}_{(j)} = \sum_{s=1, j_s \neq k}^{N_{(j)}} \widetilde{\mathbf{G}}_{(j)(k)}^{(j)(j_s)} \tag{4.90}$$

in which the matrices $\mathbf{H}_{(k)(j)}^{(j)(j_s)}$ and $\mathbf{H}_{(k)(j)}^{(k)(k_r)}$ are defined as:

$$\mathbf{H}_{(k)(j)}^{(j)(j_s)} = \widehat{\mu}_{(k)(j)}\widehat{\mu}_{(j)(j_s)} \begin{bmatrix} \left((\lambda+2\mu) g_{(j)(j_s)}^{10} g_{(k)(j)}^{10} + \mu g_{(j)(j_s)}^{01} g_{(k)(j)}^{01}\right) & \left(\lambda g_{(j)(j_s)}^{01} g_{(k)(j)}^{10} + \mu g_{(j)(j_s)}^{10} g_{(k)(j)}^{01}\right) \\ \left(\lambda g_{(j)(j_s)}^{10} g_{(k)(j)}^{01} + \mu g_{(j)(j_s)}^{01} g_{(k)(j)}^{10}\right) & \left(\mu g_{(j)(j_s)}^{10} g_{(k)(j)}^{10} + (\lambda+2\mu) g_{(j)(j_s)}^{01} g_{(k)(j)}^{01}\right) \end{bmatrix} V_{(j_s)} \tag{4.91}$$

and

$$\mathbf{H}_{(k)(j)}^{(k)(k_r)} = \widehat{\mu}_{(k)(j)}\widehat{\mu}_{(k)(k_r)} \begin{bmatrix} \left((\lambda+2\mu)g_{(k)(k_r)}^{10}g_{(k)(j)}^{10} + \mu g_{(k)(k_r)}^{01}g_{(k)(j)}^{01}\right) & \left(\lambda g_{(k)(k_r)}^{01}g_{(k)(j)}^{10} + \mu g_{(k)(k_r)}^{10}g_{(k)(j)}^{01}\right) \\ \left(\lambda g_{(k)(k_r)}^{10}g_{(k)(j)}^{01} + \mu g_{(k)(k_r)}^{01}g_{(k)(j)}^{10}\right) & \left(\mu g_{(k)(k_r)}^{10}g_{(k)(j)}^{10} + (\lambda+2\mu)g_{(k)(k_r)}^{01}g_{(k)(j)}^{01}\right) \end{bmatrix} V_{(k_r)} \tag{4.92}$$

The force vector $\mathbf{F}_{(k)}^{PDDO}$ in component form can be written as:

$$\begin{Bmatrix} F_{1(k)}^{PDDO} \\ F_{2(k)}^{PDDO} \\ 0 \\ 0 \end{Bmatrix} = \sum_{j\in H_{(k)}} \left(\mathbf{k}_{(k)(j)}^{PDDO} \begin{Bmatrix} u_{1(k)} \\ u_{2(k)} \\ u_{1(j)} \\ u_{2(j)} \end{Bmatrix} + \sum_{r=1,\, k_r\neq j}^{N_{(k)}} \mathbf{k}_{(k)(j)}^{PDDO(k)(k_r)} \begin{Bmatrix} u_{1(k)} \\ u_{2(k)} \\ u_{1(k_r)} \\ u_{2(k_r)} \end{Bmatrix} + \sum_{s=1,\, j_s\neq k}^{N_{(j)}} \mathbf{k}_{(k)(j)}^{PDDO(j)(j_s)} \begin{Bmatrix} u_{1(k)} \\ u_{2(k)} \\ u_{1(j_s)} \\ u_{2(j_s)} \end{Bmatrix} \right) \tag{4.93}$$

with

$$\mathbf{k}_{(k)(j)}^{PDDO} = \begin{bmatrix} \Phi_{11(k)} + G_{11(k)(j)}^{(k)(j)} + \widetilde{G}_{11(k)(j)}^{(j)(k)} & \Phi_{12(k)} + G_{12(k)(j)}^{(k)(j)} + \widetilde{G}_{12(k)(j)}^{(j)(k)} & -G_{11(k)(j)}^{(k)(j)} - \widetilde{G}_{11(k)(j)}^{(j)(k)} - \widetilde{\Phi}_{12(j)} & -G_{12(k)(j)}^{(k)(j)} - \widetilde{G}_{12(k)(j)}^{(j)(k)} - \widetilde{\Phi}_{12(j)} \\ \Phi_{21(k)} + G_{21(k)(j)}^{(k)(j)} + \widetilde{G}_{21(k)(j)}^{(j)(k)} & \Phi_{22(k)} + G_{22(k)(j)}^{(k)(j)} + \widetilde{G}_{22(k)(j)}^{(j)(k)} & -G_{21(k)(j)}^{(21)} - \widetilde{G}_{21(k)(j)}^{(j)(k)} - \widetilde{\Phi}_{21(j)} & -G_{22(k)(j)}^{(k)(j)} - \widetilde{G}_{22(k)(j)}^{(j)(k)} - \widetilde{\Phi}_{22(j)} \\ 0 & 0 & 0 & 0 \\ 0 & 0 & 0 & 0 \end{bmatrix} \tag{4.94}$$

$$\mathbf{k}_{(k)(j)}^{PDDO(k)(k_r)} = \begin{bmatrix} 0 & 0 & -G_{11(k)(j)}^{(k)(k_r)} & -G_{12(k)(j)}^{(k)(k_r)} \\ 0 & 0 & -G_{21(k)(j)}^{(k)(k_r)} & -G_{22(k)(j)}^{(k)(k_r)} \\ 0 & 0 & 0 & 0 \\ 0 & 0 & 0 & 0 \end{bmatrix} \tag{4.95}$$

and

$$\tilde{\mathbf{k}}_{(k)(j)}^{PDDO(j)(j_s)} = \begin{bmatrix} 0 & 0 & \tilde{G}_{11(k)(j)}^{(j)(j_s)} & \tilde{G}_{12(k)(j)}^{(j)(j_s)} \\ 0 & 0 & \tilde{G}_{21(k)(j)}^{(j)(j_s)} & \tilde{G}_{22(k)(j)}^{(j)(j_s)} \\ 0 & 0 & 0 & 0 \\ 0 & 0 & 0 & 0 \end{bmatrix} \tag{4.96}$$

4.6.5 PD Internal Traction

For the closed boundary of the PD region, Γ is split into $N_{\mathscr{B}}$ segments with $\Gamma_{(\ell)}$ representing the length of ℓ-th boundary segment. The numerical evaluation of the line integral in Eq. (4.7) is achieved by dividing the boundary segments $\Gamma_{\mathscr{B}_\sigma}$ and $\Gamma_{\mathscr{B}_\mathbf{u}}$ into $N_{\mathscr{B}_\sigma}$ and $N_{\mathscr{B}_\mathbf{u}}$ line segments, respectively. As shown in Fig. 4.7, s is the natural (local) coordinate between points ℓ_1 and ℓ_2 of the ℓ-th line segment with a unit normal vector, $\mathbf{n}^{(\ell)}$. The differential line element, $d\Gamma$, along the boundary can be related to the natural coordinate as $d\Gamma = \Gamma_{(\ell)} ds$.

As suggested by Madenci et al. (2018), it is assumed that the displacement vector, $\mathbf{u}_{(\ell)}$, and the PD form of the traction vector, $\mathbf{T}_{(\ell)}^{PD}$, vary linearly between the beginning and end points ℓ_1 and ℓ_2, respectively, of the ℓ-th line segment as:

$$\mathbf{u}_{(\ell)} = (1-s)\mathbf{u}_{(\ell_1)} + s\mathbf{u}_{(\ell_2)}, \tag{4.97}$$

and

$$\mathbf{T}_{(\ell)}^{PD} = (1-s)\mathbf{T}_{(\ell_1)}^{PD} + s\mathbf{T}_{(\ell_2)}^{PD}, \tag{4.98}$$

The PD form of traction vector $\mathbf{T}^{PD}$ at the beginning and end points ℓ_1 and ℓ_2 can be expressed as:

$$\mathbf{T}_{(\ell_1)}^{PD} = \boldsymbol{\sigma}_{(\ell_1)}^{PD}\mathbf{n}_{(\ell)} = \sum_{p=1}^{N_{(\ell_1)}} \mathbf{H}_{(\ell_1)(\ell_{1_p})}\left(\mathbf{u}_{(\ell_{1_p})} - \mathbf{u}_{(\ell_1)}\right) \tag{4.99}$$

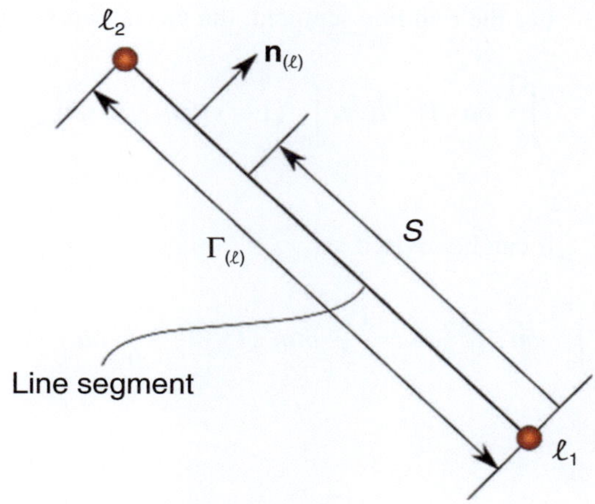

Fig. 4.7 The local coordinate in the ℓ-th line segment along the boundary

and

$$\mathbf{T}^{PD}_{(\ell_2)} = \boldsymbol{\sigma}^{PD}_{(\ell_2)}\mathbf{n}_{(\ell)} = \sum_{r=1}^{N_{(\ell_2)}} \mathbf{H}_{(\ell_2)(\ell_{2_r})}\left(\mathbf{u}_{(\ell_{2_r})} - \mathbf{u}_{(\ell_2)}\right) \tag{4.100}$$

with the matrices $\mathbf{H}_{(\ell_1)(\ell_{1_p})}$ and $\mathbf{H}_{(\ell_2)(\ell_{2_r})}$ defined as:

$$\mathbf{H}_{(\ell_1)(\ell_{1_p})} = \widehat{\mu}_{(\ell_1)(\ell_{1_p})}\begin{bmatrix} \left((\lambda+2\mu)g^{10}_{(\ell_1)(\ell_{1_p})}n^{(\ell)}_{(1)} + \mu g^{01}_{(\ell_1)(\ell_{1_p})}n^{(\ell)}_{(2)}\right) & \left(\lambda g^{01}_{(\ell_1)(\ell_{1_p})}n^{(\ell)}_{(1)} + \mu g^{10}_{(\ell_1)(\ell_{1_p})}n^{(\ell)}_{(2)}\right) \\ \left(\lambda g^{10}_{(\ell_1)(\ell_{1_p})}n^{(\ell)}_{(2)} + \mu g^{01}_{(\ell_1)(\ell_{1_p})}n^{(\ell)}_{(1)}\right) & \left(\mu g^{10}_{(\ell_1)(\ell_{1_p})}n^{(\ell)}_{(1)} + (\lambda+2\mu)g^{01}_{(\ell_1)(\ell_{1_p})}n^{(\ell)}_{(2)}\right) \end{bmatrix} V_{(j)} \tag{4.101}$$

and

$$\mathbf{H}_{(\ell_2)(\ell_{2_r})} = \widehat{\mu}_{(\ell_2)(\ell_{2_r})}\begin{bmatrix} \left((\lambda+2\mu)g^{10}_{(\ell_2)(\ell_{2_r})}n^{(\ell)}_{(1)} + \mu g^{01}_{(\ell_2)(\ell_{2_r})}n^{(\ell)}_{(2)}\right) & \left(\lambda g^{01}_{(\ell_2)(\ell_{2_r})}n^{(\ell)}_{(1)} + \mu g^{10}_{(\ell_2)(\ell_{2_r})}n^{(\ell)}_{(2)}\right) \\ \left(\lambda g^{10}_{(\ell_2)(\ell_{2_r})}n^{(\ell)}_{(2)} + \mu g^{01}_{(\ell_2)(\ell_{2_r})}n^{(\ell)}_{(1)}\right) & \left(\mu g^{10}_{(\ell_2)(\ell_{2_r})}n^{(\ell)}_{(1)} + (\lambda+2\mu)g^{01}_{(\ell_2)(\ell_{2_r})}n^{(\ell)}_{(2)}\right) \end{bmatrix} V_{(j)} \tag{4.102}$$

For the ℓ-th line segment, the virtual work can be evaluated as:

$$\int_0^{\Gamma_{(\ell)}} \delta\mathbf{u}\cdot\mathbf{T}^{PD}d\Gamma = \int_0^1 \left[(1-s)\delta\mathbf{u}_{(\ell_1)} + s\delta\mathbf{u}_{(\ell_2)}\right]\cdot\left[(1-s)\mathbf{T}^{PD}_{(\ell_1)} + s\mathbf{T}^{PD}_{(\ell_2)}\right]\Gamma_{(\ell)}ds. \tag{4.103}$$

It can be revised as:

$$\int_0^{\Gamma_{(\ell)}} \delta\mathbf{u}\cdot\mathbf{T}^{PD}d\Gamma = \frac{\Gamma_{(\ell)}}{3}\delta\mathbf{u}_{(\ell_1)}\mathbf{T}^{PD}_{(\ell_1)} + \frac{\Gamma_{(\ell)}}{6}\delta\mathbf{u}_{(\ell_1)}\mathbf{T}^{PD}_{(\ell_2)} + \frac{\Gamma_{(\ell)}}{6}\mathbf{u}_{(\ell_2)}\mathbf{T}^{PD}_{(\ell_1)} + \frac{\Gamma_{(\ell)}}{3}\mathbf{u}_{(\ell_2)}\mathbf{T}^{PD}_{(\ell_2)} \tag{4.104}$$

or

$$\int_0^{\Gamma_{(\ell)}} \delta\mathbf{u}\cdot\mathbf{T}^{PD}d\Gamma = \left\{ \delta\mathbf{u}^T_{(\ell_1)} \quad \delta\mathbf{u}^T_{(\ell_2)} \right\} \left\{ \begin{array}{c} \mathbf{F}^{PD}_{(\ell_1)} \\ \mathbf{F}^{PD}_{(\ell_2)} \end{array} \right\} \tag{4.105}$$

with the force vectors $\mathbf{F}^{PD}_{(\ell_1)}$ and $\mathbf{F}^{PD}_{(\ell_2)}$ at beginning and end points ℓ_1 and ℓ_2 defined as:

$$\left\{ \begin{array}{c} \mathbf{F}^{PD}_{(\ell_1)} \\ \mathbf{F}^{PD}_{(\ell_2)} \end{array} \right\} = \begin{bmatrix} \frac{\Gamma_{(\ell)}}{3} & \frac{\Gamma_{(\ell)}}{6} \\ \frac{\Gamma_{(\ell)}}{6} & \frac{\Gamma_{(\ell)}}{3} \end{bmatrix} \left\{ \begin{array}{c} \mathbf{T}^{PD}_{(\ell_1)} \\ \mathbf{T}^{PD}_{(\ell_2)} \end{array} \right\} \tag{4.106}$$

or

$$\left\{ \begin{array}{c} \mathbf{F}^{PD}_{(\ell_1)} \\ \mathbf{F}^{PD}_{(\ell_2)} \end{array} \right\} = \left\{ \begin{array}{c} \sum\limits_{p=1}^{N_{(\ell_1)}} \frac{\Gamma_{(\ell)}\mathbf{H}_{(\ell_1)(\ell_{1_p})}}{3}\left(\mathbf{u}_{(\ell_{1_p})} - \mathbf{u}_{(\ell_1)}\right) + \sum\limits_{r=1}^{N_{(\ell_2)}} \frac{\Gamma_{(\ell)}\mathbf{H}_{(\ell_2)(\ell_{2_r})}}{6}\left(\mathbf{u}_{(\ell_{2_r})} - \mathbf{u}_{(\ell_2)}\right) \\ \sum\limits_{r=1}^{N_{(\ell_2)}} \frac{\Gamma_{(\ell)}\mathbf{H}_{(\ell_2)(\ell_{2_r})}}{3}\left(\mathbf{u}_{(\ell_{2_r})} - \mathbf{u}_{(\ell_2)}\right) + \sum\limits_{p=1}^{N_{(\ell_1)}} \frac{\Gamma_{(\ell)}\mathbf{H}_{(\ell_1)(\ell_{1_p})}}{6}\left(\mathbf{u}_{(\ell_{1_p})} - \mathbf{u}_{(\ell_1)}\right) \end{array} \right\} \tag{4.107}$$

Also, it can be recast in component form as:

$$\begin{Bmatrix} F^{PD}_{1(\ell_1)} \\ F^{PD}_{2(\ell_1)} \\ F^{PD}_{1(\ell_2)} \\ F^{PD}_{1(\ell_2)} \end{Bmatrix} = \sum_{p=1,}^{N_{(\ell_1)}} \mathbf{k}^{\boldsymbol{\sigma}}_{(\ell_1)\left(\ell_{1_p}\right)} \begin{Bmatrix} u_{1(\ell_1)} \\ u_{2(\ell_1)} \\ u_{1\left(\ell_{1_p}\right)} \\ u_{2\left(\ell_{1_p}\right)} \end{Bmatrix} + \mathbf{k}^{\boldsymbol{\sigma}}_{(\ell_1)(\ell_2)} \begin{Bmatrix} u_{1(\ell_1)} \\ u_{2(\ell_1)} \\ u_{1(\ell_2)} \\ u_{2(\ell_2)} \end{Bmatrix}$$
$$+ \sum_{r=1,\, \ell_{2_r} \neq \ell_1}^{N_{(\ell_2)}} \widehat{\mathbf{k}}^{\boldsymbol{\sigma}}_{(\ell_2)(\ell_{2_r})} \begin{Bmatrix} u_{1(\ell_1)} \\ u_{2(\ell_1)} \\ u_{(\ell_{2_r})} \\ u_{(\ell_{2_r})} \end{Bmatrix} \tag{4.108}$$

and

$$\begin{Bmatrix} F^{PD}_{1(\ell_2)} \\ F^{PD}_{2(\ell_2)} \\ F^{PD}_{1(\ell_1)} \\ F^{PD}_{1(\ell_1)} \end{Bmatrix} = \sum_{r=1}^{N_{(\ell_2)}} \mathbf{k}^{\boldsymbol{\sigma}}_{(\ell_2)(\ell_{2_r})} \begin{Bmatrix} u_{1(\ell_2)} \\ u_{2(\ell_2)} \\ u_{1(\ell_{2_r})} \\ u_{2(\ell_{2_r})} \end{Bmatrix} + \mathbf{k}^{\boldsymbol{\sigma}}_{(\ell_2)(\ell_1)} \begin{Bmatrix} u_{1(\ell_2)} \\ u_{2(\ell_2)} \\ u_{1(\ell_1)} \\ u_{2(\ell_1)} \end{Bmatrix}$$
$$+ \sum_{p=1,\, \ell_{1_p} \neq \ell_2}^{N_{(\ell_1)}} \widehat{\mathbf{k}}^{\boldsymbol{\sigma}}_{(\ell_1)\left(\ell_{1_p}\right)} \begin{Bmatrix} u_{1(\ell_2)} \\ u_{2(\ell_2)} \\ u_{1\left(\ell_{1_p}\right)} \\ u_{2\left(\ell_{1_p}\right)} \end{Bmatrix} \tag{4.109}$$

in which the matrices $\mathbf{k}^{\boldsymbol{\sigma}}_{(\ell_1)\left(\ell_{1_p}\right)}$, $\mathbf{k}^{\boldsymbol{\sigma}}_{(\ell_1)(\ell_2)}$, $\widehat{\mathbf{k}}^{\boldsymbol{\sigma}}_{(\ell_2)(\ell_{2_r})}$ for $\ell_1 - \ell_2$ interactions are defined as:

$$\mathbf{k}^{\boldsymbol{\sigma}}_{(\ell_1)\left(\ell_{1_p}\right)} = \begin{bmatrix} -\dfrac{\Gamma_{(\ell)} H^{(11)}_{(\ell_1)\left(\ell_{1_p}\right)}}{3} & -\dfrac{\Gamma_{(\ell)} H^{(12)}_{(\ell_1)\left(\ell_{1_p}\right)}}{3} & \dfrac{\Gamma_{(\ell)} H^{(11)}_{(\ell_1)\left(\ell_{1_p}\right)}}{3} & \dfrac{\Gamma_{(\ell)} H^{(12)}_{(\ell_1)\left(\ell_{1_p}\right)}}{3} \\ -\dfrac{\Gamma_{(\ell)} H^{(21)}_{(\ell_1)\left(\ell_{1_p}\right)}}{3} & -\dfrac{\Gamma_{(\ell)} H^{(22)}_{(\ell_1)\left(\ell_{1_p}\right)}}{3} & \dfrac{\Gamma_{(\ell)} H^{(21)}_{(\ell_1)\left(\ell_{1_p}\right)}}{3} & \dfrac{\Gamma_{(\ell)} H^{(22)}_{(\ell_1)\left(\ell_{1_p}\right)}}{3} \\ 0 & 0 & 0 & 0 \\ 0 & 0 & 0 & 0 \end{bmatrix} \tag{4.110}$$

$$\mathbf{k}^{\sigma}_{(\ell_1)(\ell_2)} = \begin{bmatrix} \frac{\Gamma_{(\ell)}H^{(11)}_{(\ell_2)(\ell_1)}}{6} & \frac{\Gamma_{(\ell)}H^{(12)}_{(\ell_2)(\ell_1)}}{6} & -\sum_{r=1}^{N_{(\ell_2)}}\frac{\Gamma_{(\ell)}H^{(11)}_{(\ell_2)(\ell_{2_r})}}{6} & -\sum_{r=1}^{N_{(\ell_2)}}\frac{\Gamma_{(\ell)}H^{(12)}_{(\ell_2)(\ell_{2_r})}}{6} \\ \frac{\Gamma_{(\ell)}H^{(21)}_{(\ell_2)(\ell_1)}}{6} & \frac{\Gamma_{(\ell)}H^{(22)}_{(\ell_2)(\ell_1)}}{6} & -\sum_{r=1}^{N_{(\ell_2)}}\frac{\Gamma_{(\ell)}H^{(21)}_{(\ell_2)(\ell_{2_r})}}{6} & -\sum_{r=1}^{N_{(\ell_2)}}\frac{\Gamma_{(\ell)}H^{(22)}_{(\ell_2)(\ell_{2_r})}}{6} \\ 0 & 0 & 0 & 0 \\ 0 & 0 & 0 & 0 \end{bmatrix} \quad (4.111)$$

$$\widehat{\mathbf{k}}^{\sigma}_{(\ell_2)(\ell_{2_r})} = \begin{bmatrix} 0 & 0 & \frac{\Gamma_{(\ell)}H^{(11)}_{(\ell_2)(\ell_{2_r})}}{6} & \frac{\Gamma_{(\ell)}H^{(12)}_{(\ell_2)(\ell_{2_r})}}{6} \\ 0 & 0 & \frac{\Gamma_{(\ell)}H^{(21)}_{(\ell_2)(\ell_{2_r})}}{6} & \frac{\Gamma_{(\ell)}H^{(22)}_{(\ell_2)(\ell_{2_r})}}{6} \\ 0 & 0 & 0 & 0 \\ 0 & 0 & 0 & 0 \end{bmatrix} \quad (4.112)$$

The stiffness matrices $\mathbf{k}^{\sigma}_{(\ell_2)(\ell_{2_r})}$, $\mathbf{k}^{\sigma}_{(\ell_2)(\ell_1)}$, $\widehat{\mathbf{k}}^{\sigma}_{(\ell_1)(\ell_{1_p})}$ for $\ell_2 - \ell_1$ interactions are defined as:

$$\mathbf{k}^{\sigma}_{(\ell_2)(\ell_{2_r})} = \begin{bmatrix} -\frac{\Gamma_{(\ell)}H^{(11)}_{(\ell_2)(\ell_{2_r})}}{3} & -\frac{\Gamma_{(\ell)}H^{(11)}_{(\ell_2)(\ell_{2_r})}}{3} & \frac{\Gamma_{(\ell)}H^{(11)}_{(\ell_2)(\ell_{2_r})}}{3} & \frac{\Gamma_{(\ell)}H^{(11)}_{(\ell_2)(\ell_{2_r})}}{3} \\ -\frac{\Gamma_{(\ell)}H^{(11)}_{(\ell_2)(\ell_{2_r})}}{3} & -\frac{\Gamma_{(\ell)}H^{(11)}_{(\ell_2)(\ell_{2_r})}}{3} & \frac{\Gamma_{(\ell)}H^{(11)}_{(\ell_2)(\ell_{2_r})}}{3} & \frac{\Gamma_{(\ell)}H^{(11)}_{(\ell_2)(\ell_{2_r})}}{3} \\ 0 & 0 & 0 & 0 \\ 0 & 0 & 0 & 0 \end{bmatrix} \quad (4.113)$$

$$\mathbf{k}^{\sigma}_{(\ell_2)(\ell_1)} = \begin{bmatrix} \frac{\Gamma_{(\ell)}H^{(11)}_{(\ell_1)(\ell_2)}}{6} & \frac{\Gamma_{(\ell)}H^{(12)}_{(\ell_1)(\ell_2)}}{6} & -\sum_{p=1}^{N_{(\ell_1)}}\frac{\Gamma_{(\ell)}H^{(11)}_{(\ell_1)(\ell_{1_p})}}{6} & -\sum_{j=1}^{N_{(\ell_1)}}\frac{\Gamma_{(\ell)}H^{(12)}_{(\ell_1)(\ell_{1_p})}}{6} \\ \frac{\Gamma_{(\ell)}H^{(21)}_{(\ell_1)(\ell_2)}}{6} & \frac{\Gamma_{(\ell)}H^{(22)}_{(\ell_1)(\ell_2)}}{6} & -\sum_{p=1}^{N_{(\ell_1)}}\frac{\Gamma_{(\ell)}H^{(21)}_{(\ell_1)(\ell_{1_p})}}{6} & -\sum_{j=1}^{N_{(\ell_1)}}\frac{\Gamma_{(\ell)}H^{(22)}_{(\ell_1)(\ell_{1_p})}}{6} \\ 0 & 0 & 0 & 0 \\ 0 & 0 & 0 & 0 \end{bmatrix} \quad (4.114)$$

and

$$\widehat{\mathbf{k}}^{\boldsymbol{\sigma}}_{(\ell_1)(\ell_{1_p})} = \begin{bmatrix} 0 & 0 & \dfrac{\Gamma_{(\ell)} H^{(11)}_{(\ell_1)(\ell_{1_p})}}{6} & \dfrac{\Gamma_{(\ell)} H^{(12)}_{(\ell_1)(\ell_{1_p})}}{6} \\ 0 & 0 & \dfrac{\Gamma_{(\ell)} H^{(21)}_{(\ell_1)(\ell_{1_p})}}{6} & \dfrac{\Gamma_{(\ell)} H^{(22)}_{(\ell_1)(\ell_{1_p})}}{6} \\ 0 & 0 & 0 & 0 \\ 0 & 0 & 0 & 0 \end{bmatrix} \tag{4.115}$$

As shown in Figs. 4.8 and 4.9, the forces due to traction at points ℓ_1 and ℓ_2 of the ℓ-th line segment include (1) the interactions shown as blue bonds with stiffness matrices $\mathbf{k}^{\sigma}_{(\ell_1)(\ell_2)}$ and $\mathbf{k}^{\sigma}_{(\ell_2)(\ell_1)}$, (2) network of interactions around points ℓ_1 and point ℓ_2 shown as red bonds with stiffness matrices $\mathbf{k}^{\boldsymbol{\sigma}}_{(\ell_1)(\ell_{1_p})}$ and $\mathbf{k}^{\boldsymbol{\sigma}}_{(\ell_2)(\ell_{2_r})}$, respectively, due to direct interactions, and (3) network of interactions around points ℓ_1 and ℓ_2 shown as green bonds with stiffness matrices, $\widehat{\mathbf{k}}^{\boldsymbol{\sigma}}_{(\ell_1)(\ell_{1_p})}$ and $\widehat{\mathbf{k}}^{\boldsymbol{\sigma}}_{(\ell_2)(\ell_{2_r})}$, respectively, due to indirect interactions.

Along the boundary, the traction vector, $\mathbf{T}^*_{(\ell)}$, represents the applied values of the traction vector in the ℓ-th line segment. It is also assumed to vary linearly between the beginning and end points ℓ_1 and ℓ_2, respectively, of the ℓ-th line segment as:

$$\mathbf{T}^*_{(\ell)} = (1-s)\mathbf{T}^*_{(\ell_1)} + s\mathbf{T}^*_{(\ell_2)} \tag{4.116}$$

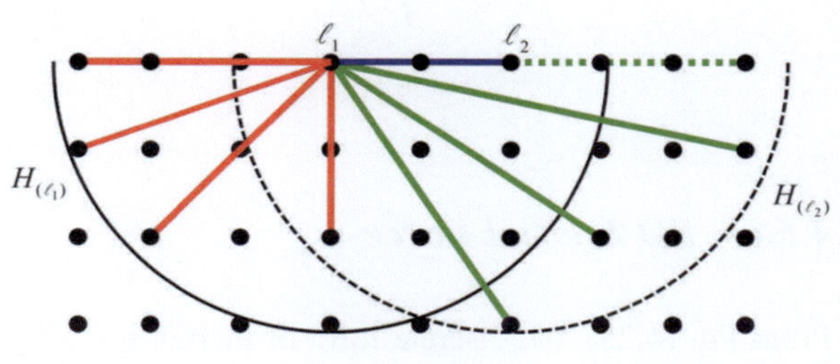

Fig. 4.8 Element (bond) connectivity between ℓ_1 and ℓ_2 for direct and indirect tractions at point ℓ_1

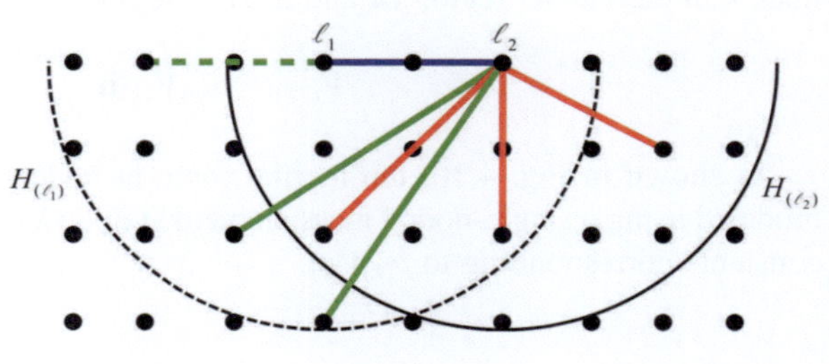

Fig. 4.9 Element (bond) connectivity between ℓ_2 and ℓ_1 for direct and indirect tractions at point ℓ_2

For the ℓ-th line segment, the virtual work due to applied traction can be evaluated as:

$$\int_0^{\Gamma_{(\ell)}} \delta\mathbf{u} \cdot \mathbf{T}^* d\Gamma = \int_0^1 \left[(1-s)\delta\mathbf{u}_{(\ell_1)} + s\delta\mathbf{u}_{(\ell_2)}\right] \cdot \left[(1-s)\mathbf{T}^*_{(\ell_1)} + s\mathbf{T}^*_{(\ell_2)}\right] \Gamma_{(\ell)} ds. \tag{4.117}$$

It can be revised as:

$$\int_0^{\Gamma_{(\ell)}} \delta\mathbf{u} \cdot \mathbf{T}^* d\Gamma = \frac{\Gamma_{(\ell)}}{3} \delta\mathbf{u}_{(\ell_1)} \mathbf{T}^*_{(\ell_1)} + \frac{\Gamma_{(\ell)}}{6} \delta\mathbf{u}_{(\ell_1)} \mathbf{T}^*_{(\ell_2)} + \frac{\Gamma_{(\ell)}}{6} \mathbf{u}_{(\ell_2)} \mathbf{T}^*_{(\ell_1)} + \frac{\Gamma_{(\ell)}}{3} \mathbf{u}_{(\ell_2)} \mathbf{T}^*_{(\ell_2)} \tag{4.118}$$

or

$$\int_0^{\Gamma_{(\ell)}} \delta\mathbf{u} \cdot \mathbf{T}^* d\Gamma = \left\{ \delta\mathbf{u}^T_{(\ell_1)} \quad \delta\mathbf{u}^T_{(\ell_2)} \right\} \left\{ \begin{array}{c} \mathbf{f}^*_{(\ell_1)} \\ \mathbf{f}^*_{(\ell_2)} \end{array} \right\} \tag{4.119}$$

where the force vectors $\mathbf{f}^*_{(\ell_1)}$ and $\mathbf{f}^*_{(\ell_2)}$ are evaluated as:

$$\mathbf{f}^*_{(\ell_1)} = \frac{\Gamma_{(\ell)}}{3} \mathbf{T}^*_{(\ell_1)} + \frac{\Gamma_{(\ell)}}{6} \mathbf{T}^*_{(\ell_2)} \tag{4.120}$$

and

$$\mathbf{f}^*_{(\ell_2)} = \frac{\Gamma_{(\ell)}}{6} \mathbf{T}^*_{(\ell_1)} + \frac{\Gamma_{(\ell)}}{3} \mathbf{T}^*_{(\ell_2)} \tag{4.121}$$

4.6.6 PD Inertial Force

From Eq. (4.23), the discrete form of inertial force $\mathbf{F}_{\ddot{\mathbf{u}}(k)}$ due to the acceleration of node k in the PDDO region $\mathscr{R}$ and interior region $\mathcal{D}$ can be expressed as:

$$\mathbf{F}_{\ddot{\mathbf{u}}(k)} = \rho_{(k)} V_{(k)} \ddot{\mathbf{u}}_{(k)} \tag{4.122}$$

As shown in Fig. 4.10, the inertial force at node k due to acceleration can be modeled using a single-noded mass element at node k (shown in red) with the “mass constant” corresponding to $\rho_{(k)} V_{(k)}$.

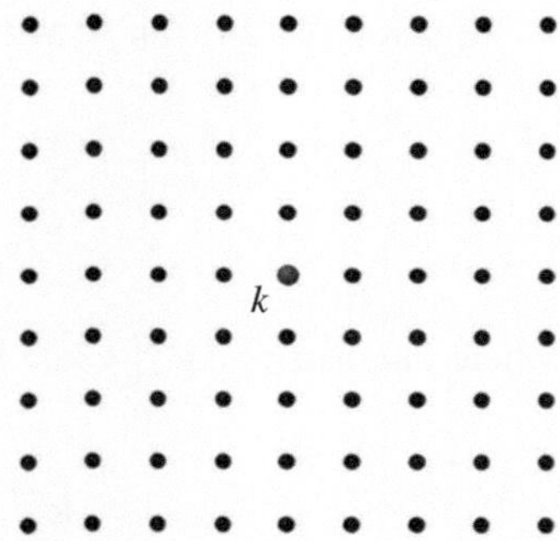

Fig. 4.10 Single-noded mass element with mass constant $\rho_{(k)}V_{(k)}$ at the node k for modeling the inertial force due to acceleration

4.7 Discrete Form of PD Internal Net Field Flow

For the interior region $\mathcal{D}$ and outer region $\mathcal{R}$, the discrete form of the virtual field variation at point $\mathbf{x}_{(k)}$ due to the internal net field flow can be expressed as:

$$\sum_{k=1}^{N_{\mathcal{D}}+N_{\mathcal{R}}} \delta\Theta_{(k)}^T \left(\nabla \cdot \mathbf{q}_{(k)}^{\alpha}\right) V_{(k)} = \sum_{k=1}^{N_{\mathcal{D}}+N_{\mathcal{R}}} \delta\Theta_{(k)}^T L_{(k)}^{\alpha} V_{(k)} = \sum_{k=1}^{N_{\mathcal{D}}+N_{\mathcal{R}}} \delta\Theta_{(k)}^T F_{(k)}^{\alpha} \tag{4.123}$$

in which the $\alpha = BB$ and $PDDO$.

4.7.1 BB-PD Interactions

As derived in Sect. 3.13.1 of Chap. 3, the discrete form of internal net field flow $F_{(k)}^{BB}$ after multiplying with the volume, $V_{(k)}$, at node k can be obtained as:

$$F_{(k)}^{BB} = L_{(k)}^{BB} V_{(k)} = \sum_{j \in H_{(k)}} \Phi_{(k)(j)}^{BB} \left(\Theta_{(j)} - \Theta_{(k)}\right) + \sum_{p \in \widetilde{H}_{(k)}} \widetilde{\Phi}_{(p)(k)}^{BB} \left(\Theta_{(p)} - \Theta_{(k)}\right) \tag{4.124}$$

with

$$\Phi_{(k)(j)}^{BB} = \widehat{\mu}_{(k)(j)} \frac{\kappa_{(k)}}{2\left|\boldsymbol{\xi}_{(k)(j)}\right|} V_{(j)} V_{(k)} \tag{4.125}$$

and

$$\widetilde{\Phi}_{(p)(k)}^{BB} = \widehat{\mu}_{(p)(k)} \frac{\kappa_{(p)}}{2\left|\boldsymbol{\xi}_{(p)(k)}\right|} V_{(p)} V_{(k)} \tag{4.126}$$

Equation (4.124) can be written in component form as:

$$\left\{ \begin{array}{c} F_{(k)}^{BB} \\ 0 \end{array} \right\} = \sum_{j=1}^{N_{(k)}} \mathbf{k}_{(k)(j)}^{BB} \left\{ \begin{array}{c} \Theta_{(k)} \\ \Theta_{(j)} \end{array} \right\} + \sum_{p=1}^{\widetilde{N}_{(k)}} \widetilde{\mathbf{k}}_{(p)(k)}^{BB} \left\{ \begin{array}{c} \Theta_{(k)} \\ \Theta_{(p)} \end{array} \right\} \tag{4.127}$$

where the matrices, $\mathbf{k}_{(k)(j)}^{BB}$ and $\widetilde{\mathbf{k}}_{(p)(k)}^{BB}$ are defined as:

$$\mathbf{k}_{(k)(j)}^{BB} = \left[\begin{array}{cc} -\Phi_{(k)(j)}^{BB)} & \Phi_{(k)(j)}^{BB} \\ 0 & 0 \end{array} \right] \tag{4.128}$$

and

$$\widetilde{\mathbf{k}}_{(p)(k)}^{BB} = \left[\begin{array}{cc} -\widetilde{\Phi}_{(p)(k)}^{BB} & \widetilde{\Phi}_{(p)(k)}^{BB} \\ 0 & 0 \end{array} \right] \tag{4.129}$$

4.7.2 PDDO Interactions

As derived in Sect. 3.13.2 of Chap. 3, the discrete form of divergence of the field flux $F_{(k)}^{PDDO}$ at point k can be written as:

$$F_{(k)}^{PDDO} = \sum_{j \in H_{(k)}} \left(\begin{array}{c} -\widetilde{\Phi}_{(j)}\Theta_{(j)} + \sum\limits_{s=1, j_s \neq k}^{N_{(j)}} \widetilde{G}_{(k)(j)}^{(j)(j_s)}\Theta_{(j_s)} + \widetilde{G}_{(k)(j)}^{(j)(k)}\left(\Theta_{(k)} - \Theta_{(j)}\right) \\ -\sum\limits_{r=1, k_r \neq j}^{N_{(k)}} G_{(k)(j)}^{(k)(k_r)}\Theta_{(k_r)} + \Phi_{(k)}\Theta_{(k)} + G_{(k)(j)}^{(k)(j)}\left(\Theta_{(k)} - \Theta_{(j)}\right) \end{array} \right) \tag{4.130}$$

with

$$G_{(k)(j)}^{(k)(k_r)} = H_{(k)(j)}^{(k)(k_r)} V_{(j)} V_{(k)} \tag{4.131}$$

$$\Phi_{(k)} = \sum_{r=1, k_r \neq j}^{N_{(k)}} G_{(k)(j)}^{(k)(k_r)} \tag{4.132}$$

and

$$\widetilde{G}_{(k)(j)}^{(j)(j_s)} = H_{(k)(j)}^{(j)(j_s)} V_{(j)} V_{(k)} \tag{4.133}$$

$$\widetilde{\Phi}_{(j)} = \sum_{s=1, j_s \neq k}^{N_{(j)}} \widetilde{G}_{(j)(k)}^{(j)(j_s)} \tag{4.134}$$

in which the coefficients $H_{(k)(j)}^{(j)(j_s)}$ and $H_{(k)(j)}^{(k)(k_r)}$ are defined as:

$$H_{(k)(j)}^{(j)(j_s)} = \widehat{\mu}_{(k)(j)} \widehat{\mu}_{(j)(j_s)} m_{(j)} \left(g_{(j)(j_s)}^{10} g_{(k)(j)}^{10} + g_{(j)(j_s)}^{01} g_{(k)(j)}^{01} \right) V_{(j_s)} \tag{4.135}$$

and

$$H_{(k)(j)}^{(k)(k_r)} = \widehat{\mu}_{(k)(j)} \widehat{\mu}_{(k)(k_r)} m_{(k)} \left(g_{(k)(k_r)}^{10} g_{(k)(j)}^{10} + g_{(k)(k_r)}^{01} g_{(k)(j)}^{01} \right) V_{(k_r)} \tag{4.136}$$

The divergence of the field flux $F_{(k)}^{PDDO}$ in component form can be written as:

$$\left\{ \begin{array}{c} F_{(k)}^{PDDO} \\ 0 \end{array} \right\} = \begin{array}{l} \displaystyle\sum_{j \in H_{(k)}} \left(\mathbf{k}_{(k)(j)}^{PDDO} \left\{ \begin{array}{c} \Theta_{(k)} \\ \Theta_{(j)} \end{array} \right\} + \sum_{r=1, k_r \neq j}^{N_{(k)}} \mathbf{k}_{(k)(j)}^{PDDO(k)(k_r)} \left\{ \begin{array}{c} \Theta_{(k)} \\ \Theta_{(k_r)} \end{array} \right\} \right. \\ \displaystyle \left. + \sum_{s=1, j_s \neq k}^{N_{(j)}} \widetilde{\mathbf{k}}_{(k)(j)}^{PDDO(j)(j_s)} \left\{ \begin{array}{c} \Theta_{(k)} \\ \Theta_{(j_s)} \end{array} \right\} \right) \end{array} \tag{4.137}$$

with

$$\mathbf{k}_{(k)(j)}^{PDDO} = \left[\begin{array}{cc} \left(\Phi_{(k)} + G_{(k)(j)}^{(k)(j)} + \widetilde{G}_{(k)(j)}^{(j)(k)} \right) & \left(\Phi_{(k)} + G_{(k)(j)}^{(k)(j)} + \widetilde{G}_{(k)(j)}^{(j)(k)} \right) \\ 0 & 0 \end{array} \right. \\ \left. \begin{array}{cc} \left(-G_{(k)(j)}^{(k)(j)} - \widetilde{G}_{11(k)(j)}^{(j)(k)} - \widetilde{\Phi}_{(j)} \right) & \left(-G_{(k)(j)}^{(k)(j)} - \widetilde{G}_{(k)(j)}^{(j)(k)} - \widetilde{\Phi}_{(j)} \right) \\ 0 & 0 \end{array} \right] \tag{4.138}$$

$$\mathbf{k}_{(k)(j)}^{PDDO(k)(k_r)} = \begin{bmatrix} 0 & 0 & -G_{(k)(j)}^{(k)(k_r)} & -G_{(k)(j)}^{(k)(k_r)} \\ 0 & 0 & 0 & 0 \end{bmatrix} \tag{4.139}$$

and

$$\widetilde{\mathbf{k}}_{(k)(j)}^{PDDO(j)(j_s)} = \begin{bmatrix} 0 & 0 & \widetilde{G}_{(k)(j)}^{(j)(j_s)} & \widetilde{G}_{(k)(j)}^{(j)(j_s)} \\ 0 & 0 & 0 & 0 \end{bmatrix} \tag{4.140}$$

4.7.3 PD Internal Field Flux

For the closed boundary of the PD region, Γ is split into $N_{\mathscr{B}}$ segments with $\Gamma_{(\ell)}$ representing the length of ℓ-th boundary segment. The numerical evaluation of the line integrals in Eq. (4.16) is achieved by dividing the boundary segments $\Gamma_{\mathscr{B}_{\mathbf{q}}}$ and $\Gamma_{\mathscr{B}_{\Theta}}$ into $N_{\mathscr{B}_{\mathbf{q}}}$ and $N_{\mathscr{B}_{\Theta}}$ line segments, respectively. As shown in Fig. 4.7, s is the natural (local) coordinate between points ℓ_1 and ℓ_2 of the ℓ-th line segment with a unit normal vector, $\mathbf{n}^{(\ell)}$. The differential line element, $d\Gamma$ along the boundary can be related to the natural coordinate as $d\Gamma = \Gamma_{(\ell)} ds$.

As suggested by Madenci et al. (2019), it is assumed that the field, $\Theta_{(\ell)}$, and the PD form of field flux, $\mathbf{q}_{(\ell)}^{PD}$, vary linearly between the beginning and end points ℓ_1 and ℓ_2, respectively, of the ℓ-th line segment as:

$$\Theta_{(\ell)} = (1-s)\Theta_{(\ell_1)} + s\Theta_{(\ell_2)} \tag{4.141}$$

and

$$\mathbf{q}_{(\ell)}^{PD} = (1-s)\mathbf{q}_{(\ell_1)}^{PD} + s\mathbf{q}_{(\ell_2)}^{PD} \tag{4.142}$$

The PD form of field flux vector $\mathbf{q}^{PD}$ at the beginning and end points ℓ_1 and ℓ_2 can be expressed as:

$$\mathbf{q}_{(\ell_1)}^{PD} = \sum_{p=1}^{N_{(\ell_1)}} \mathbf{H}_{(\ell_1)(\ell_{1_p})}\left(\Theta_{(\ell_{1_p})} - \Theta_{(\ell_1)}\right) \tag{4.143}$$

and

$$\mathbf{q}_{(\ell_2)}^{PD} = \sum_{r=1}^{N_{(\ell_2)}} \mathbf{H}_{(\ell_2)(\ell_{2_r})}\left(\Theta_{(\ell_{2_r})} - \Theta_{(\ell_2)}\right) \tag{4.144}$$

in which the vectors $\mathbf{H}_{(\ell_1)(\ell_{1_p})}$ and $\mathbf{H}_{(\ell_2)(\ell_{2_r})}$ are defined as

$$\mathbf{H}_{(\ell_1)(\ell_{1_p})} = \widehat{\mu}_{(\ell_1)(\ell_{1_p})}\left\{ m_{(\ell_1)} g_{(\ell_1)(\ell_{1_p})}^{10} \quad - m_{(\ell_1)} g_{(\ell_1)(\ell_{1_p})}^{01} \right\}^T V_{(j)} \tag{4.145}$$

and

$$\mathbf{H}_{(\ell_2)(\ell_{2_r})} = \widehat{\mu}_{(\ell_2)(\ell_{2_r})}\left\{ m_{(\ell_2)} g_{(\ell_2)(\ell_{2_r})}^{10} \quad - m_{(\ell_2)} g_{(\ell_2)(\ell_{2_r})}^{01} \right\}^T V_{(p)} \tag{4.146}$$

For the ℓ-th line segment, the virtual field variation due to field flux can be evaluated as:

$$\int_0^{\Gamma_{(\ell)}} \delta\Theta \mathbf{q}^{PD} \cdot \mathbf{n} d\Gamma = \int_0^1 \left[(1-s)\delta\Theta_{(\ell_1)} + s\delta\Theta_{(\ell_2)}\right] \cdot \left[(1-s)\mathbf{q}_{(\ell_1)} + s\mathbf{q}_{(\ell_2)}\right] \mathbf{n}_{(\ell)} \Gamma_{(\ell)} ds \tag{4.147}$$

It can be revised as:

$$\begin{aligned} \int_0^{\Gamma_{(\ell)}} \delta\Theta \mathbf{q}^{PD} \cdot \mathbf{n} d\Gamma = & \frac{\mathbf{n}_{(\ell)}\Gamma_{(\ell)}}{3} \delta\Theta_{(\ell_1)} \mathbf{q}_{(\ell_1)}^{PD} + \frac{\mathbf{n}_{(\ell)}\Gamma_{(\ell)}}{6} \delta\Theta_{(\ell_1)} \mathbf{q}_{(\ell_2)}^{PD} \\ & + \frac{\mathbf{n}_{(\ell)}\Gamma_{(\ell)}}{6} \delta\Theta_{(\ell_2)} \mathbf{q}_{(\ell_1)}^{PD} + \frac{\mathbf{n}_{(\ell)}\Gamma_{(\ell)}}{3} \delta\Theta_{(\ell_2)} \mathbf{q}_{(\ell_2)}^{PD} \end{aligned} \tag{4.148}$$

or

$$\int_0^{\Gamma_{(\ell)}} \delta\Theta \mathbf{q}^{PD} \cdot \mathbf{n}_{(\ell)} d\Gamma = \left\{ \delta\Theta_{(\ell_1)} \quad \delta\Theta_{(\ell_2)} \right\} \left\{ \begin{array}{c} F_{(\ell_1)}^{PD} \\ F_{(\ell_2)}^{PD} \end{array} \right\} \tag{4.149}$$

where $F_{(\ell_1)}^{PD}$ and $F_{(\ell_2)}^{PD}$ at beginning and end points ℓ_1 and ℓ_2 are defined as:

$$\left\{ \begin{array}{c} F_{(\ell_1)}^{PD} \\ F_{(\ell_2)}^{PD} \end{array} \right\} = \begin{bmatrix} \frac{\Gamma_{(\ell)}}{3} & \frac{\Gamma_{(\ell)}}{6} \\ \frac{\Gamma_{(\ell)}}{6} & \frac{\Gamma_{(\ell)}}{3} \end{bmatrix} \left\{ \begin{array}{c} \mathbf{n}_{(\ell)} \cdot \mathbf{q}_{(\ell_1)}^{PD} \\ \mathbf{n}_{(\ell)} \cdot \mathbf{q}_{(\ell_2)}^{PD} \end{array} \right\} \tag{4.150}$$

or

$$\left\{ \begin{array}{c} F_{(\ell_1)}^{PD} \\ F_{(\ell_2)}^{PD} \end{array} \right\} = \left\{ \begin{array}{c} \sum\limits_{p=1}^{N_{(\ell_1)}} \frac{\Gamma_{(\ell)}}{3} \mathbf{n}_{(\ell)} \cdot \mathbf{H}_{(\ell_1)(\ell_{1_p})} \left(\Theta_{(\ell_{1_p})} - \Theta_{(\ell_1)}\right) + \sum\limits_{r=1}^{N_{(\ell_2)}} \frac{\Gamma_{(\ell)}}{6} \mathbf{n}_{(\ell)} \cdot \mathbf{H}_{(\ell_2)(\ell_{2_r})} \left(\Theta_{(\ell_{2_r})} - \Theta_{(\ell_2)}\right) \\ \sum\limits_{r=1}^{N_{(\ell_2)}} \frac{\Gamma_{(\ell)}}{3} \mathbf{n}_{(\ell)} \cdot \mathbf{H}_{(\ell_2)(\ell_{2_r})} \left(\Theta_{(\ell_{2_r})} - \Theta_{(\ell_2)}\right) + \sum\limits_{p=1}^{N_{(\ell_1)}} \frac{\Gamma_{(\ell)}}{6} \mathbf{n}_{(\ell)} \cdot \mathbf{H}_{(\ell_1)(\ell_{1_p})} \left(\Theta_{(\ell_{1_p})} - \Theta_{(\ell_1)}\right) \end{array} \right\} \tag{4.151}$$

It also can be revised in component form as:

$$\begin{aligned}\left\{\begin{array}{c}F_{1(\ell_1)}^{PD}\\0\end{array}\right\} &= \sum_{p=1}^{N_{(\ell_1)}} \mathbf{k}_{(\ell_1)(\ell_{1_p})}^{\mathbf{q}} \left\{\begin{array}{c}\Theta_{(\ell_1)}\\\Theta_{(\ell_{1_p})}\end{array}\right\} + \mathbf{k}_{(\ell_1)(\ell_2)}^{\mathbf{q}} \left\{\begin{array}{c}\Theta_{(\ell_1)}\\\Theta_{(\ell_2)}\end{array}\right\}\\&+ \sum_{r=1,\,\ell_{2_r}\neq\ell_1}^{N_{(\ell_2)}} \widehat{\mathbf{k}}_{(\ell_2)(\ell_{2_r})}^{\mathbf{q}} \left\{\begin{array}{c}\Theta_{(\ell_1)}\\\Theta_{(\ell_{2_r})}\end{array}\right\}\end{aligned} \tag{4.152}$$

and

$$\begin{aligned}\left\{\begin{array}{c}F_{(\ell_2)}^{PD}\\0\end{array}\right\} &= \sum_{r=1}^{N_{(\ell_2)}} \mathbf{k}_{(\ell_2)(\ell_{2_r})}^{\mathbf{q}} \left\{\begin{array}{c}\Theta_{(\ell_2)}\\\Theta_{(\ell_{2_r})}\end{array}\right\} + \mathbf{k}_{(\ell_2)(\ell_1)}^{\mathbf{q}} \left\{\begin{array}{c}\Theta_{(\ell_2)}\\\Theta_{(\ell_1)}\end{array}\right\}\\&+ \sum_{p=1,\,\ell_{1_p}\neq\ell_2}^{N_{(\ell_1)}} \widehat{\mathbf{k}}_{(\ell_1)(\ell_{1_p})}^{\mathbf{q}} \left\{\begin{array}{c}\Theta_{(\ell_2)}\\\Theta_{(\ell_{1_p})}\end{array}\right\}\end{aligned} \tag{4.153}$$

in which the matrices $\mathbf{k}_{(\ell_1)(\ell_{1_p})}^{\mathbf{q}}$, $\mathbf{k}_{(\ell_1)(\ell_2)}^{\mathbf{q}}$, $\widehat{\mathbf{k}}_{(\ell_2)(\ell_{2_r})}^{\mathbf{q}}$ for $\ell_1 - \ell_2$ interactions are defined as:

$$\mathbf{k}_{(\ell_1)(\ell_{1_p})}^{\mathbf{q}} = \begin{bmatrix} -\dfrac{\Gamma_{(\ell)}}{3}\mathbf{n}_{(\ell)}\cdot\mathbf{H}_{(\ell_1)(\ell_{1_p})} & \dfrac{\Gamma_{(\ell)}}{3}\mathbf{n}_{(\ell)}\cdot\mathbf{H}_{(\ell_1)(\ell_{1_p})}\\ 0 & 0\end{bmatrix} \tag{4.154}$$

$$\mathbf{k}_{(\ell_1)(\ell_2)}^{\mathbf{q}} = \begin{bmatrix} \dfrac{\Gamma_{(\ell)}}{6}\mathbf{n}_{(\ell)}\cdot\mathbf{H}_{(\ell_2)(\ell_1)} & -\sum\limits_{r=1,\,\ell_{2_r}\neq\ell_1}^{N_{(\ell_2)}}\dfrac{\Gamma_{(\ell)}}{6}\mathbf{n}_{(\ell)}\cdot\mathbf{H}_{(\ell_2)(\ell_{2_r})}\\ 0 & 0\end{bmatrix} \tag{4.155}$$

$$\widehat{\mathbf{k}}_{(\ell_2)(\ell_{2_r})}^{\mathbf{q}} = \begin{bmatrix} 0 & \dfrac{\Gamma_{(\ell)}}{6}\mathbf{n}_{(\ell)}\cdot\mathbf{H}_{(\ell_2)(\ell_{2_r})}\\ 0 & 0\end{bmatrix} \tag{4.156}$$

The matrices $\mathbf{k}_{(\ell_2)(\ell_{2_r})}^{\mathbf{q}}$, $\mathbf{k}_{(\ell_2)(\ell_1)}^{\mathbf{q}}$, and $\widehat{\mathbf{k}}_{(\ell_1)(\ell_{1_p})}^{\mathbf{q}}$ for $\ell_2 - \ell_1$ interactions are defined as:

$$\mathbf{k}_{(\ell_2)(\ell_{2_r})}^{\mathbf{q}} = \begin{bmatrix} -\dfrac{\Gamma_{(\ell)}}{3}\mathbf{n}_{(\ell)}\cdot\mathbf{H}_{(\ell_2)(\ell_{2_r})} & \dfrac{\Gamma_{(\ell)}}{3}\mathbf{n}_{(\ell)}\cdot\mathbf{H}_{(\ell_2)(\ell_{2_r})}\\ 0 & 0\end{bmatrix} \tag{4.157}$$

$$\mathbf{k}_{(\ell_2)(\ell_1)}^{\mathbf{q}} = \begin{bmatrix} \dfrac{\Gamma_{(\ell)}}{6}\mathbf{n}_{(\ell)}\cdot\mathbf{H}_{(\ell_1)(\ell_2)} & -\sum\limits_{p=1,\,\ell_{1_p}\neq\ell_2}^{N_{(\ell_1)}}\dfrac{\Gamma_{(\ell)}}{6}\mathbf{n}_{(\ell)}\cdot\mathbf{H}_{(\ell_1)(\ell_{1_p})}\\ 0 & 0\end{bmatrix} \tag{4.158}$$

and

$$\widehat{\mathbf{k}}^{\mathbf{q}}_{(\ell_1)(\ell_{1_p})} = \begin{bmatrix} 0 & \frac{\Gamma_{(\ell)}}{6}\mathbf{n}_{(\ell)} \cdot \mathbf{H}_{(\ell_1)(\ell_{1_p})} \\ 0 & 0 \end{bmatrix} \tag{4.159}$$

As shown in Figs. 4.11 and 4.12, the total internal net field flow due to field interactions at points ℓ_1 and ℓ_2 of the ℓ-th line segment include (1) the interactions shown as blue bonds with stiffness matrices $\mathbf{k}^{\mathbf{q}}_{(\ell_1)(\ell_2)}$ and $\mathbf{k}^{\mathbf{q}}_{(\ell_2)(\ell_1)}$, (2) network of interactions around points ℓ_1 and point ℓ_2 shown as red bonds with stiffness matrices $\mathbf{k}^{\mathbf{q}}_{(\ell_1)(\ell_{1_p})}$ and $\mathbf{k}^{\mathbf{q}}_{(\ell_2)(\ell_{2_r})}$, respectively, due to direct interactions, and (3) network of interactions around points ℓ_1 and ℓ_2 shown as green bonds with stiffness matrices, $\widehat{\mathbf{k}}^{\mathbf{q}}_{(\ell_1)(\ell_{1_p})}$ and $\widehat{\mathbf{k}}^{\mathbf{q}}_{(\ell_2)(\ell_{2_r})}$, respectively, due to indirect interactions.

Along the boundary, the field flux, $\mathbf{q}^*_{(\ell)}$, represents the applied values of field flux vector in the ℓ-th line segment. It is also assumed to vary linearly between the beginning and end points ℓ_1 and ℓ_2, respectively, of the ℓ-th line segment as:

$$\mathbf{q}^*_{(\ell)} = (1-s)\mathbf{q}^*_{(\ell_1)} + s\mathbf{q}^*_{(\ell_2)} \tag{4.160}$$

For the ℓ-th line segment, the virtual field variation due to applied flux can be evaluated as:

$$\int_0^{\Gamma_{(\ell)}} \delta\Theta\mathbf{q}^* \cdot \mathbf{n}d\Gamma = \int_0^1 \left[(1-s)\delta\Theta_{(\ell_1)} + s\delta\Theta_{(\ell_2)}\right] \cdot \left[(1-s)\mathbf{q}^*_{(\ell_1)} + s\mathbf{q}^*_{(\ell_2)}\right]\mathbf{n}\Gamma_{(\ell)}ds \tag{4.161}$$

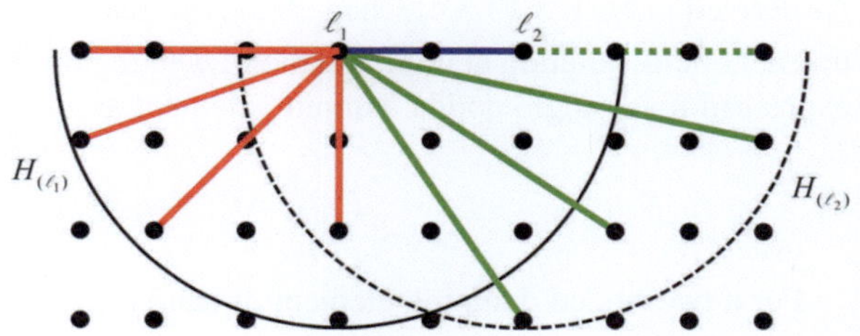

Fig. 4.11 Element (bond) connectivity between ℓ_1 and ℓ_2 for direct and indirect field flow at point ℓ_1

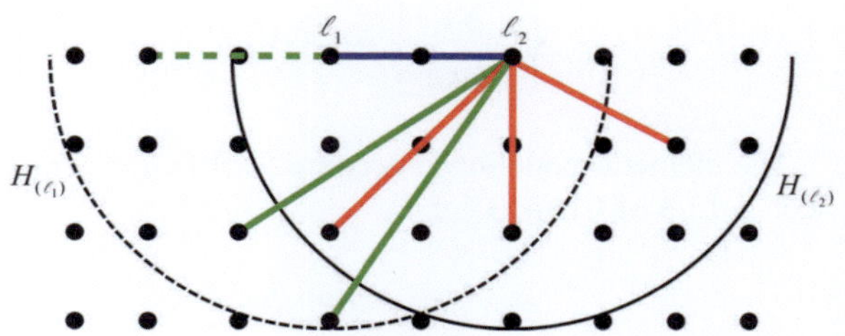

Fig. 4.12 Element (bond) connectivity between ℓ_2 and ℓ_1 for direct and indirect field flow at point ℓ_2

It can be revised as:

$$\int_0^{\Gamma_{(\ell)}} \delta\Theta \mathbf{q}^* \cdot \mathbf{n} d\Gamma = \frac{\Gamma_{(\ell)}\mathbf{n}_{(\ell)}}{3}\delta\Theta_{(\ell_1)}\mathbf{q}^*_{(\ell_1)} + \frac{\Gamma_{(\ell)}\mathbf{n}_{(\ell)}}{6}\delta\Theta_{(\ell_1)}\mathbf{q}^*_{(\ell_2)} + \frac{\Gamma_{(\ell)}\mathbf{n}_{(\ell)}}{6}\Theta_{(\ell_2)}\mathbf{q}^*_{(\ell_1)} + \frac{\Gamma_{(\ell)}\mathbf{n}_{(\ell)}}{3}\Theta_{(\ell_2)}\mathbf{q}^*_{(\ell_2)} \tag{4.162}$$

or

$$\int_0^{\Gamma_{(\ell)}} \delta\Theta \mathbf{q}^* \cdot \mathbf{n} d\Gamma = \left\{ \delta\Theta_{(\ell_1)} \quad \delta\Theta^*_{(\ell_2)} \right\} \begin{Bmatrix} f^*_{(\ell_1)} \\ f^*_{(\ell_2)} \end{Bmatrix} \tag{4.163}$$

where $f^*_{(\ell_1)}$ and $f^*_{(\ell_2)}$ are evaluated as:

$$f^*_{(\ell_1)} = \frac{\mathbf{n}_{(\ell)}\Gamma_{(\ell)}}{3}\mathbf{q}^*_{(\ell_1)} + \frac{\mathbf{n}_{(\ell)}\Gamma_{(\ell)}}{6}\mathbf{q}^*_{(\ell_2)} \tag{4.164}$$

and

$$f^*_{(\ell_2)} = \frac{\mathbf{n}_{(\ell)}\Gamma_{(\ell)}}{6}\mathbf{q}^*_{(\ell_1)} + \frac{\mathbf{n}_{(\ell)}\Gamma_{(\ell)}}{6}\mathbf{q}^*_{(\ell_2)} \tag{4.165}$$

4.7.4 PD Transient Field Variation

As derived in Sect. 3.13.3 of Chap. 3, the discrete form of field flow $F_{\dot{\Theta}(k)}$ due to the transient field variation at node k in the PDDO region $\mathcal{R}$ and interior region $\mathcal{D}$ can be expressed for a single noded damping element as:

$$F_{\dot{\Theta}(k)} = C_{(k)}V_{(k)}\dot{\Theta}_{(k)} \tag{4.166}$$

For a two-noded damping element, it can be expressed as:

$$\begin{Bmatrix} F_{\dot{\Theta}(k)} \\ F_{\dot{\Theta}(k_d)} \end{Bmatrix} = C_{(k)}V_{(k)} \begin{bmatrix} 1 & -1 \\ -1 & 1 \end{bmatrix} \begin{Bmatrix} \dot{\Theta}_{(k)} \\ \dot{\Theta}_{(k_d)} \end{Bmatrix} \tag{4.167}$$

The element modeling procedure for PD transient field variation is described in Sect. 3.13.3 of Chap. 3.

4.7.5 PD Transient Material Property

As derived in Sect. 3.13.4 of Chap. 3, the discrete form of field flow $F_{\Theta(k)}$ due to the time rate of change of material property at node k in the PDDO region $\mathcal{R}$ and interior region $\mathcal{D}$ can be expressed for a two noded element as:

$$\left\{ \begin{array}{c} F_{\Theta(k)} \\ F_{\Theta(k_d)} \end{array} \right\} = \dot{C}_{(k)} V_{(k)} \left[\begin{array}{cc} 1 & -1 \\ -1 & 1 \end{array} \right] \left\{ \begin{array}{c} \Theta_{(k)} \\ \Theta_{(k_d)} \end{array} \right\} \tag{4.168}$$

The element modeling procedure for PD transient material property variation is described in Sect. 3.13.4 of Chap. 3.

References

Anicode, S. V. K., & Madenci, E. (2022). Seamless coupling of bond- and state-based peridynamic and finite element analyses. *Mechanics and materials., 173*, 104433.

Anicode, S. V. K., & Madenci, E. (2023). Direct coupling of dual-horizon peridynamics with finite elements for irregular discretization without an overlap zone. *Engineering with Computers, 40*, 1–31.

Madenci, E., Dorduncu, M., Barut, A., & Phan, N. (2018). Weak form of peridynamics for nonlocal essential and natural boundary conditions. *Computer Methods in Applied Mechanics and Engineering, 337*, 598–631.

Madenci, E., Barut, A., & Dorduncu, M. (2019). *Peridynamic differential operator for numerical analysis*. Springer International Publishing.

Madenci, E., Roy, P., & Behera, D. (2022). *Advances in peridynamics*. Springer Nature.

Chapter 5
Coupling of Peridynamics with Finite Elements

5.1 Introduction

The finite element method (FEM) is computationally robust and very effective for modeling material response in the absence of crack initiation and growth. However, damage prediction using the FEM can be very cumbersome because the derivatives of displacements are undefined at the discontinuities. Although the capability of peridynamics (PD) is unrivaled for failure prediction, it is computationally more costly than the FEM. Hence, it is desirable to couple the dual horizon PD and FEM to take advantage of their salient features. Even in the case of a single horizon, the coupling of these two approaches is not straightforward because the PD theory is nonlocal with volume constraints, whereas the FEM is local with surface constraints. Various coupling strategies for a single-horizon PD exist through the use of an overlap region in which the PD nodes and FEM nodes coexist and enable transfer of information as discussed by Anicode and Madenci (2023).

This chapter presents a coupling approach for the solution of equilibrium and field equations with uniform and non-uniform discretization in a commercial finite element analysis framework, Ansys (Anicode & Madenci, 2022, 2023). It is based on the variational principles. The PD and FE regions share the same nodes along their interface without any constraint conditions and/or an overlap zone. This coupling approach does not suffer from any unphysical concentrations near the interface in either steady-state or transient analysis. The PD region can be completely or partially surrounded by a region of traditional finite elements. It is a general framework to perform bond-based (BB), ordinary state-based (OSB), and non-ordinary state-based (NOSB) PD analysis coupled with traditional elements. MATRIX27 elements native to Ansys are used for the PD region and the traditional elements for the remaining FE region. The applications of this coupling approach are demonstrated in Chaps. 11 and 13 for solution of equilibrium and field equations, respectively. The results from the coupled PD/FE approach agree well with those of FEM for all boundary and loading conditions.

E. Madenci et al., *Peridynamics and Its Applications Using Ansys*,
https://doi.org/10.1007/978-3-031-59896-8_5

5.2 Coupling of PD with FE Regions for Deformation Analysis

Unlike the previous coupling techniques, this approach does not require a morphing or a blending function that facilitates coupling over a transition zone. As shown in Fig. 5.1, the PD domain with an arbitrary geometry is interfaced with the FE domain. The domain, Ω, is split into FE and PD domains of Ω^{FE} and Ω^{PD}, respectively, i.e., $\Omega = \Omega^{PD} \cup \Omega^{FE}$. The domain, Ω^{PD}, is composed of $\Omega^{PD} = \mathcal{D} \cup \mathscr{R} \cup \mathscr{B}$ with an interior region, $\mathcal{D}$, outer region, $\mathscr{R}$, and a "real boundary layer" region, $\mathscr{B}$, with nonzero volume. The extent of region, $\mathscr{R}$, is defined by δ for bond-based (BB) and 2δ for ordinary state-based (OSB), and non-ordinary state-based (NOSB) PD, and the width of region $\mathscr{B}$ is equal to one PD grid spacing. The mid-surface of the boundary layer region, $\mathscr{B}$, with a narrow width, b, defines the surface of the PD region, S^{PD}. The volume of boundary layer region is $V_{\mathscr{B}} = bS^{PD}$ and it is much smaller than the volume of outer region, $V_{\mathscr{R}}$, i.e., $V_{\mathscr{B}} \ll V_{\mathscr{R}}$.

The surface defining the PD region, S^{PD}, can be subjected to displacement constraints and/or external tractions. It is split into $S_{\mathbf{u}}^{PD}$, $S_{\boldsymbol{\sigma}}^{PD}$, and S_{I}^{PD} representing parts of the boundary subjected to displacement constraints, $\mathbf{U}^{*PD}$, external tractions, $\mathbf{T}^{*PD}$, and PD/FE interface, respectively. Similar to the PD region, the surface defining the FE region, S^{FE}, is split accordingly. The superscripts appearing as PD and FE indicate quantities associated with the domains of PD and FE, respectively.

Applying the principle of virtual work for the entire domain Ω leads to

$$\delta \mathrm{W}_{\text{internal}}^{PD} + \delta W_{\text{inertial}}^{PD} + \delta W_{\text{internal}}^{FE} + \delta W_{\text{inertial}}^{FE} = \delta W_{\mathbf{u}}^{PD} + \delta W_{\mathbf{u}}^{FE} + \delta W_{\boldsymbol{\sigma}}^{PD} + \delta W_{\boldsymbol{\sigma}}^{FE} + \delta W_{\mathbf{b}}^{PD} + \delta W_{\mathbf{b}}^{FE} \tag{5.1}$$

in which $\delta(\ )$ denotes the virtual value of the quantity. The virtual work due to internal forces, inertial forces, external tractions, and body loads are denoted by

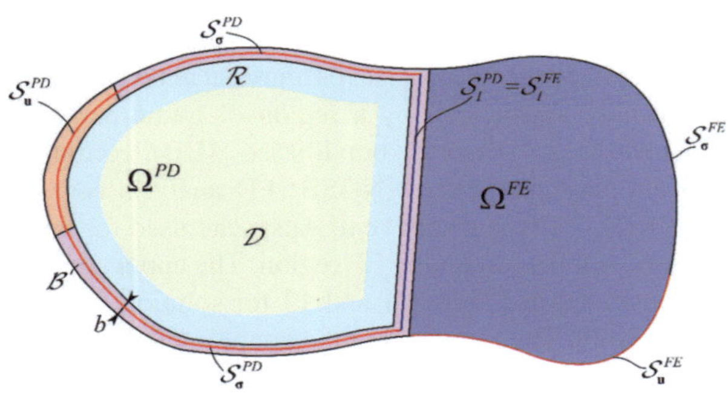

Fig. 5.1 Coupling of PD and FE regions for deformation analysis with a seamless PD/FE interface

$\delta W^{\alpha}_{\text{internal}}$, $\delta W^{\alpha}_{\text{inertial}}$, $\delta W^{\alpha}_{\boldsymbol{\sigma}}$, and $\delta W^{\alpha}_{\mathbf{b}}$, respectively, in regions of $\alpha = PD, FE$. The expression in Eq. (5.1) can be rewritten as:

$$\begin{aligned}
&\int\limits_{\Omega^{PD}} \rho^{PD}(\mathbf{x},t)\ddot{\mathbf{u}}(\mathbf{x},t)\cdot\delta\mathbf{u}(\mathbf{x})dV + \int\limits_{\Omega^{PD}} \mathbf{P}^{PD}(\mathbf{x},t) : \delta\nabla\mathbf{u}(\mathbf{x})dV \\
&+ \int\limits_{\Omega^{FE}} \rho^{FE}(\mathbf{x},t)\ddot{\mathbf{u}}(\mathbf{x},t)\cdot\delta\mathbf{u}(\mathbf{x})dV + \int\limits_{\Omega^{FE}} \mathbf{P}^{FE}(\mathbf{x},t) : \delta\nabla\mathbf{u}(\mathbf{x})dV = \\
&\int\limits_{S^{PD}_{\boldsymbol{\sigma}}} \mathbf{T}^{*PD}(\mathbf{x},t)\cdot\delta\mathbf{u}(\mathbf{x})dS + \int\limits_{S^{FE}_{\boldsymbol{\sigma}}} \mathbf{T}^{*FE}(\mathbf{x},t)\cdot\delta\mathbf{u}(\mathbf{x})dS \\
&+ \int\limits_{\Omega^{PD}} \mathbf{b}^{PD}(\mathbf{x},t)\cdot\delta\mathbf{u}(\mathbf{x})dV + \int\limits_{\Omega^{FE}} \mathbf{b}^{FE}(\mathbf{x},t)\cdot\delta\mathbf{u}(\mathbf{x})dV
\end{aligned} \tag{5.2}$$

in which $\rho^{\alpha}(\mathbf{x},t)$, $\mathbf{P}^{\alpha}(\mathbf{x},t)$, and $\mathbf{b}^{\alpha}(\mathbf{x},t)$ represent the density of material point, the first Piola-Kirchhoff stress tensor, and body force density vector, respectively, for $\alpha = PD, FE$. The displacement and acceleration vectors are denoted as $\mathbf{u}(\mathbf{x},t)$ and $\ddot{\mathbf{u}}(\mathbf{x},t)$, respectively. Prior to applying the divergence theorem, Eq. (5.2) can be recast as:

$$\begin{aligned}
&\int\limits_{\Omega^{PD}} \rho^{PD}(\mathbf{x},t)\ddot{\mathbf{u}}(\mathbf{x},t)\cdot\delta\mathbf{u}(\mathbf{x})dV + \int\limits_{\Omega^{PD}} \left[\nabla\cdot\left(\mathbf{P}^{PD}(\mathbf{x},t)\delta\mathbf{u}(\mathbf{x})\right)\right. \\
&- \left.\left(\nabla\cdot\mathbf{P}^{PD}(\mathbf{x},t)\right)\cdot\delta\mathbf{u}(\mathbf{x})\right]dV + \int\limits_{\Omega^{FE}} \rho^{FE}(\mathbf{x},t)\ddot{\mathbf{u}}(\mathbf{x},t)\cdot\delta\mathbf{u}(\mathbf{x})dV \\
&+ \int\limits_{\Omega^{FE}} \mathbf{P}^{FE}(\mathbf{x},t) : \delta\nabla\mathbf{u}(\mathbf{x})dV = \\
&\int\limits_{S^{PD}_{\boldsymbol{\sigma}}} \mathbf{T}^{*PD}(\mathbf{x},t)\cdot\delta\mathbf{u}(\mathbf{x})dS + \int\limits_{S^{FE}_{\boldsymbol{\sigma}}} \mathbf{T}^{*FE}(\mathbf{x},t)\cdot\delta\mathbf{u}(\mathbf{x})dS \\
&+ \int\limits_{\Omega^{PD}} \mathbf{b}^{PD}(\mathbf{x},t)\cdot\delta\mathbf{u}(\mathbf{x})dV + \int\limits_{\Omega^{FE}} \mathbf{b}^{FE}(\mathbf{x},t)\cdot\delta\mathbf{u}(\mathbf{x})dV
\end{aligned} \tag{5.3}$$

Invoking the divergence theorem along with $(\mathbf{P}^{\alpha}(\mathbf{x}, t)\delta\mathbf{u}(\mathbf{x})) \cdot \mathbf{n}(\mathbf{x}) = (\mathbf{P}^{\alpha\mathrm{T}}(\mathbf{x}, t)\mathbf{n}(\mathbf{x})) \cdot \delta\mathbf{u}(\mathbf{x})$ $\alpha = PD, FE$ into Eq. (5.3) results in:

$$\begin{aligned}
&\int\limits_{\Omega^{PD}} \rho^{PD}(\mathbf{x}, t)\ddot{\mathbf{u}}(\mathbf{x}, t) \cdot \delta\mathbf{u}(\mathbf{x})dV + \oint\limits_{S^{PD}} \left(\mathbf{P}^{PD\mathrm{T}}(\mathbf{x}, t)\mathbf{n}(\mathbf{x})\right) \cdot \delta\mathbf{u}(\mathbf{x})dS \\
&\quad - \int\limits_{\Omega^{PD}} \left(\nabla \cdot \mathbf{P}^{PD}(\mathbf{x}, t)\right) \cdot \delta\mathbf{u}(\mathbf{x})dV + \int\limits_{\Omega^{FE}} \rho^{FE}(\mathbf{x}, t)\ddot{\mathbf{u}}(\mathbf{x}, t) \cdot \delta\mathbf{u}(\mathbf{x})dV \\
&\quad + \int\limits_{\Omega^{FE}} \mathbf{P}^{FE}(\mathbf{x}, t) : \delta\nabla\mathbf{u}(\mathbf{x})dV = \\
&\quad \int\limits_{S_\sigma^{PD}} \mathbf{T}^{*PD}(\mathbf{x}, t) \cdot \delta\mathbf{u}(\mathbf{x})dS + \int\limits_{S_\sigma^{FE}} \mathbf{T}^{*FE}(\mathbf{x}, t) \cdot \delta\mathbf{u}(\mathbf{x})dS \\
&\quad + \int\limits_{\Omega^{PD}} \mathbf{b}^{PD}(\mathbf{x}, t) \cdot \delta\mathbf{u}(\mathbf{x})dV + \int\limits_{\Omega^{FE}} \mathbf{b}^{FE}(\mathbf{x}, t) \cdot \delta\mathbf{u}(\mathbf{x})dV
\end{aligned} \tag{5.4}$$

in which $\delta\mathbf{u}(\mathbf{x})$ is the virtual displacement vector, $\mathbf{T}^{PD}(\mathbf{x}, t) = \mathbf{P}^{PD\mathrm{T}}(\mathbf{x}, t)\mathbf{n}(\mathbf{x})$ is the PD representation of the traction vector with $\mathbf{n}(\mathbf{x})$ being the unit normal vector to the surface (Madenci et al., 2019, 2022). Replacing the divergence of stress tensor $\nabla\cdot\mathbf{P}^{PD}(\mathbf{x}, t)$ with the PD representation of the internal force vector, $\mathbf{L}^{PD}$, and splitting the volume of the domain, Ω^{PD}, into inner region, $\mathcal{D}$, outer region, $\mathcal{R}$, and the boundary layer region, $\mathcal{B}$, Eq. (5.4) can be rewritten as:

$$\begin{aligned}
&\int\limits_{\mathcal{D}\cup\mathcal{R}} \rho^{PD}(\mathbf{x}, t)\ddot{\mathbf{u}}(\mathbf{x}, t) \cdot \delta\mathbf{u}(\mathbf{x})dV + \int\limits_{\mathcal{B}} \rho^{PD}(\mathbf{x}, t)\ddot{\mathbf{u}}(\mathbf{x}, t) \cdot \delta\mathbf{u}(\mathbf{x})dV \\
&\quad - \int\limits_{\mathcal{D}} \mathbf{L}^{PD}(\mathbf{x}, t) \cdot \delta\mathbf{u}(\mathbf{x})dV - \int\limits_{\mathcal{R}} \mathbf{L}^{PDDO}(\mathbf{x}, t) \cdot \delta\mathbf{u}(\mathbf{x})dV - \int\limits_{\mathcal{B}} \mathbf{L}^{PDDO}(\mathbf{x}, t) \cdot \delta\mathbf{u}(\mathbf{x})dV \\
&\quad + \oint\limits_{S^{PD}} \left(\mathbf{P}^{PD\mathrm{T}}(\mathbf{x}, t)\mathbf{n}(\mathbf{x})\right) \cdot \delta\mathbf{u}(\mathbf{x})dS + \int\limits_{\Omega^{FE}} \rho^{FE}(\mathbf{x}, t)\ddot{\mathbf{u}}(\mathbf{x}, t) \cdot \delta\mathbf{u}(\mathbf{x})dV \\
&\quad + \int\limits_{\Omega^{FE}} \mathbf{P}^{FE}(\mathbf{x}, t) : \delta\nabla\mathbf{u}(\mathbf{x})dV = \\
&\quad \int\limits_{S_\sigma^{PD}} \mathbf{T}^{*PD}(\mathbf{x}, t) \cdot \delta\mathbf{u}(\mathbf{x})dS + \int\limits_{S_\sigma^{FE}} \mathbf{T}^{*FE}(\mathbf{x}, t) \cdot \delta\mathbf{u}(\mathbf{x})dS \\
&\quad + \int\limits_{\mathcal{D}\cup\mathcal{R}} \mathbf{b}^{PD}(\mathbf{x}, t) \cdot \delta\mathbf{u}(\mathbf{x})dV + \int\limits_{\mathcal{B}} \mathbf{b}^{PD}(\mathbf{x}, t) \cdot \delta\mathbf{u}(\mathbf{x})dV + \int\limits_{\Omega^{FE}} \mathbf{b}^{FE}(\mathbf{x}, t) \cdot \delta\mathbf{u}(\mathbf{x})dV
\end{aligned} \tag{5.5a}$$

or

$$
\begin{aligned}
&\int_{\mathcal{D}\cup\mathcal{R}} \rho^{PD}(\mathbf{x},t)\ddot{\mathbf{u}}(\mathbf{x},t)\cdot\delta\mathbf{u}(\mathbf{x})dV + b\int_{S}\rho^{PD}(\mathbf{x},t)\ddot{\mathbf{u}}(\mathbf{x},t)\cdot\delta\mathbf{u}(\mathbf{x})dS \\
&\quad - \int_{\mathcal{D}}\mathbf{L}^{PD}(\mathbf{x},t)\cdot\delta\mathbf{u}(\mathbf{x})dV - \int_{\mathcal{R}}\mathbf{L}^{PDDO}(\mathbf{x},t)\cdot\delta\mathbf{u}(\mathbf{x})dV - b\int_{S}\mathbf{L}^{PDDO}(\mathbf{x},t)\cdot\delta\mathbf{u}(\mathbf{x})dS \\
&\quad + \oint_{S^{PD}}\left(\mathbf{P}^{PDT}(\mathbf{x},t)\mathbf{n}(\mathbf{x})\right)\cdot\delta\mathbf{u}(\mathbf{x})dS + \int_{\Omega^{FE}}\rho^{FE}(\mathbf{x},t)\ddot{\mathbf{u}}(\mathbf{x},t)\cdot\delta\mathbf{u}(\mathbf{x})dV \\
&\quad + \int_{\Omega^{FE}}\mathbf{P}^{FE}(\mathbf{x},t):\delta\nabla\mathbf{u}(\mathbf{x})dV = \\
&\quad \int_{S_{\sigma}^{PD}}\mathbf{T}^{*PD}(\mathbf{x},t)\cdot\delta\mathbf{u}(\mathbf{x})dS + \int_{S_{\sigma}^{FE}}\mathbf{T}^{*FE}(\mathbf{x},t)\cdot\delta\mathbf{u}(\mathbf{x})dS \\
&\quad + \int_{\mathcal{D}\cup\mathcal{R}}\mathbf{b}^{PD}(\mathbf{x},t)\cdot\delta\mathbf{u}(\mathbf{x})dV + b\int_{S}\mathbf{b}^{PD}(\mathbf{x},t)\cdot\delta\mathbf{u}(\mathbf{x})dS + \int_{\Omega^{FE}}\mathbf{b}^{FE}(\mathbf{x},t)\cdot\delta\mathbf{u}(\mathbf{x})dV
\end{aligned}
\tag{5.5b}
$$

The boundary layer region, $\mathcal{B}$, is defined by the surface of the domain, S^{PD}, with its width, b, and for $b \ll 1$, Eq. (5.5b) reduces to:

$$
\begin{aligned}
&\int_{\mathcal{D}\cup\mathcal{R}} \rho^{PD}(\mathbf{x},t)\ddot{\mathbf{u}}(\mathbf{x},t)\cdot\delta\mathbf{u}(\mathbf{x})dV - \int_{\mathcal{D}}\mathbf{L}^{\alpha}(\mathbf{x},t)\cdot\delta\mathbf{u}(\mathbf{x})dV \\
&- \int_{\mathcal{R}}\mathbf{L}^{PDDO}(\mathbf{x},t)\cdot\delta\mathbf{u}(\mathbf{x})dV + \oint_{S^{PD}}\left(\mathbf{P}^{PDT}(\mathbf{x},t)\mathbf{n}(\mathbf{x})\right)\cdot\delta\mathbf{u}(\mathbf{x})dS \\
&+ \int_{\Omega^{FE}}\rho^{FE}(\mathbf{x},t)\ddot{\mathbf{u}}(\mathbf{x},t)\cdot\delta\mathbf{u}(\mathbf{x})dV + \int_{\Omega^{FE}}\mathbf{P}^{FE}(\mathbf{x},t):\delta\nabla\mathbf{u}(\mathbf{x})dV \quad \text{for } \alpha = BB, OSB, NOSB \\
&= \int_{S_{\sigma}^{PD}}\mathbf{T}^{*PD}(\mathbf{x},t)\cdot\delta\mathbf{u}(\mathbf{x})dS + \int_{S_{\sigma}^{FE}}\mathbf{T}^{*FE}(\mathbf{x},t)\cdot\delta\mathbf{u}(\mathbf{x})dS \\
&+ \int_{\mathcal{D}\cup\mathcal{R}}\mathbf{b}^{PD}(\mathbf{x},t)\cdot\delta\mathbf{u}(\mathbf{x})dV + \int_{\Omega^{FE}}\mathbf{b}^{FE}(\mathbf{x},t)\cdot\delta\mathbf{u}(\mathbf{x})dV
\end{aligned}
\tag{5.6}
$$

The equilibrium equation valid in the entire domain $\Omega = \Omega^{PD} \cup \Omega^{FE}$ is expressed only in terms of displacements as unknowns.

For a two-dimensional domain, Eq. (5.6) concerns only area and line integrations and simplifies to:

$$\begin{aligned}
&\int_{\mathcal{D}\cup\mathcal{R}} \rho^{PD}(\mathbf{x},t)\ddot{\mathbf{u}}(\mathbf{x},t)\cdot\delta\mathbf{u}(\mathbf{x})dA - \int_{\mathcal{D}} \mathbf{L}^{\alpha}(\mathbf{x},t)\cdot\delta\mathbf{u}(\mathbf{x})dA \\
&- \int_{\mathcal{R}} \mathbf{L}^{PDDO}(\mathbf{x},t)\cdot\delta\mathbf{u}(\mathbf{x})dA + \oint_{\Gamma^{PD}} \mathbf{T}^{PD}(\mathbf{x},t)\cdot\delta\mathbf{u}(\mathbf{x})d\Gamma \\
&+ \int_{\Omega^{FE}} \rho^{FE}(\mathbf{x},t)\ddot{\mathbf{u}}(\mathbf{x},t)\cdot\delta\mathbf{u}(\mathbf{x})dA + \int_{\Omega^{FE}} \mathbf{P}^{FE}(\mathbf{x},t) : \delta\nabla\mathbf{u}(\mathbf{x})dA \quad \text{for } \alpha = BB, OSB, NOSB \\
&= \int_{\Gamma^{PD}_{\sigma}} \mathbf{T}^{*PD}(\mathbf{x},t)\cdot\delta\mathbf{u}(\mathbf{x})d\Gamma + \int_{\Gamma^{FE}_{\sigma}} \mathbf{T}^{*FE}(\mathbf{x},t)\cdot\delta\mathbf{u}(\mathbf{x})d\Gamma \\
&+ \int_{\mathcal{D}\cup\mathcal{R}} \mathbf{b}^{PD}(\mathbf{x},t)\cdot\delta\mathbf{u}(\mathbf{x})dA + \int_{\Omega^{FE}} \mathbf{b}^{FE}(\mathbf{x},t)\cdot\delta\mathbf{u}(\mathbf{x})dA
\end{aligned} \tag{5.7}$$

in which $\mathcal{D}$ and $\mathcal{R}$ represent the area of the interior and outer regions, Γ^{PD} represents the entire boundary of the domain, and $\Gamma^{PD}_{\mathcal{B}_\sigma}$ and $\Gamma^{PD}_{\mathcal{B}_\mathbf{u}}$ represent the segment of the boundary subjected to the external tractions and displacement constraints, respectively. The region $\mathcal{B}$ with a thickness of b is referred to as the boundary layer region. As the width approaches to zero, i.e., $b \to 0$, Γ coincides with the boundary of PD domain, Ω^{PD}, as shown in Fig. 5.1.

The PD and FE domains are bounded by $\Gamma^{PD} = \Gamma^{PD}_{\mathcal{B}_\sigma} \cup \Gamma^{PD}_{\mathcal{B}_\mathbf{u}} \cup \Gamma^{PD}_{\mathcal{I}}$ and $\Gamma^{FE} = \Gamma^{FE}_{\mathcal{B}_\sigma} \cup \Gamma^{FE}_{\mathcal{B}_\mathbf{u}} \cup \Gamma^{FE}_{\mathcal{I}}$, respectively. Along the PD/FE interface, $\mathbf{x} \in \left(\Gamma^{PD}_{\mathcal{I}} = \Gamma^{FE}_{\mathcal{I}}\right)$, the displacement is continuous, i.e., $\mathbf{u}^{PD}(\mathbf{x},t) = \mathbf{u}^{FE}(\mathbf{x},t)$ along with their corresponding variations as $\delta\mathbf{u}^{PD}(\mathbf{x},t) = \delta\mathbf{u}^{FE}(\mathbf{x},t)$. The coupling approach is valid for BB-, OSB- and NOSB-PD, i.e., $\mathbf{L}^{\alpha}(\mathbf{x})$ with $\alpha = BB, OSB, NOSB$. The evaluation of each integral in Eq. (5.7) can be performed by discretizing the domain and employing numerical quadrature that requires the discrete form of PD internal force vector, $\mathbf{L}^{\alpha}(\mathbf{x})$.

5.3 Coupling of PD with FE Regions for Field Analysis

As shown in Fig. 5.2, the PD domain with an arbitrary geometry is interfaced with the FE domain. The domain, Ω, is split into FE and PD domains of Ω^{FE} and Ω^{PD}, respectively, i.e., $\Omega = \Omega^{PD} \cup \Omega^{FE}$. The domain, Ω^{PD}, is composed of $\Omega^{PD} = \mathcal{D} \cup \mathcal{R} \cup \mathcal{B}$ with an interior region, $\mathcal{D}$, outer region, $\mathcal{R}$, and a "real boundary layer" region, $\mathcal{B}$, with nonzero volume. The extent of region, $\mathcal{R}$, is defined by δ for BB and the width of region $\mathcal{B}$ is equal to one PD grid spacing. The mid-surface of the boundary layer region, $\mathcal{B}$, with a narrow width, b, defines the surface of the PD

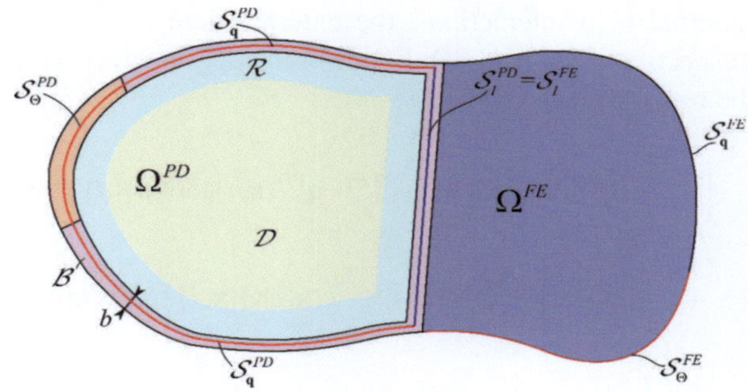

Fig. 5.2 Coupling of PD and FE regions for field analysis with a seamless PD/FE interface

region, S^{PD}. The volume of boundary layer region is $V_{\mathscr{B}} = bS^{PD}$ and it is much smaller than the volume of outer region, $V_{\mathscr{R}}$, i.e., $V_{\mathscr{B}} \ll V_{\mathscr{R}}$.

The surface defining the PD domain, S^{PD}, can be subjected to field constraints and/or external field flux. It is split into S_{Θ}^{PD}, $S_{\mathbf{q}}^{PD}$, and S_{I}^{PD} representing parts of the boundary subjected to field constraints, Θ^{*PD}, external field flux, $\mathbf{q}^{*PD}$, and PD-FE interface, respectively. Similar to the PD domain, the surface defining the FE domain, S^{FE}, is split accordingly. The superscripts appearing as PD and FE indicate quantities associated with the domains of PD and FE, respectively.

Application of the variational approach leads to:

$$\delta W_{\text{field}}^{PD} + \delta W_{\text{trans}}^{PD} + \delta W_{\text{field}}^{FE} + \delta W_{\text{trans}}^{FE} = \delta W_{\Theta}^{PD} + \delta W_{\Theta}^{FE} + \delta W_{\mathbf{q}}^{PD} + \delta W_{\mathbf{q}}^{FE} \\ + \delta W_{s}^{PD} + \delta W_{s}^{FE} \tag{5.8}$$

in which $\delta W_{\text{internal}}^{\alpha}$ and $\delta W_{\text{trans}}^{\alpha}$ represent the virtual value of the field potential and transient fluctuation due to internal field flux, respectively, and $\delta W_{\mathbf{q}}^{\alpha}$ and δW_{s}^{α} represent the virtual value of the external field interactions due to the known field flux and generated field source, respectively, in domains of $\alpha = PD, FE$. This expression can be rewritten as:

$$\begin{aligned} &\int\limits_{\Omega^{PD}} \mathbf{q}^{PD}(\mathbf{x},t)\cdot\delta\nabla\Theta(\mathbf{x})dV + \int\limits_{\Omega^{PD}} \left(C(\mathbf{x},t)\dot{\Theta}(\mathbf{x},t) + \dot{C}(\mathbf{x},t)\Theta(\mathbf{x},t)\right)\delta\Theta(\mathbf{x})dV \\ &\quad + \int\limits_{\Omega^{FE}} \mathbf{q}^{FE}(\mathbf{x},t)\cdot\delta\nabla\Theta(\mathbf{x})dV + \int\limits_{\Omega^{FE}} C^{FE}(\mathbf{x},t)\dot{\Theta}(\mathbf{x},t)\delta\Theta(\mathbf{x})dV = \\ &\quad\quad \int\limits_{S_{\mathbf{q}}^{PD}} \mathbf{q}^{*PD}(\mathbf{x},t)\cdot\mathbf{n}(\mathbf{x})\delta\Theta(\mathbf{x})dS + \int\limits_{S_{\mathbf{q}}^{FE}} \mathbf{q}^{*FE}(\mathbf{x},t)\cdot\mathbf{n}(\mathbf{x})\delta\Theta(\mathbf{x})dS \\ &\quad\quad + \int\limits_{\Omega^{PD}} s_{b}^{PD}(\mathbf{x},t)\delta\Theta(\mathbf{x})dV + \int\limits_{\Omega^{FE}} s_{b}^{FE}(\mathbf{x},t)\delta\Theta(\mathbf{x})dV \end{aligned} \tag{5.9}$$

in which $\Theta(\mathbf{x},t)$ is the unknown field variable, $C^{\alpha}(\mathbf{x},t)$, $\dot{C}^{\alpha}(\mathbf{x},t)$,$\mathbf{q}^{\alpha}(\mathbf{x},t)$, $\mathbf{q}^{*\alpha}(\mathbf{x},t)$, and $s_{b}^{\alpha}(\mathbf{x},t)$ are the physical property of material and its time variation, the flux due to internal field interactions, the external field flux, and the source generation term, respectively, for $\alpha = PD, FE$. Prior to applying the divergence theorem, Eq. (5.9) can be recast as:

$$
\begin{aligned}
&\int_{\Omega^{PD}} \left[\nabla \cdot (\mathbf{q}^{PD}(\mathbf{x},t)\delta\Theta(\mathbf{x})) - (\nabla \cdot \mathbf{q}^{PD}(\mathbf{x},t))\delta\Theta(\mathbf{x})\right] dV \\
&+ \int_{\Omega^{PD}} \left(C^{PD}(\mathbf{x},t)\dot{\Theta}(\mathbf{x},t) + \dot{C}^{PD}(\mathbf{x},t)\Theta(\mathbf{x},t)\right)\delta\Theta(\mathbf{x})dV \\
&+ \int_{\Omega^{FE}} \mathbf{q}^{FE}(\mathbf{x},t)\cdot\delta\nabla\Theta(\mathbf{x})dV + \int_{\Omega^{FE}} \left(C^{FE}(\mathbf{x},t)\dot{\Theta}(\mathbf{x},t) + \dot{C}^{FE}(\mathbf{x},t)\Theta(\mathbf{x},t)\right)\delta\Theta(\mathbf{x})dV = \\
&\int_{S_{\mathbf{q}}^{PD}} \mathbf{q}^{*PD}(\mathbf{x},t)\cdot\mathbf{n}(\mathbf{x})\delta\Theta(\mathbf{x})dS + \int_{S_{\mathbf{q}}^{FE}} \mathbf{q}^{*FE}(\mathbf{x},t)\cdot\mathbf{n}(\mathbf{x})\delta\Theta(\mathbf{x})dS \\
&+ \int_{\Omega^{PD}} s_b^{PD}(\mathbf{x},t)\delta\Theta(\mathbf{x})dV + \int_{\Omega^{FE}} s_b^{FE}(\mathbf{x},t)\delta\Theta(\mathbf{x})dV
\end{aligned}
\tag{5.10}
$$

Invoking the divergence theorem along with $(\mathbf{q}^{PD}(\mathbf{x},t)\delta\Theta(\mathbf{x}))\cdot\mathbf{n}(\mathbf{x}) = (\mathbf{q}^{PD}(\mathbf{x},t)\cdot\mathbf{n}(\mathbf{x}))\delta\Theta(\mathbf{x})$ into Eq. (5.10) results in:

$$
\begin{aligned}
&\int_{\Omega^{PD}} \left(C^{PD}(\mathbf{x},t)\dot{\Theta}(\mathbf{x},t) + \dot{C}^{PD}(\mathbf{x},t)\Theta(\mathbf{x},t)\right)\delta\Theta(\mathbf{x})dV \\
&+ \oint_{S^{PD}} (\mathbf{q}^{PD}(\mathbf{x},t)\cdot\mathbf{n}(\mathbf{x}))\delta\Theta(\mathbf{x})dS - \int_{\Omega^{PD}} (\nabla\cdot\mathbf{q}^{PD}(\mathbf{x},t))\delta\Theta(\mathbf{x})dV \\
&+ \int_{\Omega^{FE}} \left(C^{FE}(\mathbf{x},t)\dot{\Theta}(\mathbf{x},t) + \dot{C}^{FE}(\mathbf{x},t)\Theta(\mathbf{x},t)\right)\delta\Theta(\mathbf{x})dV + \int_{\Omega^{FE}} \mathbf{q}^{FE}(\mathbf{x},t)\cdot\delta\nabla\Theta(\mathbf{x})dV = \\
&\int_{S_{\mathbf{q}}^{PD}} \mathbf{q}^{*PD}(\mathbf{x},t)\cdot\mathbf{n}(\mathbf{x})\delta\Theta(\mathbf{x})dS + \int_{S_{\mathbf{q}}^{FE}} \mathbf{q}^{*FE}(\mathbf{x},t)\cdot\mathbf{n}(\mathbf{x})\delta\Theta(\mathbf{x})dS \\
&+ \int_{\Omega^{PD}} s_b^{PD}(\mathbf{x},t)\delta\Theta(\mathbf{x})dV + \int_{\Omega^{FE}} s_b^{FE}(\mathbf{x},t)\delta\Theta(\mathbf{x})dV
\end{aligned}
\tag{5.11}
$$

in which $\delta\Theta(\mathbf{x})$ is the virtual field distribution and $\mathbf{n}(\mathbf{x})$ is the unit normal vector to the surface. Replacing the divergence of field flux vector $\nabla\cdot\mathbf{q}^{PD}(\mathbf{x},t)$ with the PD representation of the internal field flux, L^{PD}, and splitting the volume of the domain, Ω^{PD}, into inner region, $\mathcal{D}$, outer region, $\mathcal{R}$, and the boundary layer region, $\mathcal{B}$, Eq. (5.11) can be rewritten as:

$$
\begin{aligned}
&\int\limits_{\mathcal{D}\cup\mathcal{R}} \left(C^{PD}(\mathbf{x},t)\dot{\Theta}(\mathbf{x},t) + \dot{C}^{PD}(\mathbf{x},t)\Theta(\mathbf{x},t)\right)\delta\Theta(\mathbf{x})dV \\
&+ \int\limits_{\mathcal{B}} \left(C^{PD}(\mathbf{x},t)\dot{\Theta}(\mathbf{x},t) + \dot{C}^{PD}(\mathbf{x},t)\Theta(\mathbf{x},t)\right)\delta\Theta(\mathbf{x})dV - \int\limits_{\mathcal{D}} L^{PD}(\mathbf{x},t)\delta\Theta(\mathbf{x})dV \\
&- \int\limits_{\mathcal{R}} L^{PDDO}(\mathbf{x},t)\delta\Theta(\mathbf{x})dV - \int\limits_{\mathcal{B}} L^{PDDO}(\mathbf{x},t)\delta\Theta(\mathbf{x})dV \\
&+ \oint\limits_{S^{PD}} \left(\mathbf{q}^{PD}(\mathbf{x},t)\cdot\mathbf{n}(\mathbf{x})\right)\delta\Theta(\mathbf{x})dS + \int\limits_{\Omega^{FE}} \left(C^{FE}(\mathbf{x},t)\dot{\Theta}(\mathbf{x},t) + \dot{C}^{FE}(\mathbf{x},t)\Theta(\mathbf{x},t)\right)\delta\Theta(\mathbf{x})dV \\
&+ \int\limits_{\Omega^{FE}} \mathbf{q}^{FE}(\mathbf{x},t)\cdot\delta\nabla\Theta(\mathbf{x})dV = \\
&\quad \int\limits_{S^{PD}_{\mathbf{q}}} \mathbf{q}^{*PD}(\mathbf{x},t)\cdot\mathbf{n}(\mathbf{x})\delta\Theta(\mathbf{x})dS + \int\limits_{S^{FE}_{\mathbf{q}}} \mathbf{q}^{*FE}(\mathbf{x},t)\cdot\mathbf{n}(\mathbf{x})\delta\Theta(\mathbf{x})dS \\
&\quad + \int\limits_{\mathcal{D}\cup\mathcal{R}} s_b^{PD}(\mathbf{x},t)\delta\Theta(\mathbf{x})dV + \int\limits_{\mathcal{B}} s_b^{PD}(\mathbf{x},t)\delta\Theta(\mathbf{x})dV + \int\limits_{\Omega^{FE}} s_b^{FE}(\mathbf{x},t)\delta\Theta(\mathbf{x})dV
\end{aligned}
\tag{5.12a}
$$

or

$$
\begin{aligned}
&\int\limits_{\mathcal{D}\cup\mathcal{R}} \left(C^{PD}(\mathbf{x},t)\dot{\Theta}(\mathbf{x},t) + \dot{C}^{PD}(\mathbf{x},t)\Theta(\mathbf{x},t)\right)\delta\Theta(\mathbf{x})dV \\
&+ b\int\limits_{S^{PD}} \left(C^{PD}(\mathbf{x},t)\dot{\Theta}(\mathbf{x},t) + \dot{C}^{PD}(\mathbf{x},t)\Theta(\mathbf{x},t)\right)\delta\Theta(\mathbf{x})dS - \int\limits_{\mathcal{D}} L^{PD}(\mathbf{x},t)\delta\Theta(\mathbf{x})dV \\
&- \int\limits_{\mathcal{R}} L^{PDDO}(\mathbf{x},t)\delta\Theta(\mathbf{x})dV - b\int\limits_{S^{PD}} L^{PDDO}(\mathbf{x},t)\delta\Theta(\mathbf{x})dS \\
&+ \oint\limits_{S^{PD}} \left(\mathbf{q}^{PD}(\mathbf{x},t)\cdot\mathbf{n}(\mathbf{x})\right)\delta\Theta(\mathbf{x})dS + \int\limits_{\Omega^{FE}} \left(C^{FE}(\mathbf{x},t)\dot{\Theta}(\mathbf{x},t) + \dot{C}^{FE}(\mathbf{x},t)\Theta(\mathbf{x},t)\right)\delta\Theta(\mathbf{x})dV \\
&+ \int\limits_{\Omega^{FE}} \mathbf{q}^{FE}(\mathbf{x},t)\cdot\delta\nabla\Theta(\mathbf{x})dV = \\
&\quad \int\limits_{S^{PD}_{\mathbf{q}}} \mathbf{q}^{*PD}(\mathbf{x},t)\cdot\mathbf{n}(\mathbf{x})\delta\Theta(\mathbf{x})dS + \int\limits_{S^{FE}_{\mathbf{q}}} \mathbf{q}^{*FE}(\mathbf{x},t)\cdot\mathbf{n}(\mathbf{x})\delta\Theta(\mathbf{x})dS \\
&\quad + \int\limits_{\mathcal{D}\cup\mathcal{R}} s_b^{PD}(\mathbf{x},t)\delta\Theta(\mathbf{x})dV + b\int\limits_{S^{PD}} s_b^{PD}(\mathbf{x},t)\delta\Theta(\mathbf{x})dS + \int\limits_{\Omega^{FE}} s_b^{FE}(\mathbf{x},t)\delta\Theta(\mathbf{x})dV
\end{aligned}
\tag{5.12b}
$$

The boundary layer region, $\mathscr{B}$, is defined by the surface of the domain, S^{PD}, with its width, b, and for $b \ll 1$, Eq. (5.12b) reduces to:

$$\begin{aligned}
&\int\limits_{\mathcal{D}\cup\mathscr{R}} \left(C^{PD}(\mathbf{x},t)\dot{\Theta}(\mathbf{x},t) + \dot{C}^{PD}(\mathbf{x},t)\Theta(\mathbf{x},t)\right)\delta\Theta(\mathbf{x})dV - \int\limits_{\mathcal{D}} L^{PD}(\mathbf{x},t)\delta\Theta(\mathbf{x})dV \\
&- \int\limits_{\mathscr{R}} L^{PDDO}(\mathbf{x},t)\delta\Theta(\mathbf{x})dV + \oint\limits_{S^{PD}} \left(\mathbf{q}^{PD}(\mathbf{x},t)\cdot\mathbf{n}(\mathbf{x})\right)\delta\Theta(\mathbf{x})dS \\
&+ \int\limits_{\Omega^{FE}} \left(C^{FE}(\mathbf{x},t)\dot{\Theta}(\mathbf{x},t) + \dot{C}^{FE}(\mathbf{x},t)\Theta(\mathbf{x},t)\right)\delta\Theta(\mathbf{x})dV + \int\limits_{\Omega^{FE}} \mathbf{q}^{FE}(\mathbf{x},t)\cdot\delta\nabla\Theta(\mathbf{x})dV = \\
&\int\limits_{S^{PD}_{\mathbf{q}}} \mathbf{q}^{*PD}(\mathbf{x},t)\cdot\mathbf{n}(\mathbf{x})\delta\Theta(\mathbf{x})dS + \int\limits_{S^{FE}_{\mathbf{q}}} \mathbf{q}^{*FE}(\mathbf{x},t)\cdot\mathbf{n}(\mathbf{x})\delta\Theta(\mathbf{x})dS \\
&+ \int\limits_{\mathcal{D}\cup\mathscr{R}} s_b^{PD}(\mathbf{x},t)\delta\Theta(\mathbf{x})dV + \int\limits_{\Omega^{FE}} s_b^{FE}(\mathbf{x},t)\delta\Theta(\mathbf{x})dV
\end{aligned} \tag{5.13}$$

The field equation valid in the entire domain $\Omega = \Omega^{PD} \cup \Omega^{FE}$ is expressed only in terms of the unknown field variable. For a two-dimensional domain, Eq. (5.13) concerns only area and line integrations and simplifies to:

$$\begin{aligned}
&\int\limits_{\mathcal{D}\cup\mathscr{R}} \left(C^{PD}(\mathbf{x},t)\dot{\Theta}(\mathbf{x},t) + \dot{C}^{PD}(\mathbf{x},t)\Theta(\mathbf{x},t)\right)\delta\Theta(\mathbf{x})dA - \int\limits_{\mathcal{D}} L^{PD}(\mathbf{x},t)\delta\Theta(\mathbf{x})dA \\
&- \int\limits_{\mathscr{R}} L^{PDDO}(\mathbf{x},t)\delta\Theta(\mathbf{x})dA + \oint\limits_{\Gamma^{PD}} \left(\mathbf{q}^{PD}(\mathbf{x},t)\cdot\mathbf{n}(\mathbf{x})\right)\delta\Theta(\mathbf{x})d\Gamma \\
&+ \int\limits_{\Omega^{FE}} \left(C^{FE}(\mathbf{x},t)\dot{\Theta}(\mathbf{x},t) + \dot{C}^{FE}(\mathbf{x},t)\Theta(\mathbf{x},t)\right)\delta\Theta(\mathbf{x})dA + \int\limits_{\Omega^{FE}} \mathbf{q}^{FE}(\mathbf{x},t)\cdot\delta\nabla\Theta(\mathbf{x})dA = \\
&\int\limits_{\Gamma^{PD}_{\mathbf{q}}} \left(\mathbf{q}^{*PD}(\mathbf{x},t)\cdot\mathbf{n}(\mathbf{x})\right)\delta\Theta(\mathbf{x})d\Gamma + \int\limits_{\Gamma^{FE}_{\mathbf{q}}} \left(\mathbf{q}^{*FE}(\mathbf{x},t)\cdot\mathbf{n}(\mathbf{x})\right)\delta\Theta(\mathbf{x})d\Gamma \\
&+ \int\limits_{\mathcal{D}\cup\mathscr{R}} s_b^{PD}(\mathbf{x},t)\delta\Theta(\mathbf{x})dA + \int\limits_{\Omega^{FE}} s_b^{FE}(\mathbf{x},t)\delta\Theta(\mathbf{x})dA
\end{aligned} \tag{5.14}$$

in which $\mathcal{D}$ and $\mathscr{R}$ represent the area of the interior and outer regions, Γ^{PD} represents the entire boundary of the domain, $\Gamma^{PD}_{\mathscr{B}_{\mathbf{q}}}$ and $\Gamma^{PD}_{\mathscr{B}_{\Theta}}$ represent the segment of the boundary subjected to the external field flux and field constraints, respectively. The region, $\mathscr{B}$, with a thickness of b is referred to as the boundary layer region. As the width approaches zero, i.e., $b \to 0$, Γ coincides with the boundary of PD domain as shown in Fig. 5.2.

The PD and FE domains are bounded by $\Gamma^{PD} = \Gamma^{PD}_{\mathscr{B}_\mathbf{q}} \cup \Gamma^{PD}_{\mathscr{B}_\Theta} \cup \Gamma^{PD}_{\mathcal{I}}$ and $\Gamma^{FE} = \Gamma^{FE}_{\mathscr{B}_\mathbf{q}} \cup \Gamma^{FE}_{\mathscr{B}_\Theta} \cup \Gamma^{FE}_{\mathcal{I}}$, respectively. Along the PD/FE interface, $\mathbf{x} \in \left(\Gamma^{PD}_{\mathcal{I}} = \Gamma^{FE}_{\mathcal{I}}\right)$, the field variable is continuous, i.e., $\Theta^{PD}(\mathbf{x}, t) = \Theta^{FE}(\mathbf{x}, t)$ along with their corresponding variations as $\delta\Theta^{PD}(\mathbf{x}, t) = \delta\Theta^{FE}(\mathbf{x}, t)$. The evaluation of each integral in Eq. (5.14) can be performed by discretizing the domain and employing numerical quadrature that requires the discrete form of PD internal field flux, $L^{BB}(\mathbf{x})$.

5.4 Discretization for PD/FEM Coupling

For two-dimensional analysis, the material point k in the PD domain has an incremental volume, boundary area, and length denoted as $V_{(k)} = hA_{(k)}$, $S_{(k)} = h\Gamma_{(k)}$, and $\Gamma_{(k)}$, respectively, where $h = 1$ is the thickness of the domain.

5.4.1 *Deformation Analysis*

As shown in Fig. 5.3, the domain is divided into PD and FE regions. The PD and FE regions are bounded by $\Gamma^{PD} = \Gamma^{PD}_{\mathscr{B}_\sigma} \cup \Gamma^{PD}_{\mathscr{B}_\mathbf{u}} \cup \Gamma^{PD}_{\mathcal{I}}$ and $\Gamma^{FE} = \Gamma^{FE}_{\mathscr{B}_\sigma} \cup \Gamma^{FE}_{\mathscr{B}_\mathbf{u}} \cup \Gamma^{FE}_{\mathcal{I}}$, respectively. The PD region $\Omega^{PD} = \mathcal{D}^{PD} \cup \mathscr{R}^{PD} \cup \mathscr{B}^{PD} \cup \mathcal{I}^{PD}$ is discretized with $N^{PD} = N^{PD}_{\mathcal{D}} + N^{PD}_{\mathscr{R}} + N^{PD}_{\mathscr{B}} + N^{PD}_{\mathcal{I}}$ PD points, where $N^{PD}_{\mathcal{D}}$, $N^{PD}_{\mathscr{R}}$, $N^{PD}_{\mathscr{B}}$, and $N^{PD}_{\mathcal{I}}$ represent the number of points associated with regions, $\mathcal{D}^{PD}$, $\mathscr{R}^{PD}$, $\mathscr{B}^{PD}$, and $\mathcal{I}^{PD}$, respectively. For the FE domain, $\Omega^{FE} = \mathcal{D}^{FE} \cup \mathscr{B}^{FE}$ is discretized with E elements. Along the PD/FE interface, $\mathbf{x} \in \left(\Gamma^{PD}_{\mathcal{I}} = \Gamma^{FE}_{\mathcal{I}}\right)$, the displacement is continuous, i.e., $\mathbf{u}^{PD}(\mathbf{x}, t) = \mathbf{u}^{FE}(\mathbf{x}, t)$ with their corresponding variations as $\delta\mathbf{u}^{PD}(\mathbf{x}, t) = \delta\mathbf{u}^{FE}(\mathbf{x}, t)$. The coupling approach is valid for BB-, OSB-, and NOSB-PD, i.e., $\mathbf{L}^{\alpha}(\mathbf{x})$ with $\alpha = BB, OSB, NOSB$.

Under small deformation assumptions, the discrete form of Eq. (5.7) for virtual work can be expressed as:

$$
\begin{aligned}
&\sum_{k=1}^{N^{PD}_{\mathcal{D}}+N^{PD}_{\mathscr{R}}} \rho^{PD}_{(k)} \delta\mathbf{u}^{PD}_{(k)} \cdot \ddot{\mathbf{u}}^{PD}_{(k)} V_{(k)} - \sum_{k=1}^{N^{PD}_{\mathcal{D}}} \delta\mathbf{u}^{PD}_{(k)} \cdot \left(\sum_{j=1}^{N_{(k)}} \mathbf{t}^{\alpha}_{(k)(j)} V_{(j)} - \sum_{p=1}^{\widetilde{N}_{(k)}} \mathbf{t}^{\alpha}_{(p)(k)} V_{(p)} \right) V_{(k)} \\
&- \sum_{k=1}^{N^{PD}_{\mathscr{R}}} \delta\mathbf{u}^{PD}_{(k)} \cdot \left(\sum_{j=1}^{N_{(k)}} \left(\mathbf{t}^{PDDO}_{(j)(k)} - \mathbf{t}^{PDDO}_{(k)(j)} \right) V_{(j)} \right) V_{(k)} + \sum_{k=1}^{N^{PD}_{\mathscr{B}_\sigma}+N^{PD}_{\mathscr{B}_\mathbf{u}}+N^{PD}_{\mathcal{I}}} \delta\mathbf{u}^{PD}_{(k)} \cdot \left(\boldsymbol{\sigma}^{PD}_{(k)} \mathbf{n}_{(k)} \right) \Gamma_{(k)} \\
&+ \sum_{e=1}^{E} \rho^{FE}_{(k)} \delta\mathbf{u}^{FE}_{(e)} \cdot \ddot{\mathbf{u}}^{FE}_{(k)} V_{(k)} + \sum_{e=1}^{E} \operatorname{tr}\left(\left(\nabla \delta\mathbf{u}^{FE}_{(e)} \right)^{T} \boldsymbol{\sigma}^{FE}_{(e)} \right) V_{(e)} - \sum_{k=1}^{N^{PD}_{\mathscr{B}_\sigma}} \delta\mathbf{u}^{PD}_{(k)} \cdot \mathbf{T}^{*PD} \Gamma_{(k)} \\
&- \sum_{k=1}^{N^{PD}_{\mathcal{D}}+N^{PD}_{\mathscr{R}}} \delta\mathbf{u}^{PD}_{(k)} \cdot \mathbf{b}^{PD}_{(k)} V_{(k)} - \sum_{e=1}^{E_{\mathscr{B}_\sigma}} \delta\mathbf{u}^{FE}_{(e)} \cdot \mathbf{T}^{*FE}_{(e)} \Gamma_{(e)} - \sum_{e=1}^{E} \delta\mathbf{u}^{FE}_{(e)} \cdot \mathbf{b}^{FE}_{(e)} V_{(e)} = 0
\end{aligned}
\tag{5.15}
$$

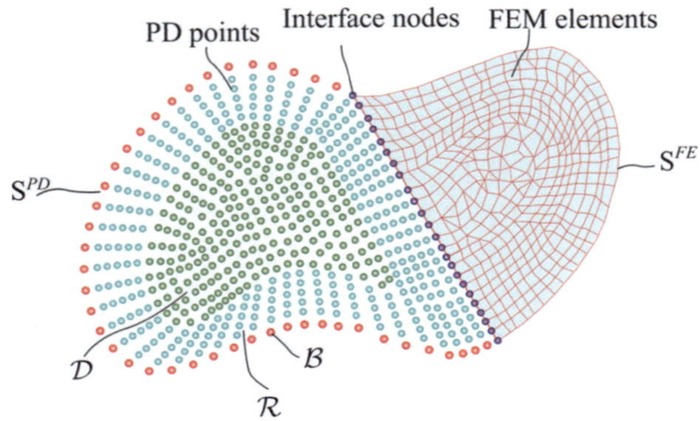

Fig. 5.3 Numerical discretization of the domain for a PD equilibrium equation

In the interior PD region, $\mathcal{D}^{PD}$**,** the virtual work due to internal force vector can be evaluated as:

$$\begin{aligned}&\sum_{k=1}^{N_{\mathcal{D}}^{PD}} \delta\mathbf{u}_{(k)}^{PD} \cdot \left(\sum_{j=1}^{N_{(k)}} \mathbf{t}_{(k)(j)}^{\alpha} V_{(j)} - \sum_{p=1}^{\widetilde{N}_{(k)}} \mathbf{t}_{(p)(k)}^{\alpha} V_{(p)} \right) V_{(k)} = \\ &\left\{\delta\mathbf{V}_{\mathcal{D}}^{PD}\right\}^T \sum_{k=1}^{N_{\mathcal{D}}^{PD}} \left(\sum_{j=1}^{N_{(k)}} \mathbf{k}_{(k)(j)}^{\alpha} - \sum_{p=1}^{\widetilde{N}_{(k)}} \mathbf{k}_{(p)(k)}^{\alpha} \right) \left\{\mathbf{V}^{PD}\right\}\end{aligned} \tag{5.16}$$

where $\mathbf{V}^{PD}$ is the unknown displacement vector. It can be partitioned as:

$$\mathbf{V}^{PD} = \left\{ \mathbf{V}_{\mathcal{D}}^{PD} \;\; \mathbf{V}_{\mathscr{R}}^{PD} \;\; \mathbf{V}_{\mathscr{B}_{\sigma}}^{PD} \;\; \mathbf{V}_{\mathscr{B}_{\mathbf{u}}}^{PD} \;\; \mathbf{V}_{\mathcal{I}}^{PD} \right\}^T \tag{5.17}$$

in which the vectors, $\mathbf{V}_{\mathcal{D}}^{PD}$, $\mathbf{V}_{\mathscr{R}}^{PD}$, $\mathbf{V}_{\mathscr{B}_{\sigma}}^{PD}$, $\mathbf{V}_{\mathscr{B}_{\mathbf{u}}}^{PD}$, and $\mathbf{V}_{\mathcal{I}}^{PD}$ include the displacement components at points located in regions $\mathcal{D}^{PD}$, $\mathscr{R}^{PD}$, $\Gamma_{\mathscr{B}_{\sigma}}^{PD}$, $\Gamma_{\mathscr{B}_{\mathbf{u}}}^{PD}$, and $\Gamma_{\mathcal{I}}^{FE}$, respectively. Subsequently, the virtual work due to the PD internal force vector ,$\mathbf{F}_{\mathcal{D}}$, in region, $\mathcal{D}^{PD}$, can be evaluated as:

$$\begin{aligned}&\sum_{k=1}^{N_{\mathcal{D}}^{PD}} \delta\mathbf{u}_{(k)}^{PD} \cdot \left(\sum_{j=1}^{N_{(k)}} \mathbf{t}_{(k)(j)}^{\alpha} V_{(j)} - \sum_{p=1}^{\widetilde{N}_{(k)}} \mathbf{t}_{(p)(k)}^{\alpha} V_{(p)} \right) V_{(k)} \\ &= \left\{\delta\mathbf{V}_{\mathcal{D}}^{PD}\right\}^T \left\{\mathbf{F}_{\mathcal{D}}\right\} \\ &= \left\{\delta\mathbf{V}_{\mathcal{D}}^{PD}\right\}^T \left[\mathbf{K}_{\mathcal{D}\mathcal{D}}^{PD} \quad \mathbf{K}_{\mathcal{D}\mathscr{R}}^{PD} \quad \mathbf{0} \quad \mathbf{0} \quad \mathbf{0} \right] \begin{Bmatrix} \mathbf{V}_{\mathcal{D}}^{PD} \\ \mathbf{V}_{\mathscr{R}}^{PD} \\ \mathbf{V}_{\mathscr{B}_{\sigma}}^{PD} \\ \mathbf{V}_{\mathscr{B}_{\mathbf{u}}}^{PD} \\ \mathbf{V}_{\mathcal{I}}^{PD} \end{Bmatrix}\end{aligned} \tag{5.18}$$

where the submatrices $\mathbf{K}_{\mathscr{D}\mathscr{D}}^{PD}$ and $\mathbf{K}_{\mathscr{D}\mathscr{R}}^{PD}$ with dimensions $2N_{\mathscr{D}}^{PD} \times 2N_{\mathscr{D}}^{PD}$ and $2N_{\mathscr{D}}^{PD} \times 2N_{\mathscr{R}}^{PD}$, respectively, represent the self-stiffness and coupling stiffness matrices between the unknown vectors of $\mathbf{V}_{\mathscr{D}}^{PD}$ and $\mathbf{V}_{\mathscr{R}}^{PD}$. The virtual work due to the internal force vector, $\mathbf{F}_{\mathscr{R}}$, in the PDDO region, $\mathscr{R}^{PD}$, and the virtual work due to internal traction, $\mathbf{F}_{\mathscr{B}_\sigma}$, around the PD boundary, $\mathscr{B}^{PD}$, and interface, $\mathcal{I}^{PD}$, can be evaluated as:

$$\begin{aligned}
&\sum_{k=1}^{N_{\mathscr{R}}^{PD}} \delta\mathbf{u}_{(k)}^{PD} \cdot \left(\sum_{j=1}^{N_{(k)}} \left(\mathbf{t}_{(j)(k)}^{PDDO} - \mathbf{t}_{(k)(j)}^{PDDO} \right) V_{(j)} \right) V_{(k)} \\
&= \left\{ \delta\mathbf{V}_{\mathscr{R}}^{PD} \right\}^T \sum_{k=1}^{N_{\mathscr{R}}^{PD}} \left(\sum_{j=1}^{N_{(k)}} \left(\mathbf{k}_{(j)(k)}^{\alpha} - \mathbf{k}_{(k)(j)}^{\alpha} \right) \right) \left\{ \mathbf{V}^{PD} \right\} \\
&= \left\{ \delta\mathbf{V}_{\mathscr{R}}^{PD} \right\}^T \left\{ \mathbf{F}_{\mathscr{R}} \right\} \\
&= \left\{ \delta\mathbf{V}_{\mathscr{R}}^{PD} \right\}^T \begin{bmatrix} \mathbf{K}_{\mathscr{R}\mathscr{D}}^{PD} & \mathbf{K}_{\mathscr{R}\mathscr{R}}^{PD} & \mathbf{K}_{\mathscr{R}\mathscr{B}_\sigma}^{PD} & \mathbf{K}_{\mathscr{R}\mathscr{B}_\mathbf{u}}^{PD} & \mathbf{K}_{\mathscr{R}\mathcal{I}}^{PD} \end{bmatrix} \begin{Bmatrix} \mathbf{V}_{\mathscr{D}}^{PD} \\ \mathbf{V}_{\mathscr{R}}^{PD} \\ \mathbf{V}_{\mathscr{B}_\sigma}^{PD} \\ \mathbf{V}_{\mathscr{B}_\mathbf{u}}^{PD} \\ \mathbf{V}_{\mathcal{I}}^{PD} \end{Bmatrix}
\end{aligned} \tag{5.19}$$

and

$$\begin{aligned}
&\sum_{k=1}^{N_{\mathscr{B}_\sigma}^{PD}+N_{\mathscr{B}_\mathbf{u}}^{PD}+N_{\mathcal{I}}^{PD}} \delta\mathbf{u}_{(k)}^{PD} \cdot \left(\boldsymbol{\sigma}_{(k)}^{PD} \mathbf{n}_{(k)} \right) \Gamma_{(k)} = \begin{Bmatrix} \delta\mathbf{V}_{\mathscr{B}_\sigma}^{PD} \\ \delta\mathbf{V}_{\mathcal{I}}^{PD} \end{Bmatrix}^T \begin{Bmatrix} \mathbf{F}_{\mathscr{B}_\sigma} \\ \mathbf{F}_{\mathcal{I}} \end{Bmatrix} \begin{Bmatrix} \mathbf{V}_{\mathscr{B}_\sigma}^{PD} \\ \mathbf{V}_{\mathscr{B}_\mathbf{u}}^{PD} \\ \mathbf{V}_{\mathcal{I}}^{PD} \end{Bmatrix} \\
&= \begin{Bmatrix} \delta\mathbf{V}_{\mathscr{B}_\sigma}^{PD} \\ \delta\mathbf{V}_{\mathscr{B}_\mathbf{u}}^{PD} \\ \delta\mathbf{V}_{\mathcal{I}}^{PD} \end{Bmatrix} \begin{bmatrix} \mathbf{0} & \mathbf{K}_{\mathscr{B}_\sigma\mathscr{R}}^{PD} & \mathbf{K}_{\mathscr{B}_\sigma\mathscr{B}_\sigma}^{PD} & \mathbf{K}_{\mathscr{B}_\sigma\mathscr{B}_\mathbf{u}}^{PD} & \mathbf{K}_{\mathscr{B}_\sigma\mathcal{I}}^{PD} \\ \mathbf{0} & \mathbf{K}_{\mathscr{B}_\mathbf{u}\mathscr{R}}^{PD} & \mathbf{K}_{\mathscr{B}_\mathbf{u}\mathscr{B}_\sigma}^{PD} & \mathbf{K}_{\mathscr{B}_\mathbf{u}\mathscr{B}_\mathbf{u}}^{PD} & \mathbf{K}_{\mathscr{B}_\mathbf{u}\mathcal{I}}^{PD} \\ \mathbf{0} & \mathbf{K}_{\mathcal{I}\mathscr{R}}^{PD} & \mathbf{K}_{\mathcal{I}\mathscr{B}_\sigma}^{PD} & \mathbf{K}_{\mathcal{I}\mathscr{B}_\mathbf{u}}^{PD} & \mathbf{K}_{\mathcal{I}\mathcal{I}}^{PD} \end{bmatrix} \begin{Bmatrix} \mathbf{V}_{\mathscr{D}}^{PD} \\ \mathbf{V}_{\mathscr{R}}^{PD} \\ \mathbf{V}_{\mathscr{B}_\sigma}^{PD} \\ \mathbf{V}_{\mathscr{B}_\mathbf{u}}^{PD} \\ \mathbf{V}_{\mathcal{I}}^{PD} \end{Bmatrix}
\end{aligned} \tag{5.20}$$

where the submatrices $\mathbf{K}_{\mathscr{R}\mathscr{R}}^{PD}$ and $\mathbf{K}_{\mathscr{R}\mathscr{D}}^{PD}$, $\mathbf{K}_{\mathscr{R}\mathscr{B}_\sigma}^{PD}$, $\mathbf{K}_{\mathscr{R}\mathscr{B}_\mathbf{u}}^{PD}$ $\mathbf{K}_{\mathscr{R}\mathcal{I}}^{PD}$ represent the self-stiffness and coupling stiffness matrices between the unknown vectors of $\mathbf{V}_{\mathscr{D}}^{PD}$, $\mathbf{V}_{\mathscr{R}}^{PD}$, $\mathbf{V}_{\mathscr{B}_\sigma}^{PD}$, $\mathbf{V}_{\mathscr{B}_\mathbf{u}}^{PD}$, and $\mathbf{V}_{\mathcal{I}}^{PD}$. Their dimensions are $2N_{\mathscr{R}}^{PD} \times 2N_{\mathscr{R}}^{PD}$, $2N_{\mathscr{R}}^{PD} \times 2N_{\mathscr{D}}^{PD}$, $2N_{\mathscr{R}}^{PD} \times 2N_{\mathscr{B}_\sigma}^{PD}$, $2N_{\mathscr{R}}^{PD} \times 2N_{\mathscr{B}_\sigma}^{PD}$ and $2N_{\mathscr{R}}^{PD} \times 2N_{\mathcal{I}}^{PD}$, respectively. The submatrices $\mathbf{K}_{\mathscr{B}_\sigma\mathscr{B}_\sigma}^{PD}$, $\mathbf{K}_{\mathscr{B}_\mathbf{u}\mathscr{B}_\mathbf{u}}^{PD}$, $\mathbf{K}_{\mathcal{I}\mathcal{I}}^{PD}$ and $\mathbf{K}_{\mathscr{B}_\sigma\mathscr{R}}^{PD}$, $\mathbf{K}_{\mathscr{B}_\sigma\mathscr{B}_\mathbf{u}}^{PD}$, $\mathbf{K}_{\mathscr{B}_\sigma\mathcal{I}}^{PD}$, $\mathbf{K}_{\mathscr{B}_\mathbf{u}\mathscr{R}}^{PD}$, $\mathbf{K}_{\mathscr{B}_\mathbf{u}\mathscr{B}_\sigma}^{PD}$, $\mathbf{K}_{\mathscr{B}_\mathbf{u}\mathcal{I}}^{PD}$, $\mathbf{K}_{\mathcal{I}\mathscr{R}}^{PD}$, $\mathbf{K}_{\mathcal{I}\mathscr{B}_\sigma}^{PD}$,$\mathbf{K}_{\mathcal{I}\mathscr{B}_\mathbf{u}}^{PD}$ represent the self-stiffness and coupling stiffness matrices between the unknown vectors of $\mathbf{V}_{\mathscr{R}}^{PD}$, $\mathbf{V}_{\mathscr{B}_\sigma}^{PD}$, $\mathbf{V}_{\mathscr{B}_\mathbf{u}}^{PD}$, and $\mathbf{V}_{\mathcal{I}}^{PD}$. Their dimensions are

$2N^{PD}_{\mathscr{B}_\sigma} \times 2N^{PD}_{\mathscr{B}_\sigma}$, $2N^{PD}_{\mathscr{B}_\mathbf{u}} \times 2N^{PD}_{\mathscr{B}_\mathbf{u}}$, $2N^{PD}_{\mathcal{I}} \times 2N^{PD}_{\mathcal{I}}$, $2N^{PD}_{\mathscr{B}_\sigma} \times 2N^{PD}_{\mathscr{R}}$, $2N^{PD}_{\mathscr{B}_\sigma} \times 2N^{PD}_{\mathscr{B}_\mathbf{u}}$, $2N^{PD}_{\mathscr{B}_\sigma} \times 2N^{PD}_{\mathcal{I}}$, $2N^{PD}_{\mathscr{B}_\mathbf{u}} \times 2N^{PD}_{\mathscr{R}}$, $2N^{PD}_{\mathscr{B}_\mathbf{u}} \times 2N^{PD}_{\mathscr{B}_\sigma}$, $2N^{PD}_{\mathscr{B}_\mathbf{u}} \times 2N^{PD}_{\mathcal{I}}$, $2N^{PD}_{\mathcal{I}} \times 2N^{PD}_{\mathscr{R}}$, $2N^{PD}_{\mathcal{I}} \times 2N^{PD}_{\mathscr{B}_\sigma}$ and $2N^{PD}_{\mathcal{I}} \times 2N^{PD}_{\mathscr{B}_\mathbf{u}}$, respectively.

The virtual work due to the inertial force vectors $\mathbf{F}_{\ddot{\mathbf{u}}\mathcal{D}}$ and $\mathbf{F}_{\ddot{\mathbf{u}}\mathscr{R}}$ in the PD domain of $\mathcal{D} \cup \mathscr{R}$ can be evaluated as:

$$\sum_{k=1}^{N^{PD}_{\mathcal{D}}+N^{PD}_{\mathscr{R}}} \rho^{PD}_{(k)} \delta\mathbf{u}^{PD}_{(k)} \cdot \ddot{\mathbf{u}}^{PD}_{(k)} V_{(k)} = \begin{Bmatrix} \delta\mathbf{V}^{PD}_{\mathcal{D}} \\ \delta\mathbf{V}^{PD}_{\mathscr{R}} \end{Bmatrix}^T \begin{Bmatrix} \mathbf{F}_{\ddot{\mathbf{u}}\mathcal{D}} \\ \mathbf{F}_{\ddot{\mathbf{u}}\mathscr{R}} \end{Bmatrix}$$
$$= \begin{Bmatrix} \delta\mathbf{V}^{PD}_{\mathcal{D}} \\ \delta\mathbf{V}^{PD}_{\mathscr{R}} \end{Bmatrix}^T \begin{bmatrix} \mathbf{M}^{PD}_{\mathcal{D}\mathcal{D}} & \mathbf{0} & \mathbf{0} & \mathbf{0} & \mathbf{0} \\ \mathbf{0} & \mathbf{M}^{PD}_{\mathscr{R}\mathscr{R}} & \mathbf{0} & \mathbf{0} & \mathbf{0} \end{bmatrix} \begin{Bmatrix} \ddot{\mathbf{V}}^{PD}_{\mathcal{D}} \\ \ddot{\mathbf{V}}^{PD}_{\mathscr{R}} \\ \ddot{\mathbf{V}}^{PD}_{\mathscr{B}_\sigma} \\ \ddot{\mathbf{V}}^{PD}_{\mathscr{B}_\mathbf{u}} \\ \ddot{\mathbf{V}}^{PD}_{\mathcal{I}} \end{Bmatrix} \tag{5.21}$$

in which the submatrices $\mathbf{M}^{PD}_{\mathcal{D}\mathcal{D}}$ and $\mathbf{M}^{PD}_{\mathscr{R}\mathscr{R}}$ are diagonal mass matrices in PD regions $\mathcal{D}^{PD}$ and $\mathscr{R}^{PD}$ with dimensions $2N^{PD}_{\mathcal{D}} \times 2N^{PD}_{\mathcal{D}}$ and $2N^{PD}_{\mathscr{R}} \times 2N^{PD}_{\mathscr{R}}$, respectively.

Due to the applied external tractions, $\mathbf{T}^{*PD}$ and body loads, $\mathbf{b}^{*PD}$, the virtual work can be evaluated as:

$$\sum_{k=1}^{N^{PD}_{\mathscr{B}_\sigma}} \delta\mathbf{u}^{PD}_{(k)} \cdot \mathbf{T}^{*PD} \Gamma_{(k)} = \left\{ \delta\mathbf{V}^{PD}_{\mathscr{B}_\sigma} \right\}^T \left\{ \mathbf{F}^{*PD}_{\mathscr{B}_\sigma} \right\} \tag{5.22}$$

and

$$\sum_{k=1}^{N^{PD}_{\mathcal{D}}+N^{PD}_{\mathscr{R}}} \delta\mathbf{u}^{PD}_{(k)} \cdot \mathbf{b}^{PD}_{(k)} V_{(k)} = \left\{ \delta\mathbf{V}^{PD}_{\mathcal{D}} \quad \delta\mathbf{V}^{PD}_{\mathscr{R}} \right\}^T \begin{Bmatrix} \mathbf{F}^{*PD}_{\mathbf{b}_\mathcal{D}} \\ \mathbf{F}^{*PD}_{\mathbf{b}_\mathscr{R}} \end{Bmatrix} \tag{5.23}$$

in which the force vectors $\mathbf{F}^{*PD}_{\mathscr{B}_\sigma}$, $\mathbf{F}^{*PD}_{\mathbf{b}_\mathcal{D}}$, and $\mathbf{F}^{*PD}_{\mathbf{b}_\mathscr{R}}$ are defined as:

$$\mathbf{F}^{*PD}_{\mathscr{B}_\sigma} = \left\{ \mathbf{T}^{*PD}_{(1)} S_{(1)}, \mathbf{T}^{*PD}_{(2)} S_{(2)}, \mathbf{T}^{*PD}_{(3)} S_{(3)}, \ldots, \mathbf{T}^{*PD}_{(N_{\mathscr{B}_\sigma})} S_{(N_{\mathscr{B}_\sigma})} \right\}^T \tag{5.24}$$

$$\mathbf{F}^{*PD}_{\mathbf{b}_\mathcal{D}} = \left\{ \mathbf{b}^{PD}_{(1)} V_{(1)}, \mathbf{b}^{PD}_{(2)} V_{(2)}, \mathbf{b}^{PD}_{(3)} V_{(3)}, \ldots, \mathbf{b}^{PD}_{(N_\mathcal{D})} V_{(N_\mathcal{D})} \right\}^T \tag{5.25}$$

and

$$\mathbf{F}^{*PD}_{\mathbf{b}_\mathscr{R}} = \left\{ \mathbf{b}^{PD}_{(1)} V_{(1)}, \mathbf{b}^{PD}_{(2)} V_{(2)}, \mathbf{b}^{PD}_{(3)} V_{(3)}, \ldots, \mathbf{b}^{PD}_{(N_\mathscr{R})} V_{(N_\mathscr{R})} \right\}^T \tag{5.26}$$

In the FE domain, the vector of unknown displacements, $\mathbf{V}^{FE}$, and its variation, $\delta\mathbf{V}^{FE}$, are defined as:

$$\mathbf{V}^{FE} = \left\{ \mathbf{V}_{\mathcal{I}}^{FE} \quad \mathbf{V}_{\mathcal{D}}^{FE} \quad \mathbf{V}_{\mathcal{B}_\sigma}^{FE} \quad \mathbf{V}_{\mathcal{B}_\mathbf{u}}^{FE} \right\}^T \tag{5.27}$$

and

$$\delta\mathbf{V}^{FE} = \left\{ \delta\mathbf{V}_{\mathcal{I}}^{FE} \quad \delta\mathbf{V}_{\mathcal{D}}^{FE} \quad \delta\mathbf{V}_{\mathcal{B}_\sigma}^{FE} \quad \delta\mathbf{V}_{\mathcal{B}_\mathbf{u}}^{FE} \right\}^T \tag{5.28}$$

where the vectors $\mathbf{V}_{\mathcal{I}}^{FE}$, $\mathbf{V}_{\mathcal{D}}^{FE}$, $\mathbf{V}_{\mathcal{B}_\sigma}^{FE}$ and $\mathbf{V}_{\mathcal{B}_\mathbf{u}}^{FE}$ contain the unknown displacement components at nodal points of PD/FE interface, $\Gamma_{\mathcal{I}}^{FE}$, in the FE region Ω^{FE}, and along the boundary segments, $\Gamma_{\mathcal{B}_\sigma}^{FE}$ and $\Gamma_{\mathcal{B}_\mathbf{u}}^{FE}$, respectively. The virtual work due to internal forces in the FE domain can be expressed as:

$$\sum_{e=1}^{E} \mathrm{tr}\left(\left(\nabla\delta\mathbf{u}_{(e)}^{FE}\right)^T \boldsymbol{\sigma}_{(e)}^{FE}\right) V_{(e)} = \sum_{e=1}^{E} \left\{\delta\mathbf{v}_{(e)}^{FE}\right\}^T \mathbf{k}_{(e)}^{FE} \left\{\mathbf{v}_{(e)}^{FE}\right\}$$
$$= \left\{\delta\mathbf{V}^{FE}\right\}^T \mathbf{K}^{FE} \left\{\mathbf{V}^{FE}\right\} \tag{5.29}$$

where $\mathbf{K}^{FE}$ represents the stiffness matrix in the FE region. It can be partitioned as:

$$\mathbf{K}^{FE} = \begin{bmatrix} \mathbf{K}_{\mathcal{I}\mathcal{I}}^{FE} & \mathbf{K}_{\mathcal{I}\mathcal{D}}^{FE} & \mathbf{K}_{\mathcal{I}\mathcal{B}_\sigma}^{FE} & \mathbf{K}_{\mathcal{I}\mathcal{B}_\mathbf{u}}^{FE} \\ \mathbf{K}_{\mathcal{D}\mathcal{I}}^{FE} & \mathbf{K}_{\mathcal{D}\mathcal{D}}^{FE} & \mathbf{K}_{\mathcal{D}\mathcal{B}_\sigma}^{FE} & \mathbf{K}_{\mathcal{D}\mathcal{B}_\mathbf{u}}^{FE} \\ \mathbf{K}_{\mathcal{B}_\sigma\mathcal{I}}^{FE} & \mathbf{K}_{\mathcal{B}_\sigma\mathcal{D}}^{FE} & \mathbf{K}_{\mathcal{B}_\sigma\mathcal{B}_\sigma}^{FE} & \mathbf{K}_{\mathcal{B}_\sigma\mathcal{B}_\mathbf{u}}^{FE} \\ \mathbf{K}_{\mathcal{B}_\mathbf{u}\mathcal{I}}^{FE} & \mathbf{K}_{\mathcal{B}_\mathbf{u}\mathcal{D}}^{FE} & \mathbf{K}_{\mathcal{B}_\mathbf{u}\mathcal{B}_\sigma}^{FE} & \mathbf{K}_{\mathcal{B}_\mathbf{u}\mathcal{B}_\mathbf{u}}^{FE} \end{bmatrix} \tag{5.30}$$

in which $\mathbf{K}_{\mathcal{I}\mathcal{I}}^{FE}$, $\mathbf{K}_{\mathcal{D}\mathcal{D}}^{FE}$, $\mathbf{K}_{\mathcal{B}_\sigma\mathcal{B}_\sigma}^{FE}$, and $\mathbf{K}_{\mathcal{B}_\mathbf{u}\mathcal{B}_\mathbf{u}}^{FE}$ represent the self-stiffness submatrices that are only associated with the displacement vectors $\mathbf{V}_{\mathcal{I}}^{FE}$, $\mathbf{V}_{\mathcal{D}}^{FE}$, $\mathbf{V}_{\mathcal{B}_\sigma}^{FE}$, and $\mathbf{V}_{\mathcal{B}_\mathbf{u}}^{FE}$, respectively. The off-diagonal submatrices represent the coupling stiffness matrices among these displacement vectors.

The virtual work due to inertial forces in the FE region can be evaluated as:

$$\sum_{e=1}^{E} \rho_{(e)}^{FE} \delta\mathbf{u}_{(e)}^{FE} \cdot \ddot{\mathbf{u}}_{(e)}^{FE} V_{(e)}$$
$$= \begin{Bmatrix} \delta\mathbf{V}_{\mathcal{I}}^{FE} \\ \delta\mathbf{V}_{\mathcal{D}}^{FE} \\ \delta\mathbf{V}_{\mathcal{B}_\sigma}^{FE} \\ \delta\mathbf{V}_{\mathcal{B}_\mathbf{u}}^{FE} \end{Bmatrix}^T \begin{bmatrix} \mathbf{M}_{\mathcal{I}\mathcal{I}}^{FE} & \mathbf{0} & \mathbf{0} & \mathbf{0} \\ \mathbf{0} & \mathbf{M}_{\mathcal{D}\mathcal{D}}^{FE} & \mathbf{0} & \mathbf{0} \\ \mathbf{0} & \mathbf{0} & \mathbf{M}_{\mathcal{B}_\sigma\mathcal{B}_\sigma}^{FE} & \mathbf{0} \\ \mathbf{0} & \mathbf{0} & \mathbf{0} & \mathbf{M}_{\mathcal{B}_\mathbf{u}\mathcal{B}_\mathbf{u}}^{FE} \end{bmatrix} \begin{Bmatrix} \ddot{\mathbf{V}}_{\mathcal{I}}^{FE} \\ \ddot{\mathbf{V}}_{\mathcal{D}}^{FE} \\ \ddot{\mathbf{V}}_{\mathcal{B}_\sigma}^{FE} \\ \ddot{\mathbf{V}}_{\mathcal{B}_\mathbf{u}}^{FE} \end{Bmatrix} \tag{5.31}$$

in which the submatrices $\mathbf{M}_{\mathcal{I}\mathcal{I}}^{FE}$, $\mathbf{M}_{\mathcal{D}\mathcal{D}}^{FE}$, $\mathbf{M}_{\mathscr{B}_\sigma\mathscr{B}_\sigma}^{FE}$, and $\mathbf{M}_{\mathscr{B}_\mathbf{u}\mathscr{B}_\mathbf{u}}^{FE}$ are diagonal matrices along the PD/FE interface, $\Gamma_{\mathcal{I}}^{FE}$, in the FE region Ω^{FE}, and along the boundary segments, $\Gamma_{\mathscr{B}_\sigma}^{FE}$ and $\Gamma_{\mathscr{B}_\mathbf{u}}^{FE}$ with dimensions $2N_{\mathcal{I}}^{FE} \times 2N_{\mathcal{I}}^{FE}$, $2N_{\mathcal{D}}^{FE} \times 2N_{\mathcal{D}}^{FE}$, $2N_{\mathscr{B}_\sigma}^{FE} \times 2N_{\mathscr{B}_\sigma}^{FE}$, and $2N_{\mathscr{B}_\mathbf{u}}^{FE} \times 2N_{\mathscr{B}_\mathbf{u}}^{FE}$, respectively.

Due to the applied external tractions and body loads, the virtual work can be evaluated as:

$$\sum_{e=1}^{E_{\mathscr{B}_\sigma}} \delta\mathbf{u}_{(e)}^{FE} \cdot \mathbf{T}_{(e)}^{*FE} \Gamma_{(e)} = \left\{ \delta\mathbf{V}_{\mathscr{B}_\sigma}^{FE} \right\}^T \left\{ \mathbf{F}_{\mathscr{B}_\sigma}^{*FE} \right\} \tag{5.32}$$

and

$$\sum_{e=1}^{E} \delta\mathbf{u}_{(e)}^{FE} \cdot \mathbf{b}_{(e)}^{FE} V_{(e)} = \left\{ \delta\mathbf{V}_{\mathcal{I}}^{FE} \quad \delta\mathbf{V}_{\mathcal{D}}^{FE} \quad \delta\mathbf{V}_{\mathscr{B}_\sigma}^{FE} \quad \delta\mathbf{V}_{\mathscr{B}_\mathbf{u}}^{FE} \right\}^T \begin{Bmatrix} \mathbf{F}_{\mathbf{b}_{\mathcal{I}}}^{*FE} \\ \mathbf{F}_{\mathbf{b}_{\mathcal{D}}}^{*FE} \\ \mathbf{F}_{\mathbf{b}_{\mathscr{B}_\sigma}}^{*FE} \\ \mathbf{F}_{\mathbf{b}_{\mathscr{B}_\mathbf{u}}}^{*FE} \end{Bmatrix} \tag{5.33}$$

where the vectors $\mathbf{F}_{\mathscr{B}_\sigma}^{*FE}$, $\mathbf{F}_{\mathbf{b}_{\mathcal{I}}}^{*FE}$, $\mathbf{F}_{\mathbf{b}_{\mathcal{D}}}^{*FE}$, $\mathbf{F}_{\mathbf{b}_{\mathscr{B}_\sigma}}^{*FE}$, and $\mathbf{F}_{\mathbf{b}_{\mathscr{B}_\mathbf{u}}}^{*FE}$ are defined as:

$$\mathbf{F}_{\mathscr{B}_\sigma}^{*FE} = \left\{ \mathbf{T}_{(1)}^{*FE} S_{(1)}, \mathbf{T}_{(2)}^{*FE} S_{(2)}, \mathbf{T}_{(3)}^{*FE} S_{(3)}, \ldots, \mathbf{T}_{\left(N_{\mathscr{B}_\sigma}^{FE}\right)}^{*FE} S_{\left(N_{\mathscr{B}_\sigma}^{FE}\right)} \right\}^T \tag{5.34}$$

$$\mathbf{F}_{\mathbf{b}_{\mathcal{I}}}^{*FE} = \left\{ \mathbf{b}_{(1)}^{FE} V_{(1)}, \mathbf{b}_{(2)}^{FE} V_{(2)}, \mathbf{b}_{(3)}^{FE} V_{(3)}, \ldots, \mathbf{b}_{\left(N_{\mathcal{I}}^{PD}\right)}^{FE} V_{\left(N_{\mathcal{I}}^{PD}\right)} \right\}^T \tag{5.35}$$

$$\mathbf{F}_{\mathbf{b}_{\mathcal{D}}}^{*FE} = \left\{ \mathbf{b}_{(1)}^{FE} V_{(1)}, \mathbf{b}_{(2)}^{FE} V_{(2)}, \mathbf{b}_{(3)}^{FE} V_{(3)}, \ldots, \mathbf{b}_{\left(N_{\mathcal{D}}^{PD}\right)}^{FE} V_{\left(N_{\mathcal{D}}^{PD}\right)} \right\}^T \tag{5.36}$$

$$\mathbf{F}_{\mathbf{b}_{\mathscr{B}_\sigma}}^{*FE} = \left\{ \mathbf{b}_{(1)}^{FE} V_{(1)}, \mathbf{b}_{(2)}^{FE} V_{(2)}, \mathbf{b}_{(3)}^{FE} V_{(3)}, \ldots, \mathbf{b}_{\left(N_{\mathscr{B}_\sigma}^{PD}\right)}^{FE} V_{\left(N_{\mathscr{B}_\sigma}^{PD}\right)} \right\}^T \tag{5.37}$$

and

$$\mathbf{F}_{\mathbf{b}_{\mathscr{B}_\mathbf{u}}}^{*FE} = \left\{ \mathbf{b}_{(1)}^{FE} V_{(1)}, \mathbf{b}_{(2)}^{FE} V_{(2)}, \mathbf{b}_{(3)}^{FE} V_{(3)}, \ldots, \mathbf{b}_{\left(N_{\mathscr{B}_\mathbf{u}}^{PD}\right)}^{FE} V_{\left(N_{\mathscr{B}_\mathbf{u}}^{PD}\right)} \right\}^T \tag{5.38}$$

After combining Eqs. (5.18) to (5.26) and (5.31) to (5.38), the virtual work expression can be cast in a matrix form as:

$$\delta\mathbf{V}^T \left(\mathbf{F}_{\ddot{\mathbf{u}}} + \mathbf{F} - \mathbf{F}_{\mathbf{b}}^* \right) = \delta\mathbf{V}^T \left(\mathbf{M}\ddot{\mathbf{V}} + \mathbf{K}\mathbf{V} - \mathbf{F} \right) = 0 \tag{5.39}$$

in which the mass and stiffness matrices and force vector can be expressed as:

$$\mathbf{M}=\begin{bmatrix} \mathbf{M}^{PD}_{\mathcal{D}\mathcal{D}} & \mathbf{0} & \mathbf{0} & \mathbf{0} & \mathbf{0} & \mathbf{0} & \mathbf{0} & \mathbf{0} \\ \mathbf{0} & \mathbf{M}^{PD}_{\mathcal{R}\mathcal{R}} & \mathbf{0} & \mathbf{0} & \mathbf{0} & \mathbf{0} & \mathbf{0} & \mathbf{0} \\ \mathbf{0} & \mathbf{0} & \mathbf{0} & \mathbf{0} & \mathbf{0} & \mathbf{0} & \mathbf{0} & \mathbf{0} \\ \mathbf{0} & \mathbf{0} & \mathbf{0} & \mathbf{0} & \mathbf{0} & \mathbf{0} & \mathbf{0} & \mathbf{0} \\ \mathbf{0} & \mathbf{0} & \mathbf{0} & \mathbf{0} & \mathbf{M}^{FE}_{\mathcal{I}\mathcal{I}} & \mathbf{0} & \mathbf{0} & \mathbf{0} \\ \mathbf{0} & \mathbf{0} & \mathbf{0} & \mathbf{0} & \mathbf{0} & \mathbf{M}^{PD}_{\mathcal{D}\mathcal{D}} & \mathbf{0} & \mathbf{0} \\ \mathbf{0} & \mathbf{0} & \mathbf{0} & \mathbf{0} & \mathbf{0} & \mathbf{0} & \mathbf{M}^{FE}_{\mathcal{B}_\sigma\mathcal{B}_\sigma} & \mathbf{0} \\ \mathbf{0} & \mathbf{0} & \mathbf{0} & \mathbf{0} & \mathbf{0} & \mathbf{0} & \mathbf{0} & \mathbf{M}^{FE}_{\mathcal{B}_\mathbf{u}\mathcal{B}_\mathbf{u}} \end{bmatrix} \tag{5.40}$$

$$\mathbf{K}=\begin{bmatrix} \mathbf{K}^{PD}_{\mathcal{D}\mathcal{D}} & \mathbf{K}^{PD}_{\mathcal{D}\mathcal{R}} & \mathbf{0} & \mathbf{0} & \mathbf{0} & \mathbf{0} & \mathbf{0} & \mathbf{0} \\ \mathbf{K}^{PD}_{\mathcal{R}\mathcal{D}} & \mathbf{K}^{PD}_{\mathcal{R}\mathcal{R}} & \mathbf{K}^{PD}_{\mathcal{R}\mathcal{B}_\sigma} & \mathbf{K}^{PD}_{\mathcal{R}\mathcal{B}_\mathbf{u}} & \mathbf{K}^{PD}_{\mathcal{R}\mathcal{I}} & \mathbf{0} & \mathbf{0} & \mathbf{0} \\ \mathbf{0} & \mathbf{K}^{PD}_{\mathcal{B}_\sigma\mathcal{R}} & \mathbf{K}^{PD}_{\mathcal{B}_\sigma\mathcal{B}_\sigma} & \mathbf{K}^{PD}_{\mathcal{B}_\sigma\mathcal{B}_\mathbf{u}} & \mathbf{K}^{PD}_{\mathcal{B}_\sigma\mathcal{I}} & \mathbf{0} & \mathbf{0} & \mathbf{0} \\ \mathbf{0} & \mathbf{K}^{PD}_{\mathcal{B}_\mathbf{u}\mathcal{R}} & \mathbf{K}^{PD}_{\mathcal{B}_\mathbf{u}\mathcal{B}_\sigma} & \mathbf{K}^{PD}_{\mathcal{B}_\mathbf{u}\mathcal{B}_\mathbf{u}} & \mathbf{K}^{PD}_{\mathcal{B}_\mathbf{u}\mathcal{I}} & \mathbf{0} & \mathbf{0} & \mathbf{0} \\ \mathbf{0} & \mathbf{K}^{PD}_{\mathcal{I}\mathcal{R}} & \mathbf{K}^{PD}_{\mathcal{I}\mathcal{B}_\sigma} & \mathbf{K}^{PD}_{\mathcal{I}\mathcal{B}_\mathbf{u}} & \mathbf{K}^{PD}_{\mathcal{I}\mathcal{I}}+\mathbf{K}^{FE}_{\mathcal{I}\mathcal{I}} & \mathbf{K}^{FE}_{\mathcal{I}\mathcal{D}} & \mathbf{K}^{FE}_{\mathcal{I}\mathcal{B}_\sigma} & \mathbf{K}^{FE}_{\mathcal{I}\mathcal{B}_\mathbf{u}} \\ \mathbf{0} & \mathbf{0} & \mathbf{0} & \mathbf{0} & \mathbf{K}^{FE}_{\mathcal{D}\mathcal{I}} & \mathbf{K}^{FE}_{\mathcal{D}\mathcal{D}} & \mathbf{K}^{FE}_{\mathcal{D}\mathcal{B}_\sigma} & \mathbf{K}^{FE}_{\mathcal{D}\mathcal{B}_\mathbf{u}} \\ \mathbf{0} & \mathbf{0} & \mathbf{0} & \mathbf{0} & \mathbf{K}^{FE}_{\mathcal{B}_\sigma\mathcal{I}} & \mathbf{K}^{FE}_{\mathcal{B}_\sigma\mathcal{D}} & \mathbf{K}^{FE}_{\mathcal{B}_\sigma\mathcal{B}_\sigma} & \mathbf{K}^{FE}_{\mathcal{B}_\sigma\mathcal{B}_\mathbf{u}} \\ \mathbf{0} & \mathbf{0} & \mathbf{0} & \mathbf{0} & \mathbf{K}^{FE}_{\mathcal{B}_\mathbf{u}\mathcal{I}} & \mathbf{K}^{FE}_{\mathcal{B}_\mathbf{u}\mathcal{D}} & \mathbf{K}^{FE}_{\mathcal{B}_\mathbf{u}\mathcal{B}_\sigma} & \mathbf{K}^{FE}_{\mathcal{B}_\mathbf{u}\mathcal{B}_\mathbf{u}} \end{bmatrix} \tag{5.41}$$

$$\begin{aligned} \mathbf{V} &= \left\{ \mathbf{V}^{PD} \quad \mathbf{V}^{FE} \right\}^{\mathbf{T}} \\ &= \left\{ \mathbf{V}^{PD}_{\mathcal{D}} \quad \mathbf{V}^{PD}_{\mathcal{R}} \quad \mathbf{V}^{PD}_{\mathcal{B}_\sigma} \quad \mathbf{V}^{PD}_{\mathcal{B}_\mathbf{u}} \quad \mathbf{V}^{PD}_{\mathcal{I}}=\mathbf{V}^{FE}_{\mathcal{I}} \quad \mathbf{V}^{FE}_{\mathcal{D}} \quad \mathbf{V}^{FE}_{\mathcal{B}_\sigma} \quad \mathbf{V}^{FE}_{\mathcal{B}_\mathbf{u}} \right\}^{\mathbf{T}} \end{aligned} \tag{5.42}$$

and

$$\mathbf{F}=\left\{ \mathbf{F}^{*PD}_{\mathbf{b}_\mathcal{D}} \quad \mathbf{F}^{*PD}_{\mathbf{b}_\mathcal{R}} \quad \mathbf{F}^{*PD}_{\mathcal{B}_\sigma} \quad \mathbf{0} \quad \mathbf{F}^{*FE}_{\mathbf{b}_\mathcal{I}} \quad \mathbf{F}^{*FE}_{\mathbf{b}_\mathcal{D}} \quad \mathbf{F}^{*FE}_{\mathcal{B}_\sigma}+\mathbf{F}^{*FE}_{\mathbf{b}_{\mathcal{B}_\sigma}} \quad \mathbf{F}^{*FE}_{\mathbf{b}_{\mathcal{B}_\mathbf{u}}} \right\}^{\mathbf{T}} \tag{5.43}$$

The quantities necessary to construct the submatrices and subvectors associated with the PD region in Eqs. (5.41) and (5.43) are derived in Chaps. 3 and 4.

5.4.2 Field Analysis

Similar to the discretization of PD/FEM coupling for deformation analysis, the domain for the field analysis is divided into PD and FE regions, as shown in Fig. 5.4. The PD and FE regions are bounded by $\Gamma^{PD}=\Gamma^{PD}_{\mathcal{B}_\mathbf{q}}\cup\Gamma^{PD}_{\mathcal{B}_\Theta}\cup\Gamma^{PD}_{\mathcal{I}}$ and $\Gamma^{FE}=\Gamma^{FE}_{\mathcal{B}_\mathbf{q}}\cup\Gamma^{FE}_{\mathcal{B}_\Theta}\cup\Gamma^{FE}_{\mathcal{I}}$, respectively. The PD domain $\Omega^{PD}=\mathcal{D}^{PD}\cup\mathcal{R}^{PD}\cup\mathcal{B}^{PD}\cup\mathcal{I}^{PD}$ is discretized with $N^{PD}=N^{PD}_{\mathcal{D}}+N^{PD}_{\mathcal{R}}+N^{PD}_{\mathcal{B}}+N^{PD}_{\mathcal{I}}$ PD points, where $N^{PD}_{\mathcal{D}}$, $N^{PD}_{\mathcal{R}}$, $N^{PD}_{\mathcal{B}}$, and $N^{PD}_{\mathcal{I}}$ represent the number of points associated with regions,

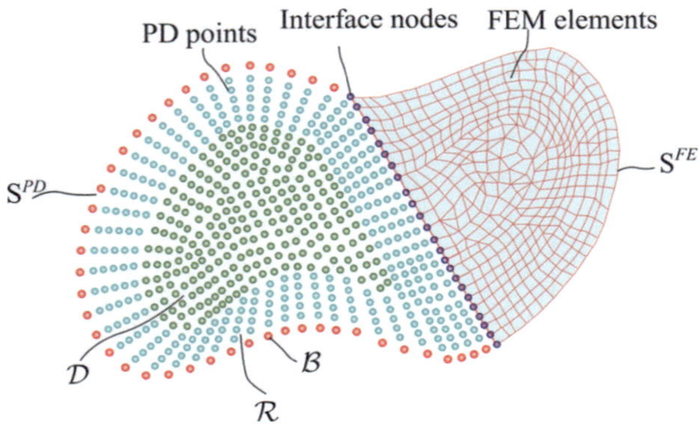

Fig. 5.4 Numerical discretization of the domain for the PD field equation

$\mathcal{D}^{PD}$, $\mathcal{R}^{PD}$, $\mathcal{B}^{PD}$, and $\mathcal{I}^{PD}$, respectively. The FE domain $\Omega^{FE} = \mathcal{D}^{FE} \cup \mathcal{B}^{FE}$ is discretized with E elements. Along the PD/FE interface, $\mathbf{x} \in \left(\Gamma_{\mathcal{I}}^{PD} = \Gamma_{\mathcal{I}}^{FE}\right)$, the field is continuous, i.e., $\Theta^{PD}(\mathbf{x}, t) = \Theta^{FE}(\mathbf{x}, t)$ with their corresponding variations as $\delta\Theta^{PD}(\mathbf{x}, t) = \delta\Theta^{FE}(\mathbf{x}, t)$.

The discrete form of Eq. (5.14) can be expressed as:

$$
\begin{aligned}
&\sum_{k=1}^{N_{\mathcal{D}}^{PD}+N_{\mathcal{R}}^{PD}} \left(C_{(k)}^{PD}\dot{\Theta}_{(k)}^{PD} + \dot{C}_{(k)}^{PD}\Theta_{(k)}^{PD}\right)V_{(k)}\delta\Theta_{(k)}^{PD} - \sum_{k=1}^{N_{\mathcal{D}}^{PD}} \delta\Theta_{(k)}^{PD}\left(\sum_{j=1}^{N_{(k)}} h_{(k)(j)}^{BB}V_{(j)} - \sum_{p=1}^{\widetilde{N}_{(k)}} (h_{(p)(k)}^{BB})V_{(p)}\right)V_{(k)} \\
&- \sum_{k=1}^{N_{\mathcal{R}}^{PD}} \delta\Theta_{(k)}^{PD}\left(\sum_{j=1}^{N_{(k)}} (h_{(j)(k)}^{PDDO} - h_{(k)(j)}^{PDDO})V_{(j)}\right)V_{(k)} \\
&+ \sum_{k=1}^{N_{\mathcal{B}\mathbf{q}}^{PD}+N_{\mathcal{B}\Theta}^{PD}+N_{I}^{PD}} \delta\Theta_{(k)}^{PD}\left(\mathbf{q}_{(k)}^{PD}\cdot\mathbf{n}_{(k)}\right)S_{(k)} + \sum_{e=1}^{E}\left(C_{(k)}^{FE}\dot{\Theta}_{(k)}^{FE} + \dot{C}_{(k)}^{FE}\Theta_{(k)}^{FE}\right)V_{(k)}\delta\Theta_{(k)}^{PD} \\
&+ \sum_{e=1}^{E} \nabla\delta\Theta_{(e)}^{FE}\cdot\mathbf{q}_{(e)}^{FE}V_{(e)} - \sum_{k=1}^{N_{\mathcal{B}\mathbf{q}}^{PD}} \delta\Theta_{(k)}^{PD}\left(\mathbf{q}_{(k)}^{PD*}\cdot\mathbf{n}_{(k)}\right)\Gamma_{(k)} - \sum_{k=1}^{N_{\mathcal{D}}^{PD}+N_{\mathcal{R}}^{PD}} \delta\Theta_{(k)}^{PD}s_{b(k)}^{PD}V_{(k)} \\
&- \sum_{e=1}^{E_{\mathcal{B}\mathbf{q}}} \delta\Theta_{(e)}^{FE}\left(\mathbf{q}_{(e)}^{*FE}\cdot\mathbf{n}_{(e)}\right)\Gamma_{(e)} - \sum_{e=1}^{E} \delta\Theta_{(e)}^{FE}s_{b(e)}^{FE}V_{(e)} = 0
\end{aligned}
\tag{5.44}
$$

The virtual field variation due to internal net field flow in Eq. (5.44) in the interior PD region, $\mathcal{D}^{PD}$, can be evaluated as:

$$
\begin{aligned}
&\sum_{k=1}^{N_{\mathcal{D}}^{PD}} \delta\Theta_{(k)}^{PD}\left(\sum_{j=1}^{N_{(k)}} h_{(k)(j)}^{BB}V_{(j)} - \sum_{p=1}^{\widetilde{N}_{(k)}}\left(h_{(p)(k)}^{BB}\right)V_{(p)}\right)V_{(k)} = \\
&\left\{\delta\mathbf{V}_{\mathcal{D}}^{PD}\right\}^{T}\sum_{k=1}^{N_{\mathcal{D}}^{PD}}\left(\sum_{j=1}^{N_{(k)}}\mathbf{k}_{(k)(j)}^{BB} - \sum_{p=1}^{\widetilde{N}_{(k)}}\mathbf{k}_{(p)(k)}^{BB}\right)\left\{\mathbf{V}^{PD}\right\}
\end{aligned}
\tag{5.45}
$$

where the unknown vector,$\mathbf{V}$, consists of the field variable. It can be partitioned as:

$$\mathbf{V}^{PD} = \left\{ \mathbf{V}_{\mathcal{D}}^{PD} \quad \mathbf{V}_{\mathcal{R}}^{PD} \quad \mathbf{V}_{\mathcal{B}_{\mathbf{q}}}^{PD} \quad \mathbf{V}_{\mathcal{B}_{\Theta}}^{PD} \quad \mathbf{V}_{\mathcal{I}}^{PD} \right\}^{T} \tag{5.46}$$

In which the vectors $\mathbf{V}_{\mathcal{D}}^{PD}$, $\mathbf{V}_{\mathcal{R}}^{PD}$, $\mathbf{V}_{\mathcal{B}_{\mathbf{q}}}^{PD}$, $\mathbf{V}_{\mathcal{B}_{\Theta}}^{PD}$, and $\mathbf{V}_{\mathcal{I}}^{PD}$ include the field values at points located in regions $\mathcal{D}^{PD}$, $\mathcal{R}^{PD}$, $\Gamma_{\mathcal{B}_{\mathbf{q}}}^{PD}$, $\Gamma_{\mathcal{B}_{\Theta}}^{PD}$, and $\Gamma_{\mathcal{I}}^{PD}$, respectively. The virtual field variation due to internal net field flow $\mathbf{F}_{\mathcal{D}}$ in the interior PD region, $\mathcal{D}^{PD}$, can be recast as:

$$\begin{aligned}
&\sum_{k=1}^{N_{\mathcal{D}}^{PD}} \delta\Theta_{(k)}^{PD} \left(\sum_{j=1}^{N_{(k)}} h_{(k)(j)}^{BB} V_{(j)} - \sum_{p=1}^{\widetilde{N}_{(k)}} \left(h_{(p)(k)}^{BB} \right) V_{(p)} \right) V_{(k)} \\
&= \left\{ \delta\mathbf{V}_{\mathcal{D}}^{PD} \right\}^{T} \left\{ \mathbf{F}_{\mathcal{D}} \right\} \\
&= \left\{ \delta\mathbf{V}_{\mathcal{D}}^{PD} \right\}^{T} \left[\mathbf{K}_{\mathcal{D}\mathcal{D}}^{PD} \quad \mathbf{K}_{\mathcal{D}\mathcal{R}}^{PD} \quad \mathbf{0} \quad \mathbf{0} \quad \mathbf{0} \right] \left\{ \begin{array}{c} \mathbf{V}_{\mathcal{D}}^{PD} \\ \mathbf{V}_{\mathcal{R}}^{PD} \\ \mathbf{V}_{\mathcal{B}_{\mathbf{q}}}^{PD} \\ \mathbf{V}_{\mathcal{B}_{\Theta}}^{PD} \\ \mathbf{V}_{\mathcal{I}}^{PD} \end{array} \right\}
\end{aligned} \tag{5.47}$$

where the submatrices $\mathbf{K}_{\mathcal{D}\mathcal{D}}$ and $\mathbf{K}_{\mathcal{D}\mathcal{R}}$ with dimensions $N_{\mathcal{D}}^{PD} \times N_{\mathcal{D}}^{PD}$ and $N_{\mathcal{D}}^{PD} \times N_{\mathcal{R}}^{PD}$, respectively, represent the self-stiffness and coupling stiffness matrices between the unknown vectors of $\mathbf{V}_{\mathcal{D}}^{PD}$ and $\mathbf{V}_{\mathcal{R}}^{PD}$. The virtual field variation due to internal net field flow $\mathbf{F}_{\mathcal{R}}$ in the exterior region $\mathcal{R}^{PD}$ and due to internal field flux $\mathbf{F}_{\mathcal{B}}$ around the PD boundary $\mathcal{B}^{PD}$ and interface $\mathcal{I}^{PD}$ can be evaluated as:

$$\begin{aligned}
&\sum_{k=1}^{N_{\mathcal{R}}^{PD}} \delta\Theta_{(k)}^{PD} \left(\sum_{j=1}^{N_{(k)}} \left(h_{(j)(k)}^{PDDO} - h_{(k)(j)}^{PDDO} \right) V_{(j)} \right) V_{(k)} \\
&= \left\{ \delta\mathbf{V}_{\mathcal{R}}^{PD} \right\}^{T} \sum_{k=1}^{N_{\mathcal{R}}^{PD}} \left(\sum_{j=1}^{N_{(k)}} \left(\mathbf{k}_{(j)(k)}^{PDDO} - \mathbf{k}_{(k)(j)}^{PDDO} \right) \right) \left\{ \mathbf{V}^{PD} \right\} = \left\{ \delta\mathbf{V}_{\mathcal{R}}^{PD} \right\}^{T} \left\{ \mathbf{F}_{\mathcal{R}} \right\} \\
&= \left\{ \delta\mathbf{V}_{\mathcal{R}}^{PD} \right\}^{T} \left[\mathbf{K}_{\mathcal{R}\mathcal{D}}^{PD} \quad \mathbf{K}_{\mathcal{R}\mathcal{R}}^{PD} \quad \mathbf{K}_{\mathcal{R}\mathcal{B}_{\mathbf{q}}}^{PD} \quad \mathbf{K}_{\mathcal{R}\mathcal{B}_{\Theta}}^{PD} \quad \mathbf{K}_{\mathcal{R}\mathcal{I}}^{PD} \right] \left\{ \begin{array}{c} \mathbf{V}_{\mathcal{D}}^{PD} \\ \mathbf{V}_{\mathcal{R}}^{PD} \\ \mathbf{V}_{\mathcal{B}_{\mathbf{q}}}^{PD} \\ \mathbf{V}_{\mathcal{B}_{\Theta}}^{PD} \\ \mathbf{V}_{\mathcal{I}}^{PD} \end{array} \right\}
\end{aligned} \tag{5.48}$$

and

$$\sum_{k=1}^{N^{PD}_{\mathscr{B}_{\mathbf{q}}}+N^{PD}_{\mathscr{B}_{\Theta}}+N^{PD}_{\mathcal{I}}} \delta\Theta^{PD}_{(k)}\left(\mathbf{q}^{PD}_{(k)}\cdot\mathbf{n}_{(k)}\right)S_{(k)}$$

$$= \begin{Bmatrix} \delta\mathbf{V}^{PD}_{\mathscr{B}_{\mathbf{q}}} \\ \delta\mathbf{V}^{PD}_{\mathscr{B}_{\Theta}} \\ \delta\mathbf{V}^{PD}_{\mathcal{I}} \end{Bmatrix}^T \begin{bmatrix} \mathbf{0} & \mathbf{K}^{PD}_{\mathscr{B}_{\mathbf{q}}\mathscr{R}} & \mathbf{K}^{PD}_{\mathscr{B}_{\mathbf{q}}\mathscr{B}_{\mathbf{q}}} & \mathbf{K}^{PD}_{\mathscr{B}_{\mathbf{q}}\mathscr{B}_{\Theta}} & \mathbf{K}^{PD}_{\mathscr{B}_{\mathbf{q}}\mathcal{I}} \\ \mathbf{0} & \mathbf{K}^{PD}_{\mathscr{B}_{\Theta}\mathscr{R}} & \mathbf{K}^{PD}_{\mathscr{B}_{\Theta}\mathscr{B}_{\mathbf{q}}} & \mathbf{K}^{PD}_{\mathscr{B}_{\Theta}\mathscr{B}_{\Theta}} & \mathbf{K}^{PD}_{\mathscr{B}_{\Theta}\mathcal{I}} \\ \mathbf{0} & \mathbf{K}^{PD}_{\mathcal{I}\mathscr{R}} & \mathbf{K}^{PD}_{\mathcal{I}\mathscr{B}_{\mathbf{q}}} & \mathbf{K}^{PD}_{\mathcal{I}\mathscr{B}_{\Theta}} & \mathbf{K}^{PD}_{\mathcal{I}\mathcal{I}} \end{bmatrix} \begin{Bmatrix} \mathbf{V}^{PD}_{\mathcal{D}} \\ \mathbf{V}^{PD}_{\mathscr{R}} \\ \mathbf{V}^{PD}_{\mathscr{B}_{\mathbf{q}}} \\ \mathbf{V}^{PD}_{\mathscr{B}_{\Theta}} \\ \mathbf{V}^{PD}_{\mathcal{I}} \end{Bmatrix} \tag{5.49}$$

$$= \begin{Bmatrix} \delta\mathbf{V}^{PD}_{\mathscr{B}_{\mathbf{q}}} \\ \delta\mathbf{V}^{PD}_{\mathcal{I}} \end{Bmatrix}^T \begin{Bmatrix} \mathbf{F}_{\mathscr{B}_{\mathbf{q}}} \\ \mathbf{F}_{\mathcal{I}} \end{Bmatrix}$$

where the submatrices $\mathbf{K}^{PD}_{\mathscr{R}\mathscr{R}}$ and $\mathbf{K}^{PD}_{\mathscr{R}\mathcal{D}}$, $\mathbf{K}^{PD}_{\mathscr{R}\mathscr{B}_{\mathbf{q}}}$, $\mathbf{K}^{PD}_{\mathscr{R}\mathscr{B}_{\Theta}}$, $\mathbf{K}^{PD}_{\mathscr{R}\mathcal{I}}$ represent the self-stiffness and coupling stiffness matrices between the unknown vectors of $\mathbf{V}^{PD}_{\mathcal{D}}$, $\mathbf{V}^{PD}_{\mathscr{R}}$, $\mathbf{V}^{PD}_{\mathscr{B}_{\mathbf{q}}}$, $\mathbf{V}^{PD}_{\mathscr{B}_{\Theta}}$, and $\mathbf{V}^{PD}_{\mathcal{I}}$. Their dimensions are $N^{PD}_{\mathscr{R}}\times N^{PD}_{\mathscr{R}}$, $N^{PD}_{\mathscr{R}}\times N^{PD}_{\mathcal{D}}$, $N^{PD}_{\mathscr{R}}\times N^{PD}_{\mathscr{B}_{\mathbf{q}}}$, $N^{PD}_{\mathscr{R}}\times N^{PD}_{\mathscr{B}_{\Theta}}$ and $N^{PD}_{\mathscr{R}}\times N^{PD}_{\mathcal{I}}$, respectively. The submatrices $\mathbf{K}^{PD}_{\mathscr{B}_{\mathbf{q}}\mathscr{B}_{\mathbf{q}}}$, $\mathbf{K}^{PD}_{\mathscr{B}_{\Theta}\mathscr{B}_{\Theta}}$, $\mathbf{K}^{PD}_{\mathcal{I}\mathcal{I}}$ and $\mathbf{K}^{PD}_{\mathscr{B}_{\mathbf{q}}\mathscr{R}}$, $\mathbf{K}^{PD}_{\mathscr{B}_{\mathbf{q}}\mathscr{B}_{\Theta}}$, $\mathbf{K}^{PD}_{\mathscr{B}_{\mathbf{q}}\mathcal{I}}$, $\mathbf{K}^{PD}_{\mathscr{B}_{\Theta}\mathscr{R}}$, $\mathbf{K}^{PD}_{\mathscr{B}_{\Theta}\mathscr{B}_{\mathbf{q}}}$, $\mathbf{K}^{PD}_{\mathscr{B}_{\Theta}\mathcal{I}}$, $\mathbf{K}^{PD}_{\mathcal{I}\mathscr{R}}$, $\mathbf{K}^{PD}_{\mathcal{I}\mathscr{B}_{\mathbf{q}}}$, $\mathbf{K}^{PD}_{\mathcal{I}\mathscr{B}_{\Theta}}$ represent the self-stiffness and coupling stiffness matrices between the unknown vectors of $\mathbf{V}^{PD}_{\mathscr{R}}$, $\mathbf{V}^{PD}_{\mathscr{B}_{\mathbf{q}}}$, $\mathbf{V}^{PD}_{\mathscr{B}_{\Theta}}$, and $\mathbf{V}^{PD}_{\mathcal{I}}$. Their dimensions are $N^{PD}_{\mathscr{B}_{\mathbf{q}}}\times N^{PD}_{\mathscr{B}_{\mathbf{q}}}$, $N^{PD}_{\mathscr{B}_{\Theta}}\times N^{PD}_{\mathscr{B}_{\Theta}}$, $N^{PD}_{\mathcal{I}}\times N^{PD}_{\mathcal{I}}$, $N^{PD}_{\mathscr{B}_{\mathbf{q}}}\times N^{PD}_{\mathscr{R}}$, $N^{PD}_{\mathscr{B}_{\mathbf{q}}}\times N^{PD}_{\mathscr{B}_{\Theta}}$, $N^{PD}_{\mathscr{B}_{\mathbf{q}}}\times N^{PD}_{\mathcal{I}}$, $N^{PD}_{\mathscr{B}_{\Theta}}\times N^{PD}_{\mathscr{R}}$, $N^{PD}_{\mathscr{B}_{\Theta}}\times N^{PD}_{\mathscr{B}_{\mathbf{q}}}$, $N^{PD}_{\mathscr{B}_{\Theta}}\times N^{PD}_{\mathcal{I}}$, $N^{PD}_{\mathcal{I}}\times N^{PD}_{\mathscr{R}}$, $N^{PD}_{\mathcal{I}}\times N^{PD}_{\mathscr{B}_{\mathbf{q}}}$, and $N^{PD}_{\mathcal{I}}\times N^{PD}_{\mathscr{B}_{\Theta}}$, respectively.

In the domain of $\mathcal{D}^{PD}\cup\mathscr{R}^{PD}$, the virtual field variation due to transient field and time rate of change material property in terms of vectors, $\mathbf{F}_{\Theta\mathscr{R}}, \mathbf{F}_{\dot{\Theta}\mathscr{R}}$, and $\mathbf{F}_{\Theta\mathcal{D}}, \mathbf{F}_{\dot{\Theta}\mathcal{D}}$ can be evaluated as:

$$\sum_{k=1}^{N_{\mathcal{D}}+N_{\mathscr{R}}} \left(C_{(k)}\dot{\Theta}_{(k)}V_{(k)} + \dot{C}_{(k)}\Theta_{(k)}V_{(k)}\right)\delta\Theta^T_{(k)} = \begin{Bmatrix} \delta\mathbf{V}^{PD}_{\mathcal{D}} \\ \delta\mathbf{V}^{PD}_{\mathscr{R}} \end{Bmatrix}^T \begin{Bmatrix} \mathbf{F}_{\dot{\Theta}\mathcal{D}} + \mathbf{F}_{\Theta\mathcal{D}} \\ \mathbf{F}_{\dot{\Theta}\mathscr{R}} + \mathbf{F}_{\Theta\mathscr{R}} \end{Bmatrix}$$

$$= \begin{Bmatrix} \delta\mathbf{V}^{PD}_{\mathcal{D}} \\ \delta\mathbf{V}^{PD}_{\mathscr{R}} \end{Bmatrix}^T \begin{bmatrix} \mathbf{C}_{\mathcal{D}\mathcal{D}} & \mathbf{0} & \mathbf{0} & \mathbf{0} \\ \mathbf{0} & \mathbf{C}_{\mathscr{R}\mathscr{R}} & \mathbf{0} & \mathbf{0} \end{bmatrix} \begin{Bmatrix} \dot{\mathbf{V}}_{\mathcal{D}} \\ \dot{\mathbf{V}}_{\mathscr{R}} \\ \dot{\mathbf{V}}_{\mathscr{B}_{\mathbf{q}}} \\ \dot{\mathbf{V}}_{\mathscr{B}_{\Theta}} \end{Bmatrix} \tag{5.50}$$

$$+ \begin{Bmatrix} \delta\mathbf{V}^{PD}_{\mathcal{D}} \\ \delta\mathbf{V}^{PD}_{\mathscr{R}} \end{Bmatrix}^T \begin{bmatrix} \dot{\mathbf{C}}_{\mathcal{D}\mathcal{D}} & \mathbf{0} & \mathbf{0} & \mathbf{0} \\ \mathbf{0} & \dot{\mathbf{C}}_{\mathscr{R}\mathscr{R}} & \mathbf{0} & \mathbf{0} \end{bmatrix} \begin{Bmatrix} \mathbf{V}_{\mathcal{D}} \\ \mathbf{V}_{\mathscr{R}} \\ \mathbf{V}_{\mathscr{B}_{\mathbf{q}}} \\ \mathbf{V}_{\mathscr{B}_{\Theta}} \end{Bmatrix}$$

in which submatrices $\mathbf{C}_{\mathcal{DD}}$, $\mathbf{C}_{\mathcal{RR}}$ represent diagonal field damping and $\dot{\mathbf{C}}_{\mathcal{DD}}$, $\dot{\mathbf{C}}_{\mathcal{RR}}$ represent the diagonal transient material property submatrices for regions $\mathcal{D}$ and $\mathcal{R}$ with dimensions $N_{\mathcal{D}} \times N_{\mathcal{D}}$ and $N_{\mathcal{R}} \times N_{\mathcal{R}}$, respectively.

Due to the applied external field flux and applied field source, the virtual field variation can be evaluated as:

$$\sum_{k=1}^{N_{\mathcal{B}_{\mathbf{q}}}^{PD}} \delta\Theta_{(k)}^{PD}\left(\mathbf{q}_{(k)}^{*PD}\cdot\mathbf{n}_{(k)}\right)S_{(k)} = \left\{\delta\mathbf{V}_{\mathcal{B}_{\mathbf{q}}}^{PD}\right\}^{T}\left\{\mathbf{F}_{\mathcal{B}_{\mathbf{q}}}^{*PD}\right\} \tag{5.51}$$

and

$$\sum_{k=1}^{N_{\mathcal{D}}^{PD}+N_{\mathcal{R}}^{PD}} \delta\Theta_{(k)}^{PD} s_{b(k)}^{PD} V_{(k)} = \left\{\delta\mathbf{V}_{\mathcal{D}}^{PD} \quad \delta\mathbf{V}_{\mathcal{R}}^{PD}\right\}^{T}\begin{Bmatrix}\mathbf{F}_{s_{\mathcal{D}}}^{*PD}\\ \mathbf{F}_{s_{\mathcal{R}}}^{*PD}\end{Bmatrix} \tag{5.52}$$

in which the vectors $\mathbf{F}_{\mathcal{B}_{\mathbf{q}}}^{*PD}$, $\mathbf{F}_{s_{\mathcal{D}}}^{*PD}$, and $\mathbf{F}_{s_{\mathcal{R}}}^{*PD}$ are defined as:

$$\mathbf{F}_{\mathcal{B}_{\mathbf{q}}}^{*PD} = \left\{\mathbf{q}_{(1)}^{*PD}\cdot\mathbf{n}_{(1)}S_{(1)}, \mathbf{q}_{(2)}^{*PD}\cdot\mathbf{n}_{(2)}S_{(2)}, \mathbf{q}_{(3)}^{*PD}\cdot\mathbf{n}_{(3)}S_{(3)}, \ldots, \mathbf{q}_{(N_{\mathcal{B}_{\mathbf{q}}})}^{*PD}\cdot\mathbf{n}_{(N_{\mathcal{B}_{\mathbf{q}}})}S_{(N_{\mathcal{B}_{\mathbf{q}}})}\right\}^{T} \tag{5.53}$$

$$\mathbf{F}_{s_{\mathcal{D}}}^{*PD} = \left\{s_{b(1)}^{*PD}V_{(1)}, s_{b(2)}^{*PD}V_{(2)}, s_{b(3)}^{*PD}V_{(3)}, \ldots, s_{b(N_{\mathcal{D}})}^{*PD}V_{(N_{\mathcal{D}})}\right\}^{T} \tag{5.54}$$

and

$$\mathbf{F}_{s_{\mathcal{R}}}^{*PD} = \left\{s_{b(1)}^{*PD}V_{(1)}, s_{b(2)}^{*PD}V_{(2)}, s_{b(3)}^{*PD}V_{(3)}, \ldots, s_{b(N_{\mathcal{D}})}^{*PD}V_{(N_{\mathcal{R}})}\right\}^{T} \tag{5.55}$$

In the FE domain, the vector of the unknown field, $\mathbf{V}^{FE}$, and its variation, $\delta\mathbf{V}^{FE}$, are defined as:

$$\mathbf{V}^{FE} = \left\{\mathbf{V}_{\mathcal{I}}^{FE} \quad \mathbf{V}_{\mathcal{D}}^{FE} \quad \mathbf{V}_{\mathcal{B}_{\mathbf{q}}}^{FE} \quad \mathbf{V}_{\mathcal{B}_{\Theta}}^{FE}\right\}^{T} \tag{5.56}$$

and

$$\delta\mathbf{V}^{FE} = \left\{\delta\mathbf{V}_{\mathcal{I}}^{FE} \quad \delta\mathbf{V}_{\mathcal{D}}^{FE} \quad \delta\mathbf{V}_{\mathcal{B}_{\mathbf{q}}}^{FE} \quad \delta\mathbf{V}_{\mathcal{B}_{\Theta}}^{FE}\right\}^{T} \tag{5.57}$$

where the vectors $\mathbf{V}_{\mathcal{I}}^{FE}$, $\mathbf{V}_{\mathcal{D}}^{FE}$, $\mathbf{V}_{\mathcal{B}_{\mathbf{q}}}^{FE}$, and $\mathbf{V}_{\mathcal{B}_{\Theta}}^{FE}$ contain the unknown field values at the nodal points of PD/FE interface $\Gamma_{\mathcal{I}}^{FE}$, in the FE region Ω^{FE}, and along the boundary segments, $\Gamma_{\mathcal{B}_{\mathbf{q}}}^{FE}$ and $\Gamma_{\mathcal{B}_{\Theta}}^{FE}$, respectively.

The virtual field variation due to the external flux in the FE domain can be expressed as:

$$\sum_{e=1}^{E} \nabla \delta \Theta_{(e)}^{FE} \cdot \mathbf{q}_{(e)}^{FE} V_{(e)} = \sum_{e=1}^{E} \left\{ \delta \mathbf{v}_{(e)}^{FE} \right\}^T \mathbf{k}_{(e)}^{FE} \left\{ \mathbf{v}_{(e)}^{FE} \right\} = \{ \delta \mathbf{V}^{FE} \}^T \mathbf{K}^{FE} \{ \mathbf{V}^{FE} \} \tag{5.58}$$

where $\mathbf{K}^{FE}$ represents the stiffness matrix in the FE region. It can be partitioned as:

$$\mathbf{K}^{FE} = \begin{bmatrix} \mathbf{K}_{\mathcal{I}\mathcal{I}}^{FE} & \mathbf{K}_{\mathcal{I}\mathcal{D}}^{FE} & \mathbf{K}_{\mathcal{I}\mathcal{B}_\mathbf{q}}^{FE} & \mathbf{K}_{\mathcal{I}\mathcal{B}_\Theta}^{FE} \\ \mathbf{K}_{\mathcal{D}\mathcal{I}}^{FE} & \mathbf{K}_{\mathcal{D}\mathcal{D}}^{FE} & \mathbf{K}_{\mathcal{D}\mathcal{B}_\mathbf{q}}^{FE} & \mathbf{K}_{\mathcal{D}\mathcal{B}_\Theta}^{FE} \\ \mathbf{K}_{\mathcal{B}_\mathbf{q}\mathcal{I}}^{FE} & \mathbf{K}_{\mathcal{B}_\mathbf{q}\mathcal{D}}^{FE} & \mathbf{K}_{\mathcal{B}_\mathbf{q}\mathcal{B}_\mathbf{q}}^{FE} & \mathbf{K}_{\mathcal{B}_\mathbf{q}\mathcal{B}_\Theta}^{FE} \\ \mathbf{K}_{\mathcal{B}_\Theta\mathcal{I}}^{FE} & \mathbf{K}_{\mathcal{B}_\Theta\mathcal{D}}^{FE} & \mathbf{K}_{\mathcal{B}_\Theta\mathcal{B}_\mathbf{q}}^{FE} & \mathbf{K}_{\mathcal{B}_\Theta\mathcal{B}_\Theta}^{FE} \end{bmatrix} \tag{5.59}$$

in which $\mathbf{K}_{\mathcal{I}\mathcal{I}}^{FE}$, $\mathbf{K}_{\mathcal{D}\mathcal{D}}^{FE}$, $\mathbf{K}_{\mathcal{B}_\mathbf{q}\mathcal{B}_\mathbf{q}}^{FE}$, and $\mathbf{K}_{\mathcal{B}_\Theta\mathcal{B}_\Theta}^{FE}$ represent the self-stiffness submatrices that are only associated with the field vectors $\mathbf{V}_{\mathcal{I}}^{FE}$, $\mathbf{V}_{\mathcal{D}}^{FE}$, $\mathbf{V}_{\mathcal{B}_\mathbf{q}}^{FE}$, and $\mathbf{V}_{\mathcal{B}_\Theta}^{FE}$, respectively. The off-diagonal submatrices represent the coupling stiffness matrices among these field vectors.

The virtual field variation due to the transient field and time rate of change of material property in the FE domain can be evaluated as:

$$\begin{aligned} &\sum_{e=1}^{E} \left(C_{(e)}^{FE} \dot{\Theta}_{(e)}^{FE} + \dot{C}_{(e)}^{FE} \Theta_{(e)}^{FE} \right) V_{(e)} \delta \Theta_{(e)}^{FE} = \\ &\begin{Bmatrix} \delta \mathbf{V}_{\mathcal{I}}^{FE} \\ \delta \mathbf{V}_{\mathcal{D}}^{FE} \\ \delta \mathbf{V}_{\mathcal{B}_\mathbf{q}}^{FE} \\ \delta \mathbf{V}_{\mathcal{B}_\Theta}^{FE} \end{Bmatrix}^T \begin{bmatrix} \mathbf{C}_{\mathcal{I}\mathcal{I}}^{FE} & \mathbf{0} & \mathbf{0} & \mathbf{0} \\ \mathbf{0} & \mathbf{C}_{\mathcal{D}\mathcal{D}}^{FE} & \mathbf{0} & \mathbf{0} \\ \mathbf{0} & \mathbf{0} & \mathbf{C}_{\mathcal{B}_\mathbf{q}\mathcal{B}_\mathbf{q}}^{FE} & \mathbf{0} \\ \mathbf{0} & \mathbf{0} & \mathbf{0} & \mathbf{C}_{\mathcal{B}_\Theta\mathcal{B}_\Theta}^{FE} \end{bmatrix} \begin{Bmatrix} \dot{\mathbf{V}}_{\mathcal{I}}^{FE} \\ \dot{\mathbf{V}}_{\mathcal{D}}^{FE} \\ \dot{\mathbf{V}}_{\mathcal{B}_\mathbf{q}}^{FE} \\ \dot{\mathbf{V}}_{\mathcal{B}_\Theta}^{FE} \end{Bmatrix} \\ &+ \begin{Bmatrix} \delta \mathbf{V}_{\mathcal{I}}^{FE} \\ \delta \mathbf{V}_{\mathcal{D}}^{FE} \\ \delta \mathbf{V}_{\mathcal{B}_\mathbf{q}}^{FE} \\ \delta \mathbf{V}_{\mathcal{B}_\Theta}^{FE} \end{Bmatrix}^T \begin{bmatrix} \dot{\mathbf{C}}_{\mathcal{I}\mathcal{I}}^{FE} & \mathbf{0} & \mathbf{0} & \mathbf{0} \\ \mathbf{0} & \dot{\mathbf{C}}_{\mathcal{D}\mathcal{D}}^{FE} & \mathbf{0} & \mathbf{0} \\ \mathbf{0} & \mathbf{0} & \dot{\mathbf{C}}_{\mathcal{B}_\mathbf{q}\mathcal{B}_\mathbf{q}}^{FE} & \mathbf{0} \\ \mathbf{0} & \mathbf{0} & \mathbf{0} & \dot{\mathbf{C}}_{\mathcal{B}_\Theta\mathcal{B}_\Theta}^{FE} \end{bmatrix} \begin{Bmatrix} \mathbf{V}_{\mathcal{I}}^{FE} \\ \mathbf{V}_{\mathcal{D}}^{FE} \\ \mathbf{V}_{\mathcal{B}_\mathbf{q}}^{FE} \\ \mathbf{V}_{\mathcal{B}_\Theta}^{FE} \end{Bmatrix} \end{aligned} \tag{5.60}$$

in which the submatrices $\mathbf{C}_{\mathcal{I}\mathcal{I}}^{FE}$, $\mathbf{C}_{\mathcal{D}\mathcal{D}}^{FE}$, $\mathbf{C}_{\mathcal{B}_\mathbf{q}\mathcal{B}_\mathbf{q}}^{FE}$, and $\mathbf{C}_{\mathcal{B}_\Theta\mathcal{B}_\Theta}^{FE}$ are diagonal field damping matrices along the PD/FE interface, $\Gamma_{\mathcal{I}}^{FE}$, inside FE region Ω^{FE}, and along the boundary segments, $\Gamma_{\mathscr{B}_{\mathbf{q}}}^{FE}$ and $\Gamma_{\mathscr{B}_{\Theta}}^{FE}$, with dimensions $N_{\mathcal{I}}^{FE} \times N_{\mathcal{I}}^{FE}$, $N_{\mathcal{D}}^{FE} \times N_{\mathcal{D}}^{FE}$, $N_{\mathscr{B}_{\mathbf{q}}}^{FE} \times N_{\mathscr{B}_{\mathbf{q}}}^{FE}$, and $N_{\mathscr{B}_{\Theta}}^{FE} \times N_{\mathscr{B}_{\Theta}}^{FE}$, respectively. Similarly, the submatrices $\dot{\mathbf{C}}_{\mathcal{I}\mathcal{I}}^{FE}$, $\dot{\mathbf{C}}_{\mathcal{D}\mathcal{D}}^{FE}$, $\dot{\mathbf{C}}_{\mathscr{B}_{\mathbf{q}}\mathscr{B}_{\mathbf{q}}}^{FE}$, and $\dot{\mathbf{C}}_{\mathscr{B}_{\Theta}\mathscr{B}_{\Theta}}^{FE}$ are diagonal transient material property matrices along the PD/FE interface, $\Gamma_{\mathcal{I}}^{FE}$, inside FE region Ω^{FE}, and along the boundary segments, $\Gamma_{\mathscr{B}_{\mathbf{q}}}^{FE}$ and $\Gamma_{\mathscr{B}_{\Theta}}^{FE}$, with dimensions $N_{\mathcal{I}}^{FE} \times N_{\mathcal{I}}^{FE}$, $N_{\mathcal{D}}^{FE} \times N_{\mathcal{D}}^{FE}$, $N_{\mathscr{B}_{\mathbf{q}}}^{FE} \times N_{\mathscr{B}_{\mathbf{q}}}^{FE}$, and $N_{\mathscr{B}_{\Theta}}^{FE} \times N_{\mathscr{B}_{\Theta}}^{FE}$, respectively.

Due to the applied field flux and field source, the virtual field variation can be evaluated as:

$$\sum_{e=1}^{E_{\mathscr{B}_{\mathbf{q}}}} \delta\Theta_{(e)}^{FE}\left(\mathbf{q}_{(e)}^{*FE} \cdot \mathbf{n}_{(e)}\right)\Gamma_{(e)} = \left\{ \delta\mathbf{V}_{\mathscr{B}_{\mathbf{q}}}^{FE} \right\}^{T} \left\{ \mathbf{F}_{\mathscr{B}_{\mathbf{q}}}^{*FE} \right\} \tag{5.61}$$

and

$$\sum_{e=1}^{E} \delta\Theta_{(e)}^{FE} s_{b(e)}^{FE} V_{(e)} = \left\{ \delta\mathbf{V}_{\mathcal{I}}^{FE} \quad \delta\mathbf{V}_{\mathcal{D}}^{FE} \quad \delta\mathbf{V}_{\mathscr{B}_{\mathbf{q}}}^{FE} \quad \delta\mathbf{V}_{\mathscr{B}_{\Theta}}^{FE} \right\}^{T} \begin{Bmatrix} \mathbf{F}_{s_{\mathcal{I}}}^{*FE} \\ \mathbf{F}_{s_{\mathcal{D}}}^{*FE} \\ \mathbf{F}_{s_{\mathscr{B}_{\mathbf{q}}}}^{*FE} \\ \mathbf{F}_{s_{\mathscr{B}_{\Theta}}}^{*FE} \end{Bmatrix} \tag{5.62}$$

where the vectors $\mathbf{F}_{\mathscr{B}_{\mathbf{q}}}^{*FE}$, $\mathbf{F}_{s_{\mathcal{I}}}^{*FE}$, $\mathbf{F}_{s_{\mathcal{D}}}^{*FE}$, $\mathbf{F}_{s_{\mathscr{B}_{\mathbf{q}}}}^{*FE}$, and $\mathbf{F}_{s_{\mathscr{B}_{\Theta}}}^{*FE}$ are defined as:

$$\begin{aligned} \mathbf{F}_{\mathscr{B}_{\mathbf{q}}}^{*FE} = \Big\{ &\left(\mathbf{q}_{(1)}^{*FE} \cdot \mathbf{n}_{(1)}\right) S_{(1)}, \left(\mathbf{q}_{(2)}^{*FE} \cdot \mathbf{n}_{(2)}\right) \Gamma_{(2)}, \left(\mathbf{q}_{(3)}^{*FE} \cdot \mathbf{n}_{(3)}\right) S_{(3)}, \ldots, \\ &\left(\mathbf{q}_{\left(N_{\mathscr{B}_{\mathbf{q}}}^{PD}\right)}^{*FE} \cdot \mathbf{n}_{\left(N_{\mathscr{B}_{\mathbf{q}}}^{PD}\right)}\right) S_{\left(N_{\mathscr{B}_{\mathbf{q}}}^{PD}\right)} \Big\}^{T} \end{aligned} \tag{5.63}$$

$$\mathbf{F}_{s_{\mathcal{I}}}^{*FE} = \left\{ s_{b(1)} V_{(1)}, s_{b(2)} V_{(2)}, s_{b(3)} V_{(3)}, \ldots, s_{b\left(N_{\mathcal{I}}^{PD}\right)} V_{\left(N_{\mathcal{I}}^{PD}\right)} \right\}^{T} \tag{5.64}$$

$$\mathbf{F}_{s_{\mathcal{D}}}^{*FE} = \left\{ s_{b(1)} V_{(1)}, s_{b(2)} V_{(2)}, s_{b(3)} V_{(3)}, \ldots, s_{b\left(N_{\mathcal{D}}^{PD}\right)} V_{\left(N_{\mathcal{D}}^{PD}\right)} \right\}^{T} \tag{5.65}$$

$$\mathbf{F}_{s_{\mathscr{B}_{\mathbf{q}}}}^{*FE} = \left\{ s_{b(1)} V_{(1)}, s_{b(2)} V_{(2)}, s_{b(3)} V_{(3)}, \ldots, s_{b\left(N_{\mathscr{B}_{\mathbf{q}}}^{PD}\right)} V_{\left(N_{\mathscr{B}_{\mathbf{q}}}^{PD}\right)} \right\}^{T} \tag{5.66}$$

and

$$\mathbf{F}_{s_{\mathscr{B}_{\Theta}}}^{*FE} = \left\{ s_{b(1)} V_{(1)}, s_{b(2)} V_{(2)}, s_{b(3)} V_{(3)}, \ldots, s_{b\left(N_{\mathscr{B}_{\Theta}}^{PD}\right)} V_{\left(N_{\mathscr{B}_{\Theta}}^{PD}\right)} \right\}^{T} \tag{5.67}$$

After combining Eqs. (5.47) to (5.52), (5.58) and (5.60) to (5.62), the variational expression can be cast in a matrix form as:

$$\delta\mathbf{V}^T\left(\mathbf{F}_{\dot{\Theta}}+\mathbf{F}+\mathbf{F}_{\Theta}-\mathbf{F}_s^*\right)=\delta\mathbf{V}^T\left(\mathbf{C}\dot{\mathbf{V}}+\mathbf{K}\mathbf{V}-\mathbf{F}\right)=0 \tag{5.68}$$

in which the damping and stiffness matrices and the force vector can be expressed as:

$$\mathbf{C}=\begin{bmatrix} \mathbf{C}_{\mathcal{D}\mathcal{D}} & \mathbf{0} & \mathbf{0} & \mathbf{0} & \mathbf{0} & \mathbf{0} & \mathbf{0} & \mathbf{0} \\ \mathbf{0} & \mathbf{C}_{\mathscr{R}\mathscr{R}} & \mathbf{0} & \mathbf{0} & \mathbf{0} & \mathbf{0} & \mathbf{0} & \mathbf{0} \\ \mathbf{0} & \mathbf{0} & \mathbf{0} & \mathbf{0} & \mathbf{0} & \mathbf{0} & \mathbf{0} & \mathbf{0} \\ \mathbf{0} & \mathbf{0} & \mathbf{0} & \mathbf{0} & \mathbf{0} & \mathbf{0} & \mathbf{0} & \mathbf{0} \\ \mathbf{0} & \mathbf{0} & \mathbf{0} & \mathbf{0} & \mathbf{C}^{FE}_{\mathcal{I}\mathcal{I}} & \mathbf{0} & \mathbf{0} & \mathbf{0} \\ \mathbf{0} & \mathbf{0} & \mathbf{0} & \mathbf{0} & \mathbf{0} & \mathbf{C}^{FE}_{\mathcal{D}\mathcal{D}} & \mathbf{0} & \mathbf{0} \\ \mathbf{0} & \mathbf{0} & \mathbf{0} & \mathbf{0} & \mathbf{0} & \mathbf{0} & \mathbf{C}^{FE}_{\mathscr{B}_\mathbf{q}\mathscr{B}_\mathbf{q}} & \mathbf{0} \\ \mathbf{0} & \mathbf{0} & \mathbf{0} & \mathbf{0} & \mathbf{0} & \mathbf{0} & \mathbf{0} & \mathbf{C}^{FE}_{\mathscr{B}_\Theta\mathscr{B}_\Theta} \end{bmatrix} \tag{5.69}$$

$$\mathbf{K}=\begin{bmatrix} \mathbf{K}^{PD}_{\mathcal{D}\mathcal{D}}+\dot{\mathbf{C}}^{PD}_{\mathcal{D}\mathcal{D}} & \mathbf{K}^{PD}_{\mathcal{D}\mathscr{R}} & \mathbf{0} & \mathbf{0} & \mathbf{0} & \mathbf{0} & \mathbf{0} & \mathbf{0} \\ \mathbf{K}^{PD}_{\mathscr{R}\mathcal{D}} & \mathbf{K}^{PD}_{\mathscr{R}\mathscr{R}}+\dot{\mathbf{C}}^{PD}_{\mathscr{R}\mathscr{R}} & \mathbf{K}^{PD}_{\mathscr{R}\mathscr{B}_\mathbf{q}} & \mathbf{K}^{PD}_{\mathscr{R}\mathscr{B}_\Theta} & \mathbf{K}^{PD}_{\mathscr{R}I} & \mathbf{0} & \mathbf{0} & \mathbf{0} \\ \mathbf{0} & \mathbf{K}^{PD}_{\mathscr{B}_\mathbf{q}\mathscr{R}} & \mathbf{K}^{PD}_{\mathscr{B}_\mathbf{q}\mathscr{B}_\mathbf{q}} & \mathbf{K}^{PD}_{\mathscr{B}_\mathbf{q}\mathscr{B}_\Theta} & \mathbf{K}^{PD}_{\mathscr{B}_\mathbf{q}I} & \mathbf{0} & \mathbf{0} & \mathbf{0} \\ \mathbf{0} & \mathbf{K}^{PD}_{\mathscr{B}_\Theta\mathscr{R}} & \mathbf{K}^{PD}_{\mathscr{B}_\Theta\mathscr{B}_\mathbf{q}} & \mathbf{K}^{PD}_{\mathscr{B}_\Theta\mathscr{B}_\Theta} & \mathbf{K}^{PD}_{\mathscr{B}_\Theta I} & \mathbf{0} & \mathbf{0} & \mathbf{0} \\ \mathbf{0} & \mathbf{K}^{PD}_{\mathcal{I}\mathscr{R}} & \mathbf{K}^{PD}_{\mathcal{I}\mathscr{B}_\mathbf{q}} & \mathbf{K}^{PD}_{\mathcal{I}\Theta} & \mathbf{K}^{PD}_{\mathcal{I}\mathcal{I}}+\mathbf{K}^{FE}_{\mathcal{I}\mathcal{I}}+\dot{\mathbf{C}}^{FE}_{\mathcal{I}\mathcal{I}} & \mathbf{K}^{FE}_{\mathcal{I}\mathcal{D}} & \mathbf{K}^{FE}_{\mathcal{I}\mathscr{B}_\mathbf{q}} & \mathbf{K}^{FE}_{\mathcal{I}\mathscr{B}_\Theta} \\ \mathbf{0} & \mathbf{0} & \mathbf{0} & \mathbf{0} & \mathbf{K}^{FE}_{\mathcal{D}\mathcal{I}} & \mathbf{K}^{FE}_{\mathcal{D}\mathcal{D}}+\mathbf{C}^{FE}_{\mathcal{D}\mathcal{D}} & \mathbf{K}^{FE}_{\mathcal{D}\mathscr{B}_\mathbf{q}} & \mathbf{K}^{FE}_{\mathcal{D}\mathscr{B}_\Theta} \\ \mathbf{0} & \mathbf{0} & \mathbf{0} & \mathbf{0} & \mathbf{K}^{FE}_{\mathscr{B}_\mathbf{q}\mathcal{I}} & \mathbf{K}^{FE}_{\mathscr{B}_\mathbf{q}\mathcal{D}} & \mathbf{K}^{FE}_{\mathscr{B}_\mathbf{q}\mathscr{B}_\mathbf{q}}+\mathbf{C}^{FE}_{\mathscr{B}_\mathbf{q}\mathscr{B}_\mathbf{q}} & \mathbf{K}^{FE}_{\mathscr{B}_\mathbf{q}\mathscr{B}_\Theta} \\ \mathbf{0} & \mathbf{0} & \mathbf{0} & \mathbf{0} & \mathbf{K}^{FE}_{\mathscr{B}_\Theta\mathcal{I}} & \mathbf{K}^{FE}_{\mathscr{B}_\Theta\mathcal{D}} & \mathbf{K}^{FE}_{\mathscr{B}_\Theta\mathscr{B}_\mathbf{q}} & \mathbf{K}^{FE}_{\mathscr{B}_\Theta\mathscr{B}_\Theta}+\mathbf{C}^{FE}_{\mathscr{B}_\Theta\mathscr{B}_\Theta} \end{bmatrix} \tag{5.70}$$

$$\begin{aligned}\mathbf{V}&=\left\{\mathbf{V}^{PD}\quad \mathbf{V}^{FE}\right\}^{\mathbf{T}}\\ &=\left\{\mathbf{V}^{PD}_{\mathcal{D}}\quad \mathbf{V}^{PD}_{\mathscr{R}}\quad \mathbf{V}^{PD}_{\mathscr{B}_\mathbf{q}}\quad \mathbf{V}^{PD}_{\mathscr{B}_\Theta}\quad \mathbf{V}^{PD}_{\mathcal{I}}=\mathbf{V}^{FE}_{\mathcal{I}}\quad \mathbf{V}^{FE}_{\mathcal{D}}\quad \mathbf{V}^{FE}_{\mathscr{B}_\mathbf{q}}\quad \mathbf{V}^{FE}_{\mathscr{B}_\Theta}\right\}^{\mathbf{T}}\end{aligned} \tag{5.71}$$

and

$$\mathbf{F} = \left\{ \mathbf{F}_{s\mathcal{D}}^{PD*} \quad \mathbf{F}_{s\mathcal{R}}^{PD*} \quad \mathbf{F}_{\mathbf{q}}^{PD*} \quad \mathbf{0} \quad \mathbf{F}_{s_{\mathcal{I}}}^{*FE} \quad \mathbf{F}_{s_{\mathcal{D}}}^{*FE} \quad \mathbf{F}_{\mathcal{B}_{\mathbf{q}}}^{*FE} + \mathbf{F}_{s_{\mathcal{B}_{\mathbf{q}}}}^{*FE} \quad \mathbf{F}_{s_{\mathcal{B}_{\Theta}}}^{*FE} \right\}^{\mathbf{T}} \tag{5.72}$$

The quantities necessary to construct the submatrices and subvectors associated with the PD region in Eqs. (5.69) and (5.71) are derived in Chaps. 3 and 4.

References

Anicode, S. V. K., & Madenci, E. (2023). Direct coupling of dual-horizon peridynamics with finite elements for irregular discretization without an overlap zone. *Engineering with Computers, 40*, 1–31.

Anicode, S. V. K., & Madenci, E. (2022). Seamless coupling of bond- and state-based peridynamic and finite element analyses. *Mechanics and Materials., 173*, 104433.

Madenci, E., Barut, A., & Dorduncu, M. (2019). *Peridynamic differential operator for numerical analysis*. Springer International Publishing.

Madenci, E., Roy, P., & Behera, D. (2022). *Advances in peridynamics*. Springer International Publishing.

Chapter 6
Coupled Peridynamic and Finite Element Analysis in the Presence of Initial Strain

6.1 Introduction

Initial strain due to thermal expansion, shrinkage, oxidation, and other physical phenomena can be a critical consideration in engineering and materials science. Understanding how material responds to initial strain and incorporating this knowledge into design and analysis processes are essential for ensuring the reliability and safety of engineering systems.

Kilic and Madenci (2009) presented a bond-based (BB) peridynamic (PD) model with initial strain arising from temperature variation. The PD force density vector is evaluated by removing the thermal stretch from the total stretch. Oterkus et al. (2014) derived a general PD framework for a fully coupled thermomechanical analysis. Oterkus and Madenci (2017) employed the BB-PD to investigate cracking of UO_2 nuclear fuel pellet during the operation of a power plant. Wang et al. (2018) studied thermal cracking in rocks using BB-PD.

Hu et al. (2018) evaluated the elastic dilatation by removing thermal dilatation from the total dilatation and determined the force density vector for the ordinary state-based (OSB) PD thermomechanical model. Gao and Oterkus (2019) presented a fully coupled OSB-PD thermomechanical model for isotropic and composite materials. D'Antuono and Morandini (2017) analyzed thermal shock response in brittle materials with OSB-PD by weakly coupling it with thermal diffusion.

Shou and Zhou (2020) presented a non-ordinary state-based (NOSB) thermomechanical model to study the formation of cracks in rocks due to temperature change. Pathrikar et al. (2021) introduced a thermomechanical model to study damage in brittle materials. In NOSB-PD, the thermal strains are subtracted from the total strain in the evaluation of the force density vectors. Recently, Liu et al. (2022) applied the OSB-PD thermomechanical model to investigate the fracture mechanisms for three-dimensional analysis of fuel pellets with non-uniform discretization. Sun et al. (2024) provided an extensive review of previous PD modeling of thermomechanical cracking in both homogeneous and heterogeneous materials.

E. Madenci et al., *Peridynamics and Its Applications Using Ansys*,
https://doi.org/10.1007/978-3-031-59896-8_6

This chapter presents an extension of the approach described in Chap. 5 to perform bond-based (BB), ordinary state-based (OSB), and non-ordinary state-based (NOSB) peridynamic (PD) analysis coupled with finite element (FE) analysis for deformation due to initial strain field. Both the PD and FE regions can be modeled with uniform or nonuniform discretization. The PD and FE regions share the same nodes along their interface without a transition zone and constraint conditions. The resulting global system of equations includes the contributions arising from the PD nodes and FE nodes. These equations are solved simultaneously without requiring an iterative procedure through implicit methods.

6.2 PD Internal Force Vectors in the Presence of Initial Deformation

The PD force density vector arises only due to mechanical deformation; thus, it requires the removal of the initial deformation from the total deformation. The corresponding initial stretch, $\overline{s}_0(\mathbf{x}, \mathbf{x}')$ of a bond, $\boldsymbol{\xi} = \mathbf{x}' - \mathbf{x}$ between points $\mathbf{x}$ and $\mathbf{x}'$ can be approximated as:

$$\overline{s}_0(\mathbf{x}, \mathbf{x}') = \frac{1}{2}\left(s_0(\mathbf{x}, \boldsymbol{\xi}) + s_0(\mathbf{x}', \boldsymbol{\xi}')\right) \tag{6.1}$$

where $s_0(\mathbf{x}, \boldsymbol{\xi})$ and $s_0(\mathbf{x}', \boldsymbol{\xi}')$ can be evaluated as:

$$s_0(\mathbf{x}, \boldsymbol{\xi}) = \mathbf{n}^T(\mathbf{x}, \boldsymbol{\xi})\boldsymbol{\varepsilon}_0(\mathbf{x})\mathbf{n}(\mathbf{x}, \boldsymbol{\xi}) \tag{6.2}$$

and

$$s_0(\mathbf{x}', \boldsymbol{\xi}') = \mathbf{n}^T(\mathbf{x}', \boldsymbol{\xi}')\boldsymbol{\varepsilon}_0(\mathbf{x}')\mathbf{n}(\mathbf{x}', \boldsymbol{\xi}') \tag{6.3}$$

in which $\boldsymbol{\varepsilon}_0(\mathbf{x})$ represents the known strain tensor arising from the initial deformation. Aligned with the bond, the unit vectors $\mathbf{n}(\mathbf{x}, \boldsymbol{\xi})$ and $\mathbf{n}(\mathbf{x}', \boldsymbol{\xi}')$ are defined as:

$$\mathbf{n}(\mathbf{x}, \boldsymbol{\xi}) = \frac{(\mathbf{x}' - \mathbf{x})}{|\mathbf{x}' - \mathbf{x}|} \tag{6.4}$$

and

$$\mathbf{n}(\mathbf{x}', \boldsymbol{\xi}') = \frac{(\mathbf{x} - \mathbf{x}')}{|\mathbf{x}' - \mathbf{x}|} \tag{6.5}$$

In accordance with Eq. (3.2) in Chap. 3, the internal force vectors due to mechanical deformation, $\mathbf{L}_m^{\alpha}(\mathbf{x}, t)$ can be expressed as:

$$\mathbf{L}_m^{\alpha}(\mathbf{x},t)=\int\limits_{H_{\mathbf{x}}}\mathbf{t}_m^{\alpha}(\mathbf{x}'-\mathbf{x},t)dV_{\mathbf{x}'}-\int\limits_{\widetilde{H}_{\mathbf{x}}}\mathbf{t}_m^{\alpha}(\mathbf{x}-\mathbf{x}',t)dV_{\mathbf{x}'}.\quad \alpha=BB,OSB,NOSB,PDDO \tag{6.6}$$

in which the force density vectors, $\mathbf{t}_m^{\alpha}(\mathbf{x}'-\mathbf{x},t)$ and $\mathbf{t}_m^{\alpha}(\mathbf{x}-\mathbf{x}',t)$ are associated with primary and secondary bonds, respectively. For uniform discretization, this equation reduces to:

$$\mathbf{L}_m^{\alpha}(\mathbf{x},t)=\int\limits_{H_{\mathbf{x}}}\left(\mathbf{t}_m^{\alpha}(\mathbf{x}'-\mathbf{x},t)-\mathbf{t}_m^{\alpha}(\mathbf{x}-\mathbf{x}',t)\right)dV_{\mathbf{x}'} \tag{6.7}$$

The PD force density vectors due to mechanical deformation can be obtained by removing the initial deformation from the total deformation. The BB-PD force density vectors in Eqs. (3.7) and (3.8) for the primary and secondary bonds, respectively, can be modified as:

$$\mathbf{t}_m^{BB}(\mathbf{x}'-\mathbf{x},t)=\widehat{\mu}(\mathbf{x}'-\mathbf{x})\frac{c(\delta_{\mathbf{x}})}{2}(s-\bar{s}_0)\frac{\mathbf{y}'-\mathbf{y}}{|\mathbf{y}'-\mathbf{y}|}\quad\text{for }\mathbf{x}'\in H_{\mathbf{x}} \tag{6.8}$$

and

$$\mathbf{t}_m^{BB}(\mathbf{x}-\mathbf{x}',t)=\widehat{\mu}(\mathbf{x}-\mathbf{x}')\frac{c(\delta_{\mathbf{x}'})}{2}(s-\bar{s}_0)\frac{\mathbf{y}-\mathbf{y}'}{|\mathbf{y}-\mathbf{y}'|}\quad\text{for }\mathbf{x}'\in \widetilde{H}_{\mathbf{x}} \tag{6.9}$$

The OSB-PD force density vectors in Eqs. (3.30) and (3.31) for the primary and secondary bonds, respectively, can be modified as:

$$\mathbf{t}_m^{OSB}(\mathbf{x}'-\mathbf{x},t)=\widehat{\mu}(\mathbf{x}'-\mathbf{x})\left\{ad(\delta_{\mathbf{x}})\frac{1}{|\boldsymbol{\xi}|}(\theta(\mathbf{x})-\theta_0(\mathbf{x}))+b(\delta_{\mathbf{x}})\frac{1}{|\boldsymbol{\xi}|}(s-\bar{s}_0)\right\}\frac{\mathbf{y}'-\mathbf{y}}{|\mathbf{y}'-\mathbf{y}|}\quad\text{for }\mathbf{x}'\in H_{\mathbf{x}} \tag{6.10}$$

and

$$\mathbf{t}_m^{OSB}(\mathbf{x}-\mathbf{x}',t)=\widehat{\mu}(\mathbf{x}-\mathbf{x}')\left\{ad(\delta_{\mathbf{x}'})\frac{1}{|\boldsymbol{\xi}|}(\theta(\mathbf{x}')-\theta_0(\mathbf{x}'))+b(\delta_{\mathbf{x}'})\frac{1}{|\boldsymbol{\xi}|}(s-\bar{s}_0)\right\}\frac{\mathbf{y}-\mathbf{y}'}{|\mathbf{y}-\mathbf{y}'|}\quad\text{for }\mathbf{x}'\in \widetilde{H}_{\mathbf{x}} \tag{6.11}$$

where $\theta_0(\mathbf{x},t)$ and $\theta_0(\mathbf{x}',t)$ represent the known initial dilatation at points $\mathbf{x}$ and $\mathbf{x}'$. The total dilatation in Eq. (3.32) and (3.33) can be expressed as:

$$\theta(\mathbf{x},t)=d(\delta_{\mathbf{x}})\int\limits_{H_{\mathbf{x}}}\widehat{\mu}(\mathbf{x}''-\mathbf{x})(s-\bar{s}_0)dV_{\mathbf{x}''}+\theta_0(\mathbf{x},t)\quad\text{for }\mathbf{x}'\in H_{\mathbf{x}} \tag{6.12}$$

and

$$\theta(\mathbf{x}',t) = d(\delta_{\mathbf{x}'}) \int\limits_{H_{\mathbf{x}'}} \widehat{\mu}(\mathbf{x}''' - \mathbf{x}')(s - \bar{s}_0) dV_{\mathbf{x}'''} + \theta_0(\mathbf{x}',t) \quad \text{for } \mathbf{x}' \in \widetilde{H}_{\mathbf{x}} \tag{6.13}$$

The NOSB-PD force density vectors in Eqs. (3.62) and (3.63) for the primary and secondary bonds, respectively, can be modified as:

$$\mathbf{t}_m^{NOSB}(\mathbf{x}' - \mathbf{x}, t) = \widehat{\mu}(\mathbf{x}' - \mathbf{x}) \mathbf{F}_{\boldsymbol{\xi}}^{PD}(\mathbf{x}, t) \left(\mathbf{C} : \left(\mathbf{E}_{\boldsymbol{\xi}}^{PD}(\mathbf{x}, t) - \boldsymbol{\varepsilon}_{\boldsymbol{\xi}0}^{PD}(\mathbf{x}) \right) \right) \mathbf{g}(\mathbf{x}, \boldsymbol{\xi}) \quad \text{for } \mathbf{x}' \in H_{\mathbf{x}} \tag{6.14a}$$

and

$$\mathbf{t}_m^{NOSB}(\mathbf{x} - \mathbf{x}', t) = \widehat{\mu}(\mathbf{x} - \mathbf{x}') \mathbf{F}_{\boldsymbol{\xi}'}^{PD}(\mathbf{x}', t) \left(\mathbf{C} : \left(\mathbf{E}_{\boldsymbol{\xi}}^{PD}(\mathbf{x}', t) - \boldsymbol{\varepsilon}_{\boldsymbol{\xi}0}^{PD}(\mathbf{x}') \right) \right) \mathbf{g}(\mathbf{x}', \boldsymbol{\xi}') \quad \text{for } \mathbf{x}' \in \widetilde{H}_{\mathbf{x}} \tag{6.14b}$$

where $\mathbf{F}_{\boldsymbol{\xi}}^{PD}$, $\mathbf{E}_{\boldsymbol{\xi}}^{PD}$, and $\mathbf{C}$ represent the bond-associated (BA) deformation gradient tensor, BA green strain tensor, and fourth-order material property tensor, respectively, and the BA initial strain tensor is defined as:

$$\boldsymbol{\varepsilon}_{\boldsymbol{\xi}0}^{PD}(\mathbf{x}) = \boldsymbol{\varepsilon}_{\boldsymbol{\xi}'0}^{PD}(\mathbf{x}') = \frac{s_0(\mathbf{x}, \boldsymbol{\xi})(\mathbf{n}(\mathbf{x}, \boldsymbol{\xi}) \otimes \mathbf{n}(\mathbf{x}, \boldsymbol{\xi})) + s_0(\mathbf{x}', \boldsymbol{\xi}')(\mathbf{n}(\mathbf{x}', \boldsymbol{\xi}') \otimes \mathbf{n}(\mathbf{x}', \boldsymbol{\xi}'))}{2}. \tag{6.15}$$

These force density vectors $\mathbf{t}_m^{\alpha}$ with $\alpha = BB, OSB, NOSB$ for the primary and secondary bonds can be recast as:

$$\mathbf{t}_m^{\alpha}(\mathbf{x}' - \mathbf{x}, t) = \mathbf{t}^{\alpha}(\mathbf{x}' - \mathbf{x}, t) + \mathbf{t}_0^{\alpha}(\mathbf{x}' - \mathbf{x}, t) \quad \text{for } \mathbf{x}' \in H_{\mathbf{x}} \tag{6.16}$$

and

$$\mathbf{t}_m^{\alpha}(\mathbf{x} - \mathbf{x}', t) = \mathbf{t}^{\alpha}(\mathbf{x} - \mathbf{x}', t) + \mathbf{t}_0^{\alpha}(\mathbf{x} - \mathbf{x}', t) \quad \text{for } \mathbf{x}' \in \widetilde{H}_{\mathbf{x}} \tag{6.17}$$

in which $\mathbf{t}^{\alpha}$ and $\mathbf{t}_0^{\alpha}$ are the force density vectors arising from total and initial deformations, respectively. The explicit forms of force density vectors $\mathbf{t}^{\alpha}(\mathbf{x}' - \mathbf{x}, t)$ and $\mathbf{t}^{\alpha}(\mathbf{x} - \mathbf{x}', t)$ are derived in Chap. 3. Arising from the initial deformation, the force density vectors, $\mathbf{t}_0^{\alpha}$, of the primary bonds for $\alpha = BB, OSB, NOSB$ can be expressed as:

$$\mathbf{t}_0^{BB}(\mathbf{x}' - \mathbf{x}, t) = -\widehat{\mu}(\mathbf{x}' - \mathbf{x}) \frac{c(\delta_{\mathbf{x}})}{2} \bar{s}_0 \frac{\mathbf{y}' - \mathbf{y}}{|\mathbf{y}' - \mathbf{y}|} \quad \text{for } \mathbf{x}' \in H_{\mathbf{x}} \tag{6.18}$$

$$\mathbf{t}_0^{OSB}(\mathbf{x}' - \mathbf{x}, t) = -\widehat{\mu}(\mathbf{x}' - \mathbf{x}) \begin{pmatrix} ad^2(\delta_{\mathbf{x}}) \dfrac{1}{|\boldsymbol{\xi}|} \displaystyle\int_{H_{\mathbf{x}}} \widehat{\mu}(\mathbf{x}'' - \mathbf{x}) \overline{s}_0 dV_{\mathbf{x}''} \\ + b(\delta_{\mathbf{x}}) \dfrac{1}{|\boldsymbol{\xi}|} \overline{s}_0 \end{pmatrix} \frac{\mathbf{y}' - \mathbf{y}}{|\mathbf{y}' - \mathbf{y}|} \quad \text{for } \mathbf{x}' \in H_{\mathbf{x}} \tag{6.19}$$

and

$$\mathbf{t}_0^{NOSB}(\mathbf{x}' - \mathbf{x}, t) = -\widehat{\mu}(\mathbf{x}' - \mathbf{x}) \mathbf{F}_{\boldsymbol{\xi}}^{PD}(\mathbf{x}, t) \left(\mathbf{C} : \boldsymbol{\varepsilon}_{\boldsymbol{\xi}0}^{PD}(\mathbf{x}) \right) \mathbf{g}(\mathbf{x}, \boldsymbol{\xi}) \quad \text{for } \mathbf{x}' \in H_{\mathbf{x}} \tag{6.20}$$

For the secondary bonds, they become:

$$\mathbf{t}_0^{BB}(\mathbf{x} - \mathbf{x}', t) = -\widehat{\mu}(\mathbf{x} - \mathbf{x}') \frac{c(\delta_{\mathbf{x}'})}{2} \overline{s}_0 \frac{\mathbf{y} - \mathbf{y}'}{|\mathbf{y} - \mathbf{y}'|} \quad \text{for } \mathbf{x}' \in \widetilde{H}_{\mathbf{x}} \tag{6.21}$$

$$\mathbf{t}_0^{OSB}(\mathbf{x} - \mathbf{x}', t) = -\widehat{\mu}(\mathbf{x} - \mathbf{x}') \begin{pmatrix} ad^2(\delta_{\mathbf{x}'}) \dfrac{1}{|\boldsymbol{\xi}'|} \displaystyle\int_{H_{\mathbf{x}'}} \widehat{\mu}(\mathbf{x}''' - \mathbf{x}') \overline{s}_0 dV_{\mathbf{x}'''} \\ + b(\delta_{\mathbf{x}'}) \dfrac{1}{|\boldsymbol{\xi}'|} \overline{s}_0 \end{pmatrix} \frac{\mathbf{y} - \mathbf{y}'}{|\mathbf{y} - \mathbf{y}'|} \quad \text{for } \mathbf{x}' \in \widetilde{H}_{\mathbf{x}} \tag{6.22}$$

and

$$\mathbf{t}_0^{NOSB}(\mathbf{x} - \mathbf{x}', t) = -\widehat{\mu}(\mathbf{x} - \mathbf{x}') \mathbf{F}_{\boldsymbol{\xi}'}^{PD}(\mathbf{x}', t) \left(\mathbf{C} : \boldsymbol{\varepsilon}_{\boldsymbol{\xi}'0}^{PD}(\mathbf{x}') \right) \mathbf{g}(\mathbf{x}', \boldsymbol{\xi}') \quad \text{for } \mathbf{x}' \in \widetilde{H}_{\mathbf{x}} \tag{6.23}$$

With these force density vectors, Eq. (6.6) can be rewritten as:

$$\mathbf{L}_m^{\alpha}(\mathbf{x}, t) = \mathbf{L}^{\alpha}(\mathbf{x}, t) + \mathbf{L}_0^{\alpha}(\mathbf{x}, t) \quad \alpha = BB, OSB, NOSB \tag{6.24}$$

where

$$\mathbf{L}^{\alpha}(\mathbf{x}, t) = \int_{H_{\mathbf{x}}} \mathbf{t}^{\alpha}(\mathbf{x}' - \mathbf{x}, t) dV_{\mathbf{x}'} - \int_{\widetilde{H}_{\mathbf{x}}} \mathbf{t}^{\alpha}(\mathbf{x} - \mathbf{x}', t) dV_{\mathbf{x}'} \tag{6.25}$$

and

$$\mathbf{L}_0^{\alpha}(\mathbf{x}, t) = \int_{H_{\mathbf{x}}} \mathbf{t}_0^{\alpha}(\mathbf{x}' - \mathbf{x}, t) dV_{\mathbf{x}'} - \int_{\widetilde{H}_{\mathbf{x}}} \mathbf{t}_0^{\alpha}(\mathbf{x} - \mathbf{x}', t) dV_{\mathbf{x}'} \tag{6.26}$$

For uniform discretization, they reduce to:

$$\mathbf{L}^{\alpha}(\mathbf{x},t)=\int\limits_{H_{\mathbf{x}}}(\mathbf{t}^{\alpha}(\mathbf{x}'-\mathbf{x},t)-\mathbf{t}^{\alpha}(\mathbf{x}-\mathbf{x}',t))dV_{\mathbf{x}'} \tag{6.27}$$

and

$$\mathbf{L}_0^{\alpha}(\mathbf{x},t)=\int\limits_{H_{\mathbf{x}}}\left(\mathbf{t}_0^{\alpha}(\mathbf{x}'-\mathbf{x},t)-\mathbf{t}_0^{\alpha}(\mathbf{x}-\mathbf{x}',t)\right)dV_{\mathbf{x}'} \tag{6.28}$$

For a single or a variable horizon, the PDDO internal force vector due to mechanical deformation can be expressed as:

$$\mathbf{L}_m^{PDDO}(\mathbf{x},t)=\int\limits_{H_{\mathbf{x}}}\left(\mathbf{t}_m^{PDDO}(\mathbf{x}',\mathbf{x}'-\mathbf{x},t)-\mathbf{t}_m^{PDDO}(\mathbf{x},\mathbf{x}'-\mathbf{x},t)\right)dV_{\mathbf{x}'} \tag{6.29}$$

In accordance with Eq. (3.71) and (3.72) along with Eq. (3.73), the force density vectors $\mathbf{t}_m^{PDDO}(\mathbf{x},\mathbf{x}'-\mathbf{x},t)$ and $\mathbf{t}_m^{PDDO}(\mathbf{x}',\mathbf{x}'-\mathbf{x},t)$ can be modified as:

$$\mathbf{t}_m^{PDDO}(\mathbf{x},\mathbf{x}'-\mathbf{x},t)=\widehat{\mu}(\mathbf{x}'-\mathbf{x})\mathbf{F}^{PD}(\mathbf{x},t)\left(\mathbf{C}:\left(\mathbf{E}^{PD}(\mathbf{x},t)-\boldsymbol{\varepsilon}_0(\mathbf{x})\right)\right)\mathbf{g}(\mathbf{x},\boldsymbol{\xi}) \tag{6.30}$$

and

$$\mathbf{t}_m^{PDDO}(\mathbf{x}',\mathbf{x}'-\mathbf{x},t)=\widehat{\mu}(\mathbf{x}'-\mathbf{x})\mathbf{F}^{PD}(\mathbf{x}',t)\left(\mathbf{C}:\left(\mathbf{E}^{PD}(\mathbf{x}',t)-\boldsymbol{\varepsilon}_0(\mathbf{x}')\right)\right)\mathbf{g}(\mathbf{x},\boldsymbol{\xi}) \tag{6.31}$$

They can be rewritten as:

$$\mathbf{t}_m^{PDDO}(\mathbf{x}'-\mathbf{x},t)=\mathbf{t}^{PDDO}(\mathbf{x}'-\mathbf{x},t)+\mathbf{t}_0^{PDDO}(\mathbf{x}'-\mathbf{x},t) \tag{6.32}$$

and

$$\mathbf{t}_m^{PDDO}(\mathbf{x}-\mathbf{x}',t)=\mathbf{t}^{PDDO}(\mathbf{x}-\mathbf{x}',t)+\mathbf{t}_0^{PDDO}(\mathbf{x}-\mathbf{x}',t) \tag{6.33}$$

Their explicit forms are derived in Chap. 3. Arising from the initial deformation, the force density vectors $\mathbf{t}_0^{PDDO}(\mathbf{x}'-\mathbf{x},t)$ and $\mathbf{t}_0^{PDDO}(\mathbf{x}-\mathbf{x}',t)$ can be expressed as:

$$\mathbf{t}_0^{PDDO}(\mathbf{x},\mathbf{x}'-\mathbf{x},t)=-\widehat{\mu}(\mathbf{x}'-\mathbf{x})\mathbf{F}^{PD}(\mathbf{x},t)(\mathbf{C}:\boldsymbol{\varepsilon}_0(\mathbf{x}))\mathbf{g}(\mathbf{x},\boldsymbol{\xi}) \tag{6.34}$$

and

$$\mathbf{t}_0^{PDDO}(\mathbf{x}', \mathbf{x}' - \mathbf{x}, t) = -\hat{\mu}(\mathbf{x}' - \mathbf{x})\mathbf{F}^{PD}(\mathbf{x}', t)(\mathbf{C} : \boldsymbol{\varepsilon}_0(\mathbf{x}'))\mathbf{g}(\mathbf{x}, \boldsymbol{\xi}) \tag{6.35}$$

The PDDO internal force vector due to mechanical deformation becomes:

$$\mathbf{L}_m^{PDDO}(\mathbf{x}, t) = \mathbf{L}^{PDDO}(\mathbf{x}, t) + \mathbf{L}_0^{PDDO}(\mathbf{x}, t) \tag{6.36}$$

where

$$\mathbf{L}^{PDDO}(\mathbf{x}, t) = \int_{H_\mathbf{x}} \left(\mathbf{t}^{PDDO}(\mathbf{x}', \mathbf{x}' - \mathbf{x}, t) - \mathbf{t}^{PDDO}(\mathbf{x}, \mathbf{x}' - \mathbf{x}, t)\right) dV_{\mathbf{x}'} \tag{6.37}$$

and

$$\mathbf{L}_0^{PDDO}(\mathbf{x}, t) = \int_{H_\mathbf{x}} \left(\mathbf{t}_0^{PDDO}(\mathbf{x}', \mathbf{x}' - \mathbf{x}, t) - \mathbf{t}_0^{PDDO}(\mathbf{x}, \mathbf{x}' - \mathbf{x}, t)\right) dV_{\mathbf{x}'} \tag{6.38}$$

6.3 PD Internal Force Vectors for Small Deformation Kinematics

As explained in Chap. 3, the bond stretch for small deformation can be approximated as:

$$s = \frac{|\mathbf{y}' - \mathbf{y}| - |\boldsymbol{\xi}|}{|\boldsymbol{\xi}|} \approx \frac{(\mathbf{u}(\mathbf{x}') - \mathbf{u}(\mathbf{x})) \cdot \mathbf{n}(\mathbf{x}, \boldsymbol{\xi})}{|\boldsymbol{\xi}|} \tag{6.39}$$

Also, the assumption of "small deformation" permits

$$\frac{(\mathbf{y}' - \mathbf{y})}{|\mathbf{y}' - \mathbf{y}|} \approx \frac{\boldsymbol{\xi}}{|\boldsymbol{\xi}|} = \frac{(\mathbf{x}' - \mathbf{x})}{|\mathbf{x}' - \mathbf{x}|} = \mathbf{n}(\mathbf{x}, \boldsymbol{\xi}) \tag{6.40}$$

and

$$\frac{(\mathbf{y} - \mathbf{y}')}{|\mathbf{y} - \mathbf{y}'|} \approx \frac{\boldsymbol{\xi}'}{|\boldsymbol{\xi}'|} = \frac{(\mathbf{x} - \mathbf{x}')}{|\mathbf{x} - \mathbf{x}'|} = \mathbf{n}(\mathbf{x}', \boldsymbol{\xi}') \tag{6.41}$$

6.3.1 BB Interactions

As derived in Sect. 3.3.1 of Chap. 3, Eqs. (3.25) and (3.26) provide the expressions for the force density vectors of the primary and secondary bonds arising from total deformation under small deformation kinematics in the form:

$$\mathbf{t}^{BB}(\mathbf{x}'-\mathbf{x},t)=\widehat{\mu}(\mathbf{x}'-\mathbf{x})\frac{\mu}{2}\mathbf{S}(\mathbf{x},\boldsymbol{\xi})(\mathbf{u}(\mathbf{x}')-\mathbf{u}(\mathbf{x}))\ \text{ for }\ \mathbf{x}'\in H_{\mathbf{x}}, \tag{6.42}$$

and

$$\mathbf{t}^{BB}(\mathbf{x}-\mathbf{x}',t)=\widehat{\mu}(\mathbf{x}-\mathbf{x}')\frac{\mu}{2}\mathbf{S}(\mathbf{x}',\boldsymbol{\xi}')(\mathbf{u}(\mathbf{x})-\mathbf{u}(\mathbf{x}'))\ \text{ for }\ \mathbf{x}'\in \widetilde{H}_{\mathbf{x}} \tag{6.43}$$

After invoking Eqs. (6.1), (6.18), and (6.21), the force density vectors arising from the initial deformation $\mathbf{t}_0^{BB}(\mathbf{x}'-\mathbf{x},t)$ and $\mathbf{t}_0^{BB}(\mathbf{x}-\mathbf{x}',t)$ can be expressed as:

$$\mathbf{t}_0^{BB}(\mathbf{x}'-\mathbf{x},t)=-\widehat{\mu}(\mathbf{x}'-\mathbf{x})\frac{c(\delta_{\mathbf{x}})}{4}(s_0(\mathbf{x},\boldsymbol{\xi})\mathbf{n}(\mathbf{x},\boldsymbol{\xi})-s_0(\mathbf{x}',\boldsymbol{\xi}')\mathbf{n}(\mathbf{x}',\boldsymbol{\xi}'))\ \text{ for }\ \mathbf{x}'\in H_{\mathbf{x}} \tag{6.44}$$

and

$$\mathbf{t}_0^{BB}(\mathbf{x}-\mathbf{x}',t)=-\widehat{\mu}(\mathbf{x}-\mathbf{x}')\frac{c(\delta_{\mathbf{x}'})}{4}(s_0(\mathbf{x}',\boldsymbol{\xi}')\mathbf{n}(\mathbf{x}',\boldsymbol{\xi}')-s_0(\mathbf{x},\boldsymbol{\xi})\mathbf{n}(\mathbf{x},\boldsymbol{\xi}))\ \text{ for }\ \mathbf{x}'\in \widetilde{H}_{\mathbf{x}} \tag{6.45}$$

Noting that $\mathbf{n}(\alpha,\beta)=(\mathbf{n}(\alpha,\beta)\otimes\mathbf{n}(\alpha,\beta))\mathbf{n}(\alpha,\beta)$, with $\alpha=\mathbf{x},\mathbf{x}'$ and $\beta=\boldsymbol{\xi},\boldsymbol{\xi}'$, these force density vectors can be recast as:

$$\mathbf{t}_0^{BB}(\mathbf{x}'-\mathbf{x},t)=-\widehat{\mu}(\mathbf{x}'-\mathbf{x})\frac{c(\delta_{\mathbf{x}})}{4}\begin{pmatrix}s_0(\mathbf{x},\boldsymbol{\xi})(\mathbf{n}(\mathbf{x},\boldsymbol{\xi})\otimes\mathbf{n}(\mathbf{x},\boldsymbol{\xi}))\mathbf{n}(\mathbf{x},\boldsymbol{\xi})\\ -s_0(\mathbf{x}',\boldsymbol{\xi}')(\mathbf{n}(\mathbf{x}',\boldsymbol{\xi}')\otimes\mathbf{n}(\mathbf{x}',\boldsymbol{\xi}'))\mathbf{n}(\mathbf{x}',\boldsymbol{\xi}')\end{pmatrix}$$
$$\text{for }\ \mathbf{x}'\in H_{\mathbf{x}} \tag{6.46}$$

and

$$\mathbf{t}_0^{BB}(\mathbf{x}-\mathbf{x}',t)=-\widehat{\mu}(\mathbf{x}-\mathbf{x}')\frac{c(\delta_{\mathbf{x}'})}{4}\begin{pmatrix}s_0(\mathbf{x}',\boldsymbol{\xi}')(\mathbf{n}(\mathbf{x}',\boldsymbol{\xi}')\otimes\mathbf{n}(\mathbf{x}',\boldsymbol{\xi}'))\mathbf{n}(\mathbf{x}',\boldsymbol{\xi}')\\ -s_0(\mathbf{x},\boldsymbol{\xi})(\mathbf{n}(\mathbf{x},\boldsymbol{\xi})\otimes\mathbf{n}(\mathbf{x},\boldsymbol{\xi}))\mathbf{n}(\mathbf{x},\boldsymbol{\xi})\end{pmatrix}$$
$$\text{for }\ \mathbf{x}'\in \widetilde{H}_{\mathbf{x}} \tag{6.47}$$

After substituting from Eq. (3.10) along with the nonlocal representation of $\mathbf{n}(\mathbf{x}, \boldsymbol{\xi}) \otimes \mathbf{n}(\mathbf{x}, \boldsymbol{\xi})$ from Eqs. (3.20) and (3.21), the initial bond forces can be rewritten as:

$$\mathbf{t}_0^{BB}(\mathbf{x}' - \mathbf{x}, t) = -\widehat{\mu}(\mathbf{x}' - \mathbf{x})\frac{\mu}{2}\mathbf{h}_1{}^{BB}(\mathbf{x}, \mathbf{x}') \ \text{ for } \mathbf{x}' \in H_{\mathbf{x}} \tag{6.48}$$

and

$$\mathbf{t}_0^{BB}(\mathbf{x} - \mathbf{x}', t) = -\widehat{\mu}(\mathbf{x} - \mathbf{x}')\frac{\mu}{2}\mathbf{h}_1{}^{BB}(\mathbf{x}', \mathbf{x}) \ \text{ for } \mathbf{x}' \in \widetilde{H}_{\mathbf{x}} \tag{6.49}$$

where the vector $\mathbf{h}_1{}^{BB}(\mathbf{x}, \mathbf{x}')$ is defined as:

$$\mathbf{h}_1(\mathbf{x}, \mathbf{x}') = \frac{s_0(\mathbf{x}, \xi)\mathbf{S}(\mathbf{x}, \xi)\xi - s_0(\mathbf{x}', \xi')\mathbf{S}(\mathbf{x}', \xi')\xi'}{2} \tag{6.50}$$

The matrix $\mathbf{S}(\mathbf{x}, \boldsymbol{\xi})$ is defined in Eq. (3.22). These force density vectors do not require any surface correction. It is worth noting that $\mathbf{S}(\mathbf{x}, \boldsymbol{\xi}) = \mathbf{S}(\mathbf{x}', \boldsymbol{\xi}')$ for material points $\mathbf{x}$ and $\mathbf{x}'$ with identical horizons.

6.3.2 OSB Interactions

As derived in Sect. 3.3.2 of Chap. 3, Eqs. (3.46) and (3.47) provide the expressions for the force density vectors of the primary and secondary bonds arising from total deformation under small deformation kinematics in the form:

$$\mathbf{t}^{OSB}(\mathbf{x}' - \mathbf{x}, t) = \widehat{\mu}(\mathbf{x}' - \mathbf{x})\left\{ \begin{array}{l} a \displaystyle\int_{H_{\mathbf{x}}} \widehat{\mu}(\mathbf{x}'' - \mathbf{x})(\widehat{\mathbf{g}}(\mathbf{x}, \xi) \otimes \mathbf{g}(\mathbf{x}, \mathbf{x}'' - \mathbf{x}))(\mathbf{u}(\mathbf{x}'') - \mathbf{u}(\mathbf{x}))dV_{\mathbf{x}''} \\ + \dfrac{n}{2(n+1)}\mu\dfrac{\delta_{\mathbf{x}}}{|\xi|}\mathbf{S}(\mathbf{x}, \xi)(\mathbf{u}(\mathbf{x}') - \mathbf{u}(\mathbf{x})) \end{array} \right\} \tag{6.51}$$

and

$$\mathbf{t}^{OSB}(\mathbf{x} - \mathbf{x}', t) = \widehat{\mu}(\mathbf{x} - \mathbf{x}')\left\{ \begin{array}{l} a \displaystyle\int_{H_{\mathbf{x}'}} \widehat{\mu}(\mathbf{x}''' - \mathbf{x}')(\widehat{\mathbf{g}}(\mathbf{x}', \xi') \otimes \mathbf{g}(\mathbf{x}', \mathbf{x}''' - \mathbf{x}'))(\mathbf{u}(\mathbf{x}''') - \mathbf{u}(\mathbf{x}'))dV_{\mathbf{x}'''} \\ + \dfrac{n}{2(n+1)}\mu\dfrac{\delta_{\mathbf{x}'}}{|\xi'|}\mathbf{S}(\mathbf{x}', \xi')(\mathbf{u}(\mathbf{x}) - \mathbf{u}(\mathbf{x}')) \end{array} \right\} \tag{6.52}$$

The force density vectors arising from the initial deformation, $\mathbf{t}_0^{BB}(\mathbf{x}' - \mathbf{x}, t)$ and $\mathbf{t}_0^{BB}(\mathbf{x} - \mathbf{x}', t)$, can be expressed as:

$$\mathbf{t}_0^{OSB}(\mathbf{x}'-\mathbf{x},t)=-\widehat{\mu}(\mathbf{x}'-\mathbf{x})\begin{pmatrix} ad(\delta_{\mathbf{x}})\dfrac{1}{|\xi|}\theta_0(\mathbf{x},t)\mathbf{n}(\mathbf{x},\xi) \\ +b(\delta_{\mathbf{x}})\dfrac{1}{|\xi|}\dfrac{1}{2}(s_0(\mathbf{x},\xi)\mathbf{n}(\mathbf{x},\xi)-s_0(\mathbf{x}',\xi')\mathbf{n}(\mathbf{x}',\xi')) \end{pmatrix}$$

$$\text{for } \mathbf{x}'\in H_{\mathbf{x}} \tag{6.53}$$

and

$$\mathbf{t}_0^{OSB}(\mathbf{x}-\mathbf{x}',t)=-\widehat{\mu}(\mathbf{x}-\mathbf{x}')\begin{pmatrix} ad(\delta_{\mathbf{x}'})\dfrac{1}{|\xi|}\theta_0(\mathbf{x}',t)\mathbf{n}(\mathbf{x}',\xi') \\ +b(\delta_{\mathbf{x}'})\dfrac{1}{|\xi|}\dfrac{1}{2}(s_0(\mathbf{x}',\xi')\mathbf{n}(\mathbf{x}',\xi')-s_0(\mathbf{x},\xi)\mathbf{n}(\mathbf{x},\xi)) \end{pmatrix}$$

$$\text{for } \mathbf{x}'\in \widetilde{H}_{\mathbf{x}} \tag{6.54}$$

The dilatation arising from the initial deformation, $\theta_0^{OSB}(\mathbf{x},t)$, can be expressed as:

$$\theta_0^{OSB}(\mathbf{x},t)=d(\delta_{\mathbf{x}})\int\limits_{H_{\mathbf{x}}}\widehat{\mu}(\mathbf{x}''-\mathbf{x})\bar{s}_0(\mathbf{x},\mathbf{x}'')dV_{\mathbf{x}''} \tag{6.55}$$

and

$$\theta_0^{OSB}(\mathbf{x}',t)=d(\delta_{\mathbf{x}'})\int\limits_{H_{\mathbf{x}'}}\widehat{\mu}(\mathbf{x}'''-\mathbf{x}')\bar{s}_0(\mathbf{x}',\mathbf{x}''')dV_{\mathbf{x}'''} \tag{6.56}$$

Noting that $\mathbf{n}(\alpha,\beta)=(\mathbf{n}(\alpha,\beta)\otimes\mathbf{n}(\alpha,\beta))\mathbf{n}(\alpha,\beta)$, with $\alpha=\mathbf{x},\,\mathbf{x}'$ and $\beta=\boldsymbol{\xi},\,\boldsymbol{\xi}'$, the force density vectors can be recast as:

$$\mathbf{t}_0^{OSB}(\mathbf{x}'-\mathbf{x},t)=-\widehat{\mu}(\mathbf{x}'-\mathbf{x})$$
$$\times\begin{pmatrix} a\theta_0(\mathbf{x},t)\widehat{\mathbf{g}}(\mathbf{x},\boldsymbol{\xi}) \\ +b(\delta_{\mathbf{x}})\dfrac{1}{|\boldsymbol{\xi}|}\dfrac{1}{2}\begin{pmatrix} s_0(\mathbf{x},\boldsymbol{\xi})(\mathbf{n}(\mathbf{x},\boldsymbol{\xi})\otimes\mathbf{n}(\mathbf{x},\boldsymbol{\xi}))\mathbf{n}(\mathbf{x},\boldsymbol{\xi}) \\ -s_0(\mathbf{x}',\boldsymbol{\xi}')(\mathbf{n}(\mathbf{x}',\boldsymbol{\xi}')\otimes\mathbf{n}(\mathbf{x}',\boldsymbol{\xi}'))\mathbf{n}(\mathbf{x}',\boldsymbol{\xi}') \end{pmatrix} \end{pmatrix}\text{for } \mathbf{x}'\in H_{\mathbf{x}} \tag{6.57}$$

and

$$\mathbf{t}_0^{OSB}(\mathbf{x}-\mathbf{x}',t) = -\widehat{\mu}(\mathbf{x}-\mathbf{x}')\begin{pmatrix} a\theta_0(\mathbf{x}',t)\widehat{\mathbf{g}}(\mathbf{x}',\xi') \\ +b(\delta_{\mathbf{x}'})\dfrac{1}{|\xi'|}\dfrac{1}{2}\begin{pmatrix} s_0(\mathbf{x}',\xi')(\mathbf{n}(\mathbf{x}',\xi')\otimes\mathbf{n}(\mathbf{x}',\xi'))\mathbf{n}(\mathbf{x}',\xi') \\ -s_0(\mathbf{x},\xi)(\mathbf{n}(\mathbf{x},\xi)\otimes\mathbf{n}(\mathbf{x},\xi))\mathbf{n}(\mathbf{x},\xi) \end{pmatrix} \end{pmatrix}$$
$$\text{for } \mathbf{x}' \in \widetilde{H}_{\mathbf{x}} \tag{6.58}$$

After substituting from Eq. (3.10) along with the nonlocal representation of $\mathbf{n}(\mathbf{x},\boldsymbol{\xi}) \otimes \mathbf{n}(\mathbf{x},\boldsymbol{\xi})$ from Eqs. (3.20) and (3.21), the initial bond forces can be rewritten as:

$$\mathbf{t}_0^{OSB}(\mathbf{x}'-\mathbf{x},t) = -\widehat{\mu}(\mathbf{x}'-\mathbf{x})\begin{pmatrix} a(d(\delta_{\mathbf{x}})\displaystyle\int_{H_{\mathbf{x}}}\widehat{\mu}(\mathbf{x}''-\mathbf{x})\bar{s}_0(\mathbf{x},\mathbf{x}'')dV_{\mathbf{x}''})\widehat{\mathbf{g}}(\mathbf{x},\xi) \\ +\dfrac{n}{2(n+1)}\mu\delta_{\mathbf{x}}\mathbf{h}_1(\mathbf{x},\mathbf{x}') \end{pmatrix} \quad \text{for } \mathbf{x}' \in H_{\mathbf{x}} \tag{6.59}$$

and

$$\mathbf{t}_0^{OSB}(\mathbf{x}-\mathbf{x}',t) = -\widehat{\mu}(\mathbf{x}-\mathbf{x}')\begin{pmatrix} a(d(\delta_{\mathbf{x}'})\displaystyle\int_{H_{\mathbf{x}'}}\widehat{\mu}(\mathbf{x}'''-\mathbf{x}')\bar{s}_0(\mathbf{x}',\mathbf{x}''')dV_{\mathbf{x}'''})\widehat{\mathbf{g}}(\mathbf{x}',\xi') \\ +\dfrac{n}{2(n+1)}\mu\delta_{\mathbf{x}'}\mathbf{h}_1(\mathbf{x}',\mathbf{x}) \end{pmatrix} \quad \text{for } \mathbf{x}' \in \widetilde{H}_{\mathbf{x}} \tag{6.60}$$

with $n = 2$ and $n = 3$ for 2D and 3D analysis, respectively.

6.3.3 NOSB Interactions

As derived in Sect. 3.3.3 of Chap. 3, Eqs. (3.67) and (3.68) along with Eq. (3.64) provide the expressions for the force density vectors of the primary and secondary bonds arising from total deformation under small deformation kinematics in the form:

$$\mathbf{t}^{NOSB}(\mathbf{x}'-\mathbf{x},t) = \widehat{\mu}(\mathbf{x}'-\mathbf{x})\boldsymbol{\sigma}_{\boldsymbol{\xi}}^{PD}(\mathbf{x},t)\mathbf{g}(\mathbf{x},\boldsymbol{\xi}) = \widehat{\mu}(\mathbf{x}'-\mathbf{x})\left(\mathbf{C}:\boldsymbol{\varepsilon}_{\boldsymbol{\xi}}^{PD}(\mathbf{x})\right)\mathbf{g}(\mathbf{x},\boldsymbol{\xi}) \tag{6.61}$$

and

$$\mathbf{t}^{NOSB}(\mathbf{x}-\mathbf{x}',t) = \widehat{\mu}(\mathbf{x}-\mathbf{x}')\boldsymbol{\sigma}_{\boldsymbol{\xi}'}^{PD}(\mathbf{x}',t)\mathbf{g}(\mathbf{x}',\boldsymbol{\xi}') = \widehat{\mu}(\mathbf{x}-\mathbf{x}')\left(\mathbf{C}:\boldsymbol{\varepsilon}_{\boldsymbol{\xi}'}^{PD}(\mathbf{x}')\right)\mathbf{g}(\mathbf{x}',\boldsymbol{\xi}') \tag{6.62}$$

By modifying Eqs. (3.67) and (3.68) along with Eq. (3.64), the NOSB force density vectors arising from the initial deformation, $\mathbf{t}_0^{NOSB}(\mathbf{x}' - \mathbf{x}, t)$ and $\mathbf{t}_0^{NOSB}(\mathbf{x} - \mathbf{x}', t)$ can be expressed as:

$$\mathbf{t}_0^{NOSB}(\mathbf{x}' - \mathbf{x}, t) = -\widehat{\mu}(\mathbf{x}' - \mathbf{x})\left(\mathbf{C} : \boldsymbol{\varepsilon}_{\xi 0}^{PD}(\mathbf{x})\right)\mathbf{g}(\mathbf{x}, \boldsymbol{\xi}) \text{ for } \mathbf{x}' \in H_{\mathbf{x}} \tag{6.63}$$

and

$$\mathbf{t}_0^{NOSB}(\mathbf{x} - \mathbf{x}', t) = -\widehat{\mu}(\mathbf{x} - \mathbf{x}')\left(\mathbf{C} : \boldsymbol{\varepsilon}_{\xi' 0}^{PD}(\mathbf{x}')\right)\mathbf{g}(\mathbf{x}', \boldsymbol{\xi}') \text{ for } \mathbf{x}' \in \widetilde{H}_{\mathbf{x}} \tag{6.64}$$

6.3.4 PDDO Interactions

As derived in Sect. 3.3.4 of Chap. 3, Eq. (3.77) and (3.78) along with Eq. (3.79) provide the expressions for the force density vectors arising from total deformation under small deformation kinematics in the form:

$$\mathbf{t}^{PDDO}(\mathbf{x}, \mathbf{x}' - \mathbf{x}, t) = \widehat{\mu}(\mathbf{x}' - \mathbf{x})\boldsymbol{\sigma}^{PD}(\mathbf{x}, t)\mathbf{g}(\mathbf{x}, \boldsymbol{\xi}) = \widehat{\mu}(\mathbf{x}' - \mathbf{x})\left(\mathbf{C} : \boldsymbol{\varepsilon}^{PD}(\mathbf{x})\right)\mathbf{g}(\mathbf{x}, \boldsymbol{\xi}) \tag{6.65}$$

and

$$\mathbf{t}^{PDDO}(\mathbf{x}', \mathbf{x}' - \mathbf{x}, t) = \widehat{\mu}(\mathbf{x}' - \mathbf{x})\boldsymbol{\sigma}^{PD}(\mathbf{x}', t)\mathbf{g}(\mathbf{x}, \boldsymbol{\xi}) = \widehat{\mu}(\mathbf{x}' - \mathbf{x})\left(\mathbf{C} : \boldsymbol{\varepsilon}^{PD}(\mathbf{x}')\right)\mathbf{g}(\mathbf{x}, \boldsymbol{\xi}) \tag{6.66}$$

By modifying Eqs. (3.77) and (3.78) along with Eq. (3.79), the PDDO force density vectors arising from the initial deformation, $\mathbf{t}_0^{PDDO}(\mathbf{x}' - \mathbf{x}, t)$ and $\mathbf{t}_0^{PDDO}(\mathbf{x} - \mathbf{x}', t)$ can be expressed as:

$$\mathbf{t}_0^{PDDO}(\mathbf{x}, \mathbf{x}' - \mathbf{x}, t) = -\widehat{\mu}(\mathbf{x}' - \mathbf{x})(\mathbf{C} : \boldsymbol{\varepsilon}_0(\mathbf{x}))\mathbf{g}(\mathbf{x}, \boldsymbol{\xi}) \tag{6.67}$$

and

$$\mathbf{t}_0^{PDDO}(\mathbf{x}', \mathbf{x}' - \mathbf{x}, t) = -\widehat{\mu}(\mathbf{x}' - \mathbf{x})(\mathbf{C} : \boldsymbol{\varepsilon}_0(\mathbf{x}'))\mathbf{g}(\mathbf{x}, \boldsymbol{\xi}) \tag{6.68}$$

6.4 PD Internal Force Vectors for Uniform Discretization

The explicit expressions for the force density vectors enable the evaluation of internal force vectors, $\mathbf{L}^{\alpha}(\mathbf{x}, t)$ and $\mathbf{L}_0^{\alpha}(\mathbf{x}, t)$, arising from total and initial deformations, respectively, in Eqs. (6.25) and (6.26) for $\alpha = BB, OSB, NOSB, PDDO$. For uniform discretization, i.e., $\delta_{\mathbf{x}} = \delta_{\mathbf{x}'} = \delta$, the BB and OSB internal force vectors can be expressed as:

$$\mathbf{L}^{BB}(\mathbf{x}, t) = \int_{H_{\mathbf{x}}} \widehat{\mu}(\mathbf{x}' - \mathbf{x})\mu\overline{\mathbf{S}}(\mathbf{x}, \boldsymbol{\xi})(\mathbf{u}(\mathbf{x}') - \mathbf{u}(\mathbf{x}))dV_{\mathbf{x}'} \tag{6.69}$$

$$\mathbf{L}_0^{BB}(\mathbf{x}, t) = \int_{H_{\mathbf{x}}} \widehat{\mu}(\mathbf{x}' - \mathbf{x})\mu\mathbf{h}_1(\mathbf{x}, \mathbf{x}')dV_{\mathbf{x}'} \tag{6.70}$$

$$\mathbf{L}^{OSB}(\mathbf{x}, t) = \widehat{\mu}(\mathbf{x}' - \mathbf{x})\left\{\begin{array}{l} a\displaystyle\int_{H_{\mathbf{x}}} \widehat{\mu}(\mathbf{x}'' - \mathbf{x})(\widehat{\mathbf{g}}(\mathbf{x}, \boldsymbol{\xi}) \otimes \mathbf{g}(\mathbf{x}, \mathbf{x}'' - \mathbf{x}))(\mathbf{u}(\mathbf{x}'') - \mathbf{u}(\mathbf{x}))dV_{\mathbf{x}''} \\ -a\displaystyle\int_{H_{\mathbf{x}'}} \widehat{\mu}(\mathbf{x}''' - \mathbf{x}')(\widehat{\mathbf{g}}(\mathbf{x}', \boldsymbol{\xi}') \otimes \mathbf{g}(\mathbf{x}', \mathbf{x}''' - \mathbf{x}'))(\mathbf{u}(\mathbf{x}''') - \mathbf{u}(\mathbf{x}'))dV_{\mathbf{x}'''} \\ +\dfrac{n}{(n+1)}\mu\dfrac{\delta}{|\boldsymbol{\xi}|}\bar{\mathbf{S}}(\mathbf{x}, \boldsymbol{\xi})(\mathbf{u}(\mathbf{x}') - \mathbf{u}(\mathbf{x})) \end{array}\right\} \tag{6.71}$$

$$\mathbf{L}_0^{OSB}(\mathbf{x}, t) = \widehat{\mu}(\mathbf{x}' - \mathbf{x})\left(ad\mathbf{h}_2(\mathbf{x}, \mathbf{x}') + \frac{n}{(n+1)}\mu\delta\mathbf{h}_1(\mathbf{x}, \mathbf{x}')\right) \tag{6.72}$$

with

$$\begin{aligned}\mathbf{h}_2(\mathbf{x}, \mathbf{x}') = &\left(\int_{H_{\mathbf{x}}} \widehat{\mu}(\mathbf{x}'' - \mathbf{x})\overline{s}_0(\mathbf{x}, \mathbf{x}'')dV_{\mathbf{x}''}\right)\widehat{\mathbf{g}}(\mathbf{x}, \boldsymbol{\xi}) \\ &-\left(\int_{H_{\mathbf{x}'}} \widehat{\mu}(\mathbf{x}''' - \mathbf{x}')\overline{s}_0(\mathbf{x}', \mathbf{x}''')dV_{\mathbf{x}'''}\right)\widehat{\mathbf{g}}(\mathbf{x}', \boldsymbol{\xi}')\end{aligned} \tag{6.73}$$

Similarly, the NOSB and PPDO internal force vectors can be expressed as:

$$\mathbf{L}^{NOSB}(\mathbf{x}, t) = \widehat{\mu}(\mathbf{x}' - \mathbf{x})((\mathbf{C} : \boldsymbol{\varepsilon}_{\boldsymbol{\xi}}^{PD}(\mathbf{x}))\mathbf{g}(\mathbf{x}, \boldsymbol{\xi}) - (\mathbf{C} : \boldsymbol{\varepsilon}_{\boldsymbol{\xi}'}^{PD}(\mathbf{x}'))\mathbf{g}(\mathbf{x}', \boldsymbol{\xi}')) \tag{6.74}$$

$$\mathbf{L}_0^{NOSB}(\mathbf{x}, t) = \widehat{\mu}(\mathbf{x}' - \mathbf{x})((\mathbf{C} : \boldsymbol{\varepsilon}_{\boldsymbol{\xi}'0}^{PD}(\mathbf{x}'))\mathbf{g}(\mathbf{x}', \boldsymbol{\xi}') - (\mathbf{C} : \boldsymbol{\varepsilon}_{\boldsymbol{\xi}0}^{PD}(\mathbf{x}))\mathbf{g}(\mathbf{x}, \boldsymbol{\xi})) \tag{6.75}$$

$$\mathbf{L}^{PDDO}(\mathbf{x}, t) = \widehat{\mu}(\mathbf{x}' - \mathbf{x})\{(\mathbf{C} : \boldsymbol{\varepsilon}^{PD}(\mathbf{x}')) - (\mathbf{C} : \boldsymbol{\varepsilon}^{PD}(\mathbf{x}))\}\mathbf{g}(\mathbf{x}, \boldsymbol{\xi}) \tag{6.76}$$

and

$$\mathbf{L}_0^{PDDO}(\mathbf{x},t) = \widehat{\mu}(\mathbf{x}' - \mathbf{x})((\mathbf{C} : \boldsymbol{\varepsilon}_0(\mathbf{x})) - (\mathbf{C} : \boldsymbol{\varepsilon}_0(\mathbf{x}')))\mathbf{g}(\mathbf{x}, \boldsymbol{\xi}) \tag{6.77}$$

The PD form of the traction vector, $\mathbf{T}_m^{PD}(\mathbf{x})$, arising from mechanical stresses can be written as:

$$\mathbf{T}_m^{PD}(\mathbf{x}) = \left(\boldsymbol{\sigma}^{PD}(\mathbf{x}) - (\mathbf{C} : \boldsymbol{\varepsilon}_0(\mathbf{x}))\right)\mathbf{n} = \mathbf{T}^* \text{ for } \mathbf{x} \in \mathscr{B}_{\boldsymbol{\sigma}} \tag{6.78}$$

The PD form of the total traction vector, $\mathbf{T}^{PD}(\mathbf{x})$, arising from the mechanical and initial strain can be expressed as:

$$\mathbf{T}^{PD}(\mathbf{x}) = \boldsymbol{\sigma}^{PD}(\mathbf{x})\mathbf{n} = \mathbf{T}^* + (\mathbf{C} : \boldsymbol{\varepsilon}_0(\mathbf{x}))\mathbf{n} = \mathbf{T}^{**} \text{ for } \mathbf{x} \in \mathscr{B}_{\boldsymbol{\sigma}} \tag{6.79}$$

and

$$\mathbf{u} = \mathbf{U}^* \text{ for } \mathbf{x} \in \mathscr{B}_{\mathbf{u}} \tag{6.80}$$

where $\mathbf{T}^*$ and $\mathbf{U}^*$ represent the applied zero and nonzero tractions on $\mathscr{B}_{\boldsymbol{\sigma}}$ and displacement constraints on $\mathscr{B}_{\mathbf{u}}$, respectively, in the "actual boundary layer" region of the domain.

6.5 Governing Equations for a Coupled PD and FE Analysis

A domain composed of PD and FE regions shown in Fig. 6.1 can be subjected to a known initial strain field, $\boldsymbol{\varepsilon}_0(\mathbf{x})$. The surface defining the PD region, S^{PD}, can be subjected to displacement constraints and/or external tractions. It is split into $S_{\mathbf{u}}^{PD}$, $S_{\boldsymbol{\sigma}}^{PD}$, and S_I^{PD} representing parts of the boundary subjected to displacement constraints, $\mathbf{U}^{*PD}$, external tractions, $\mathbf{T}^{*PD}$, and PD/FE interface, respectively. Similar to the PD region, the surface defining the FE region, S^{FE}, is split accordingly. The superscripts appearing as PD and FE indicate quantities associated with the domains of PD and FE, respectively.

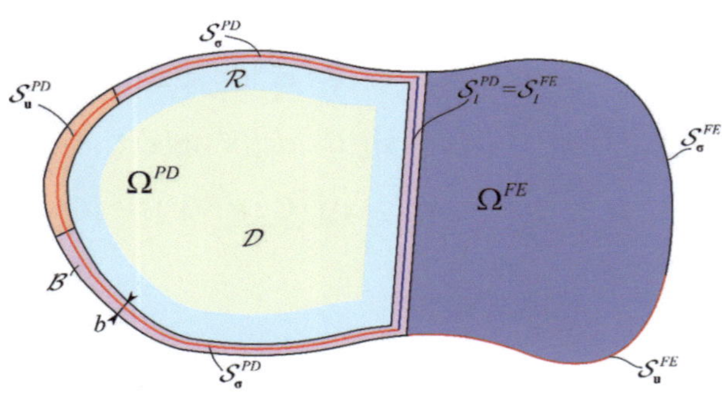

Fig. 6.1 Coupling of PD and FE regions for deformation analysis due to initial strain

The governing coupled PD with FE equations can be derived by modifying Eq. (5.7) from Chap. 5 as:

$$\begin{aligned}
&\int_{\mathcal{D}\cup\mathcal{R}} \rho^{PD}(\mathbf{x},t)\ddot{\mathbf{u}}(\mathbf{x},t)\cdot\delta\mathbf{u}(\mathbf{x})dA - \int_{\mathcal{D}} \mathbf{L}_m^{\alpha}(\mathbf{x},t)\cdot\delta\mathbf{u}(\mathbf{x})dA \\
&- \int_{\mathcal{R}} \mathbf{L}_m^{PDDO}(\mathbf{x},t)\cdot\delta\mathbf{u}(\mathbf{x})dA + \oint_{\Gamma^{PD}} \mathbf{T}_m^{PDDO}(\mathbf{x},t)\cdot\delta\mathbf{u}(\mathbf{x})d\Gamma \\
&+ \int_{\Omega^{FE}} \rho^{FE}(\mathbf{x},t)\ddot{\mathbf{u}}(\mathbf{x},t)\cdot\delta\mathbf{u}(\mathbf{x})dA + \int_{\Omega^{FE}} \mathbf{P}_m^{FE}(\mathbf{x},t) : \delta\nabla\mathbf{u}(\mathbf{x})dA \\
&= \int_{\Gamma_\sigma^{PD}} \mathbf{T}^{*PD}(\mathbf{x},t)\cdot\delta\mathbf{u}(\mathbf{x})d\Gamma + \int_{\Gamma_\sigma^{FE}} \mathbf{T}^{*FE}(\mathbf{x},t)\cdot\delta\mathbf{u}(\mathbf{x})d\Gamma \\
&+ \int_{\mathcal{D}\cup\mathcal{R}} \mathbf{b}^{PD}(\mathbf{x},t)\cdot\delta\mathbf{u}(\mathbf{x})dA + \int_{\Omega^{FE}} \mathbf{b}^{FE}(\mathbf{x},t)\cdot\delta\mathbf{u}(\mathbf{x})dA \\
&\alpha = BB, OSB, NOSB
\end{aligned} \tag{6.81}$$

Substituting from Eqs. (6.36), (6.37), (6.38), and (6.78) for the internal and boundary force vectors arising from mechanical deformation into this equation leads to:

$$\begin{aligned}
&\int_{\mathcal{D}\cup\mathcal{R}} \rho^{PD}(\mathbf{x},t)\ddot{\mathbf{u}}(\mathbf{x},t)\cdot\delta\mathbf{u}(\mathbf{x})dA - \int_{\mathcal{D}} \mathbf{L}^{\alpha}(\mathbf{x},t)\cdot\delta\mathbf{u}(\mathbf{x})dA \\
&- \int_{\mathcal{R}} \mathbf{L}^{PDDO}(\mathbf{x},t)\cdot\delta\mathbf{u}(\mathbf{x})dA + \oint_{\Gamma^{PD}} \mathbf{T}^{PDDO}(\mathbf{x},t)\cdot\delta\mathbf{u}(\mathbf{x})d\Gamma \\
&+ \int_{\Omega^{FE}} \rho^{FE}(\mathbf{x},t)\ddot{\mathbf{u}}(\mathbf{x},t)\cdot\delta\mathbf{u}(\mathbf{x})dA + \int_{\Omega^{FE}} \boldsymbol{\sigma}^{FE}(\mathbf{x},t) : \delta\nabla\mathbf{u}(\mathbf{x})dA + \\
&= \int_{\Gamma_{\boldsymbol{\sigma}}^{PD}} \mathbf{T}^{**PD}(\mathbf{x},t)\cdot\delta\mathbf{u}(\mathbf{x})d\Gamma + \int_{\Gamma_{\boldsymbol{\sigma}}^{FE}} \mathbf{T}^{*FE}(\mathbf{x},t)\cdot\delta\mathbf{u}(\mathbf{x})d\Gamma \\
&+ \int_{\mathcal{D}\cup\mathcal{R}} \mathbf{b}^{PD}(\mathbf{x},t)\cdot\delta\mathbf{u}(\mathbf{x})dA + \int_{\Omega^{FE}} \mathbf{b}^{FE}(\mathbf{x},t)\cdot\delta\mathbf{u}(\mathbf{x})dA \\
&+ \int_{\mathcal{D}} \mathbf{L}_0^{\alpha}(\mathbf{x},t)\cdot\delta\mathbf{u}(\mathbf{x})dA + \int_{\mathcal{R}} \mathbf{L}_0^{PDDO}(\mathbf{x},t)\cdot\delta\mathbf{u}(\mathbf{x})dA \\
&\alpha = BB, OSB, NOSB
\end{aligned} \tag{6.82}$$

As explained in detail in Chap. 5, the discrete form of this equation can be expressed as:

$$
\begin{aligned}
&\sum_{k=1}^{N_{\mathcal{D}}^{PD}+N_{\mathcal{R}}^{PD}} \rho_{(k)}^{PD} \boldsymbol{\delta u}_{(k)}^{PD} \cdot \ddot{\mathbf{u}}_{(k)}^{PD} V_{(k)} - \sum_{k=1}^{N_{\mathcal{D}}^{PD}} \boldsymbol{\delta u}_{(k)}^{PD} \cdot \left(\sum_{j=1}^{N_{(k)}} \mathbf{t}_{(k)(j)}^{\alpha} V_{(j)} - \sum_{p=1}^{\widetilde{N}_{(k)}} \mathbf{t}_{(p)(k)}^{\alpha} V_{(p)} \right) V_{(k)} \\
&- \sum_{k=1}^{N_{\mathcal{R}}^{PD}} \boldsymbol{\delta u}_{(k)}^{PD} \cdot \left(\sum_{j=1}^{N_{(k)}} \left(\mathbf{t}_{(j)(k)}^{PDDO} - \mathbf{t}_{(k)(j)}^{PDDO} \right) V_{(j)} \right) V_{(k)} + \sum_{k=1}^{N_{\mathcal{B}\boldsymbol{\sigma}}^{PD}+N_{\mathcal{B}\mathbf{u}}^{PD}+N_{\mathcal{I}}^{PD}} \boldsymbol{\delta u}_{(k)}^{PD} \cdot \left(\boldsymbol{\sigma}_{(k)}^{PD} \mathbf{n}_{(k)} \right) \Gamma_{(k)} \\
&+ \sum_{e=1}^{E} \rho_{(k)}^{FE} \boldsymbol{\delta u}_{(e)}^{FE} \cdot \ddot{\mathbf{u}}_{(k)}^{FE} V_{(k)} + \sum_{e=1}^{E} \operatorname{tr}\left(\left(\nabla \boldsymbol{\delta u}_{(e)}^{FE} \right)^{T} \boldsymbol{\sigma}_{(e)}^{FE} \right) V_{(e)} - \sum_{k=1}^{N_{\mathcal{B}\boldsymbol{\sigma}}^{PD}} \boldsymbol{\delta u}_{(k)}^{PD} \cdot \mathbf{T}^{**PD} \Gamma_{(k)} \\
&- \sum_{k=1}^{N_{\mathcal{D}}^{PD}+N_{\mathcal{R}}^{PD}} \boldsymbol{\delta u}_{(k)}^{PD} \cdot \mathbf{b}_{(k)}^{PD} V_{(k)} - \sum_{e=1}^{E_{\mathcal{B}\boldsymbol{\sigma}}} \boldsymbol{\delta u}_{(e)}^{FE} \cdot \mathbf{T}_{(e)}^{*FE} \Gamma_{(e)} - \sum_{e=1}^{E} \boldsymbol{\delta u}_{(e)}^{FE} \cdot \mathbf{b}_{(e)}^{FE} V_{(e)} \\
&- \sum_{k=1}^{N_{\mathcal{D}}^{PD}} \boldsymbol{\delta u}_{(k)}^{PD} \cdot \left(\sum_{j=1}^{N_{(k)}} \mathbf{t}_{0(k)(j)}^{\alpha} V_{(j)} - \sum_{p=1}^{\widetilde{N}_{(k)}} \mathbf{t}_{0(p)(k)}^{\alpha} V_{(p)} \right) V_{(k)} \\
&- \sum_{k=1}^{N_{\mathcal{R}}^{PD}} \boldsymbol{\delta u}_{(k)}^{PD} \cdot \left(\sum_{j=1}^{N_{(k)}} \left(\mathbf{t}_{0(j)(k)}^{PDDO} - \mathbf{t}_{0(k)(j)}^{PDDO} \right) V_{(j)} \right) V_{(k)} = 0
\end{aligned}
\tag{6.83}
$$

The evaluation of each of these terms is described in detail in Sect. 5.3.1 in Chap. 5 except for the last two terms arising from the presence of initial strain field. They can be evaluated as:

$$
\begin{aligned}
&\sum_{k=1}^{N_{\mathcal{D}}^{PD}} \boldsymbol{\delta u}_{(k)}^{PD} \cdot \left(\sum_{j=1}^{N_{(k)}} \mathbf{t}_{0(k)(j)}^{\alpha} V_{(j)} - \sum_{p=1}^{\widetilde{N}_{(k)}} \mathbf{t}_{0(p)(k)}^{\alpha} V_{(p)} \right) V_{(k)} \\
&= \left\{ \delta \mathbf{V}_{\mathcal{D}}^{PD} \right\}^{T} \sum_{k=1}^{N_{\mathcal{D}}^{PD}} \mathbf{L}_{0(k)}^{\alpha} V_{(k)}
\end{aligned}
\tag{6.84}
$$

and

$$
\sum_{k=1}^{N_{\mathcal{R}}^{PD}} \boldsymbol{\delta u}_{(k)}^{PD} \cdot \left(\sum_{j=1}^{N_{(k)}} \left(\mathbf{t}_{0(j)(k)}^{PDDO} - \mathbf{t}_{0(k)(j)}^{PDDO} \right) V_{(j)} \right) V_{(k)} = \left\{ \delta \mathbf{V}_{\mathcal{R}}^{PD} \right\}^{T} \sum_{k=1}^{N_{\mathcal{R}}^{PD}} \mathbf{L}_{0(k)}^{PDDO} V_{(k)} \tag{6.85}
$$

in which the $\mathbf{L}_{0(k)}^{\alpha}$ and $\mathbf{L}_{0(k)}^{PDDO}$ can be expressed as:

$$\mathbf{L}_{0(k)}^{\alpha} = \sum_{j=1}^{N_{(k)}} \mathbf{t}_{0(k)(j)}^{\alpha} V_{(j)} - \sum_{p=1}^{\widetilde{N}_{(k)}} \mathbf{t}_{0(p)(k)}^{\alpha} V_{(p)} \tag{6.86}$$

and

$$\mathbf{L}_{0(k)}^{PDDO} = \sum_{j=1}^{N_{(k)}} \left(\mathbf{t}_{0(j)(k)}^{\alpha} - \mathbf{t}_{0(k)(j)}^{\alpha} \right) V_{(j)} \tag{6.87}$$

Their summation can also be expressed in matrix form as:

$$\begin{aligned} &\sum_{k=1}^{N_{\mathcal{D}}^{PD}} \delta\mathbf{u}_{(k)}^{PD} \cdot \left(\sum_{j=1}^{N_{(k)}} \mathbf{t}_{0(k)(j)}^{\alpha} V_{(j)} - \sum_{p=1}^{\widetilde{N}_{(k)}} \mathbf{t}_{0(p)(k)}^{\alpha} V_{(p)} \right) V_{(k)} \\ &+ \sum_{k=1}^{N_{\mathcal{R}}^{PD}} \delta\mathbf{u}_{(k)}^{PD} \cdot \left(\sum_{j=1}^{N_{(k)}} \left(\mathbf{t}_{0(j)(k)}^{PDDO} - \mathbf{t}_{0(k)(j)}^{PDDO} \right) V_{(j)} \right) V_{(k)} \\ &= \left\{ \delta\mathbf{V}_{\mathcal{D}}^{PD} \quad \delta\mathbf{V}_{\mathcal{R}}^{PD} \right\}^{T} \begin{Bmatrix} \mathbf{F}_{0_{\mathcal{D}}}^{*PD} \\ \mathbf{F}_{0_{\mathcal{R}}}^{*PD} \end{Bmatrix} \end{aligned} \tag{6.88}$$

in which the vectors $\mathbf{F}_{0_{\mathcal{D}}}^{*PD}$ and $\mathbf{F}_{0_{\mathcal{R}}}^{*PD}$ represent the initial strain load vectors acting on the material points in regions $\mathcal{D}$ and $\mathcal{R}$. They are defined as:

$$\mathbf{F}_{0_{\mathcal{D}}}^{*PD} = \left\{ \mathbf{L}_{0(1)}^{\alpha} V_{(1)}, \mathbf{L}_{0(1)}^{\alpha} V_{(2)}, \mathbf{L}_{0(1)}^{\alpha} V_{(3)}, \ldots, \mathbf{L}_{0\left(N_{\mathcal{D}}^{PD}\right)}^{\alpha} V_{\left(N_{\mathcal{D}}^{PD}\right)} \right\}^{T} \tag{6.89}$$

and

$$\mathbf{F}_{0_{\mathcal{R}}}^{*PD} = \left\{ \mathbf{L}_{0(1)}^{PDDO} V_{(1)}, \mathbf{L}_{0(1)}^{PDDO} V_{(2)}, \mathbf{L}_{0(1)}^{PDDO} V_{(3)}, \ldots, \mathbf{L}_{0\left(N_{\mathcal{R}}^{PD}\right)}^{PDDO} V_{\left(N_{\mathcal{R}}^{PD}\right)} \right\}^{T} \tag{6.90}$$

The variational expression of the coupled PD with FE equations with initial strain is the same as that of Eq. (5.39) with a modified form of force vector $\mathbf{F}$ including the effect of initial strain. It is modified as:

$$\mathbf{F} = \left\{ \mathbf{F}_{\mathbf{b}_{\mathcal{D}}}^{*PD} + \mathbf{F}_{0_{\mathcal{D}}}^{*PD} \quad \mathbf{F}_{\mathbf{b}_{\mathcal{R}}}^{*PD} + \mathbf{F}_{0_{\mathcal{R}}}^{*PD} \quad \mathbf{F}_{\mathcal{B}_{\sigma}}^{*PD} \quad \mathbf{0} \quad \mathbf{F}_{\mathbf{b}_{\mathcal{I}}}^{*FE} \quad \mathbf{F}_{\mathbf{b}_{\mathcal{D}}}^{*FE} \quad \mathbf{F}_{\mathcal{B}_{\sigma}}^{*FE} + \mathbf{F}_{\mathbf{b}_{\mathcal{B}_{\sigma}}}^{*FE} \quad \mathbf{F}_{\mathbf{b}_{\mathcal{B}_{\mathbf{u}}}}^{*FE} \right\}^{\mathbf{T}} \tag{6.91}$$

References

D'Antuono, P., & Morandini, M. (2017). Thermal shock response via weakly coupled peridynamic thermo-mechanics. *International Journal of Solids and Structures, 129*, 74–89.

Gao, Y., & Oterkus, S. (2019). Ordinary state-based peridynamic modelling for fully coupled thermoelastic problems. *Continuum Mechanics and Thermodynamics, 31*, 907–937.

Hu, Y., Chen, H., Spencer, B. W., & Madenci, E. (2018). Thermomechanical peridynamic analysis with irregular non-uniform domain discretization. *Engineering Fracture Mechanics, 197*, 92–113.

Kilic, B., & Madenci, E. (2009). Peridynamic theory for thermomechanical analysis. *IEEE Transactions on Advanced Packaging, 33*, 97–105.

Liu, Q. Q., Wu, D., Madenci, E., Yu, Y., & Hu, Y. L. (2022). State-Based peridynamics for thermomechanical modeling of fracture mechanisms in nuclear fuel pellets. *Engineering Fracture Mechanics, 276*, 108917.

Oterkus, S., & Madenci, E. (2017). Peridynamic modeling of fuel pellet cracking. *Engineering Fracture Mechanics, 176*, 23–37.

Oterkus, S., Madenci, E., & Agwai, A. (2014). Fully coupled peridynamic thermomechanics. *Journal of the Mechanics and Physics of Solids, 64*, 1–23.

Pathrikar, A., Tiwari, S. B., Arayil, P., & Roy, D. (2021). Thermomechanics of damage in brittle solids: A peridynamics model. *Theoretical and Applied Fracture Mechanics, 112*, 102880.

Shou, Y., & Zhou, X. (2020). A coupled thermomechanical nonordinary state-based peridynamics for thermally induced cracking of rocks. *Fatigue & Fracture of Engineering Materials & Structures, 43*, 371–386.

Sun, W. K., Yin, B. B., Akbar, A., Kodur, V. K. R., & Liew, K. M. (2024). A coupled 3D thermo-mechanical peridynamic model for cracking analysis of homogeneous and heterogeneous materials. *Computer Methods in Applied Mechanics and Engineering, 418*, 116577.

Wang, Y., Zhou, X., & Kou, M. (2018). A coupled thermo-mechanical bond-based peridynamics for simulating thermal cracking in rocks. *International Journal of Fracture, 211*, 13–42.

Chapter 7
Peridynamics for Multiphysics Analysis

7.1 Introduction

Although peridynamics (PD) is extremely suitable to model the response of structures involving discontinuities, fractures, and other complex phenomena, it is also suitable to model multiphysics problems where different physical fields are coupled among themselves or coupled with the deformation field (Madenci & Oterkus, 2016). Within the last decade, PD theory has been applied to coupled fields representing: (1) moisture concentration, temperature, and deformation (hygro-thermo-mechanics); (2) fluid flow in porous medium and deformation (hydro-mechanics); (3) electrical potential, chemical reactions, and deformation (corrosion and electrodeposition); (4) electrical current density or atomic concentration, temperature, and deformation (electromigration); and (5) non-Fourier heat conduction and non-Fickian moisture concentration.

Kilic and Madenci (2010) included the effect of thermal field on deformation, and performed a thermo-mechanical analysis. Agwai (2011) developed the PD solution for one-dimensional fully coupled thermo-mechanics. Oterkus et al. (2014a) extended it to three-dimensional analysis to capture the thermal field induced due to dynamic loading. Also, Oterkus and Madenci (2014) extended this approach to consider heat conduction and fully coupled thermo-mechanical response of fiber-reinforced composites. Oterkus et al. (2014c) coupled moisture, temperature, vapor pressure, and deformation field equations to predict damage initiation and its propagation in electronic packages. Furthermore, Oterkus and Madenci (2017) presented the coupled PD analysis of temperature and oxygen diffusions and deformation to simulate thermal cracking behavior in uranium dioxide, UO_2 fuel pellets.

Wang and Zhou (2019) developed a coupled PD thermo-mechanical model and simulated crack propagation in rocks subjected to thermal loading. Chen et al. (2021) refined the equations of motion and thermal conduction and simulated crack propagation process in concrete. Yang et al. (2021) simulated the thermo-mechanical properties and fracture characteristics of granite under different heat treatments

E. Madenci et al., *Peridynamics and Its Applications Using Ansys*,
https://doi.org/10.1007/978-3-031-59896-8_7

through a fully coupled PD thermo-mechanical model. Wu et al. (2022) simulated crack propagation in concrete due to freeze-thaw effect in a PD framework by imposing frost heave pressure on the fracture surface. Zhang et al. (2023a) presented a PD thermo-mechanical model for simulating the thermal behavior and crack propagation in saturated rocks due to freezing. The nonlinear thermal conduction equation includes the change in thermal property and the latent water-ice phase transition at low-temperature conditions. The mechanical response includes the effects of pore ice pressure and thermal expansion. Also, Zhang et al. (2023b) presented a 3D PD model to simulate the pyrolysis process and crack propagation in charring materials due to aerodynamic heating. The model is based on the coupled solution of nonlinear thermal conduction with surface recession and equation of motion, while also including the effects of thermal expansion strain and shrinkage strain due to mass loss.

For saturated steady-state flow, Katiyar et al. (2014) presented a PD model of fluid transport in porous media. Later, Ni et al. (2020) combined finite element method (FEM) with PD for the simulation of hydraulic fracture propagation. Jabakhanji and Mohtar (2015) introduced a PD hygro-mechanical model of transient moisture flow in unsaturated, heterogeneous, and anisotropic soils. Oterkus et al. (2017) presented a fully coupled PD approach to simulate hydraulic fracturing processes by coupling the porous flow and deformation fields and their effects on each other. Zhou et al. (2020) presented a PD model to simulate the propagation and coalescence of two fissures under fluid injection conditions. Shou and Zhou (2021) developed a hydro-mechanical coupling model in the framework of non-ordinary state-based (NOSB) PD to simulate damage in fissured porous rock. Qin et al. (2021) developed a PD hydraulic fracturing model for layered rock and simulated the complex propagation behaviors at the interface. Zhou et al. (2022) developed a coupled thermo-hydro-chemo-mechanical PD model for fractured rock and simulated the water pressure and temperature in geothermal extraction. Zhang et al. (2023c) solved the equations of thermal conduction and flow diffusion by using the PD least squares minimization method and simulated crack propagation due to hydraulic fracturing while considering the effect of heat. Liu et al. (2023) presented a coupled PD hygro-mechanical model of soil shrinkage and desiccation cracking arising from water-soil interaction. Also, Zhang et al. (2023d) presented a coupled hydro-mechanical PD model to simulate unsaturated seepage and crack propagation due to moisture change in unsaturated expansive soils.

Zhao et al. (2021) developed a PD model of galvanic corrosion; the electro-potential is solved by employing the bond-based PD along with the modified PD corrosion dissolution (Jafarzadeh et al., 2019) for updating the concentration. Anicode et al. (2023) presented a unified PD modeling approach for both electro-deposition and galvanic corrosion by solving for the electro-potential and mass transport equations simultaneously in the Ansys framework.

Gerstle et al. (2008) applied the PD theory to simulate a coupled phenomenon of electromigration in a copper line. Oterkus et al. (2013) extended the one-dimensional analysis by Gerstle et al. (2008) and investigated electromigration in two-dimensional copper lines. Zhang et al. (2023e) presented a 3D PD model to predict void formation and growth in metal lines by considering the coupled governing equations of atomic concentration and the equation of motion, which include the effect of diffusion strain due to electromigration.

Recently, Anicode and Madenci (2023) introduced a coupled PD model for non-Fourier heat conduction and non-Fickian moisture concentration. The thermal and moisture fields are coupled in the form of Dual-Phase-Lag (DPL) models to resolve the issue of the propagation of thermal wave and mass of moisture with infinite speed. The nonlocal effects arising from thermal inertia, moisture inertia, and microstructural interaction are included through the PD form of coupled thermal and moisture field equations.

This chapter presents a peridynamic framework to solve multiphysics problems by considering different field equations coupled among themselves or coupled with the deformation field.

7.2 Coupling of Different Field and Equilibrium Equations

A general form of coupling among three different field equations and the displacement equilibrium equation can be expressed in matrix form as

$$\mathbf{A}(\mathbf{x},t)\begin{Bmatrix}\ddot{\mathbf{u}}(\mathbf{x},t)\\ \ddot{\phi}(\mathbf{x},t)\\ \ddot{\varphi}(\mathbf{x},t)\\ \ddot{\psi}(\mathbf{x},t)\end{Bmatrix}+\mathbf{B}(\mathbf{x},t)\begin{Bmatrix}\dot{\mathbf{u}}(\mathbf{x},t)\\ \dot{\phi}(\mathbf{x},t)\\ \dot{\varphi}(\mathbf{x},t)\\ \dot{\psi}(\mathbf{x},t)\end{Bmatrix}+\mathbf{C}(\mathbf{x},t)\begin{Bmatrix}\mathbf{u}(\mathbf{x},t)\\ \phi(\mathbf{x},t)\\ \varphi(\mathbf{x},t)\\ \psi(\mathbf{x},t)\end{Bmatrix}=$$
$$\mathbf{D}(\mathbf{x},t)\begin{Bmatrix}\nabla^2\mathbf{u}(\mathbf{x},t)+\nabla\nabla\cdot\mathbf{u}(\mathbf{x},t)\\ \nabla^2\phi(\mathbf{x},t)\\ \nabla^2\varphi(\mathbf{x},t)\\ \nabla^2\psi(\mathbf{x},t)\end{Bmatrix}+\mathbf{P}(\mathbf{x},t)\begin{Bmatrix}\mathbf{0}\\ \nabla^2\dot{\phi}(\mathbf{x},t)\\ \nabla^2\dot{\varphi}(\mathbf{x},t)\\ \nabla^2\dot{\psi}(\mathbf{x},t)\end{Bmatrix}+\mathbf{Q}(\mathbf{x},t)\begin{Bmatrix}\mathbf{0}\\ \nabla^2\ddot{\phi}(\mathbf{x},t)\\ \nabla^2\ddot{\varphi}(\mathbf{x},t)\\ \nabla^2\ddot{\psi}(\mathbf{x},t)\end{Bmatrix}$$
$$+\mathbf{H}(\mathbf{x},t)\begin{Bmatrix}\mathbf{0}\\ \nabla\phi(\mathbf{x},t)\\ \nabla\varphi(\mathbf{x},t)\\ \nabla\psi(\mathbf{x},t)\end{Bmatrix}+\begin{Bmatrix}\mathbf{b}(\mathbf{x},t)\\ s_\phi(\mathbf{x},t)+\dot{s}_\phi(\mathbf{x},t)\\ s_\varphi(\mathbf{x},t)+\dot{s}_\varphi(\mathbf{x},t)\\ s_\psi(\mathbf{x},t)+\dot{s}_\psi(\mathbf{x},t)\end{Bmatrix} \tag{7.1}$$

It is worth noting that these equations involve second-order derivatives both in space and time, t at point $\mathbf{x}$ in region $\mathscr{R}$ as shown in Fig. 7.1. The variables $\phi(\mathbf{x}, t)$, $\varphi(\mathbf{x}, t)$ and $\psi(\mathbf{x}, t)$ represent three different fields, and $\mathbf{u}(\mathbf{x}, t)$ represents the displacement vector, $\mathbf{u}^T = \{u_1, u_2, u_3\}$. The dot over the field variable denotes differentiation with respect to time. The variable $s_\alpha(\mathbf{x}, t)$ for $\alpha = \phi, \varphi, \psi$ represents the source term

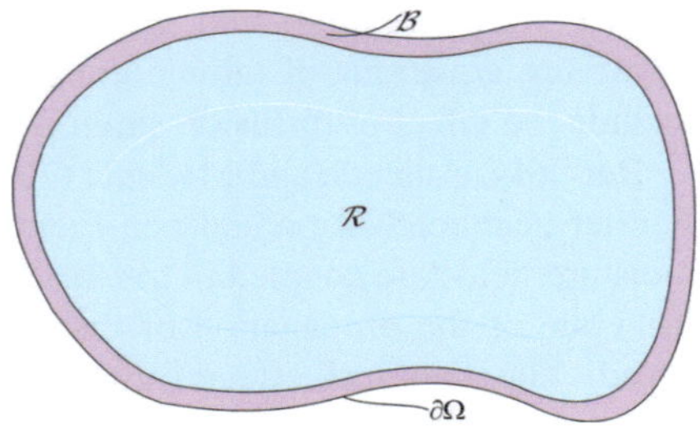

Fig. 7.1 The PDDO region, $\mathcal{R}$, of the domain with a "real" boundary layer region, $\mathcal{B}$

associated with the field equations, and $\mathbf{b}(\mathbf{x}, t)$ represents the body load vector. The known coupling coefficient matrices, $\mathbf{A}$, $\mathbf{B}$, $\mathbf{C}$, $\mathbf{D}$, $\mathbf{P}$, $\mathbf{Q}$, $\mathbf{H}$ are defined as

$$\mathbf{A}(\mathbf{x},t)=\begin{bmatrix}\mathbf{A}_{\mathbf{uu}} & \mathbf{A}_{\mathbf{u}\phi} & \mathbf{A}_{\mathbf{u}\varphi} & \mathbf{A}_{\mathbf{u}\psi}\\ \mathbf{A}_{\phi\mathbf{u}} & A_{\phi\phi} & A_{\phi\varphi} & A_{\phi\psi}\\ \mathbf{A}_{\varphi\mathbf{u}} & A_{\varphi\phi} & A_{\varphi\varphi} & A_{\varphi\psi}\\ \mathbf{A}_{\psi\mathbf{u}} & A_{\psi\phi} & A_{\psi\varphi} & A_{\psi\psi}\end{bmatrix}\mathbf{B}(\mathbf{x},t)=\begin{bmatrix}\mathbf{B}_{\mathbf{uu}} & \mathbf{B}_{\mathbf{u}\phi} & \mathbf{B}_{\mathbf{u}\varphi} & \mathbf{B}_{\mathbf{u}\psi}\\ \mathbf{B}_{\phi\mathbf{u}} & B_{\phi\phi} & B_{\phi\varphi} & B_{\phi\psi}\\ \mathbf{B}_{\varphi\mathbf{u}} & B_{\varphi\phi} & B_{\varphi\varphi} & B_{\varphi\psi}\\ \mathbf{B}_{\psi\mathbf{u}} & B_{\psi\phi} & B_{\psi\varphi} & B_{\psi\psi}\end{bmatrix}, \quad (7.2)$$

$$\mathbf{C}(\mathbf{x},t)=\begin{bmatrix}\mathbf{C}_{\mathbf{uu}} & \mathbf{C}_{\mathbf{u}\phi} & \mathbf{C}_{\mathbf{u}\varphi} & \mathbf{C}_{\mathbf{u}\psi}\\ \mathbf{C}_{\phi\mathbf{u}} & C_{\phi\phi} & C_{\phi\varphi} & C_{\phi\psi}\\ \mathbf{C}_{\varphi\mathbf{u}} & C_{\varphi\phi} & C_{\varphi\varphi} & C_{\varphi\psi}\\ \mathbf{C}_{\psi\mathbf{u}} & C_{\psi\phi} & C_{\psi\varphi} & C_{\psi\psi}\end{bmatrix}\mathbf{D}(\mathbf{x},t)=\begin{bmatrix}\mathbf{D}_{\mathbf{uu}} & \mathbf{D}_{\mathbf{u}\phi} & \mathbf{D}_{\mathbf{u}\varphi} & \mathbf{D}_{\mathbf{u}\psi}\\ \mathbf{D}_{\phi\mathbf{u}} & D_{\phi\phi} & D_{\phi\varphi} & D_{\phi\psi}\\ \mathbf{D}_{\varphi\mathbf{u}} & D_{\varphi\phi} & D_{\varphi\varphi} & D_{\varphi\psi}\\ \mathbf{D}_{\psi\mathbf{u}} & D_{\psi\phi} & D_{\psi\varphi} & D_{\psi\psi}\end{bmatrix}, \quad (7.3)$$

$$\mathbf{P}(\mathbf{x},t)=\begin{bmatrix}\mathbf{0} & \mathbf{0} & \mathbf{0} & \mathbf{0}\\ \mathbf{0} & P_{\phi\phi} & P_{\phi\varphi} & P_{\phi\psi}\\ \mathbf{0} & P_{\varphi\phi} & P_{\varphi\varphi} & P_{\varphi\psi}\\ \mathbf{0} & P_{\psi\phi} & P_{\psi\varphi} & P_{\psi\psi}\end{bmatrix}, \mathbf{Q}(\mathbf{x},t)=\begin{bmatrix}\mathbf{0} & \mathbf{0} & \mathbf{0} & \mathbf{0}\\ \mathbf{0} & Q_{\phi\phi} & Q_{\phi\varphi} & Q_{\phi\psi}\\ \mathbf{0} & Q_{\varphi\phi} & Q_{\varphi\varphi} & Q_{\varphi\psi}\\ \mathbf{0} & Q_{\psi\phi} & Q_{\psi\varphi} & Q_{\psi\psi}\end{bmatrix}, \quad (7.4)$$

and

$$\mathbf{H}(\mathbf{x},t)=\begin{bmatrix}\mathbf{0} & \mathbf{0} & \mathbf{0} & \mathbf{0}\\ \mathbf{0} & \mathbf{H}_{\phi\phi} & \mathbf{H}_{\phi\varphi} & \mathbf{H}_{\phi\psi}\\ \mathbf{0} & \mathbf{H}_{\varphi\phi} & \mathbf{H}_{\varphi\varphi} & \mathbf{H}_{\varphi\psi}\\ \mathbf{0} & \mathbf{H}_{\psi\phi} & \mathbf{H}_{\psi\varphi} & \mathbf{H}_{\psi\psi}\end{bmatrix} \quad (7.5)$$

The coefficients in these matrices facilitate the coupling among the field variables and the displacement vector. It is worth noting that the bold coefficients in these matrices represent vector quantities.

7.3 Boundary Conditions for Coupled Field Variables and Displacement

The general form of natural boundary condition at a point $\mathbf{x}$ on the boundary region $\mathscr{B}$ shown in Fig. 7.1 can be expressed in matrix form as

$$\mathbf{R}(\mathbf{x},t)\begin{Bmatrix} \nabla \mathbf{u}(\mathbf{x},t)\ \mathbf{n}(\mathbf{x}) \\ \nabla \phi(\mathbf{x},t)\cdot \mathbf{n}(\mathbf{x}) \\ \nabla \varphi(\mathbf{x},t)\cdot \mathbf{n}(\mathbf{x}) \\ \nabla \psi(\mathbf{x},t)\cdot \mathbf{n}(\mathbf{x}) \end{Bmatrix} + \mathbf{S}(\mathbf{x},t)\begin{Bmatrix} \mathbf{u}(\mathbf{x},t) \\ \phi(\mathbf{x},t) \\ \varphi(\mathbf{x},t) \\ \psi(\mathbf{x},t) \end{Bmatrix} = \begin{Bmatrix} \mathbf{T}^*_\sigma(\mathbf{x},t) \\ J^*_\phi(\mathbf{x},t) \\ J^*_\varphi(\mathbf{x},t) \\ J^*_\psi(\mathbf{x},t) \end{Bmatrix} \tag{7.6}$$

The known coupling coefficient matrices, $\mathbf{R}$ and $\mathbf{S}$ are defined as

$$\mathbf{R}(\mathbf{x},t) = \begin{bmatrix} \mathbf{R}_{\mathbf{uu}} & \mathbf{R}_{\mathbf{u}\phi} & \mathbf{R}_{\mathbf{u}\varphi} & \mathbf{R}_{\mathbf{u}\psi} \\ \mathbf{R}_{\phi\mathbf{u}} & R_{\phi\phi} & R_{\phi\varphi} & R_{\phi\psi} \\ \mathbf{R}_{\varphi\mathbf{u}} & R_{\varphi\phi} & R_{\varphi\varphi} & R_{\varphi\psi} \\ \mathbf{R}_{\psi\mathbf{u}} & R_{\psi\phi} & R_{\psi\varphi} & R_{\psi\psi} \end{bmatrix} \text{ and } \mathbf{S}(\mathbf{x},t) = \begin{bmatrix} \mathbf{S}_{\mathbf{uu}} & \mathbf{S}_{\mathbf{u}\phi} & \mathbf{S}_{\mathbf{u}\varphi} & \mathbf{S}_{\mathbf{u}\psi} \\ \mathbf{S}_{\phi\mathbf{u}} & S_{\phi\phi} & S_{\phi\varphi} & S_{\phi\psi} \\ \mathbf{S}_{\varphi\mathbf{u}} & S_{\varphi\phi} & S_{\varphi\varphi} & S_{\varphi\psi} \\ \mathbf{S}_{\psi\mathbf{u}} & S_{\psi\phi} & S_{\psi\varphi} & S_{\psi\psi} \end{bmatrix} \tag{7.7}$$

The parameters $J^*_\phi(\mathbf{x},t)$, $J^*_\varphi(\mathbf{x},t)$, and $J^*_\psi(\mathbf{x},t)$ are the imposed values of field flux on the boundary, $\mathscr{B}_{\mathbf{q}_\phi}$, $\mathscr{B}_{\mathbf{q}_\varphi}$, and $\mathscr{B}_{\mathbf{q}_\psi}$, respectively. The traction vector, $\mathbf{T}^*_{\boldsymbol{\sigma}}(\mathbf{x},t)$ represents the imposed value of traction on the boundary, $\mathscr{B}_{\boldsymbol{\sigma}}$. It is worth noting that the bold coefficients in these matrices represent vector quantities.

The essential boundary conditions on the field variables and displacement can be directly imposed as

$$\phi(\mathbf{x},t) = \phi^*(\mathbf{x},t) \qquad \forall \mathbf{x} \in \mathscr{B}_\phi \tag{7.8a}$$

$$\varphi(\mathbf{x},t) = \varphi^*(\mathbf{x},t) \qquad \forall \mathbf{x} \in \mathscr{B}_\varphi \tag{7.8b}$$

$$\psi(\mathbf{x},t) = \psi^*(\mathbf{x},t) \qquad \forall \mathbf{x} \in \mathscr{B}_\psi \tag{7.8c}$$

and

$$\mathbf{u}(\mathbf{x},t) = \mathbf{u}^*(\mathbf{x},t) \qquad \forall \mathbf{x} \in \mathscr{B}_{\mathbf{u}} \tag{7.8d}$$

The parameters $\phi^*(\mathbf{x},t)$, $\varphi^*(\mathbf{x},t)$, and $\psi^*(\mathbf{x},t)$ are the imposed values of field variables on the boundary, $\mathscr{B}_\phi$, $\mathscr{B}_\varphi$, and $\mathscr{B}_\psi$, respectively. The vector, $\mathbf{u}^*(\mathbf{x},t)$ represents the imposed value of displacement on the boundary, $\mathscr{B}_{\mathbf{u}}$.

7.4 PDDO Form of Coupled Equations

Employing the PD differential operator (PDDO), Eqs. (7.1) can be recast in integral form at point $\mathbf{x}$ with a symmetric position in its family in region $\mathscr{R}$ as (Madenci et al., 2016, 2019)

$$\mathbf{A}(\mathbf{x},t)\begin{Bmatrix}\ddot{\mathbf{u}}(\mathbf{x},t)\\ \ddot{\phi}(\mathbf{x},t)\\ \ddot{\varphi}(\mathbf{x},t)\\ \ddot{\psi}(\mathbf{x},t)\end{Bmatrix}+\mathbf{B}(\mathbf{x},t)\begin{Bmatrix}\dot{\mathbf{u}}(\mathbf{x},t)\\ \dot{\phi}(\mathbf{x},t)\\ \dot{\varphi}(\mathbf{x},t)\\ \dot{\psi}(\mathbf{x},t)\end{Bmatrix}+\mathbf{C}(\mathbf{x},t)\begin{Bmatrix}\mathbf{u}(\mathbf{x},t)\\ \phi(\mathbf{x},t)\\ \varphi(\mathbf{x},t)\\ \psi(\mathbf{x},t)\end{Bmatrix}=$$
$$\int_{H_\mathbf{x}}\begin{bmatrix}\mathbf{\Phi}_{\mathbf{uu}} & \mathbf{\Phi}_{\mathbf{u}\phi} & \mathbf{\Phi}_{\mathbf{u}\varphi} & \mathbf{\Phi}_{\mathbf{u}\psi}\\ \mathbf{\Phi}_{\phi\mathbf{u}} & \Phi_{\phi\phi} & \Phi_{\phi\varphi} & \Phi_{\phi\psi}\\ \mathbf{\Phi}_{\varphi\mathbf{u}} & \Phi_{\varphi\phi} & \Phi_{\varphi\varphi} & \Phi_{\varphi\psi}\\ \mathbf{\Phi}_{\psi\mathbf{u}} & \Phi_{\psi\varphi} & \Phi_{\psi\varphi} & \Phi_{\psi\psi}\end{bmatrix}\begin{Bmatrix}(\mathbf{u}(\mathbf{x}',t)-\mathbf{u}(\mathbf{x},t))\\ (\phi(\mathbf{x}',t)-\phi(\mathbf{x},t))\\ (\varphi(\mathbf{x}',t)-\varphi(\mathbf{x},t))\\ (\psi(\mathbf{x}',t)-\psi(\mathbf{x},t))\end{Bmatrix}dV_{\mathbf{x}'}$$
$$+\int_{H_\mathbf{x}}\begin{bmatrix}\mathbf{0}&\mathbf{0}&\mathbf{0}&\mathbf{0}\\ \mathbf{0}&\Theta_{\phi\phi}&\Theta_{\phi\varphi}&\Theta_{\phi\psi}\\ \mathbf{0}&\Theta_{\varphi\phi}&\Theta_{\varphi\varphi}&\Theta_{\varphi\psi}\\ \mathbf{0}&\Theta_{\psi\phi}&\Theta_{\psi\varphi}&\Theta_{\psi\psi}\end{bmatrix}\begin{Bmatrix}\mathbf{0}\\ (\dot{\phi}(\mathbf{x}',t)-\dot{\phi}(\mathbf{x},t))\\ (\dot{\varphi}(\mathbf{x}',t)-\dot{\varphi}(\mathbf{x},t))\\ (\dot{\psi}(\mathbf{x}',t)-\dot{\psi}(\mathbf{x},t))\end{Bmatrix}dV_{\mathbf{x}'} \tag{7.9}$$
$$+\int_{H_\mathbf{x}}\begin{bmatrix}\mathbf{0}&\mathbf{0}&\mathbf{0}&\mathbf{0}\\ \mathbf{0}&\Psi_{\phi\phi}&\Psi_{\phi\varphi}&\Psi_{\phi\psi}\\ \mathbf{0}&\Psi_{\varphi\phi}&\Psi_{\varphi\varphi}&\Psi_{\varphi\psi}\\ \mathbf{0}&\Psi_{\psi\phi}&\Psi_{\psi\varphi}&\Psi_{\psi\psi}\end{bmatrix}\begin{Bmatrix}\mathbf{0}\\ (\ddot{\phi}(\mathbf{x}',t)-\ddot{\phi}(\mathbf{x},t))\\ (\ddot{\varphi}(\mathbf{x}',t)-\ddot{\varphi}(\mathbf{x},t))\\ (\ddot{\psi}(\mathbf{x}',t)-\ddot{\psi}(\mathbf{x},t))\end{Bmatrix}dV_{\mathbf{x}'}$$
$$+\begin{Bmatrix}s_{\mathbf{u}}(\mathbf{x},t)\\ s_\phi(\mathbf{x},t)+\dot{s}_\phi(\mathbf{x},t)\\ s_\varphi(\mathbf{x},t)+\dot{s}_\varphi(\mathbf{x},t)\\ s_\psi(\mathbf{x},t)+\dot{s}_\psi(\mathbf{x},t)\end{Bmatrix}$$

where the scalar coefficients $\Phi_{\alpha\beta}=\Phi_{\alpha\beta}(\mathbf{x},\boldsymbol{\xi},t)$, $\Theta_{\alpha\beta}=\Theta_{\alpha\beta}(\mathbf{x},\boldsymbol{\xi},t)$, and $\Psi_{\alpha\beta}=\Psi_{\alpha\beta}(\mathbf{x},\boldsymbol{\xi},t)$ are defined as

$$\begin{Bmatrix}\Phi_{\alpha\beta}(\mathbf{x},\xi)\\ \Theta_{\alpha\beta}(\mathbf{x},\xi)\\ \Psi_{\alpha\beta}(\mathbf{x},\xi)\end{Bmatrix}=\omega(|\xi|)\left\{\mathrm{tr}\mathbf{G}(\mathbf{x},\xi)\begin{Bmatrix}D_{\alpha\beta}(\mathbf{x},t)\\ P_{\alpha\beta}(\mathbf{x},t)\\ Q_{\alpha\beta}(\mathbf{x},t)\end{Bmatrix}+\begin{Bmatrix}\mathbf{g}(\mathbf{x},\xi)\cdot\mathbf{H}_{\alpha\beta}(\mathbf{x},t)\\ 0\\ 0\end{Bmatrix}\right\}$$
$$\forall\alpha,\beta=\phi,\varphi,\psi \tag{7.10}$$

The vector coefficient $\mathbf{\Phi}_{\alpha\mathbf{u}}=\mathbf{\Phi}_{\alpha\mathbf{u}}(\mathbf{x},\boldsymbol{\xi},t)$ is defined as

$$\mathbf{\Phi}_{\alpha\mathbf{u}}(\mathbf{x},\boldsymbol{\xi},t)=\omega(|\boldsymbol{\xi}|)(\mathrm{tr}\mathbf{G}(\mathbf{x},\boldsymbol{\xi})\mathbf{I}+2\mathbf{G}(\mathbf{x},\boldsymbol{\xi}))\mathbf{D}_{\alpha\mathbf{u}}(\mathbf{x},t)\qquad\forall\alpha=\phi,\varphi,\psi\tag{7.11}$$

It is worth noting that the subscripts in Eqs. (7.10) and (7.11) can be interchanged. Also, the matrix $\mathbf{G}(\mathbf{x},\boldsymbol{\xi})$ and vector $\mathbf{g}(\mathbf{x},\boldsymbol{\xi})$ are derived in terms of PD functions explained in Chap. 2.

7.5 PDDO Form of Boundary Conditions

Similarly, using the PDDO, Eq. (7.6) can be recast for a point $\mathbf{x}$ on the boundary region $\mathscr{B}$ as

$$\begin{aligned}&\int\limits_{H_{\mathbf{x}}}\begin{bmatrix}\mathbf{\Omega}_{\mathbf{uu}} & \mathbf{\Omega}_{\mathbf{u}\phi} & \mathbf{\Omega}_{\mathbf{u}\varphi} & \mathbf{\Omega}_{\mathbf{u}\psi}\\ \mathbf{\Omega}_{\phi\mathbf{u}} & \Omega_{\phi\phi} & \Omega_{\phi\varphi} & \Omega_{\phi\psi}\\ \mathbf{\Omega}_{\varphi\mathbf{u}} & \Omega_{\varphi\phi} & \Omega_{\varphi\varphi} & \Omega_{\varphi\psi}\\ \mathbf{\Omega}_{\psi\mathbf{u}} & \Omega_{\psi\phi} & \Omega_{\psi\varphi} & \Omega_{\psi\psi}\end{bmatrix}\begin{Bmatrix}(\mathbf{u}(\mathbf{x}',t)-\mathbf{u}(\mathbf{x},t))\\(\phi(\mathbf{x}',t)-\phi(\mathbf{x},t))\\(\varphi(\mathbf{x}',t)-\varphi(\mathbf{x},t))\\(\psi(\mathbf{x}',t)-\psi(\mathbf{x},t))\end{Bmatrix}dV_{\mathbf{x}'}+\\ &\mathbf{S}(\mathbf{x},t)\begin{Bmatrix}\mathbf{u}(\mathbf{x},t)\\\phi(\mathbf{x},t)\\\varphi(\mathbf{x},t)\\\psi(\mathbf{x},t)\end{Bmatrix}=\begin{Bmatrix}\mathbf{T}^*_{\boldsymbol{\sigma}}(\mathbf{x},t)\\J^*_{\phi}(\mathbf{x},t)\\J^*_{\varphi}(\mathbf{x},t)\\J^*_{\psi}(\mathbf{x},t)\end{Bmatrix}\end{aligned}\tag{7.12}$$

where the scalar coefficient $\Omega_{\alpha\beta}=\Omega_{\alpha\beta}(\mathbf{x},\boldsymbol{\xi},t)$ is defined as

$$\Omega_{\alpha\beta}(\mathbf{x},\boldsymbol{\xi})=\omega(|\boldsymbol{\xi}|)\left(R_{\alpha\beta}(\mathbf{x},t)\mathbf{g}(\mathbf{x},\boldsymbol{\xi})\cdot\mathbf{n}(\mathbf{x})\right)\ \forall\alpha,\beta=\phi,\varphi\text{ and }\psi\tag{7.13}$$

and the vector coefficient is defined as

$$\mathbf{\Omega}_{\alpha\mathbf{u}}(\mathbf{x},\boldsymbol{\xi})=\omega(|\boldsymbol{\xi}|)(\mathbf{n}(\mathbf{x})\cdot\mathbf{g}(\mathbf{x},\boldsymbol{\xi}))\mathbf{R}_{\alpha\mathbf{u}}(\mathbf{x},t)\quad\forall\alpha=\phi,\varphi,\psi\text{ and }\mathbf{u}\tag{7.14}$$

where the subscripts in Eqs. (7.13) and (7.14) can be interchanged.

7.6 Numerical Discretization

The domain $\Omega=\mathscr{R}\cup\mathscr{B}$ is discretized with $N=N_{\mathscr{R}}+N_{\mathscr{B}}$ PD points as shown in Fig. 7.2. The parameters $N_{\mathscr{R}}$ and $N_{\mathscr{B}}$ represent the number of points associated with regions, $\mathscr{R}$ and $\mathscr{B}$, respectively. The discretization in region $\mathscr{R}$ can be nonuniform. The position of PD point k is denoted by $\mathbf{x}_{(k)}$ and its corresponding volume is denoted by $\Delta V_{(k)}$.

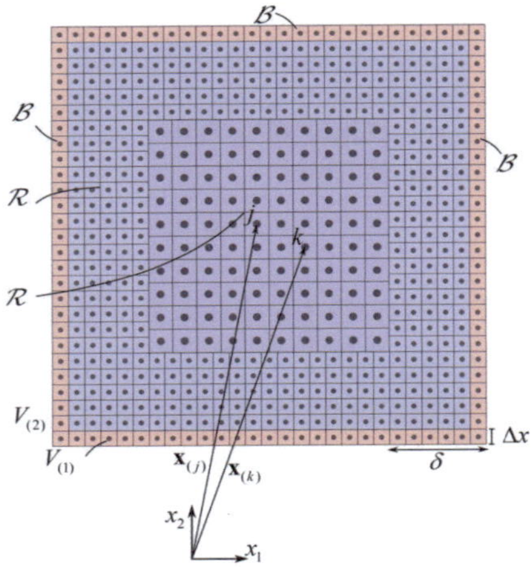

Fig. 7.2 Discretization of the domain for the PD form of coupled field and equilibrium equations with a non-uniform grid

7.6.1 *Discrete Form of Coupled Equations*

After multiplying with the volume, $\Delta V_{(k)}$ of point k, the discrete form of Eq. (7.9) can be expressed as

$$\mathbf{A}(\mathbf{x}_{(k)},t)\Delta V_{(k)}\begin{Bmatrix}\ddot{\mathbf{u}}(\mathbf{x}_{(k)},t)\\ \ddot{\phi}(\mathbf{x}_{(k)},t)\\ \ddot{\varphi}(\mathbf{x}_{(k)},t)\\ \ddot{\psi}(\mathbf{x}_{(k)},t)\end{Bmatrix}+\mathbf{B}(\mathbf{x}_{(k)},t)\Delta V_{(k)}\begin{Bmatrix}\dot{\mathbf{u}}(\mathbf{x}_{(k)},t)\\ \dot{\phi}(\mathbf{x}_{(k)},t)\\ \dot{\varphi}(\mathbf{x}_{(k)},t)\\ \dot{\psi}(\mathbf{x}_{(k)},t)\end{Bmatrix}+\mathbf{C}(\mathbf{x}_{(k)},t)\Delta V_{(k)}\begin{Bmatrix}\mathbf{u}(\mathbf{x}_{(k)},t)\\ \phi(\mathbf{x}_{(k)},t)\\ \varphi(\mathbf{x}_{(k)},t)\\ \psi(\mathbf{x}_{(k)},t)\end{Bmatrix}$$
$$=\sum_{l=1}^{N_{(k)}}\begin{bmatrix}\boldsymbol{\Phi}_{\mathbf{uu}}^{(k)(k_l)} & \boldsymbol{\Phi}_{\mathbf{u}\phi}^{(k)(k_l)} & \boldsymbol{\Phi}_{\mathbf{u}\varphi}^{(k)(k_l)} & \boldsymbol{\Phi}_{\mathbf{u}\psi}^{(k)(k_l)}\\ \boldsymbol{\Phi}_{\phi\mathbf{u}}^{(k)(k_l)} & \Phi_{\phi\phi}^{(k)(k_l)} & \Phi_{\phi\varphi}^{(k)(k_l)} & \Phi_{\phi\psi}^{(k)(k_l)}\\ \boldsymbol{\Phi}_{\varphi\mathbf{u}}^{(k)(k_l)} & \Phi_{\varphi\phi}^{(k)(k_l)} & \Phi_{\varphi\varphi}^{(k)(k_l)} & \Phi_{\varphi\psi}^{(k)(k_l)}\\ \boldsymbol{\Phi}_{\psi\mathbf{u}}^{(k)(k_l)} & \Phi_{\psi\phi}^{(k)(k_l)} & \Phi_{\psi\varphi}^{(k)(k_l)} & \Phi_{\psi\psi}^{(k)(k_l)}\end{bmatrix}\begin{Bmatrix}(\mathbf{u}(\mathbf{x}_{(k_l)},t)-\mathbf{u}(\mathbf{x}_{(k)},t))\\ (\phi(\mathbf{x}_{(k_l)},t)-\phi(\mathbf{x}_{(k)},t))\\ (\varphi(\mathbf{x}_{(k_l)},t)-\varphi(\mathbf{x}_{(k)},t))\\ (\psi(\mathbf{x}_{(k_l)},t)-\psi(\mathbf{x}_{(k)},t))\end{Bmatrix}+$$
$$\sum_{l=1}^{N_{(k)}}\begin{bmatrix}\mathbf{0} & \mathbf{0} & \mathbf{0} & \mathbf{0}\\ \mathbf{0} & \Theta_{\phi\phi}^{(k)(k_l)} & \Theta_{\phi\varphi}^{(k)(k_l)} & \Theta_{\phi\psi}^{(k)(k_l)}\\ \mathbf{0} & \Theta_{\varphi\phi}^{(k)(k_l)} & \Theta_{\varphi\varphi}^{(k)(k_l)} & \Theta_{\varphi\psi}^{(k)(k_l)}\\ \mathbf{0} & \Theta_{\psi\phi}^{(k)(k_l)} & \Theta_{\psi\varphi}^{(k)(k_l)} & \Theta_{\psi\psi}^{(k)(k_l)}\end{bmatrix}\begin{Bmatrix}\mathbf{0}\\ (\dot{\phi}(\mathbf{x}_{(k_l)},t)-\dot{\phi}(\mathbf{x}_{(k)},t))\\ (\dot{\varphi}(\mathbf{x}_{(k_l)},t)-\dot{\varphi}(\mathbf{x}_{(k)},t))\\ (\dot{\psi}(\mathbf{x}_{(k_l)},t)-\dot{\psi}(\mathbf{x}_{(k)},t))\end{Bmatrix}+$$
$$\sum_{l=1}^{N_{(k)}}\begin{bmatrix}\mathbf{0} & \mathbf{0} & \mathbf{0} & \mathbf{0}\\ \mathbf{0} & \Psi_{\phi\phi}^{(k)(k_l)} & \Psi_{\phi\varphi}^{(k)(k_l)} & \Psi_{\phi\psi}^{(k)(k_l)}\\ \mathbf{0} & \Psi_{\varphi\phi}^{(k)(k_l)} & \Psi_{\varphi\varphi}^{(k)(k_l)} & \Psi_{\varphi\psi}^{(k)(k_l)}\\ \mathbf{0} & \Psi_{\psi\phi}^{(k)(k_l)} & \Psi_{\psi\varphi}^{(k)(k_l)} & \Psi_{\psi\psi}^{(k)(k_l)}\end{bmatrix}\begin{Bmatrix}\mathbf{0}\\ (\ddot{\phi}(\mathbf{x}_{(k_l)},t)-\ddot{\phi}(\mathbf{x}_{(k)},t))\\ (\ddot{\varphi}(\mathbf{x}_{(k_l)},t)-\ddot{\varphi}(\mathbf{x}_{(k)},t))\\ (\ddot{\psi}(\mathbf{x}_{(k_l)},t)-\ddot{\psi}(\mathbf{x}_{(k)},t))\end{Bmatrix}+\begin{Bmatrix}s_{\mathbf{u}}(\mathbf{x}_{(k)},t)\\ s_{\phi}(\mathbf{x}_{(k)},t)+\dot{s}_{\phi}(\mathbf{x}_{(k)},t)\\ s_{\varphi}(\mathbf{x}_{(k)},t)+\dot{s}_{\varphi}(\mathbf{x}_{(k)},t)\\ s_{\psi}(\mathbf{x}_{(k)},t)+\dot{s}_{\psi}(\mathbf{x}_{(k)},t)\end{Bmatrix}\Delta V_{(k)}\tag{7.15}$$

where the scalar coefficients, $\boldsymbol{\Phi}_{\alpha\beta}^{(k)(k_l)} = \boldsymbol{\Phi}_{\alpha\beta}\left(\mathbf{x}_{(k)}, \boldsymbol{\xi}_{(k)(k_l)}, t\right)$, $\boldsymbol{\Theta}_{\alpha\beta}^{(k)(k_l)} = \boldsymbol{\Theta}_{\alpha\beta}\left(\mathbf{x}_{(k)}, \boldsymbol{\xi}_{(k)(k_l)}, t\right)$, and $\boldsymbol{\Psi}_{\alpha\beta}^{(k)(k_l)} = \boldsymbol{\Psi}_{\alpha\beta}\left(\mathbf{x}_{(k)}, \boldsymbol{\xi}_{(k)(k_l)}, t\right)$ with $\alpha,\ \beta = \phi,\ \varphi,\ \psi$ are defined as

$$
\begin{Bmatrix} \boldsymbol{\Phi}_{\alpha\beta}^{(k)(k_l)}\left(\mathbf{x}_{(k)}, \xi_{(k)(k_l)}, t\right) \\ \boldsymbol{\Theta}_{\alpha\beta}^{(k)(k_l)}\left(\mathbf{x}_{(k)}, \xi_{(k)(k_l)}, t\right) \\ \boldsymbol{\Psi}_{\alpha\beta}^{(k)(k_l)}\left(\mathbf{x}_{(k)}, \xi_{(k)(k_l)}, t\right) \end{Bmatrix}
= \omega\left(\left|\xi_{(k)(k_l)}\right|\right) \left\{ \begin{array}{c} \mathrm{tr}\mathbf{G}\left(\mathbf{x}_{(k)}, \xi_{(k)(k_l)}\right) \begin{Bmatrix} D_{\alpha\beta}\left(\mathbf{x}_{(k)}, t\right) \\ P_{\alpha\beta}\left(\mathbf{x}_{(k)}, t\right) \\ Q_{\alpha\beta}\left(\mathbf{x}_{(k)}, t\right) \end{Bmatrix} \\ + \begin{Bmatrix} \mathbf{g}\left(\mathbf{x}_{(k)}, \xi_{(k)(k_l)}\right) \cdot \mathbf{H}_{\alpha\beta}\left(\mathbf{x}_{(k)}, t\right) \\ 0 \\ 0 \end{Bmatrix} \end{array} \right\} \Delta V_{(k_l)} \Delta V_{(k)} \tag{7.16}
$$

The vector coefficient, $\boldsymbol{\Phi}_{\alpha\mathbf{u}}^{(k)(k_l)} = \boldsymbol{\Phi}_{\alpha\mathbf{u}}\left(\mathbf{x}_{(k)}, \boldsymbol{\xi}_{(k)(k_l)}, t\right)$ for $\alpha = \phi,\ \varphi,\ \psi,\ \mathbf{u}$ is defined as

$$
\begin{aligned}
\boldsymbol{\Phi}_{\alpha\mathbf{u}}\left(\mathbf{x}_{(k)}, \boldsymbol{\xi}_{(k)(k_l)}, t\right) &= \omega\left(\left|\boldsymbol{\xi}_{(k)(k_l)}\right|\right) \\
&\times \left(\mathrm{tr}\mathbf{G}\left(\mathbf{x}_{(k)}, \boldsymbol{\xi}_{(k)(k_l)}\right)\mathbf{I} + 2\mathbf{G}\left(\mathbf{x}_{(k)}, \boldsymbol{\xi}_{(k)(k_l)}\right)\right)\mathbf{D}_{\alpha\mathbf{u}}\left(\mathbf{x}_{(k)}, t\right)\Delta V_{(k_l)}\Delta V_{(k)}
\end{aligned} \tag{7.17}
$$

At node k, Eq. (7.15) can be expressed in vector form as

$$
\mathbf{F}_{\mathscr{R}(k)}^{\mathrm{PDDO(Inertial)}} + \mathbf{F}_{\mathscr{R}(k)}^{\mathrm{PDDO(Damp)}} + \mathbf{F}_{\mathscr{R}(k)}^{\mathrm{PDDO(Stiff)}} = \mathbf{F}_{\mathscr{R}(k)}^{\mathrm{Ext}} \tag{7.18}
$$

The vector $\mathbf{F}_{\mathscr{R}}^{\mathrm{PDDO(Stiff)}}$ corresponding to the stiffness terms can be written in component form as

$$\mathbf{F}_{\mathcal{R}(k)}^{\mathrm{PDDO(Stiff)}} = \begin{Bmatrix} F_{1(k)}^{\mathrm{PDDO(Stiff)}} \\ F_{2(k)}^{\mathrm{PDDO(Stiff)}} \\ 0 \\ F_{\phi(k)}^{\mathrm{PDDO(Stiff)}} \\ F_{\varphi(k)}^{\mathrm{PDDO(Stiff)}} \\ F_{\psi(k)}^{\mathrm{PDDO(Stiff)}} \\ 0 \\ 0 \\ 0 \\ 0 \\ 0 \\ 0 \end{Bmatrix} = \sum_{l=1}^{N_{(k)}} \mathbf{k}_{(k)(k_l)}^{PDDO} \begin{Bmatrix} u_{1(k)} \\ u_{2(k)} \\ 0 \\ \phi_{(k)} \\ \varphi_{(k)} \\ \psi_{(k)} \\ u_{1(k_l)} \\ u_{2(k_l)} \\ 0 \\ \phi_{(k_l)} \\ \varphi_{(k_l)} \\ \psi_{(k_l)} \end{Bmatrix} \tag{7.19}$$

with

$$\mathbf{k}_{(k)(k_l)}^{\mathrm{PDDO}} = \begin{Bmatrix} \mathrm{k}_{\mathbf{uu}(1)(1)}^{(k)(k_l)} & \mathrm{k}_{\mathbf{uu}(1)(2)}^{(k)(k_l)} & 0 & \mathrm{k}_{\mathbf{u}\phi(1)(1)}^{(k)(k_l)} & \mathrm{k}_{\mathbf{u}\varphi(1)(1)}^{(k)(k_l)} & \mathrm{k}_{\mathbf{u}\psi(1)(1)}^{(k)(k_l)} \\ \mathrm{k}_{\mathbf{uu}(2)(1)}^{(k)(k_l)} & \mathrm{k}_{\mathbf{uu}(2)(2)}^{(k)(k_l)} & 0 & \mathrm{k}_{\mathbf{u}\phi(2)(1)}^{(k)(k_l)} & \mathrm{k}_{\mathbf{u}\varphi(2)(1)}^{(k)(k_l)} & \mathrm{k}_{\mathbf{u}\psi(2)(1)}^{(k)(k_l)} \\ 0 & 0 & 0 & 0 & 0 & 0 \\ \mathrm{k}_{\phi\mathbf{u}(1)(1)}^{(k)(k_l)} & \mathrm{k}_{\phi\mathbf{u}(1)(2)}^{(k)(k_l)} & 0 & \mathrm{k}_{\phi\phi}^{(k)(k_l)} & \mathrm{k}_{\phi\varphi}^{(k)(k_l)} & \mathrm{k}_{\phi\psi}^{(k)(k_l)} \\ \mathrm{k}_{\varphi\mathbf{u}(1)(1)}^{(k)(k_l)} & \mathrm{k}_{\varphi\mathbf{u}(1)(2)}^{(k)(k_l)} & 0 & \mathrm{k}_{\varphi\phi}^{(k)(k_l)} & \mathrm{k}_{\varphi\varphi}^{(k)(k_l)} & \mathrm{k}_{\varphi\psi}^{(k)(k_l)} \\ \mathrm{k}_{\psi\mathbf{u}(1)(1)}^{(k)(k_l)} & \mathrm{k}_{\psi\mathbf{u}(1)(2)}^{(k)(k_l)} & 0 & \mathrm{k}_{\psi\phi}^{(k)(k_l)} & \mathrm{k}_{\psi\varphi}^{(k)(k_l)} & \mathrm{k}_{\psi\psi}^{(k)(k_l)} \\ 0 & 0 & 0 & 0 & 0 & 0 \\ 0 & 0 & 0 & 0 & 0 & 0 \\ 0 & 0 & 0 & 0 & 0 & 0 \\ 0 & 0 & 0 & 0 & 0 & 0 \\ 0 & 0 & 0 & 0 & 0 & 0 \\ 0 & 0 & 0 & 0 & 0 & 0 \end{Bmatrix}$$

$$
\left.
\begin{array}{cccccc}
\widehat{\mathrm{k}}^{(k)(k_l)}_{\mathbf{uu}(1)(1)} & \widehat{\mathrm{k}}^{(k)(k_l)}_{\mathbf{uu}(1)(2)} & 0 & \widehat{\mathrm{k}}^{(k)(k_l)}_{\mathbf{u}\phi(1)(1)} & \widehat{\mathrm{k}}^{(k)(k_l)}_{\mathbf{u}\varphi(1)(1)} & \widehat{\mathrm{k}}^{(k)(k_l)}_{\mathbf{u}\psi(1)(1)} \\
\widehat{\mathrm{k}}^{(k)(k_l)}_{\mathbf{uu}(2)(1)} & \widehat{\mathrm{k}}^{(k)(k_l)}_{\mathbf{uu}(2)(2)} & 0 & \widehat{\mathrm{k}}^{(k)(k_l)}_{\mathbf{u}\phi(2)(1)} & \widehat{\mathrm{k}}^{(k)(k_l)}_{\mathbf{u}\varphi(2)(1)} & \widehat{\mathrm{k}}^{(k)(k_l)}_{\mathbf{u}\psi(2)(1)} \\
0 & 0 & 0 & 0 & 0 & 0 \\
\widehat{\mathrm{k}}^{(k)(k_l)}_{\phi\mathbf{u}(1)(1)} & \widehat{\mathrm{k}}^{(k)(k_l)}_{\phi\mathbf{u}(1)(2)} & 0 & \widehat{\mathrm{k}}^{(k)(k_l)}_{\phi\phi} & \widehat{\mathrm{k}}^{(k)(k_l)}_{\phi\varphi} & \widehat{\mathrm{k}}^{(k)(k_l)}_{\phi\psi} \\
\widehat{\mathrm{k}}^{(k)(k_l)}_{\varphi\mathbf{u}(1)(1)} & \widehat{\mathrm{k}}^{(k)(k_l)}_{\varphi\mathbf{u}(1)(2)} & 0 & \widehat{\mathrm{k}}^{(k)(k_l)}_{\varphi\phi} & \widehat{\mathrm{k}}^{(k)(k_l)}_{\varphi\varphi} & \widehat{\mathrm{k}}^{(k)(k_l)}_{\varphi\psi} \\
\widehat{\mathrm{k}}^{(k)(k_l)}_{\psi\mathbf{u}(1)(1)} & \widehat{\mathrm{k}}^{(k)(k_l)}_{\psi\mathbf{u}(1)(2)} & 0 & \widehat{\mathrm{k}}^{(k)(k_l)}_{\psi\phi} & \widehat{\mathrm{k}}^{(k)(k_l)}_{\psi\varphi} & \widehat{\mathrm{k}}^{(k)(k_l)}_{\psi\psi} \\
0 & 0 & 0 & 0 & 0 & 0 \\
0 & 0 & 0 & 0 & 0 & 0 \\
0 & 0 & 0 & 0 & 0 & 0 \\
0 & 0 & 0 & 0 & 0 & 0 \\
0 & 0 & 0 & 0 & 0 & 0 \\
0 & 0 & 0 & 0 & 0 & 0
\end{array}
\right\}
\tag{7.20}
$$

where

$$
\begin{aligned}
&\mathrm{k}^{(k)(k_l)}_{\mathbf{uu}(1)(1)} = \Phi^{(k)(k_l)}_{\mathbf{uu}(1)(1)} + \frac{C^{(k)}_{\mathbf{uu}(1)(1)}\Delta V_{(k)}}{N_{(k)}}, \quad \mathrm{k}^{(k)(k_l)}_{\mathbf{uu}(1)(2)} = \Phi^{(k)(k_l)}_{\mathbf{uu}(1)(2)} + \frac{C^{(k)}_{\mathbf{uu}(1)(2)}\Delta V_{(k)}}{N_{(k)}} \\
&\mathrm{k}^{(k)(k_l)}_{\mathbf{u}\phi(1)(1)} = \Phi^{(k)(k_l)}_{\mathbf{u}\phi(1)(1)} + \frac{C^{(k)}_{\mathbf{u}\phi(1)(1)}\Delta V_{(k)}}{N_{(k)}}, \quad \mathrm{k}^{(k)(k_l)}_{\mathbf{u}\varphi(1)(1)} = \Phi^{(k)(k_l)}_{\mathbf{u}\varphi(1)(1)} + \frac{C^{(k)}_{\mathbf{u}\phi(1)(1)}\Delta V_{(k)}}{N_{(k)}} \\
&\mathrm{k}^{(k)(k_l)}_{\mathbf{u}\psi(1)(1)} = \Phi^{(k)(k_l)}_{\mathbf{u}\psi(1)(1)} + \frac{C^{(k)}_{\mathbf{u}\psi(1)(1)}\Delta V_{(k)}}{N_{(k)}}, \widehat{\mathrm{k}}^{(k)(k_l)}_{\mathbf{uu}(1)(1)} = -\Phi^{(k)(k_l)}_{\mathbf{uu}(1)(1)}, \widehat{\mathrm{k}}^{(k)(k_l)}_{\mathbf{uu}(1)(2)} = -\Phi^{(k)(k_l)}_{\mathbf{uu}(1)(2)} \\
&\widehat{\mathrm{k}}^{(k)(k_l)}_{\mathbf{u}\phi(1)(1)} = -\Phi^{(k)(k_l)}_{\mathbf{u}\phi(1)(1)}, \widehat{\mathrm{k}}^{(k)(k_l)}_{\mathbf{u}\varphi(1)(1)} = -\Phi^{(k)(k_l)}_{\mathbf{u}\varphi(1)(1)}, \widehat{\mathrm{k}}^{(k)(k_l)}_{\mathbf{u}\psi(1)(1)} = -\Phi^{(k)(k_l)}_{\mathbf{u}\psi(1)(1)}
\end{aligned}
\tag{7.21a}
$$

$$
\begin{aligned}
&\mathrm{k}^{(k)(k_l)}_{\mathbf{uu}(2)(1)} = \Phi^{(k)(k_l)}_{\mathbf{uu}(2)(1)} + \frac{C^{(k)}_{\mathbf{uu}(2)(1)}\Delta V_{(k)}}{N_{(k)}}, \quad \mathrm{k}^{(k)(k_l)}_{\mathbf{uu}(2)(2)} = \Phi^{(k)(k_l)}_{\mathbf{uu}(2)(2)} + \frac{C^{(k)}_{\mathbf{uu}(2)(2)}\Delta V_{(k)}}{N_{(k)}} \\
&\mathrm{k}^{(k)(k_l)}_{\mathbf{u}\phi(2)(1)} = \Phi^{(k)(k_l)}_{\mathbf{u}\phi(2)(1)} + \frac{C^{(k)}_{\mathbf{u}\phi(2)(1)}\Delta V_{(k)}}{N_{(k)}}, \quad \mathrm{k}^{(k)(k_l)}_{\mathbf{u}\varphi(2)(1)} = \Phi^{(k)(k_l)}_{\mathbf{u}\varphi(2)(1)} + \frac{C^{(k)}_{\mathbf{u}\varphi(2)(1)}\Delta V_{(k)}}{N_{(k)}} \\
&\mathrm{k}^{(k)(k_l)}_{\mathbf{u}\psi(2)(1)} = \Phi^{(k)(k_l)}_{\mathbf{u}\psi(2)(1)} + \frac{C^{(k)}_{\mathbf{u}\psi(2)(1)}\Delta V_{(k)}}{N_{(k)}}, \widehat{\mathrm{k}}^{(k)(k_l)}_{\mathbf{uu}(2)(1)} = -\Phi^{(k)(k_l)}_{\mathbf{uu}(2)(1)}, \widehat{\mathrm{k}}^{(k)(k_l)}_{\mathbf{uu}(2)(2)} = -\Phi^{(k)(k_l)}_{\mathbf{uu}(2)(2)} \\
&\widehat{\mathrm{k}}^{(k)(k_l)}_{\mathbf{u}\phi(2)(1)} = -\Phi^{(k)(k_l)}_{\mathbf{u}\phi(2)(1)}, \widehat{\mathrm{k}}^{(k)(k_l)}_{\mathbf{u}\varphi(2)(1)} = -\Phi^{(k)(k_l)}_{\mathbf{u}\varphi(2)(1)}, \widehat{\mathrm{k}}^{(k)(k_l)}_{\mathbf{u}\psi(2)(1)} = -\Phi^{(k)(k_l)}_{\mathbf{u}\psi(2)(1)}
\end{aligned}
\tag{7.21b}
$$

$$
\begin{aligned}
&\mathrm{k}_{\phi\mathbf{u}(1)(1)}^{(k)(k_l)} = \mathbf{\Phi}_{\phi\mathbf{u}(1)(1)}^{(k)(k_l)} + \frac{C_{\phi\mathbf{u}(1)(1)}^{(k)} \Delta V_{(k)}}{N_{(k)}}, \quad \mathrm{k}_{\phi\mathbf{u}(1)(2)}^{(k)(k_l)} = \mathbf{\Phi}_{\phi\mathbf{u}(1)(2)}^{(k)(k_l)} + \frac{C_{\phi\mathbf{u}(1)(2)}^{(k)} \Delta V_{(k)}}{N_{(k)}} \\
&\mathrm{k}_{\varphi\mathbf{u}(1)(1)}^{(k)(k_l)} = \mathbf{\Phi}_{\varphi\mathbf{u}(1)(1)}^{(k)(k_l)} + \frac{C_{\varphi\mathbf{u}(1)(1)}^{(k)} \Delta V_{(k)}}{N_{(k)}}, \quad \mathrm{k}_{\varphi\mathbf{u}(1)(2)}^{(k)(k_l)} = \mathbf{\Phi}_{\varphi\mathbf{u}(1)(2)}^{(k)(k_l)} + \frac{C_{\varphi\mathbf{u}(1)(2)}^{(k)} \Delta V_{(k)}}{N_{(k)}} \\
&\mathrm{k}_{\psi\mathbf{u}(1)(1)}^{(k)(k_l)} = \mathbf{\Phi}_{\psi\mathbf{u}(1)(1)}^{(k)(k_l)} + \frac{C_{\psi\mathbf{u}(1)(1)}^{(k)} \Delta V_{(k)}}{N_{(k)}}, \mathrm{k}_{\psi\mathbf{u}(1)(2)}^{(k)(k_l)} = \mathbf{\Phi}_{\psi\mathbf{u}(1)(2)}^{(k)(k_l)} + \frac{C_{\psi\mathbf{u}(1)(2)}^{(k)} \Delta V_{(k)}}{N_{(k)}} \\
&\widehat{\mathrm{k}}_{\phi\mathbf{u}(1)(1)}^{(k)(k_l)} = -\mathbf{\Phi}_{\phi\mathbf{u}(1)(1)}^{(k)(k_l)}, \quad \widehat{\mathrm{k}}_{\phi\mathbf{u}(1)(2)}^{(k)(k_l)} = -\mathbf{\Phi}_{\phi\mathbf{u}(1)(2)}^{(k)(k_l)}, \quad \widehat{\mathrm{k}}_{\varphi\mathbf{u}(1)(1)}^{(k)(k_l)} = -\mathbf{\Phi}_{\varphi\mathbf{u}(1)(1)}^{(k)(k_l)} \\
&\widehat{\mathrm{k}}_{\varphi\mathbf{u}(1)(2)}^{(k)(k_l)} = -\mathbf{\Phi}_{\varphi\mathbf{u}(1)(2)}^{(k)(k_l)}, \quad \widehat{\mathrm{k}}_{\psi\mathbf{u}(1)(1)}^{(k)(k_l)} = -\mathbf{\Phi}_{\psi\mathbf{u}(1)(1)}^{(k)(k_l)}, \quad \widehat{\mathrm{k}}_{\psi\mathbf{u}(1)(2)}^{(k)(k_l)} = -\mathbf{\Phi}_{\psi\mathbf{u}(1)(2)}^{(k)(k_l)}
\end{aligned} \tag{7.21c}
$$

and

$$
\begin{aligned}
&\mathrm{k}_{\phi\phi}^{(k)(k_l)} = \mathbf{\Phi}_{\phi\phi}^{(k)(k_l)} + \frac{C_{\phi\phi}^{(k)} \Delta V_{(k)}}{N_{(k)}}, \quad \mathrm{k}_{\phi\varphi}^{(k)(k_l)} = \mathbf{\Phi}_{\phi\varphi}^{(k)(k_l)} + \frac{C_{\phi\varphi}^{(k)} \Delta V_{(k)}}{N_{(k)}} \\
&\mathrm{k}_{\phi\psi}^{(k)(k_l)} = \mathbf{\Phi}_{\phi\psi}^{(k)(k_l)} + \frac{C_{\phi\psi}^{(k)} \Delta V_{(k)}}{N_{(k)}}, \quad \mathrm{k}_{\varphi\phi}^{(k)(k_l)} = \mathbf{\Phi}_{\varphi\phi}^{(k)(k_l)} + \frac{C_{\varphi\phi}^{(k)} \Delta V_{(k)}}{N_{(k)}} \\
&\mathrm{k}_{\varphi\varphi}^{(k)(k_l)} = \mathbf{\Phi}_{\varphi\varphi}^{(k)(k_l)} + \frac{C_{\varphi\varphi}^{(k)} \Delta V_{(k)}}{N_{(k)}}, \quad \mathrm{k}_{\varphi\psi}^{(k)(k_l)} = \mathbf{\Phi}_{\varphi\psi}^{(k)(k_l)} + \frac{C_{\varphi\psi}^{(k)} \Delta V_{(k)}}{N_{(k)}} \\
&\mathrm{k}_{\psi\phi}^{(k)(k_l)} = \mathbf{\Phi}_{\psi\phi}^{(k)(k_l)} + \frac{C_{\psi\phi}^{(k)} \Delta V_{(k)}}{N_{(k)}}, \quad \mathrm{k}_{\psi\varphi}^{(k)(k_l)} = \mathbf{\Phi}_{\psi\varphi}^{(k)(k_l)} + \frac{C_{\psi\varphi}^{(k)} \Delta V_{(k)}}{N_{(k)}} \\
&\mathrm{k}_{\psi\psi}^{(k)(k_l)} = \mathbf{\Phi}_{\psi\psi}^{(k)(k_l)} + \frac{C_{\psi\psi}^{(k)} \Delta V_{(k)}}{N_{(k)}}, \quad \widehat{\mathrm{k}}_{\phi\phi}^{(k)(k_l)} = -\mathbf{\Phi}_{\phi\phi}^{(k)(k_l)}, \quad \widehat{\mathrm{k}}_{\phi\varphi}^{(k)(k_l)} = -\mathbf{\Phi}_{\phi\varphi}^{(k)(k_l)} \\
&\widehat{\mathrm{k}}_{\phi\psi}^{(k)(k_l)} = -\mathbf{\Phi}_{\phi\psi}^{(k)(k_l)}, \quad \widehat{\mathrm{k}}_{\varphi\phi}^{(k)(k_l)} = -\mathbf{\Phi}_{\varphi\phi}^{(k)(k_l)}, \quad \widehat{\mathrm{k}}_{\varphi\varphi}^{(k)(k_l)} = -\mathbf{\Phi}_{\varphi\varphi}^{(k)(k_l)}, \quad \widehat{\mathrm{k}}_{\varphi\psi}^{(k)(k_l)} = -\mathbf{\Phi}_{\varphi\psi}^{(k)(k_l)} \\
&\widehat{\mathrm{k}}_{\psi\phi}^{(k)(k_l)} = -\mathbf{\Phi}_{\psi\phi}^{(k)(k_l)}, \quad \widehat{\mathrm{k}}_{\psi\varphi}^{(k)(k_l)} = -\mathbf{\Phi}_{\psi\varphi}^{(k)(k_l)}, \quad \widehat{\mathrm{k}}_{\psi\psi}^{(k)(k_l)} = \mathbf{\Phi}_{\psi\psi}^{(k)(k_l)}
\end{aligned} \tag{7.21d}
$$

The vector $\mathbf{F}_{\mathscr{R}(k)}^{\text{PDDO(Inertial)}}$ corresponding to inertial terms in Eq. (7.18) can be written in component form as

$$\mathbf{F}_{\mathcal{R}(k)}^{\text{PDDO(Inertial)}} = \begin{Bmatrix} F_{1(k)}^{\text{PDDO(Inertial)}} \\ F_{2(k)}^{\text{PDDO(Inertial)}} \\ 0 \\ F_{\phi(k)}^{\text{PDDO(Inertial)}} \\ F_{\varphi(k)}^{\text{PDDO(Inertial)}} \\ F_{\psi(k)}^{\text{PDDO(Inertial)}} \\ 0 \\ 0 \\ 0 \\ 0 \\ 0 \\ 0 \end{Bmatrix} = \sum_{l=1}^{N_{(k)}} \mathbf{m}_{(k)(k_l)}^{\text{PDDO}} \begin{Bmatrix} \ddot{u}_{1(k)} \\ \ddot{u}_{2(k)} \\ 0 \\ \ddot{\phi}_{(k)} \\ \ddot{\varphi}_{(k)} \\ \ddot{\psi}_{(k)} \\ \ddot{u}_{1(k_l)} \\ \ddot{u}_{2(k_l)} \\ 0 \\ \ddot{\phi}_{(k_l)} \\ \ddot{\varphi}_{(k_l)} \\ \ddot{\psi}_{(k_l)} \end{Bmatrix} \tag{7.22}$$

with

$$\mathbf{m}_{(k)(k_l)}^{\text{PDDO}} = \left\{ \begin{matrix} \mathrm{m}_{\mathbf{uu}(1)(1)}^{(k)(k_l)} & \mathrm{m}_{\mathbf{uu}(1)(2)}^{(k)(k_l)} & 0 & \mathrm{m}_{\mathbf{u}\phi(1)(1)}^{(k)(k_l)} & \mathrm{m}_{\mathbf{u}\varphi(1)(1)}^{(k)(k_l)} & \mathrm{m}_{\mathbf{u}\psi(1)(1)}^{(k)(k_l)} \\ \mathrm{m}_{\mathbf{uu}(2)(1)}^{(k)(k_l)} & \mathrm{m}_{\mathbf{uu}(2)(2)}^{(k)(k_l)} & 0 & \mathrm{m}_{\mathbf{u}\phi(2)(1)}^{(k)(k_l)} & \mathrm{m}_{\mathbf{u}\varphi(2)(1)}^{(k)(k_l)} & \mathrm{m}_{\mathbf{u}\psi(2)(1)}^{(k)(k_l)} \\ 0 & 0 & 0 & 0 & 0 & 0 \\ \mathrm{m}_{\phi\mathbf{u}(1)(1)}^{(k)(k_l)} & \mathrm{m}_{\phi\mathbf{u}(1)(2)}^{(k)(k_l)} & 0 & \mathrm{m}_{\phi\phi}^{(k)(k_l)} & \mathrm{m}_{\phi\varphi}^{(k)(k_l)} & \mathrm{m}_{\phi\psi}^{(k)(k_l)} \\ \mathrm{m}_{\varphi\mathbf{u}(1)(1)}^{(k)(k_l)} & \mathrm{m}_{\varphi\mathbf{u}(1)(2)}^{(k)(k_l)} & 0 & \mathrm{m}_{\varphi\phi}^{(k)(k_l)} & \mathrm{m}_{\varphi\varphi}^{(k)(k_l)} & \mathrm{m}_{\varphi\psi}^{(k)(k_l)} \\ \mathrm{m}_{\psi\mathbf{u}(1)(1)}^{(k)(k_l)} & \mathrm{m}_{\psi\mathbf{u}(1)(2)}^{(k)(k_l)} & 0 & \mathrm{m}_{\psi\phi}^{(k)(k_l)} & \mathrm{m}_{\psi\varphi}^{(k)(k_l)} & \mathrm{m}_{\psi\psi}^{(k)(k_l)} \\ 0 & 0 & 0 & 0 & 0 & 0 \\ 0 & 0 & 0 & 0 & 0 & 0 \\ 0 & 0 & 0 & 0 & 0 & 0 \\ 0 & 0 & 0 & 0 & 0 & 0 \\ 0 & 0 & 0 & 0 & 0 & 0 \\ 0 & 0 & 0 & 0 & 0 & 0 \end{matrix} \right.$$

$$
\left.\begin{array}{cccccc}
\widehat{\mathrm{m}}_{\mathbf{uu}(1)(1)}^{(k)(k_l)} & \widehat{\mathrm{m}}_{\mathbf{uu}(1)(2)}^{(k)(k_l)} & 0 & 0 & 0 & 0 \\
\widehat{\mathrm{m}}_{\mathbf{uu}(2)(1)}^{(k)(k_l)} & \widehat{\mathrm{m}}_{\mathbf{uu}(2)(2)}^{(k)(k_l)} & 0 & 0 & 0 & 0 \\
0 & 0 & 0 & 0 & 0 & 0 \\
0 & 0 & 0 & \widehat{\mathrm{m}}_{\phi\phi}^{(k)(k_l)} & \widehat{\mathrm{m}}_{\phi\varphi}^{(k)(k_l)} & \widehat{\mathrm{m}}_{\phi\psi}^{(k)(k_l)} \\
0 & 0 & 0 & \widehat{\mathrm{m}}_{\varphi\phi}^{(k)(k_l)} & \widehat{\mathrm{m}}_{\varphi\varphi}^{(k)(k_l)} & \widehat{\mathrm{m}}_{\varphi\psi}^{(k)(k_l)} \\
0 & 0 & 0 & \widehat{\mathrm{m}}_{\psi\phi}^{(k)(k_l)} & \widehat{\mathrm{m}}_{\psi\varphi}^{(k)(k_l)} & \widehat{\mathrm{m}}_{\psi\psi}^{(k)(k_l)} \\
0 & 0 & 0 & 0 & 0 & 0 \\
0 & 0 & 0 & 0 & 0 & 0 \\
0 & 0 & 0 & 0 & 0 & 0 \\
0 & 0 & 0 & 0 & 0 & 0 \\
0 & 0 & 0 & 0 & 0 & 0 \\
0 & 0 & 0 & 0 & 0 & 0
\end{array}\right\} \tag{7.23}
$$

where

$$
\begin{array}{l}
\mathrm{m}_{\mathbf{uu}(1)(1)}^{(k)(k_l)} = \dfrac{A_{\mathbf{uu}(1)(1)}^{(k)} \Delta V_{(k)}}{N_{(k)}}, \quad \mathrm{m}_{\mathbf{uu}(1)(2)}^{(k)(k_l)} = \dfrac{A_{\mathbf{uu}(1)(2)}^{(k)} \Delta V_{(k)}}{N_{(k)}}, \quad \mathrm{m}_{\mathbf{u}\phi(1)(1)}^{(k)(k_l)} = \dfrac{A_{\mathbf{u}\phi(1)(1)}^{(k)} \Delta V_{(k)}}{N_{(k)}} \\
\mathrm{m}_{\mathbf{u}\varphi(1)(1)}^{(k)(k_l)} = \dfrac{A_{\mathbf{u}\phi(1)(1)}^{(k)} \Delta V_{(k)}}{N_{(k)}}, \quad \mathrm{m}_{\mathbf{u}\psi(1)(1)}^{(k)(k_l)} = \dfrac{A_{\mathbf{u}\psi(1)(1)}^{(k)} \Delta V_{(k)}}{N_{(k)}}, \mathrm{m}_{\mathbf{uu}(2)(1)}^{(k)(k_l)} = \dfrac{A_{\mathbf{uu}(2)(1)}^{(k)} \Delta V_{(k)}}{N_{(k)}} \\
\mathrm{m}_{\mathbf{uu}(2)(2)}^{(k)(k_l)} = \dfrac{A_{\mathbf{uu}(2)(2)}^{(k)} \Delta V_{(k)}}{N_{(k)}}, \quad \mathrm{m}_{\mathbf{u}\phi(2)(1)}^{(k)(k_l)} = \dfrac{A_{\mathbf{u}\phi(2)(1)}^{(k)} \Delta V_{(k)}}{N_{(k)}}, \mathrm{m}_{\mathbf{u}\varphi(2)(1)}^{(k)(k_l)} = \dfrac{A_{\mathbf{u}\varphi(2)(1)}^{(k)} \Delta V_{(k)}}{N_{(k)}} \\
\mathrm{m}_{\mathbf{u}\psi(2)(1)}^{(k)(k_l)} = \dfrac{A_{\mathbf{u}\psi(2)(1)}^{(k)} \Delta V_{(k)}}{N_{(k)}}
\end{array} \tag{7.24a}
$$

$$
\begin{aligned}
&\mathrm{m}_{\phi\phi}^{(k)(k_l)} = \boldsymbol{\Psi}_{\phi\phi}^{(k)(k_l)} + \frac{A_{\phi\phi}^{(k)}\Delta V_{(k)}}{N_{(k)}}, \quad \mathrm{m}_{\phi\varphi}^{(k)(k_l)} = \boldsymbol{\Psi}_{\phi\varphi}^{(k)(k_l)} + \frac{A_{\phi\varphi}^{(k)}\Delta V_{(k)}}{N_{(k)}} \\
&\mathrm{m}_{\phi\psi}^{(k)(k_l)} = \boldsymbol{\Psi}_{\phi\psi}^{(k)(k_l)} + \frac{A_{\phi\psi}^{(k)}\Delta V_{(k)}}{N_{(k)}}, \quad \mathrm{m}_{\varphi\phi}^{(k)(k_l)} = \boldsymbol{\Psi}_{\varphi\phi}^{(k)(k_l)} + \frac{A_{\varphi\phi}^{(k)}\Delta V_{(k)}}{N_{(k)}} \\
&\mathrm{m}_{\varphi\varphi}^{(k)(k_l)} = \boldsymbol{\Psi}_{\varphi\varphi}^{(k)(k_l)} + \frac{A_{\varphi\varphi}^{(k)}\Delta V_{(k)}}{N_{(k)}}, \quad \mathrm{m}_{\varphi\psi}^{(k)(k_l)} = \boldsymbol{\Psi}_{\varphi\psi}^{(k)(k_l)} + \frac{A_{\varphi\psi}^{(k)}\Delta V_{(k)}}{N_{(k)}} \\
&\mathrm{m}_{\psi\phi}^{(k)(k_l)} = \boldsymbol{\Psi}_{\psi\phi}^{(k)(k_l)} + \frac{A_{\psi\phi}^{(k)}\Delta V_{(k)}}{N_{(k)}}, \quad \mathrm{m}_{\psi\varphi}^{(k)(k_l)} = \boldsymbol{\Psi}_{\psi\varphi}^{(k)(k_l)} + \frac{A_{\psi\varphi}^{(k)}\Delta V_{(k)}}{N_{(k)}} \\
&\mathrm{m}_{\psi\psi}^{(k)(k_l)} = \boldsymbol{\Psi}_{\psi\psi}^{(k)(k_l)} + \frac{A_{\psi\psi}^{(k)}\Delta V_{(k)}}{N_{(k)}}, \quad \widehat{\mathrm{m}}_{\phi\phi}^{(k)(k_l)} = -\boldsymbol{\Psi}_{\phi\phi}^{(k)(k_l)}, \quad \widehat{\mathrm{m}}_{\phi\varphi}^{(k)(k_l)} = -\boldsymbol{\Psi}_{\phi\varphi}^{(k)(k_l)} \\
&\widehat{\mathrm{m}}_{\phi\psi}^{(k)(k_l)} = -\boldsymbol{\Psi}_{\phi\psi}^{(k)(k_l)}, \quad \widehat{\mathrm{m}}_{\varphi\phi}^{(k)(k_l)} = -\boldsymbol{\Psi}_{\varphi\phi}^{(k)(k_l)}, \quad \widehat{\mathrm{m}}_{\varphi\varphi}^{(k)(k_l)} = -\boldsymbol{\Psi}_{\varphi\varphi}^{(k)(k_l)}, \quad \widehat{\mathrm{m}}_{\varphi\psi}^{(k)(k_l)} = -\boldsymbol{\Psi}_{\varphi\psi}^{(k)(k_l)} \\
&\widehat{\mathrm{m}}_{\psi\phi}^{(k)(k_l)} = -\boldsymbol{\Psi}_{\psi\phi}^{(k)(k_l)}, \quad \widehat{\mathrm{m}}_{\psi\varphi}^{(k)(k_l)} = -\boldsymbol{\Psi}_{\psi\varphi}^{(k)(k_l)}, \quad \widehat{\mathrm{m}}_{\psi\psi}^{(k)(k_l)} = -\boldsymbol{\Psi}_{\psi\psi}^{(k)(k_l)}
\end{aligned}
\tag{7.24b}
$$

The vector $\mathbf{F}_{\mathscr{R}(k)}^{\mathrm{PDDO(Damp)}}$ corresponding to damping terms in Eq. (7.18) can be written in component form as

$$
\mathbf{F}_{\mathcal{R}(k)}^{\mathrm{PDDO(Damp)}} = \begin{Bmatrix} F_{1(k)}^{\mathrm{PDDO(Damp)}} \\ F_{2(k)}^{\mathrm{PDDO(Damp)}} \\ 0 \\ F_{\phi(k)}^{\mathrm{PDDO(Damp)}} \\ F_{\varphi(k)}^{\mathrm{PDDO(Damp)}} \\ F_{\psi(k)}^{\mathrm{PDDO(Damp)}} \\ 0 \\ 0 \\ 0 \\ 0 \\ 0 \\ 0 \end{Bmatrix} = \sum_{l=1}^{N_{(k)}} \mathbf{c}_{(k)(k_l)}^{\mathrm{PDDO}} \begin{Bmatrix} \dot{u}_{1(k)} \\ \dot{u}_{2(k)} \\ 0 \\ \dot{\phi}_{(k)} \\ \dot{\varphi}_{(k)} \\ \dot{\psi}_{(k)} \\ \dot{u}_{1(k_l)} \\ \dot{u}_{2(k_l)} \\ 0 \\ \dot{\phi}_{(k_l)} \\ \dot{\varphi}_{(k_l)} \\ \dot{\psi}_{(k_l)} \end{Bmatrix}
\tag{7.25}
$$

with

$$
\mathbf{c}_{(k)(k_l)}^{\mathrm{PDDO}} = \left\{ \begin{array}{cccccccccccc}
c_{\mathbf{uu}(1)(1)}^{(k)(k_l)} & c_{\mathbf{uu}(1)(2)}^{(k)(k_l)} & 0 & c_{\mathbf{u}\phi(1)(1)}^{(k)(k_l)} & c_{\mathbf{u}\varphi(1)(1)}^{(k)(k_l)} & c_{\mathbf{u}\psi(1)(1)}^{(k)(k_l)} & \widehat{c}_{\mathbf{uu}(1)(1)}^{(k)(k_l)} & \widehat{c}_{\mathbf{uu}(1)(2)}^{(k)(k_l)} & 0 & 0 & 0 & 0 \\
c_{\mathbf{uu}(2)(1)}^{(k)(k_l)} & c_{\mathbf{uu}(2)(2)}^{(k)(k_l)} & 0 & c_{\mathbf{u}\phi(2)(1)}^{(k)(k_l)} & c_{\mathbf{u}\varphi(2)(1)}^{(k)(k_l)} & c_{\mathbf{u}\psi(2)(1)}^{(k)(k_l)} & \widehat{c}_{\mathbf{uu}(2)(1)}^{(k)(k_l)} & \widehat{c}_{\mathbf{uu}(2)(2)}^{(k)(k_l)} & 0 & 0 & 0 & 0 \\
0 & 0 & 0 & 0 & 0 & 0 & 0 & 0 & 0 & 0 & 0 & 0 \\
c_{\phi\mathbf{u}(1)(1)}^{(k)(k_l)} & c_{\phi\mathbf{u}(1)(2)}^{(k)(k_l)} & 0 & c_{\phi\phi}^{(k)(k_l)} & c_{\phi\varphi}^{(k)(k_l)} & c_{\phi\psi}^{(k)(k_l)} & 0 & 0 & 0 & \widehat{c}_{\phi\phi}^{(k)(k_l)} & \widehat{c}_{\phi\varphi}^{(k)(k_l)} & \widehat{c}_{\phi\psi}^{(k)(k_l)} \\
c_{\varphi\mathbf{u}(1)(1)}^{(k)(k_l)} & c_{\varphi\mathbf{u}(1)(2)}^{(k)(k_l)} & 0 & c_{\varphi\phi}^{(k)(k_l)} & c_{\varphi\varphi}^{(k)(k_l)} & c_{\varphi\psi}^{(k)(k_l)} & 0 & 0 & 0 & \widehat{c}_{\varphi\phi}^{(k)(k_l)} & \widehat{c}_{\varphi\varphi}^{(k)(k_l)} & \widehat{c}_{\varphi\psi}^{(k)(k_l)} \\
c_{\psi\mathbf{u}(1)(1)}^{(k)(k_l)} & c_{\psi\mathbf{u}(1)(2)}^{(k)(k_l)} & 0 & c_{\psi\phi}^{(k)(k_l)} & c_{\psi\varphi}^{(k)(k_l)} & c_{\psi\psi}^{(k)(k_l)} & 0 & 0 & 0 & \widehat{c}_{\psi\phi}^{(k)(k_l)} & \widehat{c}_{\psi\varphi}^{(k)(k_l)} & \widehat{c}_{\psi\psi}^{(k)(k_l)} \\
0 & 0 & 0 & 0 & 0 & 0 & 0 & 0 & 0 & 0 & 0 & 0 \\
0 & 0 & 0 & 0 & 0 & 0 & 0 & 0 & 0 & 0 & 0 & 0 \\
0 & 0 & 0 & 0 & 0 & 0 & 0 & 0 & 0 & 0 & 0 & 0 \\
0 & 0 & 0 & 0 & 0 & 0 & 0 & 0 & 0 & 0 & 0 & 0 \\
0 & 0 & 0 & 0 & 0 & 0 & 0 & 0 & 0 & 0 & 0 & 0 \\
0 & 0 & 0 & 0 & 0 & 0 & 0 & 0 & 0 & 0 & 0 & 0
\end{array} \right\} \tag{7.26}
$$

where

$$\mathrm{c}_{\mathbf{uu}(1)(1)}^{(k)(k_l)} = \frac{B_{\mathbf{uu}(1)(1)}^{(k)}\Delta V_{(k)}}{N_{(k)}}, \quad \mathrm{c}_{\mathbf{uu}(1)(2)}^{(k)(k_l)} = \frac{B_{\mathbf{uu}(1)(2)}^{(k)}\Delta V_{(k)}}{N_{(k)}}, \quad \mathrm{c}_{\mathbf{u}\phi(1)(1)}^{(k)(k_l)} = \frac{B_{\mathbf{u}\phi(1)(1)}^{(k)}\Delta V_{(k)}}{N_{(k)}}$$

$$\mathrm{c}_{\mathbf{u}\varphi(1)(1)}^{(k)(k_l)} = \frac{B_{\mathbf{u}\phi(1)(1)}^{(k)}\Delta V_{(k)}}{N_{(k)}}, \quad \mathrm{c}_{\mathbf{u}\psi(1)(1)}^{(k)(k_l)} = \frac{B_{\mathbf{u}\psi(1)(1)}^{(k)}\Delta V_{(k)}}{N_{(k)}}, \mathrm{c}_{\mathbf{uu}(2)(1)}^{(k)(k_l)} = \frac{B_{\mathbf{uu}(2)(1)}^{(k)}\Delta V_{(k)}}{N_{(k)}}$$

$$\mathrm{c}_{\mathbf{uu}(2)(2)}^{(k)(k_l)} = \frac{B_{\mathbf{uu}(2)(2)}^{(k)}\Delta V_{(k)}}{N_{(k)}}, \quad \mathrm{c}_{\mathbf{u}\phi(2)(1)}^{(k)(k_l)} = \frac{B_{\mathbf{u}\phi(2)(1)}^{(k)}\Delta V_{(k)}}{N_{(k)}}, \mathrm{c}_{\mathbf{u}\varphi(2)(1)}^{(k)(k_l)} = \frac{B_{\mathbf{u}\varphi(2)(1)}^{(k)}\Delta V_{(k)}}{N_{(k)}}$$

$$\mathrm{c}_{\mathbf{u}\psi(2)(1)}^{(k)(k_l)} = \frac{B_{\mathbf{u}\psi(2)(1)}^{(k)}\Delta V_{(k)}}{N_{(k)}} \tag{7.27a}$$

$$\mathrm{c}_{\phi\phi}^{(k)(k_l)} = \Theta_{\phi\phi}^{(k)(k_l)} + \frac{B_{\phi\phi}^{(k)}\Delta V_{(k)}}{N_{(k)}}, \quad \mathrm{c}_{\phi\varphi}^{(k)(k_l)} = \Theta_{\phi\varphi}^{(k)(k_l)} + \frac{B_{\phi\varphi}^{(k)}\Delta V_{(k)}}{N_{(k)}}$$

$$\mathrm{c}_{\phi\psi}^{(k)(k_l)} = \Theta_{\phi\psi}^{(k)(k_l)} + \frac{B_{\phi\psi}^{(k)}\Delta V_{(k)}}{N_{(k)}}, \quad \mathrm{c}_{\varphi\phi}^{(k)(k_l)} = \Theta_{\varphi\phi}^{(k)(k_l)} + \frac{B_{\varphi\phi}^{(k)}\Delta V_{(k)}}{N_{(k)}}$$

$$\mathrm{c}_{\varphi\varphi}^{(k)(k_l)} = \Theta_{\varphi\varphi}^{(k)(k_l)} + \frac{B_{\varphi\varphi}^{(k)}\Delta V_{(k)}}{N_{(k)}}, \quad \mathrm{c}_{\varphi\psi}^{(k)(k_l)} = \Theta_{\varphi\psi}^{(k)(k_l)} + \frac{B_{\varphi\psi}^{(k)}\Delta V_{(k)}}{N_{(k)}}$$

$$\mathrm{c}_{\psi\phi}^{(k)(k_l)} = \Theta_{\psi\phi}^{(k)(k_l)} + \frac{B_{\psi\phi}^{(k)}\Delta V_{(k)}}{N_{(k)}}, \quad \mathrm{c}_{\psi\varphi}^{(k)(k_l)} = \Theta_{\psi\varphi}^{(k)(k_l)} + \frac{B_{\psi\varphi}^{(k)}\Delta V_{(k)}}{N_{(k)}}$$

$$\mathrm{c}_{\psi\psi}^{(k)(k_l)} = \Theta_{\psi\psi}^{(k)(k_l)} + \frac{B_{\psi\psi}^{(k)}\Delta V_{(k)}}{N_{(k)}}, \quad \widehat{\mathrm{c}}_{\phi\phi}^{(k)(k_l)} = -\Theta_{\phi\phi}^{(k)(k_l)}, \quad \widehat{\mathrm{c}}_{\phi\varphi}^{(k)(k_l)} = -\Theta_{\phi\varphi}^{(k)(k_l)}$$

$$\widehat{\mathrm{c}}_{\phi\psi}^{(k)(k_l)} = -\Theta_{\phi\psi}^{(k)(k_l)}, \quad \widehat{\mathrm{c}}_{\varphi\phi}^{(k)(k_l)} = -\Theta_{\varphi\phi}^{(k)(k_l)}, \quad \widehat{\mathrm{c}}_{\varphi\varphi}^{(k)(k_l)} = -\Theta_{\varphi\varphi}^{(k)(k_l)}, \quad \widehat{\mathrm{c}}_{\varphi\psi}^{(k)(k_l)} = -\Theta_{\varphi\psi}^{(k)(k_l)}$$

$$\widehat{\mathrm{c}}_{\psi\phi}^{(k)(k_l)} = -\Theta_{\psi\phi}^{(k)(k_l)}, \quad \widehat{\mathrm{c}}_{\psi\varphi}^{(k)(k_l)} = -\Theta_{\psi\varphi}^{(k)(k_l)}, \quad \widehat{\mathrm{c}}_{\psi\psi}^{(k)(k_l)} = -\Theta_{\psi\psi}^{(k)(k_l)} \tag{7.27b}$$

The vectors at node k, $\mathbf{F}_{\mathscr{R}(k)}^{\text{PDDO(Stiff)}}$, $\mathbf{F}_{\mathscr{R}(k)}^{\text{PDDO(Mass)}}$, and $\mathbf{F}_{\mathscr{R}(k)}^{\text{PDDO(Damp)}}$ are constructed using a network of elements around node k, shown in red, green, and blue in Fig. 7.3 associated with stiffness, $\mathbf{k}_{(k)(k_l)}^{PDDO}$, mass, $\mathbf{m}_{(k)(k_l)}^{PDDO}$, and damping, $\mathbf{c}_{(k)(k_r)}^{PDDO}$ matrices, respectively.

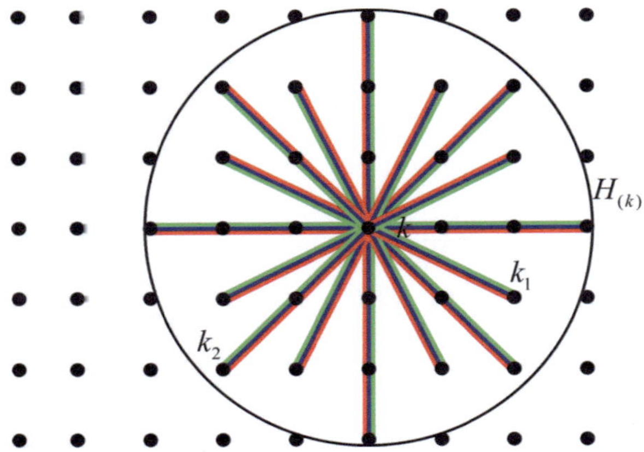

Fig. 7.3 Element connectivity (red for stiffness, green for mass, and blue for damping matrices) for a PDDO interaction of a point k with its family members in the region, $\mathscr{R}$

The external vector $\mathbf{F}^{\text{Ext}}_{\mathscr{R}(k)}$ in Eq. (7.18) can be written in component form as

$$
\mathbf{F}^{\text{Ext}}_{\mathscr{R}(k)} = \left\{ \begin{array}{c} s_{1\mathbf{u}(k)}\Delta V_{(k)} \\ s_{2\mathbf{u}(k)}\Delta V_{(k)} \\ 0 \\ s_{\phi(k)}\Delta V_{(k)} + \dot{s}_{\phi(k)}\Delta V_{(k)} \\ s_{\varphi(k)}\Delta V_{(k)} + \dot{s}_{\varphi(k)}\Delta V_{(k)} \\ s_{\psi(k)}\Delta V_{(k)} + \dot{s}_{\psi(k)}\Delta V_{(k)} \\ 0 \\ 0 \\ 0 \\ 0 \\ 0 \\ 0 \end{array} \right\} \tag{7.28}
$$

7.6.2 Discrete Form of Boundary Conditions

The discrete form of natural boundary conditions, Eq. (7.12), at point k in region $\mathscr{B}$ can be written as

$$\sum_{l=1}^{N_{(k)}} \begin{bmatrix} \boldsymbol{\Omega}_{\mathbf{uu}} & \boldsymbol{\Omega}_{\mathbf{u}\phi} & \boldsymbol{\Omega}_{\mathbf{u}\varphi} & \boldsymbol{\Omega}_{\mathbf{u}\psi} \\ \boldsymbol{\Omega}_{\phi\mathbf{u}} & \Omega_{\phi\phi} & \Omega_{\phi\varphi} & \Omega_{\phi\psi} \\ \boldsymbol{\Omega}_{\varphi\mathbf{u}} & \Omega_{\varphi\phi} & \Omega_{\varphi\varphi} & \Omega_{\varphi\psi} \\ \boldsymbol{\Omega}_{\psi\mathbf{u}} & \Omega_{\psi\phi} & \Omega_{\psi\varphi} & \Omega_{\psi\psi} \end{bmatrix} \begin{Bmatrix} \left(\mathbf{u}\left(\mathbf{x}_{(k_l)}, t\right) - \mathbf{u}\left(\mathbf{x}_{(k)}, t\right)\right) \\ \left(\phi\left(\mathbf{x}_{(k_l)}, t\right) - \phi\left(\mathbf{x}_{(k)}, t\right)\right) \\ \left(\varphi\left(\mathbf{x}_{(k_l)}, t\right) - \varphi\left(\mathbf{x}_{(k)}, t\right)\right) \\ \left(\psi\left(\mathbf{x}_{(k_l)}, t\right) - \psi\left(\mathbf{x}_{(k)}, t\right)\right) \end{Bmatrix}$$
$$+\mathbf{S}\left(\mathbf{x}_{(k)}, t\right) \begin{Bmatrix} \mathbf{u}\left(\mathbf{x}_{(k)}, t\right) \\ \phi\left(\mathbf{x}_{(k)}, t\right) \\ \varphi\left(\mathbf{x}_{(k)}, t\right) \\ \psi\left(\mathbf{x}_{(k)}, t\right) \end{Bmatrix} = \begin{Bmatrix} \mathbf{T}_{\boldsymbol{\sigma}}^{*}\left(\mathbf{x}_{(k)}, t\right) \\ J_{\phi}^{*}\left(\mathbf{x}_{(k)}, t\right) \\ J_{\varphi}^{*}\left(\mathbf{x}_{(k)}, t\right) \\ J_{\psi}^{*}\left(\mathbf{x}_{(k)}, t\right) \end{Bmatrix} \tag{7.29}$$

where the scalar coefficients are defined as

$$\begin{gathered} \Omega_{\alpha\beta}(\mathbf{x}_{(k)}, \xi_{(k)(k_l)}, t) = \omega(|\xi_{(k)(k_l)}|)(R_{\alpha\beta}(\mathbf{x}_{(k)}, t)\mathbf{g}(\mathbf{x}_{(k)}, \xi_{(k)(k_l)}) \cdot \mathbf{n}(\mathbf{x}_{(k)}))\Delta V_{(k_l)} \\ \forall \alpha, \beta = \phi, \varphi, \psi \end{gathered} \tag{7.30}$$

and the vector coefficients are defined as

$$\begin{aligned} &\boldsymbol{\Omega}_{\alpha\mathbf{u}}(\mathbf{x}_{(k)}, \xi_{(k)(k_l)}, t) \\ &\quad = \omega(|\xi_{(k)(k_l)}|)(\mathbf{n}(\mathbf{x}_{(k)}) \cdot \mathbf{g}(\mathbf{x}_{(k)}, \xi_{(k)(k_l)}))\mathbf{R}_{\alpha\mathbf{u}}(\mathbf{x}_{(k)}, t)\Delta V_{(k_l)} \forall \alpha = \phi, \varphi, \psi, \mathbf{u} \end{aligned} \tag{7.31}$$

Equation (7.29) can be simplified in vector form as

$$\mathbf{F}_{\mathscr{B}(k)}^{\text{PDDO(Stiff)}} = \mathbf{F}_{\mathscr{B}(k)}^{\text{Ext}} \tag{7.32}$$

The vector $\mathbf{F}_{\mathscr{B}(k)}^{\text{PDDO(Stiff)}}$ corresponding to the stiffness terms can be written in component form as

$$\mathbf{F}_{\mathcal{B}(k)}^{\text{PDDO(Stiff)}} = \begin{Bmatrix} F_{1(k)}^{\text{PDDO(Stiff)}} \\ F_{2(k)}^{\text{PDDO(Stiff)}} \\ 0 \\ F_{\phi(k)}^{\text{PDDO(Stiff)}} \\ F_{\varphi(k)}^{\text{PDDO(Stiff)}} \\ F_{\psi(k)}^{\text{PDDO(Stiff)}} \\ 0 \\ 0 \\ 0 \\ 0 \\ 0 \\ 0 \end{Bmatrix} = \sum_{l=1}^{N_{(k)}} \mathbf{k}_{(k)(k_l)}^{PDDO} \begin{Bmatrix} u_{1(k)} \\ u_{2(k)} \\ 0 \\ \phi_{(k)} \\ \varphi_{(k)} \\ \psi_{(k)} \\ u_{1(k_l)} \\ u_{2(k_l)} \\ 0 \\ \phi_{(k_l)} \\ \varphi_{(k_l)} \\ \psi_{(k_l)} \end{Bmatrix} \tag{7.33}$$

with

$$\mathbf{k}_{(k)(k_l)}^{\text{PDDO}} = \left\{ \begin{matrix} \mathrm{k}_{\mathbf{uu}(1)(1)}^{(k)(k_l)} & \mathrm{k}_{\mathbf{uu}(1)(2)}^{(k)(k_l)} & 0 & \mathrm{k}_{\mathbf{u}\phi(1)(1)}^{(k)(k_l)} & \mathrm{k}_{\mathbf{u}\varphi(1)(1)}^{(k)(k_l)} & \mathrm{k}_{\mathbf{u}\psi(1)(1)}^{(k)(k_l)} \\ \mathrm{k}_{\mathbf{uu}(2)(1)}^{(k)(k_l)} & \mathrm{k}_{\mathbf{uu}(2)(2)}^{(k)(k_l)} & 0 & \mathrm{k}_{\mathbf{u}\phi(2)(1)}^{(k)(k_l)} & \mathrm{k}_{\mathbf{u}\varphi(2)(1)}^{(k)(k_l)} & \mathrm{k}_{\mathbf{u}\psi(2)(1)}^{(k)(k_l)} \\ 0 & 0 & 0 & 0 & 0 & 0 \\ \mathrm{k}_{\phi\mathbf{u}(1)(1)}^{(k)(k_l)} & \mathrm{k}_{\phi\mathbf{u}(1)(2)}^{(k)(k_l)} & 0 & \mathrm{k}_{\phi\phi}^{(k)(k_l)} & \mathrm{k}_{\phi\varphi}^{(k)(k_l)} & \mathrm{k}_{\phi\psi}^{(k)(k_l)} \\ \mathrm{k}_{\varphi\mathbf{u}(1)(1)}^{(k)(k_l)} & \mathrm{k}_{\varphi\mathbf{u}(1)(2)}^{(k)(k_l)} & 0 & \mathrm{k}_{\varphi\phi}^{(k)(k_l)} & \mathrm{k}_{\varphi\varphi}^{(k)(k_l)} & \mathrm{k}_{\varphi\psi}^{(k)(k_l)} \\ \mathrm{k}_{\psi\mathbf{u}(1)(1)}^{(k)(k_l)} & \mathrm{k}_{\psi\mathbf{u}(1)(2)}^{(k)(k_l)} & 0 & \mathrm{k}_{\psi\phi}^{(k)(k_l)} & \mathrm{k}_{\psi\varphi}^{(k)(k_l)} & \mathrm{k}_{\psi\psi}^{(k)(k_l)} \\ 0 & 0 & 0 & 0 & 0 & 0 \\ 0 & 0 & 0 & 0 & 0 & 0 \\ 0 & 0 & 0 & 0 & 0 & 0 \\ 0 & 0 & 0 & 0 & 0 & 0 \\ 0 & 0 & 0 & 0 & 0 & 0 \\ 0 & 0 & 0 & 0 & 0 & 0 \end{matrix} \right.$$

$$
\left.\begin{array}{cccccc}
\widehat{\mathrm{k}}^{(k)(k_l)}_{\mathbf{uu}(1)(1)} & \widehat{\mathrm{k}}^{(k)(k_l)}_{\mathbf{uu}(1)(2)} & 0 & \widehat{\mathrm{k}}^{(k)(k_l)}_{\mathbf{u}\phi(1)(1)} & \widehat{\mathrm{k}}^{(k)(k_l)}_{\mathbf{u}\varphi(1)(1)} & \widehat{\mathrm{k}}^{(k)(k_l)}_{\mathbf{u}\psi(1)(1)} \\
\widehat{\mathrm{k}}^{(k)(k_l)}_{\mathbf{uu}(2)(1)} & \widehat{\mathrm{k}}^{(k)(k_l)}_{\mathbf{uu}(2)(2)} & 0 & \widehat{\mathrm{k}}^{(k)(k_l)}_{\mathbf{u}\phi(2)(1)} & \widehat{\mathrm{k}}^{(k)(k_l)}_{\mathbf{u}\varphi(2)(1)} & \widehat{\mathrm{k}}^{(k)(k_l)}_{\mathbf{u}\psi(2)(1)} \\
0 & 0 & 0 & 0 & 0 & 0 \\
\widehat{\mathrm{k}}^{(k)(k_l)}_{\phi\mathbf{u}(1)(1)} & \widehat{\mathrm{k}}^{(k)(k_l)}_{\phi\mathbf{u}(1)(2)} & 0 & \widehat{\mathrm{k}}^{(k)(k_l)}_{\phi\phi} & \widehat{\mathrm{k}}^{(k)(k_l)}_{\phi\varphi} & \widehat{\mathrm{k}}^{(k)(k_l)}_{\phi\psi} \\
\widehat{\mathrm{k}}^{(k)(k_l)}_{\varphi\mathbf{u}(1)(1)} & \widehat{\mathrm{k}}^{(k)(k_l)}_{\varphi\mathbf{u}(1)(2)} & 0 & \widehat{\mathrm{k}}^{(k)(k_l)}_{\varphi\phi} & \widehat{\mathrm{k}}^{(k)(k_l)}_{\varphi\varphi} & \widehat{\mathrm{k}}^{(k)(k_l)}_{\varphi\psi} \\
\widehat{\mathrm{k}}^{(k)(k_l)}_{\psi\mathbf{u}(1)(1)} & \widehat{\mathrm{k}}^{(k)(k_l)}_{\psi\mathbf{u}(1)(2)} & 0 & \widehat{\mathrm{k}}^{(k)(k_l)}_{\psi\phi} & \widehat{\mathrm{k}}^{(k)(k_l)}_{\psi\varphi} & \widehat{\mathrm{k}}^{(k)(k_l)}_{\psi\psi} \\
0 & 0 & 0 & 0 & 0 & 0 \\
0 & 0 & 0 & 0 & 0 & 0 \\
0 & 0 & 0 & 0 & 0 & 0 \\
0 & 0 & 0 & 0 & 0 & 0 \\
0 & 0 & 0 & 0 & 0 & 0 \\
0 & 0 & 0 & 0 & 0 & 0
\end{array}\right\} \tag{7.34}
$$

where

$$
\begin{aligned}
&\mathrm{k}^{(k)(k_l)}_{\mathbf{uu}(1)(1)} = \boldsymbol{\Omega}^{(k)(k_l)}_{\mathbf{uu}(1)(1)} + \frac{S^{(k)}_{\mathbf{uu}(1)(1)}}{N_{(k)}}, \quad \mathrm{k}^{(k)(k_l)}_{\mathbf{uu}(1)(2)} = \boldsymbol{\Omega}^{(k)(k_l)}_{\mathbf{uu}(1)(2)} + \frac{S^{(k)}_{\mathbf{uu}(1)(2)}}{N_{(k)}} \\
&\mathrm{k}^{(k)(k_l)}_{\mathbf{u}\phi(1)(1)} = \boldsymbol{\Omega}^{(k)(k_l)}_{\mathbf{u}\phi(1)(1)} + \frac{S^{(k)}_{\mathbf{u}\phi(1)(1)}}{N_{(k)}}, \quad \mathrm{k}^{(k)(k_l)}_{\mathbf{u}\varphi(1)(1)} = \boldsymbol{\Omega}^{(k)(k_l)}_{\mathbf{u}\varphi(1)(1)} + \frac{S^{(k)}_{\mathbf{u}\phi(1)(1)}}{N_{(k)}} \\
&\mathrm{k}^{(k)(k_l)}_{\mathbf{u}\psi(1)(1)} = \boldsymbol{\Omega}^{(k)(k_l)}_{\mathbf{u}\psi(1)(1)} + \frac{S^{(k)}_{\mathbf{u}\psi(1)(1)}}{N_{(k)}}, \widehat{\mathrm{k}}^{(k)(k_l)}_{\mathbf{uu}(1)(1)} = -\boldsymbol{\Omega}^{(k)(k_l)}_{\mathbf{uu}(1)(1)}, \widehat{\mathrm{k}}^{(k)(k_l)}_{\mathbf{uu}(1)(2)} = -\boldsymbol{\Omega}^{(k)(k_l)}_{\mathbf{uu}(1)(2)} \\
&\widehat{\mathrm{k}}^{(k)(k_l)}_{\mathbf{u}\phi(1)(1)} = -\boldsymbol{\Omega}^{(k)(k_l)}_{\mathbf{u}\phi(1)(1)}, \widehat{\mathrm{k}}^{(k)(k_l)}_{\mathbf{u}\varphi(1)(1)} = -\boldsymbol{\Omega}^{(k)(k_l)}_{\mathbf{u}\varphi(1)(1)}, \widehat{\mathrm{k}}^{(k)(k_l)}_{\mathbf{u}\psi(1)(1)} = -\boldsymbol{\Omega}^{(k)(k_l)}_{\mathbf{u}\psi(1)(1)}
\end{aligned} \tag{7.35a}
$$

$$
\begin{aligned}
&\mathrm{k}^{(k)(k_l)}_{\mathbf{uu}(2)(1)} = \boldsymbol{\Omega}^{(k)(k_l)}_{\mathbf{uu}(2)(1)} + \frac{S^{(k)}_{\mathbf{uu}(2)(1)}}{N_{(k)}}, \quad \mathrm{k}^{(k)(k_l)}_{\mathbf{uu}(2)(2)} = \boldsymbol{\Omega}^{(k)(k_l)}_{\mathbf{uu}(2)(2)} + \frac{S^{(k)}_{\mathbf{uu}(2)(2)}}{N_{(k)}} \\
&\mathrm{k}^{(k)(k_l)}_{\mathbf{u}\phi(2)(1)} = \boldsymbol{\Omega}^{(k)(k_l)}_{\mathbf{u}\phi(2)(1)} + \frac{S^{(k)}_{\mathbf{u}\phi(2)(1)}}{N_{(k)}}, \quad \mathrm{k}^{(k)(k_l)}_{\mathbf{u}\varphi(2)(1)} = \boldsymbol{\Omega}^{(k)(k_l)}_{\mathbf{u}\varphi(2)(1)} + \frac{S^{(k)}_{\mathbf{u}\varphi(2)(1)}}{N_{(k)}} \\
&\mathrm{k}^{(k)(k_l)}_{\mathbf{u}\psi(2)(1)} = \boldsymbol{\Omega}^{(k)(k_l)}_{\mathbf{u}\psi(2)(1)} + \frac{S^{(k)}_{\mathbf{u}\psi(2)(1)}}{N_{(k)}}, \widehat{\mathrm{k}}^{(k)(k_l)}_{\mathbf{uu}(2)(1)} = -\boldsymbol{\Omega}^{(k)(k_l)}_{\mathbf{uu}(2)(1)}, \widehat{\mathrm{k}}^{(k)(k_l)}_{\mathbf{uu}(2)(2)} = -\boldsymbol{\Omega}^{(k)(k_l)}_{\mathbf{uu}(2)(2)} \\
&\widehat{\mathrm{k}}^{(k)(k_l)}_{\mathbf{u}\phi(2)(1)} = -\boldsymbol{\Omega}^{(k)(k_l)}_{\mathbf{u}\phi(2)(1)}, \widehat{\mathrm{k}}^{(k)(k_l)}_{\mathbf{u}\varphi(2)(1)} = -\boldsymbol{\Omega}^{(k)(k_l)}_{\mathbf{u}\varphi(2)(1)}, \widehat{\mathrm{k}}^{(k)(k_l)}_{\mathbf{u}\psi(2)(1)} = -\boldsymbol{\Omega}^{(k)(k_l)}_{\mathbf{u}\psi(2)(1)}
\end{aligned} \tag{7.35b}
$$

$$
\begin{aligned}
&\mathrm{k}_{\phi\mathbf{u}(1)(1)}^{(k)(k_l)} = \Phi_{\phi\mathbf{u}(1)(1)}^{(k)(k_l)} + \frac{C_{\phi\mathbf{u}(1)(1)}^{(k)}}{N_{(k)}}, \quad \mathrm{k}_{\phi\mathbf{u}(1)(2)}^{(k)(k_l)} = \Phi_{\phi\mathbf{u}(1)(2)}^{(k)(k_l)} + \frac{C_{\phi\mathbf{u}(1)(2)}^{(k)}}{N_{(k)}} \\
&\mathrm{k}_{\varphi\mathbf{u}(1)(1)}^{(k)(k_l)} = \Phi_{\varphi\mathbf{u}(1)(1)}^{(k)(k_l)} + \frac{C_{\varphi\mathbf{u}(1)(1)}^{(k)}}{N_{(k)}}, \quad \mathrm{k}_{\varphi\mathbf{u}(1)(2)}^{(k)(k_l)} = \Phi_{\varphi\mathbf{u}(1)(2)}^{(k)(k_l)} + \frac{C_{\varphi\mathbf{u}(1)(2)}^{(k)}}{N_{(k)}} \\
&\mathrm{k}_{\psi\mathbf{u}(1)(1)}^{(k)(k_l)} = \Phi_{\psi\mathbf{u}(1)(1)}^{(k)(k_l)} + \frac{C_{\psi\mathbf{u}(1)(1)}^{(k)}}{N_{(k)}}, \mathrm{k}_{\psi\mathbf{u}(1)(2)}^{(k)(k_l)} = \Phi_{\psi\mathbf{u}(1)(2)}^{(k)(k_l)} + \frac{C_{\psi\mathbf{u}(1)(2)}^{(k)}}{N_{(k)}} \\
&\widehat{\mathrm{k}}_{\phi\mathbf{u}(1)(1)}^{(k)(k_l)} = -\Phi_{\phi\mathbf{u}(1)(1)}^{(k)(k_l)}, \quad \widehat{\mathrm{k}}_{\phi\mathbf{u}(1)(2)}^{(k)(k_l)} = -\Phi_{\phi\mathbf{u}(1)(2)}^{(k)(k_l)}, \quad \widehat{\mathrm{k}}_{\varphi\mathbf{u}(1)(1)}^{(k)(k_l)} = -\Phi_{\varphi\mathbf{u}(1)(1)}^{(k)(k_l)} \\
&\widehat{\mathrm{k}}_{\varphi\mathbf{u}(1)(2)}^{(k)(k_l)} = -\Phi_{\varphi\mathbf{u}(1)(2)}^{(k)(k_l)}, \quad \widehat{\mathrm{k}}_{\psi\mathbf{u}(1)(1)}^{(k)(k_l)} = -\Phi_{\psi\mathbf{u}(1)(1)}^{(k)(k_l)}, \quad \widehat{\mathrm{k}}_{\psi\mathbf{u}(1)(2)}^{(k)(k_l)} = -\Phi_{\psi\mathbf{u}(1)(2)}^{(k)(k_l)}
\end{aligned}
\tag{7.35c}
$$

and

$$
\begin{aligned}
&\mathrm{k}_{\phi\phi}^{(k)(k_l)} = \Omega_{\phi\phi}^{(k)(k_l)} + \frac{S_{\phi\phi}^{(k)}}{N_{(k)}}, \quad \mathrm{k}_{\phi\varphi}^{(k)(k_l)} = \Omega_{\phi\varphi}^{(k)(k_l)} + \frac{S_{\phi\varphi}^{(k)}}{N_{(k)}} \\
&\mathrm{k}_{\phi\psi}^{(k)(k_l)} = \Omega_{\phi\psi}^{(k)(k_l)} + \frac{S_{\phi\psi}^{(k)}}{N_{(k)}}, \quad \mathrm{k}_{\varphi\phi}^{(k)(k_l)} = \Omega_{\varphi\phi}^{(k)(k_l)} + \frac{S_{\varphi\phi}^{(k)}}{N_{(k)}} \\
&\mathrm{k}_{\varphi\varphi}^{(k)(k_l)} = \Omega_{\varphi\varphi}^{(k)(k_l)} + \frac{S_{\varphi\varphi}^{(k)}}{N_{(k)}}, \quad \mathrm{k}_{\varphi\psi}^{(k)(k_l)} = \Omega_{\varphi\psi}^{(k)(k_l)} + \frac{S_{\varphi\psi}^{(k)}}{N_{(k)}} \\
&\mathrm{k}_{\psi\phi}^{(k)(k_l)} = \Omega_{\psi\phi}^{(k)(k_l)} + \frac{S_{\psi\phi}^{(k)}}{N_{(k)}}, \quad \mathrm{k}_{\psi\varphi}^{(k)(k_l)} = \Omega_{\psi\varphi}^{(k)(k_l)} + \frac{S_{\psi\varphi}^{(k)}}{N_{(k)}} \\
&\mathrm{k}_{\psi\psi}^{(k)(k_l)} = \Omega_{\psi\psi}^{(k)(k_l)} + \frac{S_{\psi\psi}^{(k)}}{N_{(k)}}, \quad \widehat{\mathrm{k}}_{\phi\phi}^{(k)(k_l)} = -\Omega_{\phi\phi}^{(k)(k_l)}, \quad \widehat{\mathrm{k}}_{\phi\varphi}^{(k)(k_l)} = -\Omega_{\phi\varphi}^{(k)(k_l)} \\
&\widehat{\mathrm{k}}_{\phi\psi}^{(k)(k_l)} = -\Omega_{\phi\psi}^{(k)(k_l)}, \quad \widehat{\mathrm{k}}_{\varphi\phi}^{(k)(k_l)} = -\Omega_{\varphi\phi}^{(k)(k_l)}, \quad \widehat{\mathrm{k}}_{\varphi\varphi}^{(k)(k_l)} = -\Omega_{\varphi\varphi}^{(k)(k_l)}, \quad \widehat{\mathrm{k}}_{\varphi\psi}^{(k)(k_l)} = -\Omega_{\varphi\psi}^{(k)(k_l)} \\
&\widehat{\mathrm{k}}_{\psi\phi}^{(k)(k_l)} = -\Omega_{\psi\phi}^{(k)(k_l)}, \quad \widehat{\mathrm{k}}_{\psi\varphi}^{(k)(k_l)} = -\Omega_{\psi\varphi}^{(k)(k_l)}, \quad \widehat{\mathrm{k}}_{\psi\psi}^{(k)(k_l)} = \Omega_{\psi\psi}^{(k)(k_l)}
\end{aligned}
\tag{7.35d}
$$

The vector $\mathbf{F}_{\mathscr{B}(k)}^{\mathrm{PDDO(Stiff)}}$, at node k, is constructed using a network of elements around node, k, shown in red in Fig. 7.4 associated with the stiffness matrix, $\mathbf{k}_{(k)(k_l)}^{PDDO}$ defined in Eq. (7.34).

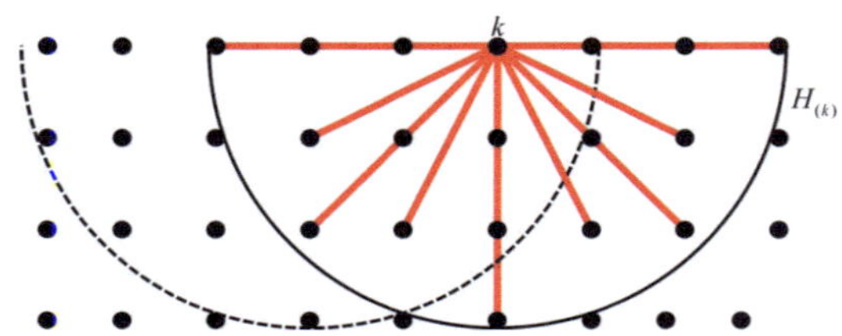

Fig. 7.4 Element connectivity for a PDDO interaction of a point k with its family members in the boundary region $\mathscr{B}$

7.6.3 Global Form of Coupled Equations

Applying Eq. (7.18) at all points $k = 1, \ldots, N_{\mathscr{R}}$ in region $\mathscr{R}$ and Eq. (7.32) at all points $k = 1, \ldots, N_{\mathscr{B}}$ in region $\mathscr{B}$ results in a system of equations as

$$\mathbf{M}\ddot{\mathbf{U}} + \mathbf{C}\dot{\mathbf{U}} + \mathbf{K}\mathbf{U} = \mathbf{F} \tag{7.36}$$

where the unknown vector $\mathbf{U}$ is defined as

$$\mathbf{U} = \left\{ \mathbf{u}_{(1)} \ \phi_{(1)} \ \varphi_{(1)} \ \psi_{(1)} \ \cdot \ \mathbf{u}_{(k)} \ \phi_{(k)} \ \varphi_{(k)} \ \psi_{(k)} \ \cdot \ \mathbf{u}_{(N)} \ \phi_{(N)} \ \varphi_{(N)} \ \psi_{(N)} \right\}^T \tag{7.37a}$$

Its time derivatives become

$$\dot{\mathbf{U}} = \left\{ \dot{\mathbf{u}}_{(1)} \ \dot{\phi}_{(1)} \ \dot{\varphi}_{(1)} \ \dot{\psi}_{(1)} \ \cdot \ \dot{\mathbf{u}}_{(k)} \ \dot{\phi}_{(k)} \ \dot{\varphi}_{(k)} \ \dot{\psi}_{(k)} \ \cdot \ \dot{\mathbf{u}}_{(N)} \ \dot{\phi}_{(N)} \ \dot{\varphi}_{(N)} \ \dot{\psi}_{(N)} \right\}^T \tag{7.37b}$$

and

$$\ddot{\mathbf{U}} = \left\{ \ddot{\mathbf{u}}_{(1)} \ \ddot{\phi}_{(1)} \ \ddot{\varphi}_{(1)} \ \ddot{\psi}_{(1)} \ \cdot \ \ddot{\mathbf{u}}_{(k)} \ \ddot{\phi}_{(k)} \ \ddot{\varphi}_{(k)} \ \ddot{\psi}_{(k)} \ \cdot \ \ddot{\mathbf{u}}_{(N)} \ \ddot{\phi}_{(N)} \ \ddot{\varphi}_{(N)} \ \ddot{\psi}_{(N)} \right\}^T \tag{7.37c}$$

The external forcing function consisting of the volumetric field generation and body loads and their rates is of the form

$$\mathbf{F} = \mathbf{F}_B + \mathbf{F}_R \tag{7.38}$$

where

$$\mathbf{F}_B = \left\{ s_{\mathbf{u}(1)} \ s_{\phi(1)} \ s_{\varphi(1)} \ s_{\psi(1)} \ \cdot \ s_{\mathbf{u}(k)} \ s_{\phi(k)} \ s_{\varphi(k)} \ s_{\psi(k)} \ \cdot \ s_{\mathbf{u}(N)} \ s_{\phi(N)} \ s_{\varphi(N)} \ s_{\psi(N)} \right\}^T \tag{7.39a}$$

and

$$\mathbf{F}_R = \left\{ \mathbf{0} \ \dot{s}_{\phi(1)} \ \dot{s}_{\varphi(1)} \ \dot{s}_{\psi(1)} \ \cdot \ \mathbf{0} \ \dot{s}_{\phi(k)} \ \dot{s}_{\varphi(k)} \ \dot{s}_{\psi(k)} \ \cdot \ \mathbf{0} \ \dot{s}_{\phi(N)} \ \dot{s}_{\varphi(N)} \ \dot{s}_{\psi(N)} \right\}^T \tag{7.39b}$$

The unknown vectors of $\mathbf{U}$, $\dot{\mathbf{U}}$ and $\ddot{\mathbf{U}}$ corresponding to the structural and field degrees of freedom in Eq. (7.36) can be grouped as $\mathbf{V}_{\mathscr{R}}$ and $\mathbf{V}_{\mathscr{B}}$ for the points in region $\mathscr{R}$ and boundary $\mathscr{B}$, respectively, along with their respective time derivatives. The inertial term in Eq. (7.36) can be written as

$$\mathbf{M}\ddot{\mathbf{U}} = \begin{bmatrix} \mathbf{M}_{\mathscr{R}\mathscr{R}}^{\mathbf{uu}} & \mathbf{0} & \mathbf{0} & \mathbf{0} & [\mathbf{0}]_{1\times 4} \\ \mathbf{0} & \mathbf{M}_{\mathscr{R}\mathscr{R}}^{\phi\phi} & \mathbf{M}_{\mathscr{R}\mathscr{R}}^{\phi\varphi} & \mathbf{M}_{\mathscr{R}\mathscr{R}}^{\phi\psi} & [\mathbf{0}]_{1\times 4} \\ \mathbf{0} & \mathbf{M}_{\mathscr{R}\mathscr{R}}^{\varphi\phi} & \mathbf{M}_{\mathscr{R}\mathscr{R}}^{\varphi\varphi} & \mathbf{M}_{\mathscr{R}\mathscr{R}}^{\varphi\psi} & [\mathbf{0}]_{1\times 4} \\ \mathbf{0} & \mathbf{M}_{\mathscr{R}\mathscr{R}}^{\psi\phi} & \mathbf{M}_{\mathscr{R}\mathscr{R}}^{\psi\varphi} & \mathbf{M}_{\mathscr{R}\mathscr{R}}^{\psi\psi} & [\mathbf{0}]_{1\times 4} \\ [\mathbf{0}]_{4\times 1} & [\mathbf{0}]_{4\times 1} & [\mathbf{0}]_{4\times 1} & [\mathbf{0}]_{4\times 1} & [\mathbf{0}]_{4\times 4} \end{bmatrix} \begin{Bmatrix} \ddot{\mathbf{V}}_{\mathbf{u}\mathscr{R}} \\ \ddot{\mathbf{V}}_{\phi\mathscr{R}} \\ \ddot{\mathbf{V}}_{\varphi\mathscr{R}} \\ \ddot{\mathbf{V}}_{\psi\mathscr{R}} \\ \ddot{\mathbf{V}}_{\mathbf{u}\mathscr{B}} \\ \ddot{\mathbf{V}}_{\phi\mathscr{B}} \\ \ddot{\mathbf{V}}_{\varphi\mathscr{B}} \\ \ddot{\mathbf{V}}_{\psi\mathscr{B}} \end{Bmatrix} \tag{7.40}$$

where the submatrix $\mathbf{M}_{\mathscr{R}\mathscr{R}}^{\mathbf{uu}}$ with dimension $2N_{\mathscr{R}} \times 2N_{\mathscr{R}}$ represents the diagonal mass matrix corresponding to vector $\ddot{\mathbf{V}}_{\mathbf{u}\mathscr{R}}$, while the submatrices $\mathbf{M}_{\mathscr{R}\mathscr{R}}^{\phi\phi}$, $\mathbf{M}_{\mathscr{R}\mathscr{R}}^{\varphi\varphi}$, and $\mathbf{M}_{\mathscr{R}\mathscr{R}}^{\psi\psi}$ with dimension $N_{\mathscr{R}} \times 1$ represent the field inertial matrices corresponding to vectors $\ddot{\mathbf{V}}_{\phi\mathscr{R}}$, $\ddot{\mathbf{V}}_{\phi\mathscr{R}}$, and $\ddot{\mathbf{V}}_{\phi\mathscr{R}}$, respectively. The submatrices $\mathbf{M}_{\mathscr{R}\mathscr{R}}^{\phi\varphi}$ and $\mathbf{M}_{\mathscr{R}\mathscr{R}}^{\phi\psi}$, $\mathbf{M}_{\mathscr{R}\mathscr{R}}^{\varphi\phi}$ and $\mathbf{M}_{\mathscr{R}\mathscr{R}}^{\varphi\psi}$, $\mathbf{M}_{\mathscr{R}\mathscr{R}}^{\psi\phi}$ and $\mathbf{M}_{\mathscr{R}\mathscr{R}}^{\psi\varphi}$ each of which with dimension $N_{\mathscr{R}} \times 1$ represents the field inertial coupling matrices between the unknown vectors of $\ddot{\mathbf{V}}_{\phi\mathscr{R}}$, $\ddot{\mathbf{V}}_{\phi\mathscr{R}}$, and $\ddot{\mathbf{V}}_{\phi\mathscr{R}}$.

The damping term in Eq. (7.36) can be written in the form

$$\mathbf{C}\dot{\mathbf{U}} = \begin{bmatrix} \mathbf{C}_{\mathscr{R}\mathscr{R}}^{\mathbf{uu}} & \mathbf{0} & \mathbf{0} & \mathbf{0} & [\mathbf{0}]_{1\times 4} \\ \mathbf{0} & \mathbf{C}_{\mathscr{R}\mathscr{R}}^{\phi\phi} & \mathbf{C}_{\mathscr{R}\mathscr{R}}^{\phi\varphi} & \mathbf{C}_{\mathscr{R}\mathscr{R}}^{\phi\psi} & [\mathbf{0}]_{1\times 4} \\ \mathbf{0} & \mathbf{C}_{\mathscr{R}\mathscr{R}}^{\varphi\phi} & \mathbf{C}_{\mathscr{R}\mathscr{R}}^{\varphi\varphi} & \mathbf{C}_{\mathscr{R}\mathscr{R}}^{\varphi\psi} & [\mathbf{0}]_{1\times 4} \\ \mathbf{0} & \mathbf{C}_{\mathscr{R}\mathscr{R}}^{\psi\phi} & \mathbf{C}_{\mathscr{R}\mathscr{R}}^{\psi\varphi} & \mathbf{C}_{\mathscr{R}\mathscr{R}}^{\psi\psi} & [\mathbf{0}]_{1\times 4} \\ [\mathbf{0}]_{4\times 1} & [\mathbf{0}]_{4\times 1} & [\mathbf{0}]_{4\times 1} & [\mathbf{0}]_{4\times 1} & [\mathbf{0}]_{4\times 4} \end{bmatrix} \begin{Bmatrix} \dot{\mathbf{V}}_{\mathbf{u}\mathscr{R}} \\ \dot{\mathbf{V}}_{\phi\mathscr{R}} \\ \dot{\mathbf{V}}_{\varphi\mathscr{R}} \\ \dot{\mathbf{V}}_{\psi\mathscr{R}} \\ \dot{\mathbf{V}}_{\mathbf{u}\mathscr{B}} \\ \dot{\mathbf{V}}_{\phi\mathscr{B}} \\ \dot{\mathbf{V}}_{\varphi\mathscr{B}} \\ \dot{\mathbf{V}}_{\psi\mathscr{B}} \end{Bmatrix} \tag{7.41}$$

where the submatrix $\mathbf{C}_{\mathscr{R}\mathscr{R}}^{\mathbf{uu}}$ with dimension $2N_{\mathscr{R}} \times 2N_{\mathscr{R}}$ represents the diagonal structural damping matrix corresponding to vector $\dot{\mathbf{V}}_{\mathbf{u}\mathscr{R}}$, while the submatrices $\mathbf{C}_{\mathscr{R}\mathscr{R}}^{\phi\phi}$, $\mathbf{C}_{\mathscr{R}\mathscr{R}}^{\varphi\varphi}$, and $\mathbf{C}_{\mathscr{R}\mathscr{R}}^{\psi\psi}$ with dimension $N_{\mathscr{R}} \times 1$ represent the field damping matrices corresponding to vectors $\dot{\mathbf{V}}_{\phi\mathscr{R}}$, $\dot{\mathbf{V}}_{\phi\mathscr{R}}$, and $\dot{\mathbf{V}}_{\phi\mathscr{R}}$, respectively. The submatrices $\mathbf{C}_{\mathscr{R}\mathscr{R}}^{\phi\varphi}$ and $\mathbf{C}_{\mathscr{R}\mathscr{R}}^{\phi\psi}$, $\mathbf{C}_{\mathscr{R}\mathscr{R}}^{\varphi\phi}$ and $\mathbf{C}_{\mathscr{R}\mathscr{R}}^{\varphi\psi}$, $\mathbf{C}_{\mathscr{R}\mathscr{R}}^{\psi\phi}$ and $\mathbf{C}_{\mathscr{R}\mathscr{R}}^{\psi\varphi}$ each with dimension $N_{\mathscr{R}} \times 1$ represent the field damping coupling matrices between the unknown vectors of $\dot{\mathbf{V}}_{\phi\mathscr{R}}$, $\dot{\mathbf{V}}_{\phi\mathscr{R}}$, and $\dot{\mathbf{V}}_{\phi\mathscr{R}}$.

The stiffness term in Eq. (7.36) can be written as

$$\mathbf{KU}=\begin{bmatrix}
\mathbf{K}^{\mathbf{uu}}_{\mathscr{R}\mathscr{R}} & \mathbf{K}^{\mathbf{u}\phi}_{\mathscr{R}\mathscr{R}} & \mathbf{K}^{\mathbf{u}\varphi}_{\mathscr{R}\mathscr{R}} & \mathbf{K}^{\mathbf{u}\psi}_{\mathscr{R}\mathscr{R}} & \mathbf{K}^{\mathbf{uu}}_{\mathscr{R}\mathscr{B}} & \mathbf{K}^{\mathbf{u}\phi}_{\mathscr{R}\mathscr{B}} & \mathbf{K}^{\mathbf{u}\varphi}_{\mathscr{R}\mathscr{B}} & \mathbf{K}^{\mathbf{u}\psi}_{\mathscr{R}\mathscr{B}} \\
\mathbf{K}^{\phi\mathbf{u}}_{\mathscr{R}\mathscr{R}} & \mathbf{K}^{\phi\phi}_{\mathscr{R}\mathscr{R}} & \mathbf{K}^{\phi\varphi}_{\mathscr{R}\mathscr{R}} & \mathbf{K}^{\phi\psi}_{\mathscr{R}\mathscr{R}} & \mathbf{K}^{\phi\mathbf{u}}_{\mathscr{R}\mathscr{B}} & \mathbf{K}^{\phi\phi}_{\mathscr{R}\mathscr{B}} & \mathbf{K}^{\phi\varphi}_{\mathscr{R}\mathscr{B}} & \mathbf{K}^{\phi\psi}_{\mathscr{R}\mathscr{B}} \\
\mathbf{K}^{\varphi\mathbf{u}}_{\mathscr{R}\mathscr{R}} & \mathbf{K}^{\varphi\phi}_{\mathscr{R}\mathscr{R}} & \mathbf{K}^{\varphi\varphi}_{\mathscr{R}\mathscr{R}} & \mathbf{K}^{\varphi\psi}_{\mathscr{R}\mathscr{R}} & \mathbf{K}^{\varphi\mathbf{u}}_{\mathscr{R}\mathscr{B}} & \mathbf{K}^{\varphi\phi}_{\mathscr{R}\mathscr{B}} & \mathbf{K}^{\varphi\varphi}_{\mathscr{R}\mathscr{B}} & \mathbf{K}^{\varphi\psi}_{\mathscr{R}\mathscr{B}} \\
\mathbf{K}^{\psi\mathbf{u}}_{\mathscr{R}\mathscr{R}} & \mathbf{K}^{\psi\phi}_{\mathscr{R}\mathscr{R}} & \mathbf{K}^{\psi\varphi}_{\mathscr{R}\mathscr{R}} & \mathbf{K}^{\psi\psi}_{\mathscr{R}\mathscr{R}} & \mathbf{K}^{\psi\mathbf{u}}_{\mathscr{R}\mathscr{B}} & \mathbf{K}^{\psi\phi}_{\mathscr{R}\mathscr{B}} & \mathbf{K}^{\psi\varphi}_{\mathscr{R}\mathscr{B}} & \mathbf{K}^{\psi\psi}_{\mathscr{R}\mathscr{B}} \\
\mathbf{K}^{\mathbf{uu}}_{\mathscr{B}\mathscr{R}} & \mathbf{K}^{\mathbf{u}\phi}_{\mathscr{B}\mathscr{R}} & \mathbf{K}^{\mathbf{u}\varphi}_{\mathscr{B}\mathscr{R}} & \mathbf{K}^{\mathbf{u}\psi}_{\mathscr{B}\mathscr{R}} & \mathbf{K}^{\mathbf{uu}}_{\mathscr{B}\mathscr{B}} & \mathbf{K}^{\mathbf{u}\phi}_{\mathscr{B}\mathscr{B}} & \mathbf{K}^{\mathbf{u}\varphi}_{\mathscr{B}\mathscr{B}} & \mathbf{K}^{\mathbf{u}\psi}_{\mathscr{B}\mathscr{B}} \\
\mathbf{K}^{\phi\mathbf{u}}_{\mathscr{B}\mathscr{R}} & \mathbf{K}^{\phi\phi}_{\mathscr{B}\mathscr{R}} & \mathbf{K}^{\phi\varphi}_{\mathscr{B}\mathscr{R}} & \mathbf{K}^{\phi\psi}_{\mathscr{B}\mathscr{R}} & \mathbf{K}^{\phi\mathbf{u}}_{\mathscr{B}\mathscr{B}} & \mathbf{K}^{\phi\phi}_{\mathscr{B}\mathscr{B}} & \mathbf{K}^{\phi\varphi}_{\mathscr{B}\mathscr{B}} & \mathbf{K}^{\phi\psi}_{\mathscr{B}\mathscr{B}} \\
\mathbf{K}^{\varphi\mathbf{u}}_{\mathscr{B}\mathscr{R}} & \mathbf{K}^{\varphi\phi}_{\mathscr{B}\mathscr{R}} & \mathbf{K}^{\varphi\varphi}_{\mathscr{B}\mathscr{R}} & \mathbf{K}^{\varphi\psi}_{\mathscr{B}\mathscr{R}} & \mathbf{K}^{\varphi\mathbf{u}}_{\mathscr{B}\mathscr{B}} & \mathbf{K}^{\varphi\phi}_{\mathscr{B}\mathscr{B}} & \mathbf{K}^{\varphi\varphi}_{\mathscr{B}\mathscr{B}} & \mathbf{K}^{\varphi\psi}_{\mathscr{B}\mathscr{B}} \\
\mathbf{K}^{\psi\mathbf{u}}_{\mathscr{B}\mathscr{R}} & \mathbf{K}^{\psi\phi}_{\mathscr{B}\mathscr{R}} & \mathbf{K}^{\psi\varphi}_{\mathscr{B}\mathscr{R}} & \mathbf{K}^{\psi\psi}_{\mathscr{B}\mathscr{R}} & \mathbf{K}^{\psi\mathbf{u}}_{\mathscr{B}\mathscr{B}} & \mathbf{K}^{\psi\phi}_{\mathscr{B}\mathscr{B}} & \mathbf{K}^{\psi\varphi}_{\mathscr{B}\mathscr{B}} & \mathbf{K}^{\psi\psi}_{\mathscr{B}\mathscr{B}}
\end{bmatrix}
\begin{Bmatrix}
\mathbf{V}_{\mathbf{u}\mathscr{R}} \\ \mathbf{V}_{\phi\mathscr{R}} \\ \mathbf{V}_{\varphi\mathscr{R}} \\ \mathbf{V}_{\psi\mathscr{R}} \\ \mathbf{V}_{\mathbf{u}\mathscr{B}} \\ \mathbf{V}_{\phi\mathscr{B}} \\ \mathbf{V}_{\varphi\mathscr{B}} \\ \mathbf{V}_{\psi\mathscr{B}}
\end{Bmatrix} \tag{7.42}$$

If the point of interest and its family member are both in region $\mathscr{R}$, the submatrix $\mathbf{K}^{\mathbf{uu}}_{\mathscr{R}\mathscr{R}}$ with dimension $2N_{\mathscr{R}} \times 2N_{\mathscr{R}}$ represents the structural self-stiffness matrix corresponding to the unknown vector of $\mathbf{V}_{\mathbf{u}\mathscr{R}}$, while the submatrices, $\mathbf{K}^{\alpha\alpha}_{\mathscr{R}\mathscr{R}}$ for $\alpha = \phi$, φ, ψ each with dimension $N_{\mathscr{R}} \times 1$ represent the field self-stiffness matrices corresponding to the unknown vectors of $\mathbf{V}_{\alpha\mathscr{R}}$. The submatrices $\mathbf{K}^{\mathbf{u}\alpha}_{\mathscr{R}\mathscr{R}}$ for $\alpha = \phi$, φ, ψ and interchangeable superscripts represent the coupled structural-field stiffness matrices each with dimension $2N_{\mathscr{R}} \times 1$, respectively, corresponding to the unknown vectors of $\mathbf{V}_{\alpha\mathscr{R}}$. The submatrices $\mathbf{K}^{\alpha\beta}_{\mathscr{R}\mathscr{R}}$ for $\alpha = \phi$, φ, ψ also with interchangeable superscripts represent the coupled field stiffness matrices each with dimension $N_{\mathscr{R}} \times 1$, respectively, between the unknown vectors of $\mathbf{V}_{\alpha\mathscr{R}}$.

If the point of interest and its family member are both in the boundary region $\mathscr{B}$, the submatrix $\mathbf{K}^{\mathbf{uu}}_{\mathscr{B}\mathscr{B}}$ with dimension $2N_{\mathscr{B}} \times 2N_{\mathscr{B}}$ represents the structural self-stiffness matrix corresponding to the unknown vector of $\mathbf{V}_{\mathbf{u}\mathscr{B}}$, while the submatrices $\mathbf{K}^{\alpha\alpha}_{\mathscr{B}\mathscr{B}}$ for $\alpha = \phi$, φ, ψ each with dimension $N_{\mathscr{B}} \times 1$, respectively, represent the field self-stiffness matrices corresponding to the unknown vectors of $\mathbf{V}_{\alpha\mathscr{B}}$. The submatrices, $\mathbf{K}^{\mathbf{u}\alpha}_{\mathscr{B}\mathscr{B}}$ for $\alpha = \phi$, φ, ψ represent the coupled structural-field stiffness matrices each with dimension $2N_{\mathscr{B}} \times 1$, respectively, corresponding to the unknown vectors of $\mathbf{V}_{\alpha\mathscr{B}}$. The submatrices, $\mathbf{K}^{\alpha\beta}_{\mathscr{B}\mathscr{B}}$ for α, $\beta = \phi$, φ, ψ with interchangeable superscripts represent the coupled field stiffness matrices each with dimension $N_{\mathscr{B}} \times 1$, respectively, between the unknown vectors of $\mathbf{V}_{\alpha\mathscr{B}}$.

If the point of interest is in region $\mathscr{R}$ and the family member is in region $\mathscr{B}$, the submatrix $\mathbf{K}^{\mathbf{uu}}_{\mathscr{R}\mathscr{B}}$ with dimension $2N_{\mathscr{R}} \times 2N_{\mathscr{R}}$ represents the structural stiffness matrix corresponding to vector of $\mathbf{V}_{\mathbf{u}\mathscr{B}}$. The submatrices $\mathbf{K}^{\mathbf{u}\alpha}_{\mathscr{R}\mathscr{B}}$ for $\alpha = \phi$, φ, ψ with interchangeable superscripts represent the structural-field coupling stiffness matrices each with dimension $2N_{\mathscr{R}} \times 1$, respectively, between the unknown vectors of $\mathbf{V}_{\alpha\mathscr{B}}$. The submatrices, $\mathbf{K}^{\alpha\beta}_{\mathscr{R}\mathscr{B}}$ for α, $\beta = \phi$, φ, ψ with interchangeable superscripts, represent the field coupling stiffness matrices each with dimension $N_{\mathscr{R}} \times 1$, respectively, between the unknown vectors of $\mathbf{V}_{\alpha\mathscr{B}}$.

If the point of interest is in region $\mathscr{B}$ and the family member is in region $\mathscr{R}$, the submatrix, $\mathbf{K}^{\mathbf{uu}}_{\mathscr{B}\mathscr{R}}$ with dimension $2N_{\mathscr{B}} \times 2N_{\mathscr{B}}$, represents the structural stiffness matrix corresponding to the vector of $\mathbf{V}_{\mathbf{u}\mathscr{R}}$. The submatrices $\mathbf{K}^{\mathbf{u}\alpha}_{\mathscr{B}\mathscr{R}}$ for $\alpha = \phi, \varphi, \psi$ with interchangeable superscripts represent the structural-field coupling stiffness matrices each with dimension $2N_{\mathscr{B}} \times 1$, respectively, between the unknown vectors of $\mathbf{V}_{\alpha\mathscr{R}}$. The submatrices $\mathbf{K}^{\alpha\beta}_{\mathscr{R}\mathscr{B}}$ for $\alpha, \beta = \phi, \varphi, \psi$ with interchangeable superscripts represent the field coupling stiffness matrices each with dimension $N_{\mathscr{B}} \times 1$, respectively, between the unknown vectors of $\mathbf{V}_{\alpha\mathscr{R}}$.

References

Agwai, A. (2011). *A peridynamic approach for coupled fields* (Ph.D. dissertation). University of Arizona.

Anicode, S. V. K., & Madenci, E. (2023). Peridynamic modeling of dual-phase-lag thermal-moisture coupling in a finite element framework. *Engineering with Computers, 39*, 911–923.

Anicode, S. V. K., Madenci, E., & Phan, N. (2023). A unified method to simulate electrodeposition and galvanic corrosion using the peridynamic differential operator. *Computer Methods in Applied Mechanics and Engineering, 408*, 115968.

Chen, W., Gu, X., Zhang, Q., & Xia, X. (2021). A refined thermo-mechanical fully coupled peridynamics with application to concrete cracking. *Engineering Fracture Mechanics, 242*, 107463.

Gerstle, W., Silling, S., Read, D., Tewary, V., & Lehoucq, R. (2008). Peridynamic simulation of electromigration. *Comput Mater Continua, 8*, 75–92.

Jabakhanji, R., & Mohtar, R. H. (2015). A peridynamic model of flow in porous media. *Advances in Water Resources, 78*, 22–35.

Jafarzadeh, S., Chen, Z., Zhao, J., & Bobaru, F. (2019). Pitting, lacy covers, and pit merger in stainless steel: 3D peridynamic models. *Corrosion Science, 150*, 17–31.

Katiyar, A., Foster, J. T., Ouchi, H., & Sharma, M. M. (2014). A peridynamic formulation of pressure driven convective fluid transport in porous media. *Journal of Computational Physics, 261*, 209–229.

Kilic, B., & Madenci, E. (2010). An adaptive dynamic relaxation method for quasi-static simulations using the peridynamic theory. *Theoretical and Applied Fracture Mechanics, 53*, 194–204.

Liu, P., Gu, X., Lu, Y., Xia, X., Madenci, E., & Zhang, Q. (2023). Peridynamics for mechanism analysis of soil desiccation cracking: Coupled hygro-mechanical model, staggered and monolithic solution. *Computer Methods in Applied Mechanics and Engineering, 406*, 115896.

Madenci, E., & Oterkus, S. (2016). Peridynamics for coupled field equations. *Handbook of Peridynamic Modeling*, 489–531.

Madenci, E., Barut, A., & Futch, M. (2016). Peridynamic differential operator and its applications. *Computer Methods in Applied Mechanics and Engineering, 304*, 408–451.

Madenci, E., Barut, A., & Dorduncu, M. (2019). *Peridynamic differential operator for numerical analysis*. Springer International Publishing.

Ni, T., Pesavento, F., Zaccariotto, M., Galvanetto, U., Zhu, Q. Z., & Schrefler, B. A. (2020). Hybrid FEM and peridynamic simulation of hydraulic fracture propagation in saturated porous media. *Computer Methods in Applied Mechanics and Engineering, 366*, 113101.

Oterkus, S., & Madenci, E. (2014). Peridynamics for Fully Coupled Thermomechanical Analysis of Fiber Reinforced Laminates. *55th AIAA/ASME/ASCE/AHS/ASC Structures, Structural Dynamics and Materials Conference*, National Harbor, Maryland, Paper No. AIAA 2014-0694

Oterkus, S., & Madenci, E. (2017). Peridynamic modeling of fuel pellet cracking. *Engineering Fracture Mechanics, 176*, 23–37.

Oterkus, S., Fox, J., & Madenci, E. (2013). Simulation of electro-migration through peridynamics. In *IEEE 63rd electronic components and technology conference* (pp. 1488–1493).

Oterkus, S., Madenci, E., & Agwai, A. (2014a). Fully coupled peridynamic thermomechanics. *Journal of the Mechanics and Physics of Solids, 64*, 1–23.

Oterkus S., Madenci E, Oterkus E, Hwang Y, Bae J, Han S. (2014c) Hygro-thermo-mechanical analysis and failure prediction in electronic packages by using peridynamics. In *64th Electronic Components & Technology Conference*, Lake Buena Vista, FL, USA.

Oterkus, S., Madenci, E., & Oterkus, E. (2017). Fully coupled poroelastic peridynamic formulation for fluid-filled fractures. *Engineering geology, 225*, 19–28.

Qin, M., Yang, D., Chen, W., & Yang, S. (2021). Hydraulic fracturing model of a layered rock mass based on peridynamics. *Engineering Fracture Mechanics, 258*, 108088.

Shou, Y., & Zhou, X. (2021). A coupled hydro-mechanical non-ordinary state-based peridynamics for the fissured porous rocks. *Engineering Analysis with Boundary Elements, 123*, 133–146.

Wang, Y. T., & Zhou, X. P. (2019). Peridynamic simulation of thermal failure behaviors in rocks subjected to heating from boreholes. *International Journal of Rock Mechanics and Mining Sciences, 117*, 31–48.

Wu, P., Liu, Y., Peng, X., & Chen, Z. (2022). Peridynamic modeling of freeze-thaw damage in concrete structures. *Mechanics of Advanced Materials and Structures*, 1–12.

Yang, Z., Yang, S. Q., & Tian, W. L. (2021). Peridynamic simulation of fracture mechanical behavior of granite specimen under real-time temperature and post-temperature treatments. *International Journal of Rock Mechanics and Mining Sciences, 138*, 104573.

Zhang, Y., Liu, C., Xiong, X., Li, J., & Madenci, E. (2023a). A coupled thermomechanical peridynamic correspondence model for damage prediction in a freezing rock. *Computers and Geotechnic, 166*, 105995.

Zhang, Y., Behera, D., & Madenci, E. (2023b). Peridynamic modeling of thermal response and cracking in charring materials due to ablation. *Engineering with Computers, 40*, 1–23.

Zhang, Y., Yu, S., & Deng, H. (2023c). Peridynamic model of deformation and failure for rock material under the coupling effect of multi-physical fields. *Theoretical and Applied Fracture Mechanics, 125*, 103912.

Zhang, Y., Madenci, E., Gu, X., & Zhang, Q. (2023d). A coupled hydro-mechanical peridynamic model for unsaturated seepage and crack propagation in unsaturated expansive soils due to moisture change. *Acta Geotechnica, 18*, 6297–6313.

Zhang, Y., Anicode, S. V. K., Fan, X., & Madenci, E. (2023e). Peridynamic modeling of void nucleation and growth in metal lines due to electromigration in a finite element framework. *Computer Methods in Applied Mechanics and Engineering, 414*, 116183.

Zhao, J., Jafarzadeh, S., Rahmani, M., Chen, Z., Kim, Y. R., & Bobaru, F. (2021). A peridynamic model for galvanic corrosion and fracture. *Electrochimica Acta, 391*, 138968.

Zhou, X. P., Wang, Y. T., & Shou, Y. D. (2020). Hydromechanical bond-based peridynamic model for pressurized and fluid-driven fracturing processes in fissured porous rocks. *International Journal of Rock Mechanics and Mining Sciences, 132*, 104383.

Zhou, X. P., Du, E. B., & Wang, Y. T. (2022). Thermo-hydro-chemo-mechanical coupling peridynamic model of fractured rock mass and its application in geothermal extraction. *Computers and Geotechnics, 148*, 104837.

Chapter 8
Ansys Elements for Peridynamic Interactions

8.1 Introduction

This chapter presents the native Ansys elements of COMBIN14, MASS21, MATRIX27, LINK31, LINK33, LINK34, and MASS71 to perform PD analysis. The structural analysis and coupled field analysis require the use of combination of COMBIN14, MASS21, and MATRIX27 elements. The field analysis requires the use of combination of LINK33 and MASS71 elements. The elements LINK31 and LINK34 are used to enforce heat convection and radiation boundary conditions when considering temperature as the field variable. A more detailed description and explanation of these elements can be found in the Ansys help manual (Ansys, 2017).

These elements contain native degrees of freedom (DOF) and allow for the material data inputs through the (Material Property) MP command. For PD analysis, certain DOFs can be tailored and the required input data are specified through the (Real Constant) R command.

8.2 Matrix Form of PD Equilibrium and Field Equations

In order to make the PD form of the equations derived in Chap. 1 for structural and field analysis, Eq. (1.15) and (1.39), consistent with the units in the finite element method, they are multiplied with the volume, $V_{(k)}$ of PD node $\mathbf{x}_{(k)}$ as

$$\rho_{(k)}\ddot{\mathbf{u}}_{(k)}V_{(k)} = \sum_{j\in H_{(k)}} \mathbf{t}_{(k)(j)}V_{(j)}V_{(k)} - \sum_{i\in \widetilde{H}_{(k)}} \mathbf{t}_{(i)(k)}V_{(i)}V_{(k)} + \mathbf{b}_{(k)}V_{(k)} \quad (8.1)$$

and

E. Madenci et al., *Peridynamics and Its Applications Using Ansys*,
https://doi.org/10.1007/978-3-031-59896-8_8

$$C(\mathbf{x}_{(k)},t)V_{(k)}\dot{\Theta}(\mathbf{x}_{(k)},t)+\dot{C}(\mathbf{x}_{(k)},t)V_{(k)}\Theta(\mathbf{x}_{(k)},t)= \\ \sum_{j\in H_{(k)}}h_{(k)(j)}V_{(j)}V_{(k)}-\sum_{i\in\widetilde{H}_{(k)}}h_{(i)(k)}V_{(i)}V_{(k)}+s_b(\mathbf{x}_{(k)},t)V_{(k)} \tag{8.2}$$

These equations can be recast in the matrix form as

$$m_{(k)}\ddot{\mathbf{u}}_{(k)}=\sum_{j\in H_{(k)}}\mathbf{k_u}\left\{\begin{array}{c}\mathbf{u}_{(k)}\\ \mathbf{u}_{(j)}\end{array}\right\}+\sum_{i\in\widetilde{H}_{(k)}}\widetilde{\mathbf{k}}_\mathbf{u}\left\{\begin{array}{c}\mathbf{u}_{(k)}\\ \mathbf{u}_{(i)}\end{array}\right\}+\mathbf{F}_{\mathbf{b}(k)} \tag{8.3}$$

and

$$\widehat{C}_{(k)}\dot{\Theta}_{(k)}+\widetilde{C}_{(k)}\Theta_{(k)}=\sum_{j\in H_{(k)}}\mathbf{k}_\Theta\left\{\begin{array}{c}\Theta_{(k)}\\ \Theta_{(j)}\end{array}\right\}+\sum_{i\in\widetilde{H}_{(k)}}\widetilde{\mathbf{k}}_\Theta\left\{\begin{array}{c}\Theta_{(k)}\\ \Theta_{(i)}\end{array}\right\}+F_{b(k)} \tag{8.4}$$

In the equation of motion, Eq. (8.3), $m_{(k)}$ represents the mass of PD node, $\mathbf{k_u}$ and $\widetilde{\mathbf{k}}_\mathbf{u}$ represent the stiffness matrices associated with the displacement vectors arising from the interactions of the node of interest with other nodes in the horizon and the dual horizon, and $\mathbf{F}_{\mathbf{b}(k)}$ is the external force acting on the PD node of interest. In the field equation, Eq. (8.4), $\widehat{C}_{(k)}$ and $\widetilde{C}_{(k)}$ are the coefficient of time derivative of the field (damping term) and time rate of change of material property, respectively. The $\mathbf{k}_\Theta$ and $\widetilde{\mathbf{k}}_\Theta$ represent the matrices associated with the field variable arising from the interactions of the PD node of interest with other nodes in the horizon and the dual horizon, and $F_{b(k)}$ is the external field source acting on the PD node of interest.

8.3 COMBIN14

The COMBIN14 element is employed to model damping effects at a PD node during the implicit transient field analysis. As shown in Fig. 8.1, the COMBIN14 element is defined by two nodes (I and J), spring constant (k), and damping coefficient (C_v). Each node has degrees of freedom (DOF) consisting of displacements (U_x, U_y, and U_z), rotations (ROT_x, ROT_y and ROT_z), pressure, and temperature.

For structural damping matrix, the damping force, **F** and torque, **T** can be evaluated as

$$\mathbf{F}=-\frac{C_v d\mathbf{u}}{dt} \tag{8.5}$$

and

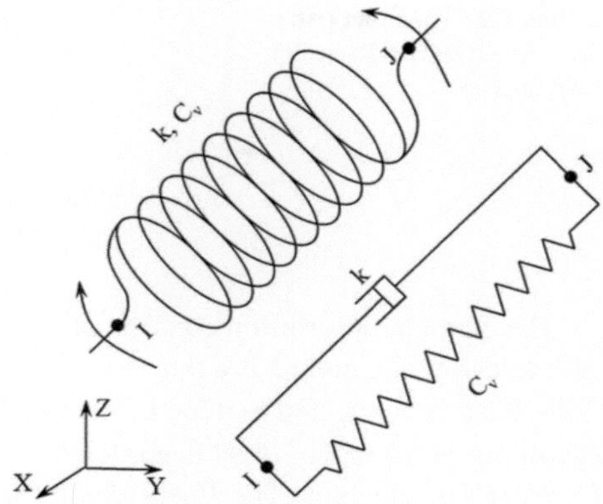

Fig. 8.1 Description of COMBIN14 element

Table 8.1 The KEYOPT command for COMBIN14 element

KEYOPT	Value	Options
KEYOPT (1)	0	Linear
	1	Nonlinear
KEYOPT (2)	0	3D - U_x, U_y, U_z
	1	1D - U_x
	2	1D - U_y
	3	1D - U_z
	4	1D - ROT_x
	5	1D - ROT_y
	6	1D - ROT_z
	7	Pressure
	8	Temperature
KEYOPT (3)	0	3D - U_x, U_y, U_z
	1	3D - ROT_x, ROT_y, ROT_z
	2	2D - U_x, U_y

$$\mathbf{T} = -\frac{C_v d\boldsymbol{\theta}}{dt} \tag{8.6}$$

with

$$C_v = C_{v1} + C_{v2}\text{v} \tag{8.7}$$

in which C_{v1} represents the linear damping and C_{v2}v facilitates the generation of nonlinear damping with v representing the velocity calculated in the previous time step. The vectors **u** and $\boldsymbol{\theta}$ include the displacement and rotation components (U_x, U_y, and U_z) and (ROT_x, ROT_y, and ROT_z), respectively.

The COMBIN14 element can be used as a longitudinal or a torsional element in 1D, 2D, or 3D applications. As specified in Ansys manual, the KEYOPT commands associated with COMBIN14 element are shown in Table 8.1.

Table. 8.2 The input parameters for COMBIN14 in PD field analysis

Input	Options	
Nodes	I, J	
KEYOPT	KEYOPT (1)	0
	KEYOPT (2)	4, 5, 8
DOFs	ROT_x, ROT_y, Temperature	
Real Constants	k-Stiffness, C_{v1} - Damping	

The element description consists of two nodes I and J. The node I is the PD node of interest, while node J is a dummy node created at the same location as that of PD node I and is connected to node I. The coefficient of time derivative of the field, $\widehat{C}_{(k)}$ (damping term) in Eq. (8.4) is modeled by using either the ROT_x or ROT_y DOF of COMBIN14 element. The linear damping for a PD node is activated through the KEYOPT command KEYOPT(1)=0. The input parameters of this element for PD field analysis are shown in Table 8.2.

To invoke this element for modeling linear damping with ROT_x DOF in the PD field analysis, Ansys APDL commands can be specified as

$$\begin{aligned}&\mathrm{ET, ETYP\_ID, COMBIN14}\\&\mathrm{KEYO, ETYP\_ID, 1, 0}\\&\mathrm{KEYO, ETYP\_ID, 2, 4}\\&\mathrm{R, REAL\_ID, , PARM\_DAMP}\\&\mathrm{E, I, I_d}\end{aligned} \tag{8.8}$$

where ETYP_ID is the element type number, REAL_ID is the real constant number, PARM _ DAMP is the value of damping parameter of the PD node I, and I_d is the dummy node created at the same location of node I. The real constants can be updated during the solution by using the RMODIF command as

$$\mathrm{RMODIF,\ \ REAL\_ID, 1,\ \ PARM\_DAMP} \tag{8.9}$$

To invoke this element for modeling transient material property of a field with TEMP DOF in the PD field analysis as described in Eq. (3.260), Ansys APDL commands can be specified as

$$\begin{aligned}&\mathrm{ET, ETYP\_ID, COMBIN14}\\&\mathrm{KEYO, ETYP\_ID, 1, 0}\\&\mathrm{KEYO, ETYP\_ID, 2, 8}\\&\mathrm{R, REAL\_ID, PARM\_STIFF}\\&\mathrm{E, I, I_d}\end{aligned} \tag{8.10}$$

$$\mathrm{RMODIF,\ \ REAL\_ID, 1,\ \ PARM\_STIFF} \tag{8.11}$$

where ETYP_ID is the element type number, REAL_ID is the real constant number, PARM_STIFF is the value of field stiffness parameter of the PD node I, and I_d is the dummy node created at the same location of node I.

8.4 MASS21

The MASS21 element is employed to include the inertial effects of a PD node during the implicit transient structural analysis. As shown in Fig. 8.2, the element is defined by one node I, which has six DOFs, i.e., displacements (U_x, U_y, and U_z) and rotations (ROT_x, ROT_y, and ROT_z) in the nodal coordinate system, (x, y, z). The element coordinate system may be initially parallel to the global or nodal or user-defined coordinate system (KEYOPT(2)) and can rotate with nodal coordinate system for large deflection analysis. In accordance with the DOF, the parameters (M_x, M_y, M_z) and (I_{xx}, I_{yy}, I_{zz}) as shown in Fig. 8.2 represent the components of mass and rotary inertia, respectively. The real constant in MASS21 element can be interpreted as mass or volume for 2D or 3D applications. As specified in Ansys manual, it can be selected by the KEYOPT command as shown in Table 8.3.

For MASS21 element, only the translational DOFs, i.e.,U_x, U_y, and U_z are employed to include the mass, $m_{(k)}$ of a PD node in Eq. (8.3). The node I is attached

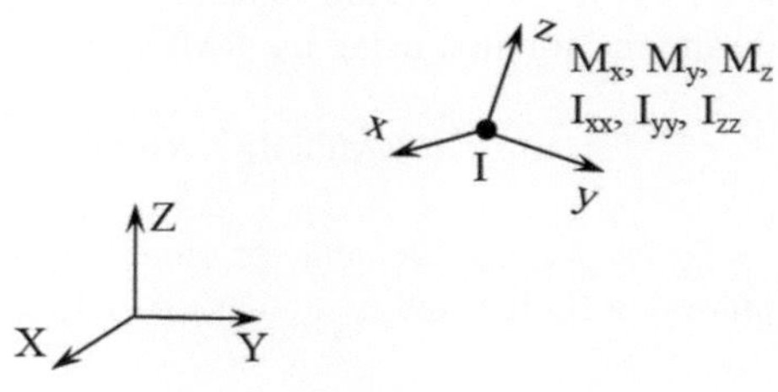

Fig. 8.2 Description of MASS21 element

Table 8.3 The KEYOPT command for MASS21 element

KEYOPT	Value	Options
KEYOPT (1)	0	Interpret real constants as mass/rotary inertia
	1	Interpret real constants as volumes and rotary inertias/density
KEYOPT (2)	0	Element coordinate system initially parallel to global coordinate system
	1	Element coordinate system initially parallel to nodal coordinate system
	4	Element coordinate system initially parallel to user-defined element coordinate system
KEYOPT (3)	0	3D Mass with rotary inertia
	2	3D Mass without rotary inertia
	3	2D Mass with rotary inertia
	4	2D Mass without rotary inertia

Table 8.4 The input parameters for MASS21 in PD structural analysis

Input	Options	
Nodes	I	
KEYOPT	KEYOPT (1)	0
	KEYOPT (3)	2 or 4
DOFs	3D or 2D	
Real Constants	MASS	

to the PD node of interest. The entry of mass corresponding to translational DOF in PD structural modeling is achieved by using KEYOPT(1)=0 and KEYOPT(3)=2 or 4. The input parameters of this element for PD structural analysis are shown in Table 8.4.

To invoke this element for 2D modeling of a PD node mass for structural analysis, Ansys APDL command can be specified as

$$\begin{aligned}&\mathrm{ET, ETYP\_ID, MASS21}\\&\mathrm{KEYO, ETYP\_ID, 1, 0}\\&\mathrm{KEYO, ETYP\_ID, 3, 4(=2\ for\ 3D)}\\&\mathrm{R, REAL\_ID, PARM\_MASS}\\&\mathrm{E, I}\end{aligned} \tag{8.12}$$

where ETYP_ID is the element type number, REAL_ID is the real constant number, and PARM_MASS is the value of the node mass. The real constants can be updated during the solution using the RMODIF command as

$$\mathrm{RMODIF,\ \ REAL\_ID, 1,\ \ PARM\_MASS} \tag{8.13}$$

To invoke this element for modeling the inertia of a PD node in thermal dual-phase-lag (DPL) analysis, Ansys APDL command can be specified as

$$\begin{aligned}&\mathrm{ET, ETYP\_ID, MASS21}\\&\mathrm{KEYO, ETYP\_ID, 1, 0}\\&\mathrm{KEYO, ETYP\_ID, 3, 0}\\&\mathrm{R, REAL\_ID, , , , PARM\_MASS}\\&\mathrm{E, I}\end{aligned} \tag{8.14}$$

where ETYP_ID is the element type number, REAL_ID is the real constant number, and PARM_MASS is the value of the DPL inertia parameter. The real constants can be updated during the solution using the RMODIF command as

$$\mathrm{RMODIF,\ \ REAL\_ID, 3,\ \ PARM\_MASS} \tag{8.15}$$

8.5 MATRIX27

The MATRIX27 element is employed for modeling the stiffness/damping/mass of a PD bond (interaction) during the implicit static and implicit transient analysis. As shown in Fig. 8.3, this element is suitable to relate the interaction of two nodes, I and J. Each node has degrees of freedom (DOF) consisting of displacements (U_x, U_y, and U_z) and rotations (ROT_x, ROT_y, and ROT_z). As specified in Ansys manual, the KEYOPT commands associated with MATRIX27 element are shown in Table 8.5. The real constants for this element support mass, damping, and stiffness coefficients in the form of symmetric, unsymmetric, and skew symmetric matrices as shown in Fig. 8.4.

PD structural analysis requires only translational DOFs. However, PD field analysis or PD coupled field-structural analysis requires combination of translational and rotational DOFs. The node I is the PD node of interest and node J is connected to either its family member in the horizon or in the dual horizon depending on the type of PD interaction. The stiffness matrices $\mathbf{k_u}$ and $\widetilde{\mathbf{k}}_{\mathbf{u}}$ in Eq. (8.3) for PD structural analysis are constructed using (U_x, U_y, and U_z) DOF of MATRIX27 element. Similarly, the matrices $\mathbf{k}_\Theta$ and $\widetilde{\mathbf{k}}_\Theta$ in Eq. (8.4) for PD field analysis are constructed using (ROT_x, ROT_y, and ROT_z) DOF of MATRIX27 element. In case of coupled field and structural analysis, the coupled element matrix involves a combination of translational and rotational DOF of MATRIX27 element. The input parameters of this element for PD analysis are shown in Table 8.6.

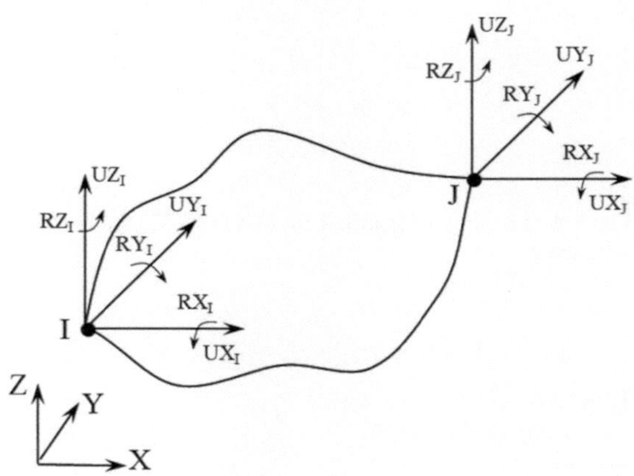

Fig. 8.3 Description of MATRIX27 element

Table 8.5 The KEYOPT command for MATRIX27 element

KEYOPT	Value	Options
KEYOPT (2)	0	Symmetric
	2	Unsymmetric
	3	Skew symmetric
KEYOPT (3)	2	Mass
	4	Stiffness
	5	Damping

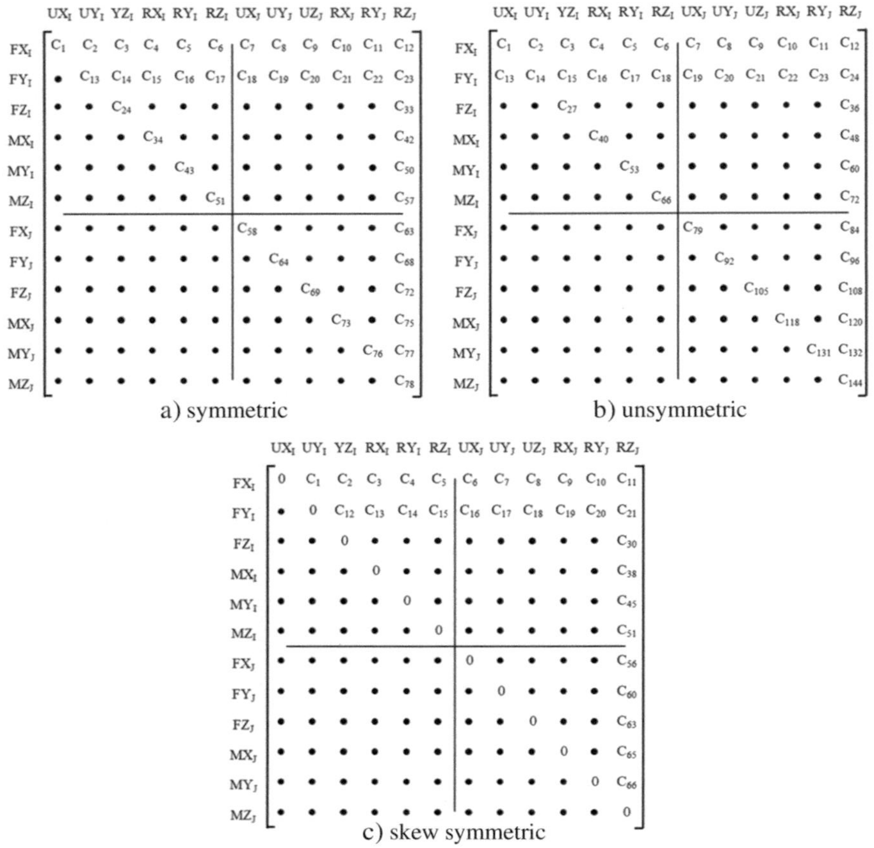

Fig. 8.4 The form of matrices for MATRIX27 element. (**a**) symmetric. (**b**) unsymmetric. (**c**) skew symmetric

Table 8.6 The input parameters for MATRIX27 in PD analysis

Input	Options
Nodes	I, J
DOFs	3D - U_x, U_y, U_z, ROT_x, ROT_y, ROT_z 2D - U_x, U_y, ROT_x, ROT_y
KEYOPTS	As summarized in Table 8.5
Real constants	$C_1, C_2, \ldots, C_{78}$ (symmetric) $C_1, C_2, \ldots, C_{144}$ (unsymmetric) $C_1, C_2, \ldots, C_{66}$ (skewsymmetric)

To invoke this element for symmetric matrix, the Ansys APDL command can be specified as

$$
\begin{aligned}
&\mathrm{ET, ETYP\_ID, MATRIX27}\\
&\mathrm{KEYO, ETYP\_ID, 2, 0}\\
&\mathrm{KEYO, ETYP\_ID, 3, 2\ or\ 4\ or\ 5}\\
&\mathrm{R, REAL\_ID,}\ \ C_1,\ \ C_2,\ \ C_3,\ \ C_4,\ \ C_5,\ \ C_6\\
&\mathrm{RMORE,}\ C_7,\ \ C_8,\ \ C_9,\ \ C_{10},\ \ C_{11},\ \ C_{12}\\
&\mathrm{RMORE,}\ C_{13},\ \ C_{14},\ \ C_{15},\ \ C_{16},\ \ C_{17},\ \ C_{18}\\
&\mathrm{RMORE,}\ C_{19},\ \ C_{20},\ \ C_{21},\ \ C_{22},\ \ C_{23},\ \ C_{24}\\
&\ldots\ldots\ldots\ldots\\
&\mathrm{RMORE,}\ C_{61},\ \ C_{62},\ \ C_{63},\ \ C_{64},\ \ C_{65},\ \ C_{66}\\
&\mathrm{RMORE,}\ C_{67},\ \ C_{68},\ \ C_{69},\ \ C_{70},\ \ C_{71},\ \ C_{72}\\
&\mathrm{RMORE,}\ C_{73},\ \ C_{74},\ \ C_{75},\ \ C_{76},\ \ C_{77},\ \ C_{78}\\
&\mathrm{E, I, J}
\end{aligned}
\tag{8.16}
$$

where ETYP_ID is the element type number; REAL_ID is the real constant number; and C_1, C_2, ..., C_{78} are the values of the mass or stiffness or damping coefficients.

The matrix coefficients can be updated during the solution using the RMODIF command as

$$
\begin{aligned}
&\mathrm{RMODIF, REAL\_ID, 1,}\ C_1,\ \ C_2,\ \ C_3,\ \ C_4,\ \ C_5,\ \ C_6\\
&\mathrm{RMODIF, REAL\_ID, 7,}\ C_7,\ \ C_8,\ \ C_9,\ \ C_{10},\ \ C_{11},\ \ C_{12}\\
&\mathrm{RMODIF, REAL\_ID, 13,}\ C_{13},\ \ C_{14},\ \ C_{15},\ \ C_{16},\ \ C_{17},\ \ C_{18}\\
&\mathrm{RMODIF, REAL\_ID, 19,}\ C_{19},\ \ C_{20},\ \ C_{21},\ \ C_{22},\ \ C_{23},\ \ C_{24}\\
&\ldots\ldots\ldots\ldots\\
&\mathrm{RMODIF, REAL\_ID, 61,}\ C_{61},\ \ C_{62},\ \ C_{63},\ \ C_{64},\ \ C_{65},\ \ C_{66}\\
&\mathrm{RMODIF, REAL\_ID, 67,}\ C_{67},\ \ C_{68},\ \ C_{69},\ \ C_{70},\ \ C_{71},\ \ C_{72}\\
&\mathrm{RMODIF, REAL\_ID, 73,}\ C_{73},\ \ C_{74},\ \ C_{75},\ \ C_{76},\ \ C_{77},\ \ C_{78}
\end{aligned}
\tag{8.17}
$$

To invoke this element for unsymmetric matrix, the Ansys APDL command can be specified as

$$
\begin{aligned}
&\mathrm{ET, ETYP\_ID, MATRIX27}\\
&\mathrm{KEYO, ETYP\_ID, 2, 2}\\
&\mathrm{KEYO, ETYP\_ID, 3, 2 \text{ or } 4 \text{ or } 5}\\
&\mathrm{R, REAL\_ID,}\ \ C_1,\ \ C_2,\ \ C_3,\ \ C_4,\ \ C_5,\ \ C_6\\
&\mathrm{RMORE,} C_7,\ \ C_8,\ \ C_9,\ \ C_{10},\ \ C_{11},\ \ C_{12}\\
&\mathrm{RMORE,} C_{13},\ \ C_{14},\ \ C_{15},\ \ C_{16},\ \ C_{17},\ \ C_{18}\\
&\ldots\ldots\ldots\ldots\\
&\mathrm{RMORE,} C_{127},\ \ C_{128},\ \ C_{129},\ \ C_{130},\ \ C_{131},\ \ C_{132}\\
&\mathrm{RMORE,} C_{133},\ \ C_{134},\ \ C_{135},\ \ C_{136},\ \ C_{137},\ \ C_{138}\\
&\mathrm{RMORE,} C_{139},\ \ C_{140},\ \ C141,\ \ C_{142},\ \ C_{143},\ \ C_{144}\\
&\mathrm{E, I, J}
\end{aligned}
\tag{8.18}
$$

where ETYP_ID is the element type number; REAL_ID is the real constant number; and C_1, C_2, …, C_{144} are the values of mass or stiffness or damping coefficients.

The matrix coefficients can be updated during the solution by using the RMODIF command as

$$
\begin{aligned}
&\mathrm{RMODIF, REAL\_ID, 1,} C_1,\ \ C_2,\ \ C_3,\ \ C_4,\ \ C_5,\ \ C_6\\
&\mathrm{RMODIF, REAL\_ID, 7,} C_7,\ \ C_8,\ \ C_9,\ \ C_{10},\ \ C_{11},\ \ C_{12}\\
&\mathrm{RMODIF, REAL\_ID, 13,} C_{13},\ \ C_{14},\ \ C_{15},\ \ C_{16},\ \ C_{17},\ \ C_{18}\\
&\ldots\ldots\ldots\ldots\\
&\mathrm{RMODIF, REAL\_ID, 127,} C_{127},\ \ C_{128},\ \ C_{129},\ \ C_{130},\ \ C_{131},\ \ C_{132}\\
&\mathrm{RMODIF, REAL\_ID, 134,} C_{133},\ \ C_{134},\ \ C_{135},\ \ C_{136},\ \ C_{137},\ \ C_{138}\\
&\mathrm{RMODIF, REAL\_ID, 139,} C_{139},\ \ C_{140},\ \ C_{141},\ \ C_{142},\ \ C_{143},\ \ C_{144}
\end{aligned}
\tag{8.19}
$$

To invoke this element for skew symmetric matrix, the Ansys APDL command can be specified as

$$
\begin{aligned}
&\mathrm{ET, ETYP\_ID, MATRIX27}\\
&\mathrm{KEYO, ETYP\_ID, 2, 3}\\
&\mathrm{KEYO, ETYP\_ID, 3, 2 \text{ or } 4 \text{ or } 5}\\
&\mathrm{R, REAL\_ID,}\ \ C_1,\ \ C_2,\ \ C_3,\ \ C_4,\ \ C_5,\ \ C_6\\
&\mathrm{RMORE,} C_7,\ \ C_8,\ \ C_9,\ \ C_{10},\ \ C_{11},\ \ C_{12}\\
&\mathrm{RMORE,} C_{13},\ \ C_{14},\ \ C_{15},\ \ C_{16},\ \ C_{17},\ \ C_{18}\\
&\ldots\ldots\ldots\ldots\\
&\mathrm{RMORE,} C_{49},\ \ C_{50},\ \ C_{51},\ \ C_{52},\ \ C_{53},\ \ C_{54}\\
&\mathrm{RMORE,} C_{55},\ \ C_{56},\ \ C_{57},\ \ C_{58},\ \ C_{59},\ \ C_{60}\\
&\mathrm{RMORE,} C_{61},\ \ C_{62},\ \ C141_{63},\ \ C_{64},\ \ C_{65},\ \ C_{66}\\
&\mathrm{E, I, J}
\end{aligned}
\tag{8.20}
$$

where ETYP_ID is the element type number; REAL_ID is the real constant number; and C_1, C_2, ..., C_{144} are the values of the mass or stiffness or damping coefficients.

The matrix coefficients can be updated during the solution by using the RMODIF command as

$$
\begin{aligned}
&\mathrm{RMODIF, REAL\_ID, 1,} C_1,\ \ C_2,\ \ C_3,\ \ C_4,\ \ C_5,\ \ C_6\\
&\mathrm{RMODIF, REAL\_ID, 7,} C_7,\ \ C_8,\ \ C_9,\ \ C_{10},\ \ C_{11},\ \ C_{12}\\
&\mathrm{RMODIF, REAL\_ID, 13,} C_{13},\ \ C_{14},\ \ C_{15},\ \ C_{16},\ \ C_{17},\ \ C_{18}\\
&\ldots\ldots\ldots\ldots\\
&\mathrm{RMODIF, REAL\_ID, 49,} C_{49},\ \ C_{50},\ \ C_{51},\ \ C_{52},\ \ C_{53},\ \ C_{54}\\
&\mathrm{RMODIF, REAL\_ID, 55,} C_{55},\ \ C_{56},\ \ C_{57},\ \ C_{58},\ \ C_{59},\ \ C_{60}\\
&\mathrm{RMODIF, REAL\_ID, 61,} C_{61},\ \ C_{62},\ \ C_{63},\ \ C_{64},\ \ C_{65},\ \ C_{66}
\end{aligned}
\tag{8.21}
$$

8.6 LINK33

The LINK33 element can be employed when performing the PD field analysis. As shown in Fig. 8.5, this element is defined by two nodes, I and J. Each node only has a single degree of freedom, i.e., temperature representing the field variable. The only real constant supported for this element is the cross-sectional area, *AREA*.

For PD field analysis, the node I is the PD node of interest and the node J is connected to its family member in the horizon or in the dual horizon. The stiffness matrices $\mathbf{k}_\Theta$ and $\widetilde{\mathbf{k}}_\Theta$ in Eq. (8.4) for PD field are modeled by using the temperature DOF of LINK33 element. Although this element supports the material data input

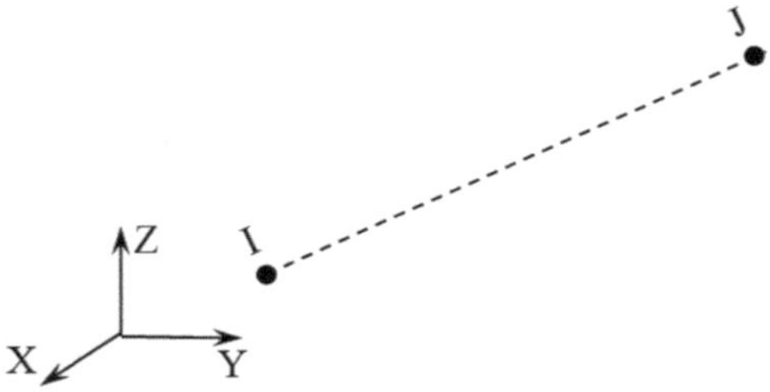

Fig. 8.5 Description of LINK33 element

Table 8.7 The input parameters for LINK33 in PD analysis

Input	Options
Nodes	I, J
Real constants	Area

through the MP command, they are not necessary for the PD analysis. The input parameters of this element for PD field analysis are shown in Table 8.7.

To invoke this element for PD field analysis, Ansys APDL command can be specified as

$$\begin{aligned}&\text{ET, ETYP\_ID, LINK33}\\&\text{R, REAL\_ID, PARM\_AREA}\\&\text{E, I, J}\end{aligned}\tag{8.22}$$

where ETYP_ID is the element type number, REAL_ID is the real constant number, and PARM_AREA is the value of the cross-sectional area. The real constants can be updated during the solution using the RMODIF command as

$$\text{RMODIF, REAL\_ID, 1, PARM\_AREA}\tag{8.23}$$

8.7 MASS71

The MASS71 element is used to model the coefficient of damping term on the left-hand side of Eq. (8.4) for a PD node during the implicit transient field analysis. It can also be used to model the source generation which can depend on the field itself in the form of a load vector. As shown in Fig. 8.6, this element is defined by one node I.

The node has one DOF, i.e., temperature. The real constants include CON1, constants A_1, A_2, A_3..., A_6 to model the temperature dependent heat generation. As specified in Ansys manual, the KEYOPT commands associated with MASS71 element are shown in Table 8.8.

To simulate the field capacitance of the PD material point, the node I is attached to the PD point of interest. The real constants for this element can be used to model the

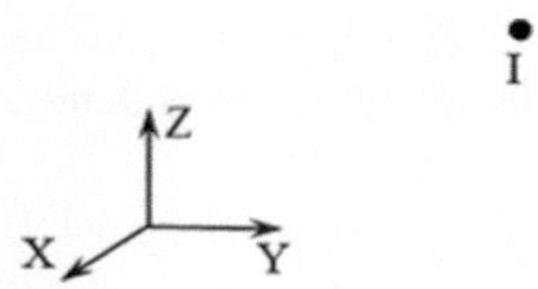

Fig. 8.6 Description of MASS71 element

Table 8.8 The KEYOPT commands for MASS71 element

KEYOPT	Value	Options
KEYOPT (3)	0	Interpret CON1 as volume (DENS and C or ENTH supplied as material properties)
	1	Interpret CON1 as thermal capacitance (DENS*C*volume)
KEYOPT (4)	0	No temperature dependent heat generation
	1	Include temperature dependent heat generation

Table 8.9 The input parameters for MASS71 in PD analysis

Input	Options	
Nodes	I	
KEYOPT	KEYOPT (3)	1
	KEYOPT (4)	0
DOFs	Temperature	
Real constants	CON1	

field capacitance or temperature dependent heat generation source. The capability of field capacitance in PD analysis is achieved by using KEYOPT(3)=1. This element supports the material data inputs through the MP command; however, this is not necessary for the PD analysis. The input parameters of this element for PD analysis are shown in Table 8.9.

To invoke this element for modeling field capacitance of a PD node, Ansys APDL commands can be specified as

$$\begin{aligned}&\text{ET, ETYP\_ID, MASS71}\\&\text{KEYO, ETYP\_ID, 3, 1}\\&\text{KEYO, ETYP\_ID, 4, 0}\\&\text{R, REAL\_ID, PARM\_CON1}\\&\text{E, I}\end{aligned}\tag{8.24}$$

where ETYP_ID is the element type number, REAL_ID is the real constant number, and PARM_CON1 is the value of the field capacitance of the PD node I. The real constants can be updated during the solution using the RMODIF command as

$$\text{RMODIF,}\quad \text{REAL\_ID, 1,}\quad \text{CON1}\tag{8.25}$$

This element can be used for modeling the field source. In the absence of temperature dependence, the following commands are used

$$\begin{aligned}&\text{ET, ETYP\_ID, MASS71}\\&\text{KEYO, ETYP\_ID, 4, 0}\\&\text{MP, QRATE, ETYP\_ID, PARM\_CON1}\\&\text{E, I}\end{aligned} \tag{8.26}$$

where ETYP_ID is the element type number and PARM_CON1 is the value of the source of the PD node I.

In the presence of temperature dependence, the following commands are used

$$\begin{aligned}&\text{ET, ETYP\_ID, MASS71}\\&\text{KEYO, ETYP\_ID, 4, 1}\\&\text{R, REAL\_ID, , A1, A2, A3, A4, A5}\\&\text{RMORE, A6}\\&\text{E, I}\end{aligned} \tag{8.27}$$

where ETYP_ID is the element type number and real constants; A1, , , , , A6 are real constants required to specify the temperature dependent source of the PD node I.

8.8 LINK31

The LINK31 element is used to model the radiation boundary condition when considering the PD thermal field equation. As shown in Fig. 8.7, this element is defined by two nodes, I and J. Each node has only a single degree of freedom, i.e., temperature. The real constants include radiating surface area A, geometric form factor F, emissivity ε, and SBC (Stefan-Boltzmann constant) σ, which appear in the expression for heat transfer due to radiation as

$$q = \sigma \varepsilon F A \left(T_{(\mathrm{I})}^4 - T_{(\mathrm{J})}^4 \right) \tag{8.28}$$

For simulating radiation boundary conditions in the PD analysis, node I is the PD node of interest, while node J is connected to a dummy node created at the same location as that of the PD node I. Although, this element supports the material data input through the MP command, they are not necessary for the PD analysis. The input parameters of this element for the PD analysis are shown in Table 8.10.

To invoke this element for modeling the PD radiation boundary condition, Ansys APDL command can be specified as

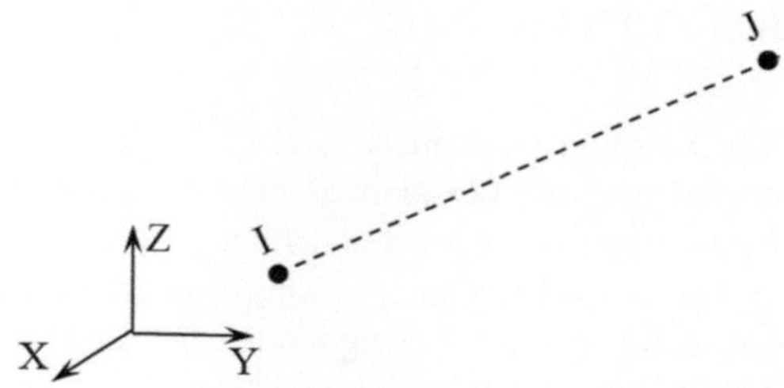

Fig. 8.7 The geometry of LINK31 element

Table 8.10 The input parameters for LINK31 in PD analysis

Input	Options
Nodes	I, J
KEYOPTS	KEYOPT (3) = 0
Real constants	Area Form factor Emissivity SBC

$$\begin{aligned}&\text{ET, ETYP\_ID, LINK31}\\&\text{KEYO, ETYP\_ID, 3, 0}\\&\text{R, REAL\_ID, PARM\_AREA, PARM\_F, PARM\_EMISSIVITY, PARM\_SBC}\\&\text{E, I, I}_{\text{d}}\end{aligned} \tag{8.29}$$

where ETYP_ID is the element type number; REAL_ID is the real constant number; and PARM_AREA, PARM_FORMFACTOR, PARM_EMISSIVITY, and PARM_SBC are the values of the cross-sectional area, form factor, emissivity, and Stefan-Boltzmann constant of the PD node I and I_d, which is the dummy node created at the same location as that of node I. The real constants can be updated during the solution using the RMODIF command as

$$\begin{aligned}&\text{RMODIF, REAL\_ID, 1, PARM\_AREA}\\&\text{RMODIF, REAL\_ID, 2, PARM\_F}\\&\text{RMODIF, REAL\_ID, 3, PARM\_EMISSIVITY}\\&\text{RMODIF, REAL\_ID, 4, PARM\_SBC}\end{aligned} \tag{8.30}$$

8.9 LINK34

The LINK34 element is used to model the convection boundary condition when considering the PD thermal field equation. As shown in Fig. 8.8, this element is defined by two nodes, I and J.

Each node has only a single degree of freedom, i.e., temperature. The real constants include convection surface area A; film coefficient h_f; and empirical convection term E, which appear in the expression for heat transfer due to convection as

$$q = h_f A E \left(T_{(\mathrm{I})} - T_{(\mathrm{J})} \right) \tag{8.31}$$

The empirical convection term E is defined as

$$E = F \left| T_{\mathrm{p(I)}} - T_{\mathrm{p(J)}} \right|^n + \frac{CC}{h_f} \tag{8.32}$$

where CC is the input constant, F is the scale factor, n is the empirical coefficient, and T_p is the temperature from the previous time step. It is to be noted that $E = F$ in Eq. (8.32) when n and CC are zero. Furthermore, the scale factor $F = 1.0$ is the default option and can be modified using KEYOPT(3)=2 option. As specified in the Ansys manual, the KEYOPT commands associated with LINK34 element are shown in Table 8.11.

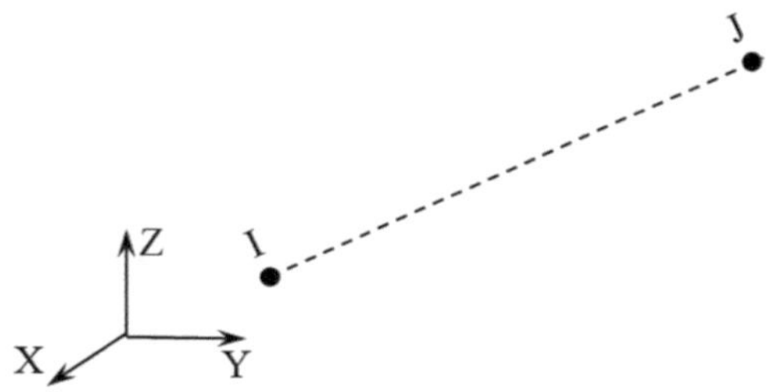

Fig. 8.8 Description of LINK34 element

Table 8.11 The KEYOPT command for LINK34 element

KEYOPT	Value	Options
KEYOPT (2)	0	Use average temperature of $T_{(\mathrm{I})}$ and $T_{(\mathrm{J})}$ to evaluate film coefficient
	1	Use greater temperature of $T_{(\mathrm{I})}$ or $T_{(\mathrm{J})}$ to evaluate film coefficient
	2	Use lesser temperature of $T_{(\mathrm{I})}$ or $T_{(\mathrm{J})}$ to evaluate film coefficient
	3	Use differential $\lvert T_{(\mathrm{I})} - T_{(\mathrm{J})} \rvert$ temperature to evaluate film coefficient
KEYOPT (3)	0	Standard element input and empirical term
	2	Use alternate input for HF and F
	3	Use discontinuous empirical term

Table 8.12 The input parameters for LINK34 in PD analysis

Input	Options
Nodes	I, J
KEYOPTS	KEYOPT (3) = 0
Real constants	Area EN-empirical coefficient (=0) CC-Input constant (=0)

For simulating convection boundary conditions in the PD analysis, node I is the PD node of interest while node J is connected to a dummy node created at the same location as that of the PD node I. The input parameters of this element for PD analysis are shown in Table 8.12.

To invoke this element for modeling the PD convection boundary condition, Ansys APDL command can be specified as

$$\begin{aligned}&\mathrm{ET, ETYP\_ID, LINK34}\\&\mathrm{KEYO, ETYP\_ID, 3, 0}\\&\mathrm{R, REAL\_ID, PARM\_AREA}\\&\mathrm{E, I, I_d}\end{aligned} \tag{8.33}$$

where ETYP_ID is the element type number, REAL_ID is the real constant number, and PARM_AREA is the value of the cross-sectional area of the PD node I and I_d, which is the dummy node created at same location as that of node I. The real constants can be updated during the solution using the RMODIF command as

$$\mathrm{RMODIF,\ \ REAL\_ID, 1,\ \ PARM\_AREA} \tag{8.34}$$

Reference

Ansys 18.2 Mechanical User's Guide. (2017)

Chapter 9
Ansys Menus for PD Analysis

9.1 Introduction

This chapter provides the procedure to create a PD model through customized Ansys submenus developed by using the Ansys UIDL (User Interface Design Language). A more detailed description of the development of customized menus can be found in the Ansys UIDL help manual (Ansys, 2017). The customized submenus are grouped under a menu named, ***PD Preprocessor***, which is located under the ***Ansys Preprocessor*** menu. The flowchart shown in Fig. 9.1 describes the general procedure to generate a PD model in the Ansys database using the menus placed under the menu, ***PD Preprocessor***. For each of the submenus in the workflow, a function block is embedded depending on the requirement to capture the user input. The function blocks include Dialog box, Picking box, and Hidden box. The subsequent sections describe in detail the functionality of each of the customized submenus.

9.2 Field Selection

As shown in Fig. 9.2, the dialog box ***Field Selection*** appears upon selecting the ***Field Selection*** submenu located under ***Peridynamic Processor*** menu. The GUI path to invoke this dialog box is shown below:

Main Menu> Preprocessor> Peridynamic Processor> Field Selection

The parameters associated with the ***Field Selection*** dialog box are ***Select Field*** and ***Select Dimension***.

Select Field: It is a drop-down menu type parameter with a list of fields for the PD analysis. The list of fields for the PD analysis is shown in Table 9.1

E. Madenci et al., *Peridynamics and Its Applications Using Ansys*,
https://doi.org/10.1007/978-3-031-59896-8_9

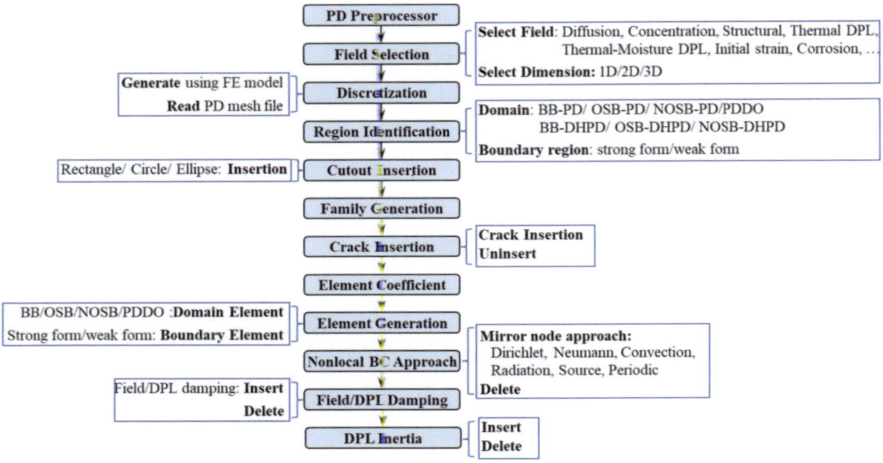

Fig. 9.1 Flowchart for PD modeling in Ansys

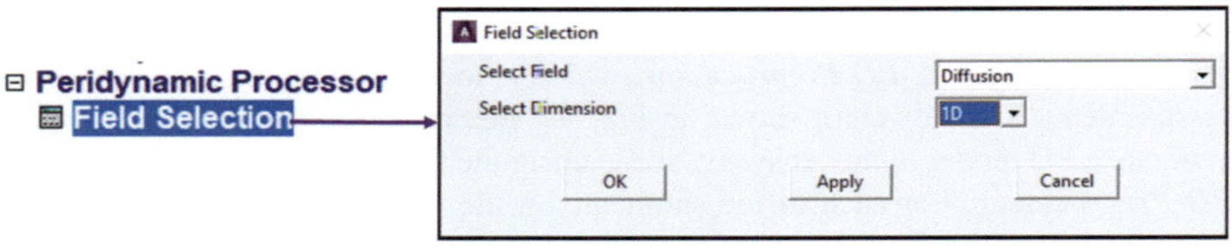

Fig. 9.2 Dialog box for selecting the field and dimension

Table 9.1 List of fields for PD analysis

Field	DOF
Diffusion	Temperature, Electric potential, Chemical concentration
Concentration	Moisture
Structural	Displacement, Velocity
Thermal DPL	Temperature
Thermal-Moisture DPL	Temperature, Moisture
Initial strain	Temperature, Displacement
Corrosion	Electric potential, Chemical concentration
Electrodeposition	Electric potential, Chemical concentration
Electromigration	Electric potential, Atomic concentration, Displacement, and Temperature
Peridynamic differential operator (PDDO)	

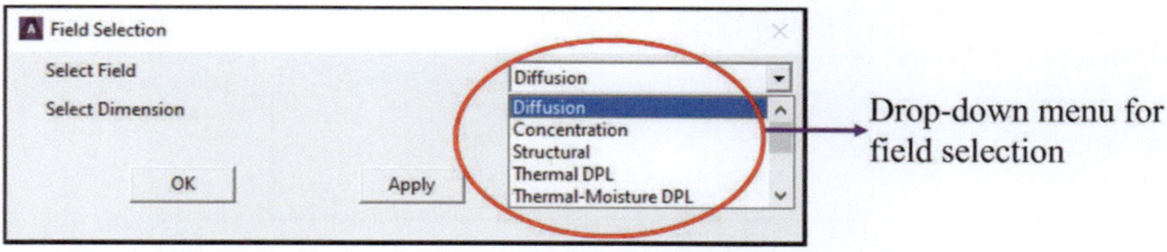

Fig. 9.3 Drop-down menu for selecting the field for PD analysis

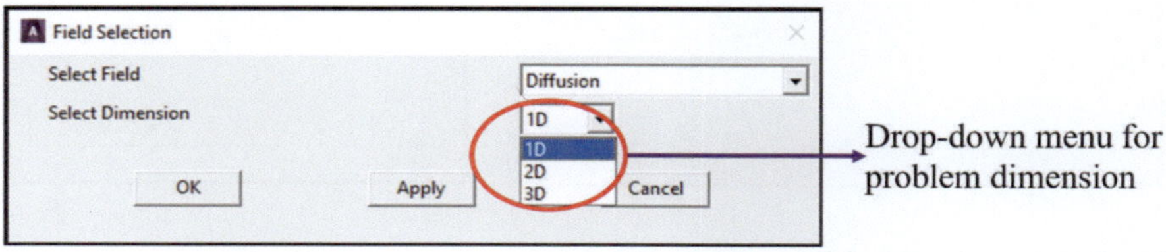

Fig. 9.4 Drop-down menu for selecting the dimension of PD analysis

The list of field types available in the GUI after expansion of the drop-down menu of the ***Field Selection*** dialog box is shown in Fig. 9.3.

Select Dimension: It is a drop-down menu type parameter with a list of dimensions available for PD analysis. It can be set to either 1D, 2D, or 3D depending on the type of analysis. The list of dimensions available is shown in Fig. 9.4 after the expansion of the drop-down menu.

The parameters in the ***Field Selection*** dialog box are set when ***OK*** or ***Apply*** button is clicked.

Note

The option for 3D is currently supported only for evaluating nonlocal derivatives using PDDO. The various field options for PD analysis are only supported for 1D and 2D.

9.3 Discretization

As shown in Fig. 9.5, the dialog box ***Discretization*** appears upon selecting the ***Discretization*** submenu located under ***Peridynamic Processor*** menu. The GUI path for this dialog box is shown below:

Main Menu> Preprocessor> Peridynamic Processor> Discretization

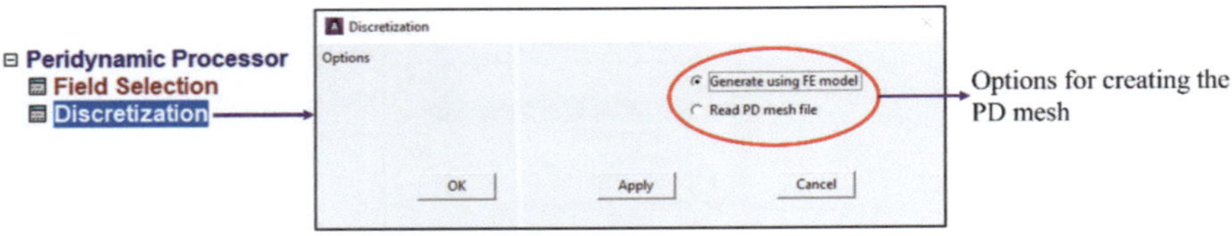

Fig. 9.5 Dialog box for PD discretization

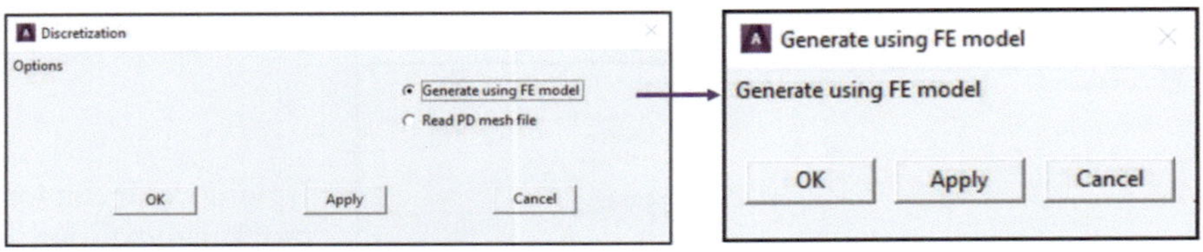

Fig. 9.6 Dialog box for confirming the PD mesh generation from the FE mesh

The PD discretization can be generated by using either the finite element mesh created natively or imported into the Ansys database. Also, a pre-existing customized PD mesh file can be imported into the Ansys database and read.

The ***Discretization*** dialog box has two options with the following radio button types: ***Generate using FE model*** and ***Read PD mesh file***:

Generate using FE model: This radio button option is selected for creating the PD mesh by using the existing finite element model. Upon selecting this radio button and clicking on the ***OK*** or ***Apply*** button, a dialog box appears as shown in Fig. 9.6 to confirm the mesh creation.

After clicking on the ***OK*** or ***Apply*** button, the existing finite elements are deleted, and the PD nodes are automatically created at the locations corresponding to the centroid of each finite element. The PD nodes are then assigned an entity depending on the geometry of the element. For example, the PD point (pink) is assigned an area of the underlying FE in a 2D analysis as shown in Fig. 9.7.

Figures 9.8, 9.9, and 9.10 provide examples of PD mesh generated in the Ansys database for a 1D, 2D, and 3D analysis, respectively, from the finite element mesh.

Read PD mesh file: This radio button option is used to import the custom PD mesh file into Ansys and to generate the PD mesh. When this option is selected, a dialog box ***Read PD Mesh File*** appears as shown in Fig. 9.11. Furthermore, upon clicking the ***Browse*** button, a file browser window shown in Fig. 9.12 appears to display the list of files in the working directory. After selecting the PD mesh file from the working directory and using the ***Open*** button, the name of the PD mesh

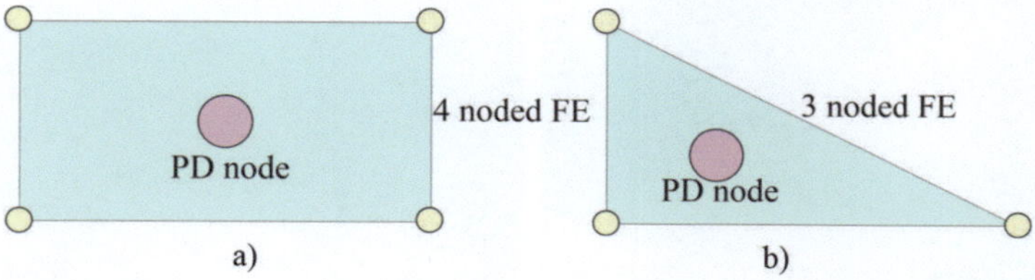

Fig. 9.7 Location of a PD point at the centroid of a finite element: (**a**) quad shape, and (**b**) triangular shape

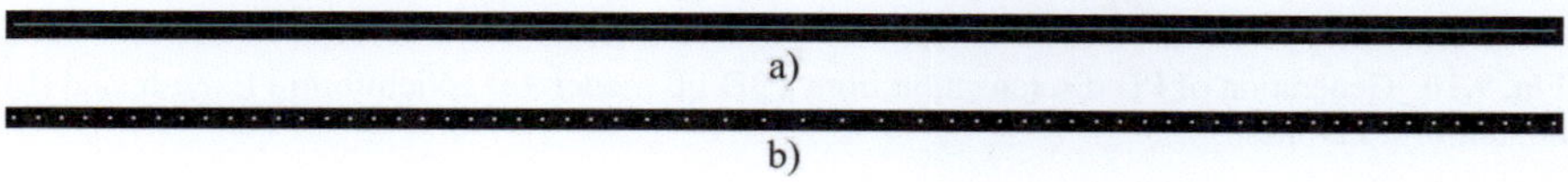

Fig. 9.8 Generation of PD discretization from a 1D FE model: (**a**) FE mesh, and (**b**) PD mesh

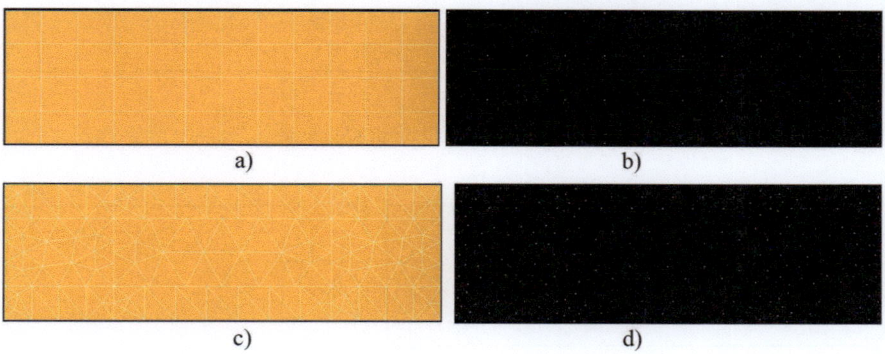

Fig. 9.9 Generation of PD discretization from a 2D FE model: (**a**) uniform FE mesh, (**b**) uniform PD mesh, (**c**) nonuniform FE mesh, and (**d**) nonuniform PD mesh

file to be read is saved in the parameter ***Filename***. Enter the number of columns to be read as a value for the parameter ***Enter the number of field variab.***

After clicking the ***OK*** or ***Apply*** button in the ***Read PD Mesh File*** dialog box, the imported PD mesh file is read, and the nodes corresponding to the PD point locations are created in the Ansys database. The specific formats of the PD mesh file to be imported using a Microsoft excel file (.csv format) or plain text file (.txt format with a comma delimiter) are shown in Figs. 9.13 and 9.14, respectively.

In the mesh file, the first line contains the number of columns, followed by the column numbering. The row or column numbering must be entered in an increasing

a) b)

Fig. 9.10 Generation of PD discretization from a 3D FE model: (**a**) nonuniform FE mesh, and (**b**) nonuniform PD mesh

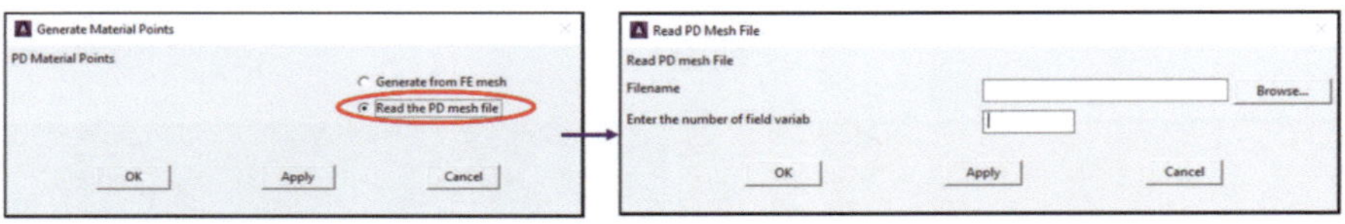

Fig. 9.11 Dialog box for selecting the existing PD mesh file name

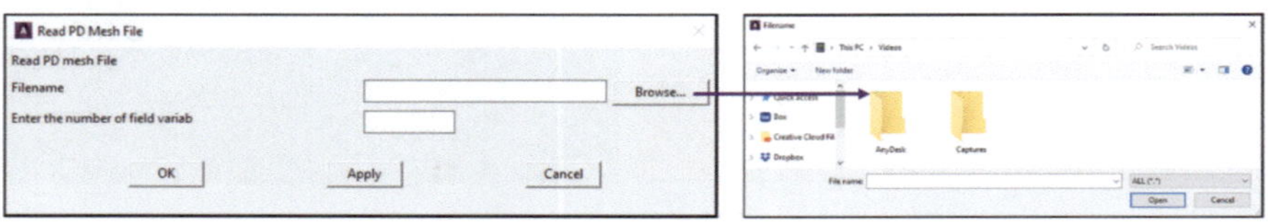

Fig. 9.12 File browser window for choosing files from the working directory

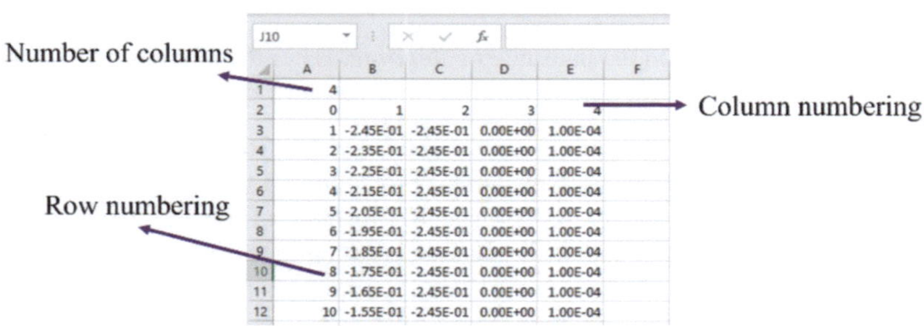

Fig. 9.13 PD mesh file in Excel (CSV) format

```
4
0,1,2,3,4
1,-2.45E-01,-2.45E-01,0.00E+00,1.00E-04
2,-2.35E-01,-2.45E-01,0.00E+00,1.00E-04
3,-2.25E-01,-2.45E-01,0.00E+00,1.00E-04
4,-2.15E-01,-2.45E-01,0.00E+00,1.00E-04
5,-2.05E-01,-2.45E-01,0.00E+00,1.00E-04
6,-1.95E-01,-2.45E-01,0.00E+00,1.00E-04
7,-1.85E-01,-2.45E-01,0.00E+00,1.00E-04
8,-1.75E-01,-2.45E-01,0.00E+00,1.00E-04
9,-1.65E-01,-2.45E-01,0.00E+00,1.00E-04
10,-1.55E-01,-2.45E-01,0.00E+00,1.00E-04
```

Fig. 9.14 PD mesh file in txt format

order beginning with a zero. Column 0 contains the numbering of PD points. Columns 1-3 correspond to the coordinates of PD points. Column 4 contains the value of the entity of PD point, i.e., length, area, or volume. For the evaluation of field derivatives using the PD mesh file as described in Chap. 10, additional columns starting from column 5 are required for the field input. The *number of columns*, i.e., the first header has to be updated accordingly for every additional column to the right.

The discretization size, Δ of a PD point is computed based on the associated entity as

$$\Delta = \begin{cases} l_e \text{ for 1D} \\ \sqrt{A_e} \text{ for 2D} \\ \sqrt[3]{V_e} \text{ for 3D} \end{cases} \tag{9.1}$$

where l_e, A_e, and V_e are the length, area, and volume of the underlying FE mesh or the values as specified in column 4 of the PD mesh file.

9.4 Region Identification

9.4.1 Domain

As shown in Fig. 9.15, the dialog box ***Domain*** appears upon selecting the ***Region Identification>Domain*** submenu located under ***Peridynamic Processor*** menu. This submenu is used for setting the PD interaction type associated with the PD mesh in the inner and outer regions as depicted in Figs. 3.11 and 3.14. The GUI path to invoke this dialog box is shown below:

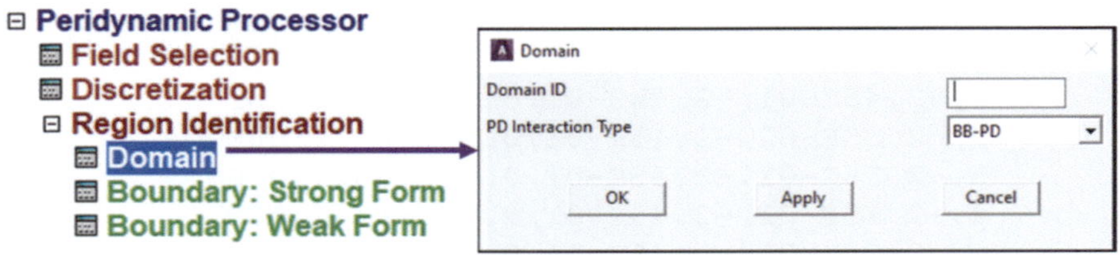

Fig. 9.15 Dialog box for identifying the specific regions of the domain

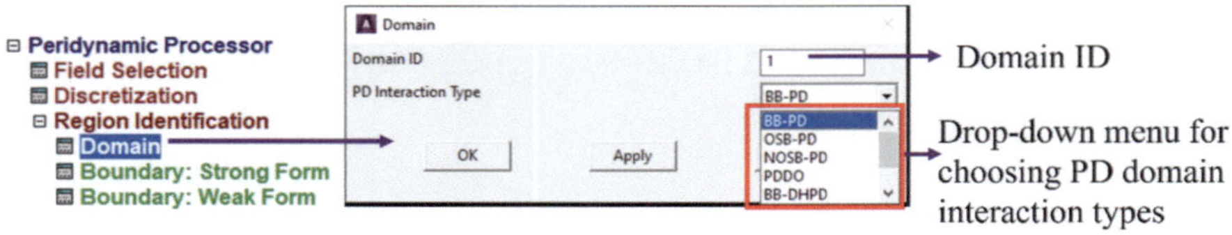

Fig. 9.16 Dialog box for selecting the specific region in domain and associated interaction type in the PD mesh

Main Menu> Preprocessor> Peridynamic Processor> Region identification> Domain

As shown in Fig 9.16, the dialog box for ***Domain*** requires the following parameters:

Domain ID: An integer parameter to identify the specific region of the domain.

PD Interaction Type: This parameter is a drop-down menu consisting of the list of PD interaction types. The supported types are Bond-Based (BB-PD), Ordinary State-Based (OSB-PD), Non-ordinary State-Based (NOSB-PD), Peridynamic Differential operator (PDDO), Bond-Based Dual Horizon (BB-DHPD), Ordinary State-Based Dual Horizon (OSB-DHPD), and Non-Ordinary State-Based Dual Horizon (NOSB-DHPD).

For modeling outer region, $\mathscr{R}$ as depicted Figs. 3.11 and 3.14, enter the ***Domain ID*** and the ***PD Interaction Type*** as PDDO. After setting the parameters of the ***Domain*** dialog box as shown in Fig 9.16 and clicking of the ***OK*** or ***Apply*** button, a node picker (Box) window appears for selecting the domain PD nodes. Use the picker option to select the outer region of the domain. Figure 9.17 shows the steps for creation of *PDDO* region in the domain.

For modeling the inner region $\mathcal{D}$, as depicted Figs. 3.1 and 3.11 after entering the Domain ID, the ***PD Interaction Type*** can be set to any of the required PD interaction type from the drop-down menu, except for PDDO. After setting the parameters of the

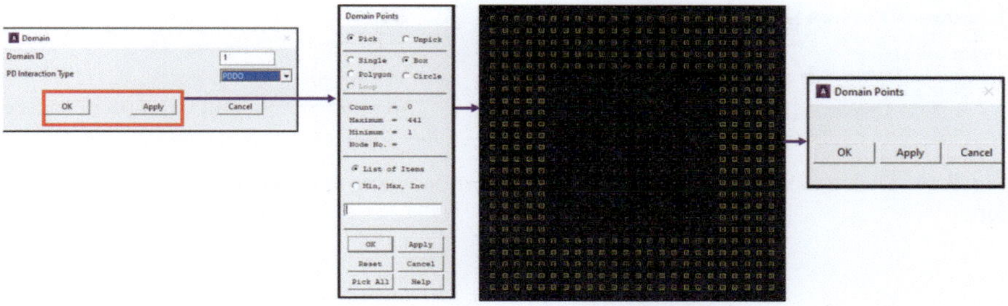

Fig. 9.17 Creation of the region in the domain corresponding to the PDDO interaction

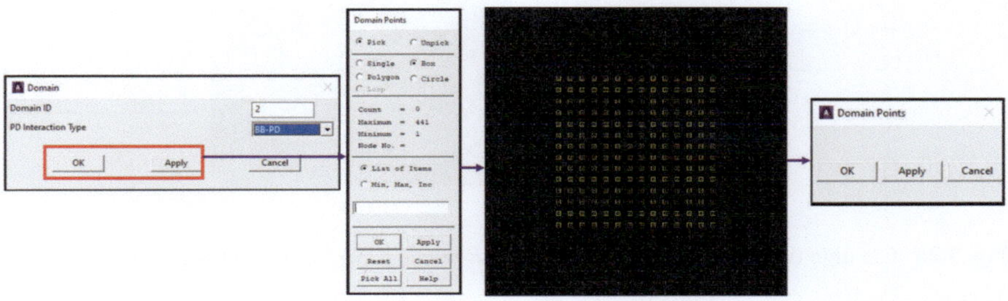

Fig. 9.18 Creation of the region in the domain corresponding to the BB-PD interaction

Domain dialog box as in Fig. 9.16 and clicking of the ***OK*** or ***Apply*** button, a node picker (Box) window appears for selecting the domain PD nodes. Use the picker option to select the inner region of the domain. Figure 9.18 shows the steps for creation of *BB-PD* region in the domain.

After selecting the nodes using the Node picker window and clicking the ***OK*** or ***Apply*** button of the node picker (Box), a dialog box labeled as ***Domain Points*** appears for confirming the selected regions. Furthermore, clicking of ***OK*** or ***Apply*** button of ***Domain Points*** dialog box creates a nodal component corresponding to the ***PD Interaction Type*** chosen. The nodal component can be viewed by using CMPLOT command by entering in Ansys command window.

9.4.2 Boundary Interactions in Strong Form

As shown in Fig. 9.19, the dialog box ***Boundary: Strong Form*** appears upon selecting the ***Region Identification>Boundary: Strong Form*** submenu located under ***Peridynamic Processor*** menu. This submenu is used for setting the PD boundary interaction type in a strong form as depicted in Figs. 3.11 and 3.14. The GUI path to invoke this dialog box is shown below:

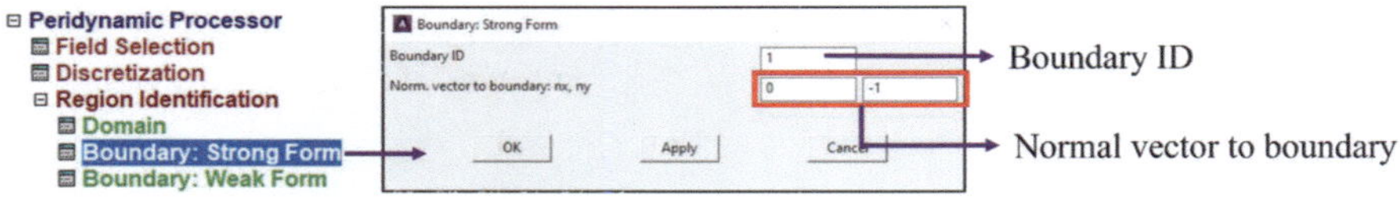

Fig. 9.19 Dialog box for entering the parameters corresponding to boundary interactions in strong form

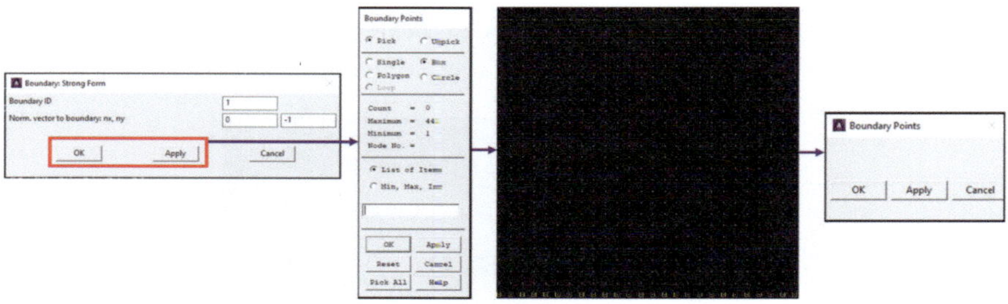

Fig. 9.20 Creation of the region corresponding to *Boundary: Strong Form* interaction

Main Menu> Preprocessor> Peridynamic Processor> Region Identification> Boundary: Strong Form

As shown in Fig. 9.19, the dialog box for a ***Boundary: Strong Form*** requires the following parameters:

Boundary ID: An integer parameter to identify the boundary region.

Norm. vector to boundary: nx, ny: The direction cosines of the unit vector normal to the selected boundary.

For modeling the boundary region, $\mathscr{B}$ as depicted Figs. 3.11 and 3.14, enter the Boundary ID and the unit normal vector of the boundary. After setting the parameters in the ***Boundary: Strong Form*** dialog box as shown in Fig. 9.19 and clicking of the ***OK*** or ***Apply*** button, a node picker (Box) window appears for selecting the domain PD nodes. Use the picker option to select the boundary region in which the boundary conditions are enforced in strong form. Figure 9.20 shows the creation of boundary region by using the node picker window corresponding to strong form of interactions. For subsequent creation of other boundary regions, a different boundary domain ID should be used.

After selecting the nodes using the Node picker window and clicking the ***OK*** or ***Apply*** button of the node picker (Box), a dialog box labeled as ***Boundary Points*** appears for confirming the selected regions. Furthermore, clicking of ***OK*** or ***Apply***

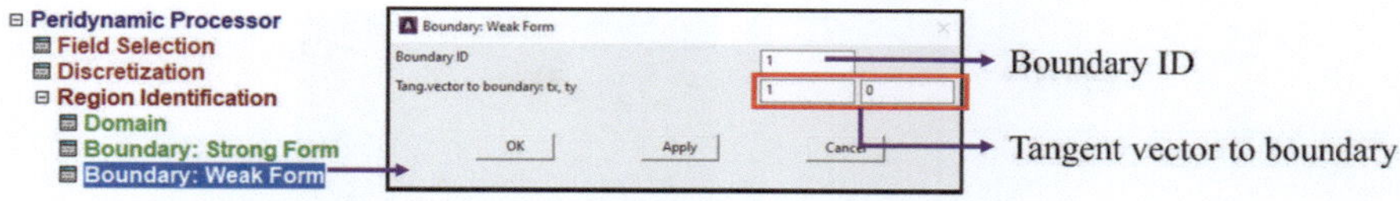

Fig. 9.21 Dialog box for entering the parameters corresponding to the weak form of boundary interactions

button of ***Boundary Points*** dialog box creates a nodal component corresponding to the ***Boundary ID***. The nodal component can be viewed by using *CMPLOT* command by entering in Ansys command window.

9.4.3 Boundary Interactions in Weak Form

As shown in Fig. 9.21, the dialog box ***Boundary: Weak Form*** appears upon selecting the ***Region Identification>Boundary: Weak Form*** submenu located under ***Peridynamic Processor*** menu. This submenu is used for setting the PD boundary interaction type in a weak form as depicted in Figs. 3.11 and 3.14. Unlike the ***Boundary: Strong Form*** case, where the unit vector normal to the boundary is required to compute the PD traction, this dialog box requires the unit vector tangent to the boundary. Additionally, the boundaries selected for the imposition of this type of boundary condition are required to be ordered in a consistent cyclic direction. The GUI path to invoke this dialog box is shown below:

Main Menu> Preprocessor> Peridynamic Processor> Region Identification> Boundary: Weak Form

As shown in Fig. 9.21, the dialog box for a ***Boundary: Weak Form*** requires the following parameters:

Boundary ID: An integer parameter to identify the boundary region.

Tang.vector to boundary: tx, ty: The direction cosines of the unit vector tangent to the selected boundary.

For modeling the boundary region, $\mathscr{B}$ as depicted Figs. 3.11 and 3.14, enter the Boundary ID and the unit tangent vector of the boundary. After setting the parameters of the ***Boundary: Weak Form*** dialog box as in Fig. 9.21 and clicking of the ***OK*** or ***Apply*** button, a node picker (Box) window appears for selecting the boundary PD nodes. Use the picker option to select the boundary region in which the boundary conditions are enforced in weak form. Figure 9.22 shows the creation of boundary region by using the node picker window corresponding to weak form of interactions.

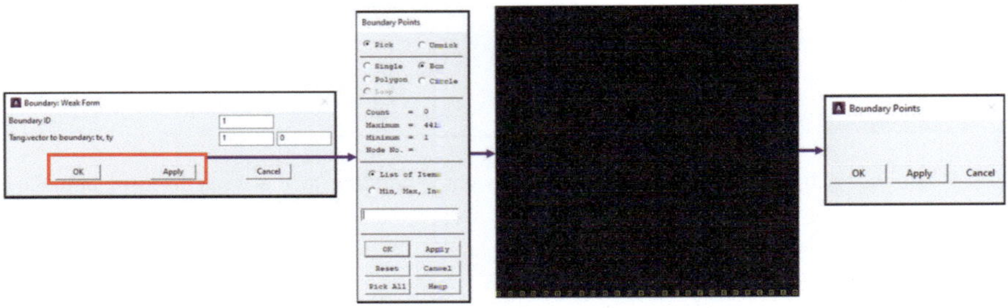

Fig. 9.22 Creation of the region corresponding to *Boundary: Weak Form* interaction

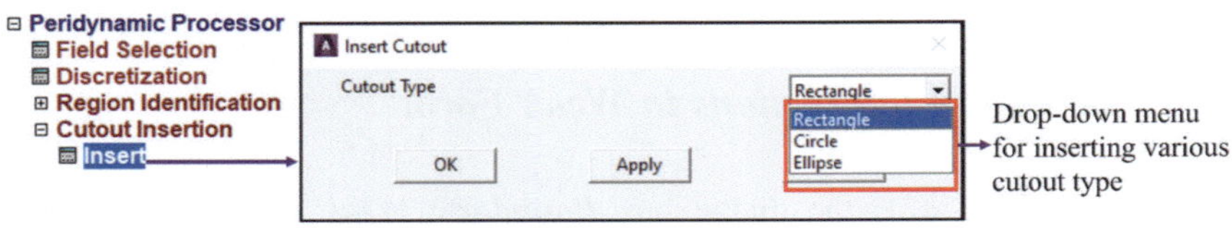

Fig. 9.23 Dialog box for inserting various cutouts in the PD mesh

For subsequent creation of other boundary regions, a different boundary domain ID should be used.

After selecting the nodes by using the Node picker window and clicking the ***OK*** or ***Apply*** button of the node picker (Box), a dialog box labeled as ***Boundary Points*** appears for confirming the selected regions. Furthermore, clicking of ***OK*** or ***Apply*** button of ***Boundary Points*** dialog box creates a nodal component corresponding to the ***Boundary ID***. The nodal component can be viewed by using *CMPLOT* command by entering in Ansys command window.

9.5 Cutout Insertion

As shown in Fig. 9.23, the dialog box ***Cutout Insertion*** appears upon selecting the ***Cutout Insertion>Insert*** submenu located under ***Peridynamic Processor*** menu. The GUI path for this dialog box is shown below:

Main Menu> Preprocessor> Peridynamic Processor> Cutout Insertion> Insert

There is only one parameter for ***Cutout Type*** in the ***Insert Cutout*** dialog box with a radio button showing the list of different cutout options.

Cutout Type: Corresponding to each cutout type in the drop-down list, a dialog box appears with an input field for cutout ID and its geometrical parameters. The input parameters for each cutout type are shown below:

Rectangle: As shown in Fig. 9.24, the dialog box for a ***Rectangular Cutout*** requires the following parameters:

Rectangular Cutout ID: An integer parameter to identify the rectangular cutout.

X1, Y1, Z1: The coordinates of the centroid of the rectangle in the global coordinate system.

Enter Length: The length of the rectangular cutout.

Enter Width: The width of the rectangular cutout.

Circle: As shown in Fig. 9.25, the dialog box for a ***Circular Cutout*** requires the following parameters:

Circular Cutout ID: An integer parameter to identify the circular cutout.

X1, Y1, Z1: The coordinates of the center of the circle in the global coordinate system.

Enter Radius: The radius of the circular cutout.

Ellipse: As shown in Fig. 9.26, the dialog box for an ***Elliptical Cutout*** requires the following parameters:

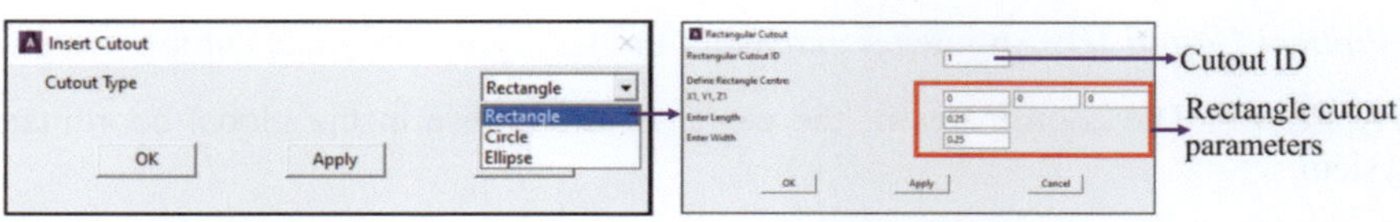

Fig. 9.24 Dialog box for entering the parameters of a rectangular cutout in the PD mesh

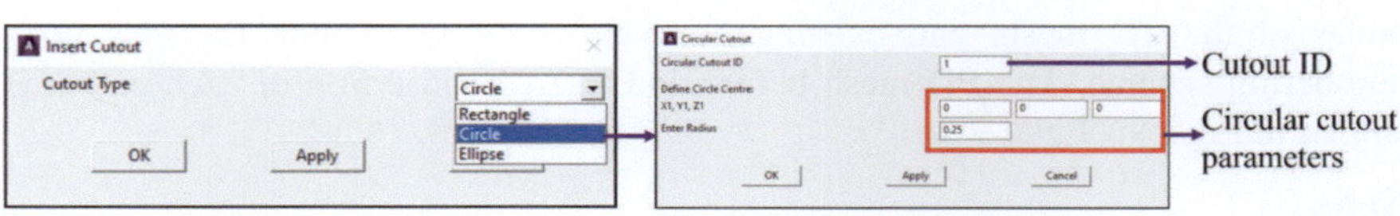

Fig. 9.25 Dialog box for entering the parameters of a circular cutout in the PD mesh

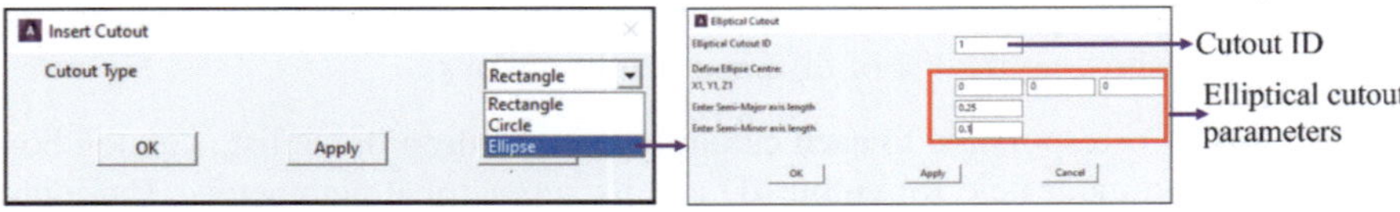

Fig. 9.26 Dialog box for entering the parameters of an elliptical cutout in the PD mesh

Fig. 9.27 The PD mesh without a cutout

Elliptical Cutout ID: An integer parameter to identify the elliptical cutout.

X1, Y1, Z1: The coordinates of the center of the ellipse in the global coordinate system.

Enter Semi-Major axis length: The length of the major axis of the elliptical cutout.

Enter Semi-Minor axis length: The length of the minor axis of the elliptical cutout.

After clicking the ***OK*** or ***Apply*** button corresponding to each cutout dialog box, the parameters are applied to the PD mesh, the cutouts are created by deleting the nodes of the PD mesh, and existing database is refreshed with the new node numbering scheme. The PD mesh before and after the insertion of each cutout is shown in Figs. 9.27 and 9.28.

Notes

The information of the nodes deleted during the cutout insertion is not saved. Hence, the original data must be restored if the cutout is to be modified.

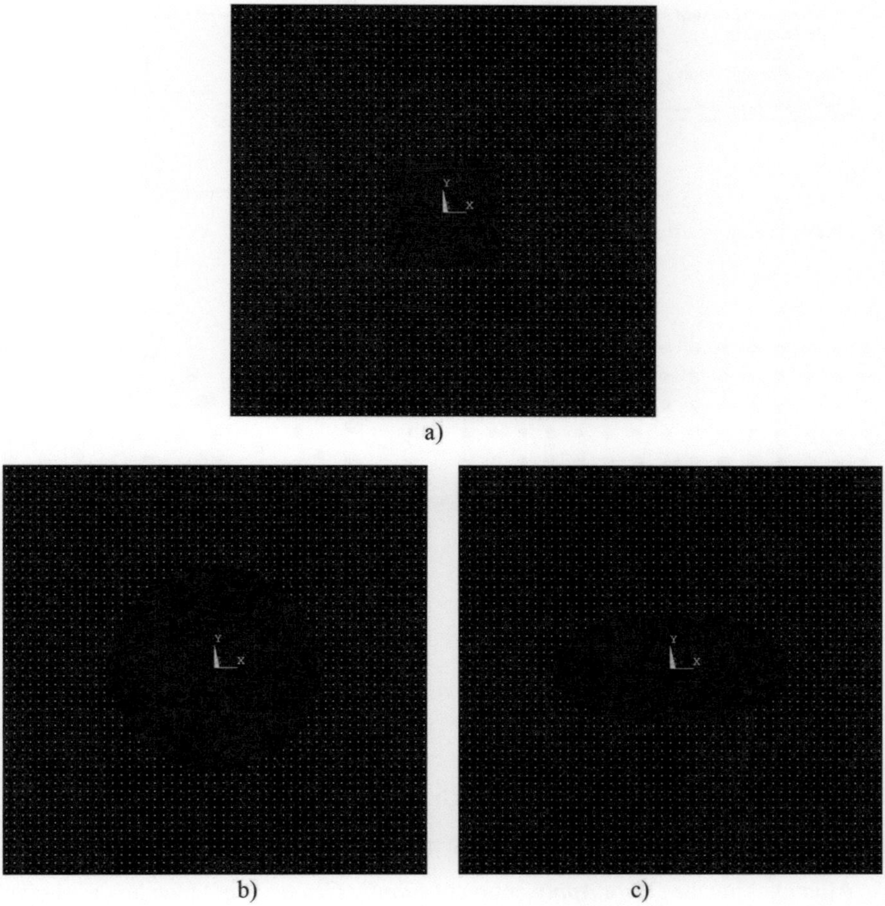

Fig. 9.28 PD mesh with various cutouts: (**a**) rectangular, (**b**) circular, and (**c**) elliptical

9.6 Family Generation

The family members of the nodes in the PD mesh are generated by using the submenu ***Family Generation*** located under ***Peridynamic Processor*** menu. As shown in Fig. 9.29, ***Family Generation*** dialog box is invoked upon clicking the ***Family Generation*** submenu. The GUI path to invoke this dialog box is shown below:

Main Menu> Preprocessor> Peridynamic Processor> Family Generation

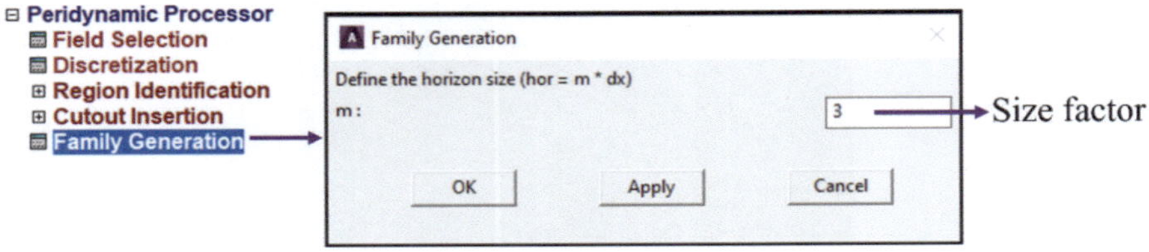

Fig. 9.29 Dialog box for family generation in PD mesh

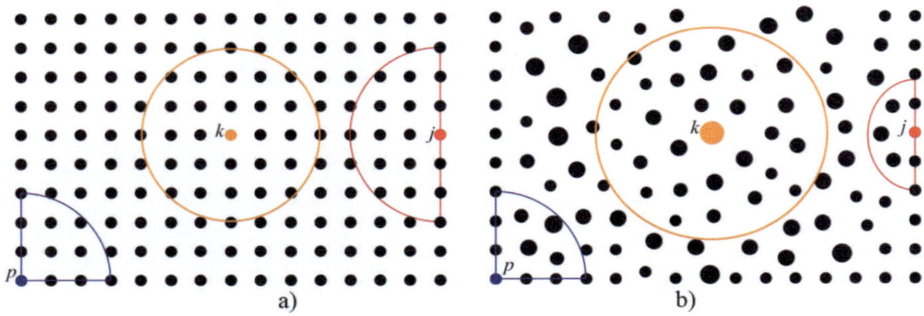

Fig. 9.30 Family members based on the search radius: (**a**) uniform PD mesh, and (**b**) nonuniform PD mesh

The parameter ***m*** used in the *Family Generation* dialog box is shown below:

m*:* Enter a parameter as a multiple of discretization size to control the horizon size.

After clicking the ***OK*** or ***Apply*** button of ***Family Generation*** dialog box as shown in Fig. 9.29, the family members for each point of the PD mesh is determined by considering the search radius (horizon), δ. The horizon of PD point is calculated by using $\delta = m\Delta$, where m is the size factor and Δ is discretization size of the PD point. The size factor $m = 3$ is set as a default option. Figure 9.30 shows a typical search area for establishing the family members of various PD points. For a specific size factor, the search radius is constant for a PD mesh with a uniform discretization. However, the search radius is dependent on the discretization size Δ for nonuniform discretization.

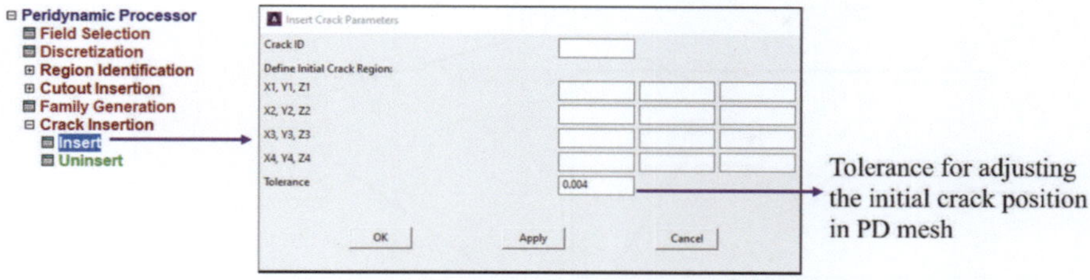

Fig. 9.31 Dialog box for inserting a crack in PD mesh

9.7 Crack Insertion

The initial crack in the PD mesh can be inserted using the submenu ***Crack Insertion>Insert*** located under ***Peridynamic Processor*** menu. As shown in Fig. 9.31, ***Insert Crack parameters*** dialog box is invoked upon clicking the ***Crack Insertion>Insert*** submenu. The GUI path to invoke this dialog box is shown below:

Main Menu> Preprocessor> Peridynamic Processor> Crack Insertion> Insert

The parameters in the ***Insert Crack parameters*** dialog box are given below:

Crack ID*:* Enter an integer parameter for crack ID.

X1, Y1, Z1*:* Enter the coordinates of the 1st end point of the crack in the global coordinates

X2, Y2, Z2*:* Enter the coordinates of the 2nd end point of the crack in the global coordinates

X3, Y3, Z3*:* Enter the coordinates of the 3rd end point of the crack in the global coordinates

X4, Y4, Z4*:* Enter the coordinates of the 4th end point of the crack in the global coordinates

Tolerance*:* Enter the tolerance value used to adjust the coordinate location of the crack in PD mesh.

In a 2D analysis, the initial crack is defined by a line. Hence, only two coordinates (*X1*, *Y1*) and (*X2*, *Y2*) corresponding to the end points of the line are used to generate the crack. In a 3D analysis, four coordinates (*X1*, *Y1, Z1*), (*X2*, *Y2, Z2*), (*X3*, *Y3, Z3*) and (*X4*, *Y4, Z4*) corresponding to the end points of the surface are used as an input to generate the crack surface. A default tolerance of $\Delta/100$ is used to adjust the coordinate locations of the crack in the PD mesh.

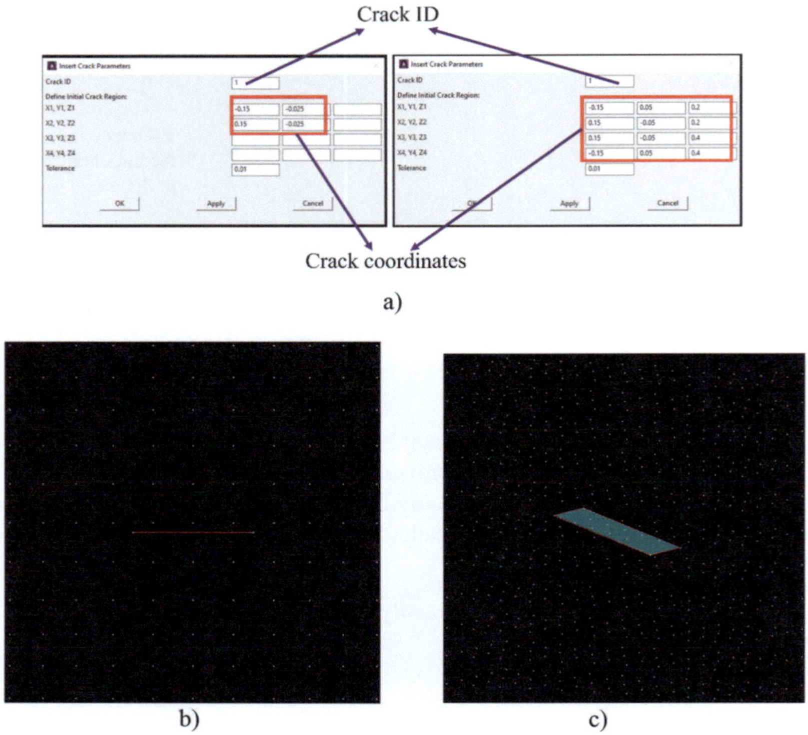

Fig. 9.32 Crack insertion in a 2D and 3D PD mesh. (**a**) Crack parameters, (**b**) line crack (2D), and (**c**) surface crack (3D)

As shown in Fig. 9.32a, after entering the crack ID and the coordinates and clicking the ***OK*** or ***Apply*** button in the ***Insert Crack parameters*** dialog box, a line or surface crack is inserted in the PD mesh with the component named as *CRACK_ID*, where *ID* is the entered input parameter.

For uninserting the existing crack in the PD mesh, the submenu ***Crack Insertion>Uninsert*** located under ***Peridynamic Processor*** menu can be used. As shown in Fig. 9.33, ***Uninsert Crack Parameters*** dialog box appears upon clicking the ***Crack Insertion>Uninsert*** submenu. The GUI path to invoke this dialog box is shown below:

Main Menu> Preprocessor> Peridynamic Processor> Crack Insertion> Uninsert

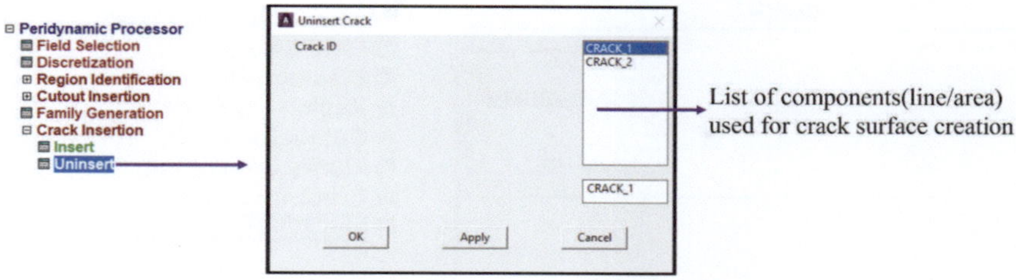

Fig. 9.33 Dialog box for uninserting the crack in PD mesh

The parameter in the ***Uninsert Crack*** dialog box shows a list of components (line/area) of the crack created with *CRACK_ID*, where *ID* is the number specified while inserting.

After clicking the ***OK*** or ***Apply*** button in ***Uninsert Crack*** dialog box, the parameters of the crack and the component (line/area) of the crack are deleted from the database and the status parameter of the bonds corresponding to the *CRACK_ID* are restored.

9.8 Element Coefficients

The coefficients of the element stiffness matrix corresponding to the PD interactions either in the domain or in the boundary of the PD mesh are generated by using the submenu ***Element Coefficients (Field)*** located under ***Peridynamic Processor*** menu. The ***Field*** refers to the field selection as previously selected in the submenu ***Field Selection.*** Figure 9.34 shows the ***Element Coefficients (Diffusion)*** *and* ***Element Coefficients (Structural)*** submenus as created based on ***Diffusion*** and ***Structural*** field selected under ***Field Selection*** submenu.

Figure 9.35 shows ***Element Coefficients (Diffusion)*** dialog box, which is invoked upon clicking the ***Element Coefficients (Diffusion)*** submenu. The GUI path to invoke this dialog box is shown below:

Main Menu> Preprocessor> Peridynamic Processor> Element Coefficients(Diffusion)

After clicking the ***OK*** or ***Apply*** button of ***Element Coefficients (Diffusion)*** dialog box as shown in Fig. 9.35, the stiffness coefficients are generated for the PD interactions in the domain and boundary for the field type ***Diffusion.***

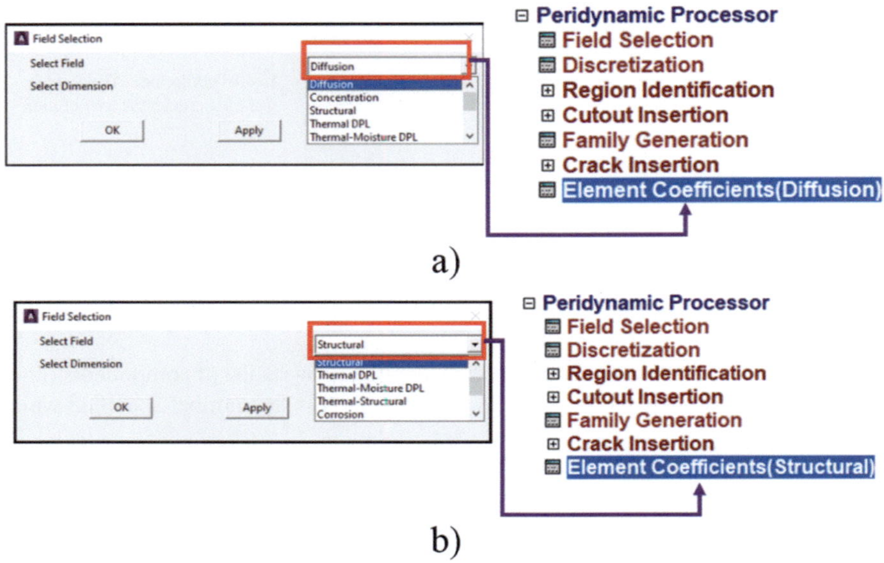

Fig. 9.34 Element Coefficients submenu created based on Field Selection submenu: (a) Diffusion, and (b) Structural

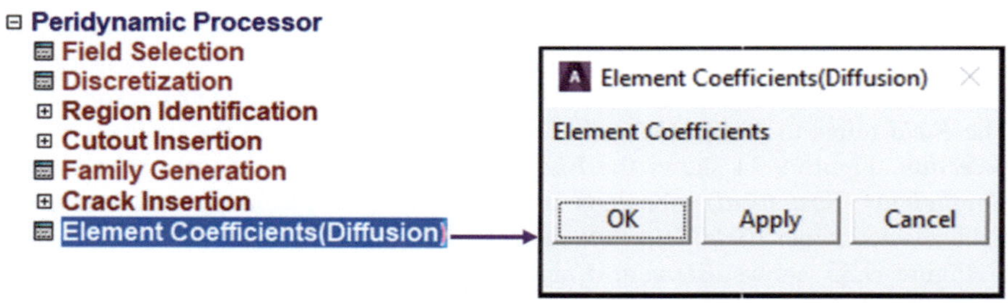

Fig. 9.35 Dialog box to generate element coefficients for diffusion field

9.9 Element Generation

The elements are generated to model the PD interactions either in the domain or in the boundary of the PD mesh by using the submenu ***Element Generation*** located under ***Peridynamic Processor*** menu.

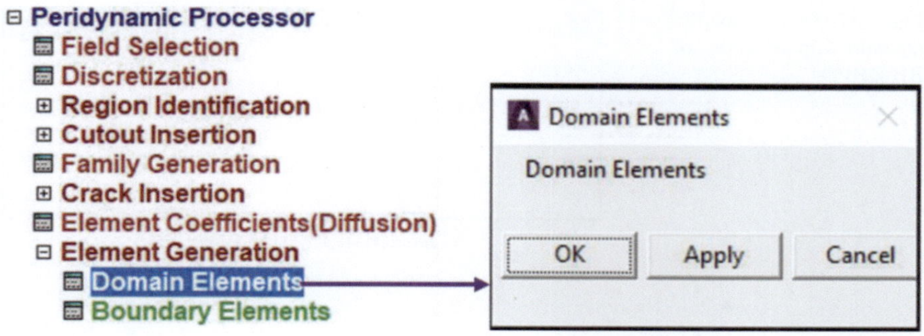

Fig. 9.36 Dialog box for element generation in the domain

9.9.1 Domain Elements

As shown in Fig. 9.36, the dialog box ***Domain Elements*** appears upon selecting the ***Element Generation>Domain Elements*** submenu located under ***Peridynamic Processor*** menu. This submenu generates the elements for the regions selected in the domain with a specific ID using the submenu ***Region Discretization>Domain***. The GUI path to invoke this dialog box is shown below:

Main Menu> Preprocessor> Peridynamic Processor> Element Generation> Domain Elements

9.9.1.1 BB-PD Elements

After clicking the ***OK*** or ***Apply*** button of ***Domain Elements*** dialog box as in Fig. 9.36, based on the options chosen in the ***Domain*** dialog box under ***Region Identification*** submenu, the element connectivity for BB-PD interactions are generated. Corresponding to the BB-DHPD interactions with dual horizon as previously specified in ***Region Identification*** submenu shown in Fig. 9.37, the element connectivity for primary and secondary bonds is shown in Fig. 9.38.

Corresponding to the BB-PD interactions with single horizon as previously specified in ***Region Identification*** submenu shown in Fig. 9.39, the element connectivity for PD bonds is shown in Fig. 9.40.

Fig. 9.37 Dialog box of ***Domain*** with selections of BB-DHPD

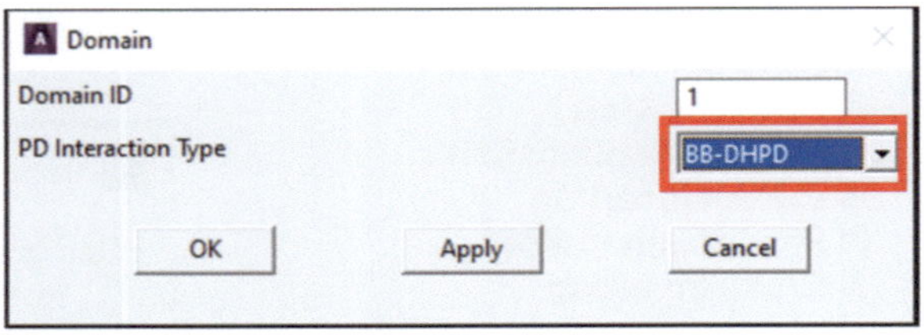

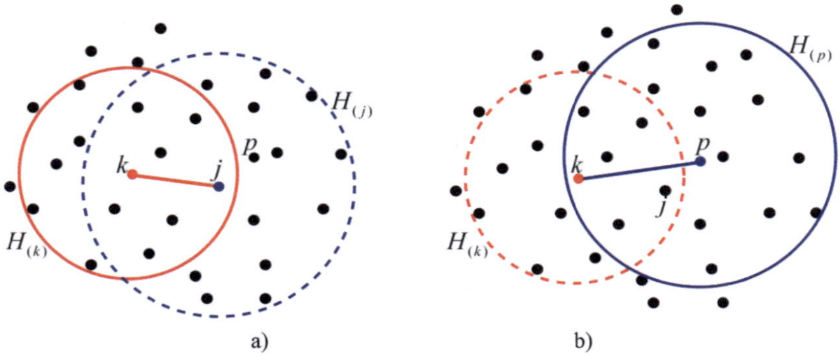

Fig. 9.38. Element connectivity for BB-DHPD interactions: (**a**) primary bond between nodes k and j, and (**b**) secondary bond between nodes k and p

Fig. 9.39 Dialog box of ***Domain*** with selections of BB-PD

Fig. 9.40 Element connectivity for BB-PD interaction between nodes k and j in $H_{(k)}$

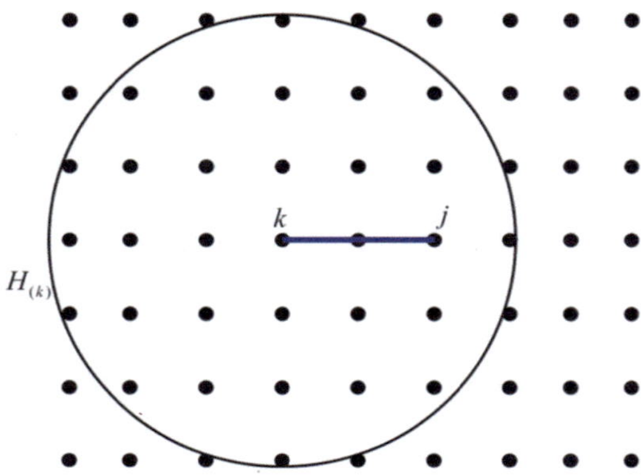

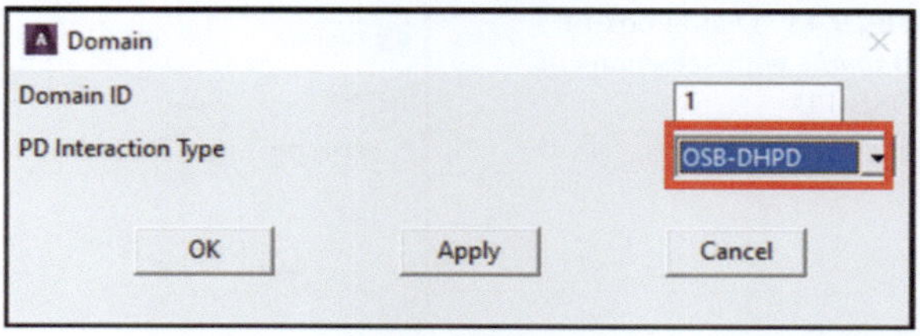

Fig. 9.41 Dialog box of ***Domain*** with selections of OSB-DHPD

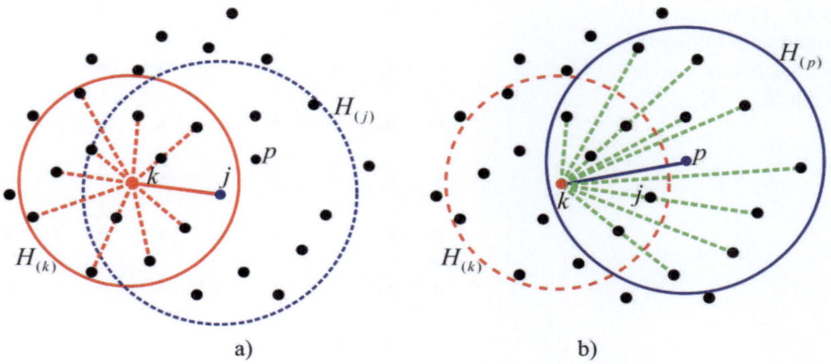

Fig. 9.42. Element connectivity for OSB-DHPD interactions: (**a**) primary bond between nodes k and j along with its direct interactions, and (**b**) secondary bond between nodes k and p along with its indirect interactions

9.9.1.2 OSB-PD Elements

After clicking the ***OK*** or ***Apply*** button of ***Domain Elements*** dialog box as in Fig. 9.36, based on the options chosen in the ***Domain*** dialog box under ***Region Identification*** submenu, the element connectivity for OSB-PD interactions are generated. Corresponding to the OSB-DHPD interactions with dual horizon as previously specified in ***Region Identification*** submenu shown in Fig. 9.41, the element connectivity for primary and secondary bonds is shown in Fig. 9.42.

Corresponding to the OSB-PD interactions with single horizon as previously specified in ***Region Identification*** submenu shown in Fig. 9.43, the element connectivity for PD bonds are shown in Fig. 9.44.

9.9.1.3 NOSB-PD Elements

After clicking the ***OK*** or ***Apply*** button of ***Domain Elements*** dialog box as in Fig. 9.36, based on the options chosen in the ***Domain*** dialog box under ***Region Identification*** submenu, the element connectivity for NOSB-PD interactions are generated. Corresponding to the NOSB-DHPD interactions with dual horizon as previously specified in ***Region Identification*** submenu shown in Fig. 9.45, the element connectivity for primary and secondary bonds are shown in Fig. 9.46.

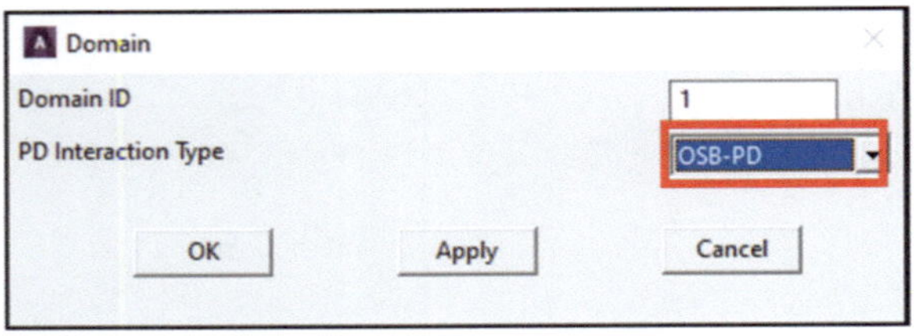

Fig. 9.43 Dialog box of ***Domain*** with selections of OSB-PD

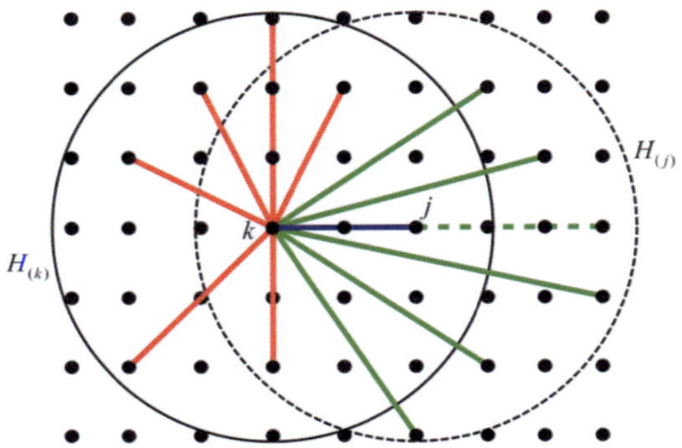

Fig. 9.44. Element connectivity for OSB-PD interactions between nodes k and j in $H_{(k)} \cup H_{(j)}$

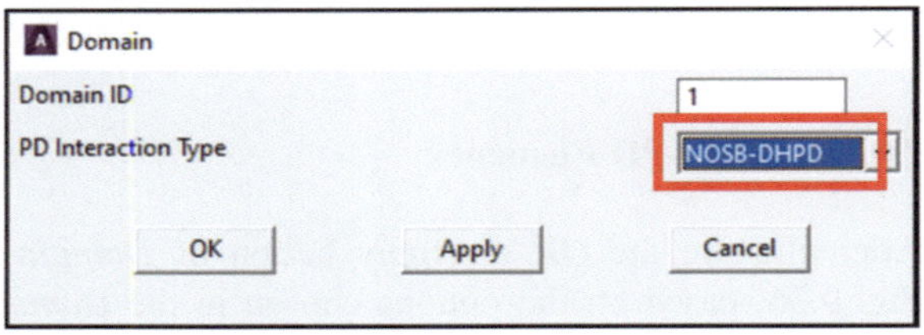

Fig. 9.45 Dialog box of ***Domain*** with selections of NOSB-DHPD

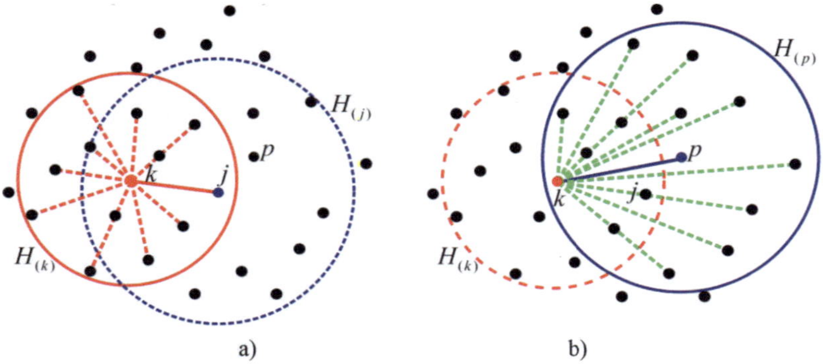

Fig. 9.46. Element connectivity for NOSB-DHPD interactions: (**a**) primary bond between nodes k and j along with its direct interactions, and (**b**) secondary bond between nodes k and p along with its indirect interactions

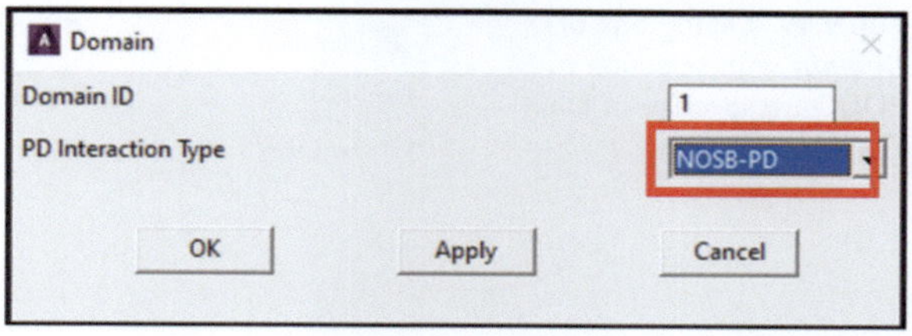

Fig. 9.47 Dialog box of ***Domain*** with selections of NOSB-PD

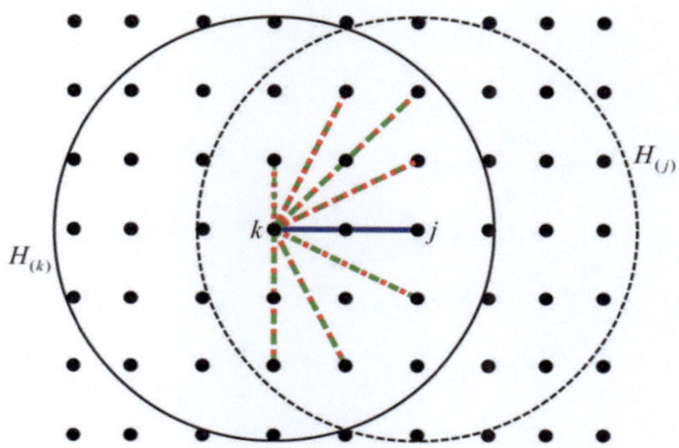

Fig. 9.48. Element connectivity for NOSB-PD interactions between nodes k and j in $H_{(k)} \cap H_{(j)}$

Corresponding to the NOSB-PD interactions with single horizon as previously specified in ***Region Identification*** submenu shown in Fig. 9.47, the element connectivity for PD bonds are shown in Fig. 9.48.

9.9.1.4 PDDO Elements

After clicking the ***OK*** or ***Apply*** button of ***Domain Elements*** dialog box as in Fig. 9.36, based on the options chosen in the ***Domain*** dialog box under ***Region Identification*** submenu, the element connectivity for PDDO interactions are generated. Corresponding to the PDDO interactions as previously specified in ***Region Identification*** submenu shown in Fig. 9.49, the element connectivity due to direct and indirect interactions at node k is shown in Figs. 9.50 and 9.51 for nonuniform and uniform mesh discretization.

9.9.2 Boundary Elements

As shown in Fig. 9.52, the dialog box ***Boundary Elements*** appears upon selecting the ***Element Generation>Boundary Elements*** submenu located under ***Peridynamic Processor*** menu. The GUI path to invoke this dialog box is shown below:

Fig. 9.49 Dialog box of ***Domain*** with selection of PDDO irrespective of Dual Horizon selection

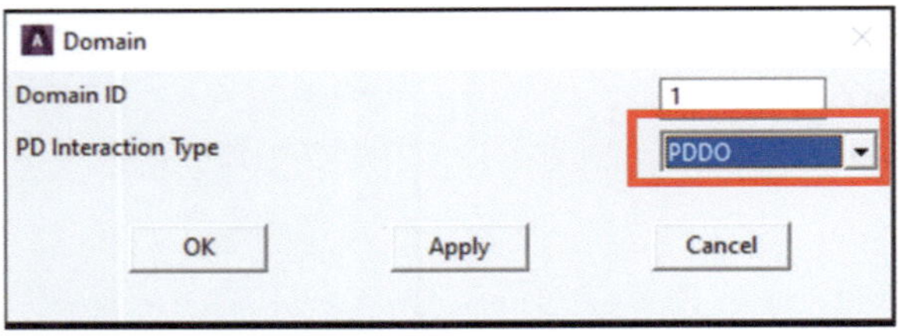

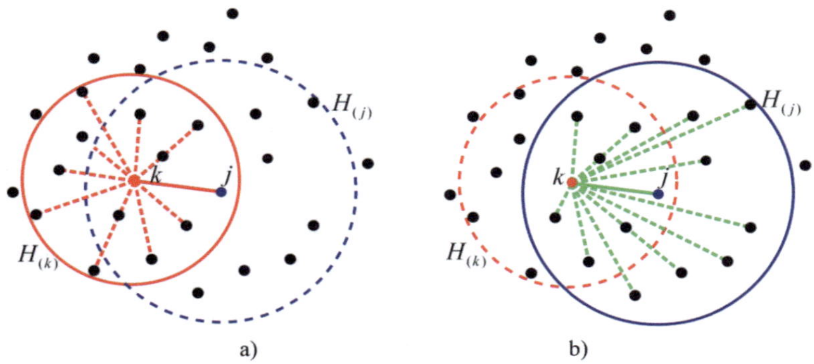

Fig. 9.50. Element connectivity for PDDO interactions between nodes k and j: (**a**) direct interaction, and (**b**) indirect interaction

Fig. 9.51. Element connectivity for constant horizon PDDO interactions between nodes k and j in $H_{(k)} \cup H_{(j)}$

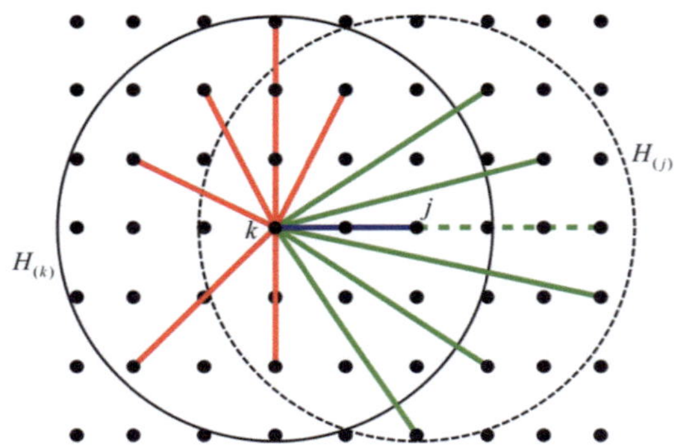

Main Menu> Preprocessor> Peridynamic Processor> Element Generation> Boundary Elements

9.9.2.1 Boundary Elements in Strong Form

After clicking the ***OK*** or ***Apply*** button of ***Boundary Elements*** dialog box as in Fig. 9.52, based on the options chosen in the ***Boundary:Strong Form*** dialog box under ***Region Identification*** submenu as in Fig. 9.53, the element connectivity for boundary interactions in strong form at node k on the boundary is generated as shown in Fig 9.54.

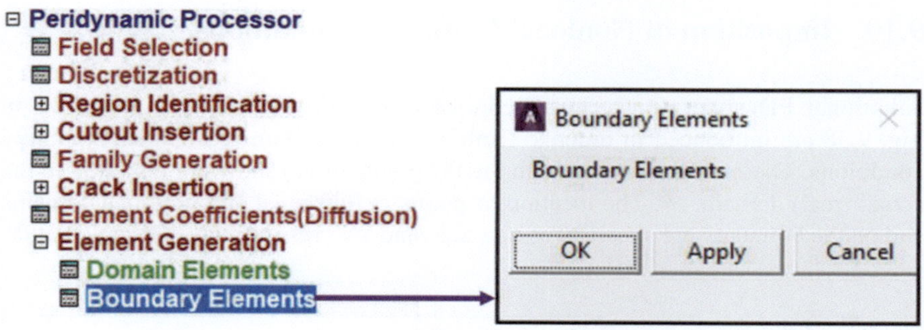

Fig. 9.52 Dialog box for boundary element generation

Fig. 9.53 Dialog box of ***Boundary: Strong Form***

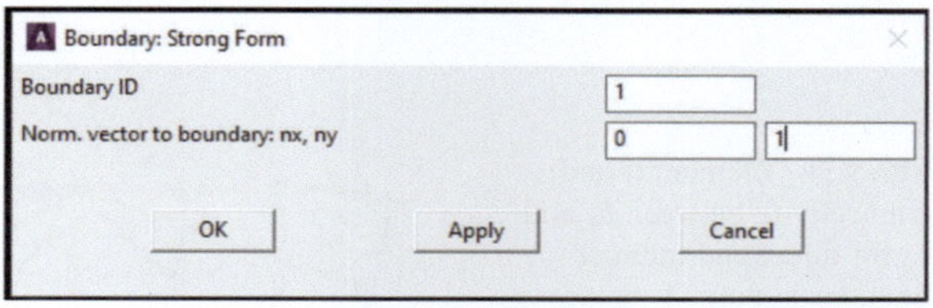

Fig. 9.54 Element connectivity between node k and other nodes in $H_{(k)}$ using strong form of PD boundary interactions

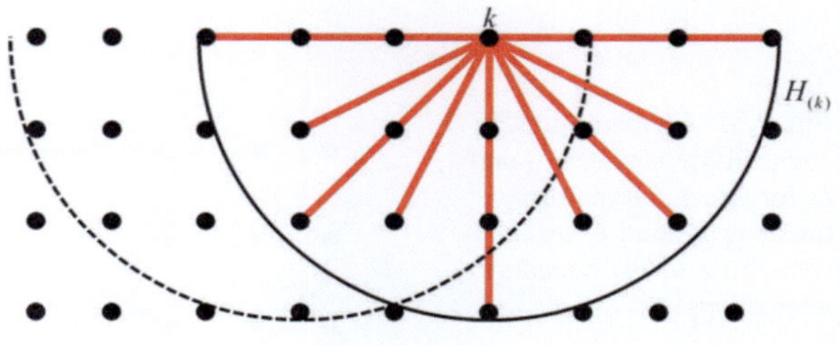

9.9.2.2 Boundary Elements in Weak Form

After clicking the ***OK*** or ***Apply*** button of ***Boundary Elements*** dialog box as in Fig. 9.52, based on the options chosen in the ***Boundary:Weak Form*** dialog box under ***Region Identification*** submenu as in Fig. 9.55, the element connectivity for boundary interactions in weak form along the boundary path $\ell_1 - \ell_2$ is shown in Figs. 9.56 and 9.57.

9.10 Imposition of Nonlocal Boundary Conditions

Additional PD nodes (pink) in a fictitious layer, $\mathscr{R}_f$ with depth δ as shown in Fig. 9.58 are introduced in order to facilitate the imposition of nonlocal boundary conditions. These points interact with the PD points (blue) shown in Fig. 9.58 in the actual (real) domain, $\mathscr{R}$. The location of points in the actual and fictitious domains and on the boundary are specified as $\mathbf{x}_{(k)}$, $\widetilde{\mathbf{x}}_{(p)}$, and $\mathbf{x}^*_{(k)}$, respectively. The field values

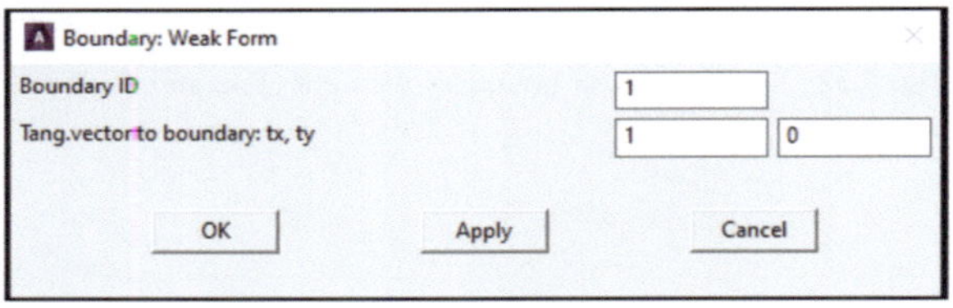

Fig. 9.55 Dialog box of ***Boundary: Weak Form***

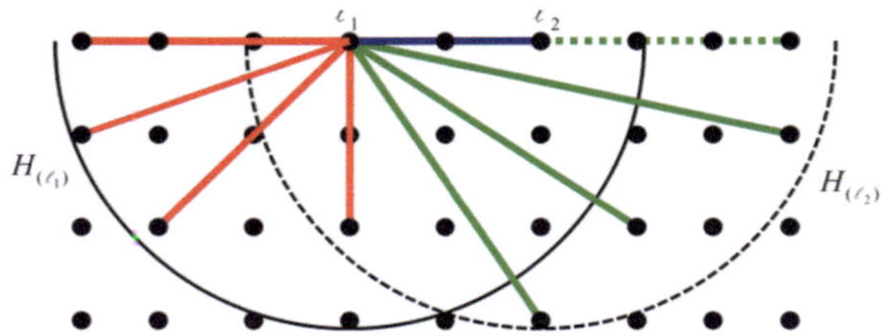

Fig. 9.56 Element (bond) connectivity between ℓ_1 and ℓ_2 for direct and indirect tractions at point ℓ_1 using weak form of PD boundary interactions

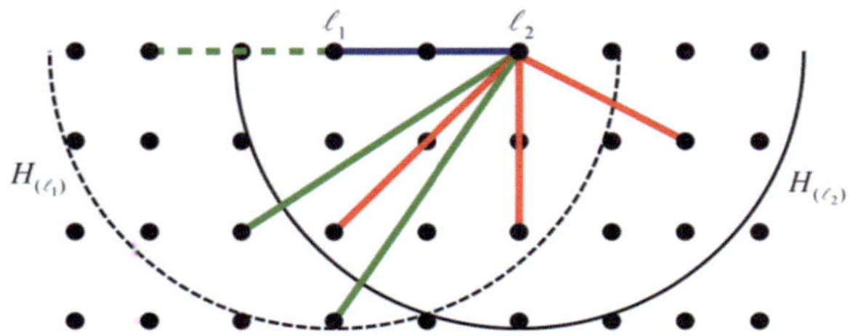

Fig. 9.57 Element (bond) connectivity between ℓ_2 and ℓ_1 for direct and indirect tractions at point ℓ_2 using weak form of PD boundary interactions

Fig. 9.58 Additional PD nodes in the fictitious region (pink colored) for the imposition of essential boundary conditions (Dirichlet)

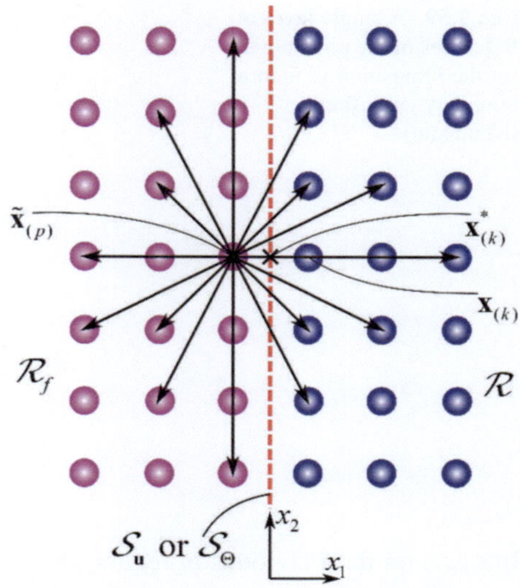

in the fictitious domain and on the boundary $\mathcal{S}_\Theta$ are denoted as $\widetilde{\Theta}_{(p)}(t) = \Theta(\widetilde{\mathbf{x}}_{(p)}, t)$ and $\Theta^*_{(k)}(t) = \Theta(\mathbf{x}^*_{(k)}, t)$, respectively. Similarly, the displacement values in the fictitious domain and on the boundary $\mathcal{S}_\mathbf{u}$ are denoted as $\widetilde{\mathbf{u}}_{(p)}(t) = \mathbf{u}(\widetilde{\mathbf{x}}_{(p)}, t)$ and $\mathbf{U}^*_{(k)}(t) = \mathbf{u}(\mathbf{x}^*_{(k)}, t)$, respectively.

9.10.1 Direct Approach

The imposition of essential (Dirichlet) boundary conditions is achieved through the PD nodes in the fictitious region, $\mathscr{R}_f$ as shown in Fig. 9.58. The known value of displacement, $\mathbf{U}^*_{(k)}(t)$ or field $\Theta^*_{(k)}(t)$ on the boundary $\mathcal{S}_\mathbf{u}$ or $\mathcal{S}_\Theta$, respectively, can be directly enforced by assigning their values to the PD nodes in the fictitious region as

$$\widetilde{\mathbf{u}}_{(p)}(t) = \mathbf{U}^*_{(k)}(t) \tag{9.2}$$

and

$$\widetilde{\Theta}_{(p)}(t) = \Theta^*_{(k)}(t) \tag{9.3}$$

The imposition of natural (Neumann) boundary conditions is achieved through a single layer of PD nodes in a "real material layer," $\mathscr{R}_r$ of width Δ as shown in Fig. 9.59. The applied nonzero traction $\mathbf{T}(\mathbf{x}^*_{(k)}, t)$, or a point force, $\mathbf{P}(\mathbf{x}^*_{(k)}, t)$, over the surface $\mathcal{S}_\boldsymbol{\sigma}$ of the boundary, can be enforced in the form of a body force density

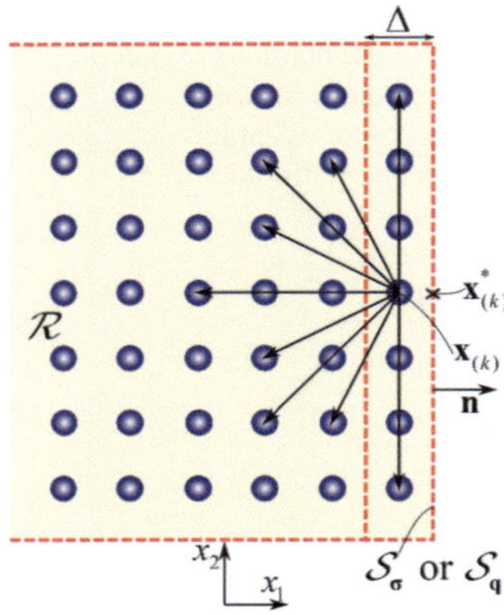

Fig. 9.59 A single layer of PD nodes in the real domain for the imposition of natural boundary conditions (Neumann)

$\mathbf{b}(\mathbf{x}_{(k)}, t)$ on the PD nodes in region, $\mathscr{R}_r$. Similarly, a specified field flux, $\mathbf{q}\left(\mathbf{x}^*_{(k)}, t\right)$, over the surface $\mathcal{S}_\mathbf{q}$ of the boundary can be enforced in the form of a field source $s_b(\mathbf{x}_{(k)}, t)$ on the PD nodes in region, $\mathscr{R}_r$.

The body force density vector, $\mathbf{b}(\mathbf{x}_{(k)}, t)$ arising from the applied nonzero traction, $\mathbf{T}\left(\mathbf{x}^*_{(k)}, t\right)$, or point force, $\mathbf{P}\left(\mathbf{x}^*_{(k)}, t\right)$ can be evaluated as

$$\mathbf{b}\left(\mathbf{x}_{(k)}, t\right) = \frac{1}{\Delta}\mathbf{T}\left(\mathbf{x}^*_{(k)}, t\right) \tag{9.4}$$

or

$$\mathbf{b}\left(\mathbf{x}_{(k)}, t\right) = \frac{1}{\mathcal{S}_\sigma \Delta}\mathbf{P}\left(\mathbf{x}^*_{(k)}, t\right) \tag{9.5}$$

The body load density is imposed as a force, which can be specified through the use Ansys F command.

The field source $s_b(\mathbf{x}_{(k)}, t)$ arising from the applied field flux, $\mathbf{q}\left(\mathbf{x}^*_{(k)}, t\right)$ can be evaluated as

$$s_b\left(\mathbf{x}_{(k)}, t\right) = -\frac{1}{\Delta}\mathbf{q}\left(\mathbf{x}^*_{(k)}, t\right) \cdot \mathbf{n}\left(\mathbf{x}^*_{(k)}, t\right) \tag{9.6}$$

where $\mathbf{n}$ is the unit normal to the boundary, $\mathcal{S}_\mathbf{q}$.

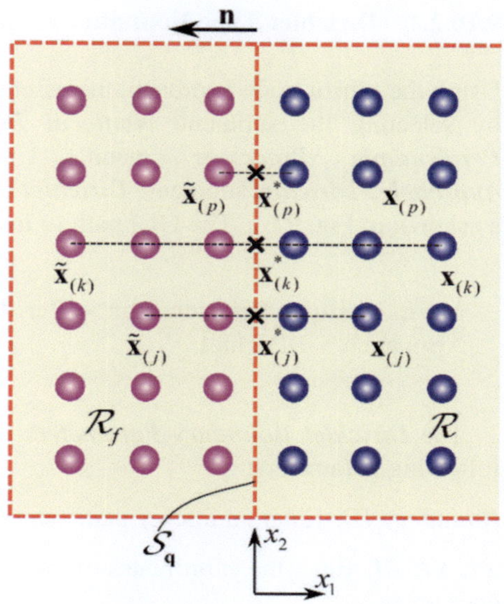

Fig. 9.60 Mirror node approach of imposition of the essential boundary conditions on PD mesh using fictitious patch

9.10.2 Mirror Node Approach

The field values on the boundary can be enforced using the mirror node approach as suggested by Madenci and Oterkus (2016). The field values on the boundary $\mathcal{S}_\Theta$ with a unit normal $\mathbf{n}$ are specified as $\Theta^*_{(\alpha)}(t) = \Theta\left(\mathbf{x}^*_{(\alpha)}, t\right)$. As shown in Fig. 9.60, the locations of PD (blue) points in the actual domain are specified as $\mathbf{x}_{(\alpha)}$ with $\alpha = k, p,$ and j. Their field values are projected to the (pink) points, $\widetilde{\mathbf{x}}_{(\alpha)}$ in the fictitious domain, $\mathcal{R}_f$ with depth δ.

The field value on the boundary $\mathcal{S}_\Theta$ can be imposed as

$$\widetilde{\Theta}_{(\alpha)}(t + \Delta t) = 2\Theta^*_{(\alpha)}(t) - \Theta_{(\alpha)}(t) \tag{9.7}$$

The known field gradient on the boundary $\mathcal{S}_\mathbf{q}$, i.e.,$\boldsymbol{\Psi}^*_{(\alpha)}(t) = \boldsymbol{\Psi}^*_{(\alpha)}\left(\mathbf{x}^*_{(\alpha)}, t\right)$ can be enforced as

$$\widetilde{\Theta}_{(\alpha)}(t + \Delta t) = \boldsymbol{\Psi}^*_{(\alpha)}(t) \cdot \left(\widetilde{\mathbf{x}}_{(\alpha)} - \mathbf{x}_{(\alpha)}\right) + \Theta_{(\alpha)}(t) \tag{9.8}$$

The relationship in Eqs. (9.7) and (9.8) can be enforced through the use of Ansys D command.

9.10.2.1 Dirichlet Type Boundary Conditions

Using the mirror node approach, the essential boundary conditions can be imposed by selecting the submenu ***Nonlocal BC Approach>Dirichlet*** located under ***Peridynamic Processor*** menu. Upon clicking the ***Nonlocal BC Approach>Dirichlet*** submenu, ***Dirichlet Boundary Parameters*** dialog box appears as shown in Fig. 9.61. The GUI path to invoke this dialog box is shown below:

Main Menu> Preprocessor> Peridynamic Processor> Nonlocal BC Approach> Dirichlet

The ***Dirichlet Boundary Parameters*** dialog box requires input values for the following parameters:

Boundary ID: Enter an integer parameter for the boundary ID.

X1, Y1, Z1: Enter the coordinates of the 1st end of the boundary point in the global coordinates.

X2, Y2, Z2: Enter the coordinates of the 2nd end of the boundary point in the global coordinates.

X3, Y3, Z3: Enter the coordinates of the 3rd end of the boundary point in the global coordinates.

After entering the boundary parameters and clicking the ***OK*** or ***Apply*** button of the dialog box in Fig. 9.61, a node picker (Box) window appears for selecting the PD nodes in the fictitious region as shown in Fig. 9.62.

Furthermore, after clicking the ***OK*** or ***Apply*** button of the node picker (Box) window, the dialog box ***Apply Boundary Condition*** appears as shown in Fig. 9.63 for choosing the parameter type of the boundary value.

The boundary value of S_Θ can be input through a constant parameter as shown in Fig. 9.63a or tabular parameter as shown in 9.63b. To specify the boundary value as a function of any input variable, a predefined tabular parameter is created using

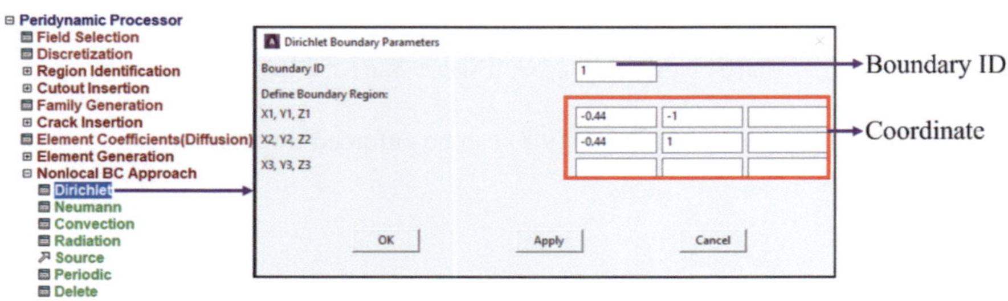

Fig. 9.61 Dialog box for entering the Dirichlet parameters in PD mesh

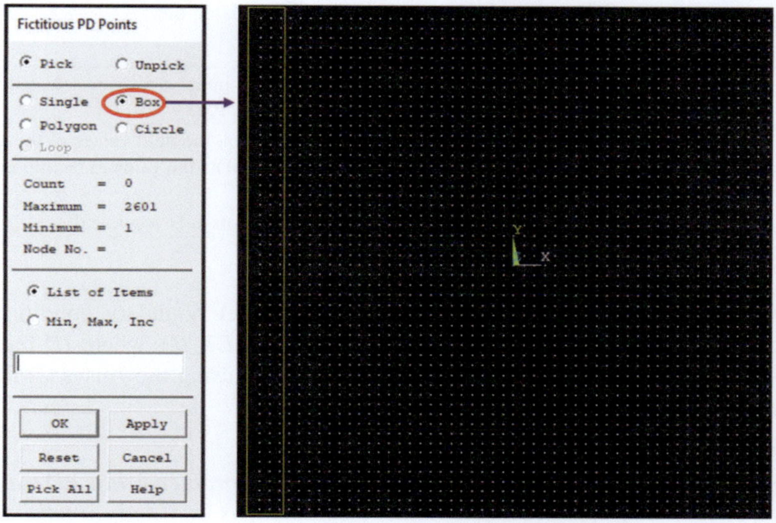

Fig. 9.62 Node picker option for selecting the set of fictitious nodes

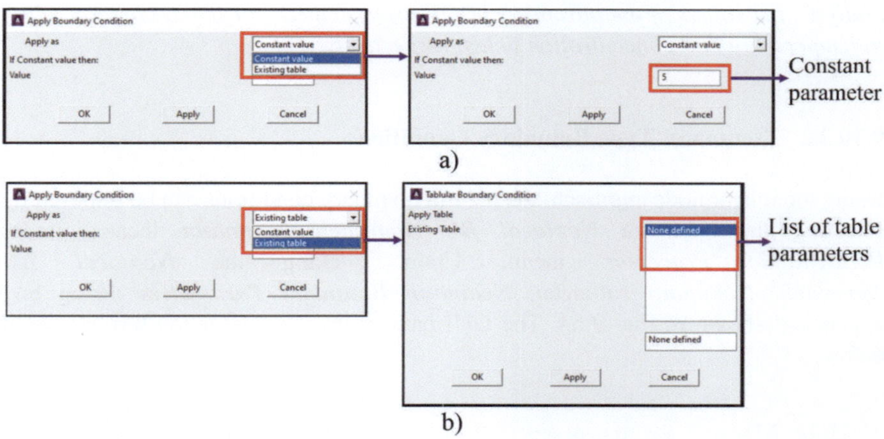

Fig. 9.63 Dialog box for entering the boundary value through different options: (**a**) Constant parameter, and (**b**) Tabular input

****DIM*** Ansys command with ***TABLE*** option and the name of the parameter as previously defined is selected as shown in Fig. 9.63b.

After clicking the ***OK*** or ***Apply*** button of the ***Apply Boundary Condition*** dialog box triggers the following actions as shown in Table 9.2.

Table 9.2 Summary of actions for ***Apply Boundary Condition*** dialog box

Step	Action
1	For each of the selected fictitious nodes, the corresponding image nodes are selected in the real domain with respect to the boundary as shown in Fig. 9.62
2	After pairing the nodes in the fictitious and the real domain, the field values of the real domain nodes and that of boundary value are used to calculate the value to be imposed on the fictitious nodes using Eq. (9.7)
3	The values of the fictitious nodes are imposed using the Ansys D command

The fictitious nodes used for creating the Dirichlet boundary condition are grouped under a nodal component named as *DIRICHLET_ID*, where *ID* refers to the ***Boundary ID*** used in Fig. 9.61 and can be viewed using the Ansys CMPLOT command.

Notes
This menu is currently supported only for transient field problems. For PD transient structural analysis, the essential boundary condition is imposed directly on the fictitious nodes as in Eq. (9.2) as an imposed displacement constraint using the D command in Ansys. For Thermal-Moisture DPL analysis, another parameter labeled as DOF is embedded in the dialog box for entering the Dirichlet boundary parameters. For imposing the Dirichlet boundary conditions for coupled field analysis, the values of the parameter, DOF are specified as 1 and 2 corresponding to temperature and concentration fields, respectively.

9.10.2.2 Neumann Type Boundary Conditions

Using the mirror node approach, the natural boundary conditions can be imposed by selecting the submenu ***Nonlocal BC Approach>Neumann*** located under ***Peridynamic Processor*** menu. Upon clicking the ***Nonlocal BC Approach>Neumann*** submenu, ***Neumann Boundary Parameters*** dialog box appears as shown in Fig. 9.64. The GUI path to invoke this dialog box is shown below:

Main Menu> Preprocessor> Peridynamic Processor> Nonlocal BC Approach> Neumann

The ***Neumann Boundary Parameters*** dialog box requires input values for the following parameters:

Boundary ID: Enter an integer parameter for the boundary ID.

X1, Y1, Z1: Enter the coordinates of the 1st end of the boundary point in the global coordinate system.

X2, Y2, Z2: Enter the coordinates of the 2nd end of the boundary point in the global coordinate system.

X3, Y3, Z3: Enter the coordinates of the 3rd end of the boundary point in the global coordinate system.

After entering the boundary parameters and clicking the ***OK*** or ***Apply*** button of the dialog box in Fig. 9.64, a node picker (Box) window appears for selecting the PD nodes in the fictitious region as shown in Fig. 9.65.

Furthermore, after clicking the ***OK*** or ***Apply*** button of the node picker (Box) window, the dialog box ***Apply Boundary Condition*** appears as shown in Fig. 9.66 for choosing the parameter type of the boundary value.

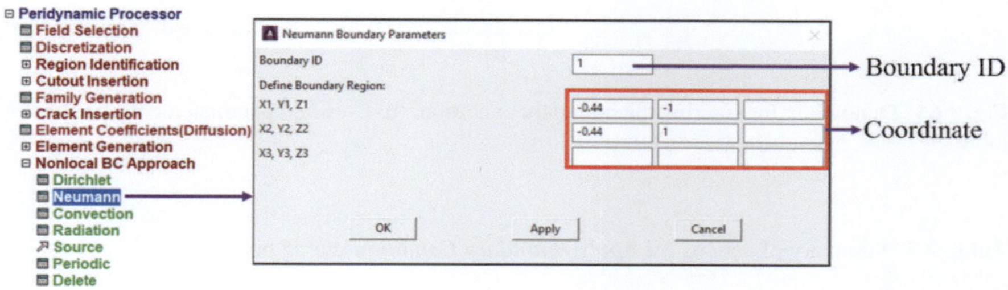

Fig. 9.64 Dialog box for entering the Neumann parameters in PD mesh

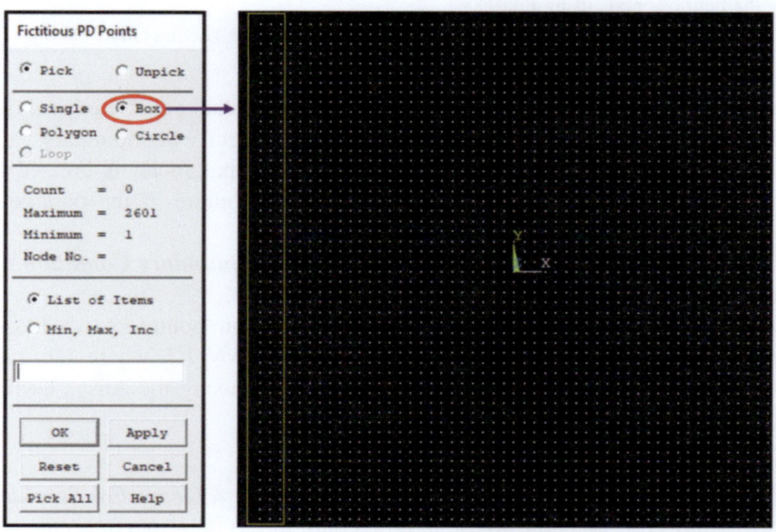

Fig. 9.65 Node picker option for selecting the set of fictitious nodes

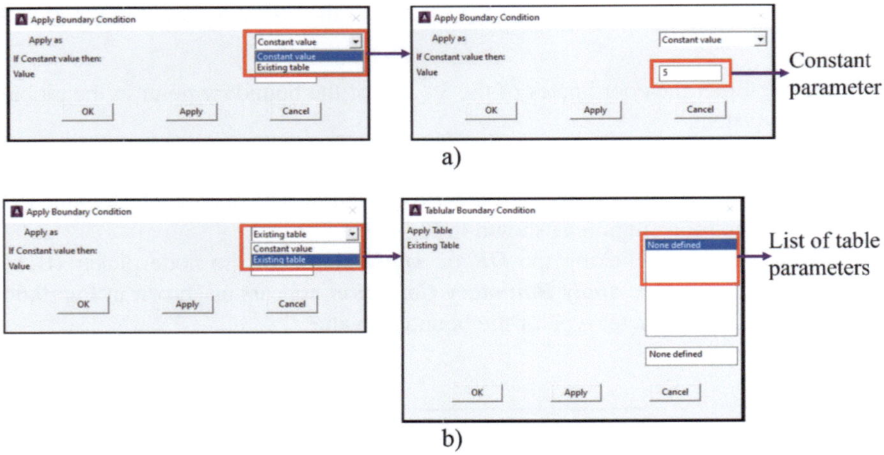

Fig. 9.66 Dialog box for entering the boundary condition: (**a**) Constant parameter, and (**b**) Tabular input

Table 9.3 Summary of actions for ***Apply Boundary Condition*** dialog box

Step	Action
1	For each of the fictitious nodes selected, the corresponding image nodes are selected in the real domain with respect to the boundary as shown in Fig 9.65
2	After pairing the nodes in the fictitious and the real domain, the field values of the real domain nodes and that of boundary value are used to calculate the value to be imposed on the fictitious node using Eq.(9.8)
3	The values of the fictitious nodes are imposed using Ansys D command

The boundary value of $\mathcal{S}_{\mathbf{q}}$ can be input through a constant parameter as shown in Fig. 9.66a or tabular parameter as shown in 9.66b. To specify the boundary value as a function of any input variable, a predefined tabular parameter is created using ****DIM*** Ansys command with ***TABLE*** option and the name of the parameter as previously defined is selected as shown in Fig. 9.66b.

After clicking the ***OK*** or ***Apply*** button of the ***Apply Boundary Condition*** dialog box triggers the following actions as shown in Table 9.3.

The fictitious nodes used for creating the Neumann boundary condition are grouped under a nodal component named as *NEUMANN_ID*, where *ID* refers to the ***Boundary ID*** used in Fig. 9.64 and can be viewed using the Ansys CMPLOT command.

Notes

This menu is currently supported only for transient field problems. For PD transient structural analysis, the natural boundary condition is imposed directly on the domain nodes in the form of a body load as in Eq. (9.4) and Eq. (9.5) using the F command in Ansys. For Thermal-Moisture DPL analysis, another parameter

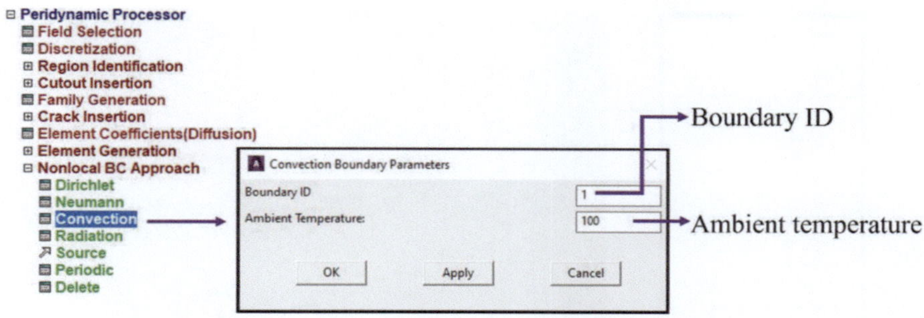

Fig. 9.67 Dialog box for entering the convection boundary condition parameters

labeled as DOF is embedded in the dialog box for entering the Neumann boundary parameters. For imposing the Neumann boundary conditions for coupled field analysis, the values of the parameter DOF are specified as 1 and 2 corresponding to temperature and concentration fields, respectively.

9.10.2.3 Convection

The convection boundary condition associated with heat conduction can be imposed by selecting the submenu ***Nonlocal BC Approach>Convection*** located under ***Peridynamic Processor*** menu. Upon clicking the ***Nonlocal BC Approach>Convection*** submenu, ***Convection Boundary Parameters*** dialog box appears as shown in Fig. 9.67. The GUI path to invoke this dialog box is shown below:

Main Menu> Preprocessor> Peridynamic Processor> Nonlocal BC Approach> Convection

The ***Convection Boundary Parameters*** dialog box requires the following parameters:

Boundary ID*:* Enter an integer parameter for the boundary ID.

Ambient Temperature*:* Enter the ambient temperature value.

After entering the boundary parameters and clicking the ***OK*** or ***Apply*** button of the dialog box in Fig 9.67, a node picker (Box) window appears to select a single layer of PD nodes close to the boundary as shown in Fig. 9.68.

Furthermore, after clicking the ***OK*** or ***Apply*** button of the node picker (Box) window, the dialog box ***Apply Convection*** shown in Fig. 9.69 appears for choosing the parameter type of the boundary value.

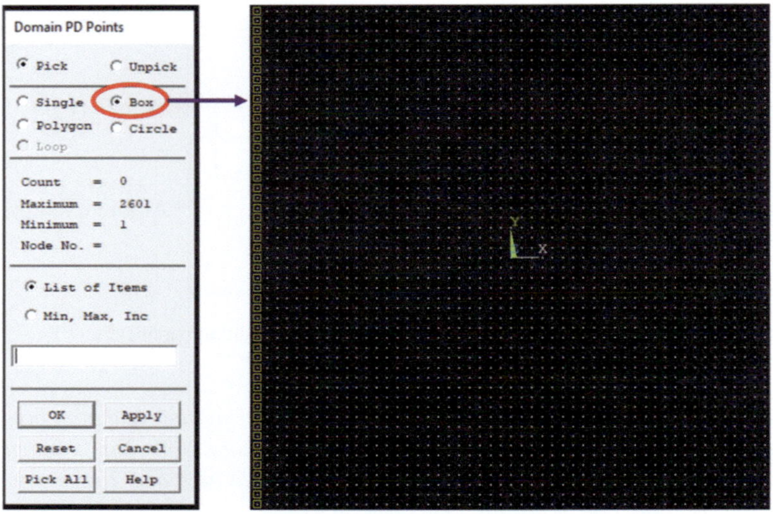

Fig. 9.68 Node picker option for selecting the set of domain nodes

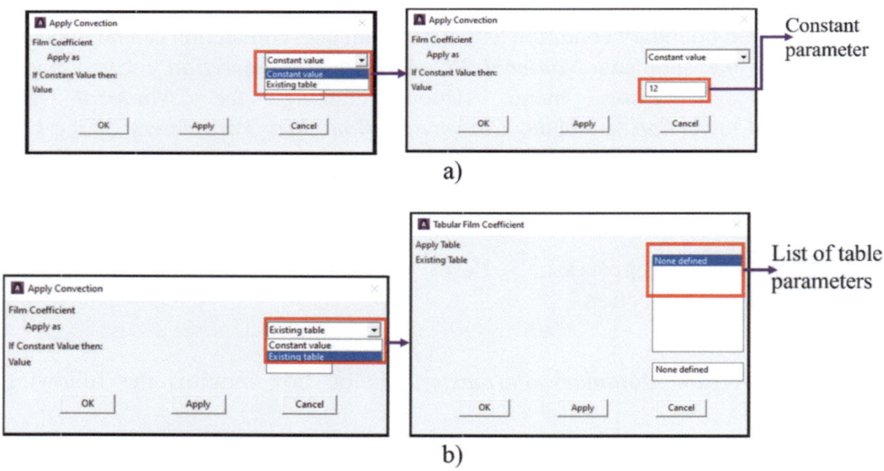

Fig. 9.69 Dialog box for entering convection film coefficient: (**a**) Constant parameter, and (**b**) Tabular input

As shown in Fig. 9.69a, b, the convection film coefficient, $\widehat{h}$, can be specified as a constant parameter or a tabular parameter. To specify the film coefficient as a function of any input variable, a predefined tabular parameter is created using ***DIM** command with ***TABLE*** option and the name of the parameter as defined is selected as shown in Fig. 9.69b

After clicking the ***OK*** or ***Apply*** button of the ***Apply Convection*** dialog box triggers the following actions as shown in Table 9.4.

The LINK34 elements using the PD domain nodes and the newly created offset nodes for modeling the convection is shown in Fig. 9.70. These elements are grouped under an element component named as *CONVECTION_ID*, where *ID* refers to the ***Boundary ID*** used in Fig. 9.67 and can be viewed using the Ansys CMPLOT command.

Notes

This menu is currently supported only for transient diffusion problems involving heat transfer.

Table 9.4 Summary of actions for ***Apply Convection*** dialog box

Step	Action
1	For the selected set of domain nodes as in Fig. 9.68, a new set of nodes with a small offset are created for generation of LINK34 elements
2	For all the new set of nodes created, the nodal temperature values are set to the ambient temperature T_∞ as defined in Fig. 9.68. This is achieved through the use of Ansys D command as *D,,TEMP,* T_∞
3	Using Eq. (8.31) and Eq. (1.43), the real constant value A is modified to include the convection film coefficient defined in Fig. 9.67 and the volume of the PD point. The modified real constant is assigned to the LINK34 elements as in Eq. (8.34) created using selected domain nodes and the newly created offset nodes

Fig. 9.70 LINK34 elements created for imposing convection boundary condition in PD mesh

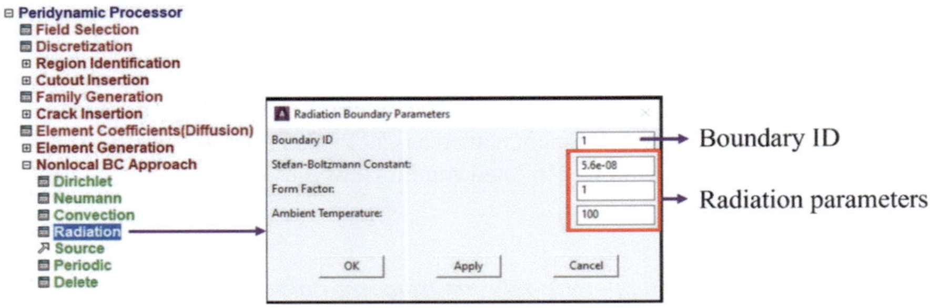

Fig. 9.71 Dialog box for entering the radiation boundary parameters in PD mesh

9.10.2.4 Radiation

The radiation boundary condition associated with heat conduction can be imposed by selecting the submenu ***Nonlocal BC Approach>Radiation*** located under ***Peridynamic Processor*** menu. As shown in Fig. 9.71, upon clicking the ***Nonlocal BC Approach>Radiation*** submenu, ***Radiation Boundary Parameters*** dialog box appears. The GUI path to invoke this dialog box is shown below:

Main Menu> Preprocessor> Peridynamic Processor> Nonlocal BC Approach> Radiation

The ***Radiation Boundary Parameters*** dialog box requires the following parameters:

Boundary ID: Enter an integer parameter for the boundary ID.

Stefan-Boltzmann Constant: Enter the value of the Stefan-Boltzmann Constant.
Form Factor: Enter the form factor value.

Ambient Temperature: Enter the ambient temperature value.

After entering the boundary parameters and clicking the ***OK*** or ***Apply*** button of the dialog box in Fig. 9.71, a node picker (Box) window appears to select a single layer of PD nodes close to the actual boundary as shown in Fig. 9.72.

Furthermore, after clicking the ***OK*** or ***Apply*** button of the node picker (Box) window, the dialog box ***Apply Radiation*** as shown in Fig. 9.73 appears for choosing the parameter type of the boundary value. The emissivity of the body, $\widehat{\varepsilon}$, can be input using constant or tabular parameter as shown in Fig. 9.73a, b. To input the emissivity as a function of any input variable, a predefined tabular parameter is created using ****DIM*** command with ***TABLE*** option and the name of the parameter as previously defined is selected as shown in Fig. 9.73b.

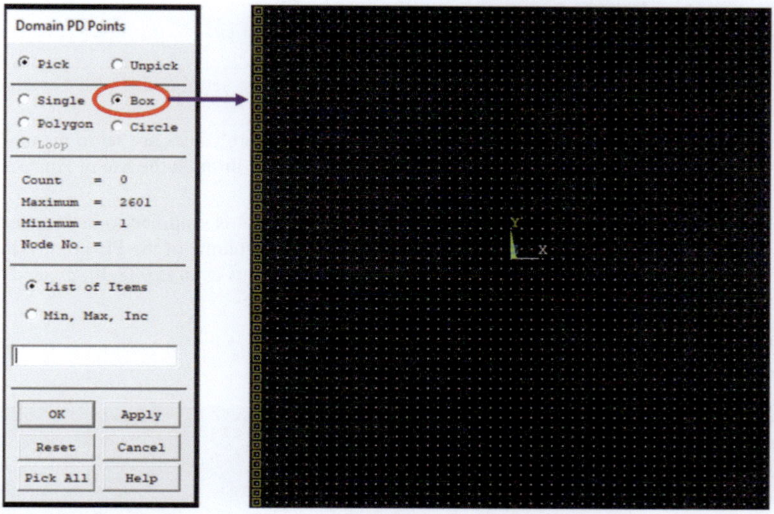

Fig. 9.72 Node picker option for selecting the set of domain nodes

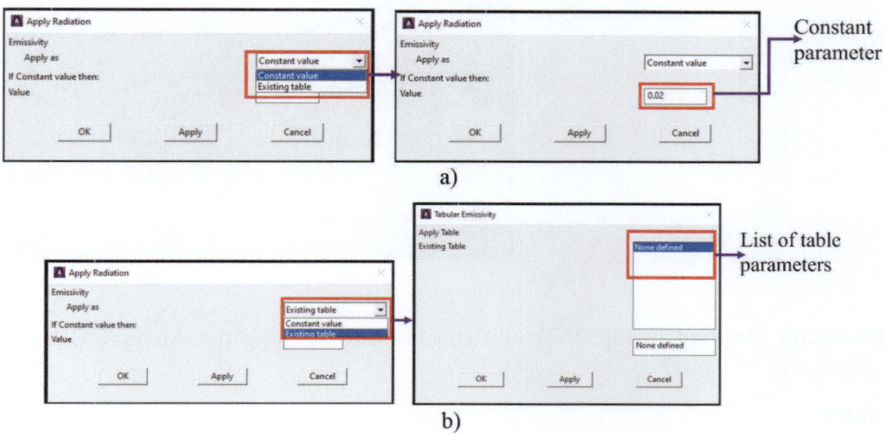

Fig. 9.73 Dialog box for entering emissivity of the radiating surface: (**a**) Constant parameter, and (**b**) Tabular input

After clicking the ***OK*** or ***Apply*** button of the ***Apply Radiation*** dialog box triggers the following actions as shown in Table 9.5.

The LINK31 elements using the PD domain nodes and the newly created offset nodes for modeling the radiation is shown in Fig. 9.74. These elements are grouped under an element component named as *RADIATION_ID*, where *ID* refers to the

Table 9.5 Summary of actions for ***Apply Radiation*** dialog box

Step	Action
1	For the selected set of domain nodes as in Fig. 9.72, a new set of nodes with a small offset are created for generation of LINK31 elements
2	For all the new set of nodes created, the nodal temperature values are set to the ambient temperature T_∞ as defined in Fig. 9.72. This is achieved through the use of Ansys D command as *D,,TEMP,* T_∞
3	Using Eq. (8.28) and Eq. (1.44), the real constant value *A* is modified to include the emissivity and form factor defined in Fig. 9.71 and the volume of the PD point. The modified real constant is assigned to the LINK31 elements as in Eq. (8.30) created using selected domain nodes and the newly created offset nodes

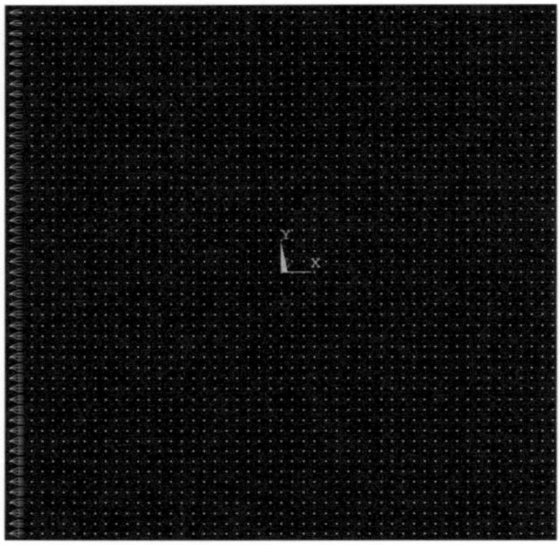

Fig. 9.74 LINK31 elements created for imposing radiation boundary condition in PD mesh

Boundary ID used in Fig. 9.71 and can be viewed using the Ansys CMPLOT command.

Notes

This menu is currently supported only for transient diffusion problems involving heat transfer.

9.10.2.5 Source

The field source is imposed by selecting the submenu ***Nonlocal BC Approach>Source*** located under ***Peridynamic Processor*** menu. Upon clicking the ***Nonlocal BC Approach>Source*** submenu, ***Domain PD points*** node picker

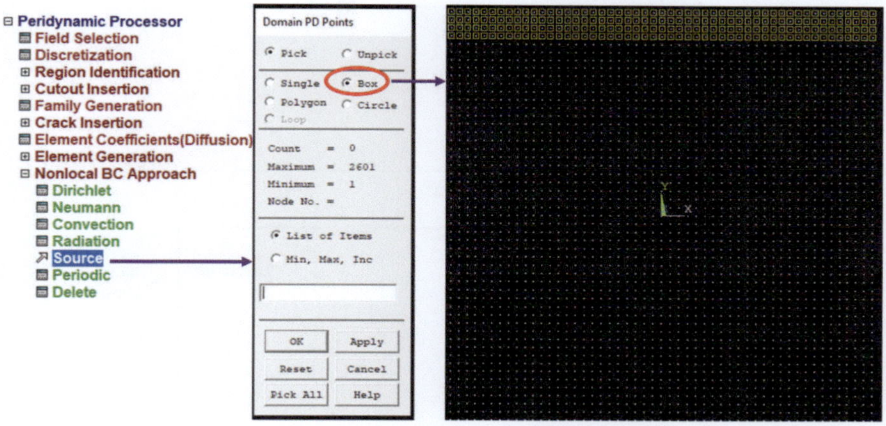

Fig. 9.75 Node picker option for selecting the set of domain nodes for applying source

option (Box) appears as shown in Fig. 9.75. The GUI path to invoke this node picker option is shown below:

> **Main Menu> Preprocessor> Peridynamic Processor> Nonlocal BC Approach> Source**

After clicking the ***OK*** or ***Apply*** button of the node picker (Box) window, the dialog box ***Apply Source*** as shown in Fig. 9.76 appears for choosing the ***Source ID*** and the parameter type of the source value. The source term, s_b as in Eq. (1.40), can be input using constant or tabular parameter as shown in Fig. 9.76. To input the source term as a function of any input variable, a predefined tabular parameter is created using ****DIM*** command with ***TABLE*** option and the name of the parameter as previously defined is selected as shown in Fig. 9.76b.

Clicking the ***OK*** or ***Apply*** button of the ***Apply Source*** dialog box triggers the following actions as shown in Table 9.6.

The MASS71 elements created using the PD domain nodes for modeling the source term is shown in Fig. 9.77. These elements are grouped under an element component named as *SRC_ID*, where *ID* refers to the ***Source ID*** used in Fig. 9.76. They can be viewed using the Ansys CMPLOT command.

Notes

This menu is currently supported only for transient field problems. Although MASS71 element has field-dependent source term, this capability is not currently supported through the GUI. The field-dependent source can be included through the Ansys commands as in Eq. (8.27).

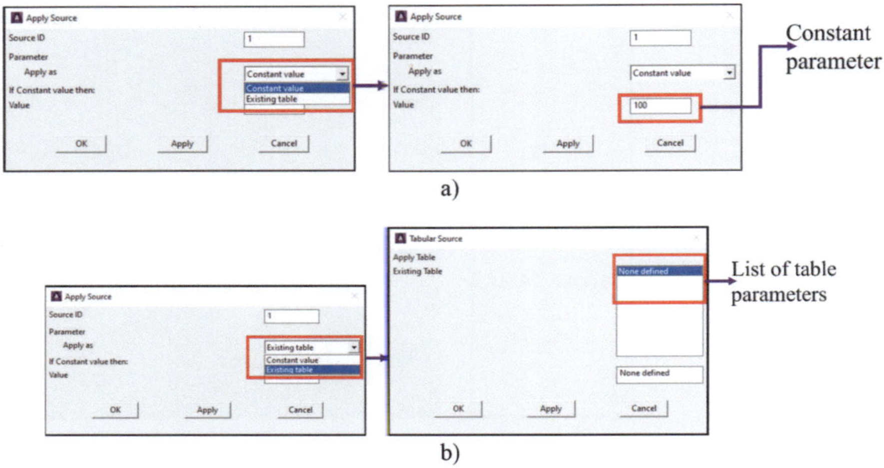

Fig. 9.76 Dialog box for entering source for PD field: (**a**) Constant parameter, and (**b**) Tabular input

Table 9.6 Summary of actions for ***Apply Source*** dialog box

Step	Action
1	For the selected set of domain nodes as in Fig. 9.75, MASS71 elements are overlaid
2	The source value PARM_CON1 is entered as in Eq. (8.26) through the MP command to include the source term defined in Fig. 9.76. The entered parameter must include the volume of the PD point

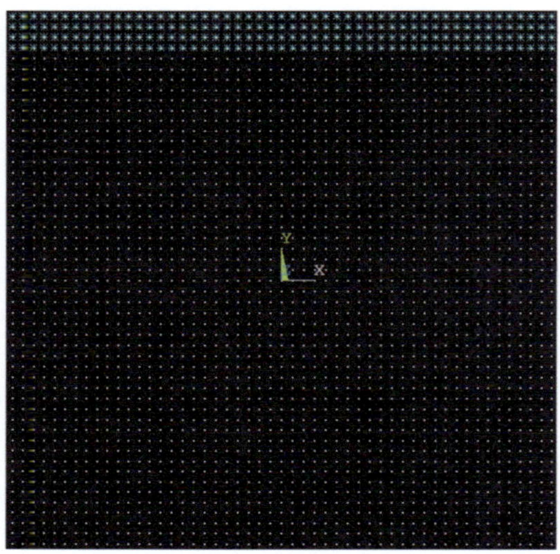

Fig. 9.77 MASS71 elements created for modeling source term for PD field analysis

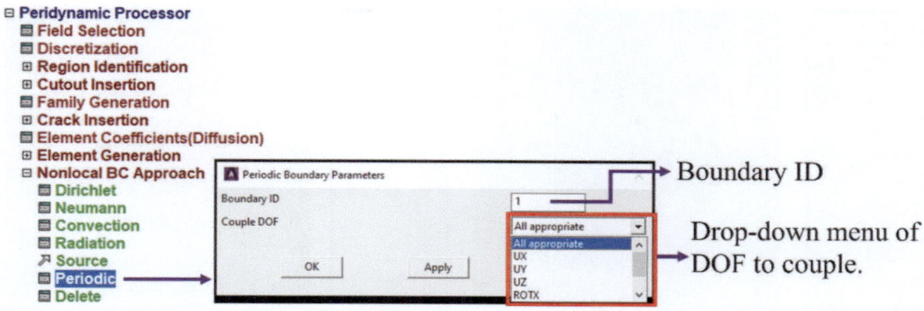

Fig. 9.78 Dialog box for creating periodic boundary conditions

9.10.2.6 Periodic Boundary Conditions

The periodic boundary condition can be imposed by selecting the submenu ***Nonlocal BC Approach>Periodic Boundary*** located under ***Peridynamic Processor*** menu. Upon clicking the ***Nonlocal BC Approach>Periodic Boundary*** submenu, ***Periodic Boundary Parameters*** dialog box appears as shown in Fig. 9.78. The GUI path to invoke this dialog box option is shown below.

Main Menu> Preprocessor> Peridynamic Processor> Nonlocal BC Approach> Periodic Boundary

Periodic boundary conditions are enforced by coupling the DOF of the master and the slave nodes by employing the Ansys CP command.

The ***Periodic Boundary Parameters*** dialog box requires the following parameter:

Boundary ID: Enter an integer parameter for the boundary ID.

Couple DOF: It is a drop-down menu type parameter with a list of DOF to couple. The supported DOF used for coupling to enforce periodic boundary conditions are UY, UY, UZ, ROTX, ROTY, and ROTZ. The default is to couple all the DOF.

After clicking the ***OK*** or ***Apply*** button of the dialog box ***Periodic Boundary Parameters***, the ***Master PD Points*** node picker option (Box) appears to select the **MASTER** nodes. Subsequent to clicking of ***OK*** or ***Apply*** button of the node picker window, another node picker option (Box) ***Slave PD Points*** appears to select the set of **SLAVE** nodes as shown in Fig. 9.79.

This menu couples DOF between the sets of master and slave nodes by looping over the master DOF node set and picking the nearest slave node.

After clicking the ***OK*** or ***Apply*** button of the ***Slave PD Points*** node picker option (Box), a dialog box ***Periodic Boundary Condition*** appears to confirm the generation of periodic BC as shown in Fig. 9.80. Furthermore, clicking ***OK*** or ***Apply*** triggers the following actions as shown in Table 9.7.

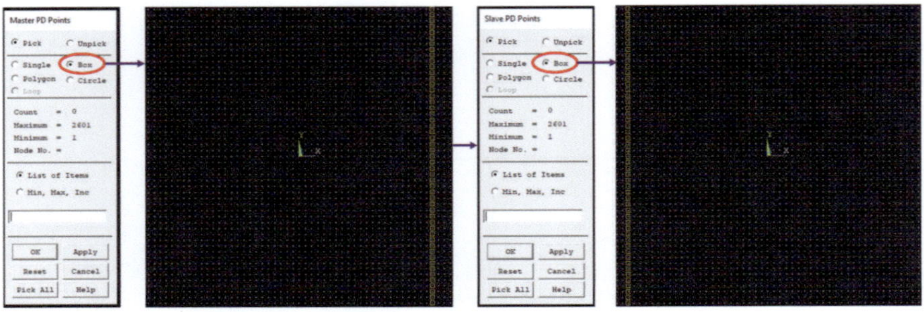

Fig. 9.79 Node picker option for selecting the sets of master and slave nodes

Table 9.7 Summary of actions for ***Periodic Boundary*** menu

Step	Action
1	For the selected set of domain and slave nodes as in Fig. 9.80, the DOF values are coupled using the Ansys CP command
2	The CP command is issued for every coupled set to enforce the condition as in Eqs. (1.48) and (1.49)

The coupled set of equations generated in Ansys through the selection of master and slave nodes is shown in Fig. 9.80. The coupled nodes are grouped under a nodal component named as *PBC_ID*, where *ID* refers to the Boundary ***ID*** used in Fig. 9.78. and can be viewed using the Ansys CMPLOT command.

9.10.2.7 Delete

The nonlocal boundary condition can be deleted by selecting the submenu ***Nonlocal BC Approach>Delete*** located under ***Peridynamic Processor*** menu. Upon clicking the ***Nonlocal BC Approach>Delete*** submenu, ***Delete*** dialog box appears as shown in Fig. 9.81. The GUI path to invoke this dialog box option is shown below:

Main Menu> Preprocessor> Peridynamic Processor> Nonlocal BC Approach> Delete

The ***Delete*** dialog box shows a list of Boundary IDs of the respective nonlocal boundary. After clicking the ***OK*** or ***Apply*** button in ***Delete*** dialog box, the parameters and components of the nonlocal boundary are deleted from the database.

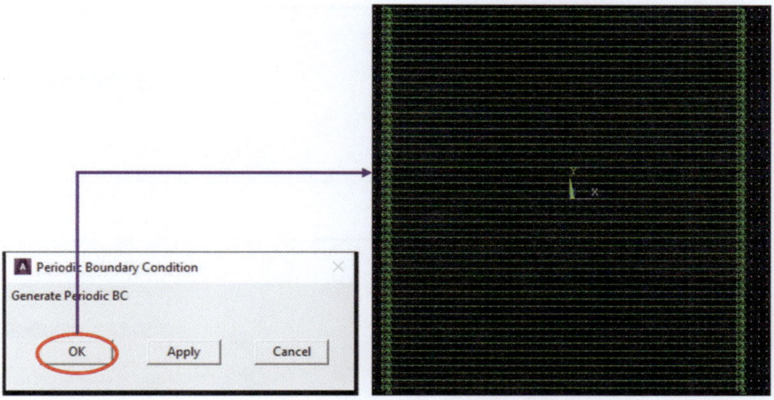

Fig. 9.80 CP equations generated by using the master and slave nodes for enforcing the periodic boundary condition

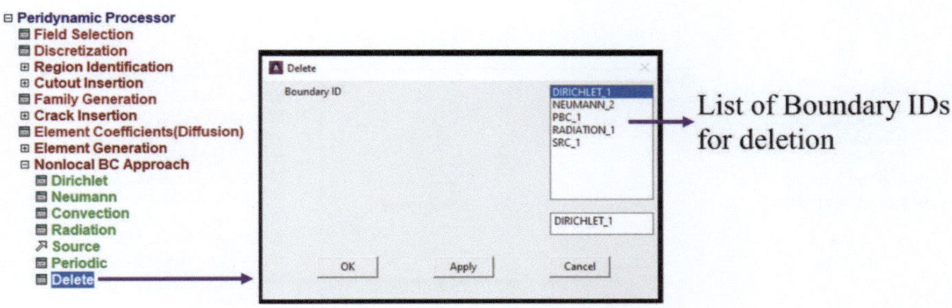

Fig. 9.81 Dialog box for deleting the nonlocal boundary conditions

9.11 Field Damping

The transient effects of the field at a node as described in Sect. 3.13.3 of Chap. 3 can be modeled by first setting the ***Select Field parameter*** in the ***Field Selection dialog box*** to ***Diffusion or Concentration*** as shown in Fig. 9.82. Also, the ANALYSIS type must be changed to transient to expose the submenu ***Field Damping*** as shown in Fig. 9.83

Upon clicking the ***Field Damping>Insert*** submenu, a node picker dialog box ***Select Domain Points*** appears, which can be used to select the region for modeling the transient effects of the field. The GUI path to invoke this node picker is shown below:

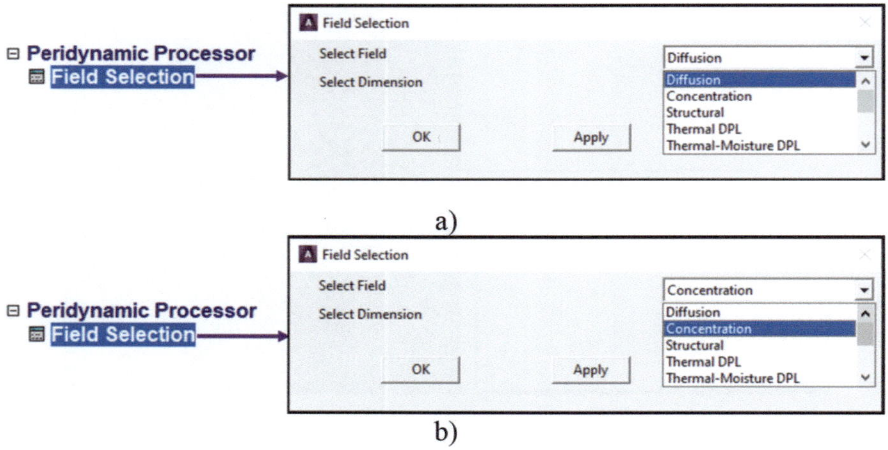

Fig. 9.82 Dialog box for selecting the fields for Field Damping: (**a**) Diffusion, and (**b**) Concentration

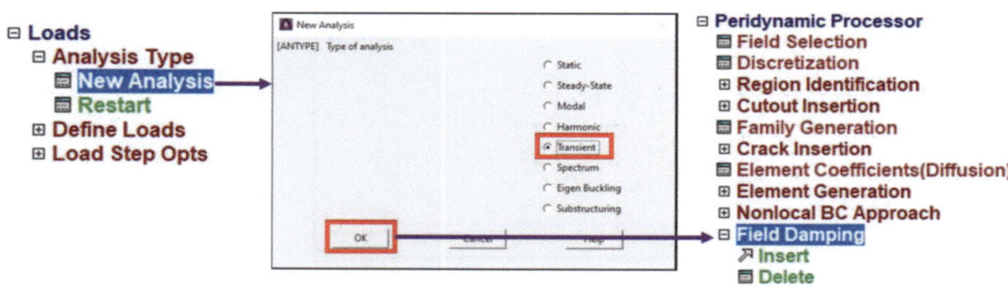

Fig. 9.83 Dialog box for selecting the analysis type for Field Damping

Main Menu> Preprocessor> Peridynamic Processor> Field Damping> Insert

After clicking the ***OK*** or ***Apply*** button of the node picker dialog box, the dialog box ***Damping Parameter*** as shown in Fig. 9.84 appears for choosing the ***Part ID*** and the parameter type value. The field damping term, $C_{(k)}V_{(k)}$ as in Eq. (3.258) can be input using constant or tabular parameter. To input the damping term as a function of any input variable, a predefined tabular parameter is created by using ****DIM*** command with ***TABLE*** option with the previously defined name of the parameter as shown in Fig. 9.84b. Clicking the ***OK*** or ***Apply*** button of the ***Damping Parameter*** dialog box triggers the following actions as shown in Table 9.8.

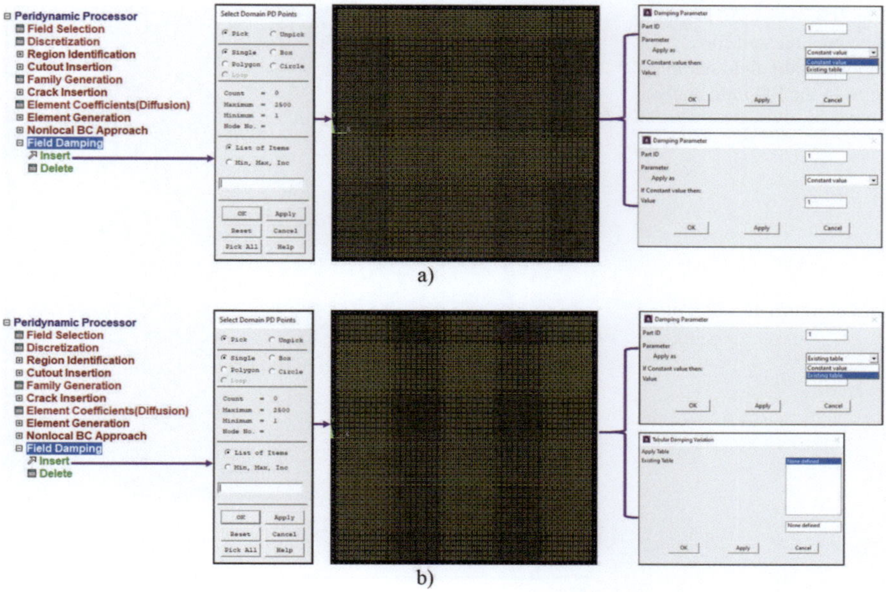

Fig. 9.84 Steps to enter the field damping values: (**a**) Constant type, and (**b**) Tabular type

Table 9.8 Summary of actions for ***Damping Parameter*** dialog box

Step	Actions
1	MASS71 elements are overlaid on the selected set of domain nodes as in Fig. 9.84
2	The real constant value CON1 is modified to include the field term defined in Fig. 9.84 and the volume of the PD point. The modified real constant is then assigned to MASS71 elements as in Eq. (8.25) created using selected domain nodes

The MASS71 elements created using the PD domain nodes for modeling the field damping term is shown in Fig. 9.85. These elements are grouped under an element component named as *MASS71_ID*, where *ID* refers to the ***Part ID*** used in Fig. 9.84. They can be viewed using the Ansys CMPLOT command.

Notes

This menu is currently supported only for transient field problems.

9.12 Thermal Dual Phase Lag (DPL) Damping

The DPL damping effects of the field at a node as described in Sect. 7.6.1 of Chap. 7 can be modeled by first setting the ***Select Field parameter*** in the ***Field Selection dialog box*** to ***Thermal DPL or Thermal-Moisture DPL*** models as shown in

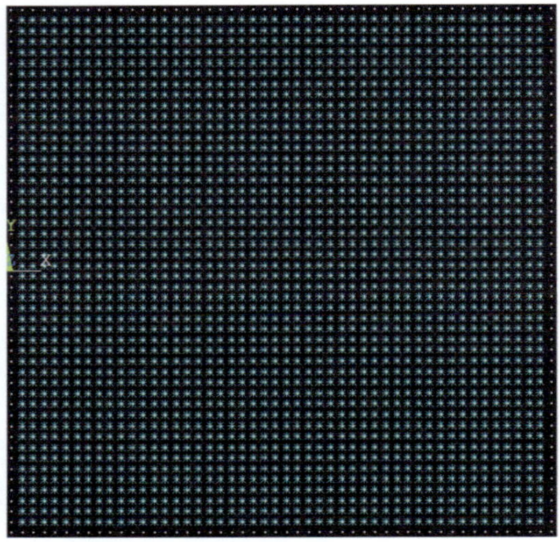

Fig. 9.85 MASS71 elements created for modeling the field damping term in the PD field analysis

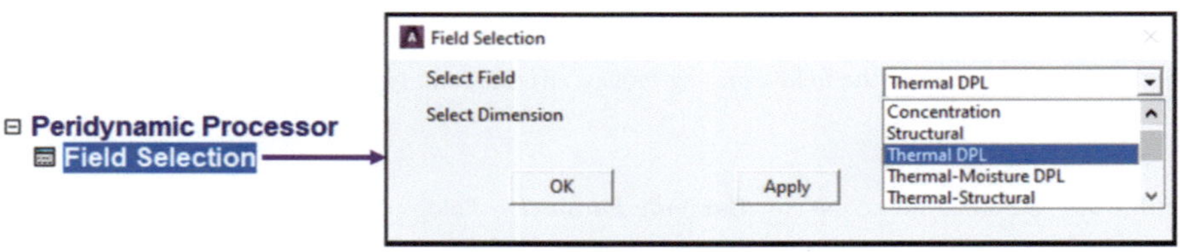

Fig. 9.86 Dialog box for selecting the field for Thermal DPL Damping

Fig. 9.86. Currently, this menu only is supported for thermal DPL transient field problems. Also, the ANALYSIS type must be changed to transient to expose the submenu ***DPL Damping*** as shown in Fig. 9.87.

Upon clicking the ***DPL Damping>Insert*** submenu, a node picker dialog box ***Select Domain Points*** appears, which can be used to select the region for modeling the DPL damping effects of field. The GUI path to invoke this node picker is shown below:

Main Menu> Preprocessor> Peridynamic Processor> DPL Damping> Insert

After clicking the ***OK*** or ***Apply*** button of the node picker dialog box, the dialog box ***DPL Damping Parameter*** as shown in Figs. 9.88 and 9.89 appears for choosing the ***Part ID*** and the parameter type value. The DPL damping term, $c_{\phi\phi}^{(k)(k_l)}$ as in Eq. (7.27b), can be input using a constant or tabular parameter. To input the damping term as a function of any input variable, a predefined tabular parameter is created

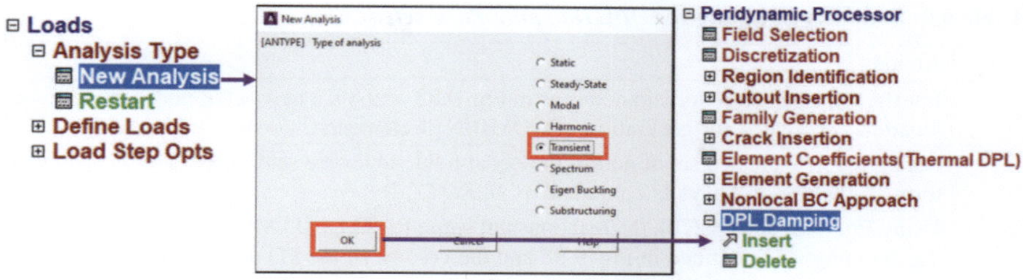

Fig. 9.87 Dialog box for selecting the analysis type for DPL Damping

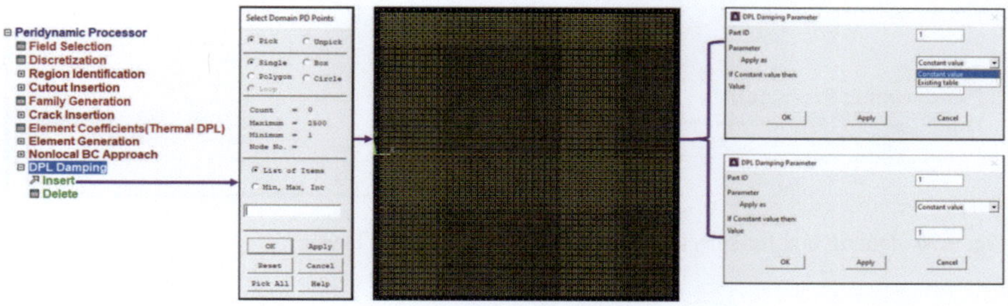

Fig. 9.88 Steps to enter the DPL damping values for Constant type

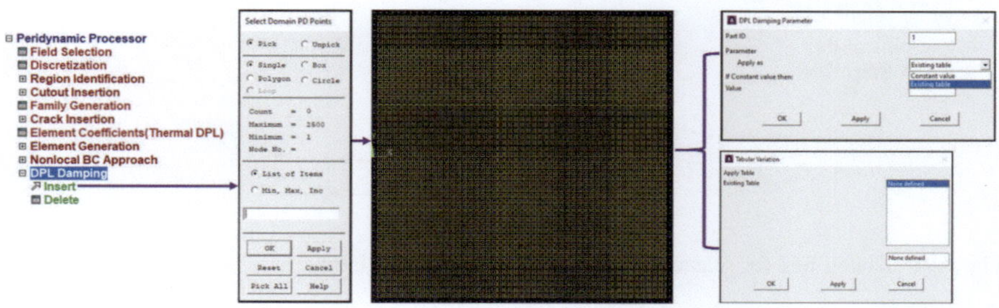

Fig. 9.89 Steps to enter the DPL damping values for Tabular type

using ****DIM*** command with ***TABLE*** option with the previously defined name of the parameter as shown in Fig. 9.89. Clicking the ***OK*** or ***Apply*** button of the ***DPL Damping Parameter*** dialog box triggers the following actions as shown in Table 9.9.

The COMBIN14 elements, created by using the PD domain nodes for modeling the DPL damping, are grouped under an element component named as *DPL_DAMP_ID*, where *ID* refers to the ***Part ID*** used in Figs. 9.88 and 9.89. They can be viewed using the Ansys CMPLOT command.

Table 9.9 Summary of actions for ***DPL Damping Parameter*** dialog box

Step	Action
1	For the selected set of domain nodes as in Fig. 9.88 or 9.89, a new set of nodes at the same location are created for generation of COMBIN14 elements
2	For all the new created set of nodes, the nodal field values are set to zero. This is achieved through the use of Ansys D command as *D,,ROTX,* 0
3	Using Eq. (8.8) and (7.27b), the real constant value PARM _ DAMP is modified to include the damping value defined in Fig. 9.88 and the volume of the PD point. The modified real constant is assigned to the COMBIN14 elements as in Eq. (8.9) created by using the selected domain nodes and the newly created offset nodes

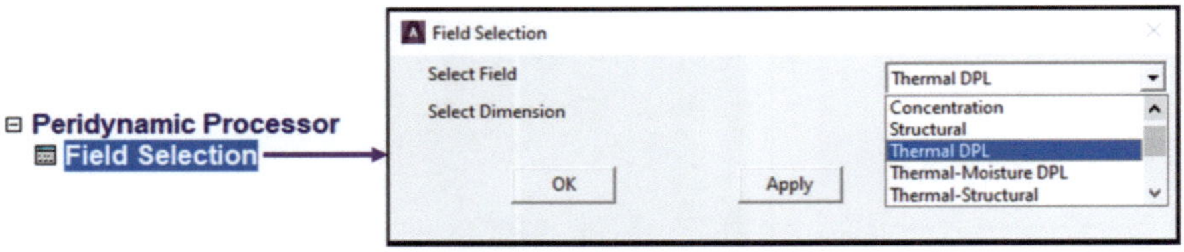

Fig. 9.90 Dialog box for selecting the field for Thermal DPL Inertia

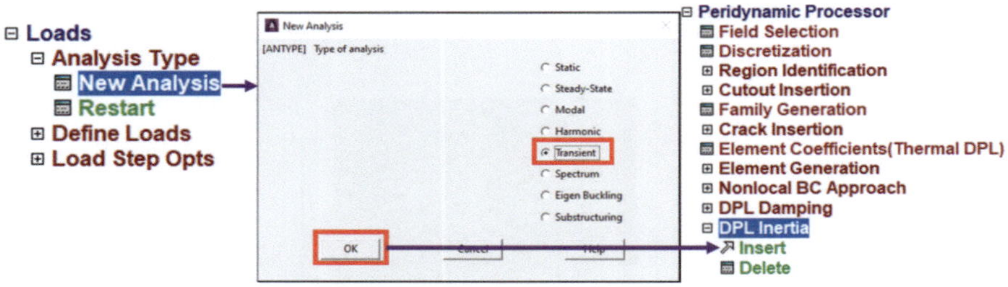

Fig. 9.91 Dialog box for selecting the analysis type for DPL Inertia

9.13 Dual Phase Lag (DPL) Inertia

The DPL inertial effects of the field at a point as described in Sect. 7.6.1 of Chap. 7 can be modeled by first setting the ***Select Field parameter*** in the ***Field Selection dialog box*** to ***Thermal DPL*** model as shown in Fig. 9.90. Currently, this menu is supported only for Thermal DPL transient field problems. Also, the ANALYSIS type must be changed to transient to expose the submenu ***DPL Inertia*** as shown in Fig. 9.91.

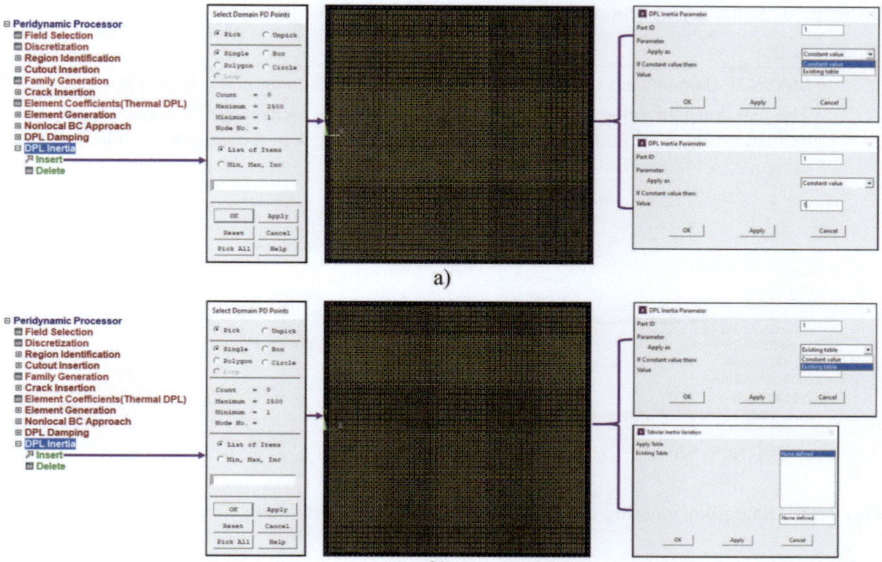

Fig. 9.92 Steps to enter the DPL inertia values: (a) Constant type, and (b) Tabular type

Upon clicking the ***DPL Inertia>Insert*** submenu, a node picker dialog box ***Select Domain Points*** appears, which can be used to select the region for modeling the DPL inertial effects of field. The GUI path to invoke this node picker is shown below:

Main Menu> Preprocessor> Peridynamic Processor> DPL Inertia> Insert

After clicking the ***OK*** or ***Apply*** button of the node picker dialog box, the dialog box ***DPL Inertia Parameter*** as shown in Fig. 9.92 appears for choosing the ***Part ID*** and the parameter type value. The DPL inertia term, $m_{\phi\phi}^{(k)(k_l)}$ as in Eq. (7.24b), can be input using constant or tabular parameter. To input the damping term as a function of any input variable, a predefined tabular parameter is created using ****DIM*** command with ***TABLE*** option with the previously defined name of the parameter as shown in Fig. 9.92b. Clicking the ***OK*** or ***Apply*** button of the ***DPL Inertia Parameter*** dialog box triggers the following actions as shown in Table 9.10.

The MASS21 elements, created using the PD domain nodes for modeling the DPL inertia, are grouped under an element component named as *DPL_INERTIA_ID*, where *ID* refers to the ***Part ID*** used in Fig. 9.92. They can be viewed using the Ansys CMPLOT command.

Table 9.10 Summary of actions for ***DPL Inertia Parameter*** dialog box

Step	Action
1	MASS21 elements are overlaid on the selected set of domain nodes as in Fig. 9.92
2	The real constant value PARM _ MASS is modified to include the DPL inertial term defined in Fig. 9.92 and the volume of the PD point. The modified real constant is then assigned to MASS21 elements as in Eq. (8.14) created using selected domain nodes

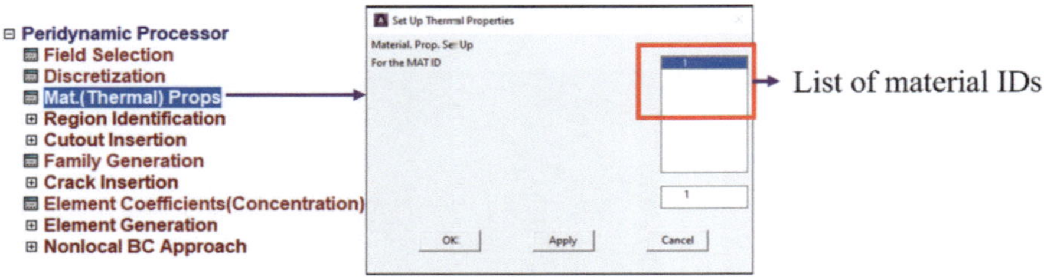

Fig. 9.93 Dialog box entering thermal material properties in PD mesh

9.14 Material Thermal Properties for Moisture Concentration

The thermal material properties required for moisture concentration analysis can be included using the submenu ***Mat. (Thermal Prop.)*** located under ***Peridynamic Processor*** menu. The GUI path to invoke this dialog box is shown below:

Main Menu> Preprocessor> Peridynamic Processor> Mat.(Thermal Props.)

As shown in Fig. 9.93, the parameter in the ***Set Up Thermal Properties*** dialog box displays a list of material IDs used in the finite model to create the PD mesh. For the input of multiple materials, different Mat IDs can be used. After selecting a specific *MAT ID* and clicking the ***OK*** or ***Apply*** button in the above dialog box, another dialog box ***Thermal Material Parameters*** appears to enter the thermal conductivity as shown in Fig. 9.94.

After clicking the ***OK*** or ***Apply*** button in ***Thermal Material Parameters*** dialog box, the value of thermal conductivity is assigned to the selected *MAT ID*.

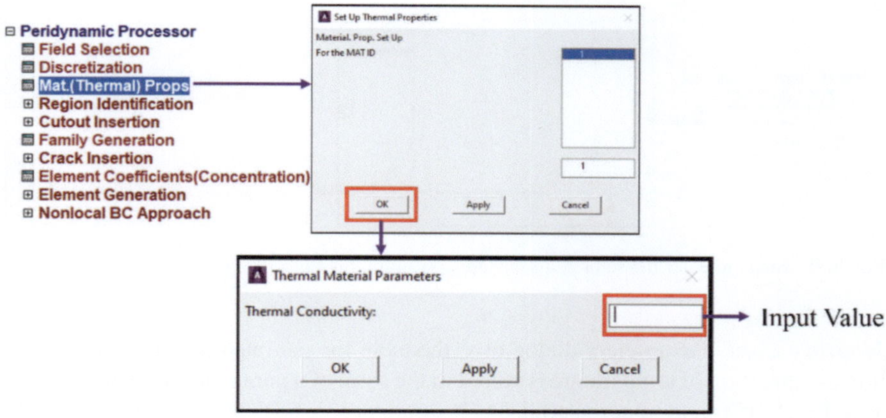

Fig. 9.94 Dialog box entering thermal conductivity in PD mesh

9.15 Peridynamic Solution

The menu ***Peridynamic Solution*** is activated under the ***Ansys Solution*** (***/SOLU***) processor specifically for the PD moisture concentration analysis, i.e., when parameter ***Select Field*** is set as ***Concentration*** in ***Field Selection*** dialog box as described in Sect. 9.2. The customized submenus grouped under this menu include ***Setup CSAT table***, ***CSAT***, and ***Time Dep. CSAT***. For each of the submenus, a function block is embedded depending on the requirement to capture the user input. The typical function blocks employed in the customization include Dialog and Picking box. Subsequent sections describe in detail the functionality of each of the customized submenus.

9.15.1 Setup CSAT Table

The material properties for saturated moisture concentration as a function of temperature is set up by using the dialog box ***Moisture Conc. Parameters***, which appears upon selecting the ***Setup CSAT table>Setup*** submenu located under ***Peridynamic Solution*** menu. The GUI path for this dialog box is shown below:

Main Menu> Solution> Peridynamic Solution> Setup CSAT table> Setup

The steps to input the values for setting up the saturated moisture concentration (CSAT) table are shown in Fig. 9.95. Upon clicking the ***OK*** or ***Apply*** button in

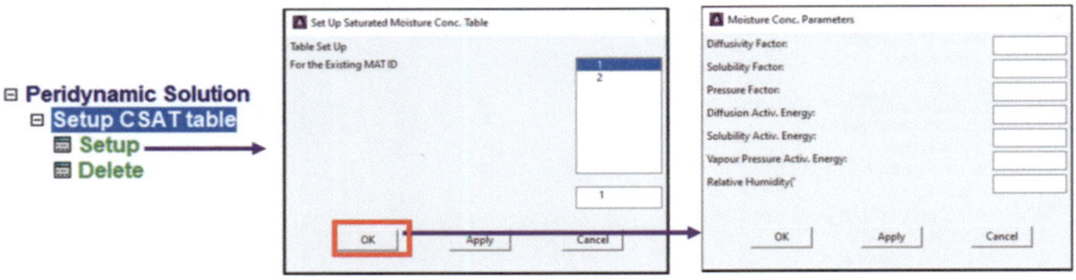

Fig. 9.95 Steps to enter the saturated moisture concentration values

Moisture Conc. Parameters dialog box, the table for saturated moisture concentration as a function of temperature is stored in the form of a parameter. Repeat the steps as in Fig. 9.95 for another material ID. The parameters *CST_MATID*, *DCS_MATID*, corresponding to saturated moisture concentration, time-dependent saturated moisture concentration, are generated, where ***MATID*** refers to the selection made in Fig. 9.95.

9.15.2 CSAT

The saturated moisture concentration as a function of temperature can be modeled as shown in Fig. 9.96. Upon clicking the ***CSAT>Apply*** submenu, a node picker dialog box ***Select Domain Points*** appears, which can be used to select the region for modeling the saturated moisture concentration properties. The GUI path to invoke this node picker is shown below:

Main Menu> Solution> Peridynamic Solution> CSAT> Apply

After clicking the ***OK*** or ***Apply*** button of the node picker dialog box, the dialog box ***CSAT Parameter*** as shown in Fig. 9.96 appears for choosing the ***MAT ID*** and the parameter type value. The CSAT defined as a field as in Eq. (3.258) can be input using constant or tabular parameter. To input the parameter as a function of any input variable, a predefined tabular parameter is created using ****DIM*** command with ***TABLE*** option with the previously defined name of the parameter as shown in Fig. 9.96b. Clicking the ***OK*** or ***Apply*** button of the ***CSAT Parameter*** dialog box triggers the following actions as shown in Table 9.11.

The COMBIN14 elements using the PD domain nodes and the newly created nodes for modeling the damping are grouped under an element component named as *CSAT_ID*, where *ID* refers to the ***MATID*** used in Fig. 9.96. They can be viewed using the Ansys CMPLOT command.

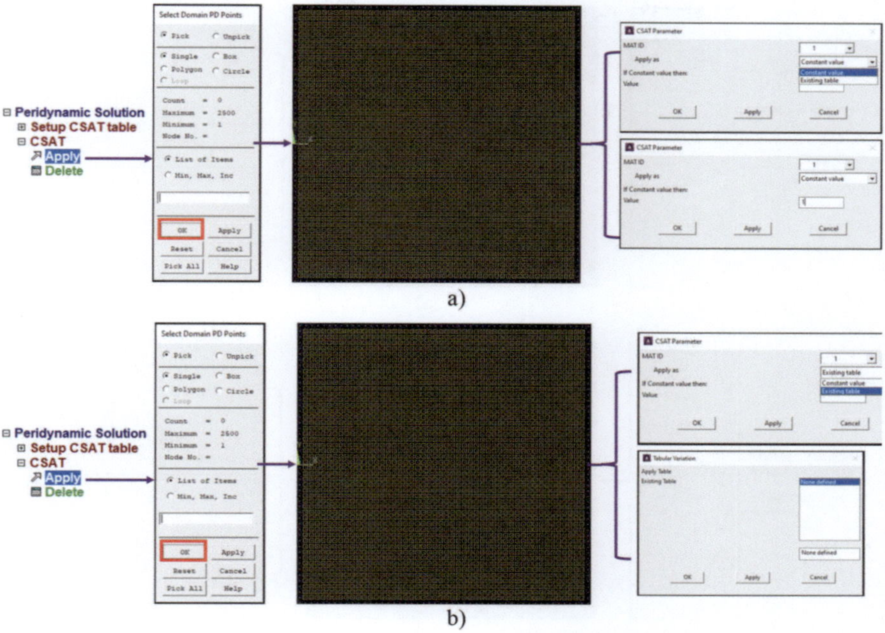

Fig. 9.96 Steps to enter the CSAT values as a function of temperature: (a) Constant type, and (b) Tabular type

Table 9.11 Summary of actions for ***CSAT Parameter*** dialog box

Step	Action
1	For the selected set of domain nodes as in Fig. 9.96, a new set of nodes at the same location are created for generation of COMBIN14 elements
2	For all the new set of nodes created, the nodal temperature values are set to the ambient temperature equal to 0 as defined in Fig. 9.96. This is achieved through the use of Ansys D command as *D,,TEMP,* 0
3	Using Eq. (8.8) and Eq. (3.258), the real constant value PARM _ DAMP is added to include the field damping factor defined in Fig. 9.96 and the volume of the PD point. The modified real constant is assigned to the COMBIN14 elements as in Eq. (8.9)

9.15.3 Time Dep. CSAT

The time-dependent saturated moisture concentration can be modeled as shown in Fig. 9.97. Upon clicking the ***Time Dep. CSAT>Apply*** submenu, a node picker dialog box ***Select Domain Points*** appears, which can be used to select the region for modeling the time-dependent saturated moisture concentration properties. The GUI path to invoke this node picker is shown below:

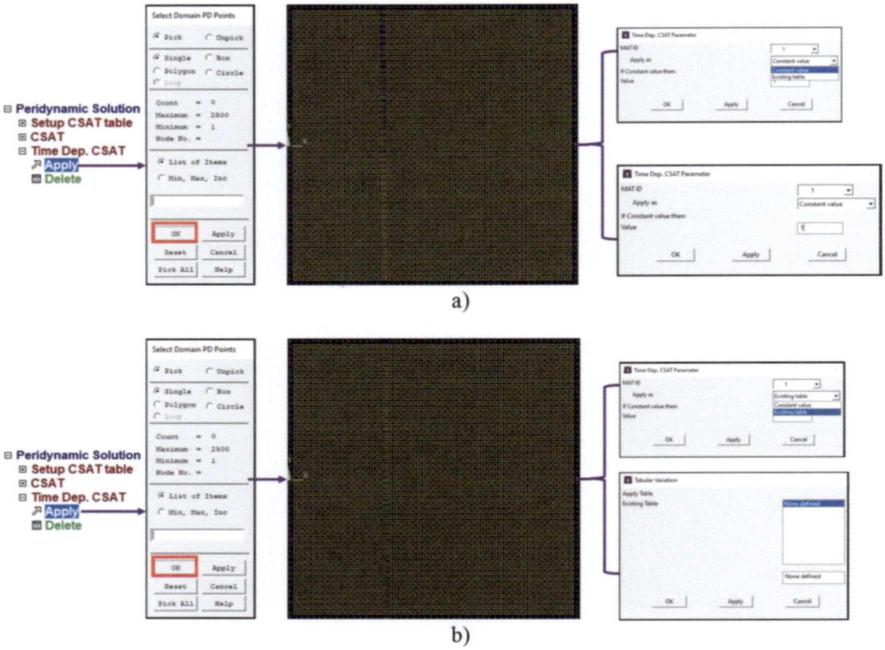

Fig. 9.97 Steps to enter the Time Dep. CSAT values as a function of temperature: (**a**) Constant type, and (**b**) Tabular type

Main Menu> Solution> Peridynamic Solution> Time Dep. CSAT> Apply

After clicking the ***OK*** or ***Apply*** button of the node picker dialog box, the dialog box ***Time Dep. CSAT Parameter*** as shown in Fig. 9.97 appears for choosing the ***MAT ID*** and the parameter type value. The time-dependent CSAT as in Eq. (3.260) can be input by using constant or tabular parameter. To input the parameter as a function of any input variable, a predefined tabular parameter is created using ****DIM*** command with ***TABLE*** option with the previously defined name of the parameter as shown in Fig. 9.97b. Clicking the ***OK*** or ***Apply*** button of the ***Time Dep. CSAT Parameter*** dialog box triggers the following actions as shown in Table 9.12.

The COMBIN14 elements using the PD domain nodes and the newly created nodes for modeling the transient material property variation using stiffness analogy are grouped under an element component named as *TCSAT_ID*, where *ID* refers to the ***MATID*** used in Fig. 9.97. They can be viewed using the Ansys CMPLOT command.

Table 9.12 Summary of actions for ***Time Dep. CSAT Parameter*** dialog box

Step	Action
1	For the selected set of domain nodes as in Fig. 9.97, a new set of nodes at the same location are created for generation of COMBIN14 elements
2	For all the new set of nodes created, the nodal temperature values are set to the ambient temperature equal to 0 as defined in Fig. 9.97. This is achieved through the use of Ansys D command as *D,,TEMP,* 0
3	Using Eq. (8.8) and Eq. (3.260), the real constant value PARM _ STIFF is added to include the field damping factor defined in Fig. 9.97 and the volume of the PD point. The modified real constant is assigned to the COMBIN14 elements as in Eq. (8.9)

References

Ansys 18.2 Mechanical User's Guide. (2017).

Madenci, E., & Oterkus, S. (2016). Ordinary state-based peridynamics for plastic deformation according to von Mises yield criteria with isotropic hardening. *Journal of the Mechanics and Physics of Solids, 86*, 192–219.

Chapter 10
Implementation of PDDO in Ansys Framework

10.1 Introduction

The peridynamic differential operator (PDDO) introduced in Chap. 2 can be applied to manipulate discrete data (Madenci et al., 2022). In real life, data sets are mostly connected with surge of irregularities, breakages, and discontinuities. It is essential that interpolation, regression, and differentiation methods do not suffer from such irregularities and scatter within a set of discrete data in a multidimensional space. Interpolation, regression, and differentiation of data play a significant role in many scientific disciplines. It is worth noting that a data point, entity, object, etc., interacts with its surrounding media. Although many methods exist for data manipulation, they tend to break down in the presence of irregularities such as a jump discontinuity and scatter. As presented by Madenci et al. (2022), the PDDO is applicable to data manipulation without any limitation. Its applicability was demonstrated by considering interpolation of real data in two and three dimensions, interpolation to approximate a three-dimensional function, adaptive data recovery in three-dimensional space, recovery of missing pixels in an image, and adaptive image compression and recovery. Also, it can be applied for model reduction by employing PD interpolation between fine and coarse grids. Furthermore, it is not limited by the number of variables without any restriction on spatial discretization.

The PD functions are the building block of the PDDO (Madenci et al., 2019). Therefore, this chapter presents the implementation of PD functions in Ansys framework. Although not limited, the applications concern the differentiation of one-, two-, and three-dimensional mathematical expressions with and without a jump discontinuity. The PD functions can be used for adaptive data recovery through regression and interpolation.

Supplementary Information The online version contains supplementary material available at https://doi.org/10.1007/978-3-031-59896-8_10.

E. Madenci et al., *Peridynamics and Its Applications Using Ansys*,
https://doi.org/10.1007/978-3-031-59896-8_10

10.2 Evaluation of PD Functions

As derived in Chap. 2, the construction of PD functions at point $\mathbf{x}_{(k)}$ shown in Fig. 10.1 requires the inverse of the shape matrix, $\mathbf{A}(\mathbf{x}_{(k)})$ in Eq. (2.14) for 2D analyis and Eq. (2.50) for 3D analysis.

For very small values of grid spacing, the condition number of the shape matrix, $\mathbf{A}(\mathbf{x}_{(k)})$, may become poor due to round-off error and its inverse may be inaccurate. Therefore, it may be advantageous to normalize the range of interaction with respect to the domain of interaction, $H_{(k)} = \{j; j \in H_{(k)}\}$ prior to evaluating the inverse of $\mathbf{A}(\mathbf{x}_{(k)})$. For the family member $\mathbf{x}_{(j)}$ of point $\mathbf{x}_{(k)}$, the relative position vector, $\boldsymbol{\xi}_{(k)(j)} = \mathbf{x}_{(j)} - \mathbf{x}_{(k)}$ can be normalized with respect to the horizon of point $\mathbf{x}_{(k)}$ as

$$\boldsymbol{\zeta}_{(k)(j)} = \boldsymbol{\xi}_{(k)(j)} / \delta_{(k)} \tag{10.1}$$

Also, the volume of point $\mathbf{x}_{(j)}$ can be normalized with respect to the volume of point $\mathbf{x}_{(k)}$ as

$$\Omega_{(j)} = V_{(j)} / V_{(k)} \tag{10.2}$$

Substituting from Eqs. (10.1) and (10.2) into Eqs. (2.52)–(2.54), the submatrices of matrix $\mathbf{A}(\mathbf{x}_{(k)})$ can be revised as

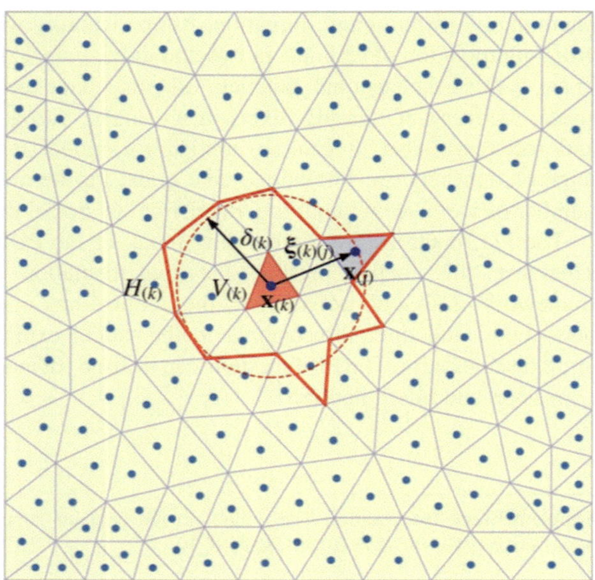

Fig. 10.1 PD point $\mathbf{x}_{(k)}$ and its family member $\mathbf{x}_{(j)}$ in its horizon, $H_{(k)} = \{j; j \in H_{(k)}\}$

$$\mathbf{A}_{11}\left(\mathbf{x}_{(k)}\right)=\delta_{(k)}^{2}V_{(k)}\sum_{j\in H_{(k)}}w_{1}\left(\left|\boldsymbol{\zeta}_{(k)(j)}\right|\right)\left(\boldsymbol{\zeta}_{(k)(j)}\otimes\boldsymbol{\zeta}_{(k)(j)}\right)\boldsymbol{\Omega}_{(j)} \tag{10.3}$$

$$\begin{aligned}\mathbf{A}_{12}\left(\mathbf{x}_{(k)}\right)&=\mathbf{A}_{21}^{T}\left(\mathbf{x}_{(k)}\right)\\&=\delta_{(k)}^{3}V_{(k)}\sum_{j\in H_{(k)}}w_{2}\left(\left|\boldsymbol{\zeta}_{(k)(j)}\right|\right)\left[\left(\boldsymbol{\zeta}_{(k)(j)}\otimes\widehat{\boldsymbol{\zeta}}_{(k)(j)}\right)\quad\left(\boldsymbol{\zeta}_{(k)(j)}\otimes\widetilde{\boldsymbol{\zeta}}_{(k)(j)}\right)\right]\boldsymbol{\Omega}_{(j)}\end{aligned} \tag{10.4}$$

$$\mathbf{A}_{22}\left(\mathbf{x}_{(k)}\right)=\delta_{(k)}^{4}\sum_{j\in H_{(k)}}w_{2}\left(\left|\boldsymbol{\zeta}_{(k)(j)}\right|\right)\begin{bmatrix}\left(\widehat{\boldsymbol{\zeta}}_{(k)(j)}\otimes\widehat{\boldsymbol{\zeta}}_{(k)(j)}\right)&\left(\widehat{\boldsymbol{\zeta}}_{(k)(j)}\otimes\widetilde{\boldsymbol{\zeta}}_{(k)(j)}\right)\\\left(\widetilde{\boldsymbol{\zeta}}_{(k)(j)}\otimes\widehat{\boldsymbol{\zeta}}_{(k)(j)}\right)&\left(\widetilde{\boldsymbol{\zeta}}_{(k)(j)}\otimes\widetilde{\boldsymbol{\zeta}}_{(k)(j)}\right)\end{bmatrix}\boldsymbol{\Omega}_{(j)} \tag{10.5}$$

in which the vectors $\boldsymbol{\zeta}$, $\widehat{\boldsymbol{\zeta}}$, and $\widetilde{\boldsymbol{\zeta}}$ are defined as

$$\boldsymbol{\zeta}_{(k)(j)}=\left\{\begin{array}{c}\zeta_{1(k)(j)}\\\zeta_{2(k)(j)}\\\zeta_{3(k)(j)}\end{array}\right\},\widehat{\boldsymbol{\zeta}}_{(k)(j)}=\left\{\begin{array}{c}\zeta_{1(k)(j)}^{2}\\\zeta_{2(k)(j)}^{2}\\\zeta_{3(k)(j)}^{2}\end{array}\right\}\text{ and }\widetilde{\boldsymbol{\zeta}}_{(k)(j)}=\left\{\begin{array}{c}\zeta_{1(k)(j)}\zeta_{2(k)(j)}\\\zeta_{1(k)(j)}\zeta_{3(k)(j)}\\\zeta_{2(k)(j)}\zeta_{3(k)(j)}\end{array}\right\} \tag{10.6}$$

The matrix $\mathbf{A}(\mathbf{x}_{(k)})$ in Eq. (2.50) can be rewritten as

$$\mathbf{A}\left(\mathbf{x}_{(k)}\right)=\mathbf{B}\overline{\mathbf{A}}\left(\mathbf{x}_{(k)}\right) \tag{10.7}$$

in which $\overline{\mathbf{A}}\left(\mathbf{x}_{(k)}\right)$ is the normalized form of the shape matrix of $\mathbf{A}(\mathbf{x}_{(k)})$. It is defined as

$$\overline{\mathbf{A}}\left(\mathbf{x}_{(k)}\right)=\begin{bmatrix}\overline{\mathbf{A}}_{11}\left(\mathbf{x}_{(k)}\right)&\overline{\mathbf{A}}_{12}\left(\mathbf{x}_{(k)}\right)\\\overline{\mathbf{A}}_{21}\left(\mathbf{x}_{(k)}\right)&\overline{\mathbf{A}}_{22}\left(\mathbf{x}_{(k)}\right)\end{bmatrix} \tag{10.8}$$

with

$$\overline{\mathbf{A}}_{11}\left(\mathbf{x}_{(k)}\right)=\sum_{j\in H_{(k)}}w_{1}\left(\left|\boldsymbol{\zeta}_{(k)(j)}\right|\right)\left(\boldsymbol{\zeta}_{(k)(j)}\otimes\boldsymbol{\zeta}_{(k)(j)}\right)\boldsymbol{\Omega}_{(j)} \tag{10.9}$$

$$\begin{aligned}\overline{\mathbf{A}}_{12}\left(\mathbf{x}_{(k)}\right)&=\overline{\mathbf{A}}_{21}^{T}\left(\mathbf{x}_{(k)}\right)\\&=\delta_{(k)}\sum_{j\in H_{(k)}}w_{2}\left(\left|\boldsymbol{\zeta}_{(k)(j)}\right|\right)\left[\left(\boldsymbol{\zeta}_{(k)(j)}\otimes\widehat{\boldsymbol{\zeta}}_{(k)(j)}\right)\quad\left(\boldsymbol{\zeta}_{(k)(j)}\otimes\widetilde{\boldsymbol{\zeta}}_{(k)(j)}\right)\right]\boldsymbol{\Omega}_{(j)}\end{aligned} \tag{10.10}$$

and

$$\overline{\mathbf{A}}_{22}\left(\mathbf{x}_{(k)}\right)=\delta_{(k)}\sum_{j\in H_{(k)}}w_2\left(\left|\boldsymbol{\zeta}_{(k)(j)}\right|\right)\begin{bmatrix}\left(\widehat{\boldsymbol{\zeta}}_{(k)(j)}\otimes\widehat{\boldsymbol{\zeta}}_{(k)(j)}\right) & \left(\widehat{\boldsymbol{\zeta}}_{(k)(j)}\otimes\widetilde{\boldsymbol{\zeta}}_{(k)(j)}\right)\\ \left(\widetilde{\boldsymbol{\zeta}}_{(k)(j)}\otimes\widehat{\boldsymbol{\zeta}}_{(k)(j)}\right) & \left(\widetilde{\boldsymbol{\zeta}}_{(k)(j)}\otimes\widetilde{\boldsymbol{\zeta}}_{(k)(j)}\right)\end{bmatrix}\Omega_{(j)} \tag{10.11}$$

The normalization matrix $\mathbf{B}(\mathbf{x}_{(k)})$ is defined as

$$\mathbf{B}_{(k)}=\begin{bmatrix}\mathbf{B}_1 & \mathbf{0}\\ \mathbf{0} & \mathbf{B}_2\end{bmatrix} \tag{10.12}$$

with

$$\mathbf{B}_1=\delta_{(k)}^2V_{(k)}\begin{bmatrix}1&0&0\\0&1&0\\0&0&1\end{bmatrix} \tag{10.13}$$

and

$$\mathbf{B}_2=\delta_{(k)}^3V_{(k)}\begin{bmatrix}1&0&0&0&0&0\\0&1&0&0&0&0\\0&0&1&0&0&0\\0&0&0&1&0&0\\0&0&0&0&1&0\\0&0&0&0&0&1\end{bmatrix} \tag{10.14}$$

Substituting from Eq. (10.7) into Eq. (2.50) leads to the system of algebraic equations for the determination of the coefficients in PD functions as

$$\mathbf{B}_{(k)}\overline{\mathbf{A}}_{(k)}\mathbf{a}_{(k)}^{p_1p_2p_3}=\mathbf{b}^{p_1p_2p_3} \tag{10.15}$$

Solution to Eq. (10.15) results in the determination of the coefficients $\mathbf{a}_{(k)}^{p_1p_2p_3}$ as

$$\mathbf{a}_{(k)}^{p_1p_2p_3}=\overline{\mathbf{A}}_{(k)}^{-1}\mathbf{B}_{(k)}^{-1}\mathbf{b}^{p_1p_2p_3} \tag{10.16}$$

The inverse of matrix **B** is expressed as

$$\mathbf{B}_{(k)}^{-1} = \begin{bmatrix} \mathbf{B}_1^{-1} & \mathbf{0} \\ \mathbf{0} & \mathbf{B}_2^{-1} \end{bmatrix} \tag{10.17}$$

with

$$\mathbf{B}_1^{-1} = \frac{1}{\delta_{(k)}^2 V_{(k)}} \begin{bmatrix} 1 & 0 & 0 \\ 0 & 1 & 0 \\ 0 & 0 & 1 \end{bmatrix} \tag{10.18}$$

and

$$\mathbf{B}_2^{-1} = \frac{1}{\delta_{(k)}^3 V_{(k)}} \begin{bmatrix} 1 & 0 & 0 & 0 & 0 & 0 \\ 0 & 1 & 0 & 0 & 0 & 0 \\ 0 & 0 & 1 & 0 & 0 & 0 \\ 0 & 0 & 0 & 1 & 0 & 0 \\ 0 & 0 & 0 & 0 & 1 & 0 \\ 0 & 0 & 0 & 0 & 0 & 1 \end{bmatrix} \tag{10.19}$$

Due to the diagonal form of matrix, $\mathbf{B}_{(k)}^{-1}$, Eq. (10.16) can be recast as

$$\mathbf{a}_{(k)}^{p_1p_2p_3} = \overline{\mathbf{A}}_{(k)}^{-1} \mathbf{b}^{p_1p_2p_3} \mathbf{B}_{(k)}^{-1} \tag{10.20}$$

After normalization, the coefficients in the PD function can be defined as

$$\overline{\mathbf{a}}_{(k)}^{p_1p_2p_3} = \overline{\mathbf{A}}_{(k)}^{-1} \mathbf{b}^{p_1p_2p_3} \tag{10.21}$$

Using Eqs. (10.17)–(10.20), the 3D PD functions after normalization can be expressed as

$$g_2^{p_1p_2p_3}\left(\boldsymbol{\zeta}_{(k)(j)}\right) = \frac{1}{\delta_{(k)} V_{(k)}} \begin{pmatrix} \overline{a}_{100}^{p_1p_2p_3} w_{100}\left(\left|\boldsymbol{\zeta}_{(k)(j)}\right|\right) \zeta_{1(k)(j)} + \overline{a}_{010}^{p_1p_2p_3} w_{010}\left(\left|\boldsymbol{\zeta}_{(k)(j)}\right|\right) \zeta_{2(k)(j)} \\ + \overline{a}_{001}^{p_1p_2p_3} w_{001}\left(\left|\boldsymbol{\zeta}_{(k)(j)}\right|\right) \zeta_{3(k)(j)} + \overline{a}_{200}^{p_1p_2p_3} w_{200}\left(\left|\boldsymbol{\zeta}_{(k)(j)}\right|\right) \zeta_{1(k)(j)}^2 \\ + \overline{a}_{020}^{p_1p_2p_3} w_{020}\left(\left|\boldsymbol{\zeta}_{(k)(j)}\right|\right) \zeta_{2(k)(j)}^2 + \overline{a}_{002}^{p_1p_2p_3} w_{002}\left(\left|\boldsymbol{\zeta}_{(k)(j)}\right|\right) \zeta_{3(k)(j)}^2 \\ + \overline{a}_{110}^{p_1p_2p_3} w_{110}\left(\left|\boldsymbol{\zeta}_{(k)(j)}\right|\right) \zeta_{1(k)(j)} \zeta_{2(k)(j)} \\ + \overline{a}_{101}^{p_1p_2p_3} w_{101}\left(\left|\boldsymbol{\zeta}_{(k)(j)}\right|\right) \zeta_{1(k)(j)} \zeta_{3(k)(j)} \\ + \overline{a}_{011}^{p_1p_2p_3} w_{011}\left(\left|\boldsymbol{\zeta}_{(k)(j)}\right|\right) \zeta_{2(k)(j)} \zeta_{3(k)(j)} \end{pmatrix} \tag{10.22}$$

Similar normalization approach can be applied to the 2D PD functions leading to

$$g_2^{p_1p_2}\left(\boldsymbol{\zeta}_{(k)(j)}\right) = \frac{1}{\delta_{(k)}V_{(k)}}\begin{pmatrix} \overline{a}_{10}^{p_1p_2} w_{10}\left(\left|\boldsymbol{\zeta}_{(k)(j)}\right|\right)\zeta_{1(k)(j)} + \overline{a}_{01}^{p_1p_2} w_{01}\left(\left|\boldsymbol{\zeta}_{(k)(j)}\right|\right)\zeta_{2(k)(j)} \\ +\overline{a}_{20}^{p_1p_2} w_{20}\left(\left|\boldsymbol{\zeta}_{(k)(j)}\right|\right)\zeta_{1(k)(j)}^2 + \overline{a}_{02}^{p_1p_2} w_{02}\left(\left|\boldsymbol{\zeta}_{(k)(j)}\right|\right)\zeta_{2(k)(j)}^2 \\ +\overline{a}_{11}^{p_1p_2} w_{11}\left(\left|\boldsymbol{\zeta}_{(k)(j)}\right|\right)\zeta_{1(k)(j)}\zeta_{2(k)(j)} \end{pmatrix} \tag{10.23}$$

10.3 Derivatives of a Function

Using the normalized PD function from Eq. (10.22), the derivatives of a function expressed in Eqs. (2.42) and (2.43) can be evaluated as

$$\begin{Bmatrix} f_{,1}\left(\mathbf{x}_{(k)}\right) \\ f_{,2}\left(\mathbf{x}_{(k)}\right) \\ f_{,3}\left(\mathbf{x}_{(k)}\right) \end{Bmatrix} = \frac{1}{\delta_{(k)}}\sum_{j\in H_{(k)}}\left(f\left(\mathbf{x}_{(j)}\right) - f\left(\mathbf{x}_{(k)}\right)\right)\begin{Bmatrix} g_2^{100}\left(\mathbf{x}_{(k)}, \boldsymbol{\zeta}_{(k)(j)}\right) \\ g_2^{010}\left(\mathbf{x}_{(k)}, \boldsymbol{\zeta}_{(k)(j)}\right) \\ g_2^{001}\left(\mathbf{x}_{(k)}, \boldsymbol{\zeta}_{(k)(j)}\right) \end{Bmatrix}\Omega_{(j)} \tag{10.24a}$$

and

$$\begin{Bmatrix} f_{,11}\left(\mathbf{x}_{(k)}\right) \\ f_{,22}\left(\mathbf{x}_{(k)}\right) \\ f_{,33}\left(\mathbf{x}_{(k)}\right) \\ f_{,12}\left(\mathbf{x}_{(k)}\right) \\ f_{,13}\left(\mathbf{x}_{(k)}\right) \\ f_{,23}\left(\mathbf{x}_{(k)}\right) \end{Bmatrix} = \frac{1}{\delta_{(k)}}\sum_{j\in H_{(k)}}\left(f\left(\mathbf{x}_{(j)}\right) - f\left(\mathbf{x}_{(k)}\right)\right)\begin{Bmatrix} g_2^{200}\left(\mathbf{x}_{(k)}, \boldsymbol{\zeta}_{(k)(j)}\right) \\ g_2^{020}\left(\mathbf{x}_{(k)}, \boldsymbol{\zeta}_{(k)(j)}\right) \\ g_2^{002}\left(\mathbf{x}_{(k)}, \boldsymbol{\zeta}_{(k)(j)}\right) \\ g_2^{110}\left(\mathbf{x}_{(k)}, \boldsymbol{\zeta}_{(k)(j)}\right) \\ g_2^{101}\left(\mathbf{x}_{(k)}, \boldsymbol{\zeta}_{(k)(j)}\right) \\ g_2^{011}\left(\mathbf{x}_{(k)}, \boldsymbol{\zeta}_{(k)(j)}\right) \end{Bmatrix}\Omega_{(j)} \tag{10.24b}$$

10.4 Applications

The applications concern the evaluation of derivatives of continuous or discontinuous discrete data. The family size of each point is specified as $\delta_{(k)} = (N+1)\Delta x$ with a grid spacing of Δx, and $N = 2$ represents the order of truncation in the TSE as explained in Chap. 2. The parameter a appearing in the Gaussian type weight function, Eq. (2.1) is specified as $a = 2$.

10.4.1 One-Dimensional Functions

The first application concerns the differentiation of a one-dimensional smooth data given in the form

$$f\left(x_{(k)}\right) = \sin\left(x_{(k)}\right) \text{ for } -\pi/2 \leq x_{(k)} \leq \pi/2 \text{ with } k = 1, .., K. \tag{10.25}$$

The uniform grid spacing between the PD points is specified as $\Delta x = 0.1$ resulting in total grid points of $K = 32$. The data set consists of points $S_1 = \{x; -0.5\pi + 0.5\Delta x \leq x \leq 0.5\pi - 0.5\Delta x\}$.

For evaluation of the derivatives of the data by using PDDO, an external PD mesh file ***ex10.4.1_sinx.csv*** consisting of grid points S_1 and the data as in Eq. (10.25) is read. The necessary format of the PD mesh file is explained in Sect. 9.2 under Chap. 9. The specific steps in Ansys framework to evaluate the derivatives of the data as described in Chap. 2 for a given PD mesh using the customized built-in menu ***Peridynamic Processor*** as shown in Fig. 10.2. The procedure for entering the parameters for each of the submenus located under this main menu is explained in Chap. 9.

Step-1 The field and dimension of the analysis are specified using the ***Field Selection*** submenu located under ***Peridynamic Processor*** menu. For evaluating the derivative of a 1D data set, the "***PDDO***" for field and "***1D***" for dimension are selected as shown in Fig. 10.2.

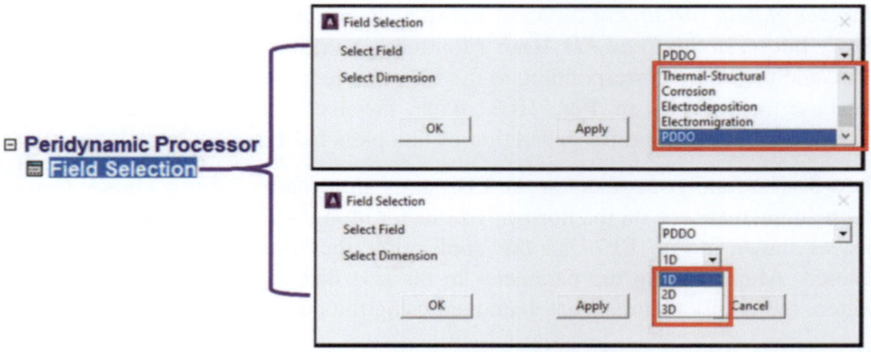

Fig. 10.2 Dialog box for selecting the field and dimension

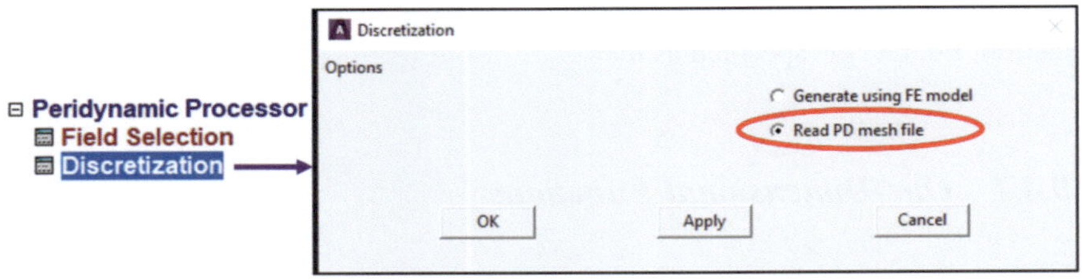

Fig. 10.3 Dialog box for selecting the external file for PD mesh generation

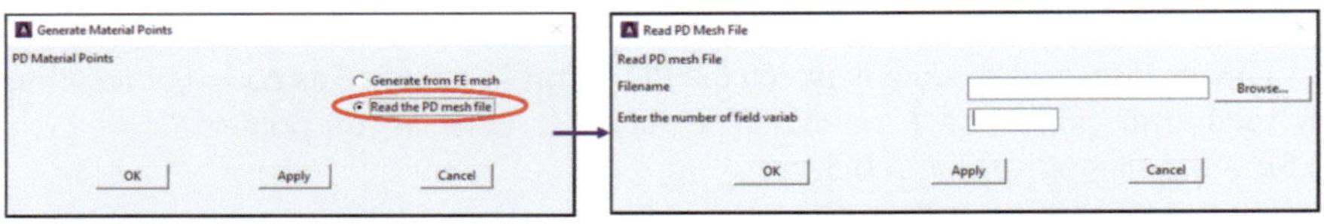

Fig. 10.4 Dialog box for selecting the existing PD mesh file name

Step-2 For the evaluation of derivatives of the PD field, PD mesh is generated in Ansys workspace by using the radio button option ***Read PD mesh file*** in ***Discretization*** submenu as shown in Fig. 10.3.

When this option is selected, a dialog box ***Read PD Mesh File*** appears as shown in Fig. 10.4.

Furthermore, upon clicking the ***Browse*** button, a file browser window shown in Fig. 10.5 appears to display the list of files in the working directory. Select the PD mesh file ***ex10.4.1_sinx.csv.***

After selecting the PD mesh file from the working directory and using the ***Open*** button, the name of the PD mesh file to be read is saved in the parameter ***Filename***. Enter the number of columns to be read as a value for the parameter ***Enter the number of field variab.*** For this case, the value to be entered is 5. Clicking the ***OK*** or ***Apply*** button in the ***Read PD Mesh File*** dialog box, the imported PD mesh file is read, and the nodes corresponding to the PD point locations as created in the Ansys database are shown in Fig. 10.6. After the mesh generation, the keyword ***_line_1d=1*** is activated for generation of line plots for 1D mesh.

Step-3 The family members of the PD points are generated using ***Family Generation*** submenu based on the horizon size factor ***m*** in the ***Family Generation*** dialog box as shown in Fig. 10.7. For this application, the horizon size factor, $m = 3$ is entered. After entering the parameter in the text box and clicking the ***OK/Apply*** button, the family members are generated using the horizon size of $\delta = 3.015\Delta x$.

Step-4 The element coefficients corresponding to the PDDO interactions around a point are generated using ***Element Coefficients (PDDO)*** submenu option as shown

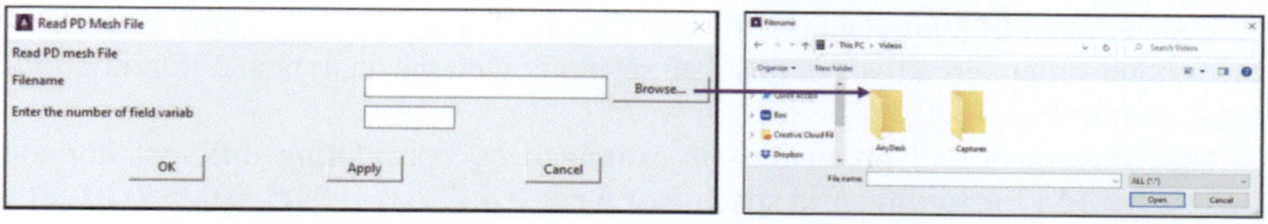

Fig. 10.5 File browser window for choosing files from the working directory

Fig. 10.6 PD nodes as created in Ansys database after reading the mesh file

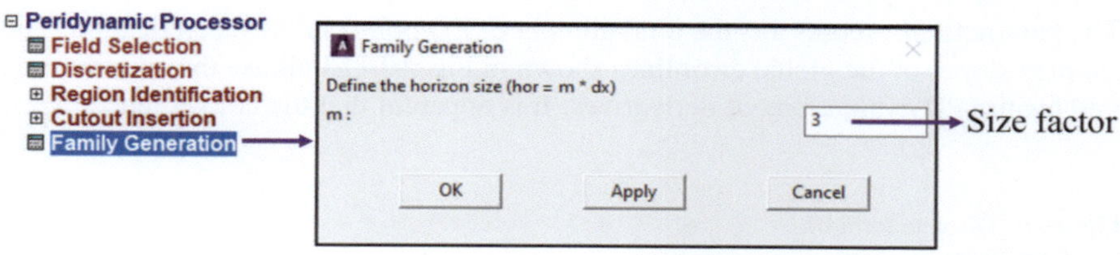

Fig. 10.7 Dialog box for defining the horizon size in the family generation

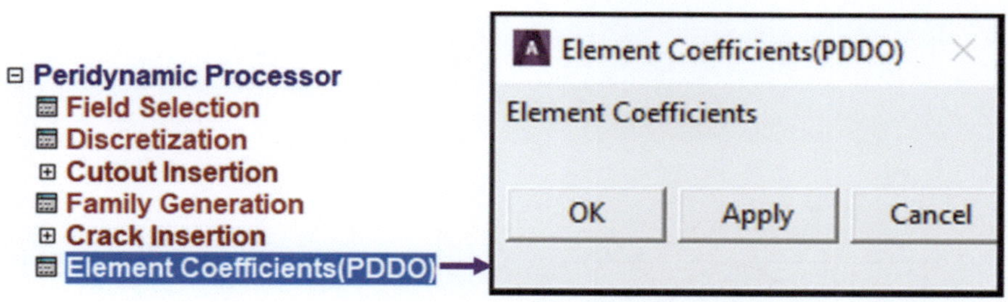

Fig. 10.8 Element Coefficients submenu for computing the nonlocal derivatives using PDDO

in Fig. 10.8. Clicking the ***OK/Apply*** button of the ***Element Coefficients (PDDO)*** dialog box generates the constants for the PDDO model. After this step, the PD predictions are generated and output to two files named as ***derv_1.dat*** and ***d1_fx.dat***, respectively. The file ***derv_1.dat*** includes the coordinates of points from columns 2 to 4, the input data in column 5, the first-order nonlocal derivatives from columns 6 to 8, and second-order nonlocal derivatives from columns 9 to 14. The file ***d1_fx.dat*** includes the data used to be plotted by using an external plotting software like python.

The variation of input data is shown in Fig. 10.9. The PD evaluations of the first- and second-order derivatives of this data set along with the analytical differentiations are shown in Fig. 10.10.

The convergence of predictions is examined by considering different horizon size, $\delta_{(k)} = 3\Delta x$ for varying grid spacing of $\Delta x = 0.4, 0.2, 0.1, 0.05, 0.025, 0.01$. The rate of convergence is computed based on a global error measure as (Mukherjee & Mukherjee, 1997)

$$\varepsilon = \frac{1}{\left|f^{(e)}\right|_{\max}} \sqrt{\frac{1}{K} \sum_{m=1}^{K} \left[f_m^{(e)} - f_m^{(c)}\right]^2} \tag{10.26}$$

in which $|f^{(e)}|_{\max}$ denotes the maximum of absolute value of the exact field variable, and the superscripts e and c denote the exact and computed solutions, respectively. The parameter, K, represents the total number of PD points in the discretization. The average slopes of the global error lines shown in Fig. 10.11 indicate the convergence rate for the PD evaluations of derivatives. It is apparent that the convergence rate is

Fig. 10.9 Discrete form of input data from function, $f(x_{(k)}) = \sin(x_{(k)})$

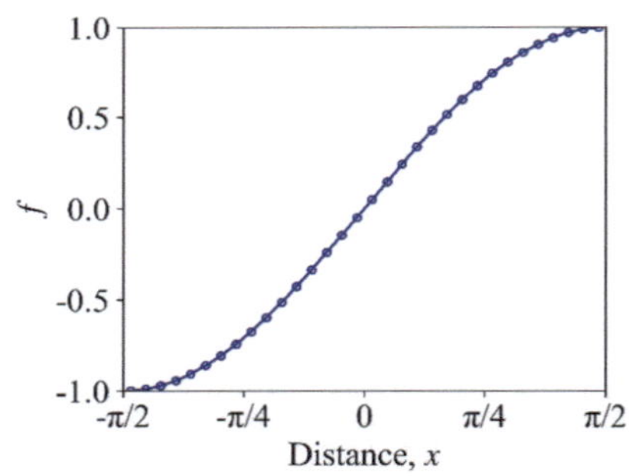

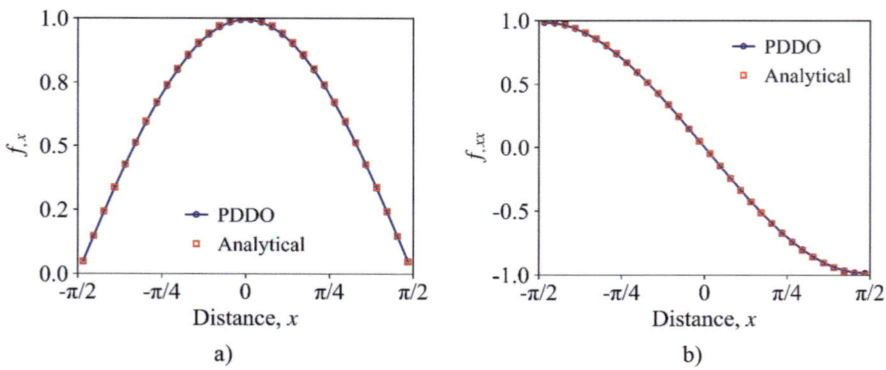

Fig. 10.10 PD and analytical evaluation of derivatives of $f(x_{(k)}) = \sin(x_{(k)})$: (**a**) $f_{,x}$ and (**b**) $f_{,xx}$

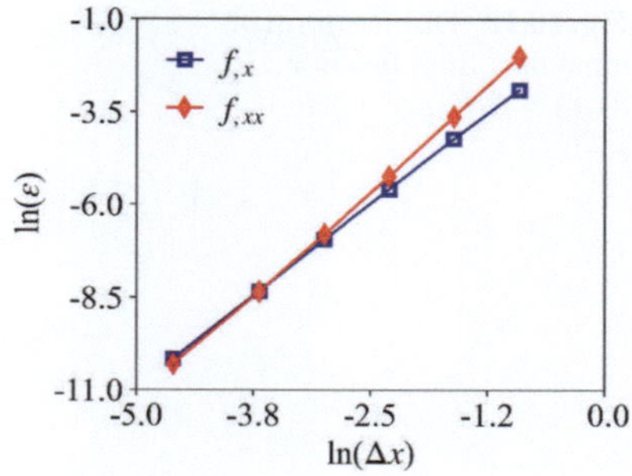

Fig. 10.11 Convergence rate for the derivatives of the function, $f(x_{(k)}) = \sin(x_{(k)})$

faster for the first derivative than that of the second derivative because of the truncation in the TSE. However, the convergence rates for the first and second derivatives become almost similar as the grid spacing decreases.

The second application concerns the differentiation of a one-dimensional data with a jump discontinuity given as

$$f\left(x_{(k)}\right) = 5H\left(x_{(k)}\right) - 5H\left(-x_{(k)}\right) \text{ for } -\pi/2 \leq x_{(k)} \leq \pi/2 \text{ with } k = 1, .., K \tag{10.27}$$

As in the case of input data in Eq. (10.27), the uniform grid spacing between the PD points is specified as $\Delta x = 0.005$ resulting in total grid points of $K = 629$. The data set consists of points $S_1 = \{x; -0.5\pi + 0.5\Delta x \leq x \leq 0.5\pi - 0.5\Delta x\}$.

For the evaluation of the derivatives of the data by using PDDO, an external PD mesh file ***ex10.4.1_sgn5.csv*** consisting of grid points S_1 and the data as in Eq. (10.27) is read. The specific steps to generate the derivatives using the ***Peridynamic Processor*** menu are similar to the preceding case. Utilizing the output files, the input data and the PD evaluation of derivatives are shown in Figs. 10.12 and 10.13. As shown in Fig. 10.13, the PD differentiation matches the analytical differentiation at all points in the set S_1 except at the origin. It captures the singularity at $x = 0$, and the strength of the singularity increases with decreasing value of Δx from $\Delta x = 0.02$ to $\Delta x = 0.005$.

10.4.2 Two-Dimensional Function

The two-dimensional data is generated from the function given as

$$f\left(x_{(k)}, y_{(k)}\right) = \frac{x_{(k)}y_{(k)}}{\sqrt{x_{(k)}^2 + y_{(k)}^2}} \text{ for } -0.5 \leq x_{(k)}, y_{(k)} \leq 0.5 \tag{10.28}$$

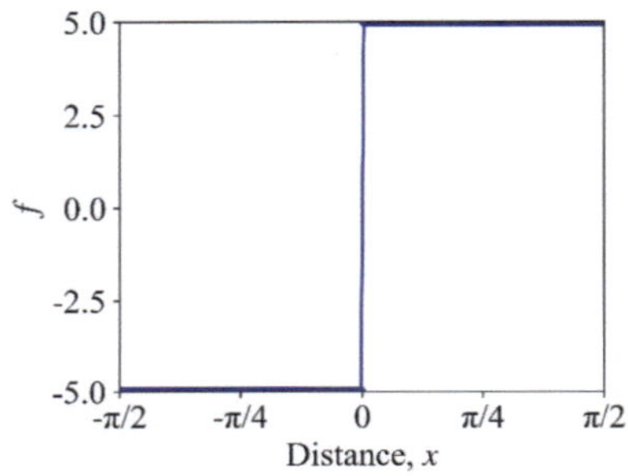

Fig. 10.12 Discrete form of input data from function, $f(x_{(k)}) = 5H(x_{(k)}) - 5H(-x_{(k)})$

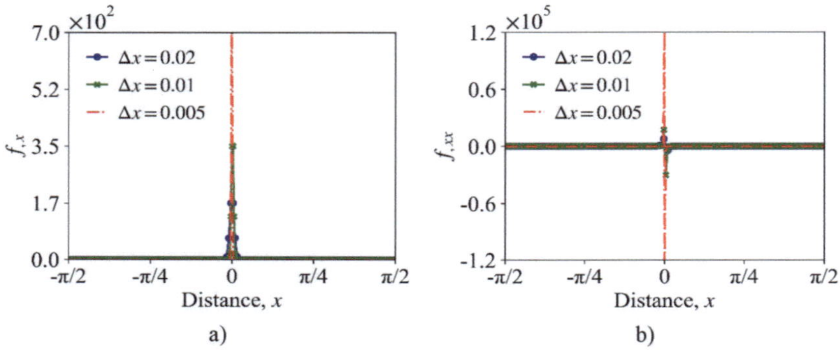

Fig. 10.13 PD and analytical differentiation of $f(x_{(k)}) = 5H(x_{(k)}) - 5H(-x_{(k)})$: **(a)** $f_{,x}$ and **(b)** $f_{,xx}$

The data is structured but nonuniform with spacing specified as

$$\Delta x = \Delta y = 0.01, \ -0.25 \leq x_{(k)}, y_{(k)} \leq 0.25 \tag{10.29}$$

and

$$\Delta x = \Delta y = 0.0147, \text{-}0.5 \leq x_{(k)}, y_{(k)} \leq -0.25 \text{ and } 0.25 \leq x_{(k)}, y_{(k)} \leq 0.5 \tag{10.30}$$

resulting in $K = 5968$ PD points. The family size of each point is specified as $\delta_{(k)} = 3\Delta x$ with an incremental area of $A_{(k)} = \Delta x^2$.

For the evaluation of derivatives of the data by using PDDO, an external PD mesh file ***ex10.4.2.csv*** consisting of 2D grid points generated with spacing as specified in Eqs. (10.29) and (10.30) along with the data in Eq. (10.28) is read.

The specific steps in Ansys framework to evaluate the derivatives of the data are similar to that of described in Sect. 10.4.1 except for the selection of the dimension in **Step-1** and discretization in **Step-2**. In **Step-1**, the "***PDDO***" for field and "***2D***' for dimension are selected as shown in Fig. 10.14.

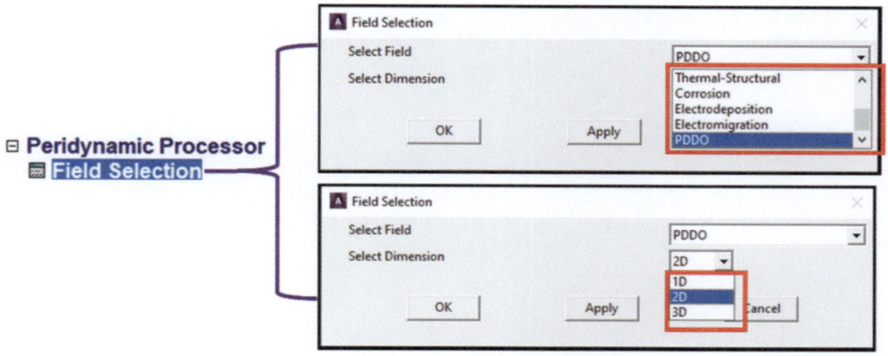

Fig. 10.14 Dialog box for selecting the field and dimension

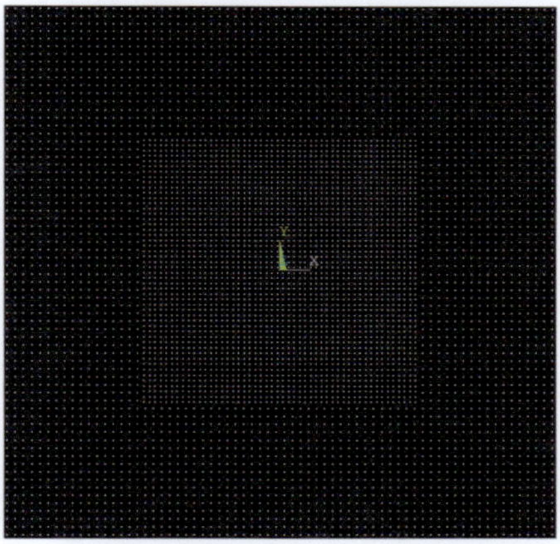

Fig. 10.15 PD nodes for the 2D data as created in Ansys database after reading of the mesh file

In **Step-2**, the PD mesh file ***ex10.4.2.csv*** is read and the value for the parameter ***Enter the number of field variab*** is set as 6. Figure 10.15 shows the nodes corresponding to the PD point locations as created in the Ansys database. After the mesh generation, the keyword ***_line_2d=1*** is activated for generation of 2D line plots.

After excuting **Steps-3** and **4**, six files named as ***derv_1.dat, line_2dx.dat***, ***line_2dy.dat, line_2dxx.dat***, ***line_2dyy.dat***, and ***line_2dxy.dat*** are created. The ***derv_1.dat*** is used to plot the input data and PD evaluation of derivatives. The remaining files are used to plot the first-order and second-order derivatives along the lines defined by set of points in S_1 and S_2. The input data with structured nonuniform grid spacing is shown in Fig. 10.16 along with the PD evaluations of the first- and second-order derivatives as shown in Figs. 10.17 and 10.18.

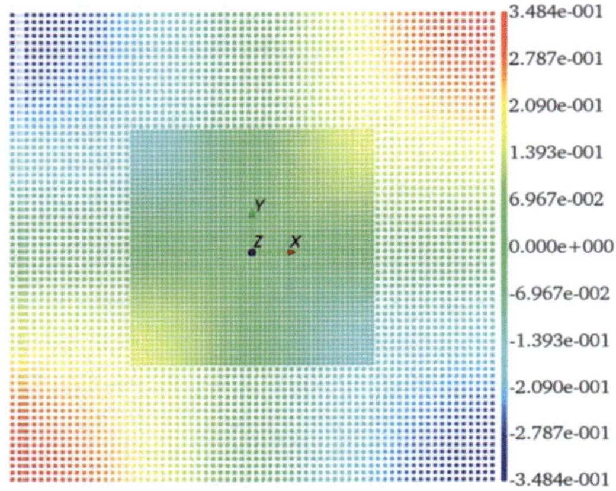

Fig. 10.16 Discrete form of 2D input data

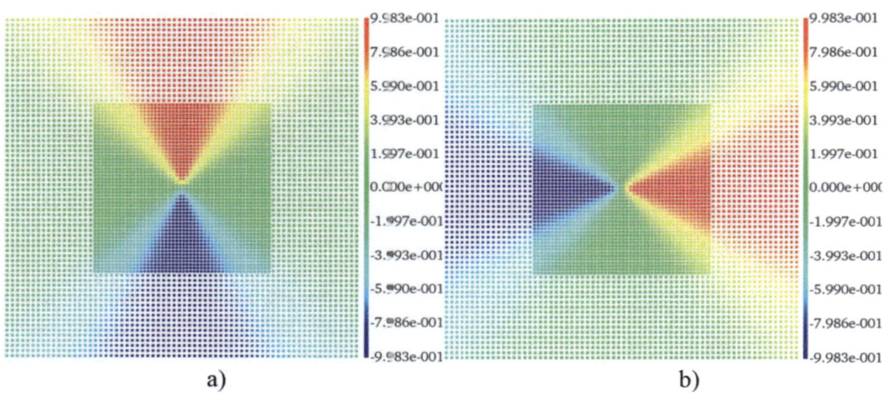

Fig. 10.17 PD evaluation of the first-order derivatives of 2D data: (**a**) $f_{,x}$, and (**b**) $f_{,y}$

The comparisons of PD and analytical evaluations of the first-order derivatives along the set of points in $S_1 = \{(x, y); 0.01 \leq x \leq 0.015\}$ and $S_2 = \{(x, y); 0.01 \leq y \leq 0.015\}$ are shown in Fig. 10.19a, b, respectively. They are in close agreement with each other at all the points except near the origin where the step discontinuity is captured approximately.

The comparisons of PD and analytical evaluations of the second-order derivatives, $f_{,xx}$ and $f_{,xy}$, along the set of points in $S_1 = \{(x, y); 0.01 \leq x \leq 0.015\}$ are shown in Fig. 10.20a, b, respectively. The second-order derivative, $f_{,yy}$, along the set of points in $S_2 = \{(x, y); 0.01 \leq y \leq 0.015\}$ is shown in Fig. 10.20c. They are in close agreement with the analytical evaluations along the points in sets S_1 and S_2; however, deviation occurs at the origin due to the singular behavior. Although not shown, this deviation reduces significantly with decreasing grid spacing or increasing the number of terms in TSE as explained by Madenci et al. (2019).

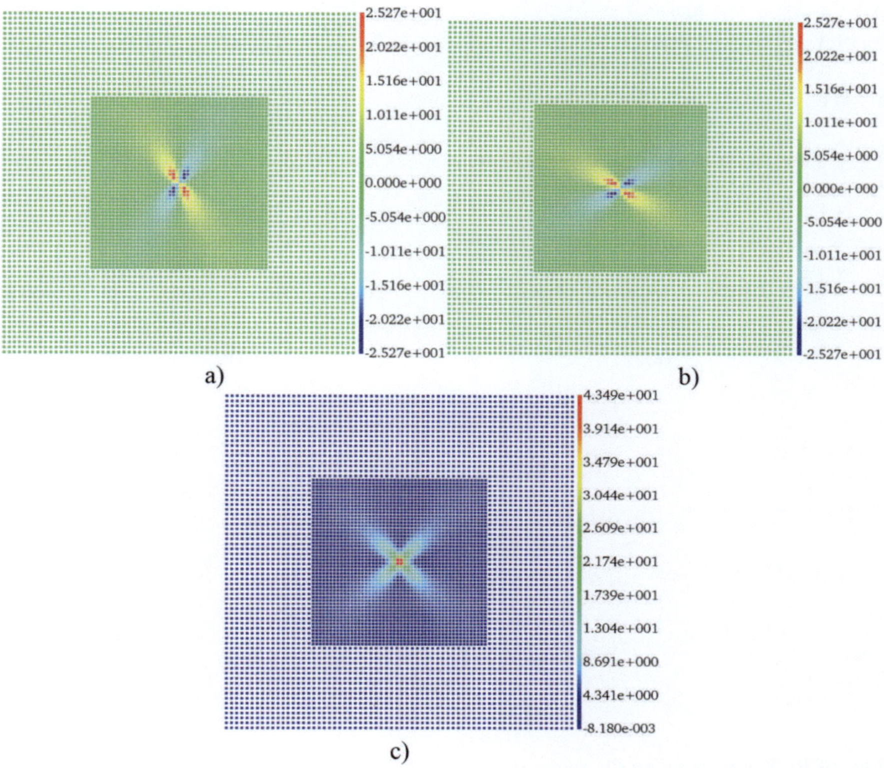

Fig. 10.18 PD evaluation of the second-order derivatives of 2D data: (**a**) $f_{,xx}$, (**b**) $f_{,yy}$, and (**c**) $f_{,xy}$

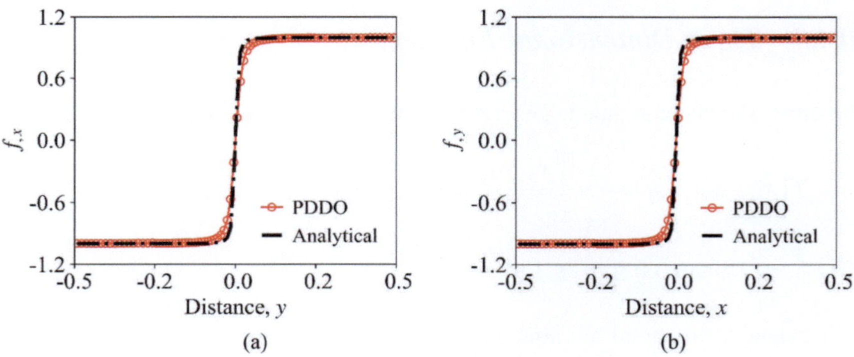

Fig. 10.19 Comparison of PD and analytical evaluation of first-order derivatives of 2D data: (**a**) $f_{,x}$ and (**b**) $f_{,y}$

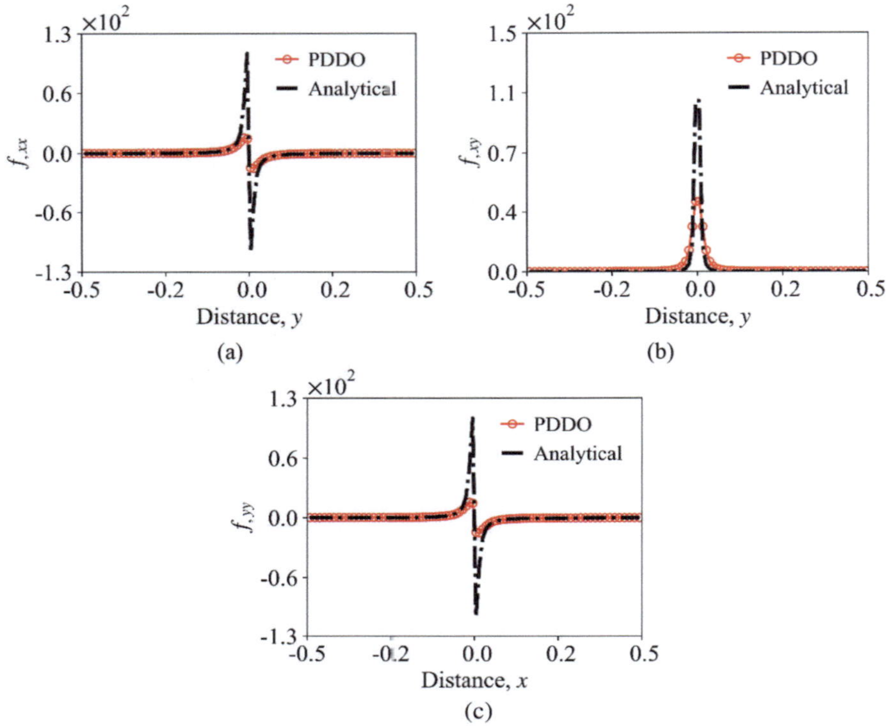

Fig. 10.20 Comparisons of PD and analytical evaluation of second-order derivatives of 2D data: **(a)** $f_{,xx}$, **(b)** $f_{,xy}$, and **(c)** $f_{,yy}$

10.4.3 Three-Dimensional Function

The three-dimensional data is generated from the function given as

$$f\left(x_{(k)}, y_{(k)}, z_{(k)}\right) = \frac{x_{(k)}y_{(k)}z_{(k)}}{\sqrt{x_{(k)}^2 + y_{(k)}^2 + z_{(k)}^2}} \quad \text{for} \quad -0.5 \leq x_{(k)}, y_{(k)} \leq 0.5 \text{ and } 0.0 \leq z_{(k)} \leq 1.0 \tag{10.31}$$

The data is structured but nonuniform with spacing specified as

$$\Delta x = \Delta y = 0.0385, \Delta z = 0.04; \; -0.25 \leq x_{(k)}, y_{(k)} \leq 0.25 \text{ and } 0.0 \leq z_{(k)} \leq 1.0 \tag{10.32}$$

and

$$\begin{aligned}&\Delta x=0.0417, \Delta y=0.0454, \Delta z=0.0476;\\&\text{-0.5} \le x_{(k)}, y_{(k)} \le -0.25, 0.25 \le x_{(k)}, y_{(k)} \le 0.5 \text{ and } 0.0 \le z_{(k)} \le 1.0\end{aligned} \tag{10.33}$$

resulting in 12793 PD points. The family size of each point is specified as $\delta_{(k)} = 3\Delta x$ with an incremental volume of $V_{(k)} = \Delta x \Delta y \Delta z$.

The specific steps in Ansys framework to evaluate the derivatives of the data for the 3D data using the customized built-in menu ***Peridynamic Processor*** is similar to the 1D and 2D data sets, except for **Steps 1** and **2**. In **Step-1**, the parameter ***Select Dimension*** is set to ***3D***. The keyword ***_line_3d=1*** is activated for generation of line plots. The PD mesh file ***ex10.4.3.csv*** is read and the value for the parameter ***Enter the number of field variab*** is set as 7 in **Step-2.** Figure 10.21 shows the nodes corresponding to the PD point locations as created in the Ansys database.

After excuting **Steps-3** and **4**, nine files named as ***derv_1.dat, line_3dx.dat***, ***line_3dy.dat***, ***line_3dz.dat, line_3dxx and line_3dxy.dat***, ***line_3dxz.dat, line_3dyy, line_3dzz.dat***, ***line_3dxz.dat***, and ***line_3dyz.dat*** are created. The ***derv_1.dat*** is used to plot the input data and PD evaluation of derivatives. The remaining files are used to plot the first-order and second-order derivatives along the lines defined by set of points in S_1 and S_2. The input data with structured nonuniform grid spacing is shown in Fig. 10.22 along with the PD evaluations of the first-order derivatives on the xy plane at $z = 1 - 0.5\Delta x$ as shown in Fig. 10.23.

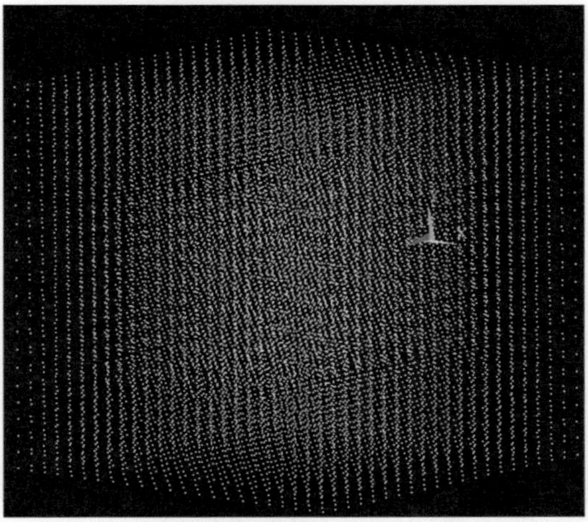

Fig. 10.21 PD nodes for a 3D data set as created in Ansys database after reading of the mesh file

Fig. 10.22 Discrete form of 3D input data

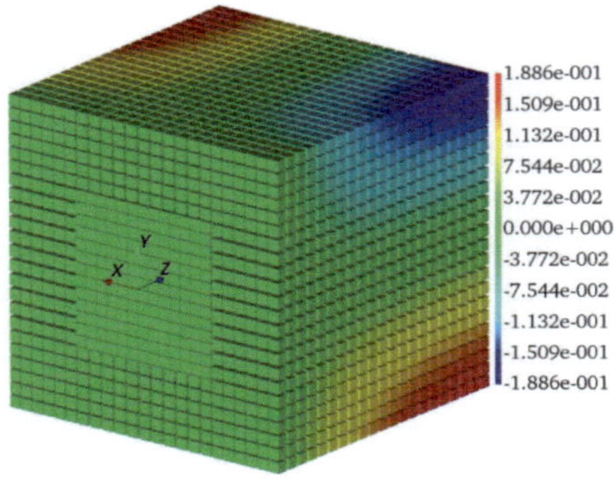

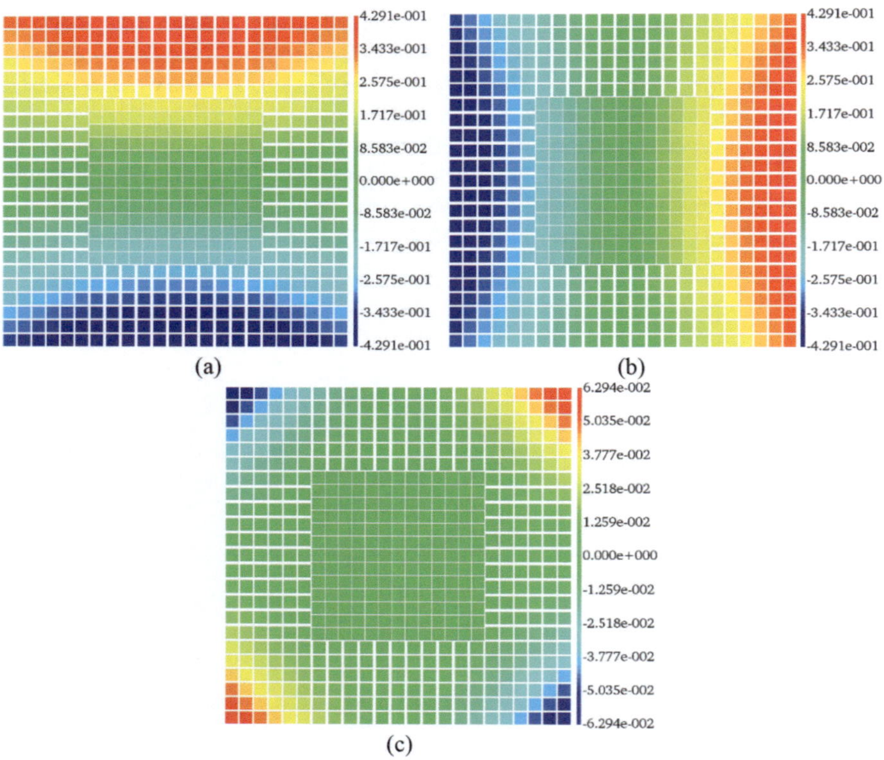

Fig. 10.23 PD evaluation of the first-order derivatives of 3D data: (**a**) $f_{,x}$, (**b**) $f_{,y}$, and (**c**) $f_{,z}$

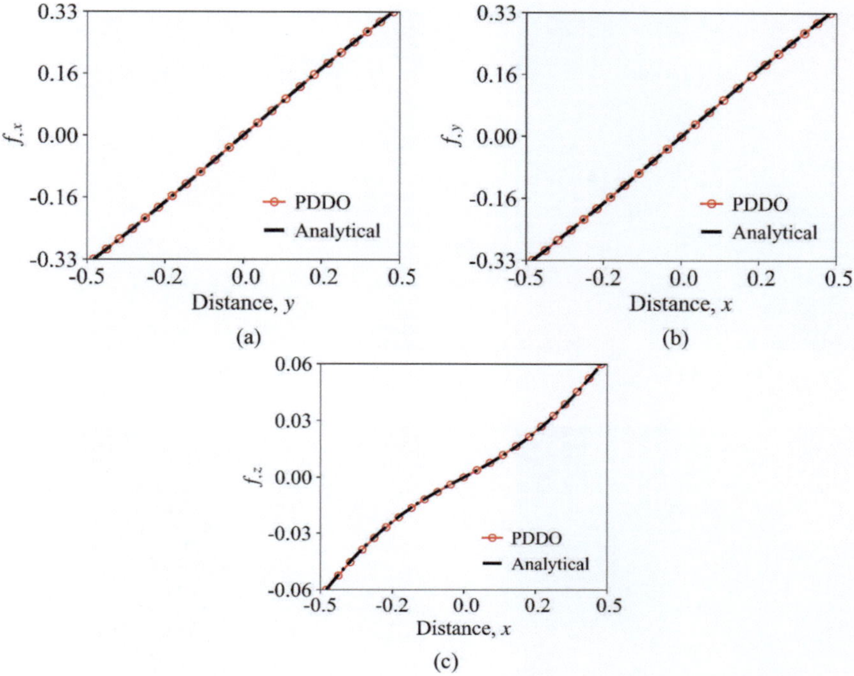

Fig. 10.24 Comparison of PD and analytical evaluation of first-order derivatives of function $f(x_{(k)}, y_{(k)}, z_{(k)})$: **(a)** $f_{,x}$, **(b)** $f_{,y}$, and **(c)** $f_{,z}$

The comparisons of PD and analytical evaluations of the first-order derivative, $f_{,x}$, for the set of points along the y-direction in $S_1 = \{(x, y, z); x = 0.5 - 0.5\Delta x, z = 1 - 0.5\Delta x\}$ is shown in Fig. 10.24a. The comparisons of PD and analytical evaluations of the first-order derivatives, $f_{,y}$ and $f_{,z}$, for the set of points along the x-direction in $S_1 = \{(x, y, z); y = 0.5 - 0.5\Delta x, z = 1 - 0.5\Delta x\}$ show close agreement as shown in Fig. 10.24b, c.

Figure 10.25 shows the PD evaluations of the second-order derivatives. The comparisons of PD and analytical evaluations of the second-order derivatives, $f_{,xx}$ and $f_{,xy}$, for the set of points along y direction in S_1 are shown in Fig. 10.26a, b, respectively. The comparisons of PD and analytical evaluations of the second-order derivatives, $f_{,yy}, f_{,zz}, f_{,xz}$, and $f_{,yz}$ for the set of points along $y$direction in S_2 are shown in Fig. 10.26c–f, respectively. They are in close agreement with each other.

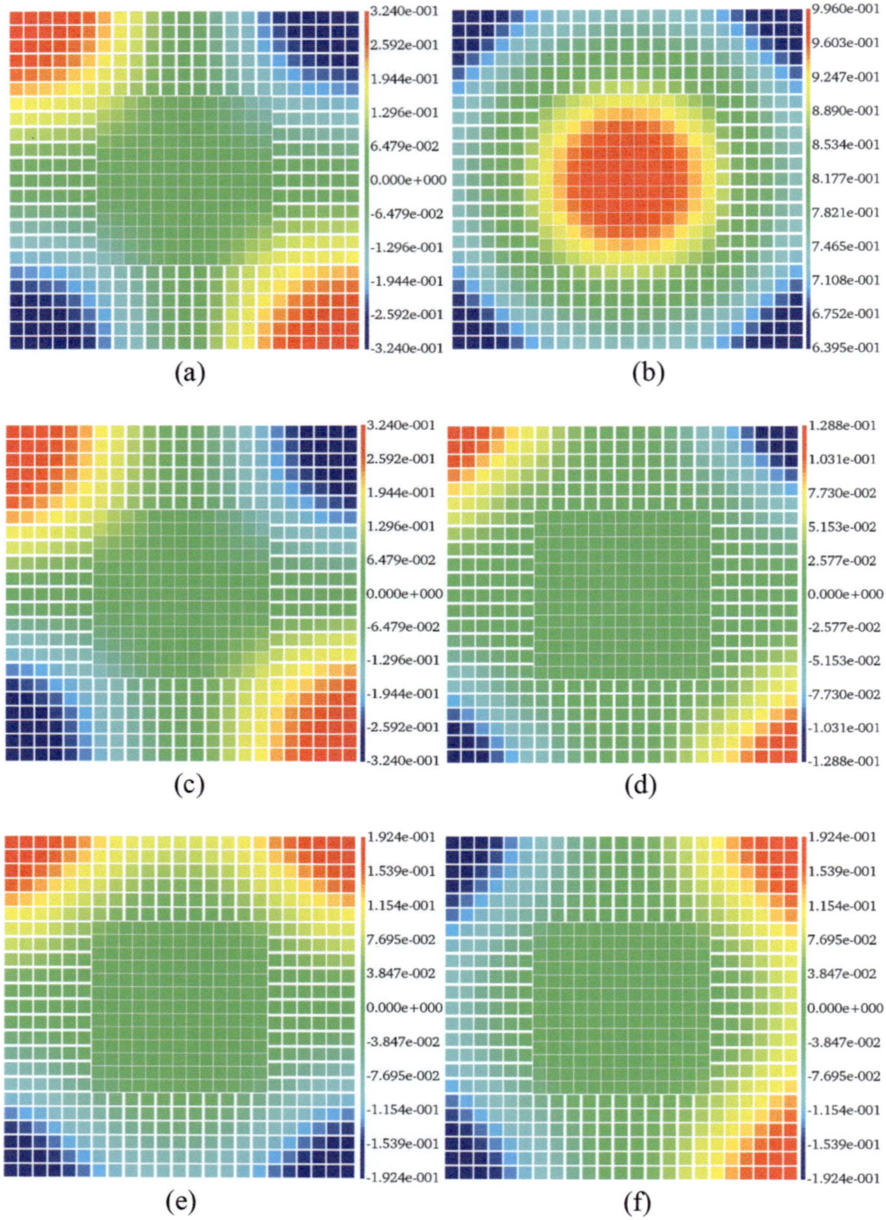

Fig. 10.25 PD evaluation of the second-order derivatives of 3D data: (**a**) $f_{,xx}$, (**b**) $f_{,xy}$, (**c**) $f_{,yy}$, (**d**) $f_{,zz}$, (**e**) $f_{,xz}$, and (**f**) $f_{,yz}$

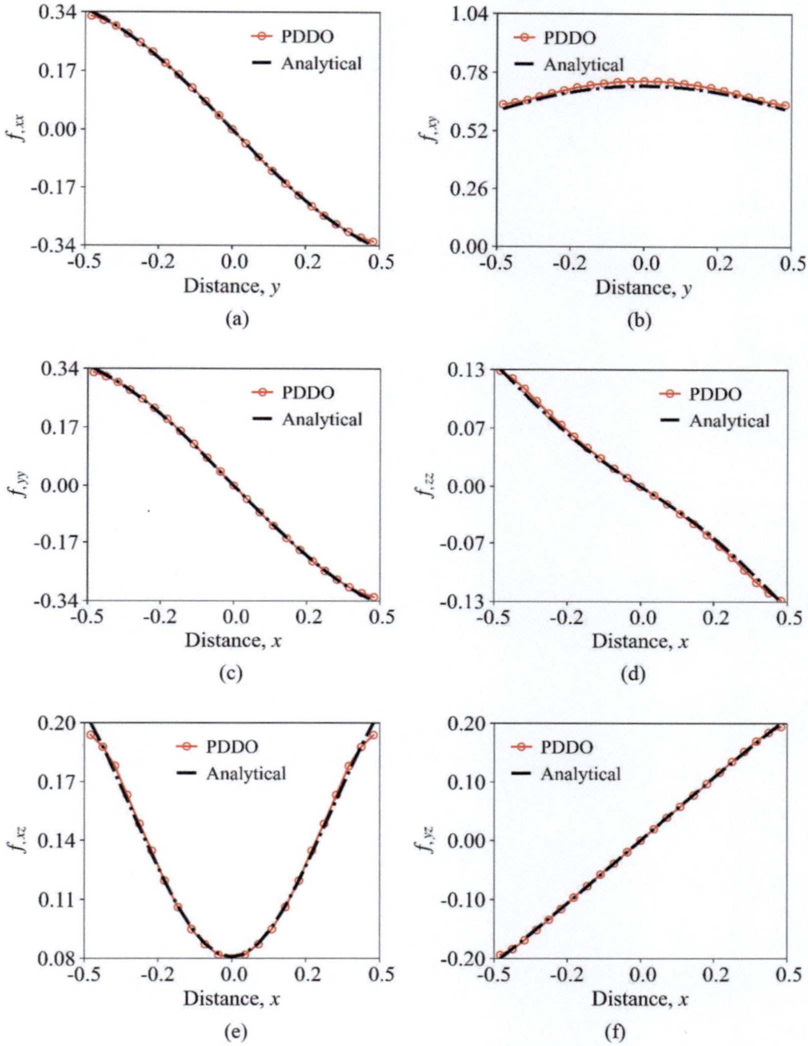

Fig. 10.26 Comparisons of PD and analytical evaluation of second-order derivatives of function $f(x_{(k)}, y_{(k)}, z_{(k)})$: **(a)** $f_{,xx}$, **(b)** $f_{,xy}$, **(c)** $f_{,yy}$, **(d)** $f_{,zz}$, **(e)** $f_{,xz}$, and **(f)** $f_{,yz}$

References

Madenci, E., Barut, A., & Dorduncu, M. (2019). *Peridynamic differential operator for numerical analysis*. Springer.

Madenci, E., Barut, A., Willmarth, E., & Phan, N. (2022). Peridynamics for data estimation, image compression/recovery, and model reduction. *Journal of Peridynamics and Nonlocal Modeling, 4*, 159–200.

Mukherjee, Y. X., & Mukherjee, S. (1997). On boundary conditions in the element-free Galerkin method. *Computational Mechanics, 19*, 264–270.

Chapter 11
Applications of Field Equations

11.1 Introduction

This chapter presents the use of Ansys macros and Graphical User Interface (GUI) capability for solving field equations such as those of thermal and moisture diffusion. The coefficients in the field equations may be time dependent. In order to demonstrate the applicability of the unified PD field equations described in Chap. 3, the thermal diffusion equation is solved by considering its strong and weak forms as explained in Chaps. 3 and 4, respectively. Also, it is solved by considering the coupled PD/FE approach as described in Chap. 5. The solution to the thermal diffusion equation demonstrates the modeling of thermal shock in a plate. The simulation of moisture diffusion concerns the desorption process from an initially fully saturated bar composed of two different materials with time-dependent saturation concentration. The solution to the moisture diffusion is constructed based on the field equation with fictitious node approach to enforce the boundary conditions as described in Sect. 1.6 of Chap. 1.

11.2 Simulation of Thermal Shock in a Plate

As shown in Fig. 11.1, the plate has a length and width of $L = W = 10\text{m}$ and a thickness of $H = 1\text{m}$. In the absence of heat source, the field equation for thermal diffusion can be expressed as

---

Supplementary Information The online version contains supplementary material available at https://doi.org/10.1007/978-3-031-59896-8_11.

E. Madenci et al., *Peridynamics and Its Applications Using Ansys*,
https://doi.org/10.1007/978-3-031-59896-8_11

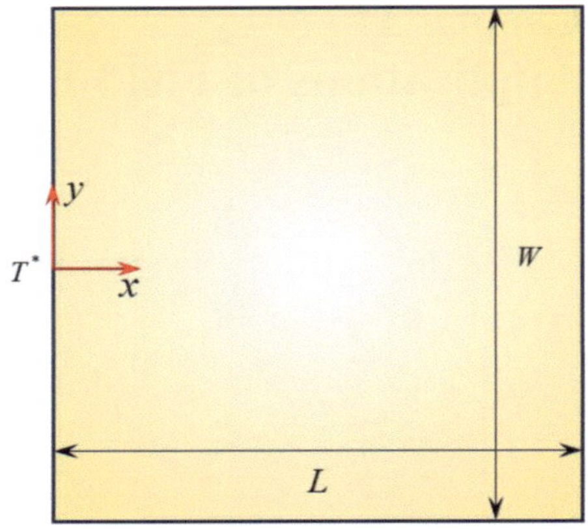

Fig. 11.1 Geometry and boundary conditions of the plate under thermal shock

$$\rho c_v \frac{\partial T}{\partial t} = k_T \nabla^2 T(\mathbf{x}, t) \tag{11.1}$$

in which $T(\mathbf{x}, t)$ represents temperature. Its specific heat capacity, thermal conductivity, and mass density are specified as $c_v = 1\text{J/kgK}$, $k_T = 1\text{W/mK}$, and $\rho = 1\ \text{kg/m}^3$, respectively. The plate has an initial temperature of

$$T(x, y, t = 0) = 0 \tag{11.2}$$

The applied temperature on the left edge is specified as

$$T(x = 0, y, t > 0) = T^*(t) = 5te^{-2t} \tag{11.3}$$

The insulated boundary conditions on the remaining edges are specified as

$$T_{,x}(x = L, y, t > 0) = 0 \text{ and } T_{,y}(x = \pm 0.5W, y, t > 0) = 0 \tag{11.4}$$

The grid spacing between the points is $\Delta x = 0.2\text{m}$ with PD horizon $\delta = 3.015\Delta x$. The simulation is performed with a time step size of $\Delta t = 5 \times 10^{-3} s$ for $t = 6s$.

11.2.1 Modeling in Ansys Framework

11.2.1.1 Assigning Degrees of Freedom

The PD model of heat conduction can be simulated by specifying the field variable, $\Theta(\mathbf{x}, t)$ as temperature $T(\mathbf{x}, t)$ in Eqs. (3.113), (4.16), and (5.14), i.e.,

$$\Theta(\mathbf{x}, t) = T(\mathbf{x}, t) \tag{11.5}$$

Also, the material property term, $C(\mathbf{x}, t)$ is set as

$$C(\mathbf{x}, t) = \rho c_v \tag{11.6}$$

In addition, the time rate of change of material property term is zero, i.e., $\dot{C}(\mathbf{x}, t) = 0$ for heat conduction.

The field flux, $\mathbf{q}(\mathbf{x}, t)$ in Eq. (3.116) is defined by the Fourier conduction law as

$$\mathbf{q}(\mathbf{x}, t) = k(\mathbf{x}) \nabla T(\mathbf{x}, t) \tag{11.7}$$

where $k(\mathbf{x})$ is the thermal conductivity.

11.2.1.2 Ansys Elements

Although, MATRIX27 and COMBIN14 have structural DOF's, they can be mimicked for transient thermal analysis by assigning the ROTX degree of freedom (DOF) as the unknown temperature field, $T(\mathbf{x}, t)$. MATRIX27 element is used for modeling the unsymmetric stiffness matrix in the PD domain, PDDO domain, and the PD boundary region. The unsymmetric stiffness coefficients in Eqs. (3.232)-(3.233), (3.249)-(3.251), and (3.257) are input as real constants to MATRIX27 element corresponding to this ROTX DOF. COMBIN14 element with ROTX DOF activated by using KEYOPT(2) = 4 is used for modeling the heat capacitance as in Eq. (11.1), and its value is input as a real constant to the element.

The supported KEYOPTS command and the procedure to input real constants applicable to these elements are described in Chap. 8. Table 11.1 shows the summary of elements and their KEYOPTS activated for modeling the simulation of thermal shock in a plate.

11.2.2 Strong Form of Thermal Diffusion Equation

This application describes the specific steps in Ansys framework to perform thermal shock simulation using the strong form of unified PD approach for the field equations as described in Chap. 3. The customized built-in menu ***Peridynamic Processor***

Table 11.1 Ansys elements used for the simulation of thermal shock in a plate

Modeling variable	Element	DOF	KEYOPTS
PD stiffness	MATRIX27	ROTX	KEYOPT (2)=2 KEYOPT (3)=4
Volumetric heat conductivity, ρc_v	COMBIN14	ROTX	KEYOPT (2)=4

is used for model generation. The procedure for entering the parameters for each of the submenus located under this main menu is explained in Chap. 9.

11.2.2.1 Preprocessing

The finite element discretization of the plate is generated using the Ansys input file: ***ex11.2.2.dat*** or imported into the Ansys database using the CDREAD command.

Step-1 The field and dimension of the analysis are specified using the ***Field Selection*** submenu located under ***Peridynamic Processor***. For the simulation of thermal shock in a plate, the "***Diffusion***" for field and "***2D***" for dimension are selected as shown in Fig. 11.2. The keyword ***_is_fld_unify*** is required and is set to 1 after the **Step-1** of the preprocessing step, i.e., ***_is_fld_unify=1*** is entered in the command window.

Step-2 The PD mesh is generated using the radio button option ***Generate using FE model*** in ***Discretization*** submenu as shown in Fig. 11.3:

Step-3 The PD regions for the field equation and boundary conditions are created using ***Region Identification*** submenu shown in Fig. 11.4. The submenu ***Region Identification>Domain*** is selected to open the dialog box ***Domain***. Two regions are

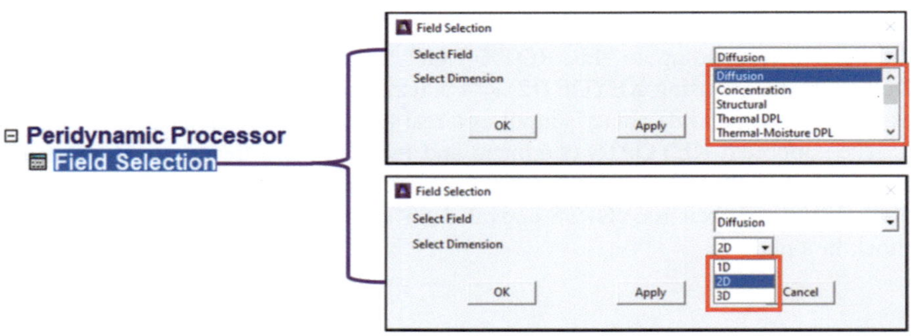

Fig. 11.2 Dialog box for selecting the field and dimension

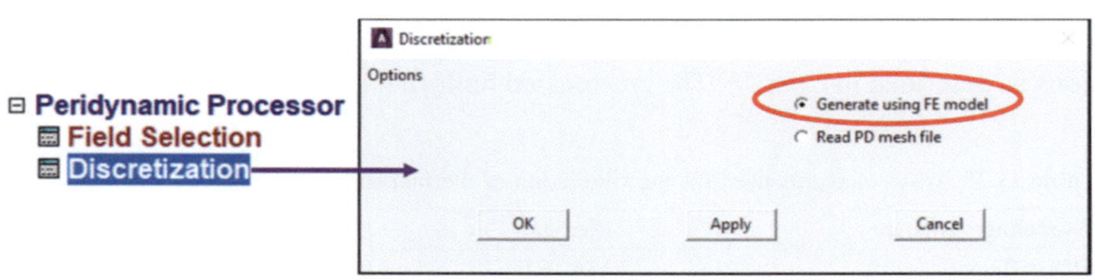

Fig. 11.3 Dialog box for PD mesh generation

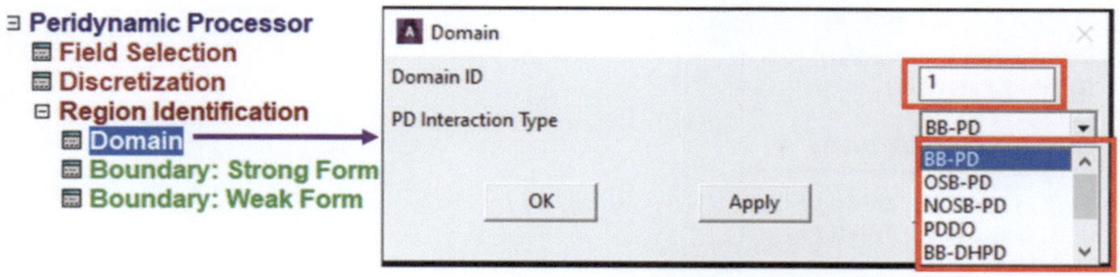

Fig. 11.4 Dialog box for selecting the domain interaction type as BB-PD

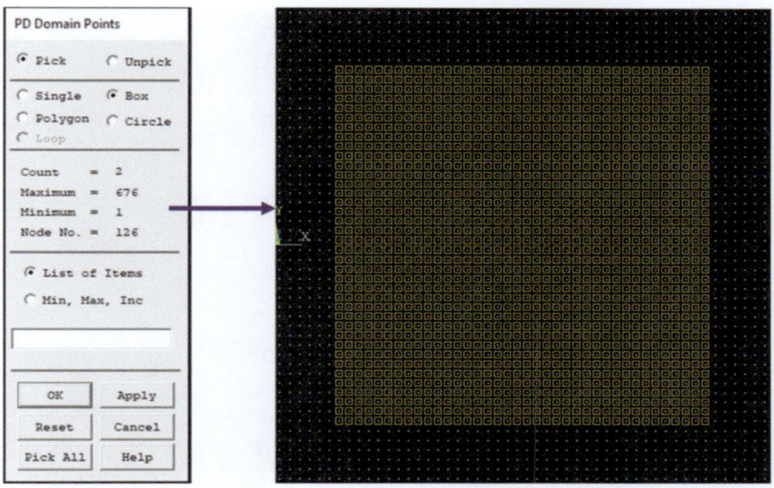

Fig. 11.5 Region identification corresponding to the BB-PD domain of the plate

created corresponding to regions of BB-PD and PDDO. For the BB-PD domain, the parameter, ***Domain ID=1*** is entered in the text box, and ***PD Interaction Type*** is selected from the drop down menu as BB-PD as shown in Fig. 11.4:

After entering the parameters and clicking the ***OK/Apply*** button, the node picker dialog box appears for selection of the desired BB-PD region as shown in Fig. 11.5. After selecting the nodes by using the Node picker window and clicking the ***OK*** or ***Apply*** button of the node picker (Box), a dialog box labeled as ***Domain Points*** appears for confirming the selected BB-PD region.

Step-4 Repeat the process for creating the PDDO region of the plate by using a different ***Domain ID*** and setting the ***PD Interaction Type*** to PDDO as shown in Fig. 11.6. The selected region for PDDO using node picker window is shown in Fig. 11.7. Clicking the ***OK*** or ***Apply*** button of the node picker (Box), a dialog box labeled as ***Domain Points*** appears for confirming the selected PDDO region.

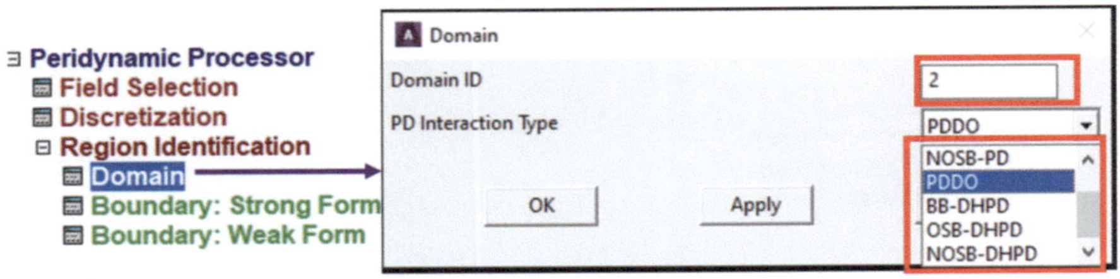

Fig. 11.6 Dialog box for selecting the domain interaction types as PDDO

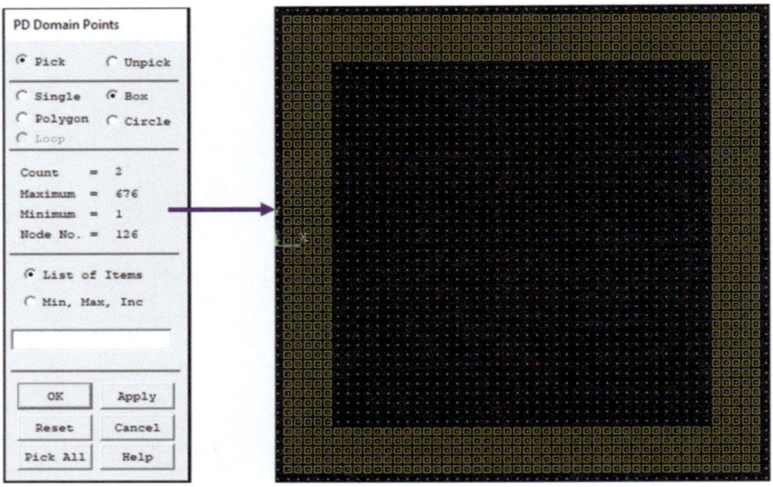

Fig. 11.7 Region identification corresponding to the PDDO domain of the plate

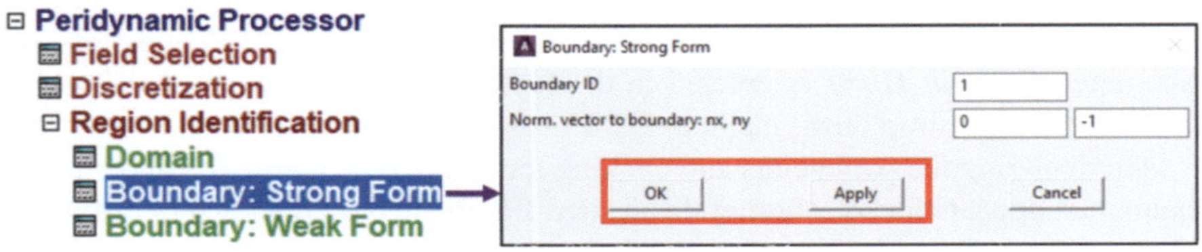

Fig. 11.8 Dialog box for selecting the boundary of the plate with strong form of interaction

Step-5 To enforce the boundary conditions in a pointwise manner, i.e., strong form, the submenu ***Region Identification>Boundary: Strong Form*** is selected to open the dialog box ***Boundary: Strong Form***. The parameters ***Boundary ID*** and ***Norm. vector to boundary: nx, ny*** are entered in the dialog box as shown in Fig. 11.8

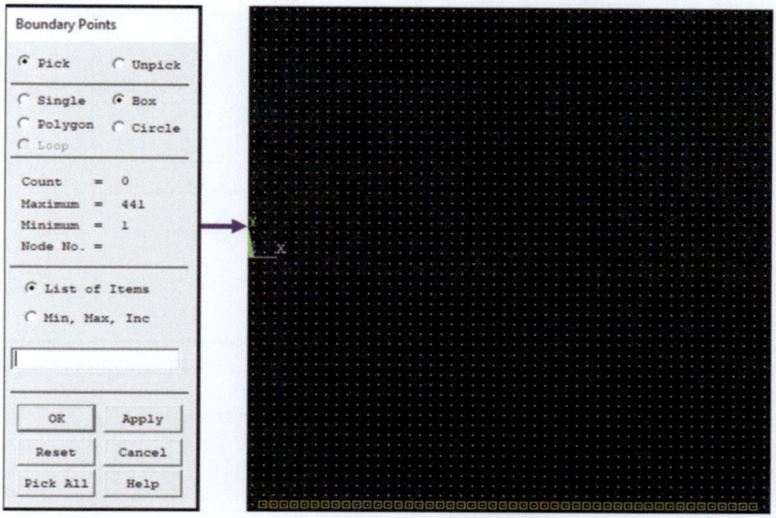

Fig. 11.9 Region identification corresponding to ***Boundary ID*=1** of the plate with strong form of interaction

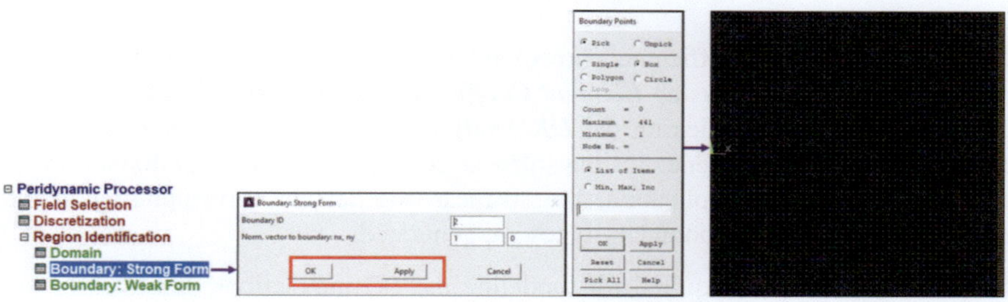

Fig. 11.10 Steps for creating the ***Boundary ID*=2** of the plate with strong form of interaction

After entering the parameters and clicking the ***OK/Apply*** button, the node picker dialog box is activated to select the desired boundary region as shown in Fig. 11.9.

Repeat the process for each boundary of the plate by entering different ***Boundary ID*** and the respective unit normal vector. Figure 11.10 shows the steps for creating the boundary region with ***Boundary ID*=2** and unit normal vector as ***nx, ny:*** 1,0.

Step-6 The family members of the PD points are generated by using ***Family Generation*** menu based on the horizon size factor m in the ***Family Generation*** dialog box as shown in Fig. 11.11. For this application, the horizon size factor, $m = 3$ is entered. After entering the parameter in the text box and clicking the ***OK/Apply*** button, the family members are generated using the horizon size of $\delta = 3.015\Delta x$.

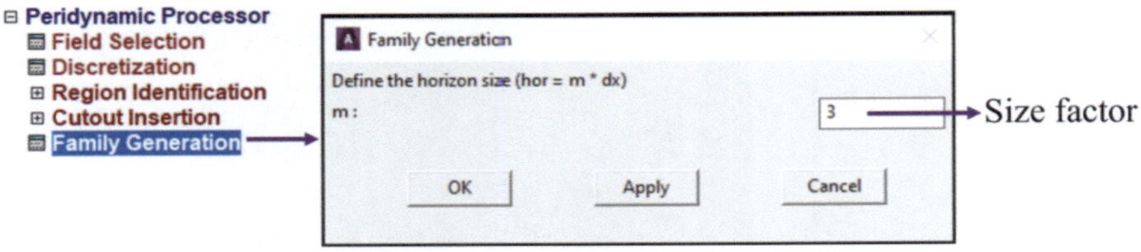

Fig. 11.11 Dialog box for defining the horizon size in the family generation

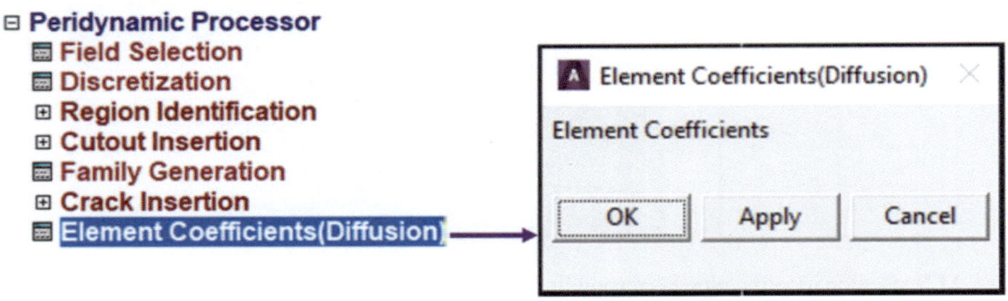

Fig. 11.12 Element coefficients submenu for Diffusion

Step-7 The element coefficients corresponding to the stiffness in the PD diffusion model are generated by using ***Element Coefficients (Diffusion)*** submenu option as shown in Fig. 11.12. Clicking the ***OK/Apply*** button of the ***Element Coefficients (Diffusion)*** dialog box generates the stiffness coefficients for the PD domain interaction types. In this application, the stiffness coefficients corresponding to the BB-PD, PDDO and PD boundary types are generated.

Step-8 The elements required for modeling the PD interactions due to diffusion both in PD and PDDO domains are created by using the dialog box ***Domain Elements***, which appears upon selecting the ***Element Generation>Domain Elements*** submenu. Clicking the ***OK/Apply*** button of the dialog box as shown in Fig. 11.13 generates the domain elements for the BB-PD and PDDO interaction types.

For generating the boundary interactions, the dialog box ***Boundary Elements***, which appears upon selecting the ***Element Generation>Boundary Elements*** submenu, is used. Clicking the ***OK/Apply*** button of the dialog box as shown in Fig. 11.14 generates the elements for the selected PD boundary interaction type.

Step-9 The transient thermal interactions are invoked by setting the ANALYSIS type to transient. The region is selected for modeling the transient thermal effects as shown in Fig. 11.15 by using the Ansys node picker dialog box or the command *NSEL,,P* and grouped under any user defined component name.

This application requires a constant value of thermal capacitance $\rho c_v = 1$J/m^{3}K. Select the nodes as grouped under the component name defined earlier and

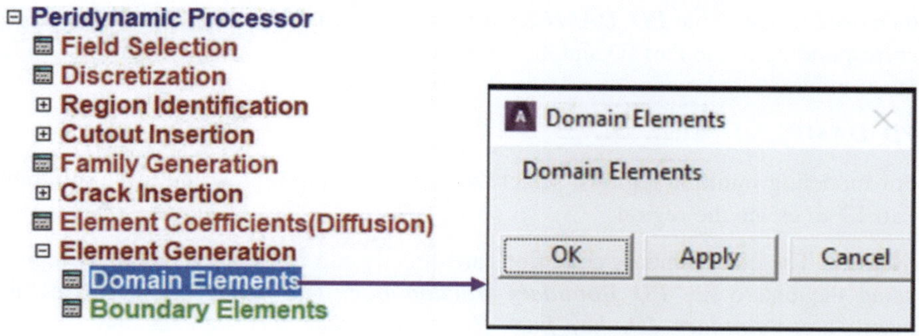

Fig. 11.13 Dialog box for domain element generation

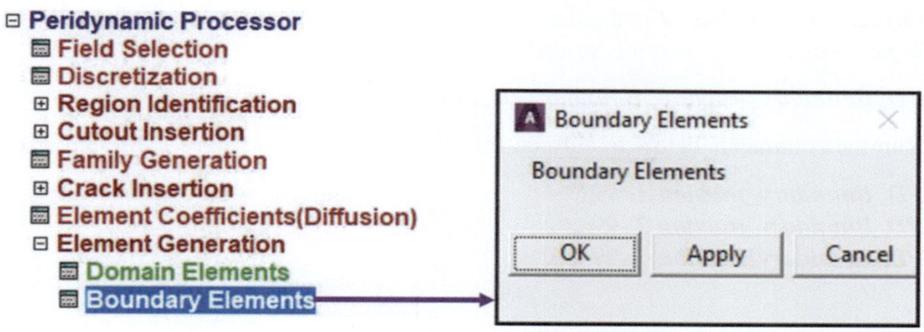

Fig. 11.14 Dialog box for boundary element generation

Fig. 11.15 Node picker option for selecting the set of domain nodes for modeling transient effects

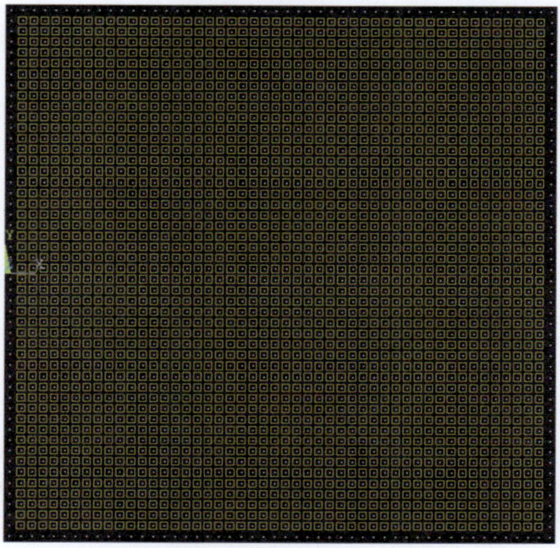

then use the macro file ***PD_DAMP2.mac*** with input arguments ***arg1***=1 and ***arg2***=1 corresponding to the Part ID and the damping constant values. Execute the macro in the command window as

PD_DAMP2,1,1

For modeling multiple regions, select the corresponding region and use a different Part ID to create the region.

Step-10 The flux boundary conditions are directly enforced on PD boundary region using the macro file ***PD_Boundary_traction.mac***. The input parameters to this macro are ***Boundary ID***, ***key_flux***, ***flux***, where ***key_flux*** denotes the key for enforcing the flux and ***flux*** denotes the value of the flux to be enforced. The parameter specified as ***key_flux=-99*** enforces zero flux condition. The parameter specified as ***key_flux=0*** does not enforce any flux condition. To enforce nonzero flux, set the ***key_flux=1*** and enter the nonzero value of the flux to be enforced. The input format of this macro is of the form

PD_Boundary_traction, ***Boundary ID***, ***key_flux***,, ***flux***,

Run the commands below to enforce zero flux conditions on edges 1 through 3.

PD_Boundary_traction,*1,-99,,,*
PD_Boundary_traction,*2,-99,,,*
PD_Boundary_traction,*3,-99,,,*

11.2.2.2 Solver

As this application requires the imposition of time-dependent boundary conditions, the analysis is performed over specified N time steps and updating the thermal boundary condition based on the current time.

Due to the presence of structural degrees of freedom in the assembled stiffness matrix, NEWMARK time integration is employed during the transient analysis. However, in order to remove the inertial effects during time integration and retain the damping effects while performing implicit transient thermal analysis, the command TINTP, QUASI is employed.

The initial condition for the transient case is specified by using the following command

```
IC, comp_ID, ROTX, val
```

where comp_ID is the nodal component containing the PD nodes, ROTX represents the activated rotational-X DOF value representing temperature, and val = 0 ° C represents the input value.

Thermal shock in Eq. (11.3) is applied as a temperature boundary condition along the left edge of the plate. The macro file ***PD_Therm_Solve.mac*** is used to solve for the total simulation time specified as input to this macro. This macro requires the input of the ***simulation time*** and ***time step size*** and the ***boundary_ID*** corresponding to which the time-dependent boundary conditions are specified.

PD_Therm_Solve,6,5e-03,4

11.2.2.3 Postprocessing

The temperature contour plots can be generated by outputting the values of ROTX DOF to the external file along with the coordinates, which can be read using in Matlab or Python. The macro file ***PD_Post_Temp.mac*** can be used to output the temperature variation along the centerline of the plate.

Arising from the strong form of boundary conditions, Fig. 11.16 shows the temperature variation predicted along the centerline $y = 0$ at $t = 3$s and $t = 6$s into the simulation and compared with the FEM solution. The FEM solution is performed by using PLANE55 elements.

11.2.3 Weak Form of Thermal Diffusion Equation

This application describes the specific steps in Ansys framework to perform thermal shock simulation using the weak form of unified PD approach for the field equations as described in Chap. 4. To generate the FE model, read the input file ***ex11.2.3.dat***. The application is same as that of Sect. 11.2.1; therefore, the steps for pre-, postprocessing, and solver remain the same except for the imposition of boundary conditions in preprocessing in **Step-6**. To enforce the boundary conditions in an average manner, i.e., the weak form submenu ***Region Identification>Boundary: Weak Form*** is selected to open the dialog box ***Boundary: Weak Form.*** The parameters ***Boundary ID=1*** and ***Tang. vector to boundary: tx, ty:1,0*** are entered in the dialog box as shown in Fig. 11.17.

After entering the parameters and clicking the ***OK/Apply*** button, the node picker dialog box appears to select the desired boundary region as shown in Fig. 11.18.

Repeat the process for each boundary of the plate by entering a different ***Boundary ID*** and the respective unit tangent vector. Figure 11.19 shows the steps in creating the boundary region with ***Boundary ID=2*** and unit tangent vector ***tx, ty: 0,1***.

The zero flux conditions on the boundary regions are satisfied naturally without any additional constraints. As a consequence, **Step-10** of the preprocessing step of Sect. 11.2.2.1 is skipped during the model generation. Sects. 11.2.2.2 and 11.2.2.3

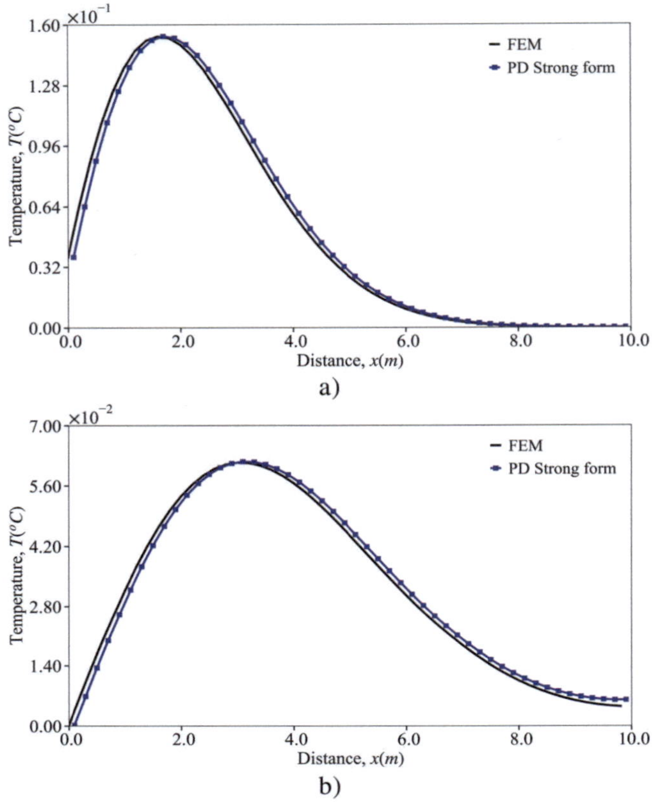

Fig. 11.16 Temperature variation from the strong form of equation along the centerline $y = 0$ at: (**a**) $t = 3$s and (**b**) $t = 6$s

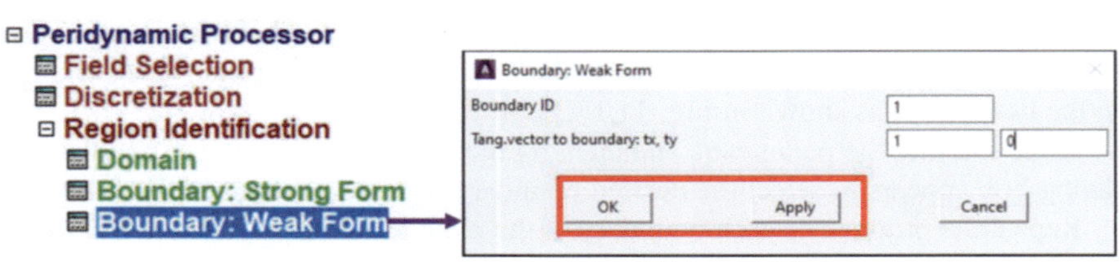

Fig. 11.17 Dialog box for selecting the boundary of the plate with weak form of interaction

corresponding to the solver and the postprocessing remain unchanged. Arising from the weak form of boundary conditions, Fig. 11.20 shows the temperature variation along the centerline $y = 0$ at $t = 3$s and $t = 6$s into the simulation and compared with the FEM solution. The FEM solution is performed by using PLANE55 elements.

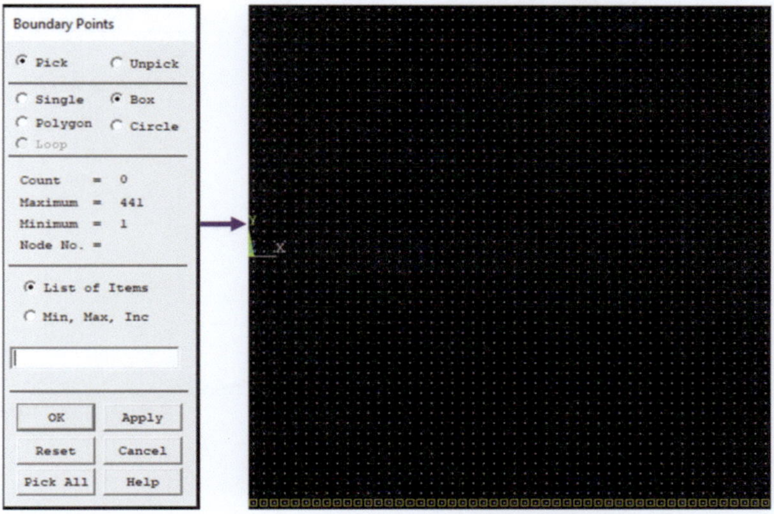

Fig. 11.18 Region identification corresponding to ***Boundary ID*=1** of the plate with weak form of interaction

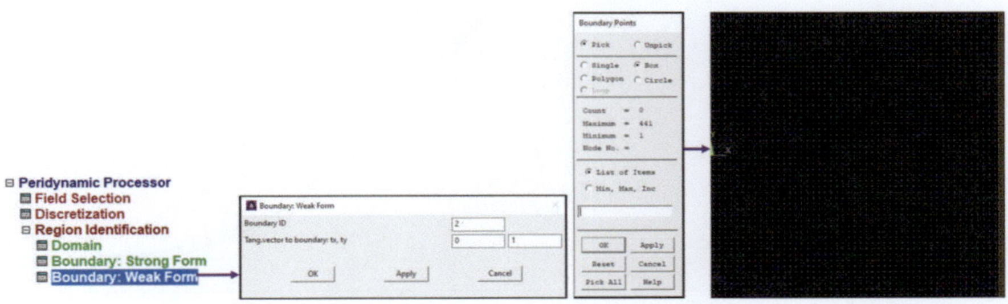

Fig. 11.19 Steps for creating the ***Boundary ID*=2** of the plate with weak form of interaction

11.2.4 Coupled PD/FE Form of Thermal Diffusion Equation

This application describes the specific steps in Ansys framework to perform thermal shock simulation using the coupled PD/FE approach for the field equations as described in Chap. 5. The application is the same as that of Sect. 11.2.1. However, the left and right halves of the plate are modeled as PD and FE regions, respectively. The PD region is discretized as described in Sect. 11.2.2.1, and the FE region is modeled using the traditional PLANE182 elements. To generate PD region using FE model, copy the lines 1 to 30 from the input file ***ex11.2.4.dat,*** paste it in the command window and follow the Steps 1 to 10. Because the PLANE182 element does not have ROTX degree of freedom, the keyword ***is_fld_pd_fe*** is required and is

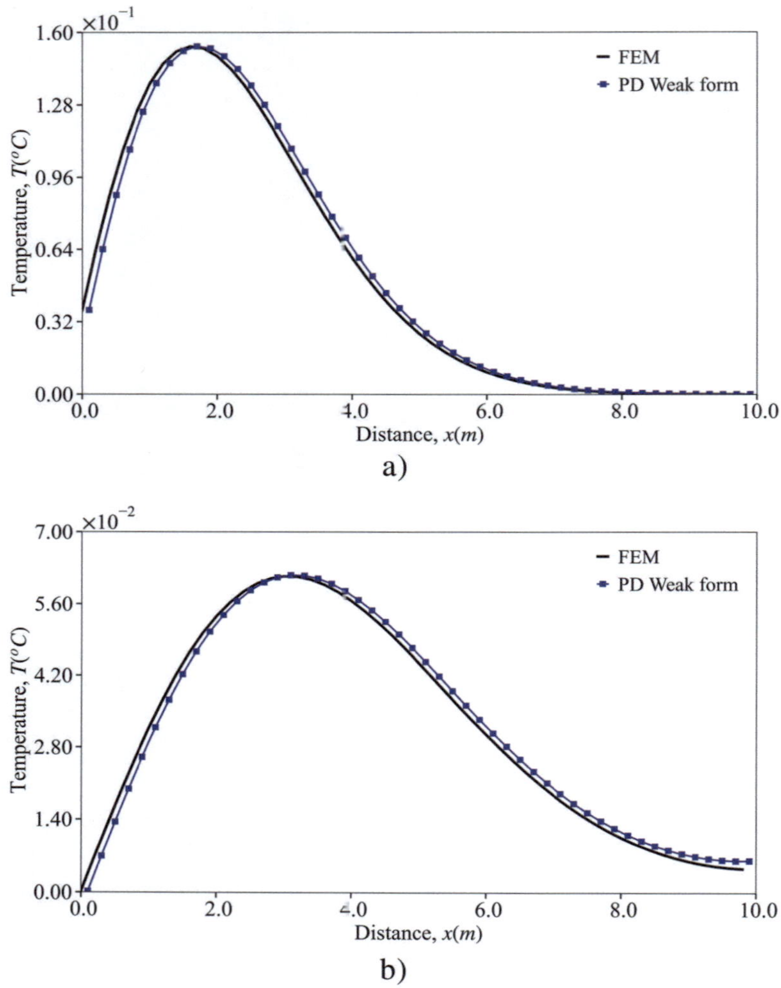

Fig. 11.20 Temperature variation from the weak form of equation along the centerline $y = 0$ at: (**a**) $t = 3$s and (**b**) $t = 6$s

set to 1 after **Step-1** during preprocessing. This operation can activate the degree of freedom UX by changing KEY options as the KEYOPT(3)=1 and KEYOPT(2)=1 for MATRIX27 and COMBIN14 elements, respectively. To generate the FE model to be coupled with the PD region, paste the lines from 32 to 49 of the input file ***ex11.2.4.dat***. The resulting PD/FE element of the plate is shown in Fig. 11.21. The steps for pre-, postprocessing, and solver described in Sect. 11.2.3 remain the same.

The interface nodes between PD and FE regions coincide, i.e., they share the same node number and location as shown in Fig. 11.22.

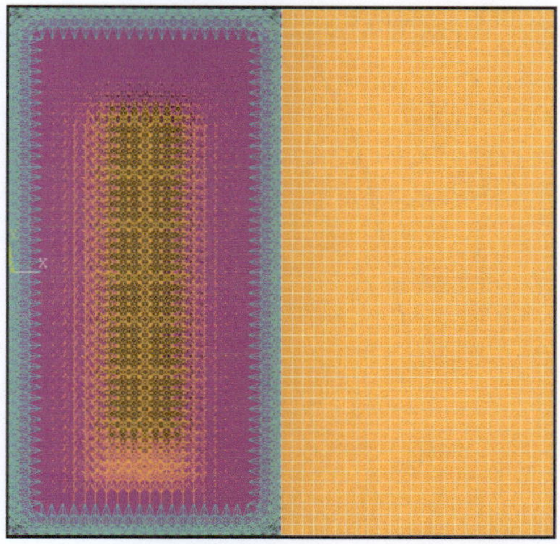

Fig. 11.21 Elements of the model with PD and FE regions

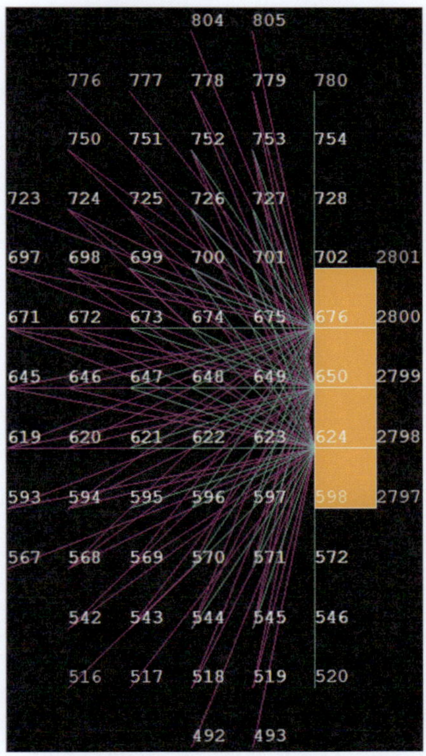

Fig. 11.22 Interface nodes between PD domain on the left and FE domain on the right

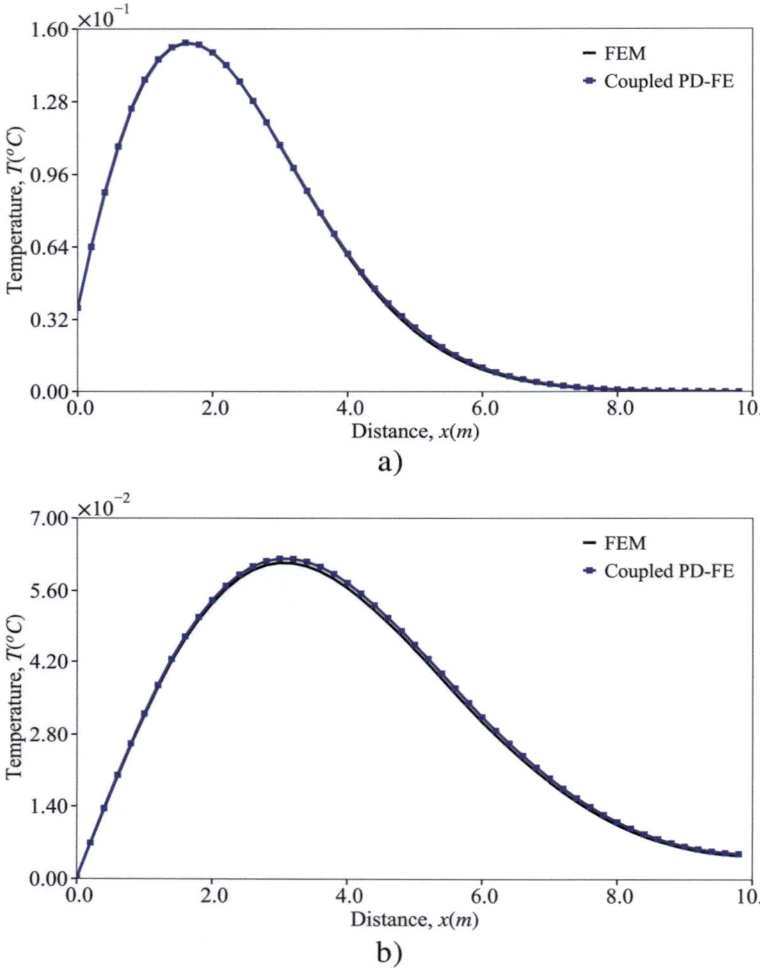

Fig. 11.23 Temperature variation from the coupled PD/FE analysis along the centerline $y = 0$ at: (**a**) $t = 3$s and (**b**) $t = 6$s

The zero flux conditions on the PD boundary regions are satisfied naturally without any additional constraints. As a consequence, **Step-10** of the preprocessing step of Sect. 11.2.2.1 is skipped during the model generation. Sects. 11.2.2.2 and 11.2.2.3 corresponding to the solver and the postprocessing remain unchanged. Figure 11.23 shows the temperature variation from the coupled PD/FE analysis along the centerline $y = 0$ at $t = 3$s and $t = 6$s into the simulation and compared with the FEM solution. The FEM solution is performed by using PLANE55 elements.

11.3 Simulation of Moisture Diffusion in a Nonhomogeneous Bar

As described in detail by Anicode and Madenci (2021), the field equation for moisture diffusion with desorption can be expressed as

$$\frac{\partial M(\mathbf{x},t)}{\partial t} = D(\mathbf{x},t)\nabla \cdot (\nabla M(\mathbf{x},t)) + G(\mathbf{x},t) \tag{11.8}$$

where $M(\mathbf{x}, t)$ is the moisture concentration, $D(\mathbf{x}, t)$ is the material diffusivity, and $G(\mathbf{x}, t)$ is the diffusing substance generation rate per unit volume. This equation is valid only in homogeneous domains. In order to consider nonhomogeneous domains, Wang et al. (2017) introduced a normalized wetness parameter, $\overline{M}(\mathbf{x},t)$ in the form of

$$\overline{M}(\mathbf{x},t) = \frac{M(\mathbf{x},t)}{M_{sat}(\mathbf{x},t)} \tag{11.9}$$

in which $M_{sat}(\mathbf{x}, t)$ is the saturated moisture concentration, which defines the degree of saturation in a material. This normalization satisfies the equality of chemical potentials at the interface of dissimilar materials. Invoking the wetness parameter, Eq. (11.9) into Eq. (11.8) leads to the wetness equation as

$$\begin{aligned} &M_{sat}(\mathbf{x},t)\frac{\partial \overline{M}(\mathbf{x},t)}{\partial t} + \overline{M}(\mathbf{x},t)\frac{\partial M_{sat}(\mathbf{x},t)}{\partial t} = \\ &D(\mathbf{x},t)\nabla \cdot \nabla \left(M_{sat}(\mathbf{x},t)\overline{M}(\mathbf{x},t)\right) + G(\mathbf{x},t) \end{aligned} \tag{11.10}$$

The diffusivity, $D(\mathbf{x}, t)$, and saturated concentration, $M_{sat}(\mathbf{x}, t)$, can be expressed as

$$D(\mathbf{x},t) = D_0 e^{\left(\frac{-E_D}{RT(\mathbf{x},t)}\right)} \tag{11.11}$$

and

$$M_{sat}(t) = S(\mathbf{x},t)P_{AVP}(\mathbf{x},t) \tag{11.12}$$

in which D_0 is the diffusivity factor, R is the universal gas constant ($R = 8.3145$ J/Kmol), E_D is the activation energy of the diffusivity and $T(t)$ is temperature. The solubility, $S(t)$, and the ambient vapor pressure, $P_{AVP}(t)$, are defined as

$$S(\mathbf{x},t) = S_0 e^{\left(\frac{E_S}{RT(\mathbf{x},t)}\right)} \tag{11.13}$$

and

$$P_{AVP}(\mathbf{x},t) = P_{SVP}(\mathbf{x},t) RH(t) \tag{11.14}$$

in which S_0 is the solubility factor, E_S is the activation energy of the solubility, and $RH(t)$ indicates the relative humidity. The solubility activation energy of dissimilar materials can be either equal or unequal at the interface. The saturated vapor pressure, $P_{SVP}(\mathbf{x}, t)$ is defined as

$$P_{SVP}(\mathbf{x},t) = P_0 e^{\left(\frac{-E_{VP}}{RT(\mathbf{x},t)}\right)} \tag{11.15}$$

in which P_0 is the pressure factor and E_{VP} is the vapor pressure activation energy. With these expressions, the saturated concentration, $M_{sat}(\mathbf{x}, t)$, can be rewritten as

$$M_{sat}(\mathbf{x},t) = S_0 P_0 e^{\left(\frac{E_S - E_{VP}}{RT(\mathbf{x},t)}\right)} \times RH(t) \tag{11.16}$$

As shown in Fig. 11.24, a 1D bar of two different materials has a length of $2L = 2$ mm and a thickness of $H = W = L/100$. Its cross-sectional area is specified as $A = H \times W$. The bar is initially fully saturated at 85 °C/100% RH with insulated lateral surfaces. The solubility activation energy of materials 1 and 2 can be either equal or unequal at the interface. The properties of each material previously considered by Wang et al. (2017) are given in Tables 11.2 and 11.3.

The initial and boundary conditions in terms of the wetness parameter can be specified as

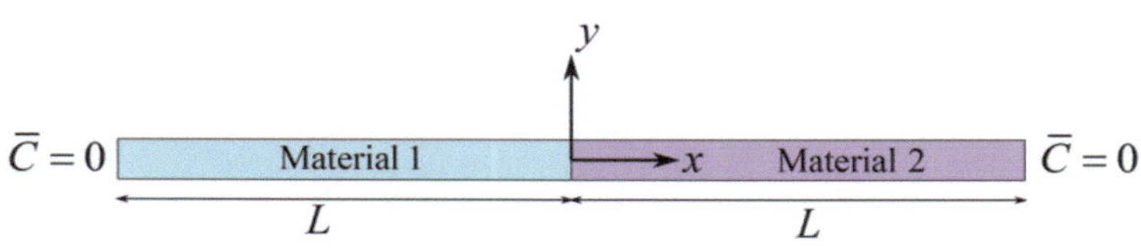

Fig. 11.24 Geometry and boundary conditions of the nonhomogeneous bar

Table 11.2 Thermal material properties

	Material 1	Material 2
Density, ρ (kg/m^3)	3×10^3	3×10^3
Specific heat, c_v (J/kgK)	1.5×10^3	1.5×10^3
Thermal conductivity, k (W/mK)	0.6	0.6

Table 11.3 Diffusion material properties

	Material 1	Material 2
Diffusivity factor, D_0 (m^2/s)	5×10^{-3}	4×10^{-3}
Solubility factor, S_0 (kg/m^3Pa)	6×10^{-10}	2×10^{-10}
Pressure factor, P_0 (Pa)	5.0492×10^{10}	5.0492×10^{10}
Diffusion activation energy, E_D (J/mol)	5×10^4	5×10^4
Solubility activation energy, E_S (J/mol)	4×10^4	4×10^4
		4.5×10^4
Vapor pressure activation energy, E_{VP} (J/mol)	4.08737×10^4	4.08737×10^4

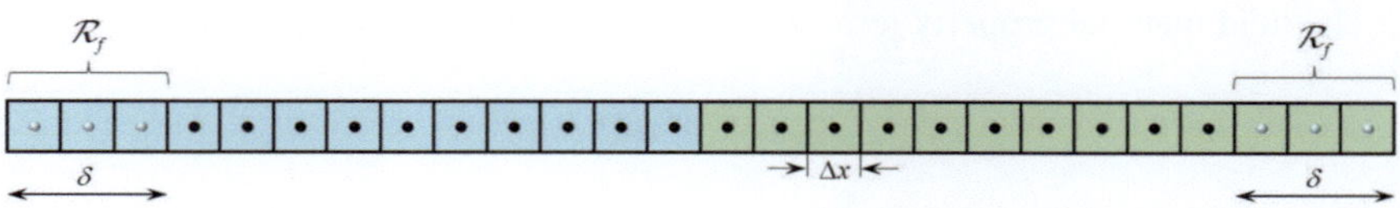

Fig. 11.25 Fictitious regions in PD discretization of the bar made of two dissimilar materials

$$\overline{M}(x, t=0) = 1 \text{ with } -L \leq x \leq L \tag{11.17}$$

and

$$\overline{M}(x = \pm L, t) = 0 \tag{11.18}$$

Both uniform and nonuniform temperature fields are considered. The uniform temperature is applied as

$$T(x, t) = (85 + 2t)^\circ\text{C} \tag{11.19}$$

The nonuniform temperature is achieved by solving the thermal conduction equation as described in Eq. (11.1) with the following boundary condition

$$T(x = \pm L, t) = (85 + 2t)^\circ\text{C} \tag{11.20}$$

The grid spacing between the nodes is specified as $\Delta x = 0.5H$ with a horizon size of $\delta = 3.015\Delta x$ resulting in 200 nodes. The domain at the left($x = -L$) and right end ($x = L$) is extended using 3 fictious nodes as shown in the Fig. 11.25.

The solution time step sizes are specified as $\Delta t = 2$s and $\Delta t = 0.1$s for uniform and nonuniform temperature boundary conditions, respectively, with a total simulation time of $t = 80$s.

11.3.1 Modeling in Ansys Framework

11.3.1.1 Assigning Degrees of Freedom

The PD model of moisture diffusion with time-varying saturated concentration can be simulated by specifying the field variable, $\Theta(\mathbf{x}, t)$, as normalized concentration $\overline{M}(\mathbf{x}, t)$ in Eqs. (3.113), (4.16), and (5.14), i.e.,

$$\Theta(\mathbf{x}, t) = \overline{M}(\mathbf{x}, t) \tag{11.21}$$

The field material property term, $C(\mathbf{x}, t)$, represents the saturated concentration as

$$C(\mathbf{x}, t) = M_{sat}(\mathbf{x}, t) \tag{11.22}$$

The time rate of change of saturated concentration is set as the time-dependent saturated concentration

$$\dot{C}(\mathbf{x}, t) = \frac{\partial M_{sat}(\mathbf{x}, t)}{\partial t} \tag{11.23}$$

The field flux, $\mathbf{q}(\mathbf{x}, t)$, as in Eq. (3.116) is defined by the Fickian law as

$$\mathbf{q}(\mathbf{x}, t) = D(\mathbf{x}, t)\nabla\left(M_{sat}(\mathbf{x}, t)\overline{M}(\mathbf{x}, t)\right) \tag{11.24}$$

where the product term $D(\mathbf{x}, t)M_{sat}(\mathbf{x}, t)$ is interpreted as the moisture conductivity.

11.3.1.2 Ansys Elements

LINK33 element is used for modeling the symmetric stiffness matrices in the PD domain and fictitious region. The symmetric stiffness coefficients appearing in Eq. (3.235) along with Eqs. (3.232) and (3.233) are input as real constants to LINK33 element.

COMBIN14 element with TEMP DOF activated using KEYOPT(2) = 8 is used for modeling the time rate of change of material property in Eq. (11.23), and its value is input as a real constant to the element.

MASS71 element with KEYOPT(3) =1 activated is used for modeling the field material property term as in Eq. (11.22), and its value is input as a real constant to the element.

The supported KEYOPTS command and the procedure to input real constant applicable to these elements are described in Chap. 8. Table 11.4 shows the summary of elements and their KEYOPTS activated for modeling the simulation of moisture diffusion with time-varying saturated concentration.

Table 11.4 Ansys element modeling summary for moisture diffusion with time-varying saturated concentration

Modeling variables	Element	DOF	KEYOPTS
PD stiffness	LINK33	TEMP	NA
Time-dependent saturated concentration, $\dot{C}_{sat}(\mathbf{x}, t)$	COMBIN14	TEMP	KEYOPT(2)=8
Saturated concentration, $C_{sat}(\mathbf{x}, t)$	MASS71	TEMP	KEYOPT(3)=1

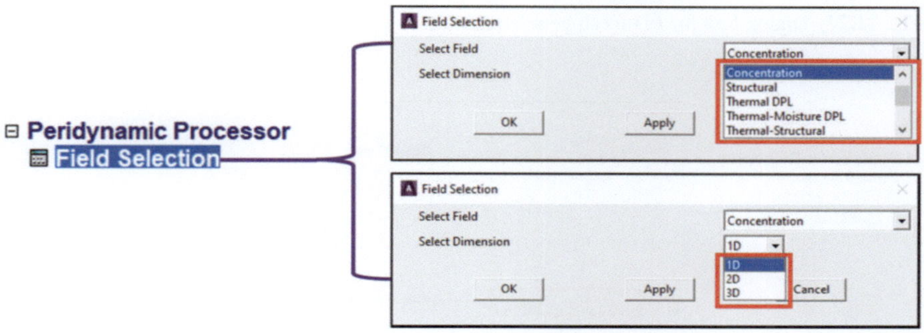

Fig. 11.26 Dialog box for selecting the field and dimension

11.3.2 *Moisture Diffusion with Time-Varying and Spatially Uniform Temperature*

11.3.2.1 Preprocessing

The finite element discretization of the 1D bar can be generated by using the Ansys input file: ***ex11.3.2.dat*** or imported into the Ansys database using the CDREAD command. As the model involves two different materials, two separate material IDs are required for specifying the moisture properties.

Step-1 The field and dimension of the analysis are specified using the ***Field Selection*** submenu located under ***Peridynamic Processor***. For the simulation of moisture concentration in a bar, the "***Concentration***" for field and "***1D***" for dimension are selected as shown in Fig. 11.26.

Step-2 Furthermore, the PD mesh is generated using the radio button option ***Generate using FE model*** in ***Discretization*** submenu as shown in Fig. 11.27.

As the current application involves the usage of fictitious nodes for imposition of boundary conditions, they are created on the right and left ends of the domain as shown in Fig. 11.25. They can be created by either extending the domain in the finite element model or after the creation of the PD mesh with additional 3 nodes having a spacing of Δx using N command at the left and right ends of the bar.

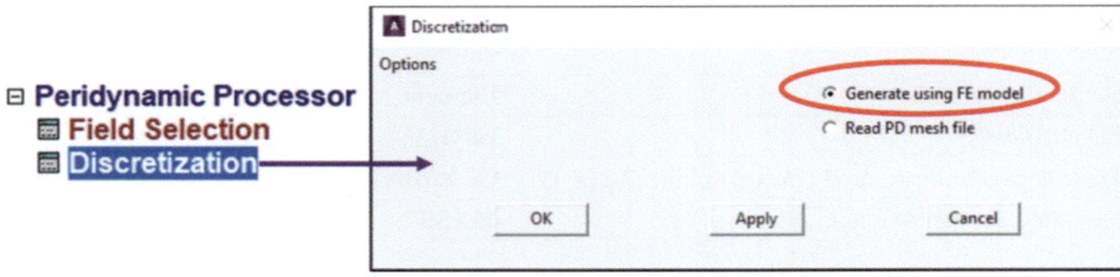

Fig. 11.27 Dialog box for PD mesh generation

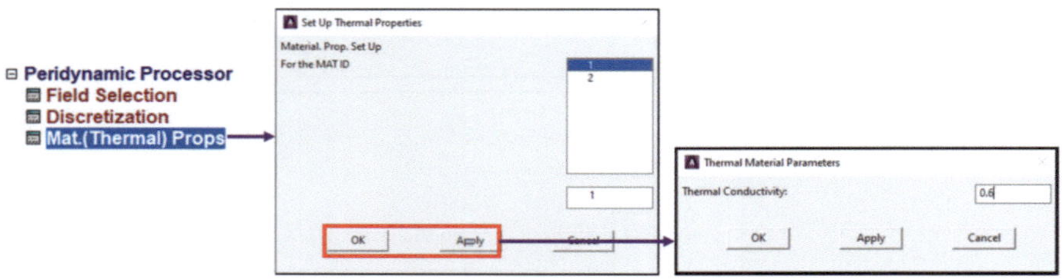

Fig. 11.28 Steps for entering the thermal properties required for moisture concentration

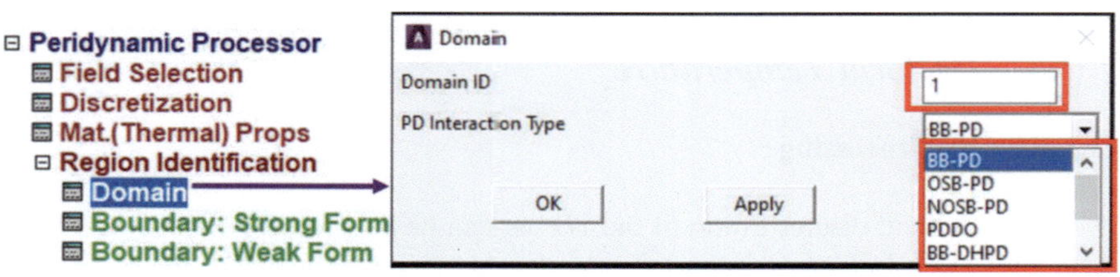

Fig. 11.29 Dialog box for selecting the domain interaction type as BB-PD

Step-3 The procedure to use this menu is explained in Sect. 9.14 in Chap. 9. As shown in Fig. 11.28, the dialog box ***Set Up Thermal Properties***, which appears after selecting the submenu ***Mat. (Thermal Props)***, displays the *MAT ID* used to create the PD mesh from the FE mesh. After selecting a specific *MAT ID* and clicking the ***OK*** or ***Apply*** button in this dialog box, another dialog box ***Thermal Material Parameters*** appears to enter the thermal conductivity. The thermal material properties listed in Table 11.2 are entered for each *MAT ID*.

Step-4 The PD regions for the field equations are created by using ***Region Identification*** submenu shown in Fig. 11.29. The submenu ***Region Identification >Domain*** is selected to open the dialog box ***Domain***. Only one region corresponding to BB-PD is created. The parameter ***Domain ID=1*** is entered in the

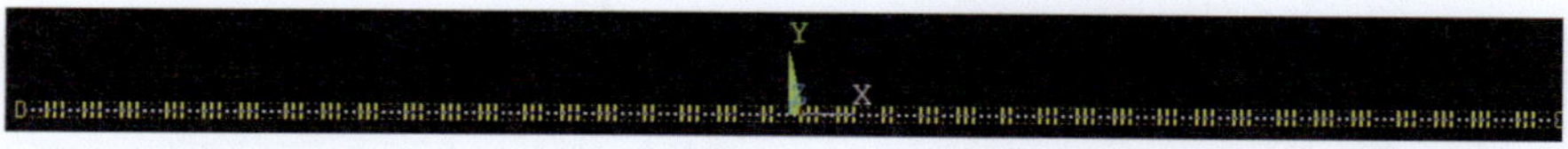

Fig. 11.30 PD discretization of the bar with fictitious regions

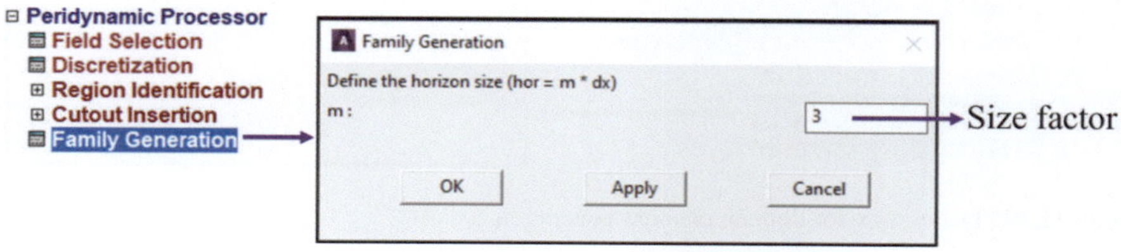

Fig. 11.31 Dialog box for defining the horizon size in the family generation

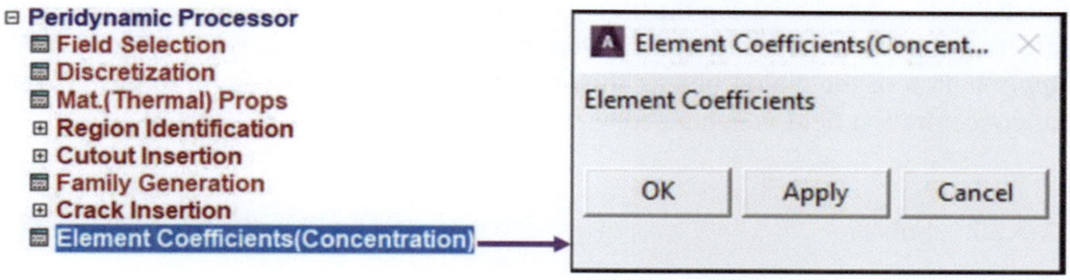

Fig. 11.32 Element Coefficients submenu for Concentration

text box and ***PD Interaction Type*** is selected from the drop-down menu as BB-PD as shown in Fig. 11.29.

After entering the parameters and clicking the ***OK/Apply*** button, the node picker dialog box appears to select the desired region. All regions including the fictitious nodes are selected as BB-PD region as shown in Fig. 11.30.

Step-5 The family members of the PD points are generated using ***Family Generation*** submenu based on the horizon size factor $\boldsymbol{m}$ in the ***Family Generation*** dialog box as shown in Fig. 11.31. For this application, the horizon size factor, $m = 3$ is entered. After entering the parameter in the text box and clicking the ***OK/Apply*** button, the family members are generated using the horizon size of $\delta = 3.015\Delta x$.

Step-6 The element coefficients corresponding to stiffness in the PD concentration model are generated using ***Element Coefficients (Concentration)*** submenu option as shown in Fig. 11.32. Clicking the ***OK/Apply*** button of the ***Element Coefficients (Concentration)*** dialog box generates the stiffness coefficients for the PD domain interaction types. In this application, the stiffness coefficients correspond to BB-PD type interactions.

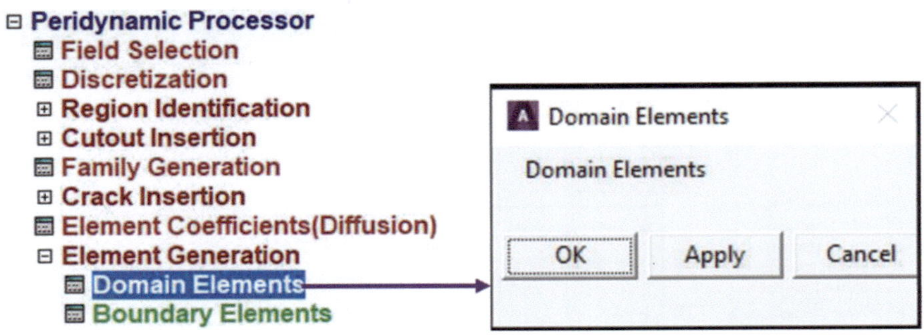

Fig. 11.33 Dialog box for domain element generation

Step-7 The elements required for modeling the PD interactions for the concentration field are created by using the dialog box ***Domain Elements***, which appears upon selecting the ***Element Generation>Domain Elements*** submenu. Clicking the ***OK/ Apply*** button of the dialog box as shown in Fig. 11.33, the BB-PD interaction type for concentration field is generated.

11.3.2.2 Solver

The menu ***Peridynamic Solution*** located under ***Solution*** processor is used for simulating the time-dependent PD moisture diffusion process.

Step-1 As this application requires the input of time-varying temperature as in Eq. (11.19) or (11.20), a table of the temperature profile is created using ****DIM*** Ansys command. For the simulation purpose, it is assumed that the temperature profile is known at every instant of the moisture simulation time step. The APDL commands can be executed to create a table type parameter for time-varying temperature as in Eq. (11.19) or (11.20) as follows:

```
totime=80 !! total simulation time
deltime=2.0 !! time step size
boundary_temp=273+85 !! boundary temperature
nsteps=totime/deltime
*dim,DRC,table,nsteps+1 !! Create table with parameter name DRC
*do,ii,1, nsteps +1
DRC (ii,0)=deltime*(ii-1)
DRC (ii,1)=boundary_temp+2*(deltime*(ii-1))
*enddo
```

The first and second columns of the table type parameter, *DRC*, correspond to the time and value of the temperature field, respectively.

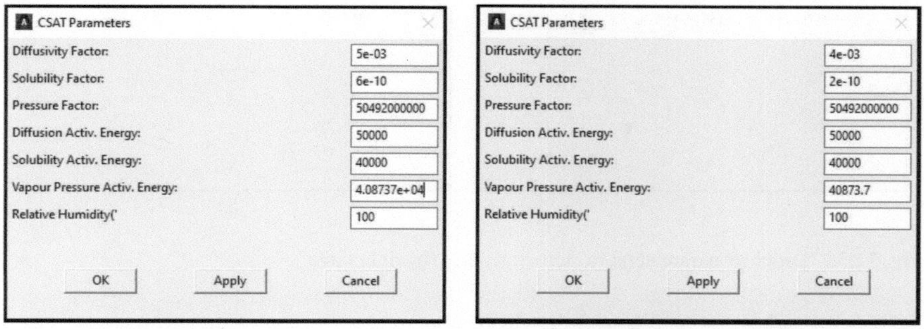

a) b)

Fig. 11.34 CSAT parameters. (**a**) MAT ID=1 and (**b**) MAT ID=2

The macro file ***PD_MOIST_TEMP.mac*** is executed for loading the temperature data and for further evaluation of the moisture properties. The input parameters to this macro are the ***simulation time, time step size***, ***the initial temperature***, and ***spatial temperature***. The ***spatial temperature*** is set to zero. After executing the command below, a table is created internally.

PD_MOIST_TEMP.mac,80,2.0,358,358,0

The table contains the spatial variation of temperature at different instants of time for evaluation of the moisture material properties as given in Table 11.3

Step-2 The parameters for computing the saturated concentration as given in Table 11.3 are entered in the dialog box ***Moisture Conc. Parameters***, which appear upon selecting the ***Setup CSAT table>Setup*** submenu. For each Mat ID, the parameters entered are shown in Fig. 11.34.

Step-3 After setting up the parameters for all material properties, the menu ***CSAT>Apply*** is used to model the transient field property, i.e., saturated concentration M_{sat} as in Eq. (11.22). The details of the procedure for using this submenu are explained in Sect. 9.15.2 in Chap. 9. Select the left region of the bar $-L < x < 0$ using the node picker window and apply the tabular parameter by selecting the table name *CST_1* as generated from **Step-2**.

Step-4 After setting up the saturated concentration, the time-dependent saturated concentration, i.e., $\dot{M}_{sat}$ as in Eq. (11.23) needs to be setup. The details of the procedure for using the ***Time Dep. CSAT>Apply*** submenu are explained in Chap. 9. Similar to **Step-3**, select the left region of the bar $-L < x < 0$ using the node picker window and apply the tabular parameter by selecting the table name *DCS_1* as generated from **Step-1**. Repeat **Steps-3** and **4** again for the right region of the bar $0 < x < L$ by selecting MAT ID=2 with tables *CST_2* and *DCS_2*.

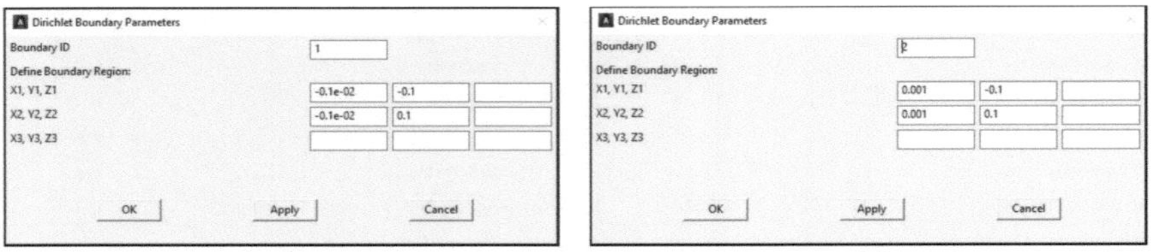

a) b)

Fig. 11.35 Dirichlet parameters: (**a**) left edge, and (**b**) right edge

Step-5 After setting up the CSAT and Time Dep. CSAT for both the left and right regions of the bar, the zero concentration boundary conditions in Eq. (11.18) are specified using the submenu:

Main Menu> Preprocessor> Peridynamic Processor> Nonlocal BC Approach> Dirichlet

Usage of this submenu requires a priori modeling of fictitious nodes around the actual domain, and the details for using this submenu are explained in Sect. 9.10.2.1 in Chap. 9. In this application, corresponding to the left and right ends of the bar, two ***Dirichlet*** boundaries are created. Figure 11.35 shows the ***Dirichlet*** Boundary parameters created with two separate IDs. Selecting the fictitious nodes for each boundary and clicking the ***OK*** or ***Apply*** button of the node picker (Box) window, the dialog box ***Apply Boundary Condition*** appears where the parameter ***Apply as*** is set to ***Constant Value*** and value entered is 0 corresponding to the zero moisture as in Eq. (11.18). For boundary ID set as 1, the fictitious nodes are chosen as 1, 2, and 3, while for the boundary ID set as 2, the fictitious nodes are chosen as 204, 205, and 206.

Step-6 After setting up of the boundary condition for moisture diffusion as in **Step-6,** enter the Ansys Solution processor (/SOLU) to set up the initial condition for the transient case using the following APDL command:

```
NSEL,ALL
CM, node_all, nodes !!!node_all is the user defined component name
IC, node_all, TEMP, val
```

where *node_all* is the nodal component containing the PD mesh nodes, *TEMP* represents the activated TEMP DOF value corresponding to the temperature, and *val* = 1 represents the input value as in Eq. (11.17).

Step-7 After setting up of the initial condition for moisture diffusion as in **Step-6**, without exiting the Ansys solution processor (/SOLU), the macro

PD_MOIST_SOLVE.mac is executed for the moisture diffusion. This macro requires the input of the ***simulation time*** and ***time step size***. After executing the command below, the temperature DOF, which is the normalized concentration, $\overline{M}$, as in Eq. (11.19), is solved.

PD_MOIST_SOLVE.mac,80,2.0

11.3.2.3 Postprocessing

The normalized concentration value, $\overline{M}$, is obtained as a nodal solution with *TEMP* DOF. The actual concentration accumulation at various times can be output to the file named as ***pd-moisture.out*** by using the macro file ***PD_Post_Moist.mac.*** This macro requires the input of the ***input time*** and ***solubility_key***. The parameter ***solubility_key=0*** refers to equal solubility case i.e., $E_{S_1} = E_{S_2} = 4e + 04$ and for the case ***solubility_key=1*** refers to the unequal solubility, ***i.e.,*** $E_{S_1} = 4e + 04; E_{S_2} = 4.5e + 04$. For the current problem, execute the macro as

PD_Post_Moist,60,0

and

PD_Post_Moist,60,1

to postprocess the results for the equal and unequal solubility cases, respectively, at time $t = 60s$. It should be noted that the model needs to be rerun with different solubility in **Step-2** of solver for unequal solubility case.

The results concern desorption due to uniform temperature with equal and unequal solubility activation energy, E_S of materials 1 and 2. The PD predictions recover the FE results of 1D, 2D, and 3D models (Diyaroglu et al. 2017a, b, 2018; Anicode & Madenci, 2021). Figure 11.36 shows the concentration results with equal

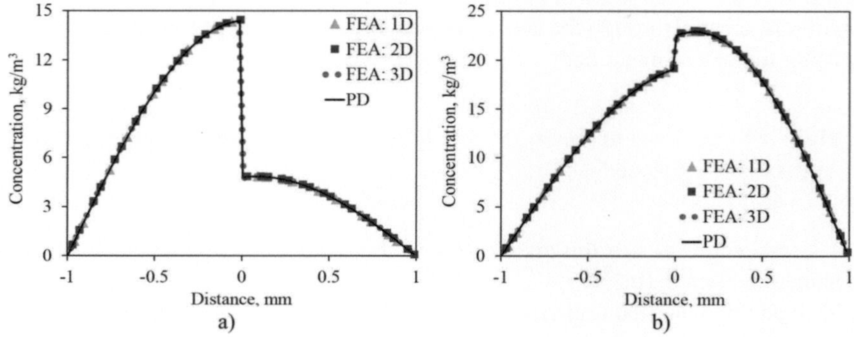

Fig. 11.36 Moisture concentration along the bar under nonuniform temperature at $t = 60s$ with: (**a**) equal E_S and (**b**) unequal E_S

and unequal values of E_S along the bar at $t = 60$s for uniform temperature. The PD predictions are in excellent agreement with those of FEA for both equal and unequal values of E_S.

11.3.3 Moisture Diffusion with Time-Varying and Spatially Nonuniform Temperature

The preprocessing operations for the moisture diffusion with time-varying and spatially nonuniform temperature are the same as those of time-varying uniform temperature. To generate PD region using FE model, read the input file ***ex11.3.3.dat***. After finishing **Step-7** of preprocessing step as in Sect. 11.3.2.1, the consideration of spatially nonuniform temperature requires the thermal analysis to be solved a priori. The thermal analysis is performed by using the same time steps as used for the moisture field. To account for time-dependent thermal boundary condition as in Eq. (11.20), a table of the temperature profile is created by using ****DIM*** Ansys command as shown below for each time step of the thermal simulation.

```
totime=80 !! total simulation time
deltime=0.1 !! time step size
boundary_temp=273+85 !! boundary temperature
nsteps=totime/deltime
*dim,DRC,table,nsteps+1 !! Create table with parameter name DRC
*do,ii,1, nsteps+1
DRC (ii,0)=deltime*(ii-1)
DRC (ii,1)=boundary_temp+2*(deltime*(ii-1))
*enddo
```

The first and second columns of the table parameter correspond to the time and the value of the temperature field, respectively. The **tabular temperature** boundary conditions, created through the above commands, are applied **at the ends of the bar** by using the below menu path:

Main Menu> Preprocessor> Peridynamic Processor> Nonlocal BC Approach> Dirichlet

The procedure to use this menu to read the tabular type parameter is already explained in Sect. 9.10.2.1.

To model the thermal transient term, the submenu ***Field Damping*** is used, and it can be accessed using the menu path:

Main Menu> Preprocessor> Peridynamic Processor> Field Damping> Insert

The procedure to input the constant damping parameter, $\rho C_v = 4.5 \times 10^6$ J/m^{3}K using this submenu, is already explained in Sect. 9.11. Select the nodes from $-L < x < 0$ and $0 < x < L$ corresponding to ***Part ID=1*** and ***Part ID=2***, respectively. As part of the solution procedure, the ***spatial temperature*** option in the macro file ***PD_MOIST_TEMP.mac*** is now set to 1 to solve for the temperature field equation. The input arguments of this macro are already explained in Sect. 11.3.2. A table is created internally after executing the command below:

PD_MOIST_TEMP.mac,80,0.1,358,358,1

The table contains the nonuniform temperature profile from the solution of the thermal problem. After the temperature table is set up, follow the steps from **Step-2** to **Step-7** as in Sect. 11.3.2.2 to set up the moisture properties. In **Step-8**, the macro ***PD_MOIST_SOLVE.mac*** is executed for the moisture diffusion. The input parameters for this macro are already explained in Sect. 11.3.2.2. The macro is executed using the below command

PD_MOIST_SOLVE.mac,80,0.1

For the postprocessing of the results, follow the similar steps as given in Sect. 11.3.2.3. The results concern desorption due to nonuniform temperature with equal and unequal solubility activation energy, E_S, of materials 1 and 2. The PD predictions from the output file ***pd-moisture.out*** recover the FE results of 1D, 2D and 3D models (Diyaroglu et al. 2017a, b, 2018; Anicode & Madenci, 2021).

Figure 11.37 shows the concentration results along the bar at $t = 60$s for nonuniform temperature. The PD predictions are in excellent agreement with those of FEA for both equal and unequal values of E_S.

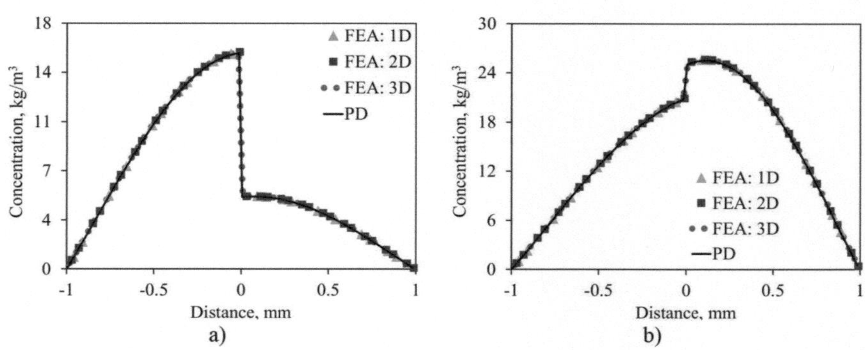

Fig. 11.37 Moisture concentration along the bar under nonuniform temperature at $t = 60$s with: (**a**) equal E_S and (**b**) unequal E_S

References

Anicode, S. V. K., & Madenci, E. (2021). Three-dimensional moisture diffusion simulation with time dependent saturated concentration in Ansys through native thermal and spring elements. *Microelectronics Reliability, 123*, 114167.

Diyaroglu, C., Madenci, E., Oterkus, S., & Oterkus, E.(2017a). Peridynamic solution of wetness equation with time dependent saturated concentration in ansys framework. In *67th electronic components & technology conference*, Orlando, FL.

Diyaroglu, C., Oterkus, S., Oterkus, E., & Madenci, E. (2017b). Peridynamic Modeling of Diffusion by Using Finite Element Analysis. *IEEE Transactions on Components, Packaging and Manufacturing Technology, 7*, 1823–1831.

Diyaroglu, C., Madenci, E., Oterkus, S., & Oterkus, E. (2018). A novel moisture diffusion modeling approach using finite element analysis. *Microelectronics and Optoelectronics, 7*, 438.

Wang, J., Liu, R., Liu, D., & Park, S. (2017). Advancement in simulating moisture diffusion in electronic packages under dynamic thermal loading conditions. *Microelectronics Reliability, 73*, 42–53.

Chapter 12
Applications of Peridynamics for Static Structural Analysis

12.1 Introduction

The applications demonstrate the strong and weak forms of single- and dual-horizon peridynamic (PD) equilibrium equations described in Chaps. 3 and 4 by considering an elastic plate with or without a crack under static loading through uniform and nonuniform discretization, as considered by Anicode and Madenci (2022, 2023). The first application demonstrates the PD modeling of a plate without a crack through uniform and nonuniform discretization under uniaxial tension. The second application demonstrates the PD modeling of a plate with a crack through uniform and nonuniform discretization under displacement constraint. The parameter a appearing in the Gaussian type weight function, Eq. (2.1) is specified as $a = 3$ in the presence of a crack, and $a = 2$ in the absence of a crack. The operations of the preprocessing, solution and postprocessing of the models are demonstrated using Ansys GUI in a step-by-step fashion.

12.2 Plate Under Tension

As shown in Fig. 12.1, the plate geometry is defined by its length, $L = 100$ mm width, $W = 100$ mm, and thickness, $h = 1$ mm. The plate is subjected to a tensile stress of $\sigma_0 = 1.0$GPa along the right edge. It has roller support conditions along the left and bottom edges. It is free of tractions along the top edge. The material is isotropic with a Young's modulus of $E = 65.8$ GPa and Poisson's ratio of either $\nu = 0.333$ or $\nu \neq 0.333$. The simulation is performed for an elastic response with small deformation under plane stress conditions.

Supplementary Information The online version contains supplementary material available at https://doi.org/10.1007/978-3-031-59896-8_12.

E. Madenci et al., *Peridynamics and Its Applications Using Ansys*,
https://doi.org/10.1007/978-3-031-59896-8_12

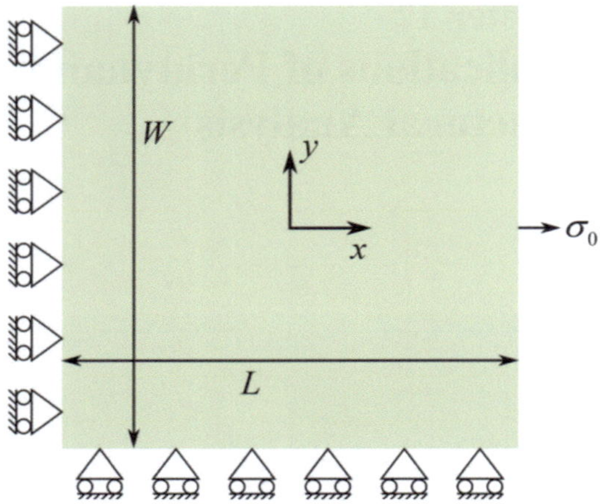

Fig. 12.1 Plate geometry and boundary conditions

Table 12.1 MATRIX27 element used for the quasi-static structural analysis

Element	DOF	KEYOPTS
MATRIX27	UX,UY	KEYOPT(2)=2 KEYOPT(3)=4

12.2.1 Modeling in Ansys Framework

In this application, the MATRIX27 element is employed for modeling the unsymmetric stiffness matrices in the PD, PDDO, and PD boundary regions. The unsymmetric stiffness coefficients corresponding to these regions as in Eqs. (3.165)–(3.166), (3.178)–(3.181), (3.198)–(3.201), and (3.218)–(3.220) and along the boundary region as in Eq. (3.226) are input as real constants to MATRIX27 element. The UX and UY degrees of freedom represent the components of displacement field $\mathbf{u}(\mathbf{x}, t)$ in Eqs. (3.98)–(3.101).

The supported KEYOPTS command and the procedure to input real constant applicable to these elements are described in Chap. 8. Table 12.1 shows the summary of MATRIX27 element and the KEYOPTS activated for quasi-static structural analysis.

12.2.2 Strong Form of BB-PD with Uniform Discretization

The plate is discretized with a uniform grid spacing of Δx and the PD horizon is specified as $\delta = 3.015\Delta x$. The grid spacing is $\Delta x = 1.25$mm with 6,084 PD points in regions $\mathcal{D}$ and $\mathcal{R}$, and 316 boundary points in region $\mathcal{B}$ as shown in Fig. 12.2. In this simulation, the Poisson's ratio is set as $\nu = 0.333$.

The boundary conditions are imposed as

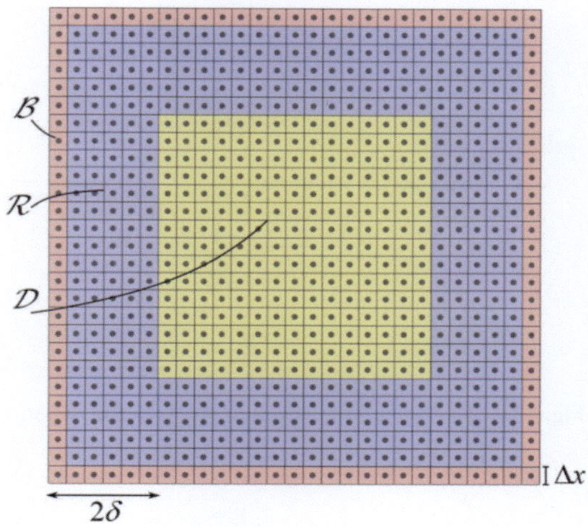

Fig. 12.2 PD regions of discretization in the plate

$$\left.\begin{array}{l} u_y(x,y)=0 \\ \sigma_{yy}^{PD}(x,y)=0 \end{array}\right\} \text{for } (-L/2+\Delta/2 \le x \le L/2-\Delta/2) \text{ and } y=(-W/2+\Delta/2) \quad (12.1)$$

$$\left.\begin{array}{l} u_x(x,y)=0 \\ \sigma_{xy}^{PD}(x,y)=0 \end{array}\right\} \text{for } x=(-L/2+\Delta/2) \text{ and } (-W/2+\Delta/2 \le y \le W/2-\Delta/2) \quad (12.2)$$

$$\left.\begin{array}{l} \sigma_{xy}^{PD}(x,y)=0 \\ \sigma_{yy}^{PD}(x,y)=0 \end{array}\right\} \text{for } (-L/2+\Delta/2 \le x \le L/2-\Delta/2) \text{ and } y=(W/2-\Delta/2) \quad (12.3)$$

$$\left.\begin{array}{l} \sigma_{xx}^{PD}(x,y)=\sigma_0 \\ \sigma_{xy}^{PD}(x,y)=0 \end{array}\right\} \text{for } x=(L/2-\Delta/2) \text{ and } (-W/2+\Delta/2 \le y \le W/2-\Delta/2) \quad (12.4)$$

12.2.2.1 Preprocessing

The finite element discretization of the plate is generated using the Ansys input file: ***ex12.2.2.dat*** or imported into the Ansys database using the CDREAD command. Subsequently, the BB-PD model of the plate is created through the following steps:

Step-1 The field and dimension of the analysis are specified using the ***Field Selection*** submenu located under ***Peridynamic Processor***. For the structural response of the plate, the "***Structural***" for field and "***2D***" for dimension are selected as shown in Fig. 12.3.

Step-2 Furthermore, the PD mesh is generated using the radio button option ***Generate using FE model*** in ***Discretization*** submenu as shown in Fig. 12.4.

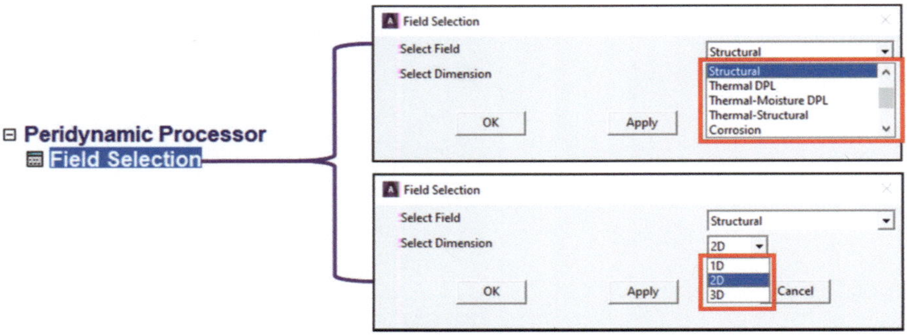

Fig. 12.3 Dialog box for selecting the field and dimension

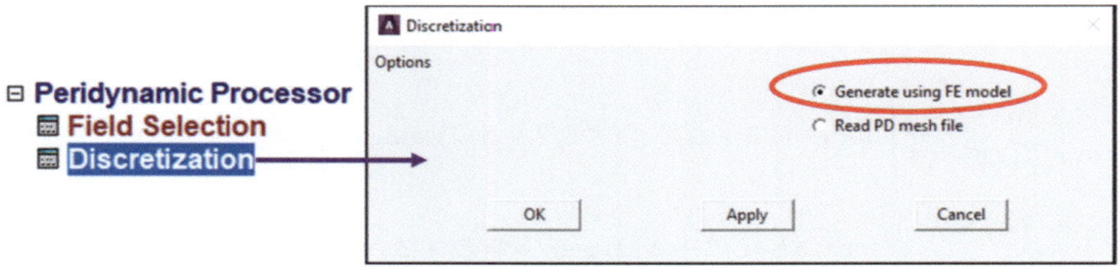

Fig. 12.4 Dialog box for PD mesh generation

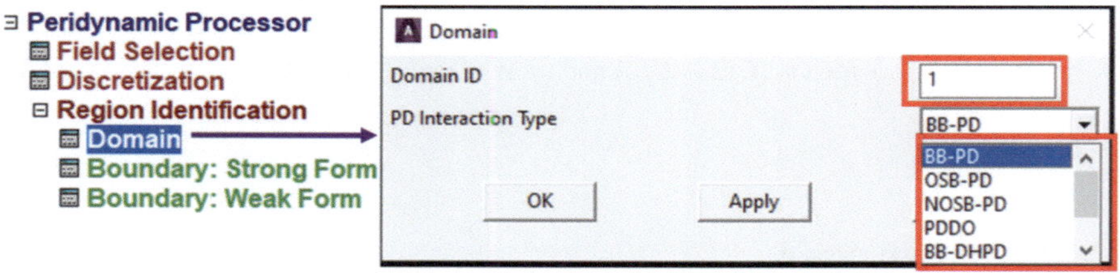

Fig. 12.5 Dialog box for selecting the domain interaction type as BB-PD

Step-3 The BB-PD and PDDO regions for the structural equations and boundary conditions are created using ***Region Identification*** submenu shown in Fig. 12.5. The submenu ***Region Identification >Domain*** is selected to open the dialog box ***Domain***. In this application, two zones are created corresponding to regions BB-PD and PDDO. For the BB-PD region, the parameter ***Domain ID=1*** is entered in the text box, and ***PD Interaction Type*** is selected from the drop down menu as BB-PD as shown in Fig. 12.5.

After entering the parameters and clicking the ***OK/Apply*** button, the node picker dialog box appears to select the desired region. After selecting the nodes by using the Node picker window and clicking the ***OK*** or ***Apply*** button of the node picker (Box), a

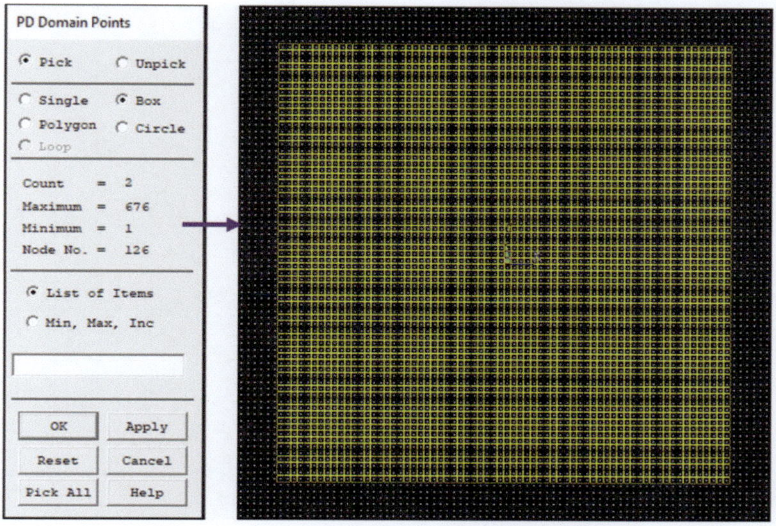

Fig. 12.6 Region identification corresponding to BB-PD domain of the plate

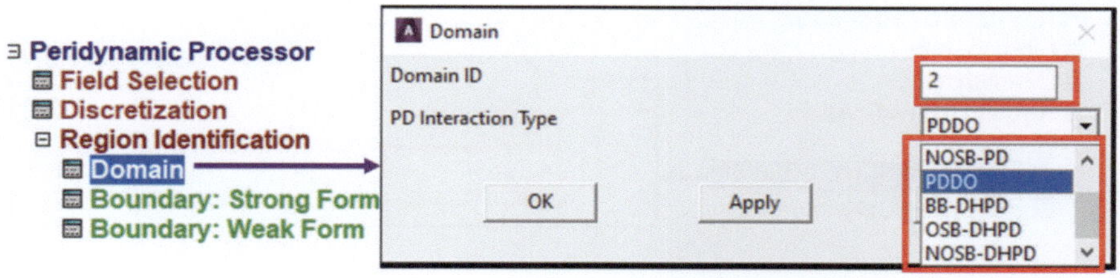

Fig. 12.7 Dialog box for selecting the domain interaction types as PDDO

dialog box labeled as ***Domain Points*** appears for confirming the selected BB-PD regions as shown in Fig. 12.6.

Step-4 Repeat the process for creating the PDDO region of the plate using a different ***Domain ID*** and setting the ***PD Interaction Type*** to PDDO as shown in Fig. 12.7. The selected region for PDDO by using node picker window is shown in Fig. 12.8. Clicking the ***OK*** or ***Apply*** button of the node picker (Box), a dialog box labeled as ***Domain Points*** appears for confirming the selected PDDO region.

Step-5 To enforce the boundary conditions in a pointwise manner (strong form), the submenu ***Region Identification >Boundary: Strong Form*** is selected to open the dialog box ***Boundary: Strong Form***. The parameters ***Boundary ID*** and ***Norm. vector to boundary: nx, ny*** are entered in the dialog box as shown in Fig. 12.9.

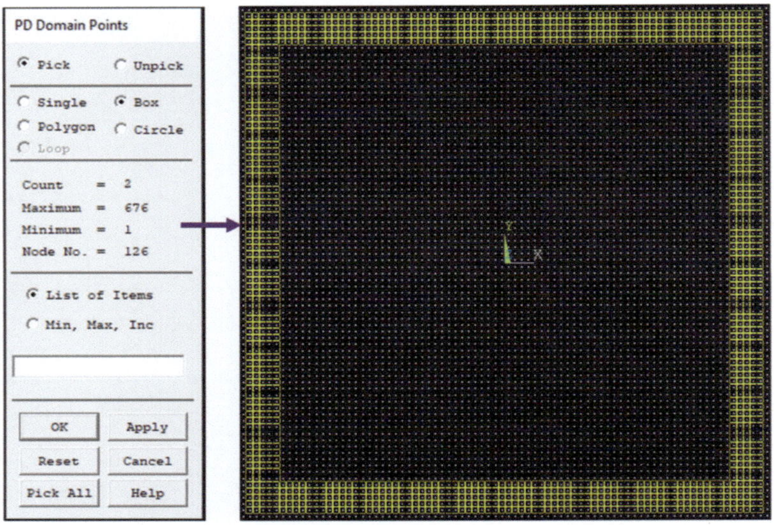

Fig. 12.8 Region identification corresponding to the PDDO domain of the plate

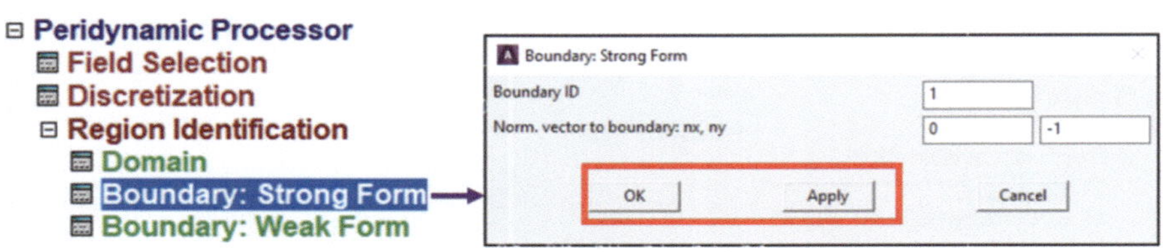

Fig. 12.9 Dialog box for selecting the boundary of the plate with strong form of interaction

Enter ***Boundary ID=1***, the unit normal vector ***nx, ny: 0, -1***. After entering the parameters and clicking the ***OK/Apply*** button, the node picker dialog box appears to select the desired boundary region as shown in Fig. 12.10. Repeat the process for each boundary of the plate by entering different ***Boundary ID*** and the respective unit normal vector. Figure 12.11 shows the steps for creating the boundary region with ***Boundary ID=2*** and unit normal vector ***nx, ny: 1, 0***.

Step-6 The family members of the PD points are generated by using ***Family Generation*** menu based on the horizon size factor ***m*** in the ***Family Generation*** dialog box as shown in Fig. 12.12. For this application, the horizon size factor, $m = 3$ is entered. After entering the parameter in the text box and clicking the ***OK/Apply*** button, the family members are generated using the horizon size of $\delta = 3.015\Delta x$.

Step-7 The element coefficients corresponding to the stiffness for PD structural model are generated using ***Element Coefficients (Structural)*** submenu option as

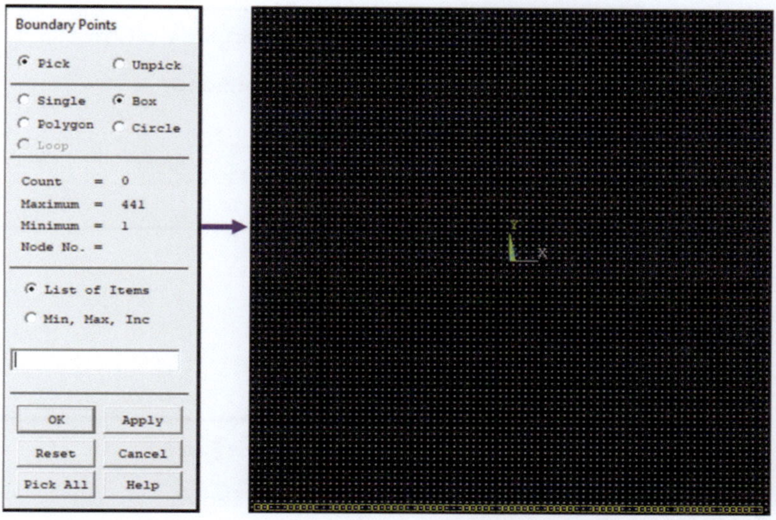

Fig. 12.10 Region identification corresponding to ***Boundary ID*=1** of the plate with strong form of interaction

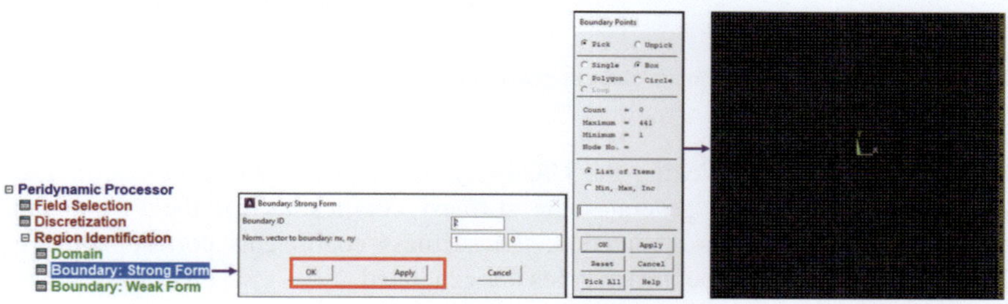

Fig. 12.11 Steps for creating the ***Boundary ID*=2** of the plate with strong form of interaction

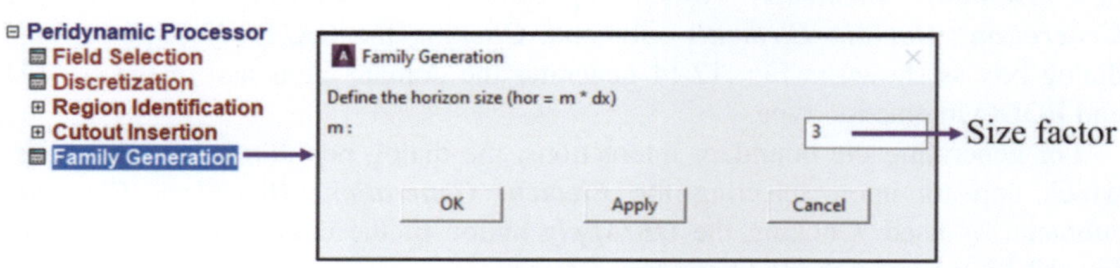

Fig. 12.12 Dialog box for defining the horizon size in the family generation

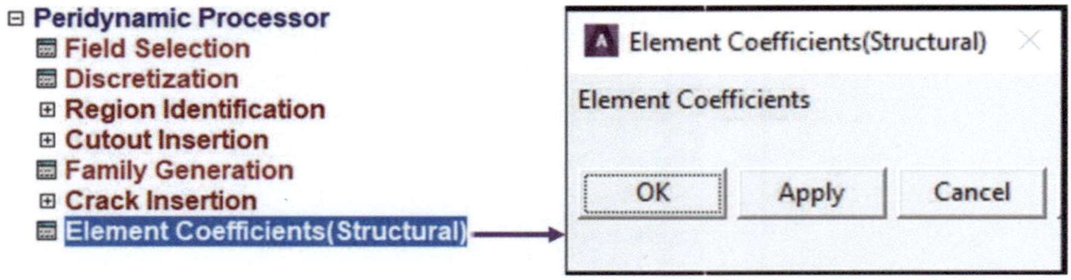

Fig. 12.13 Element Coefficients submenu for Structural analysis

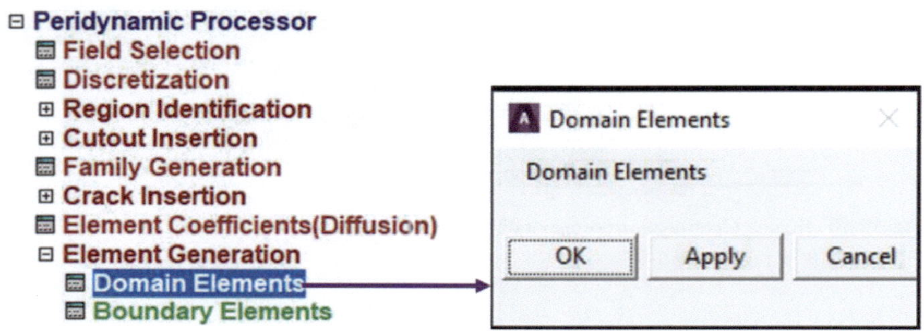

Fig. 12.14 Dialog box for domain element generation

shown in Fig. 12.13. Clicking the ***OK/Apply*** button of the ***Element Coefficients (Structural)*** dialog box generates the stiffness coefficients for the PD domain interaction types. In this application, the stiffness coefficients corresponding to BB-PD, PDDO and PD boundary types are generated.

Step-8 The elements required for modeling the PD interactions for the structural analysis both inside the BB-PD and PDDO domains are created by using the dialog box ***Domain Elements***, which appears upon selecting the ***Element Generation>Domain Elements*** submenu. Clicking the ***OK/Apply*** button of the dialog box as shown in Fig. 12.14 generates the domain elements for the BB-PD and PDDO interaction type.

For generating the boundary interactions, the dialog box ***Boundary Elements***, which appears upon selecting the ***Element Generation>Boundary Elements*** submenu, is used. Clicking the ***OK/Apply*** button of the dialog box as shown in Fig. 12.15 generates the PD boundary interactions.

Figure 12.16 shows the various Ansys elements used to model the domain and boundary regions. The boundary layer elements are shown in cyan color, the inner region resembling gold color, and the PDDO region with pink color.

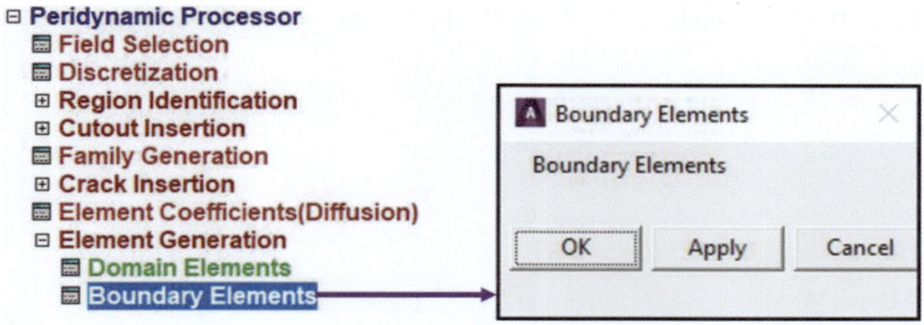

Fig. 12.15 Dialog box for boundary element generation

Fig. 12.16 PD elements in Ansys corresponding to different regions

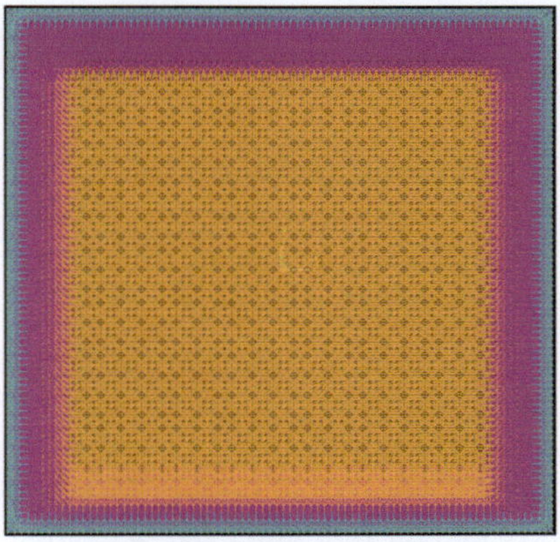

Step-9 The traction boundary conditions are directly enforced on PD boundary region using the ***PD_Boundary_traction.mac*** macro. The input parameters to this macro are ***Boundary ID***, ***key_tractionX***, ***key_tractionX***, ***tractionX***, and ***tractionY***, where ***key_tractionX***, ***key_tractionY*** denote the keys for enforcing the traction and ***tractionX***, ***tractionY*** denote the traction values to be enforced. The parameters specified as ***key_tractionX=-99/key_tractionY=-99*** enforce zero traction conditions in X and Y directions. Similarly, the parameters specified as ***key_tractionX=0/ key_tractionY=0*** do not enforce any traction conditions in X and Y directions. To enforce nonzero traction, set the ***key_tractionX=1/key_tractionY=1*** and enter the nonzero value of the traction to be enforced. The input format of this macro is of the form

PD_Boundary_traction, ***BoundaryID, key_tractionX, key_tractionY, tractionX, tractionY***

Run the below commands to enforce traction conditions on edges 1 through 4.

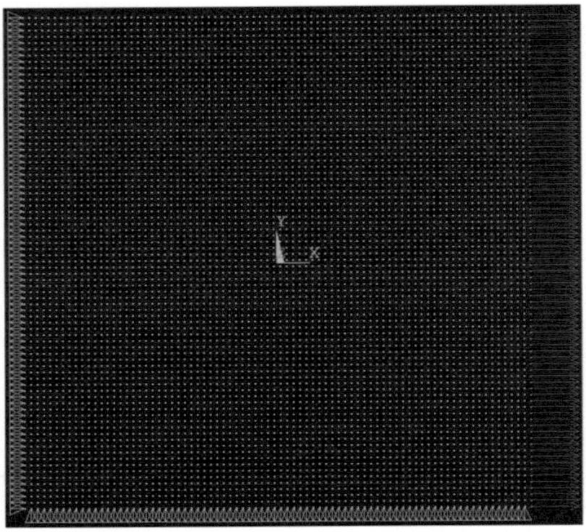

Fig. 12.17 Displacement BC directly enforced on the left and bottom PD points

PD_Boundary_traction,1,0,-99,0,0
PD_Boundary_traction,2,1,-99,1e+09,0
PD_Boundary_traction,3,-99,-99,0,0
PD_Boundary_traction,4,-99,0,0,0

The displacement boundary conditions are directly enforced using Ansys D command by selecting the PD points on the left and bottom edges, i.e. ***Boundary ID*=1, 4** as shown in Fig. 12.17 and constraining the normal directions, i.e., UY, UX respectively. For ***Boundary ID*=3,** tractions in X and Y directions are enforced as zero and for ***Boundary ID*=2** traction in X direction is set to 1e+09 while it is set to zero in Y direction.

12.2.2.2 Postprocessing

The displacement contour plots can be generated directly by using the Ansys PLNS command. To generate the line plots of displacements along the centerline of the plate, input file ***PD_Post_Disp.mac*** is used. The displacements along X and Y directions are output to the files named as ***disp_ux.out*** and ***disp_uy.out***. The PD stress plots are generated using the macro file ***PD_Post_Stress.mac***, which outputs the file *pdstress.dat*, which can be read in the VTK python module ***plot_pdstress.py***.

Figures 12.18 and 12.19 show the comparisons of BB-PD displacement predictions with those of FEM. Also, BB-PD horizontal and vertical displacements are compared with those of FEM along the centerlines of the plate in Fig. 12.20.

Figure 12.21 shows the BB-PD stress predictions for Poisson's ratio of $\nu = 0.333$. The horizontal normal stress variation is same as the applied stress, $\sigma_0 = 1.0$ GPa. The vertical normal and shear stress variation is minimal, and it can be considered almost zero throughout the plate.

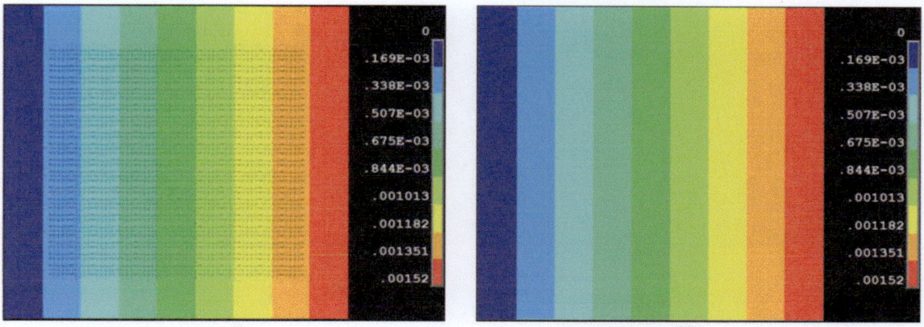

a) b)

Fig. 12.18 Comparison of horizontal displacement predictions in the plate under tension: (**a**) BB-PD, and (**b**) FEM

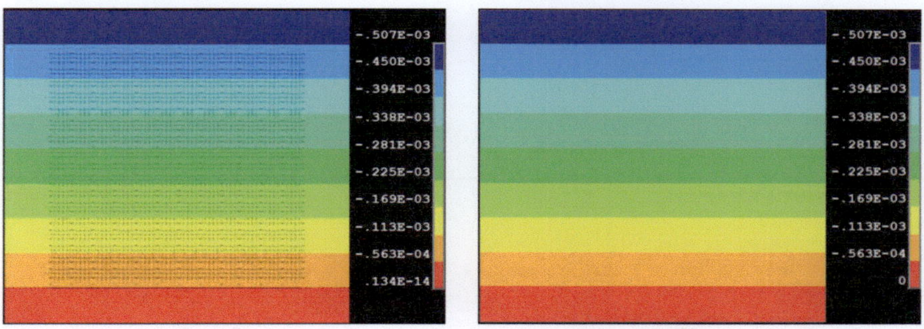

a) b)

Fig. 12.19 Comparison of vertical displacement predictions in the plate under tension: (**a**) BB-PD, and (**b**) FEM

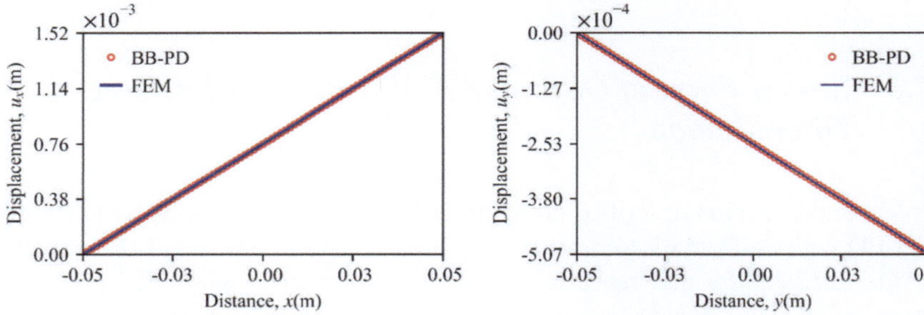

a) b)

Fig. 12.20 Comparison of BB-PD and FEM horizontal displacement predictions in the plate under tension: (**a**) $u_x(x, y = 0)$ and (**b**) $u_y(x = 0, y)$

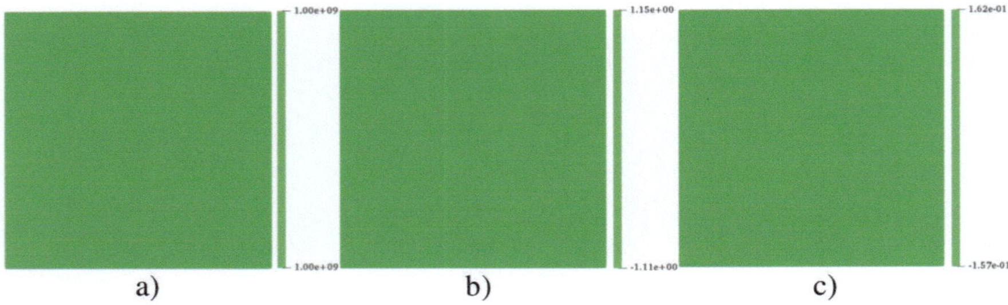

Fig. 12.21 BB-PD stress predictions in the plate under tension: (**a**) σ_{xx}, (**b**) σ_{yy}, and (**c**) σ_{xy}

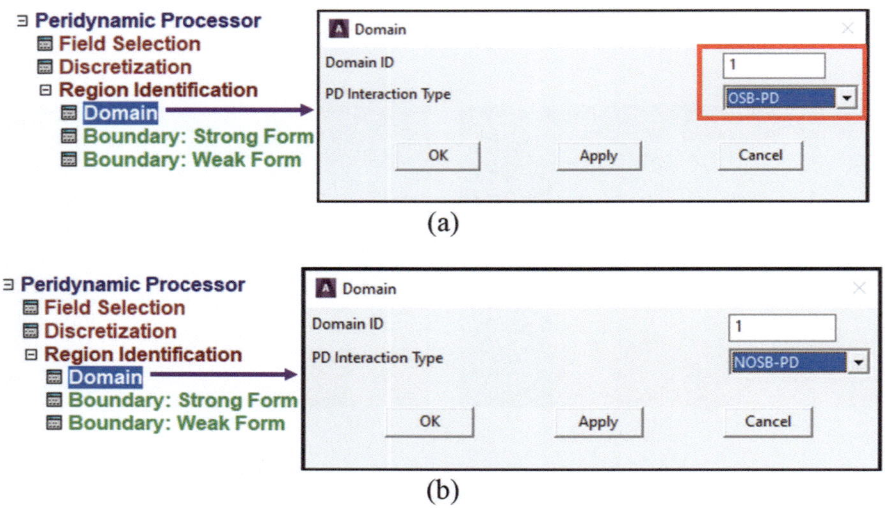

Fig. 12.22 Dialog box for selecting the domain interaction type as: (**a**) OSB-PD and (**b**) NOSB-PD

12.2.3 Strong Form of OSB- and NOSB-PD with Uniform Discretization

The plate under tension described in the preceding section is modeled by OSB- and NOSB-PD with uniform discretization for a Poisson's ratio of $\nu = 0.1$. Reading the input file ***ex12.2.3.dat*** and repeating the similar steps as in Sect. 12.2.2.1 of the **Preprocessing** stage, except for selecting the inner region of domain as either OSB or NOSB as in **Step-3**, lead to the results with NOSB and OSB models. Figure 12.22 shows the selection of OSB-PD and NOSB-PD interactions in PD region.

Following the postprocessing Sec. 12.2.2.2, Figs. 12.23, 12.24, and 12.25 show the OSB and NOSB-PD and FEM displacement predictions, respectively. Also,

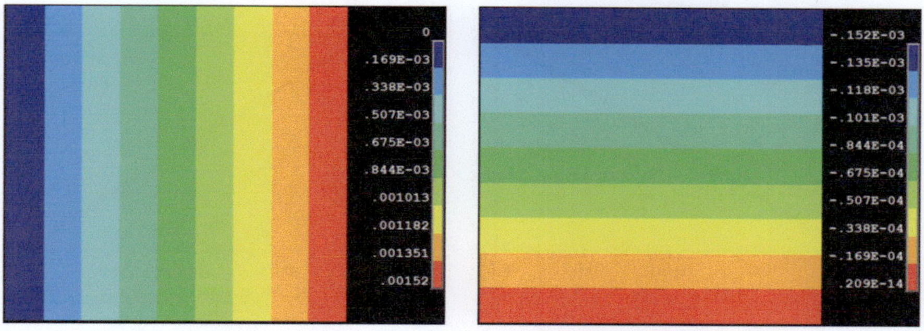

a) b)

Fig. 12.23 OSB displacement predictions in the plate under tension: (**a**) u_x and (**b**) u_y

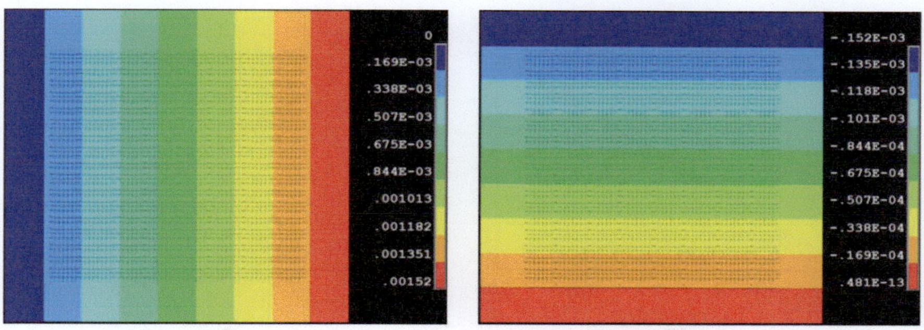

a) b)

Fig. 12.24 NOSB displacement predictions in the plate under tension: (**a**) u_x and (**b**) u_y

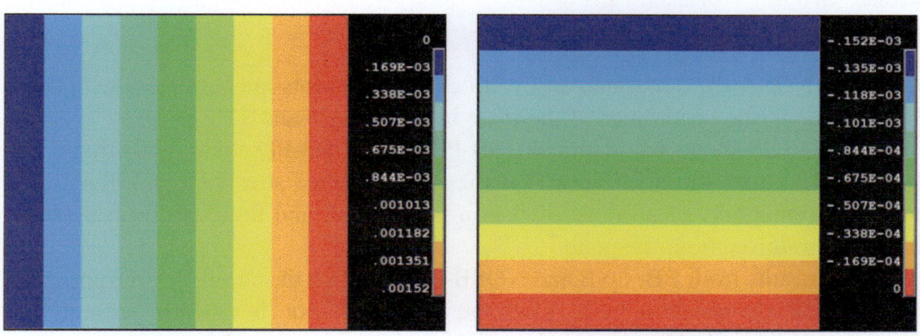

a) b)

Fig. 12.25 FEM displacement predictions in the plate under tension: (**a**) u_x and (**b**) u_y

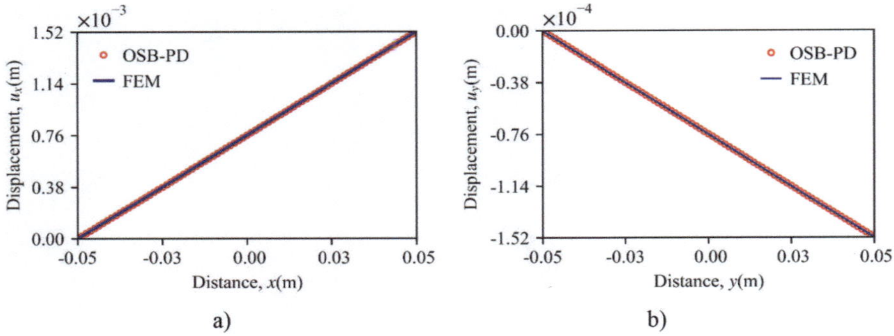

Fig. 12.26 Comparison of OSB-PD and FEM displacement predictions in the plate under tension: (**a**) $u_x(x, y = 0)$ and (**b**) $u_y(x = 0, y)$

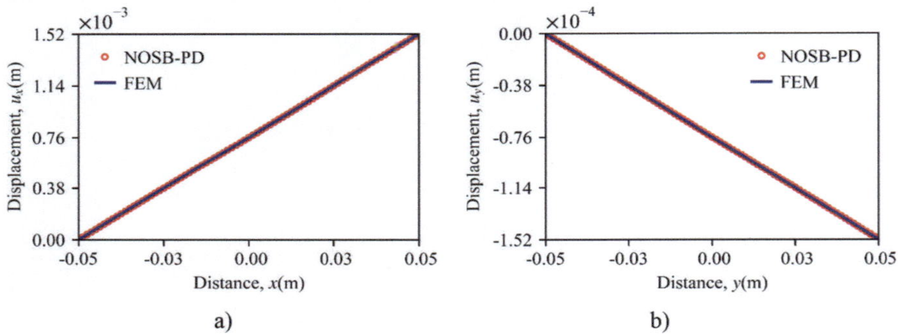

Fig. 12.27 Comparison of NOSB-PD and FEM displacement predictions in the plate under tension: (**a**) $u_x(x, y = 0)$ and (**b**) $u_y(x = 0, y)$

OSB and NOSB-PD horizontal and vertical displacements are compared with those of FEM along the centerlines of the plate in Figs. 12.26 and 12.27.

Figure 12.28 shows the identical OSB-PD and NOSB-PD stress predictions. The horizontal normal stress variation is same as the applied stress, $\sigma_0 = 1.0\text{GPa}$. The vertical normal and shear stress variation is minimal, and they can be considered almost zero throughout the plate.

These results from BB, OSB and NOSB-PD establish the validity of the approach for a simple deformation field. Also, the stress variation is uniform without any unphysical stress concentrations near the boundaries.

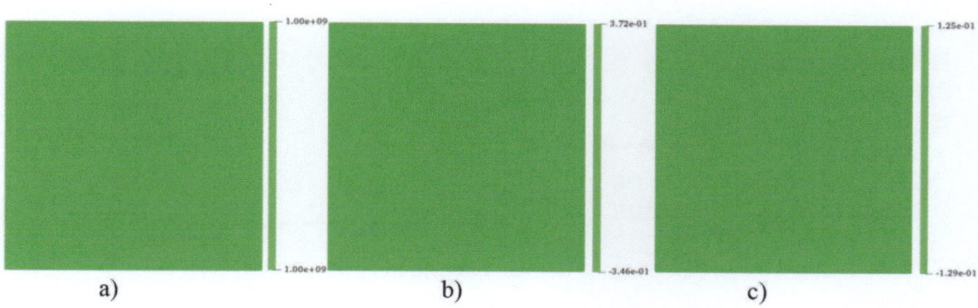

Fig. 12.28 Identical OSB-PD and NOSB-PD stress predictions in the plate under tension: (**a**) σ_{xx}, (**b**) σ_{yy}, and (**c**) σ_{xy}

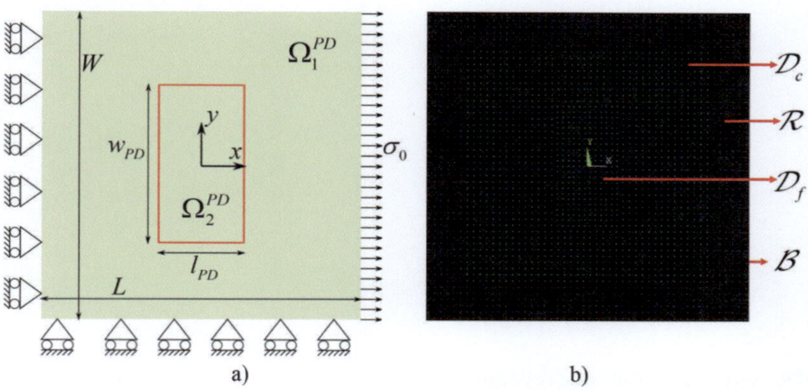

Fig. 12.29 Coarse and fine discretization regions in plate: (**a**) Geometry and boundary conditions, and (**b**) PD discretization

12.2.4 Strong Form of NOSB-PD with Nonuniform Discretization

The response of the plate under tension described in the preceding section is modeled by NOSB-DHPD with nonuniform discretization, Poisson's ratio of $\nu = 0.1$ and subjected to a tensile stress of $\sigma_0 = 0.1\text{MPa}$. As shown in Fig. 12.29a, the fine PD region, Ω_2^{PD}, has a length and width of $l_{PD} = 20\text{mm}$ and $w_{PD} = 40\text{mm}$, while the remaining region corresponds to coarse PD, Ω_1^{PD}. The coarse and fine PD regions Ω_1^{PD} and Ω_2^{PD} are discretized with a mesh size of $\Delta x = \sqrt{A_{(k)}} = 2.0\text{mm}$ and 1.8mm, respectively. As shown in Fig. 12.29b, $\mathcal{D}_f$ representing the fine PD region, Ω_2^{PD} consists of 276 nodes while $\mathcal{D}_c \cup \mathcal{R}$ representing the coarse PD region, Ω_1^{PD} consists has 2104 nodes and $\mathcal{B}$ representing the boundary region has 196 nodes.

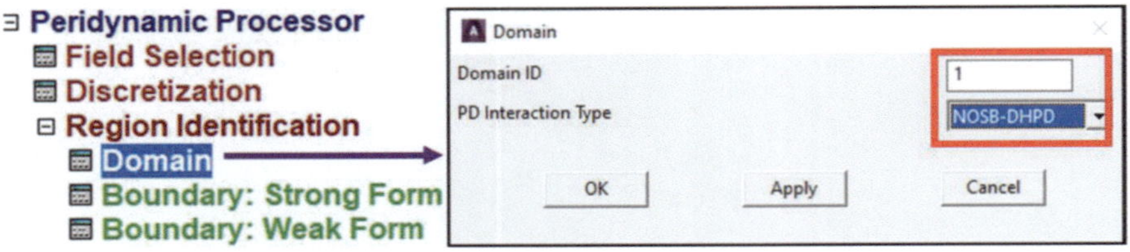

Fig. 12.30 Dialog box for selecting the domain interaction type as NOSB-DHPD

Fig. 12.31 PD elements in Ansys corresponding to different regions

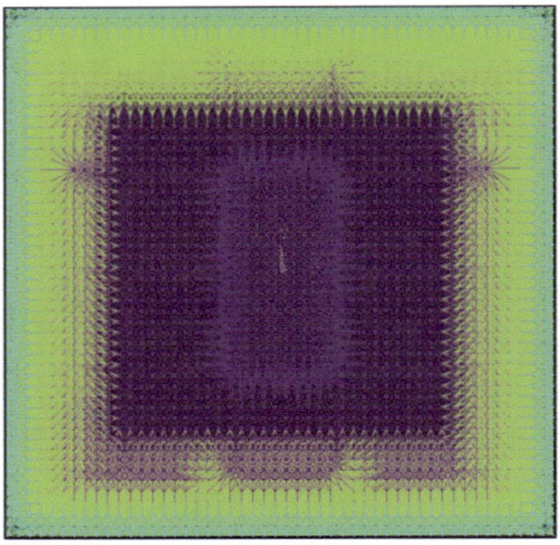

The finite element discretization of the plate is generated using the Ansys input file: ***ex12.2.4.dat*** or imported into the Ansys database using the CDREAD command. Furthermore, steps in the preprocessing are similar to those described in Sect. 12.2.2.1 for a uniform mesh, except for **Step-6** where the ***PD Interaction Type*** is changed to NOSB-DHPD, i.e., the dual-horizon NOSB is used as an interaction type for ***Domain ID=1*** as shown in Fig. 12.30.

Figure 12.31 shows PD elements with a nonuniform mesh. The boundary layer element is shown in cyan color, the inner region resembling pink color, and the PDDO region with light green color.

In **Step-9** of Sect. 12.2.2.1, run the below commands to enforce traction conditions on edges 1 through 4.

PD_Boundary_traction,1,0,-99,0,0
PD_Boundary_traction,2,1,-99,1e+05,0
PD_Boundary_traction,3,-99,-99,0,0
PD_Boundary_traction,4,-99,0,0,0

Following the postprocessing Sect. 12.2.2.2, Figs. 12.32 and 12.33 show the comparisons of NOSB-DHPD displacement predictions with those of FEM. Also, NOSB-DHPD horizontal and vertical displacements are compared with those of FEM along the centerlines of the plate in Fig. 12.34.

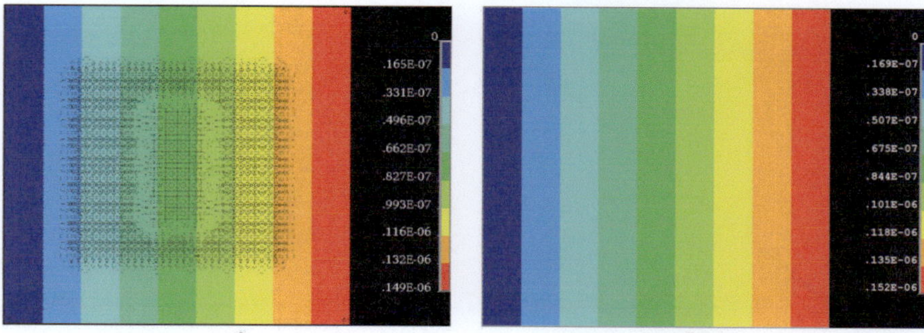

a) b)

Fig. 12.32 Comparison of horizontal displacement predictions in the plate under tension: (**a**) NOSB-DHPD and (**b**) FEM

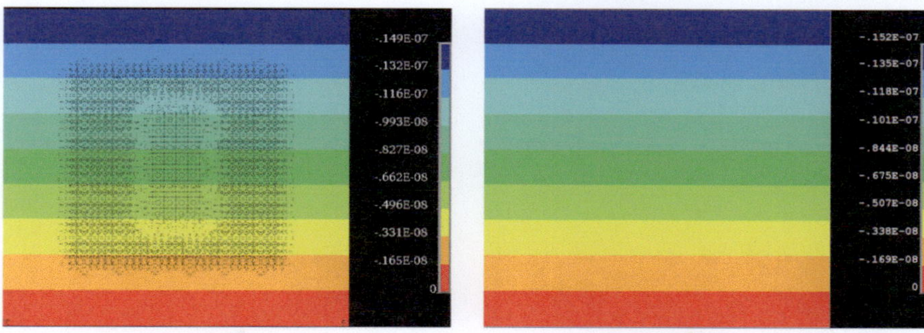

a) b)

Fig. 12.33 Comparison of vertical displacement predictions in the plate under tension: (**a**) NOSB-DHPD and (**b**) FEM

12.3 Plate with a Crack Under Tension

As shown in Fig. 12.35, the plate geometry is defined by its length, $L = 100$mm width, $W = 100$mm, and thickness, $h = 1$mm. The crack length is $2a = 20$mm. The material is isotropic with a Young's modulus of $E = 65.8$GPa and Poisson's ratio of $\nu = 0.3$. It has roller support conditions along the left and bottom edges. It is free of tractions along the top edge.

As shown in Fig. 12.36a, the plate is split into fine and coarse PD regions. The fine region, Ω_2^{PD}, has a length and width of $l_{PD} = 20$ mm and $w_{PD} = 40$ mm, while the remaining part of the plate corresponds to coarse region, Ω_1^{PD}. The coarse and fine PD regions Ω_1^{PD} and Ω_2^{PD} are discretized with a mesh size of $\Delta x = \sqrt{A_{(k)}} = 1.5$ mm and 1.2 mm, respectively. As shown in Fig. 12.36b, $\mathcal{D}_f$ representing the fine PD region, Ω_2^{PD} consists of 612 nodes while $\mathcal{D}_c \cup \mathcal{R}$

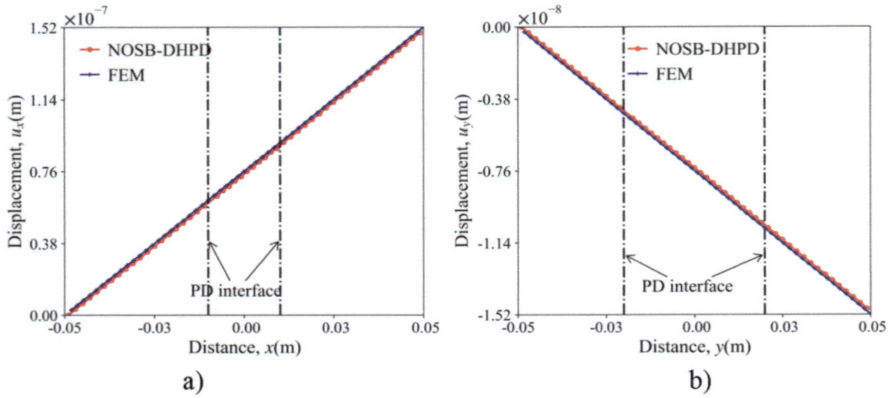

Fig. 12.34 Comparison of NOSB-DHPD and FEM horizontal displacement predictions in the plate under tension: (**a**) $u_x(x, y = 0)$ and (**b**) $u_y(x = 0, y)$

Fig. 12.35 Plate geometry and boundary conditions

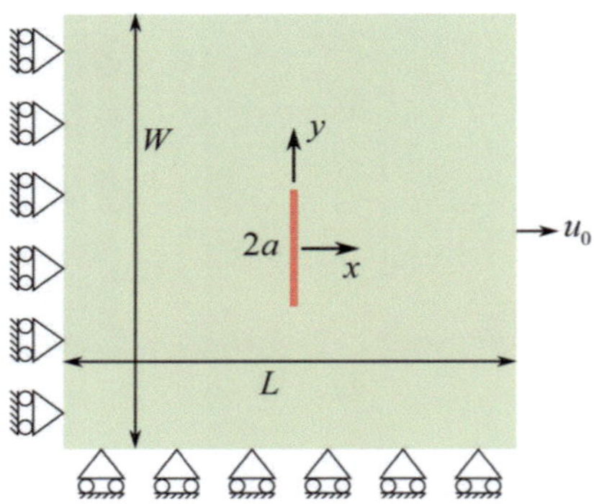

representing the coarse PD region, Ω_1^{PD} consists has 3912 nodes and $\mathscr{B}$ representing the boundary region has 266 nodes.

The plate is subjected to a displacement constraint of u_0 along the right edge. The boundary conditions are imposed as

$$\left.\begin{array}{l} u_y(x, y) = 0 \\ \sigma_{yy}^{\mathrm{PD}}(x, y) = 0 \end{array}\right\} \text{for } (-L/2 + \Delta/2 \leq x \leq L/2 - \Delta/2) \text{ and } y = (-W/2 + \Delta/2) \tag{12.5}$$

$$\left.\begin{array}{l} u_x(x, y) = 0 \\ \sigma_{xy}^{\mathrm{PD}}(x, y) = 0 \end{array}\right\} \text{for } x = (-L/2 + \Delta/2) \text{ and } (-W/2 + \Delta/2 \leq y \leq W/2 - \Delta/2) \tag{12.6}$$

$$\left.\begin{array}{l} \sigma_{xy}^{\mathrm{PD}}(x, y) = 0 \\ \sigma_{yy}^{\mathrm{PD}}(x, y) = 0 \end{array}\right\} \text{for } (-L/2 + \Delta/2 \leq x \leq L/2 - \Delta/2) \text{ and } y = (W/2 - \Delta/2) \tag{12.7}$$

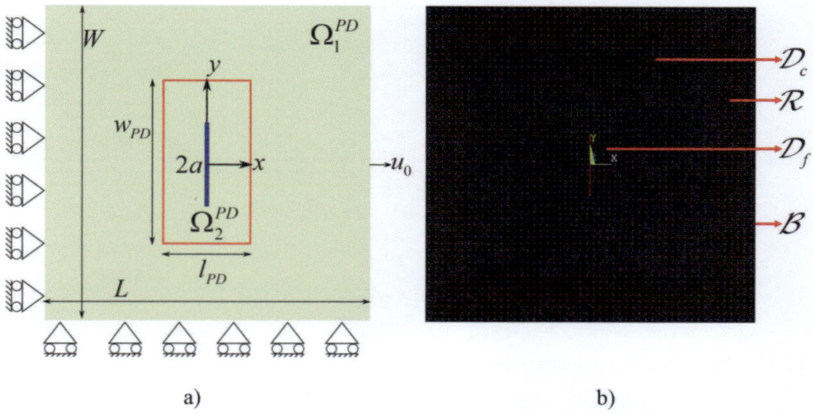

Fig. 12.36 Coarse and fine discretization regions in plate with a crack: (**a**) Geometry and boundary conditions and (**b**) PD discretization

$$\left.\begin{array}{l} u_x(x,y)=u_0 \\ \sigma_{xy}^{\mathrm{PD}}(x,y)=0 \end{array}\right\} \text{ for } x=(L/2-\Delta/2) \text{ and } (-W/2+\Delta/2\leq y\leq W/2-\Delta/2) \tag{12.8}$$

12.3.1 Strong Form of OSB-PD with Nonuniform Discretization

12.3.1.1 Preprocessing

The plate with a crack under tension is modeled by OSB-PD with nonuniform discretization. The Ansys modeling procedure is similar to that of Sect. 12.2. The finite element discretization of the plate can be generated by using the Ansys input file: ***ex12.3.1.dat*** or imported into the Ansys database using the CDREAD command.

Step-1 The field and dimension of the analysis are specified using the ***Field Selection*** submenu located under ***Peridynamic Processor***. For the structural response of the plate, the "***Structural***" for field and "***2D***" for dimension is selected as shown in Fig. 12.37.

Step-2 The PD mesh is generated using the radio button option ***Generate using FE model*** in ***Discretization*** submenu as shown in Fig. 12.38.

Step-3 The PD regions for the structural equations and boundary conditions are created using ***Region Identification*** submenu shown in Fig. 12.39. The submenu ***Region Identification >Domain*** is selected to open the dialog box ***Domain***. In this

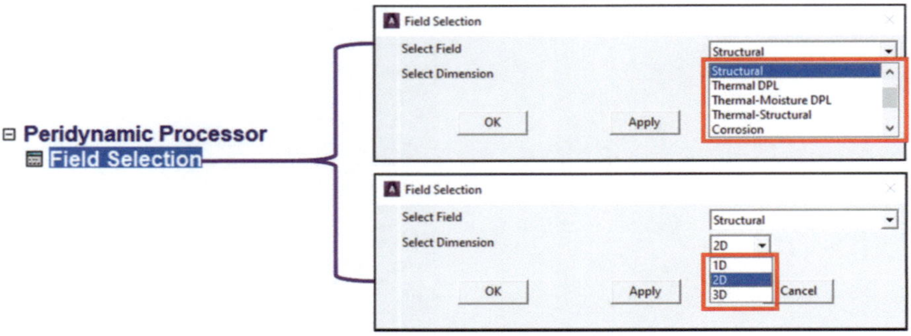

Fig. 12.37 Dialog box for selecting the field and dimension

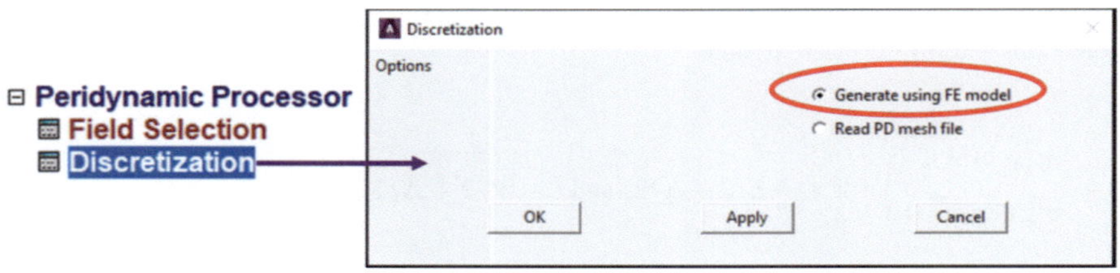

Fig. 12.38 Dialog box for PD mesh generation

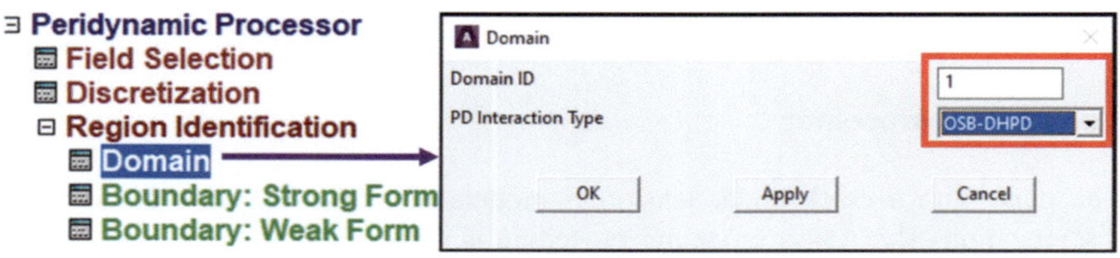

Fig. 12.39 Dialog box for selecting the domain interaction type as OSB-DHPD

application, two zones corresponding to OSB-DHPD and PDDO regions are created. For the OSB-DHPD region, the parameter, ***Domain ID=1*** is entered in the text box, and ***PD Interaction Type*** is selected from the drop-down menu as OSB-DHPD as shown in Fig. 12.39.

After entering the parameters and clicking the ***OK/Apply*** button, the node picker dialog box appears to select the desired region. After selecting the nodes using the Node picker window and clicking the ***OK*** or ***Apply*** button of the node picker (Box), a dialog box labeled as ***Domain Points*** appears for confirming the selected OSB-DHPD region as shown in Fig. 12.40.

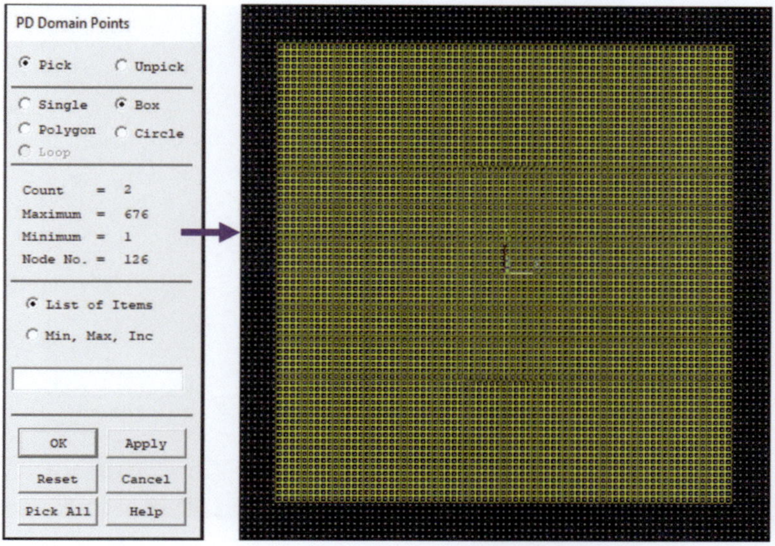

Fig. 12.40 Region identification corresponding to OSB-DHPD region of the plate

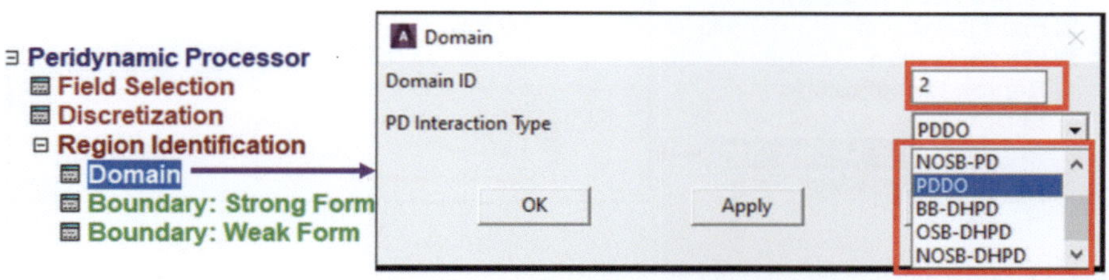

Fig. 12.41 Dialog box for selecting the domain interaction types as PDDO

Step-4 Repeat the process for creating the PDDO region of the plate using a different ***Domain ID*** and setting the ***PD Interaction Type*** to PDDO as shown in Fig. 12.41. The selected region for PDDO by using node picker window is shown in Fig. 12.42. Clicking the ***OK*** or ***Apply*** button of the node picker (Box), a dialog box labeled as ***Domain Points*** appears for confirming the selected PDDO region.

Step-5 To enforce the boundary conditions in a pointwise manner (strong form), the submenu ***Region Identification >Boundary: Strong Form*** is selected to open the dialog box ***Boundary: Strong Form.*** The parameters ***Boundary ID*** and ***Norm. vector to boundary: nx, ny*** are entered in the dialog box as shown in Fig. 12.43.

Enter ***Boundary ID=1***, and its unit normal vector ***nx, ny: 0, -1***. After entering the parameters and clicking the ***OK/Apply*** button, the node picker dialog box appears to select the desired boundary region as shown in Fig. 12.44. Repeat the process for

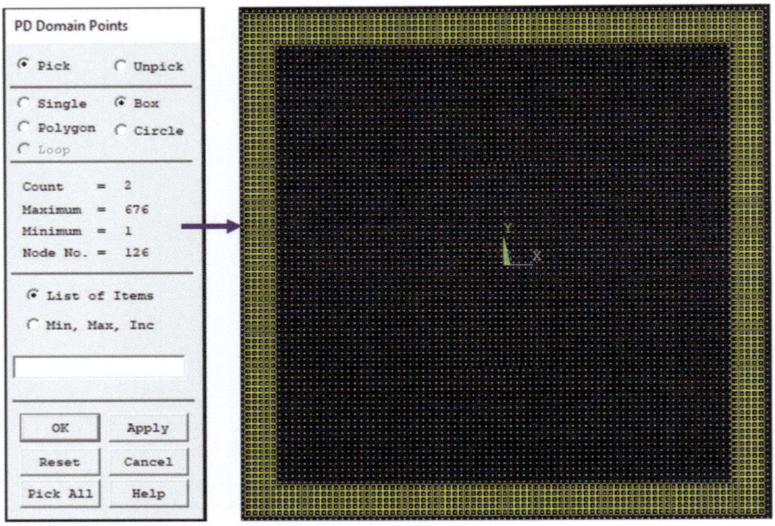

Fig. 12.42 Region identification corresponding to PDDO region of the plate

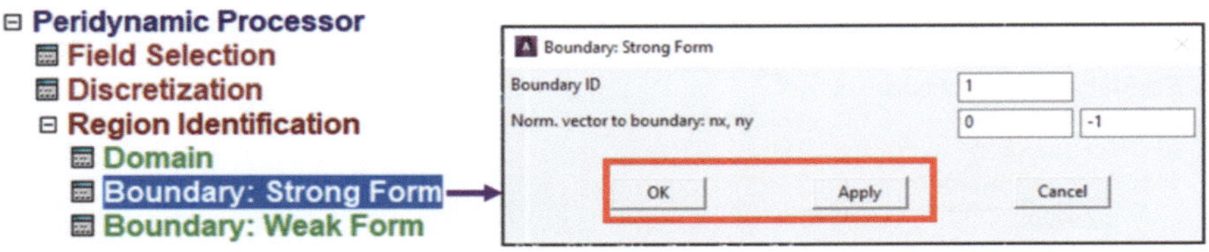

Fig. 12.43 Dialog box for selecting the boundary of the plate with strong form of interaction

each boundary of the plate by entering different ***Boundary ID*** and its unit normal vector. Figure 12.45 shows the steps in creating the boundary region with ***Boundary ID=2*** and unit normal vector, ***nx, ny:1, 0***.

Step-6 The family members of the PD points are generated using ***Family Generation*** submenu based on the horizon size factor ***m*** in the ***Family Generation*** dialog box as shown in Fig. 12.46. For this application, the horizon size factor $m = 3$ is entered. After entering the parameter in the text box and clicking the ***OK/Apply*** button, the family members are generated using the horizon size of $\delta = 3.015\Delta x$.

Step-7 The initial crack in the PD mesh is inserted using ***Crack Insertion*** submenu. Additional details of using this submenu are given in Sect. 9.7 of Chap. 9. As shown in Fig. 12.47, ***Insert Crack parameters*** dialog box appears upon clicking the ***Crack Insertion>Insert*** submenu. The parameters X1,Y1=0, -0.01 and X2,Y2=0, 0.01 denoting the crack coordinates are entered.

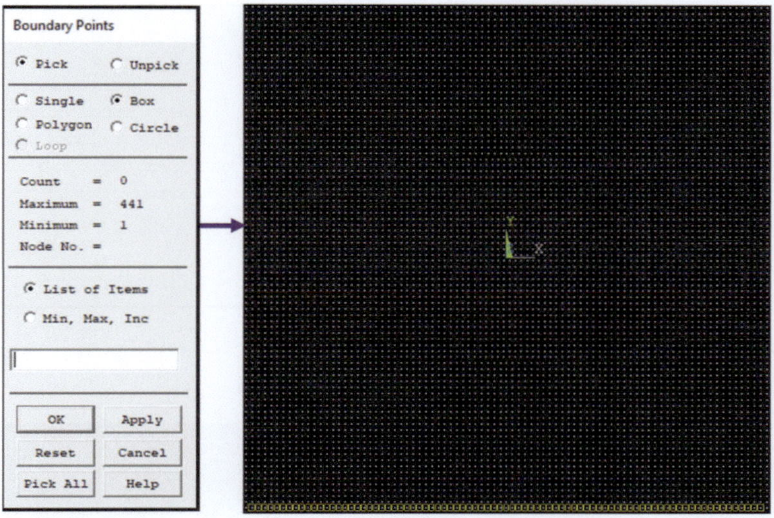

Fig. 12.44 Region identification corresponding to ***Boundary ID=1*** of the plate with strong form of interaction

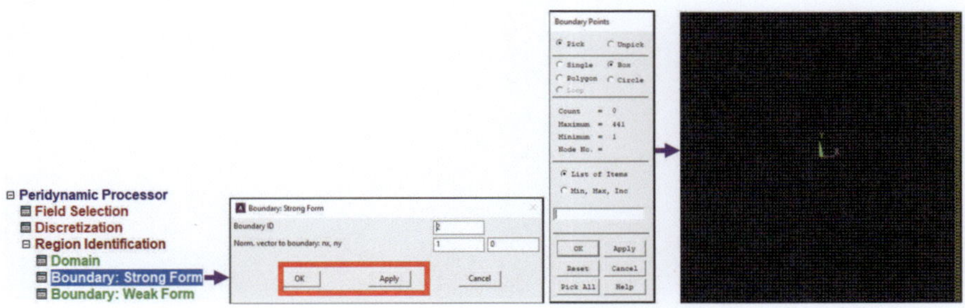

Fig. 12.45 Steps for creating the ***Boundary ID=2*** of the plate with strong form of interaction

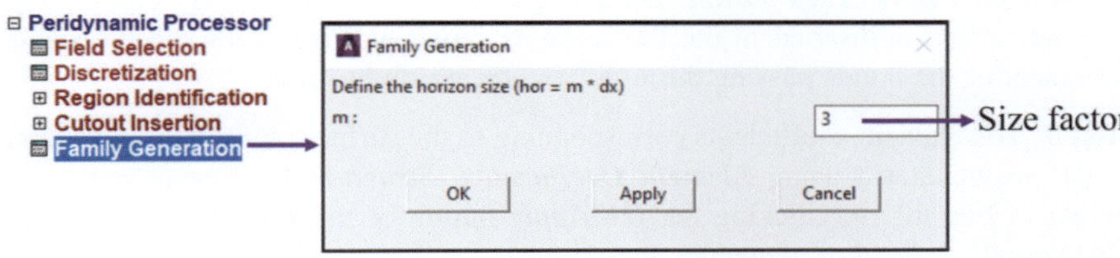

Fig. 12.46 Dialog box for defining the horizon size in the family generation

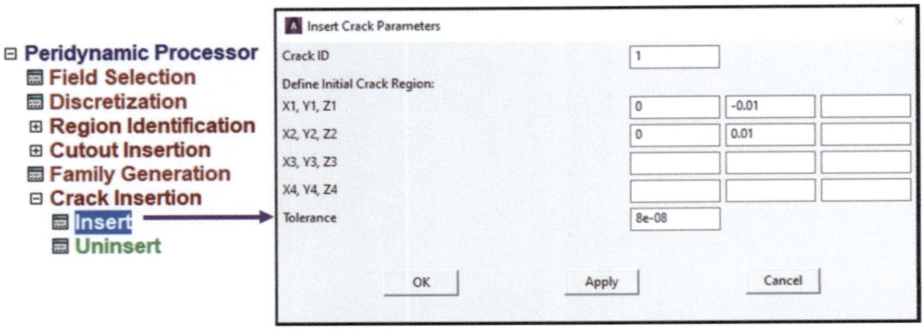

Fig. 12.47 Dialog box for inserting an initial crack

Fig. 12.48 Line crack along Y-axis of the plate

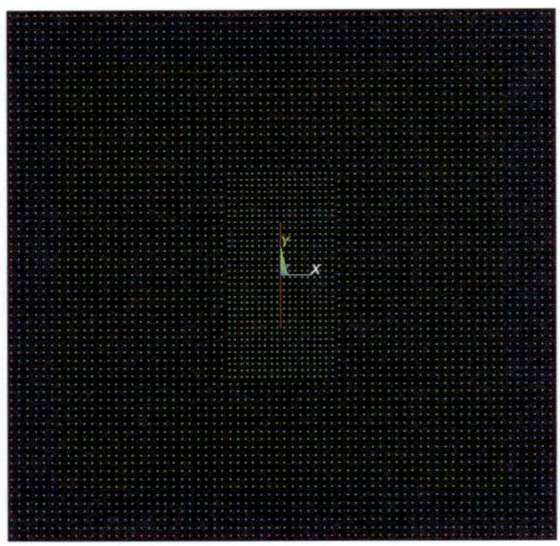

After entering the crack ID, coordinates of the crack and clicking the ***OK*** or ***Apply*** button in the ***Insert Crack parameters*** dialog box, a line (shown in red) representing the sharp crack is inserted in the PD mesh as shown in Fig. 12.48 and the status parameter of the bonds passing through this line are set to zero.

Step-8 The element coefficients corresponding to the stiffness for the PD structural model are generated using ***Element Coefficients (Structural)*** submenu option as shown in Fig. 12.49. Clicking the ***OK/Apply*** button of the ***Element Coefficients (Structural)*** dialog box generates the stiffness coefficients for the PD interaction type. In this application, the stiffness coefficients corresponding to OSB-DHPD , PDDO, and PD boundary types are generated.

Step-9 The elements required for modeling the PD interactions for the structural analysis both in the OSB-DHPD and PDDO domains are created by using the dialog

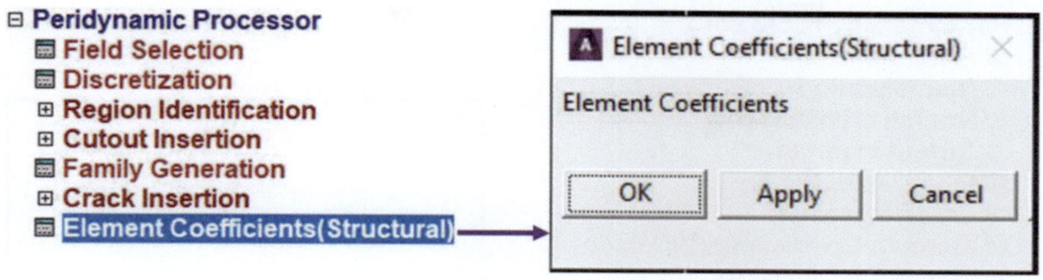

Fig. 12.49 Element Coefficients submenu for Structural

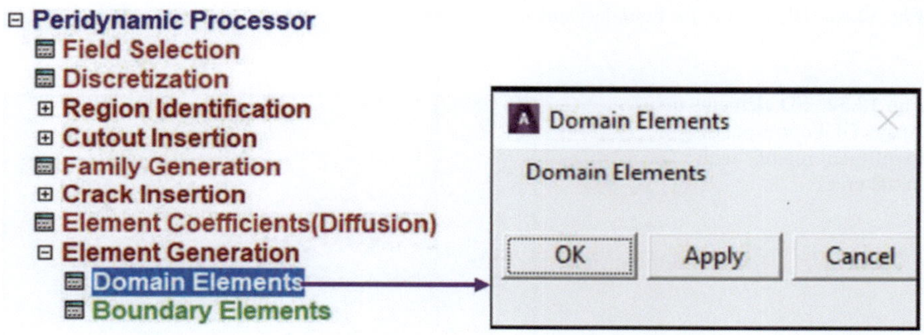

Fig. 12.50 Dialog box for domain element generation

box ***Domain Elements***, which appears upon selecting the ***Element Generation>Domain Elements*** submenu. Clicking the ***OK/Apply*** button of the dialog box as shown in Fig. 12.50 generates the domain elements for the OSB-DHPD and PDDO interaction type.

For generating the boundary interactions, the dialog box ***Boundary Elements***, which appears upon selecting the ***Element Generation>Boundary Elements*** submenu, is used. Clicking the ***OK/Apply*** button of the dialog box as shown in Fig. 12.51 generates the elements for the selected PD boundary interaction type.

Figure 12.52 shows the various Ansys elements used to model domain and boundary regions. The boundary layer element is shown in cyan color; the inner region, OSB-DHPD resembling blue color; and the PDDO region with light green color.

Step-10 The traction boundary conditions are directly enforced on PD boundary region using the ***PD_Boundary_traction.mac*** macro. The input parameters to this macro are ***Boundary ID***, ***key_tractionX***, ***key_tractionY***, ***tractionX***, and ***tractionY***, where ***key_tractionX***, ***key_tractionY*** denote the keys for enforcing the traction and ***tractionX***, ***tractionY*** denote the traction values to be enforced. The parameters specified as ***key_tractionX=-99/key_tractionY=-99*** enforce zero traction conditions in X and Y directions. Similarly, the parameters specified as ***key_tractionX=0/***

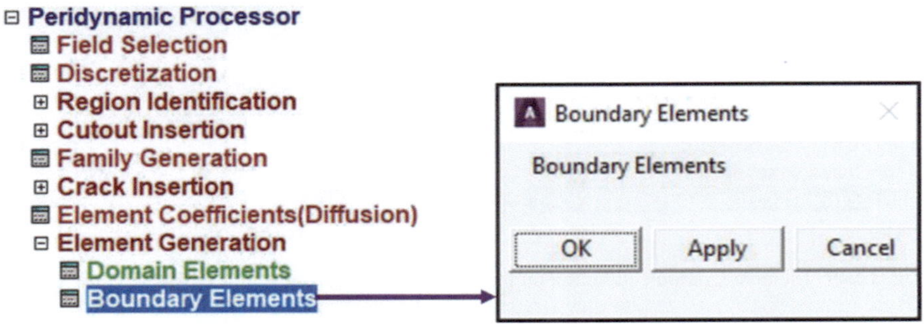

Fig. 12.51 Dialog box for boundary element generation

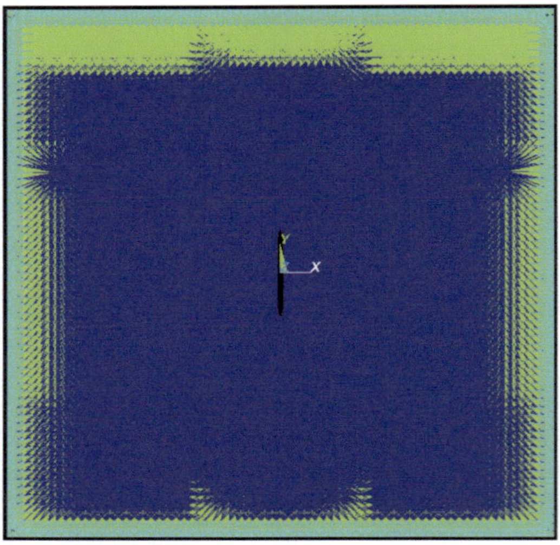

Fig. 12.52 PD elements in Ansys GUI corresponding to different regions with initial crack

key_tractionY=0 do not enforce any traction conditions in X and Y directions. To enforce nonzero traction, set the ***key_tractionX=1/key_tractionY=1*** and enter the nonzero value of the traction to be enforced. The input format of this macro is of the form

PD_Boundary_traction, BoundaryID, key_tractionX, key_tractionY, tractionX, tractionY

Run the below commands to enforce traction conditions on edges 1 through 4.

PD_Boundary_traction,1,0,-99,0,0
PD_Boundary_traction,2,-99,0,0,0
PD_Boundary_traction,3,-99,-99,0,0
PD_Boundary_traction,4,-99,0,0,0

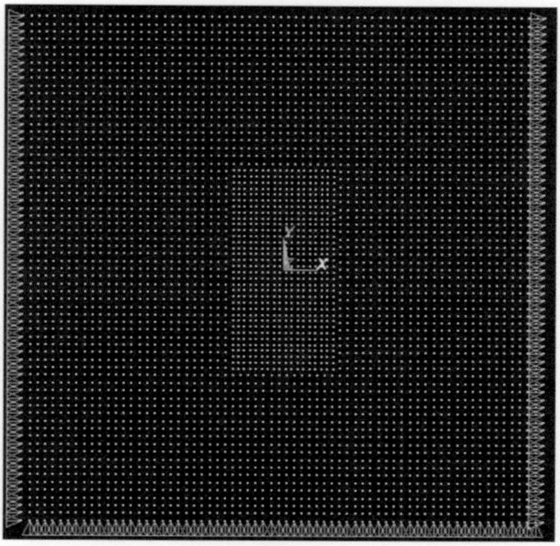

Fig. 12.53 Displacement BC directly enforced on the left, right, and bottom PD nodes

The displacement boundary conditions are directly enforced using Ansys D command by selecting the PD points on the left and bottom edges, i.e., ***Boundary ID=1, 4*** as shown in Fig. 12.53 and constraining the normal directions, i.e., u_y, u_x, respectively. For ***Boundary ID=3,*** tractions in X and Y directions are enforced as zero and for ***Boundary ID=2*** traction in Y direction is enforced as zero and the nonzero displacement constraint of $u_x = u_0 = 0.001\text{m}$ is applied.

12.3.1.2 Postprocessing

The displacement contour plots can be generated directly by using Ansys PLNS command. To generate the line plots of displacements along the centerline of the plate, input file ***PD_Post_Disp-DH.mac*** is used. Before using this macro, two sets of nodal components with names as post_nds and post_ndsy are created in the database as shown in Fig. 12.54. The displacements along X and Y are output to the files named as ***disp_crk_ux.out*** and ***disp_crk_uy.out***. The crack opening displacement is output to the file ***crkpd_open.dat.***

Figures 12.55 and 12.56 show the comparisons of OSB-DHPD displacement predictions in the plate with a crack under tension with those of FEM. Also, OSB-DHPD horizontal and vertical displacements are compared with those of FEM along the centerlines of the plate in Fig. 12.57. In addition, Fig. 12.58 shows the comparison of crack opening displacement predicted by OSB-DHPD model against the FEM solution.

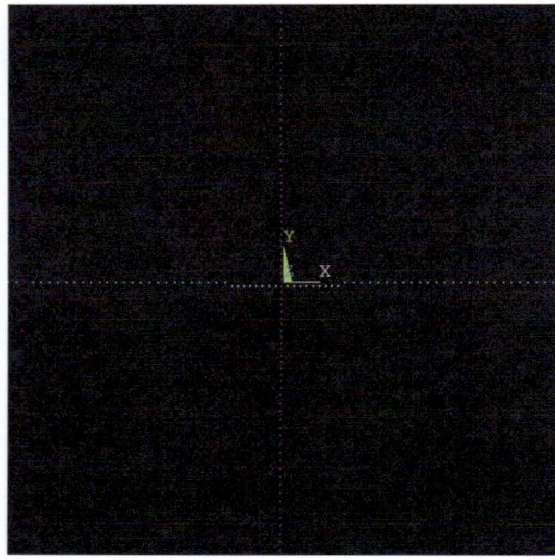

Fig. 12.54 Nodal components post_nds and post_ndsy created for viewing the displacement results along X and Y, respectively

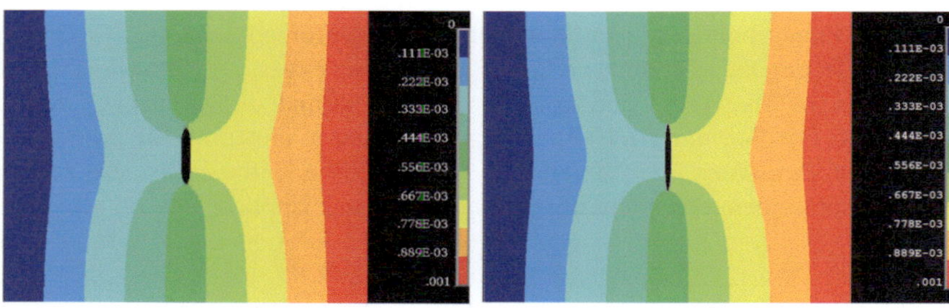

a) b)

Fig. 12.55 Comparison of horizontal displacement predictions in the plate with a crack under tension: (**a**) OSB-DHPD, and (**b**) FEM

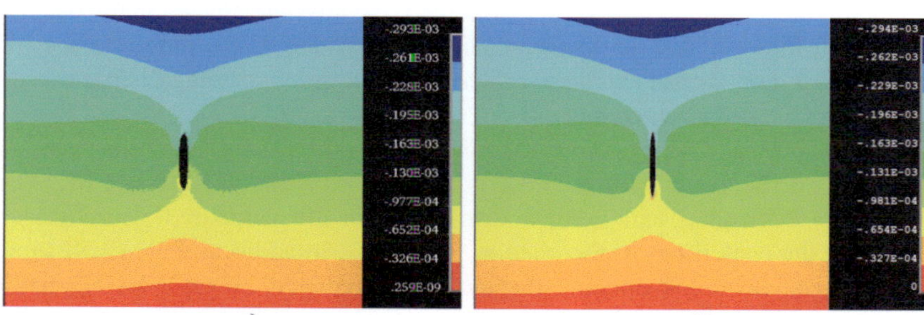

a) b)

Fig. 12.56 Comparison of vertical displacement predictions in the plate with a crack under tension: (**a**) OSB-DHPD, and (**b**) FEM

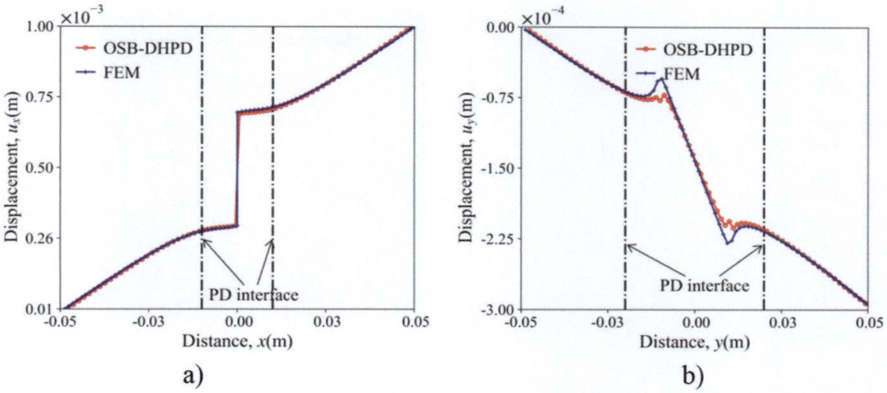

Fig. 12.57 Comparison of OSB-DHPD and FEM displacement predictions in the plate with a crack under tension: (**a**) $u_x(x, y = 0)$ and (**b**) $u_y(x = 0, y)$

Fig. 12.58 Comparison of the crack opening displacement, u_x, predicted by OSB-DHPD approach and FEM

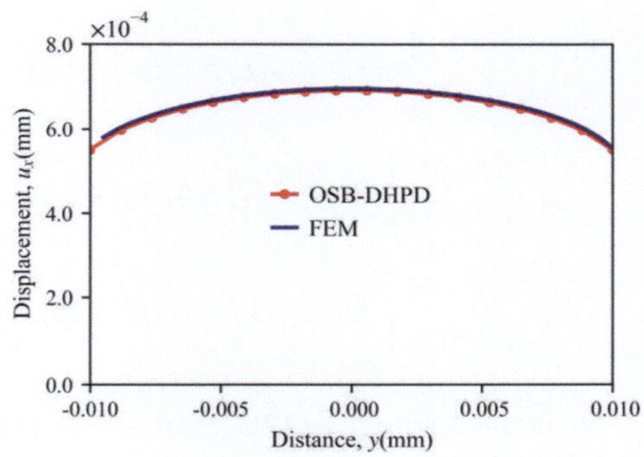

12.3.2 *Weak Form of NOSB-PD with Nonuniform Discretization*

The plate geometry, material property and boundary conditions are the same as that of Sect. 12.3.1 except for the loading on the right edge as shown in Fig. 12.59. It is subjected to a uniform traction of $\sigma_0 = 0.1$MPa along the right edge. The boundary conditions are imposed as

$$\left.\begin{array}{l} u_y(x,y)=0 \\ \sigma_{yy}^{\mathrm{PD}}(x,y)=0 \end{array}\right\} \text{ for } (-L/2+\Delta/2 \le x \le L/2-\Delta/2) \text{ and } y=(-W/2+\Delta/2) \tag{12.9}$$

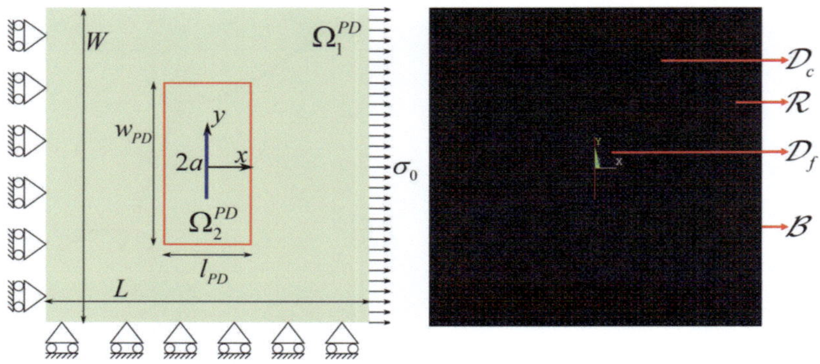

Fig. 12.59 Coarse and fine discretization regions in plate with a crack: (**a**) Geometry and boundary conditions, and (**b**) PD discretization

$$\left.\begin{array}{r} u_x(x,y)=0 \\ \sigma_{xy}^{\mathrm{PD}}(x,y)=0 \end{array}\right\} \text{for } x=(-L/2+\Delta/2) \text{ and } (-W/2+\Delta/2\le y\le W/2-\Delta/2) \quad (12.10)$$

$$\left.\begin{array}{r} \sigma_{xy}^{\mathrm{PD}}(x,y)=0 \\ \sigma_{yy}^{\mathrm{PD}}(x,y)=0 \end{array}\right\} \text{for } (-L/2+\Delta/2\le x\le L/2-\Delta/2) \text{ and } y=(W/2-\Delta/2) \quad (12.11)$$

$$\left.\begin{array}{r} \sigma_{xx}^{PD}(x,y)=\sigma_0 \\ \sigma_{xy}^{\mathrm{PD}}(x,y)=0 \end{array}\right\} \text{for } x=(L/2-\Delta/2) \text{ and } (-W/2+\Delta/2\le y\le W/2-\Delta/2) \quad (12.12)$$

Read the input file ***ex12.3.2.dat*** and repeat the similar steps as in Sect. 12.3.1.1 of the **Preprocessing** stage except for selecting the inner region of domain as NOSB-DHPD in **Step-3** and enforcing the boundary conditions in a weak form in **Step-5**. Figure 12.60 shows the selection of NOSB-DHPD interaction in PD region in **Step-3**.

In **Step-5**, to enforce the boundary conditions in a weak form, the submenu ***Region Identification >Boundary: Weak Form*** is selected to open the dialogue box ***Boundary: Weak Form.*** The parameters ***Boundary ID*** and ***Tang. vector to boundary: tx, ty*** are entered in the dialogue box as shown in Fig. 12.61.

Enter ***Boundary ID=1***, and its unit tangent vector**:** ***tx, ty: 1, 0.*** After entering the parameters and clicking the ***OK/Apply*** button, the node picker dialogue box is activated to select the desired boundary region as shown in Fig. 12.62. Repeat the process for each boundary of the plate by entering different ***Boundary ID*** and its unit tangent vector. Figure 12.63 shows the steps in creating the boundary region with ***Boundary ID=2*** and unit tangent vector, ***tx, ty:0, 1***.

Following similar steps as in preceding cases, various Ansys elements for modeling PD, PDDO and boundary regions are generated as shown in Fig. 12.64. The

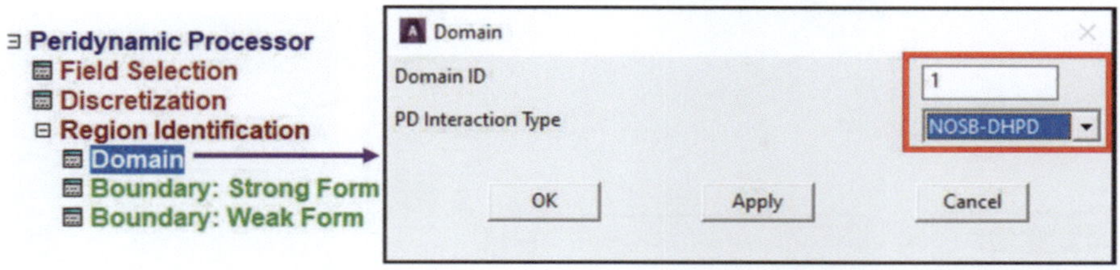

Fig. 12.60 Dialog box for selecting the domain interaction type as NOSB-DHPD

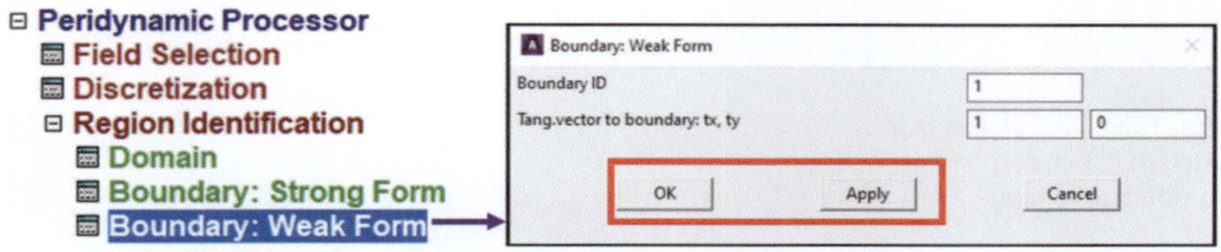

Fig. 12.61 Dialog box for selecting the boundary of the plate with weak form of interaction

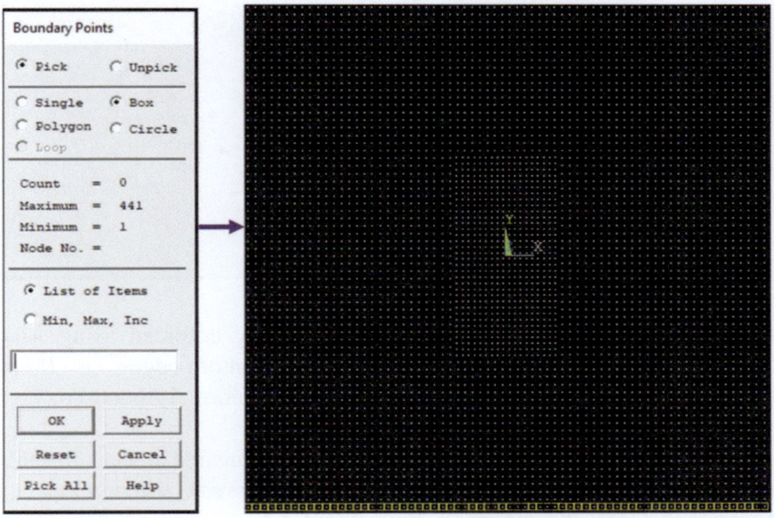

Fig. 12.62 Region identification corresponding to ***Boundary ID*=1** of the plate with weak form of interaction

boundary layer elements are shown in cyan color, the inner NOSB-DHPD region in pink color and the PDDO region resembling light green color.

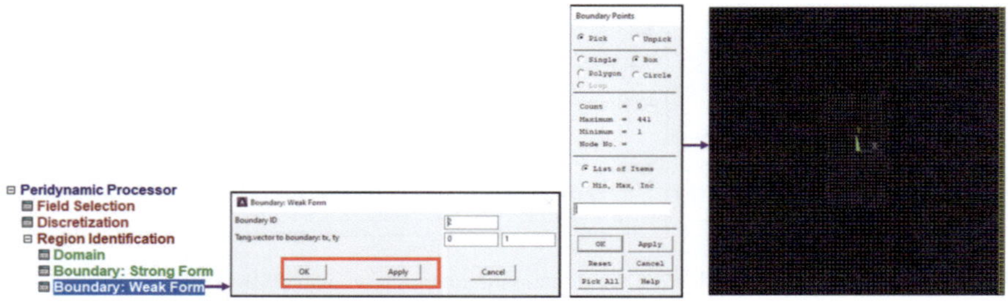

Fig. 12.63 Steps for creating the ***Boundary ID=2*** of the plate with weak form of interaction

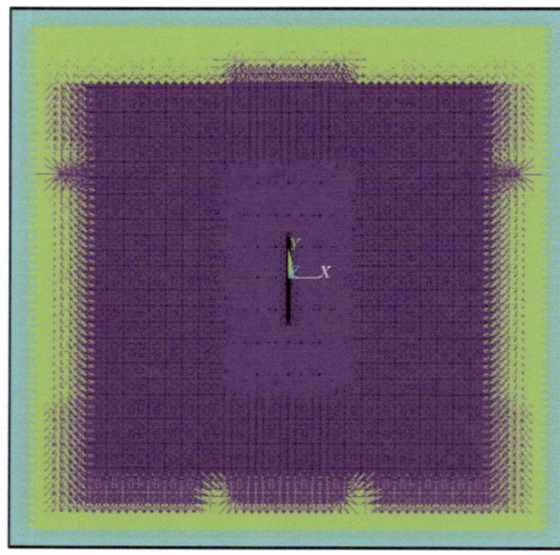

Fig. 12.64 PD elements in Ansys GUI corresponding to different regions

The displacement boundary conditions are directly enforced using Ansys D command by selecting the PD points on the left and bottom edges, i.e., ***Boundary ID=1, 4*** as shown in Fig. 12.65 and constraining the normal directions, i.e., UY, UX, respectively.

The zero traction boundary conditions are naturally satisfied in weak formulation. To apply nonzero traction, the macro file ***PD_Boundaryweak_traction.mac*** is used. The input parameters to this macro are ***Boundary ID***, ***key_tractionX***, ***key_tractionY***, ***tractionX,*** and ***tractionY,*** where ***key_tractionX***, ***key_tractionY*** denote the keys for enforcing the traction and ***tractionX***, ***tractionY*** denote the traction values to be enforced. The parameters specified as ***key_tractionX=-99/key_tractionY=-99*** enforce zero traction conditions in X and Y directions. To enforce nonzero traction, set the ***key_tractionX=1/key_tractionY=1*** and enter the nonzero value of the traction to be enforced. Run the below commands to enforce traction conditions on edges 2.

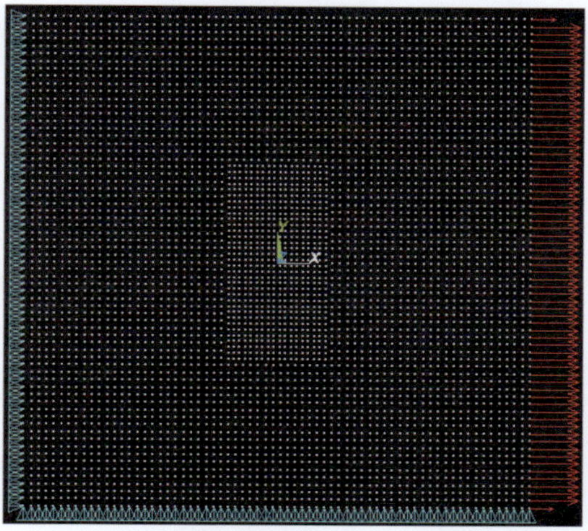

Fig. 12.65 Displacement BC directly enforced on the left and bottom PD nodes and the traction enforced on the right edge

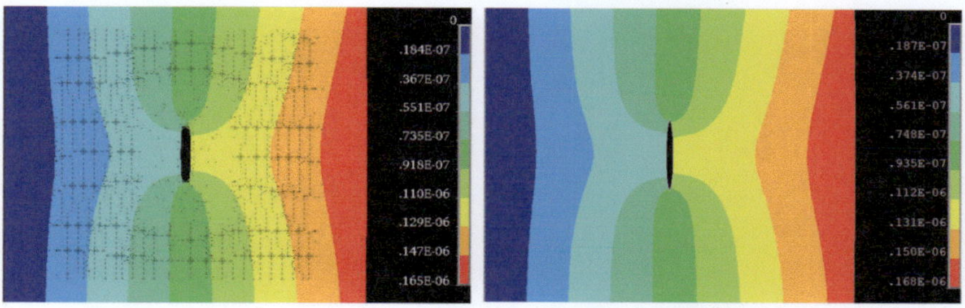

a) b)

Fig. 12.66 Comparison of horizontal displacement predictions in the plate with a crack under tension: (**a**) NOSB-DHPD and (**b**) FEM

PD_Boundaryweak_traction, 2, 1, -99, 1e+05, 0

After applying the boundary conditions and solution, the displacement contour plots can be generated directly by using Ansys PLNS command following the similar steps for postprocessing described in Sect. 12.2.1. Figures 12.66 and 12.67 show the comparisons of NOSB-DHPD horizontal and vertical displacement predictions in the plate with a crack under tension with those of FEM. Also, NOSB-DHPD horizontal and vertical displacements are compared with those of FEM along the centerlines of the plate in Fig. 12.68. The PD predictions successfully capture the displacement field around the crack tip and match the FE results. Figure 12.69 shows the comparison of crack opening displacement predicted by NOSB-DHPD model against the FEM solution.

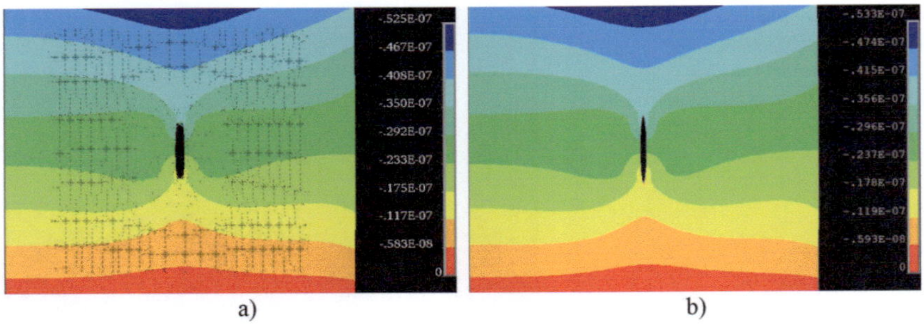

Fig. 12.67 Comparison of vertical displacement predictions in the plate with a crack under tension: (**a**) NOSB-DHPD and (**b**) FEM

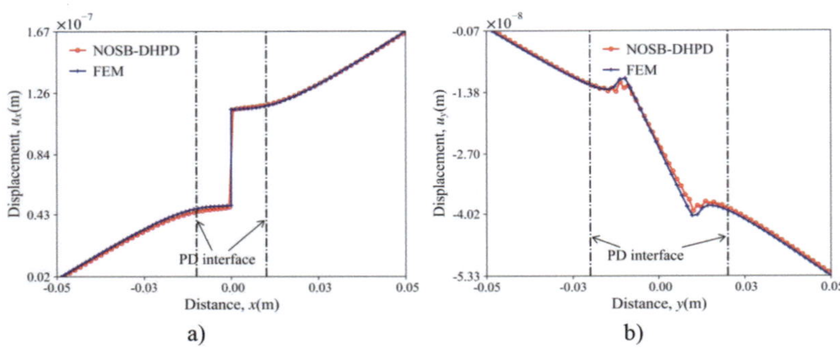

Fig. 12.68 Comparison of NOSB-DHPD and FEM displacement predictions in the plate with a crack under tension: (**a**) $u_x(x, y = 0)$ and (**b**) $u_y(x = 0, y)$

Fig. 12.69 Comparison of the crack opening displacement, u_x, predicted by NOSB-DHPD approach and FEM

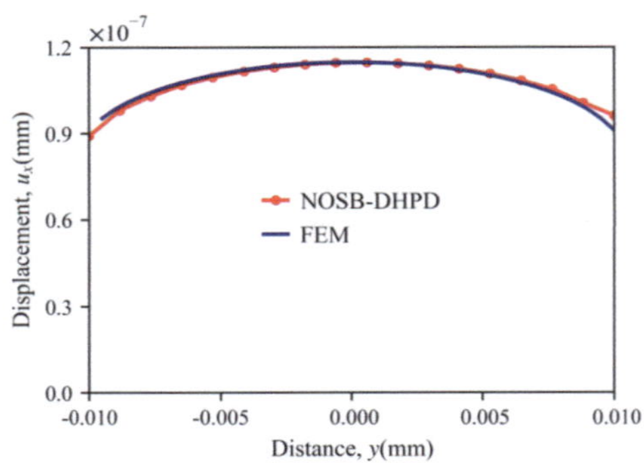

References

Anicode, S. V. K., & Madenci, E. (2022). Bond-and state-based peridynamic analysis in a commercial finite element framework with native elements. *Computer Methods in Applied Mechanics and Engineering, 398*, 115208.

Anicode, S. V. K., & Madenci, E. (2023). Direct coupling of dual-horizon peridynamics with finite elements for irregular discretization without an overlap zone. *Engineering with Computers, 40*, 1–31.

Chapter 13
Coupled Peridynamics and Finite Element Method for Static and Transient Structural Analysis

13.1 Introduction

This chapter presents the applications of coupled bond-based (BB), ordinary state-based (OSB), and non-ordinary state-based (NOSB) peridynamics (PD) and finite element method by considering an elastic plate with or without a crack under static and dynamic loading through uniform and nonuniform discretization as considered by Anicode and Madenci (2022, 2023). Also, this coupling approach is demonstrated by considering a plate with or without a crack subjected to initial thermal strain through uniform discretization. The first application demonstrates the PD/FE modeling with uniform discretization under traction. The second application demonstrates the PD/FE modeling with nonuniform discretization under displacement loading. The third application demonstrates the insertion of a crack in uniform discretization under displacement loading. Subsequent applications demonstrate the FE/PD modeling with uniform and nonuniform discretization, respectively, under transient compression. The last two applications demonstrate FE/PD modeling with uniform discretization in the presence of initial thermal strain with or without a crack.

The parameter a appearing in the Gaussian type weight function, Eq. (2.1), is specified as $a = 2$ for static loading resulting in homogeneous deformation and $a = 3$ for static loading resulting in nonhomogeneous deformation and transient loading.

Based on the approach described in Chap. 5, the solutions to all these applications are constructed by using Ansys Graphical User Interface (GUI) in a step-by-step fashion and compared with finite element analysis (FEA) solutions. The specific steps in the preprocessing, solution, and post-processing phases of the analysis are displayed through the GUI in a step-by-step fashion for each application. The PD operations of the preprocessing, solution, and post-processing for each application

Supplementary Information The online version contains supplementary material available at https://doi.org/10.1007/978-3-031-59896-8_13.

E. Madenci et al., *Peridynamics and Its Applications Using Ansys*,
https://doi.org/10.1007/978-3-031-59896-8_13

are provided in the form of Ansys input files. The PD operations are executed through the macro files. The input files containing the encrypted macro files to generate the simulations are provided as supplementary material.

13.2 Modeling in Ansys Framework

The applications in this chapter describe the specific steps in Ansys framework to perform quasi-static and transient structural simulations by using the coupled PD/FE approach as described in Chap. 5. The customized built-in menu ***Peridynamic Processor*** is used for model generation. The procedure for entering the parameters for each of the submenus located under this main menu is explained in Chap. 9.

For structural analysis, MATRIX27 element is used for modeling the unsymmetric stiffness matrices in the PD and PDDO domain along with the PD boundary region. The unsymmetric stiffness coefficients corresponding to PD and PDDO regions as in Eqs. (3.165 and 3.166), (3.178–3.181), (3.198–3.201), and (3.218–3.220) and along the boundary region as in Eqs. (4.110–4.115) are input as real constants to MATRIX27 element. The UX and UY degrees of freedom represent the components of the displacement field, $\mathbf{u}(\mathbf{x}, t)$ in Eq. (5.7). For structural transient analysis, MASS21 elements are overlaid on the nodes in PD and PDDO regions. The mass matrix with LUMP,ON command is activated in the FE region.

The supported KEYOPTS command and the procedure to input real constants applicable to these elements are described in Chap. 8. Table 13.1 shows the summary of MATRIX27 and MASS21 elements and the KEYOPTS activated for structural simulations.

13.3 Static Response of a Plate Under Traction

13.3.1 Model Setup

Figure 13.1 shows the plate geometry with a Cartesian coordinate system located at its center and the boundary conditions. It has square dimensions with an edge length of $W = L = 1$ m. The left and bottom edges of the plate are on roller supports. Its top edge is free of tractions, and the right edge is subjected to a uniform tension of

Table 13.1 Element used for the quasi-static and transient structural simulation

Modeling variable	Element	DOF	KEYOPTS
PD stiffness	MATRIX27	UX,UY	KEYOPT(2)=2 KEYOPT(3)=4
MASS matrix	MASS21	UX,UY	KEYOPT(1)=0 KEYOPT(3)=4

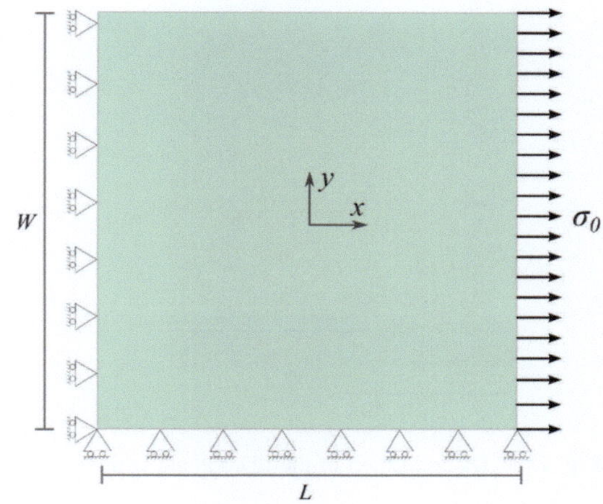

Fig. 13.1 Description of plate geometry, boundary, and loading conditions

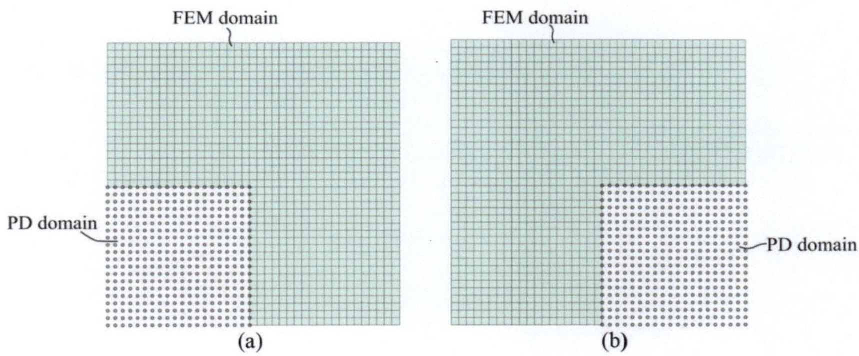

Fig. 13.2 Discretization of the plate with PD and FE regions: (**a**) Model I and (**b**) Model II

$\sigma_0 = 1.0$ MPa. It is isotropic with Young's modulus and Poisson's ratio of $E = 70$ GPa and $\nu = 0.25$, respectively. The PD and the FE regions are discretized with a uniform mesh size of $\Delta x = 0.02$ m. In this simulation, the static response of the plate is modeled by coupled OSB-PD/FE and NOSB-PD/FE, respectively.

In order to consider different combinations of boundary and applied loading conditions on the PD and FE domains, the plate is analyzed by two different models, as shown in Fig. 13.2. In the first model, Model I, shown in Fig. 13.2a, the displacement boundary conditions are enforced on PD and FE domains and loads are applied on the FE domain. In the second model, Model II, shown in Fig. 13.2b, the displacement boundary conditions are enforced on the FE domain and applied loads on the PD and FE domains. The FE domain has 1875 rectangular elements with 1976 nodes. The PD regions of $\mathcal{D} \cup \mathcal{R}$ and $\mathcal{B}$ consist of 576 and 100 points, respectively, as shown in Fig. 13.3.

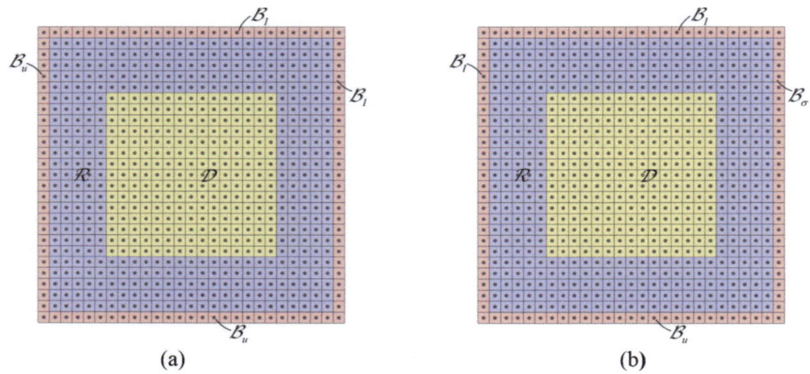

Fig. 13.3 Discretization of the PD domain with $\mathcal{D} \cup \mathcal{R}$ and $\mathcal{B}$: (**a**) Model I and (**b**) Model II

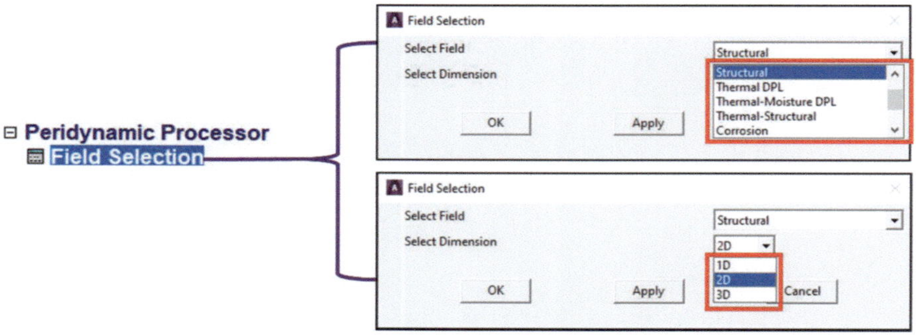

Fig. 13.4 Dialog box for selecting the field and dimension

13.3.2 Preprocessing

13.3.2.1 Preprocessing for Model I

For Model I, the plate under tension described in the preceding section is modeled by using OSB-PD. The finite element discretization of the plate is generated by using the Ansys input file: ***ex13.3.2_model1.dat*** or imported into the Ansys database using the CDREAD command. Copy the lines from 1 to 33 in the above input file and paste in the command window. This generates the finite element model for creating the PD domain.

Step-1 The field and dimension of the analysis are specified using the ***Field Selection*** submenu located under ***Peridynamic Processor***. For simulation of the structural response, the "***Structural***" for field and "***2D***" for dimension are selected as shown in Fig. 13.4.

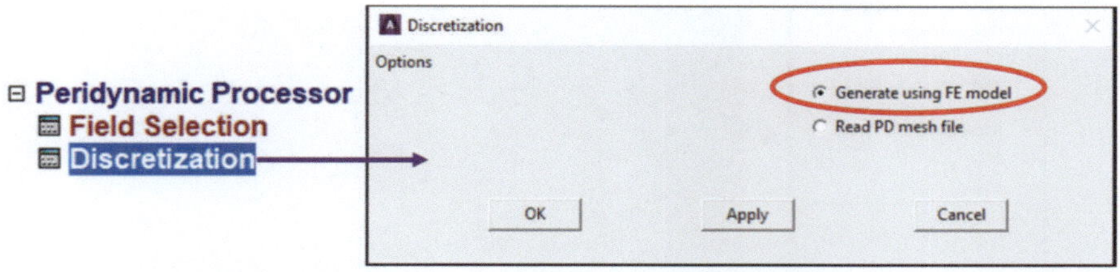

Fig. 13.5 Dialog box for PD mesh generation

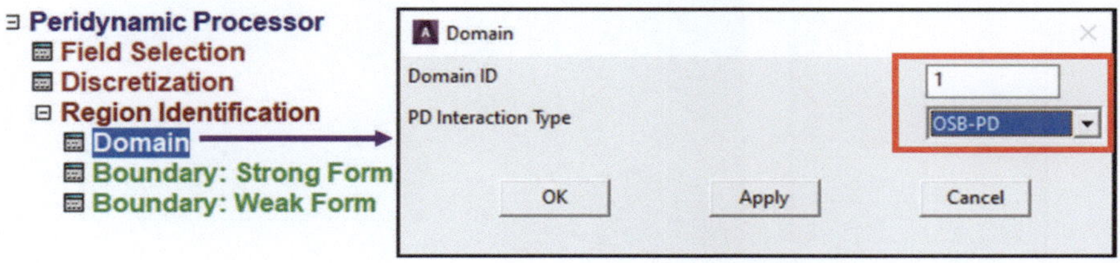

Fig. 13.6 Dialog box for selecting the domain interaction type as OSB-PD

Step-2 The PD mesh is generated using the radio button option ***Generate using FE model*** in ***Discretization*** submenu as shown in Fig. 13.5:

Step-3 The PD regions for the structural equations and boundary conditions are created using ***Region Identification*** submenu shown in Fig. 13.6. The submenu ***Region Identification>Domain*** is selected to open the dialog box ***Domain***. In this application, two regions corresponding to OSB-PD and PDDO regions are created. For the coupled OSB-PD/FE simulation, the parameter, ***Domain ID=1*** is entered in the text box and ***PD Interaction Type*** is selected from the drop-down menu as OSB-PD as shown in Fig. 13.6.

After entering the parameters and clicking the ***OK/Apply*** button, the node picker dialog box is activated to select the desired region. After selecting the nodes by using the Node Picker window and clicking the ***OK*** or ***Apply*** button of the node picker (Box), a dialog box labeled as ***Domain Points*** appears to confirm the selected OSB-PD regions as shown in Fig. 13.7.

Setp-4 Repeat the process for creating the PDDO region of the plate using a different ***Domain ID*** and setting the ***PD Interaction Type*** to PDDO as shown in Fig. 13.8. The selected region for PDDO using node picker window is shown in Fig. 13.9. Clicking the ***OK*** or ***Apply*** button of the node picker (Box), a dialog box labeled as ***Domain Points*** appears to confirm the selected PDDO region.

Setp-5 To enforce the boundary conditions in a coupled PD/FE region, the submenu ***Region Identification>Boundary: Weak Form*** is selected to open the

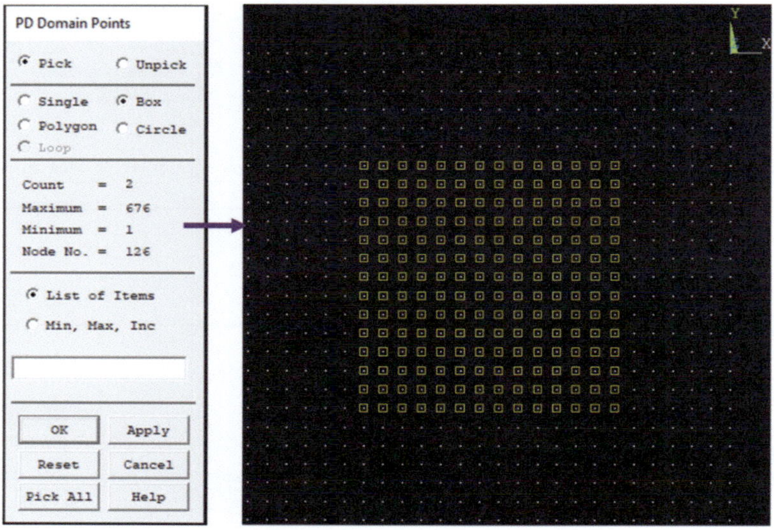

Fig. 13.7 Region identification corresponding to OSB-PD domain of the plate

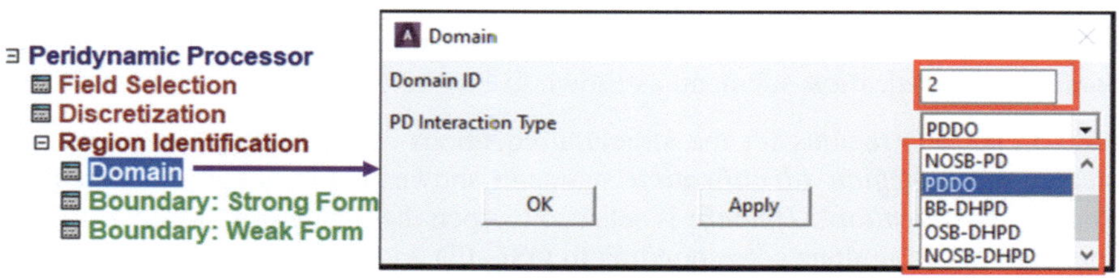

Fig. 13.8 Dialog box for selecting the domain interaction types as PDDO

dialog box ***Boundary: Weak Form.*** The parameters ***Boundary ID*** and ***Tang. vector to boundary: tx, ty*** are entered in the dialog box as shown in Fig. 13.10.

Enter ***Boundary ID=1***, and its unit tangent vector: ***tx,ty: 1,0***. After entering the parameters and clicking the ***OK/Apply*** button, the node picker dialog box appears to select the desired boundary region as shown in Fig. 13.11. Repeat the process for each boundary of the plate by entering different ***Boundary ID*** and its unit tangent vector. Fig. 13.12 shows the steps in creating the boundary region with ***Boundary ID=2*** and unit tangent vector, ***tx,ty:0,1***.

Step-6 The family members of the PD points are generated using ***Family Generation*** menu based on the horizon size factor ***m*** in the ***Family Generation*** dialog box as shown in Fig. 13.13. For this application, the horizon size factor $m = 3$ is entered. After entering the parameter in the text box and clicking the ***OK/Apply*** button, the family members are generated using the horizon size of $\delta = 3.015\Delta x$.

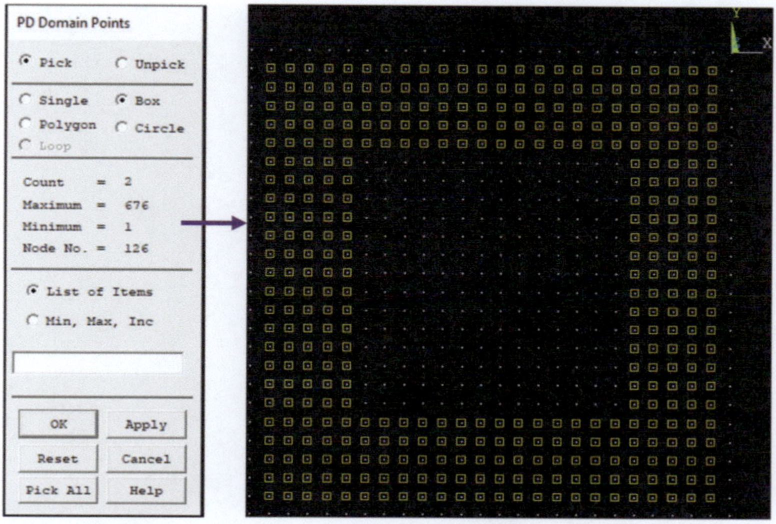

Fig. 13.9 Region identification corresponding to PDDO domain of the plate

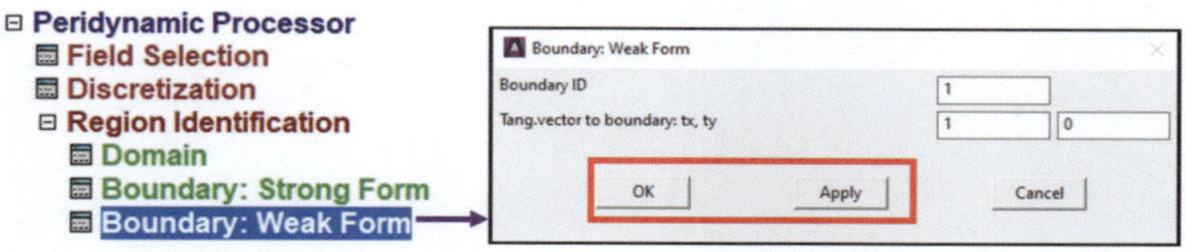

Fig. 13.10 Dialog box for selecting the boundary of the plate with weak form of interaction

Step-7 The element coefficients corresponding to the stiffness for the PD structural analysis are generated using ***Element Coefficients (Structural)*** submenu option as shown in Fig. 13.14. Clicking the ***OK/Apply*** button of the ***Element Coefficients (Structural)*** dialog box generates the stiffness coefficients for the PD interaction types. In this application, the stiffness coefficients corresponding to OSB-PD and PDDO types are generated.

Step-8 The elements required for modeling the PD interactions due to the structural analysis inside the domain are created by using the dialog box ***Domain Elements***, which appears upon selecting the ***Element Generation>Domain Elements*** submenu. Clicking the ***OK/Apply*** button of the dialog box as shown in Fig. 13.15 generates the domain elements for the OSB-PD and PDDO interaction types.

For generating the boundary interactions, the dialog box ***Boundary Elements***, which appears upon selecting the ***Element Generation>Boundary Elements***

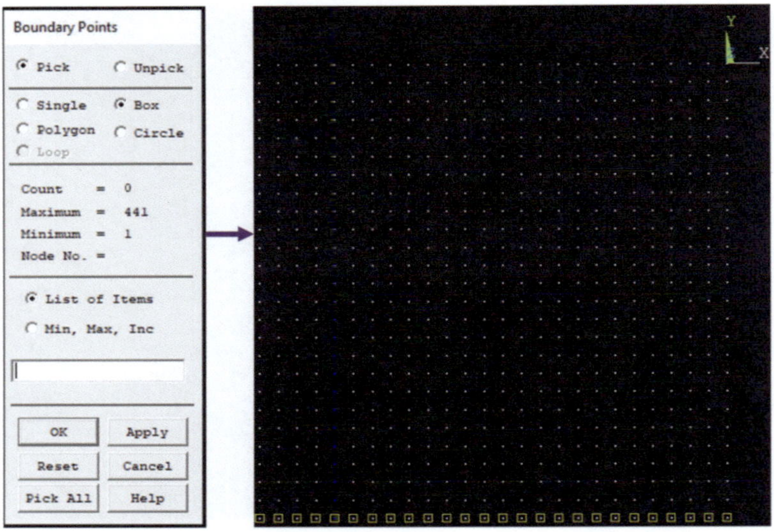

Fig. 13.11 Region identification corresponding to ***Boundary ID*=1** of the plate with weak form of interaction for coupling PD with FE

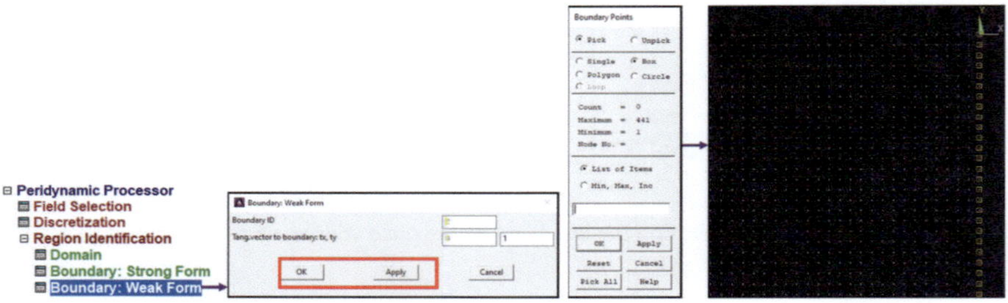

Fig. 13.12 Steps for creating the ***Boundary ID*=2** of the plate with weak form of interaction for coupling PD with FE

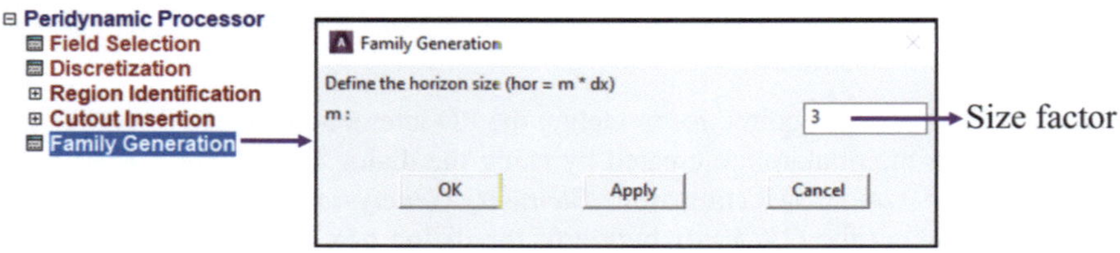

Fig. 13.13 Dialog box for defining the horizon size in the family generation

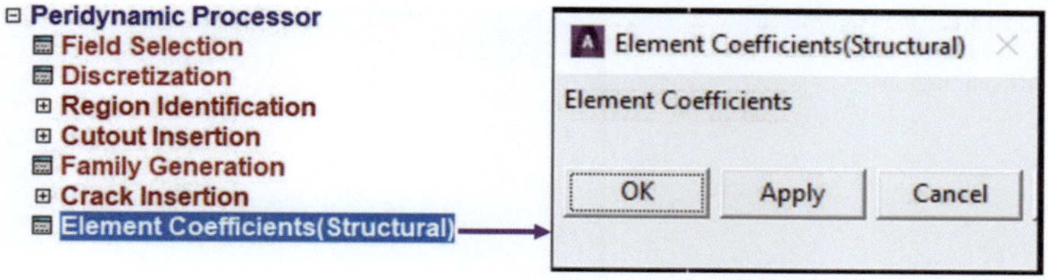

Fig. 13.14 Element coefficients submenu for structural

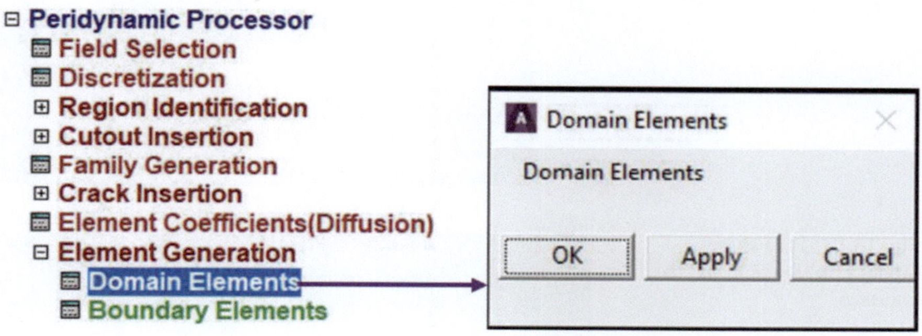

Fig. 13.15 Dialog box for domain element generation

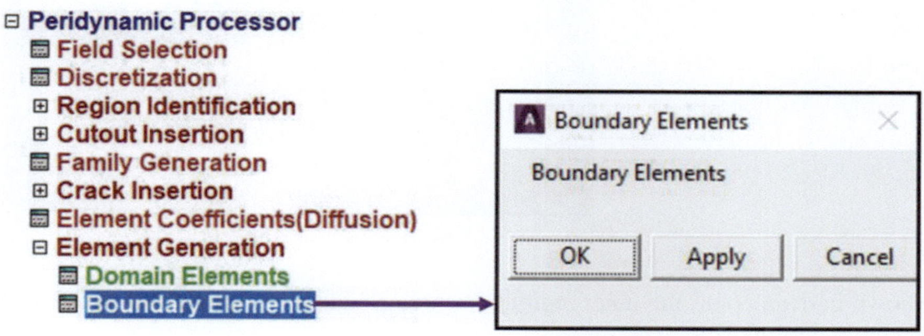

Fig. 13.16 Dialog box for boundary element generation

submenu is used. Clicking the ***OK/Apply*** button of the dialog box as shown in Fig. 13.16 generates the elements for the selected PD boundary interaction type.

Copy the lines from 39 to 56 in the input file ***ex13.3.2_model1.dat*** to create FE region surrounding the PD region. Fig. 13.17 shows the various Ansys elements used to model domain and boundary regions. The boundary layer elements are

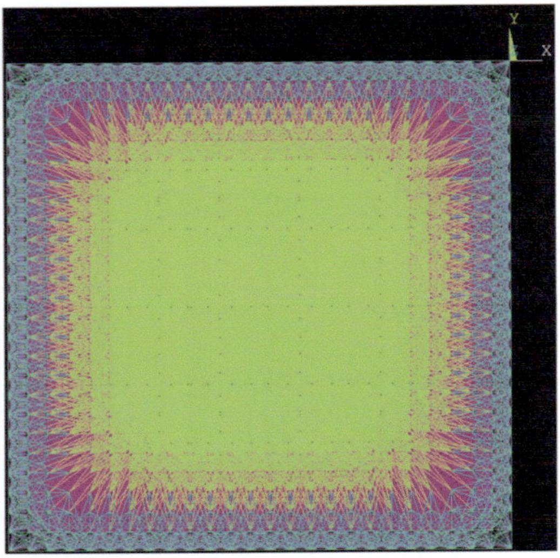

Fig. 13.17 PD elements in Ansys corresponding to different regions

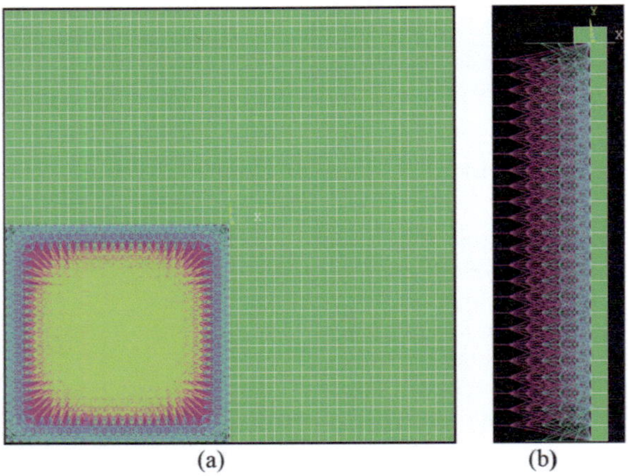

Fig. 13.18 PD/FE for Model-I: (**a**) Domain and (**b**) Interface

shown in cyan color, the inner region resembling light green color, and the PDDO region with pink color.

Step-9 The FE model is built around the PD domain such that the nodes of PD and FE region coincide at the interface, i.e., both the FE and PD regions share the same node number and location. The FE nodes near the interface can be made to coincide with the PD nodes by issuing the Ansys command NUMMRG, NODES.

Figure 13.18a shows the Ansys PLANE182 elements (shown in green) near the PD region sharing the ***Boundary ID=2,3***. Fig. 13.18b shows the elements along the interface of PD and FE regions.

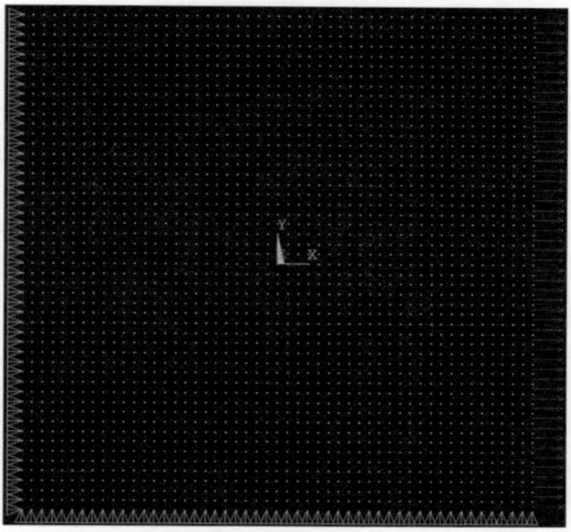

Fig. 13.19 Boundary condition set up for the coupled PD/FE model with displacement constraints on the left and bottom edges and traction on the right edge

Step-10 The traction boundary conditions are naturally satisfied without any requirement for direct enforcement. The displacement boundary conditions are directly enforced using Ansys D command by selecting the nodes on the left and bottom edges as shown in Fig. 13.19. The nodal DOF, UX and UY are constrained along the left and bottom edges, respectively. On the right edge of the plate consisting of FE nodes, the traction value of $\sigma_0 = 1.0\text{MPa}$ is applied in the X direction using the native Ansys command SF. Copy the lines from 58 to 71 in the input file ***ex13.3.2_model1.dat*** to apply the traction load on FE nodes.

13.3.2.2 Preprocessing for Model II

The plate under tension described in the preceding section is modeled as Model II by using NOSB-PD. The finite element discretization of the plate is generated by using the Ansys input file: ***ex13.3.2_model2.dat*** or imported into the Ansys database using the CDREAD command. This generates the finite element model for creating the Model II PD domain. Repeat the steps as given in Sect. 13.3.2.1, except for the region identification in **Step-3** where the domain interaction type is now set to NOSB-PD as shown in Fig. 13.20.

Copy the lines from 39 to 56 in the input file ***ex13.3.2_model2.dat*** to create FE region surrounding the PD region. After **Step-8**, various Ansys elements to model domain and boundary regions are generated as shown in Fig. 13.21. The boundary layer elements are shown in cyan color, the inner NOSB-PD region resembling light blue color and the PDDO region with pink color.

After **Step-9**, the Ansys PLANE182 elements (shown in green) near the PD region sharing the ***Boundary ID=3,4*** are shown in Fig. 13.22a. Figure 13.22b shows the elements along the interface of PD and FE regions.

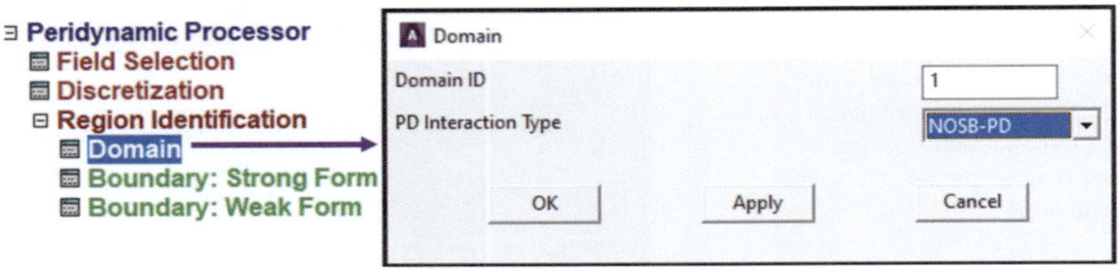

Fig. 13.20 Dialog box for selecting the domain interaction type as NOSB-PD

Fig. 13.21 PD elements in Ansys corresponding to different regions

(a) (b)

Fig. 13.22 PD and FE elements for Model II: (**a**) Domain and (**b**) Interface

The displacement boundary conditions are directly enforced using Ansys D command by selecting the nodes on the left and bottom edges, as shown in Fig. 13.19. The nodal DOF, UX and UY are constrained along the left and bottom edges, respectively. On the right edge of the plate consisting of PD and FE nodes, the traction value of $\sigma_0 = 1.0$MPa is applied in the X direction using the native Ansys SF command for FE nodes.

For the PD nodes, the traction boundary conditions are directly enforced on the PD boundary region by using the ***PD_Boundaryweak_traction.mac*** macro. The input parameters to this macro are: ***Boundary ID***, ***key_tractionX***, ***key_tractionY***, ***tractionX***, and ***tractionY***. The parameters ***key_tractionX*** and ***key_tractionY*** denote the keys for enforcing the traction conditions, and ***tractionX*** and ***tractionY*** denote the traction values to be enforced. The key parameters with ***key_tractionX=-1/ key_tractionY=-1*** do not enforce any traction condition either in X or Y direction, as they are automatically satisfied in weak form.

To enforce non-zero traction, these parameters are set as *key_tractionX=1/ key_tractionY=1*. along with the value of the traction to be enforced. Run the command given below to enforce the traction value $\sigma_0 = 1.0$ MPa on the right edge, i.e., ***Boundary ID*=2** of the PD domain:

PD_Boundaryweak_traction, 2,1,-1,1e+06,1.

Copy the lines from 58 to 73 in the input file ***ex13.3.2_model2.dat*** to apply the traction load in both FE and PD nodes.

13.3.3 Post-processing

The displacement contour plots can be generated directly by using Ansys PLNS command. To generate the line plots of displacements along the center line of the plate, first set the parameter ***_Model*=1 or 2** and run the macro file ***PD-FE_Post_Disp.mac***. The displacements along X and Y are output to the files named as ***disp_ux.out*** and ***disp_uy.out*** corresponding to each Model I or II.

Figures 13.23 and 13.24 show the displacement variations u_x and u_y along the horizontal and vertical directions from the coupled PD/FE analysis for OSB and NOSB interactions. Corresponding to each model, the results are almost identical. Also, they agree well with those of FE analysis as shown in Fig. 13.25. The black regions at the center of the NOSB-PD region and near the FE-PD interface for the Fig. 13.24 correspond to the background color and visible because of the coarsely discretized MATRIX27 elements. This plotting artifact disappears as the discretization becomes fine.

Figure 13.26 shows the comparison of PD/FE displacement predictions for OSB and NOSB interactions with those of FEM along $y = -0.25$m in both model I and II, and $x = -0.25$ m in model I, and $x = 0.25$ m in model II. The PD/FE predictions capture the FE results and do not present any kinks near the interface.

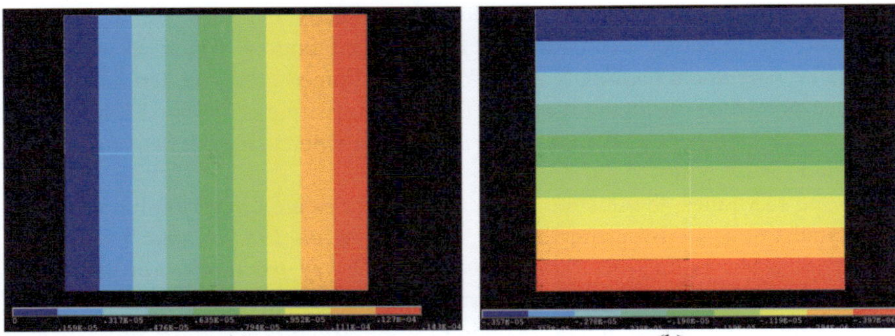

(a) (b)

Fig. 13.23 Coupled OSB-PD/FE displacement predictions for model I: (**a**) u_x and (**b**) u_y

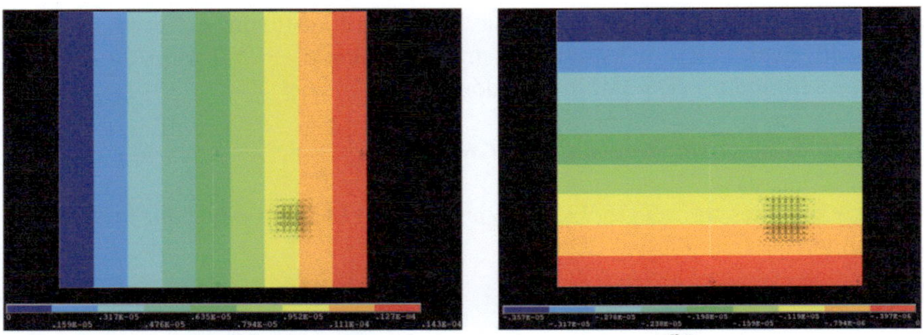

(a) (b)

Fig. 13.24 Coupled NOSB-PD/FE displacement predictions for model II: (**a**) u_x and (**b**) u_y

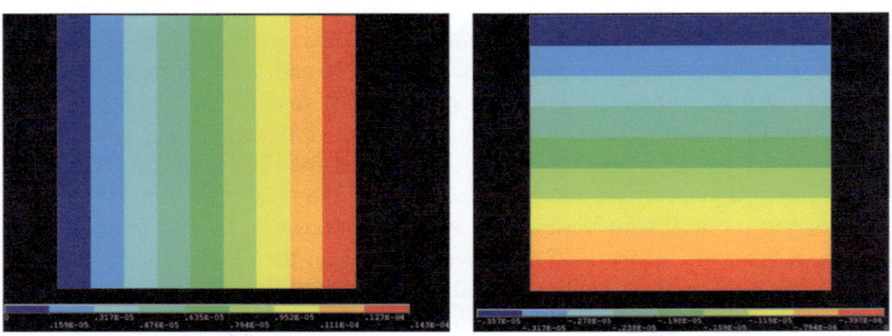

(a) (b)

Fig. 13.25 FE displacement predictions: (**a**) u_x and (**b**) u_y

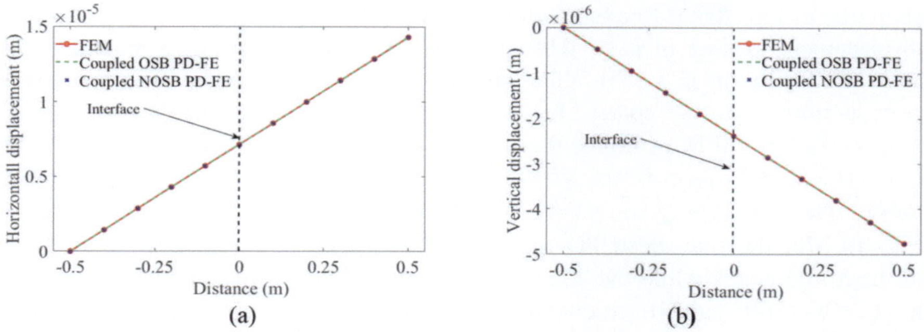

Fig. 13.26 Comparison of Model I (OSB PD-FE) and Model II (NOSB PD-FE) displacement predictions with those of FEM along particular lines: (**a**) $u_x(x, y = -0.25\text{m})$ and (**b**) $u_y(x = \pm 0.25 \text{ m}, y)$

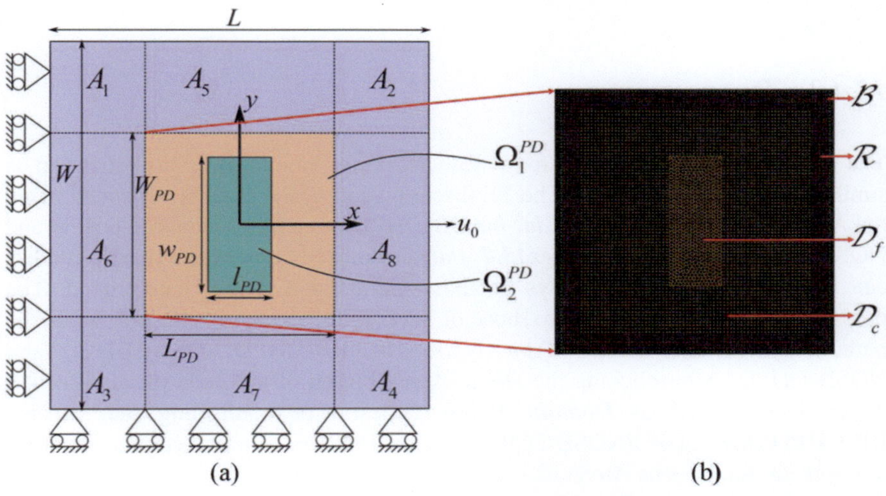

Fig. 13.27 Coupled DHPD/FE plate model: (**a**) geometry and boundary conditions and (**b**) nonuniform discretization in PD region

13.4 Static Response of a Plate Under Displacement Loading

13.4.1 Model Setup

Figure 13.27a shows the plate geometry with a Cartesian coordinate system located at its center and the boundary conditions. It has square dimensions with an edge length of $L = W = 1.5\text{m}$. The left and bottom edges of the plate are on roller supports. Its top edge is free of tractions, and the right edge is subjected to a uniform displacement loading of $u_0 = 0.001$ m. It is isotropic with Young's modulus and Poisson's ratio of $E = 70$ GPa and $\nu = 0.33$, respectively. The geometry corresponding to the coarse PD region, Ω_1^{PD} has a length and width of $L_{PD} = W_{PD} = 1.0$ m while for the fine PD region Ω_2^{PD}, the length and width are $l_{PD} = 0.2$ m and $w_{PD} = 0.8$ m. The PD regions Ω_1^{PD} and Ω_2^{PD} are discretized with a mesh size of $\Delta x = \sqrt{A_{(k)}} = 0.0125$ m and 0.01 m, respectively. As shown in Fig. 13.27b, the fine mesh PD region, Ω_2^{PD} consisting of region $\mathcal{D}_f$ has 1430 unstructured nodes while the coarse PD region Ω_1^{PD} consisting of $\mathcal{D}_c \cup \mathcal{R}$ and $\mathcal{B}$ has (3984+1460) and 316 structured nodes, respectively. The FE region as shown in Fig. 13.27a, consisting of areas $A_1 \ldots A_8$ is meshed with varying element sizes. The areas, $A_1 \ldots A_4$ are meshed with an element size of 0.05125 m × 0.05125 m resulting in 4 × 25 elements, while the areas, $A_5 \ldots A_8$ are meshed with an element size of 0.0125 m × 0.05125 m resulting in 4 × 316 elements. The solution is constructed by coupled BB-DHPD/FE, OSB-DHPD/FE and NOSB-DHPD/FE.

13.4.2 Preprocessing

The modeling procedure in Ansys framework for nonuniform discretization is similar to that of Sect. 13.3.2. The FE discretization of the plate is generated using the Ansys input file: ***ex13.4_bb.dat for*** BB-DHPD model. For the OSB-DHPD and NOSB-DHPD Ansys input file: ***ex13.4_state.dat*** is used or the respective FE models can be imported into the Ansys database using the CDREAD command. The preprocessing steps are similar to those of Sect. 13.3.2 except for **Step-3** where the parameter ***PD Interaction Type*** is set to BB-DHPD, OSB-DHPD, and NOSB-DHPD. After clicking the ***OK*** or ***Apply*** button of the node picker (Box), a dialog box labeled as ***Domain Points*** appears for confirming the selected BB-DHPD/OSB-DHPD/NOSB-DHPD regions. Following similar steps as in previous application, various Ansys elements to model domain and boundary regions are generated as shown in Fig. 13.28. The boundary layer element is shown in cyan color, the inner region resembling pink color, and the PDDO region with light green color.

The FE model is built around the PD domain such that the nodes of PD and FE regions coincide at the interface, i.e., both the FE and PD regions share the same node number and location. The FE nodes near the interface can be made to coincide with the PD nodes by using the Ansys command NUMMRG,NODES. Copy the lines from 65 to 104 in the above input file and paste in the command window to generate the FE model surrounding the PD region. Fig. 13.29a shows the Ansys PLANE182 elements (shown in orange) around the PD region sharing the ***Boundary ID=1,2,3 and 4***. Fig. 13.29b shows the elements along the interface of PD and FE regions for a ***Boundary ID=2***.

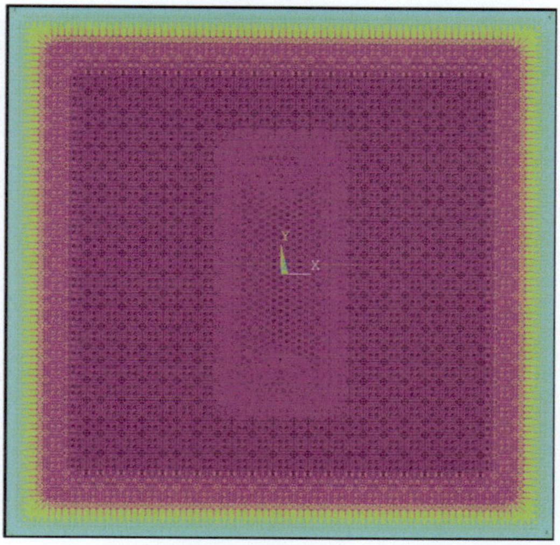

Fig. 13.28 PD elements in Ansys corresponding to different regions

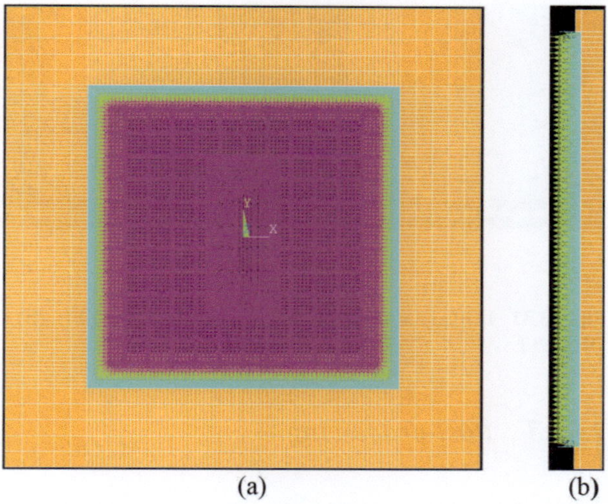

Fig. 13.29 PD-FE elements. (**a**) Domain and (**b**) Interface PD-FE

Copy lines 106 to 120 from either of the above input files and paste them in the command window to apply the displacement constraints and solve for the coupled PD-FE model. The displacement boundary conditions are directly enforced using Ansys D command by selecting the nodes on the left and bottom edges, as shown in Fig. 13.30. The nodal DOF, UX and UY are constrained along the left and bottom edges, respectively. On the right edge of the plate, a non-zero displacement value of $u_0 = 0.001\text{m}$ is applied in X direction.

Fig. 13.30 Boundary condition set up for the coupled DHPD/FE model with displacement constraints on the left bottom edges and right edge

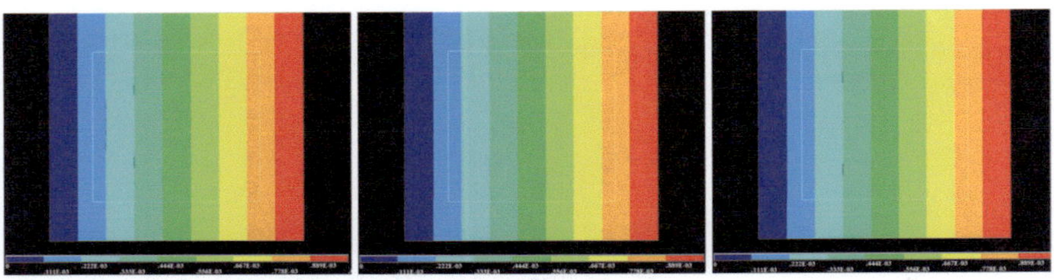

a) b) c)

Fig. 13.31 Horizontal displacement, u_x: (**a**) BB-DHPD/FE, (**b**) OSB-DHPD/FE, and (**c**) NOSB-DHPD/FE

13.4.3 Post-processing

The displacement contour plots can be generated directly by using the Ansys PLNS command. To generate the line plots of the displacements along the X and Y center lines, they are output to the files named ***disp_ux_dh.out*** and ***disp_uy_dh.out*** by running the macro file ***post_dual_hori.mac***. Figures 13.31 and 13.32 show the horizontal and vertical displacement variations, respectively, from coupled BB-DHPD/FE, OSB-DHPD/FE, and NOSB-DHPD/FE models. As expected, the displacement field is smooth and linearly varying. Figures 13.33 and 13.34 show the comparison of DHPD/FE displacement predictions along the center lines $y = 0$ and $x = 0$, with those of FEM. The coupled DHPD/FE predictions near the coarse-fine PD regions do not present any oscillations due to different horizon sizes. Also, the predictions capture the FE results without any kinks near the DHPD/FE interface.

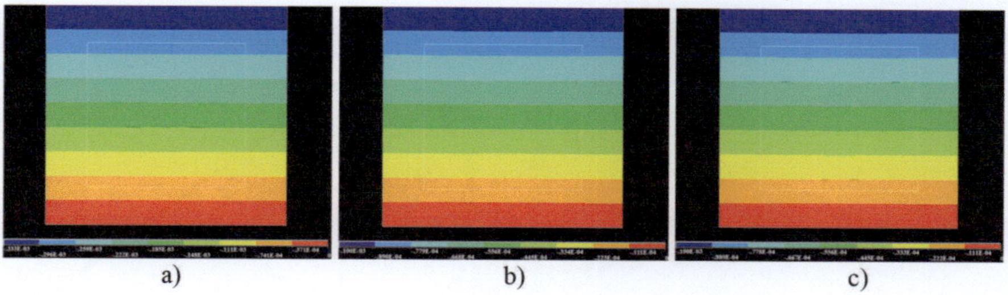

Fig. 13.32 Vertical displacement, u_y: (**a**) BB-DHPD/FE, (**b**) OSB-DHPD/FE, and (**c**) NOSB-DHPD/FE

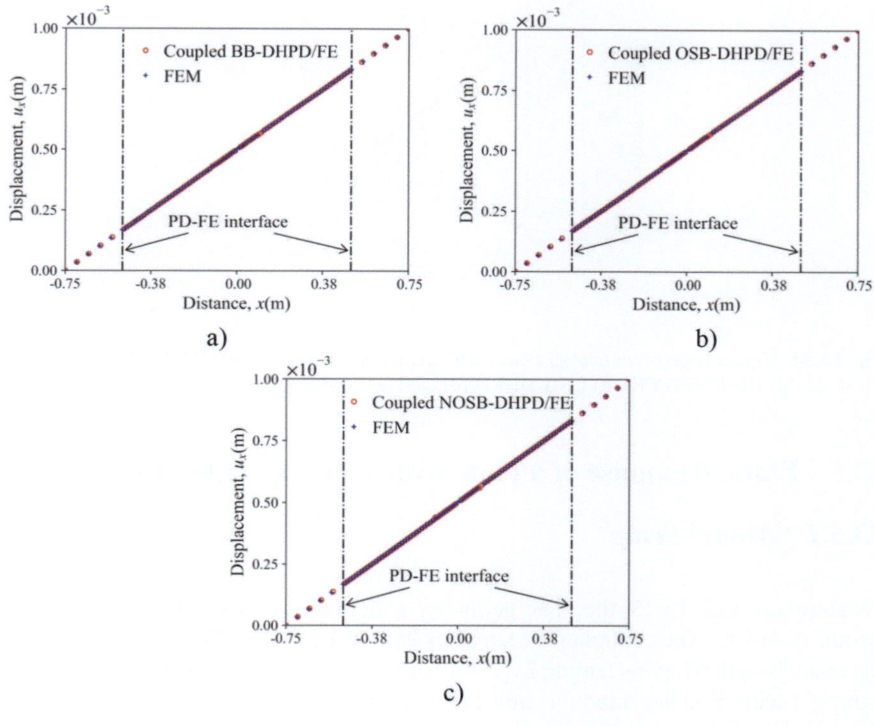

Fig. 13.33 Comparison of horizontal displacement predictions from coupled DHPD-FE with those of FEM, u_x: (**a**) BB-DHPD/FE, (**b**) OSB-DHPD/FE, and (**c**) NOSB-DHPD/FE

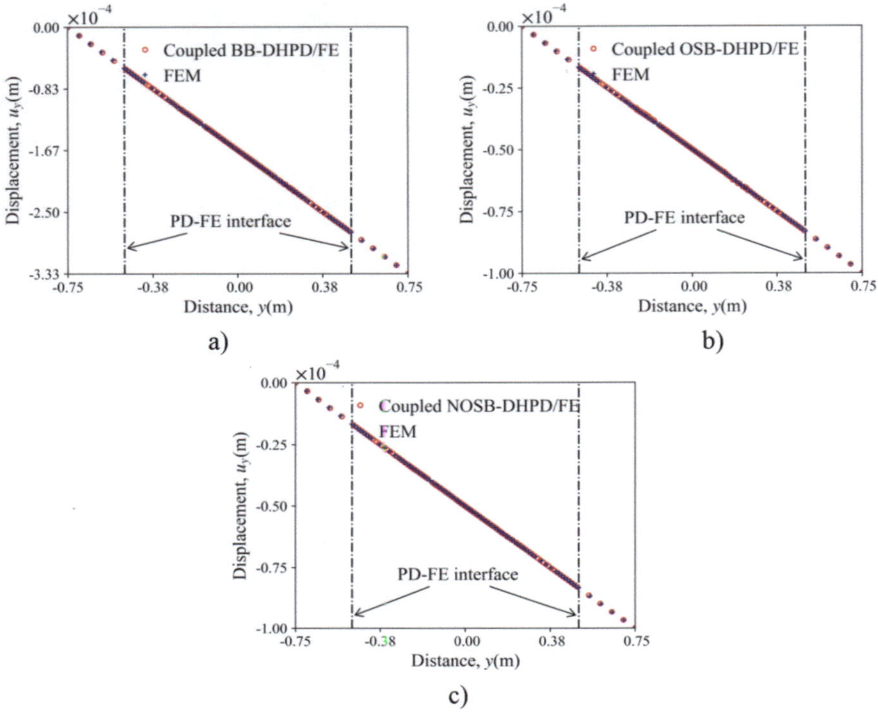

Fig. 13.34 Comparison of vertical displacement predictions from coupled DHPD-FE with those of FEM, u_y: (**a**) BB-DHPD/FE, (**b**) OSB-DHPD/FE, and (**c**) NOSB-DHPD/FE

13.5 Static Response of a Plate with a Crack Under Tension

13.5.1 Model Setup

As shown in Fig. 13.35, the plate geometry is defined by its length, $L = 1$ m, and width, $W = 1$ m. The central crack length is $2a = 0.1$ m. The PD domain containing the crack is defined by its length, $L_{PD} = 0.5$ m, and width, $W_{PD} = 0.5$ m. The plate is isotropic with Young's modulus and Poisson's ratio of $E = 50$MPa and $\nu = 0.25$, respectively. It is subjected to displacement loading of $u_0 = 0.001$ m along the right edge. Figure 13.36 shows the discretization of the PD domain. It has a uniform grid spacing of $\Delta x = 0.005$ m resulting in 9604 PD points inside $\mathcal{D} \cup \mathcal{R}$ and 396 boundary points in the region $\mathcal{B}$ along the perimeter of the PD domain. The FE domain outside the PD domain is discretized with the same grid spacing as that of the PD domain resulting in 30600 elements (PLANE182) and 31200 nodes.

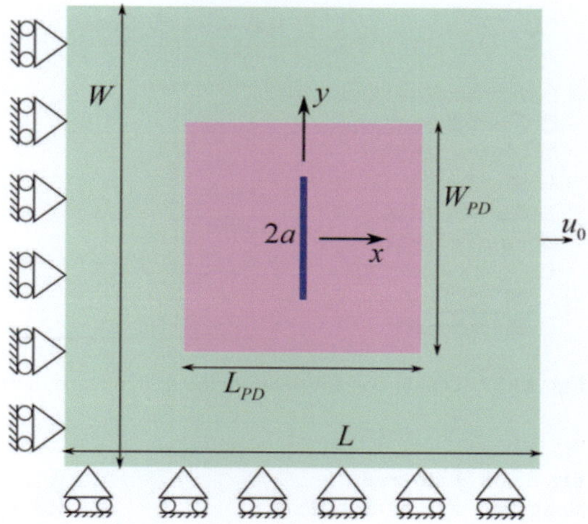

Fig. 13.35 Geometry of a plate with a crack, boundary and loading conditions

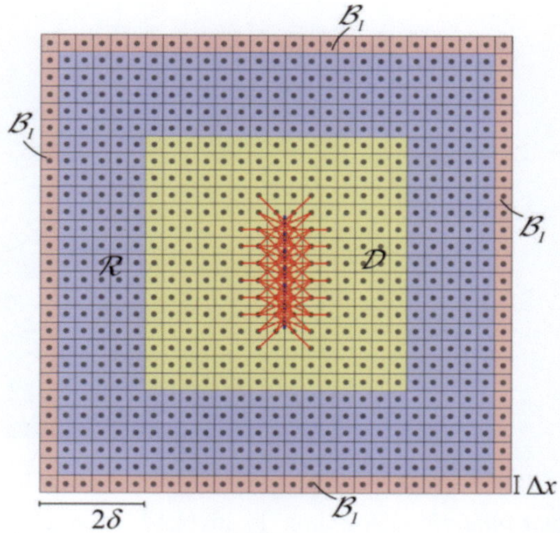

Fig. 13.36 Discretization of the PD domain in the plate with a crack

13.5.2 Preprocessing

The plate with a crack under tension described in the preceding section is modeled by using OSB-PD. The modeling procedure in the Ansys framework is similar to that of Sect. 13.3.2.1 besides adding a new step for crack insertion after **Step-6**. The finite element discretization of the plate is generated using the Ansys input file: ***ex13.5.dat*** from lines 1 to 40 or imported into the Ansys database using the CDREAD command. After generating the family members in **Step-6**, the initial crack in the PD mesh is inserted by using ***Crack Insertion*** submenu. Further details of using this submenu are given in Sect. 9.7 of Chap. 9. As shown in Fig. 13.37, ***Insert Crack***

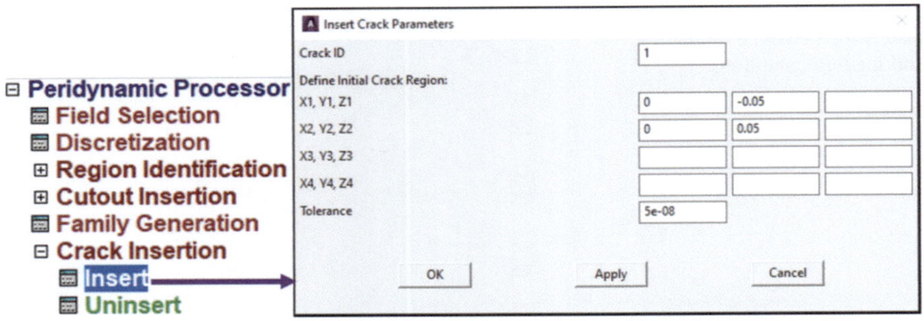

Fig. 13.37 Dialog box for defining the horizon size in the family generation

Fig. 13.38 Line crack along the Y axis of the plate

parameters dialog box is invoked upon clicking the ***Crack Insertion>Insert*** submenu. The parameters X1, Y1=0,-0.05 and X2, Y2=0,0.05 denoting the crack coordinates are entered.

After entering the crack ID and coordinates of the crack and clicking the ***OK*** or ***Apply*** button in the ***Insert Crack parameters*** dialog box, a line (shown in red) representing the sharp crack is inserted in the PD mesh as shown in Fig. 13.38 and the status parameter of the bonds passing through this line are set to zero.

Figure 13.39 shows various Ansys elements used to model OSB-PD, PDDO, and boundary regions. The boundary layer elements are shown in cyan color, the OSB-PD region resembling light green color, and the PDDO region with pink color. Due to the relatively small size of the PDDO region as compared to the OSB-PD in this model, the elements for the PDDO region as in pink color have limited visibility.

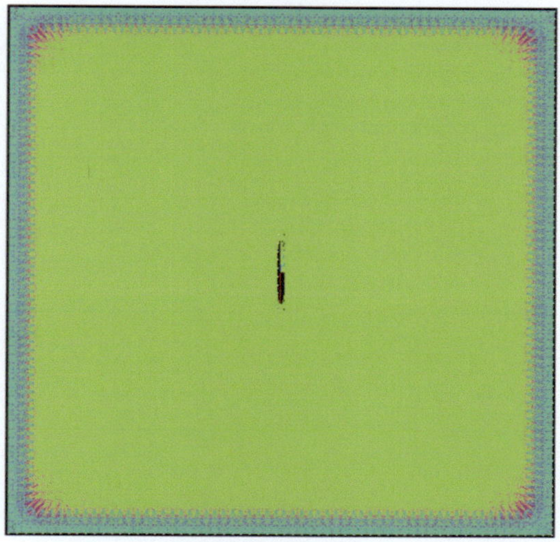

Fig. 13.39 PD elements in Ansys corresponding to different regions

(a) (b)

Fig. 13.40 PD-FE elements. (**a**) Domain and (**b**) Interface PD-FE

The FE model is built around the PD domain such that the nodes of the PD and FE regions coincide at the interface, i.e., both the FE and PD regions share the same node number and location. The FE nodes near the interface can be made to coincide with the PD nodes by issuing the Ansys command NUMMRG,NODES. Copy the lines from 45 to 66 in the above input file and paste them in the command window to generate the FE model surrounding the PD region. Figure 13.40a shows the Ansys PLANE182 finite elements (shown in light blue) surrounded by the PD regions sharing the ***Boundary ID=1,2,3 and 4***. Figure 13.40b shows the elements along the interface of PD and FE regions for a ***Boundary ID=2***.

The traction boundary conditions are naturally satisfied without any requirement for direct enforcement. The displacement boundary conditions are directly enforced using Ansys D command by selecting the nodes on the left and bottom edges, as

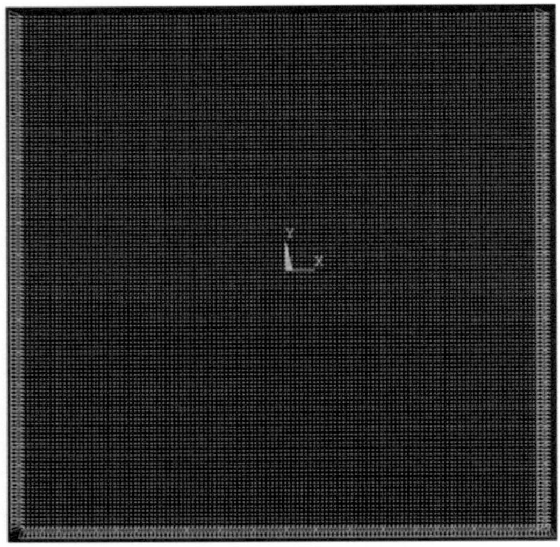

Fig. 13.41 Boundary condition set up for the coupled PD/FE model with displacement constraints on the left, bottom, and right edges

shown in Fig. 13.41. The nodal DOF, UX and UY are constrained along the left and bottom edges, respectively. On the right edge of the plate, a non-zero displacement value of $u_0 = 0.001$m is applied in X direction.

13.5.3 Post-processing

The displacement contour plots can be generated directly by using the Ansys PLNS command. To generate the line plots of displacements along the center line of the plate, the displacements along X and Y are output to the files named ***disp_ux.out*** and ***disp_uy.out*** by running the macro file ***PD_Post_Disp_crk.mac***.

Figures 13.42 and 13.43 show the comparison of the displacement variations from the FEM and coupled OSB-PD/FE models, indicating a close agreement. As shown in Fig. 13.44, the displacement variations along the center lines of the plate from the coupled OSB-PD/FE model match those of FEM.

13.6 Dynamic Response of a Plate Under Compression

13.6.1 Model Setup

The plate geometry and boundary conditions are shown in Fig. 13.45. The origin of the coordinate system coincides with the center of the plate. The plate has square dimensions with an edge length of $W = L = 8$ m. Young's modulus and Poisson's ratio of the isotropic plate are $E = 50$ MPa and $\nu = 0.333$, respectively. Its density is

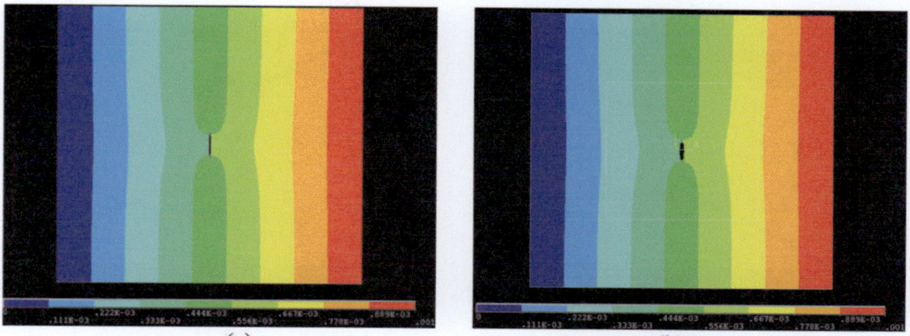

(a) (b)

Fig. 13.42 Horizontal displacement in a plate with a crack under tension: (**a**) FEM, and (**b**) coupled OSB-PD/FE model

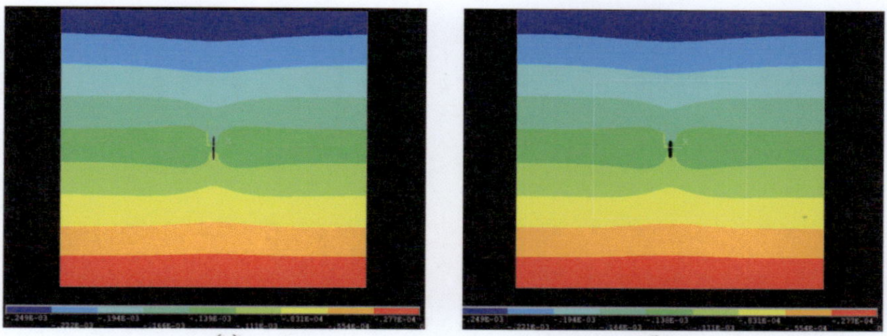

(a) (b)

Fig. 13.43 Vertical displacement in a plate with a crack under tension: (**a**) FEM, and (**b**) coupled OSB-PD/FE model

specified as $\rho = 2647$ kg/m^3. The bottom edge of the plate is fixed in both horizontal and vertical directions, the top edge of the plate is subjected to a transient pressure, $p(t)$. The vertical edges are free of tractions.

The applied pressure history for $0 < t < 0.5$s is shown in Fig. 13.46.

13.6.2 Uniform Discretization

The dynamic response of the plate is modeled by coupled NOSB-PD/FE with uniform discretization. The PD domain at the center of the plate is defined by dimensions, $L_{PD} = W_{PD} = 4$ m. The PD and FE domains are discretized with a uniform mesh size of $\Delta x = 0.1$ m. The PD regions of $\mathcal{D} \cup \mathcal{R}$ and $\mathcal{B}_I$ shown in Fig. 13.47 consist of 1521 and 160 points, respectively. The FE domain has 4800

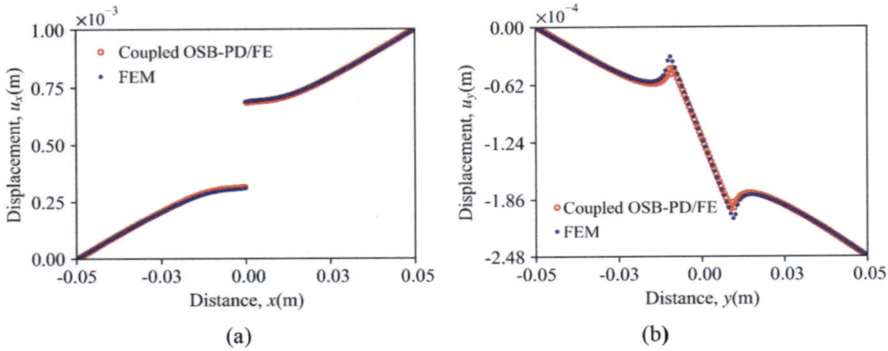

Fig. 13.44 Comparison of the displacements from the coupled OSB-PD/FE model with the FEM: (**a**)$u_x(x, y = 0)$, and (**b**) $u_y(x = 0, y)$

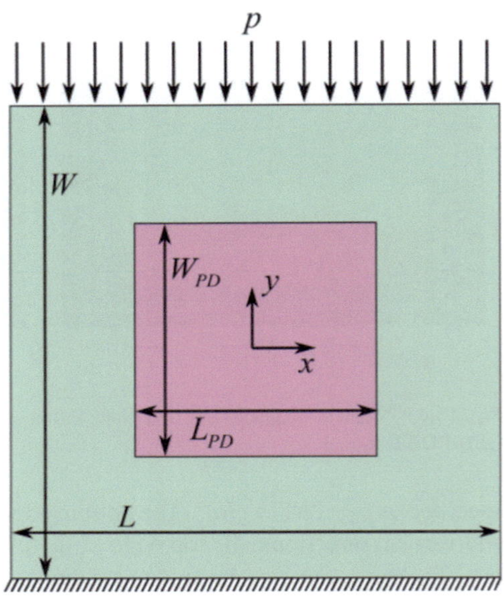

Fig. 13.45 Description of PD and FE domains of the plate, boundary, and applied loading conditions

Ansys PLANE182 elements with 5,040 nodes. The implicit analysis is performed in two load steps with a time step of $\Delta t = 10^{-3}$s. The first load step is in the form of a ramp until $0.1s$, and the second load step is constant for a total time of 0.4s.

13.6.2.1 Preprocessing

The finite element discretization of the plate is generated using the Ansys input file: ***ex13.6.2.dat*** from lines 1 to 42 or imported into the Ansys database using the

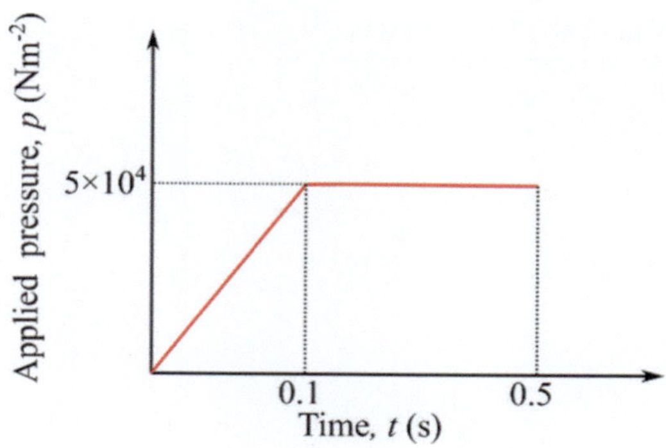

Fig. 13.46 Transient loading and unloading history

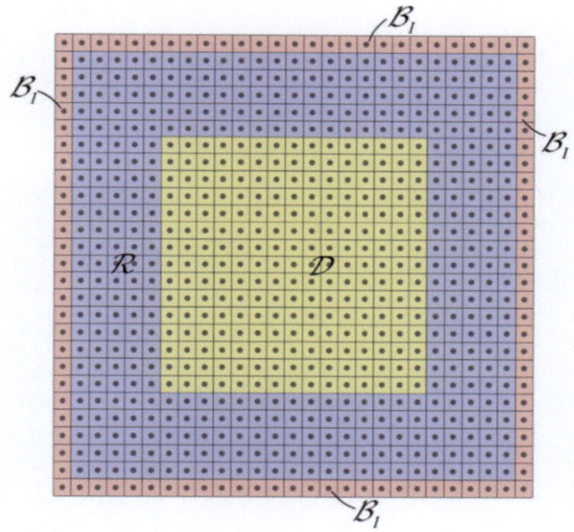

Fig. 13.47 Discretization of regions $\mathcal{D} \cup \mathcal{R}$ and $\mathcal{B}_I$ in the PD domain

CDREAD command. This generates the finite element model for creating the PD domain. The first 8 steps of preprocessing are the same as that of Sect. 13.3.2 with NOSB-PD. Fig. 13.48 shows various Ansys elements used to model NOSB-PD, PDDO, and boundary regions. The boundary layer element is shown in cyan color, the inner NOSB-PD region resembling light green color, and the PDDO region with pink color.

After generating the PD elements, the MASS21 elements are overlaid on PD and PDDO points marked with square symbols as shown in Fig. 13.49. Additional details for using this element are explained in Sect. 8.4 in Chap. 8. Copy the lines from 44 to 73 in the above input file and paste in the command window to generate the mass elements in the PD and PDDO regions.

The FE model is built around the PD domain such that the nodes of the PD and FE regions coincide at the interface, i.e., both the FE and PD region share the same node number and location. The FE nodes near the interface can be made to coincide with the PD nodes by using the Ansys command NUMMRG,NODES command. Copy

Fig. 13.48 PD elements in Ansys corresponding to different regions

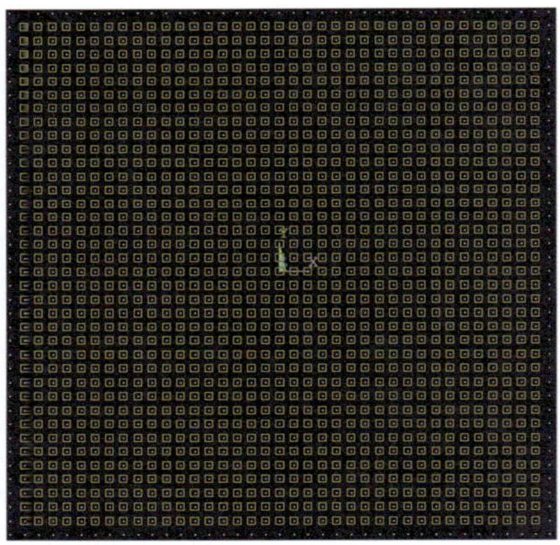

Fig. 13.49 Region corresponding to domain for overlaying the mass elements

the lines from 75 to 97 in the above input file and paste in the command window to generate the FE region surrounding the PD. Fig. 13.50a shows the Ansys PLANE182 elements (shown in light yellow) surrounding the PD region sharing the ***Boundary ID=1,2,3 and 4***. Fig. 13.50b shows the elements along the interface of PD and FE regions for a ***Boundary ID=4***.

The displacement boundary conditions are directly enforced using Ansys D command by selecting the nodes along the edge as shown below in Fig. 13.51 and constraining the displacement in all directions. The traction $\sigma_0 = 5 \times 10^4$ Pa is applied on the FE nodes using Ansys SF command, by ramping it for 1000 substeps

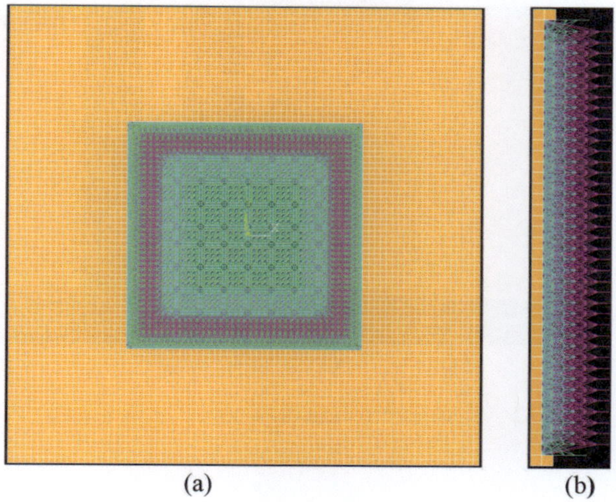

Fig. 13.50 PD-FE elements. (**a**) Domain and (**b**) Interface PD-FE

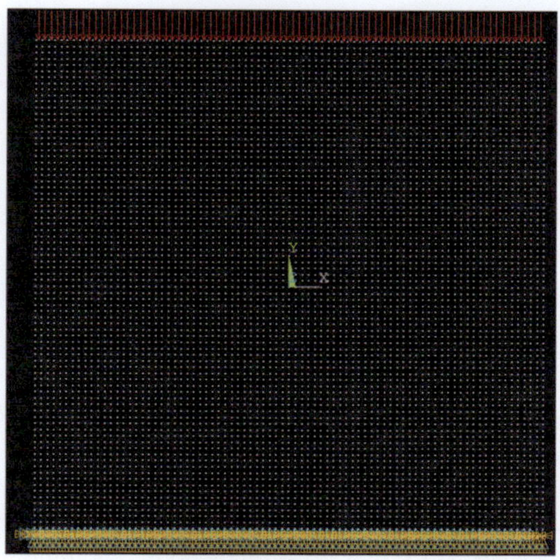

Fig. 13.51 Boundary condition set up for the coupled PD/FE model with displacement constraints on bottom edge and traction on the top edge

for a simulation time of $t = 0.1s$. Furthermore, the analysis is performed for a simulation time of $t = 0.4s$ while keeping the load as constant.

13.6.2.2 Post-processing

The displacement contour plots at various time steps can be generated directly by using Ansys PLNS command. Figures 13.52, 13.53, 13.54, 13.55, and 13.56 show the comparison of the displacement profile, u_y at time instants of $t = 10^{-3}$s,

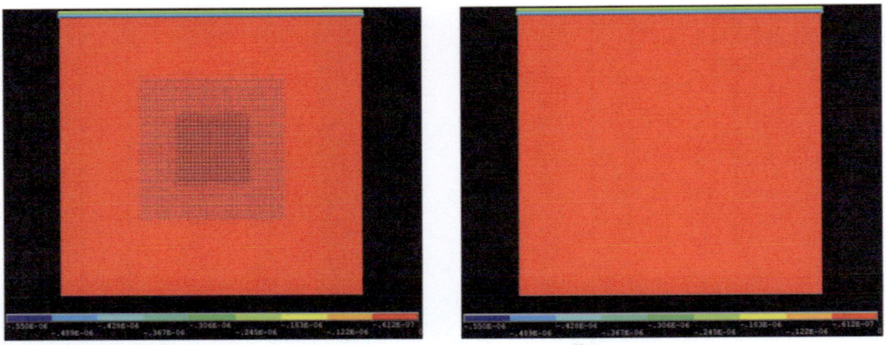

(a) (b)

Fig. 13.52 Variation of vertical displacement, u_y at $t = 10^{-3}$s: (**a**) coupled NOSB-PD/FE, and (**b**) FEM

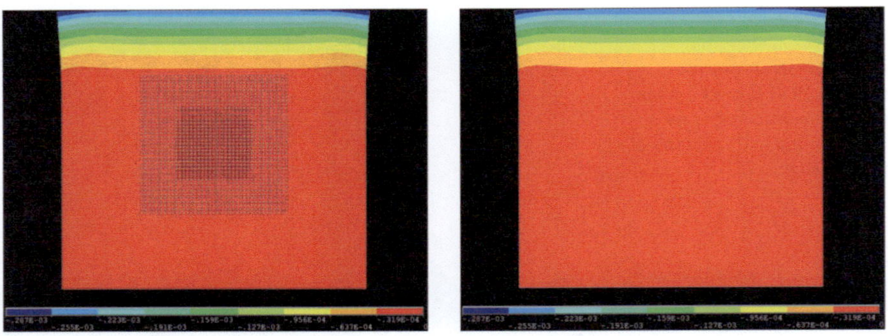

(a) (b)

Fig. 13.53 Variation of vertical displacement, u_y at $t = 2 \times 10^{-2}$s: (**a**) coupled NOSB-PD/FE, and (**b**) FEM

$t = 2 \times 10^{-2}$s, $t = 4 \times 10^{-2}$s, $t = 8 \times 10^{-2}$s, and $t = 4 \times 10^{-1}$s between coupled NOSB-PD/FE and FEM predictions. The time history displacement of the specific node can be viewed using NSOL command in /POST26 processor. Figure 13.57 shows the displacement history of the point P ($x =- 1$ m, $y =1$ m) located at the interface of PD-FE region and its comparison with the FE prediction. The coupled NOSB-PD/FE predictions capture the FE solutions without any dispersion and kinks in the solution as wave reflection and interference occur.

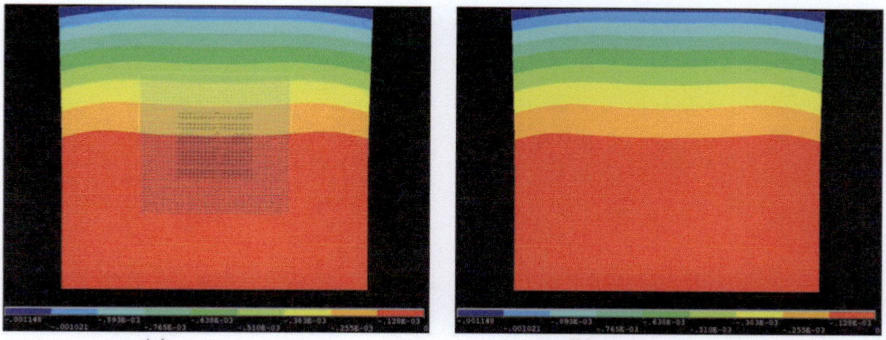

(a) (b)

Fig. 13.54 Variation of vertical displacement, u_y at $t = 4 \times 10^{-2}$s: (**a**) coupled NOSB-PD/FE, and (**b**) FEM

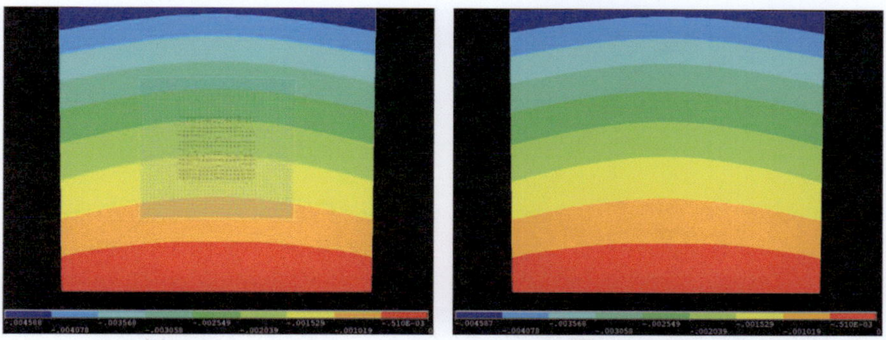

(a) (b)

Fig. 13.55 Variation of vertical displacement, u_y at $t = 8 \times 10^{-2}$s: (**a**) coupled NOSB-PD/FE, and (**b**) FEM

13.6.3 Nonuniform Discretization

As shown in Fig. 13.58a, the geometry corresponding to the coarse PD region, Ω_1^{PD} has a length and width of $L_{PD} = W_{PD} = 4$ m while for the fine PD region, Ω_2^{PD} the length and width are $l_{PD} = w_{PD} = 2$ m. The PD regions Ω_1^{PD} and Ω_2^{PD} are discretized with a mesh size of $\Delta x = 0.08$ m and 0.06 m, respectively. As shown in Fig. 13.58b, the fine mesh PD region, Ω_2^{PD} consisting of region $\mathcal{D}_f$ has 1156 structured nodes while the coarse PD region i.e., Ω_1^{PD}consisting of $\mathcal{D}_c \cup \mathcal{R}$ and $\mathcal{B}$ has (896+880) and 200 structured nodes, respectively. The FE region as shown in Fig. 13.58a, consisting of areas $A_1 \ldots A_8$ is meshed with varying sizes. The areas, $A_1 \ldots A_4$ are meshed with an element size of 0.408m × 0.408m resulting in 4 × 25 elements while

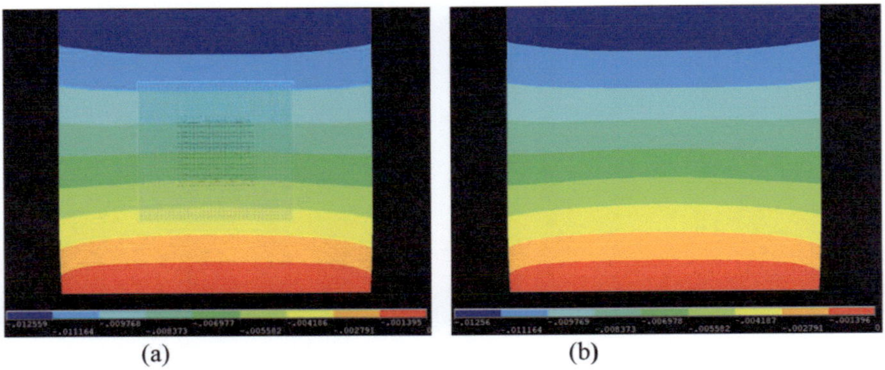

Fig. 13.56 Variation of vertical displacement, u_y at $t = 0.4$s: (**a**) coupled NOSB-PD/FE, and (**b**) FEM

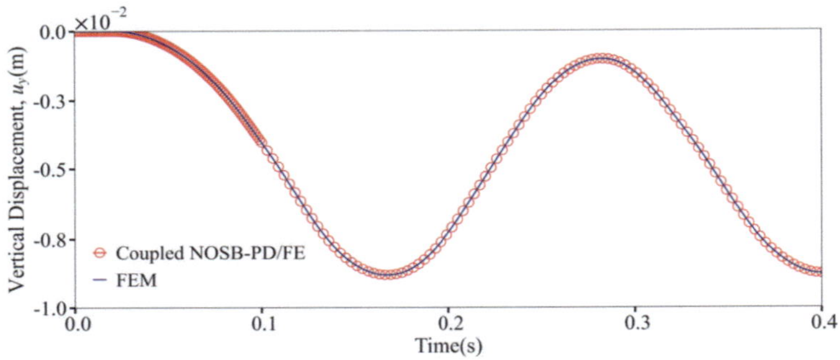

Fig. 13.57 Comparison of the vertical displacement at point P ($x = -1\text{m}, y = 1\text{m}$) from the coupled NOSB-PD/FE model with FEM as time progresses

the areas, $A_5 \ldots A_8$ are meshed with an element size of 0.0769m × 0.04065m resulting in 4 × 200 elements.

The finite element discretization of the plate is generated using the Ansys input file: ***ex13.6.3.dat*** or imported into the Ansys database using the CDREAD command. This generates the finite element model for creating the PD domain. The first 8 steps to generate the PD interaction are the same as that of Sect. 13.3.2 except for the selection of the ***PD Interaction Type***. The NOSB-DHPD of ***PD Interaction Type*** is selected as shown in Fig. 13.59.

After setting the ***PD Interaction Type*** and following similar steps, the Ansys elements used for PD, PDDO and boundary regions are generated as shown in Fig. 13.60. The boundary layer element are shown in cyan color, the inner NOSB-DHPD region resembling pink color, and the PDDO region with light green color.

Corresponding to the coarse region, $\mathcal{D}_c$ and fine region, $\mathcal{D}_f$, two MASS21 element types are defined with real constants based on the density and volume of the PD point

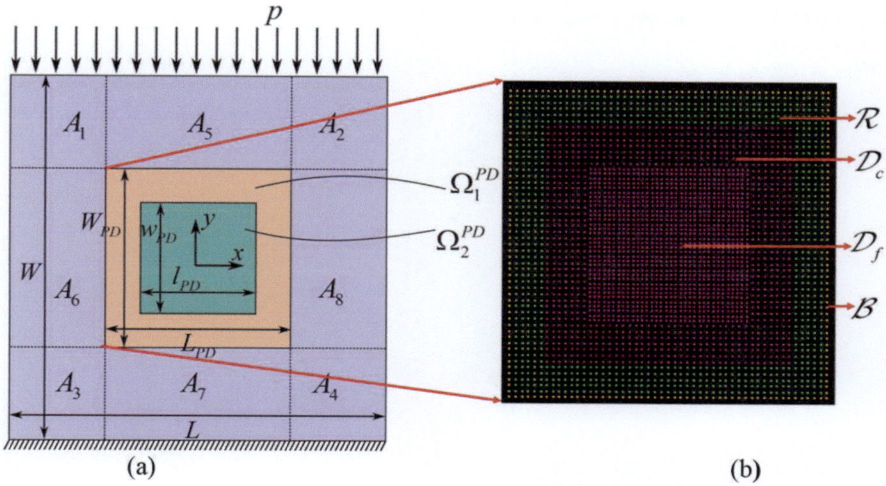

Fig. 13.58 Plate with FE and PD regions: (**a**) geometry and boundary conditions, and (**b**) nonuniform discretization in PD region

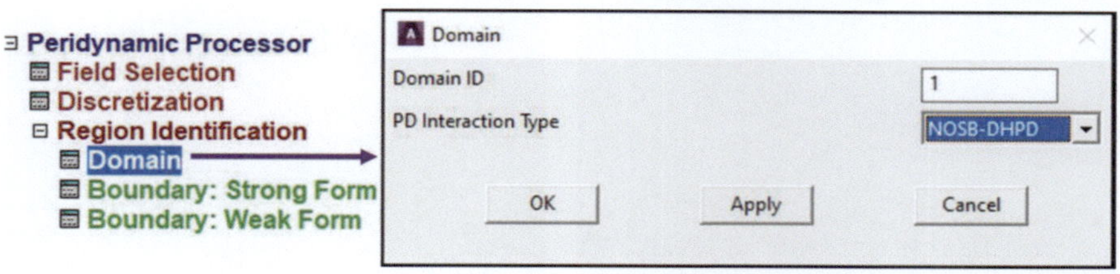

Fig. 13.59 Dialog box for selecting the domain interaction type as NOSB-DHPD

as shown in Fig. 13.61. Copy the lines from 65 to 95 in the above input file and paste in the command window to generate the mass elements in the PD and PDDO regions.

The FE model is built around the PD domain. Figure 13.62 shows the elements in the domain and the interface of PD/FE region. Copy the lines from 101 to 179 in the above input file and paste in the command window to generate the FE region surrounding the PD region. Ansys PLANE182 elements (shown in orange) surrounding the PD region sharing the ***Boundary ID=1,2,3 and 4*** are shown in Fig. 13.62a. Figure 13.62b shows the elements along the interface of PD and FE regions for a ***Boundary ID=2***.

The displacement boundary conditions are directly enforced using Ansys D command by selecting the nodes along the bottom edge as shown below in Fig. 13.63 and constraining the displacement in all directions. The traction $\sigma_0 = 5 \times 10^4$Pa is applied on the FE nodes using Ansys SF command, by ramping

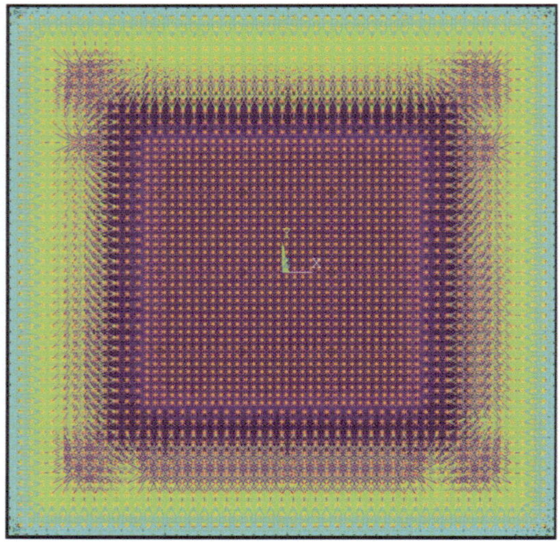

Fig. 13.60 PD elements in Ansys corresponding to different regions

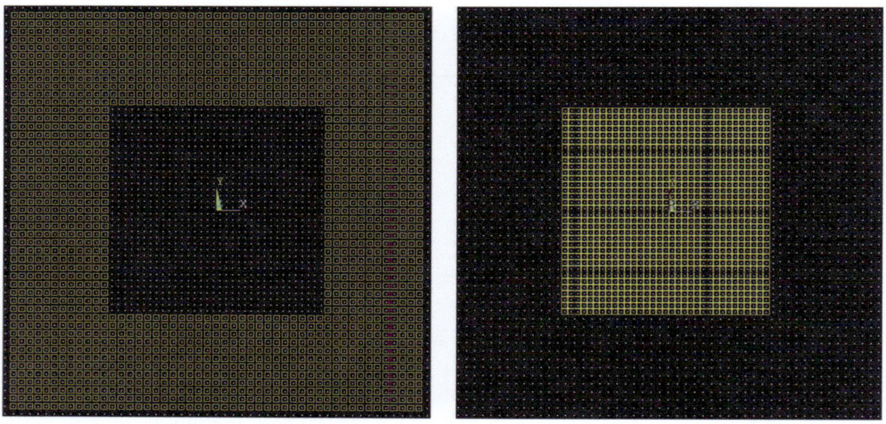

Fig. 13.61 Region corresponding to domain for overlaying the mass elements: (**a**) Coarse mesh, $\mathcal{D}_c$, and (**b**) Fine mesh, $\mathcal{D}_f$

it for 1000 substeps for a simulation time of $t = 0.1s$. Furthermore, the analysis is performed for a simulation time of $t = 0.4s$ while keeping the load as constant.

Figure 13.64 shows the coupled NOSB-DHPD/FE vertical displacement u_y predictions at instants of $t = 0.01s$, $0.02s$, $0.04s$, $0.08s$ and $0.4s$. Also, Fig. 13.65 shows the displacement history of the node located at the interface of PD-FE region and its comparison with the FE prediction. The coupled NOSB-DHPD/FE predictions capture the FE solutions without any dispersion and kinks in the solution as wave reflection and interference occur. The slight deviation between the results for finite element and coupled NOSB-DHPD/FE is due to the difference in locations of the PD

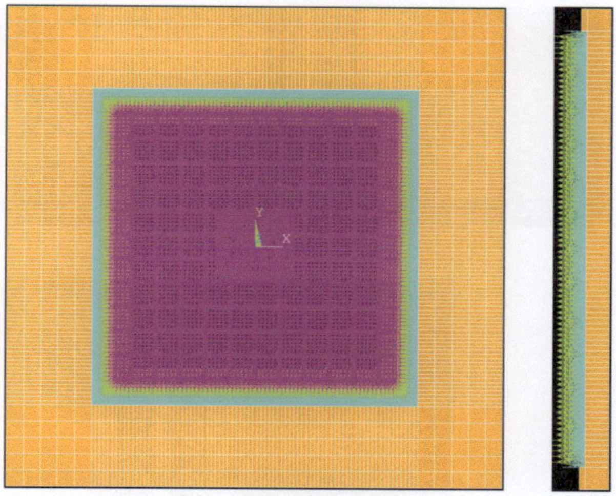

Fig. 13.62 DHPD-FE elements. (**a**) Domain and (**b**) Interface

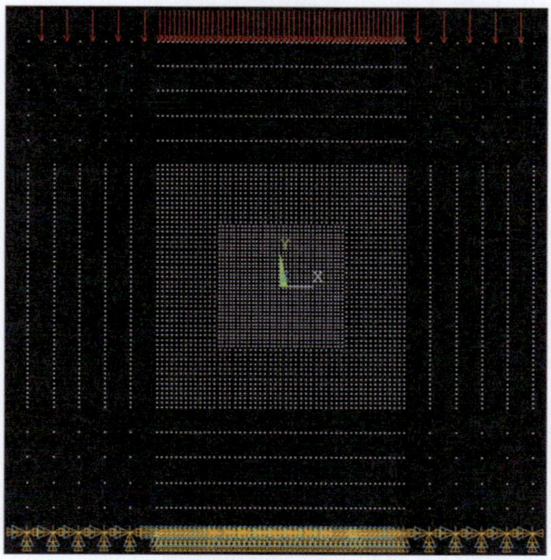

Fig. 13.63 Boundary condition set up for the coupled DHPD-FE model with displacement constraints on bottom edge and traction on the top edge

point and FE node, P_{PD-FE} ($x = 0.9507\text{m}, y = 0.9507\text{m}$) and P_{FEM}($x = 1.0\text{m}$, $y = 1.0\text{m}$), respectively.

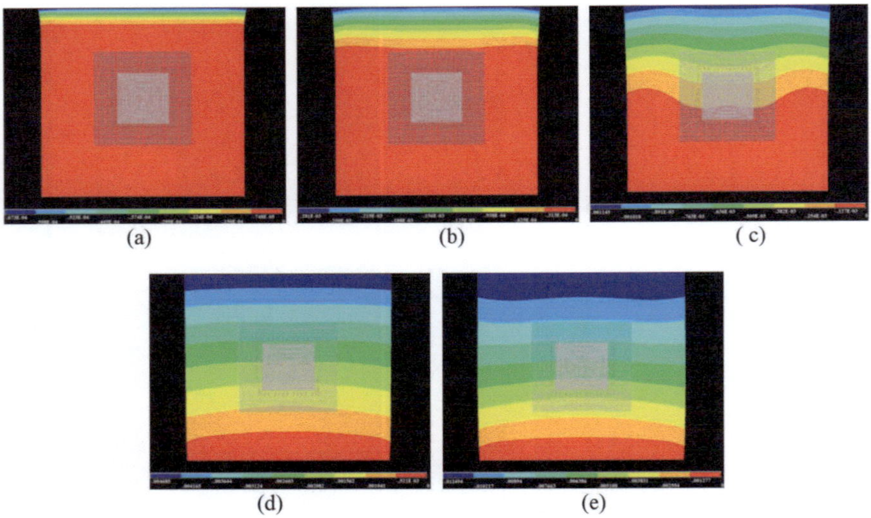

Fig. 13.64 Coupled NOSB-DHPD/FE vertical displacement, u_y predictions at time: (**a**) $t = 0.01s$, (**b**) $t = 0.02s$, (**c**) $t = 0.04s$, (**d**) $t = 0.08s$, (**e**) $t = 0.4s$)

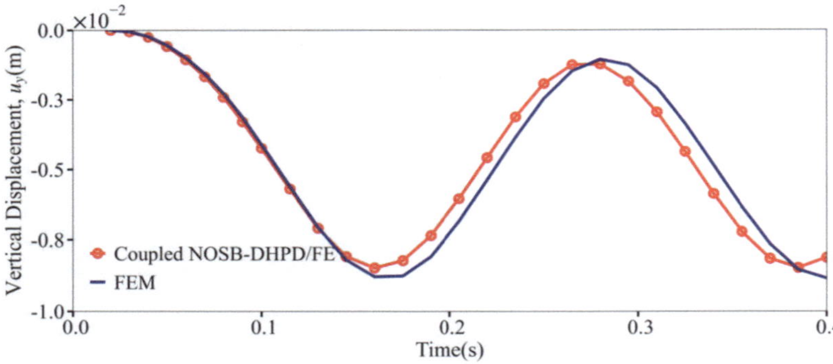

Fig. 13.65 Comparison of the vertical displacement at points P_{PD-FE} ($x = 0.9507$m, $y = 0.9507$m) and P_{FEM}($x = 1.0$m, $y = 1.0$m) from the coupled NOSB-DHPD/FE with FEM as time progresses

13.7 Static Response of a Plate Under Initial Thermal Strain

Figure 13.66 shows the plate geometry and boundary conditions. A Cartesian coordinate system is located at the center of the plate with dimensions length $L = 1$ m and width $W = 0.5$ m. The geometry corresponding to the PD region, Ω^{PD} has a length $L_{PD} = 0.25$ m and width $W_{PD} = W = 0.5$ m. It has Young's

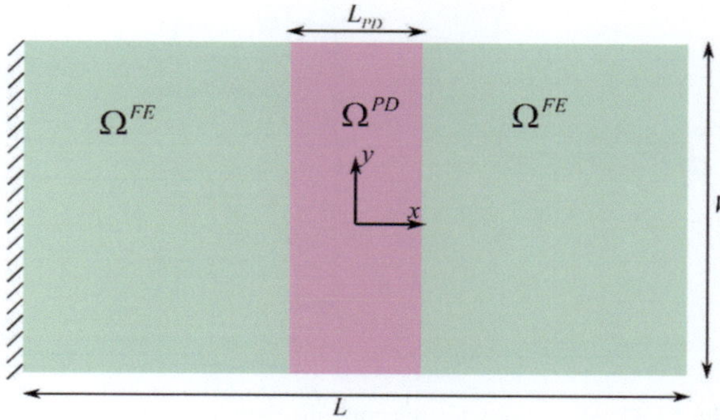

Fig. 13.66 FE and PD regions of plate: geometry and boundary conditions

modulus and Poisson's ratio of $E = 0.05$ GPa and $\nu = 0.25$ or $\nu = 0.33$, respectively, and coefficient of thermal expansion of $\alpha = 0.01\text{K}^{-1}$. The left edge of the plate is fixed, and other edges of the plate are free of tractions. It is subjected to an initial strain arising from a quadratic temperature variation $T(x) = 400x^2$. The PD and the FE regions are discretized with a uniform mesh size of $\Delta x = 0.005$ m. In this simulation, the static response of the plate under initial thermal strain is modeled by coupled BB-PD/FE, OSB-PD/FE and NOSB-PD/FE. The Poisson's ratio is set as $\nu = 0.33$ for coupled BB-PD/FE and as $\nu = 0.25$ for OSB-PD/FE and NOSB-PD/FE.

13.7.1 Preprocessing

The finite element discretization of the plate is generated using the Ansys input file, ***ex13.7_bb.dat,*** from lines 1 to 49 or imported into the Ansys database using the CDREAD command. This generates the finite element model for creating the PD domain. The coupled BB-PD/FE model is built through the following steps.

Step-1 The field and dimension of the analysis are specified using the ***Field Selection*** submenu located under ***Peridynamic Processor***. For the simulation of response of a plate under initial thermal strain, the "***Thermal-Structural***" for field and "***2D***" for dimension are selected as shown in Fig. 13.67.

Step-2 The PD mesh is generated using the radio button option ***Generate using FE model*** in ***Discretization*** submenu as shown in Fig. 13.68:

Step-3 The PD regions for the structural equations and boundary conditions are created using ***Region Identification*** submenu shown in Fig. 13.69. The submenu ***Region Identification>Domain*** is selected to open the dialog box ***Domain***. For the current application, two regions corresponding to BB-PD and PDDO are created. The parameter, ***Domain ID=1*** is entered in the text box and ***PD Interaction Type*** is selected from the drop down menu as BB-PD as shown in Fig. 13.69.

After entering the parameters and clicking the ***OK/Apply*** button, the node picker dialog box appears to select the desired region. After selecting the nodes using the

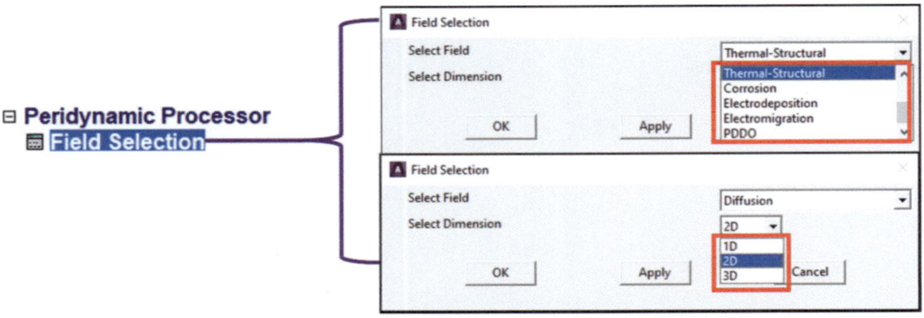

Fig. 13.67 Dialog box for selecting the field and dimension

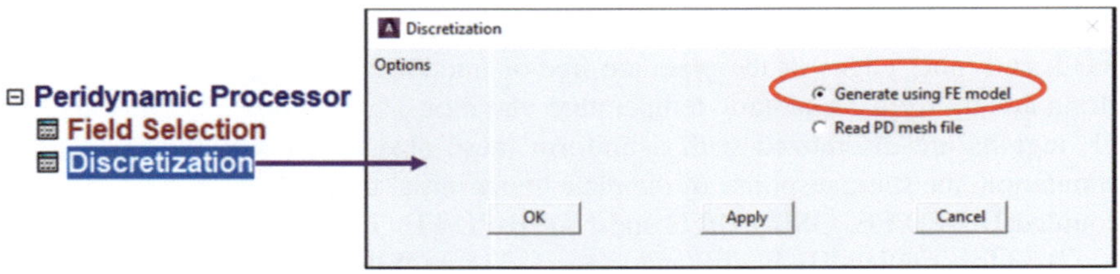

Fig. 13.68 Dialog box for PD mesh generation

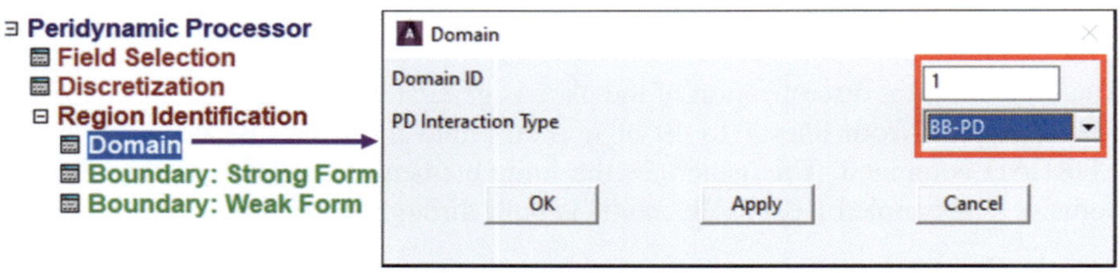

Fig. 13.69 Dialog box for selecting the domain interaction type as BB-PD

Node picker window and clicking the ***OK*** or ***Apply*** button of the node picker (Box), a dialog box labeled as ***Domain Points*** appears for confirming the selected BB-PD regions as shown in Fig. 13.70.

Setp-4 Repeat the process for creating the PDDO region of the plate using a different ***Domain ID*** and setting the ***PD Interaction Type*** to PDDO as shown in Fig. 13.71. The selected region for PDDO using node picker window is shown in Fig. 13.72. Clicking the ***OK*** or ***Apply*** button of the node picker (Box), a dialog box labeled as ***Domain Points*** appears for confirming the selected PDDO region.

Setp-5 To enforce the boundary conditions in coupled PD/FE analysis, the submenu ***Region Identification>Boundary: Weak Form*** is selected to open the

Fig. 13.70 Region identification corresponding to BB-PD domain of the plate

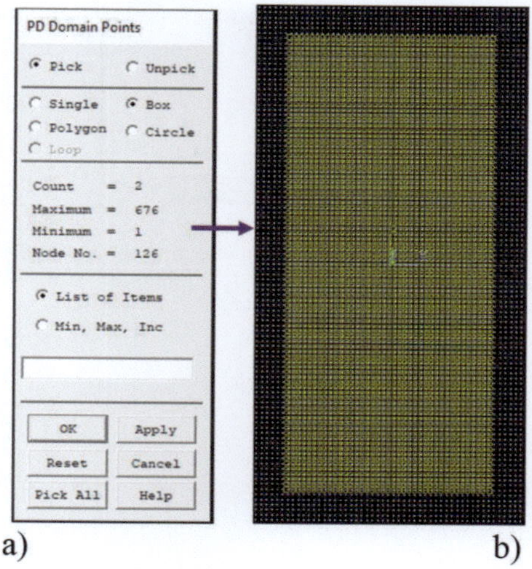

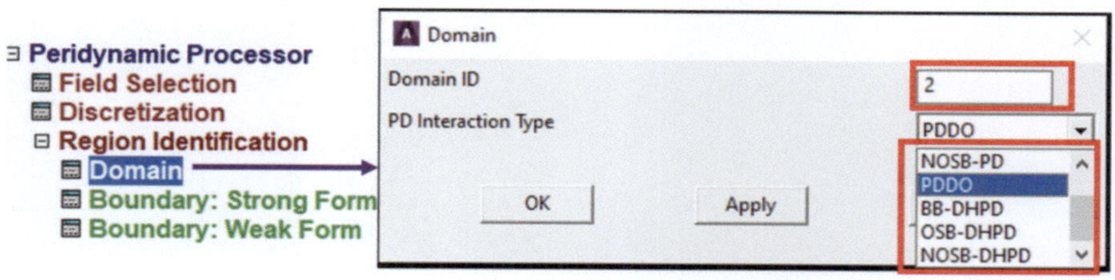

Fig. 13.71 Dialog box for selecting the domain interaction types as PDDO

Fig. 13.72 Region identification corresponding to PDDO domain of the plate

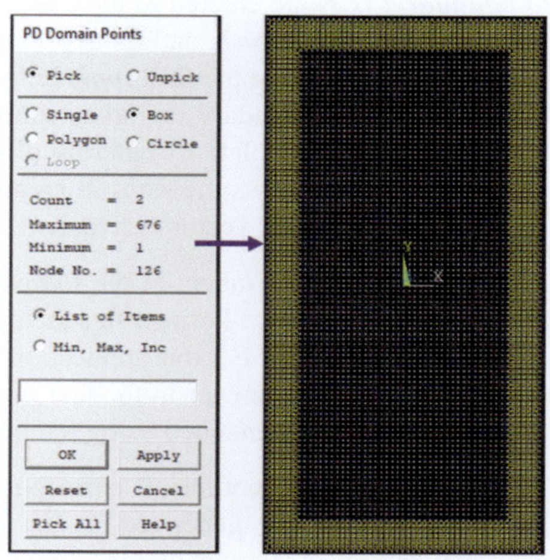

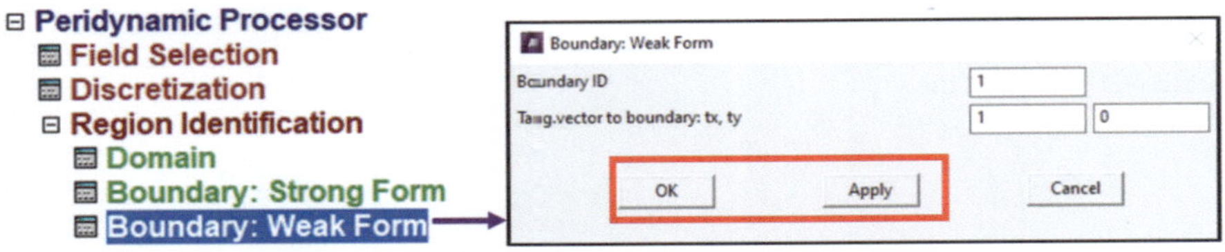

Fig. 13.73 Dialog box for selecting the boundary of the plate with weak form of interaction

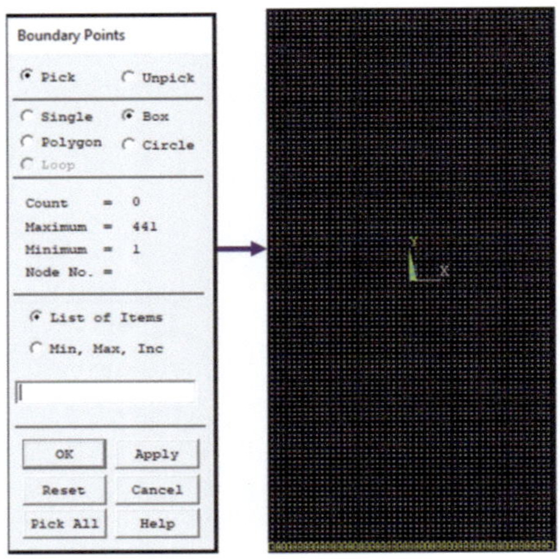

Fig. 13.74 Region identification corresponding to ***Boundary ID*=1** of the plate with weak form of interaction for coupled PD/FE analysis

dialog box ***Boundary: Weak Form.*** The parameters ***Boundary ID*** and ***Tang. vector to boundary: tx,ty*** are entered in the dialog box as shown in Fig. 13.73.

Enter ***Boundary ID=1***, and its unit tangent vector ***tx,ty: 1,0***. After entering the parameters and clicking the ***OK/Apply*** button, the node picker dialog box appears to select the desired boundary region as shown in Fig. 13.74. Repeat the process for each boundary of the plate by entering different ***Boundary ID*** and its unit tangent vector. Fig. 13.75 shows the steps in creating the boundary region with ***Boundary ID=2*** and unit tangent vector ***tx,ty:0,1***.

Step-6 The family members of the PD points are generated using ***Family Generation*** menu based on the horizon size factor ***m*** in the ***Family Generation*** dialog box as shown in Fig. 13.76. For this application, the horizon size factor, $m = 3$ is entered. After entering the parameter in the text box and clicking the ***OK/Apply*** button, the family members are generated using the horizon size of $\delta = 3.015\Delta x$.

Step-7 The element coefficients corresponding to the stiffness for the PD thermal-structural model are generated using ***Element Coefficients (Therm.-Struct)*** submenu option as shown in Fig. 13.77. Clicking the ***OK/Apply*** button of the ***Element Coefficients (Therm.-Struct)*** dialog box generates the stiffness coefficients

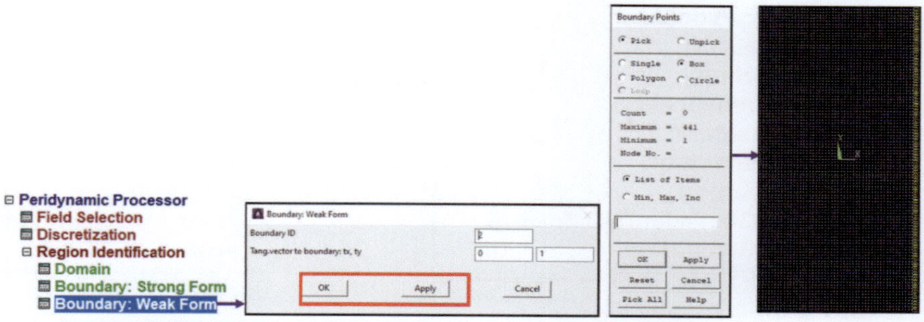

Fig. 13.75 Steps for creating the ***Boundary ID*=2** of the plate with weak form of interaction for coupled PD/FE analysis

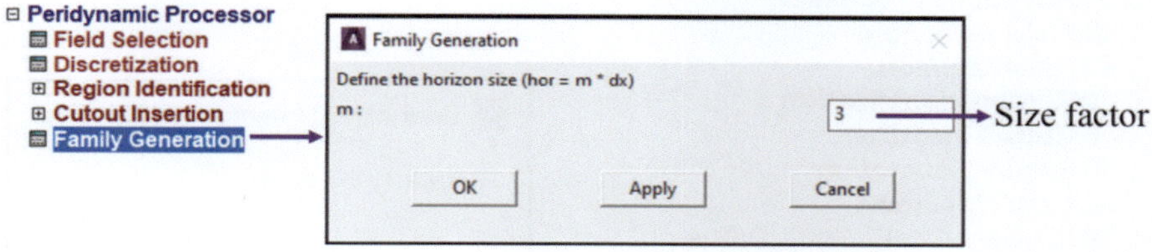

Fig. 13.76 Dialog box for defining the horizon size in the family generation

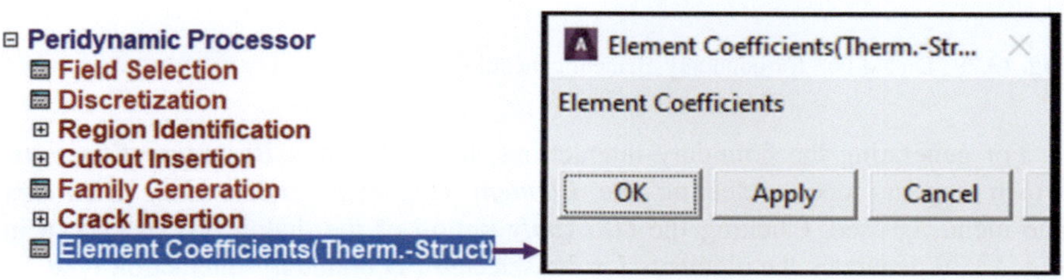

Fig. 13.77 Element Coefficients submenu for Thermal-Structural analysis

for the PD interaction types. In this application, the stiffness coefficients corresponding to BB-PD and PDDO types are generated.

Step-8 The elements required for modeling the PD interactions in the thermal-structural analysis in the domain are created by using the dialog box ***Domain Elements*** which appears upon selecting the ***Element Generation>Domain Elements*** submenu. Clicking the ***OK/Apply*** button of the dialog box as shown in Fig. 13.78 generates the domain elements for the BB-PD and PDDO interaction type.

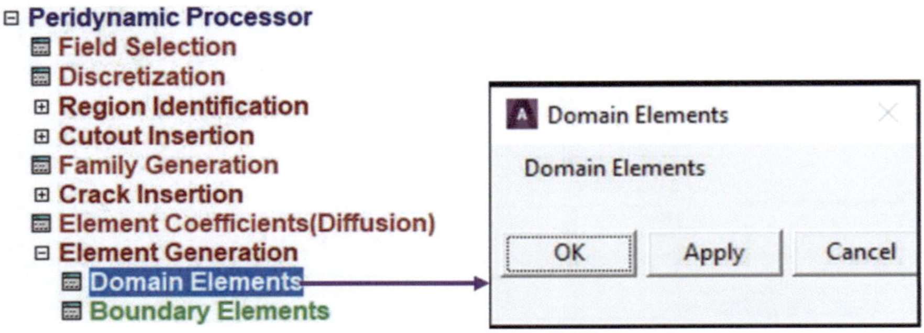

Fig. 13.78 Dialog box for domain element generation

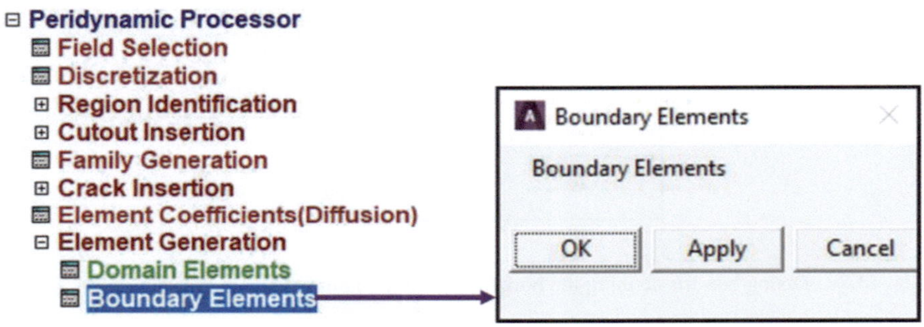

Fig. 13.79 Dialog box for boundary element generation

For generating the boundary interactions, the dialog box ***Boundary Elements,*** which appears upon selecting the ***Element Generation>Boundary Elements*** sub-menu, is used. Clicking the ***OK/Apply*** button of the dialog box as shown in Fig. 13.79 generates the elements for the selected PD boundary interaction type.

Figure 13.80 shows various Ansys elements used to model PD, PDDO, and boundary regions. The boundary layer element are shown in cyan color, the inner BB-PD region resembling golden yellow color, and the PDDO region with pink color.

Step-9 The FE model is built around the PD domain such that the nodes of the PD and FE regions coincide at the interface, i.e., both the FE and PD region share the same node number and location. The FE nodes near the interface can be made to coincide with the PD nodes by issuing the Ansys command NUMMRG, NODES. Copy the lines from 51 to 67 in the above input file and paste in the command window to generate the FE model surrounding the PD region. Figure 13.81a shows the Ansys PLANE182 finite elements (shown in green) near the PD region sharing the ***Boundary ID=2,4***. Figure 13.81b shows the elements along the interface of PD and FE regions for a ***Boundary ID=2***.

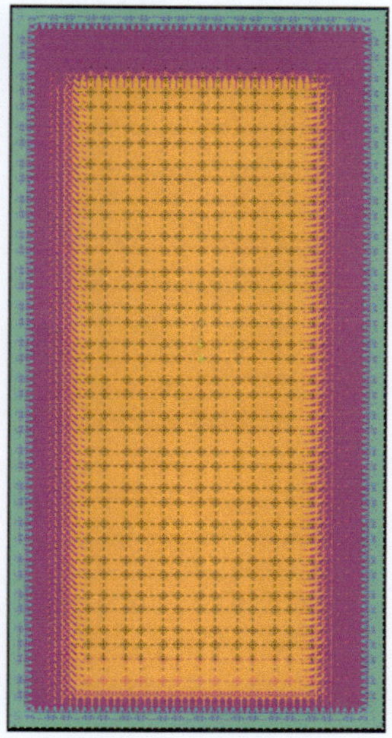

Fig. 13.80 PD elements in Ansys corresponding to different regions

Step-10 The initial thermal strain in the PD region is invoked by using the macro ***PD_inistrain.mac***. The input parameters to this macro are: ***Temperature variation type***, ***Direction***, ***Constant_1*** and ***Constant_2*** where ***Temperature variation type***, denotes the type of temperature field variation. The supported variation types include Constant, Linear and Quadratic. The second argument ***Direction*** denotes the direction of spatial variation of temperature. The supported directions are ***Direction=0,1,2*** denote the variation only in X, only in Y and both X and Y, respectively. ***Constant_1*** and ***Constant_2*** are the constants used to describe the temperature variation corresponding to the type chosen. To input linear temperature variation in X direction, of the form $Ax+b$, the arguments are specified as ***Temperature variation type=1, Direction=0, Constant_1=A*** and ***Constant_2=B***. Run the below command to enforce quadratic temperature variation only in X direction for $T(x) = 400x^2$

PD_inistrain,2,0,400,0.

Run the attached input file ***FE_inistrain.mac*** with the command, *FE_inistrain,2* to input the body loads due to quadratic thermal load variation, $T(x) = 400x^2$ for the FE domain. The traction boundary conditions are naturally satisfied without any additional requirements. The displacement boundary conditions are directly enforced using Ansys D command by selecting the nodes on the left edge as shown in Fig. 13.82. The nodal DOF, UX and UY are constrained along the left edge. The arrows shown in red color denote the traction vector due to the quadratic thermal load on the PD boundaries. The tractions due to the thermal load in the FEM

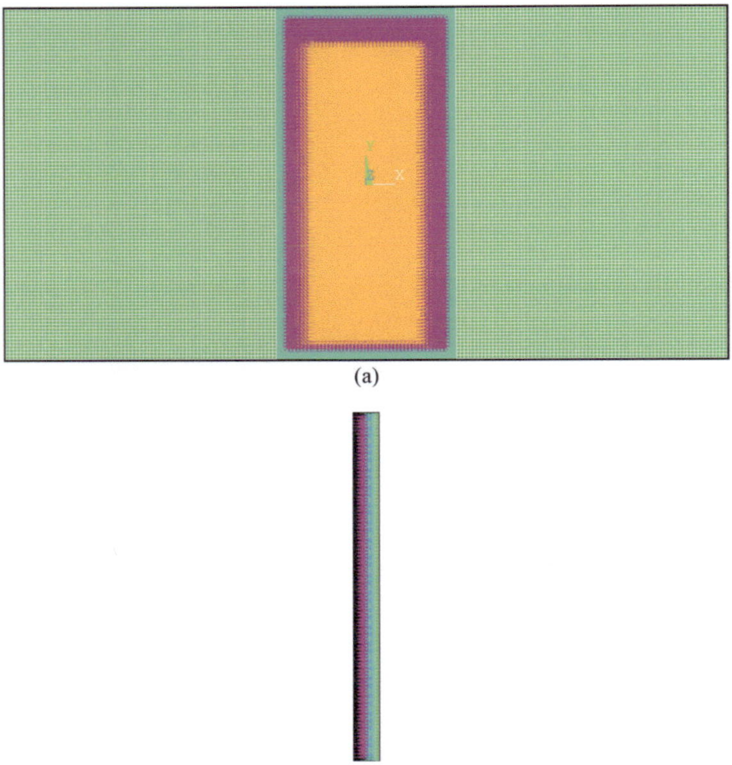

(a)

(b)

Fig. 13.81 PD-FE elements: (**a**) Domain and (**b**) Interface

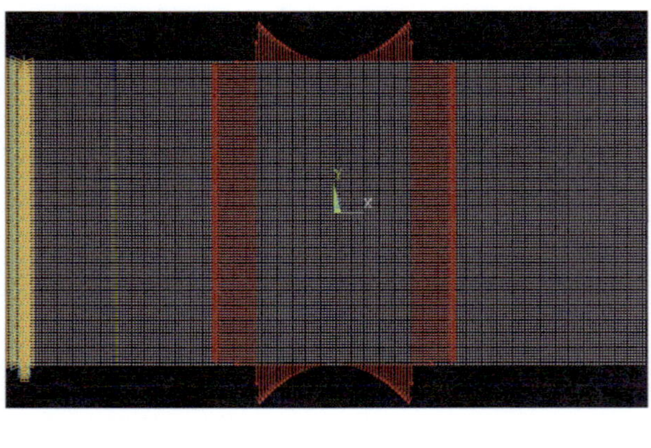

Fig. 13.82 Boundary condition set up for the coupled PD-FE model with displacement constraints on the left edge

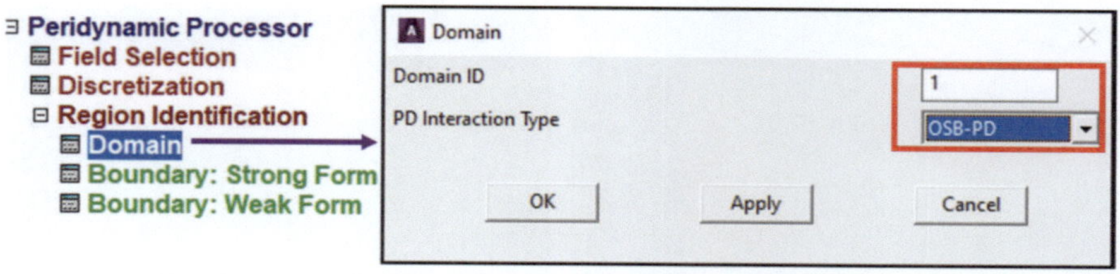

Fig. 13.83 Dialog box for selecting the domain interaction type as OSB-PD

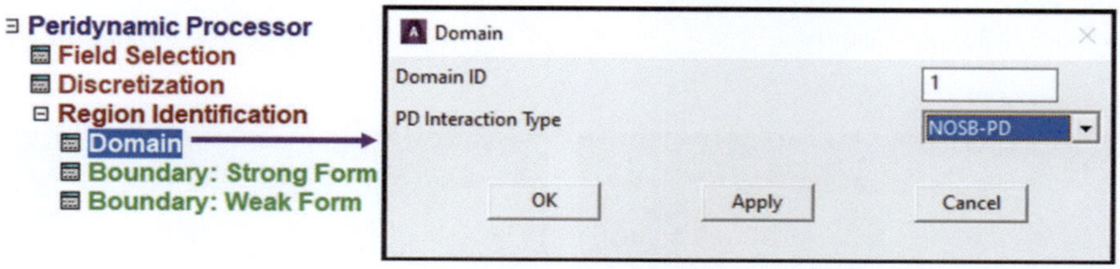

Fig. 13.84 Dialog box for selecting the domain interaction type as NOSB-PD

region are treated by using the PLANE182 elements. To apply the displacement constraints and solve the model, copy the lines from 70 to 80.

To generate the results for the coupled OSB-PD/FE and coupled NOSB-PD/FE, the finite element model is used for creating the PD domain by using the same Ansys input file: ***ex13.7_state.dat*** from lines 1 to 49. The operations for generating the PD and FEM elements and imposing the boundary conditions follow similar preprocessing steps from **Step-1** to **Step-10** as described above except for **Step-3** where the parameter ***PD interaction type*** is changed to OSB-PD and NOSB-PD as shown in Figs. 13.83 and 13.84, respectively.

13.7.2 Post-processing

The displacement contour plots can be generated directly by using Ansys PLNS command. To generate the line plots of displacements along the center line of the plate, the displacements along X and Y are output to the files named as ***disp_ux.out*** and ***disp_uy.out*** by running the mac file ***PD_Post_Disp_inistrain.mac***. Figures 13.85, 13.86, and 13.87 show the horizontal and vertical displacement variations, respectively, from coupled BB-PD/FE, OSB-PD/FE, and NOSB-PD/FE models. Figures 13.88 and 13.89 show the comparison of PD/FE displacement predictions along the center lines $y = 0$ and $x = 0$, respectively, for BB, OSB and NOSB interactions with those of FEM. The coupled PD/FE predictions agree with

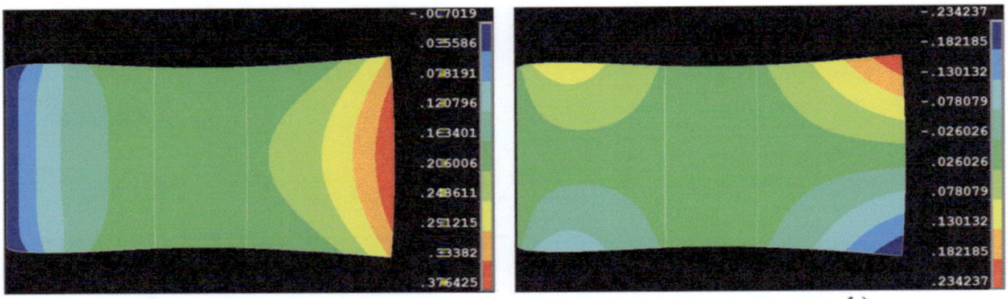

a) b)

Fig. 13.85 BB-PD/FE analysis displacement predictions due to quadratic temperature variation with $v = 0.33$: (**a**) $u_x$and (**b**) u_y

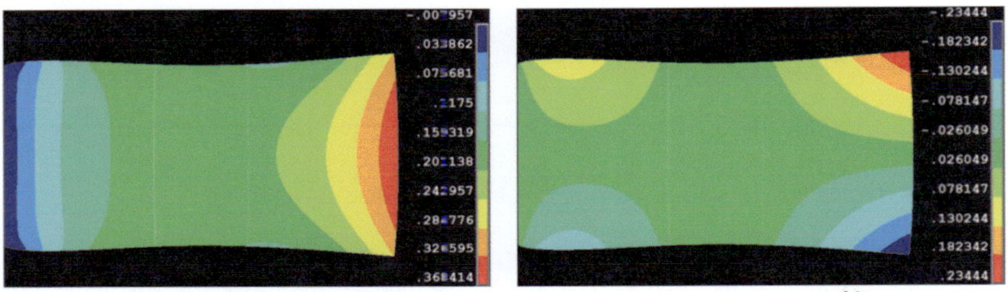

a) b)

Fig. 13.86 Coupled OSB-PD/FE analysis displacement predictions due to quadratic temperature variation with $v = 0.25$: (**a**) u_x and (**b**) u_y

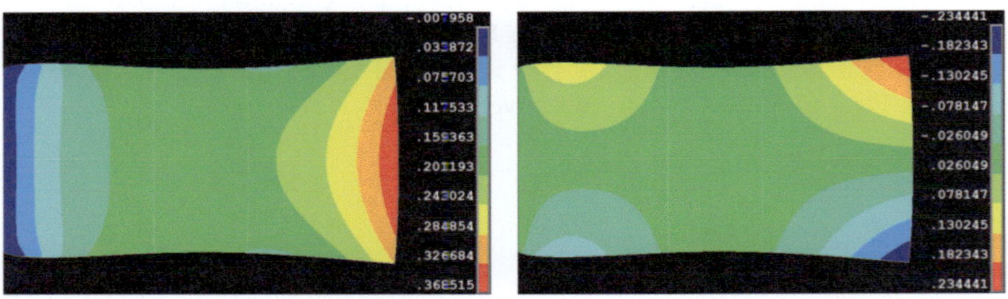

a) b)

Fig. 13.87 Coupled NOSB-PD/FE analysis displacement predictions due to quadratic temperature variation with $v = 0.25$: (**a**) u_x and (**b**) u_y

FE results. Also, the predictions capture the FE results without any kinks near the PD/FE interface.

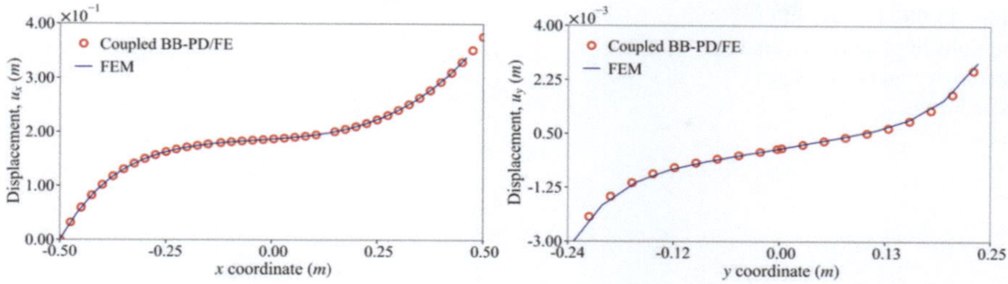

Fig. 13.88 Comparison of coupled BB-PD/FE and FEM displacement predictions along the horizontal and vertical centerlines of a plate subjected quadratic temperature variation with $\nu = 0.33$: (**a**) u_x and (**b**) u_y

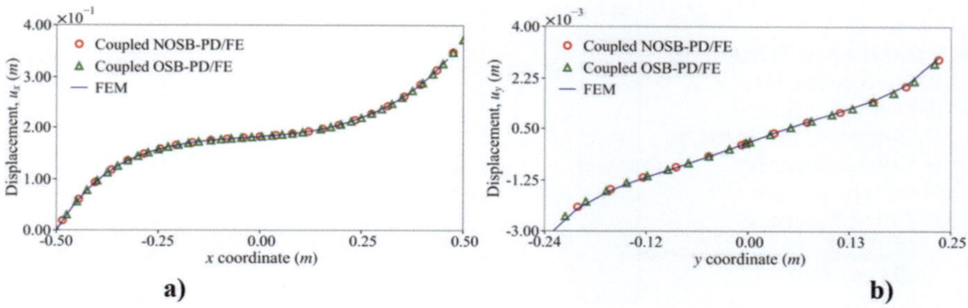

Fig. 13.89 Comparison of coupled NOSB-PD/FE, OSB-PD/FE and FEM displacement predictions along the horizontal and vertical centerlines of a plate subjected quadratic temperature load with $v = 0.25$: (**a**) u_x and (**b**) u_y

13.8 Static Response of a Plate with a Crack Due to Combined Initial Thermal Strain and Mechanical Loading

As shown in Fig. 13.90, a central crack of $2a = 0.1m$ is introduced in the plate described in Sect. 13.7. Also, the boundary conditions remain the same except for the right edge which is subjected to displacement constraint of $u(x = 0.5L) = u_0 = 0.005m$. The PD and the FE regions are discretized with a uniform mesh size of $\Delta x = 0.005$ m. It is subjected to a thermal strain arising from the linear variation of temperature as $T(x) = 5x$. The Poisson's ratio is set as $\nu = 0.33$ for coupled BB-PD/FE and as $\nu = 0.1$ for OSB-PD/FE and NOSB-PD/FE.

The FE discretization of the plate is generated using the Ansys input file: ***ex13.8_bb.dat for*** BB-PD model. For the OSB-PD and NOSB-PD Ansys input file: ***ex13.8_state.dat*** is used, or the respective FE models can be imported into the Ansys database using the CDREAD command. In both input files, the commands in lines 1 to 50 are used to discretize the PD region with traditional finite

Fig. 13.90 FE and PD regions of plate with a crack: geometry and boundary conditions

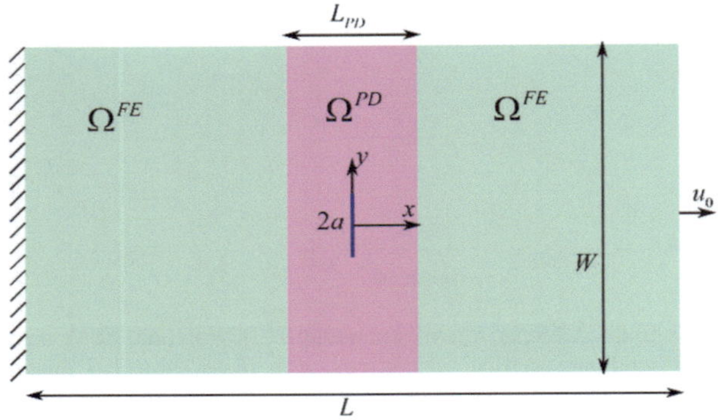

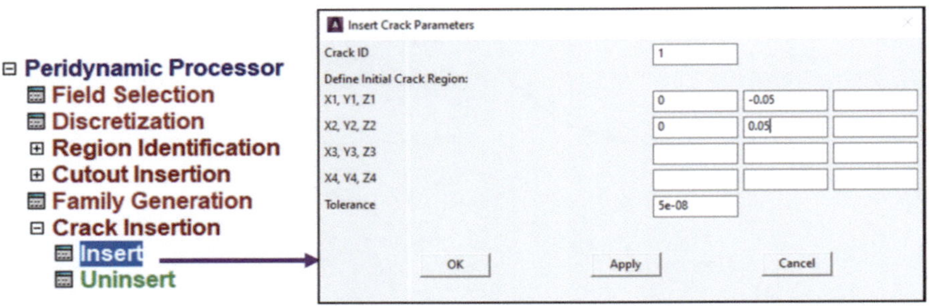

Fig. 13.91 Dialog box for entering the parameters for the crack insertion in PD mesh

elements; the commands in lines 51 to 70 are used to generate FE elements in the FE region, and the remaining lines are used for solving and plotting the displacement solution. The operations for generating the PD and FEM elements and imposing the boundary conditions are the same as those in Sect. 13.7.1. After generation of the family members in **Step-6**, the initial crack in the PD mesh is inserted using ***Crack Insertion*** submenu. As shown in Fig. 13.91, ***Insert Crack parameters*** dialog box appears upon clicking the ***Crack Insertion>Insert*** submenu. The parameters X1, Y1=0,-0.05, and X2, Y2=0,0.05 denoting the crack coordinates are entered.

After entering the crack ID, coordinates of the crack and clicking the ***OK*** or ***Apply*** button in the ***Insert Crack parameters*** dialog box, a line (shown in red) representing the sharp crack is inserted in the PD mesh as shown in Fig. 13.92 and the status parameter of the bonds passing through this line are set to zero.

After the crack insertion step, follow steps from **Step 7** to **Step 9** as discussed in the previous case to generate the coupled PD/FE model for initial thermal strain. Copy the lines from 51 to 70 in the above input file and paste in the command window to generate the FE model surrounding the PD region. After the model setup at the end of **Step-9**, save the model database. The coupled PD/FE model with an initial crack in PD region is shown in Fig. 13.93. The PD region sharing the

Fig. 13.92 Line crack along Y axis of the plate

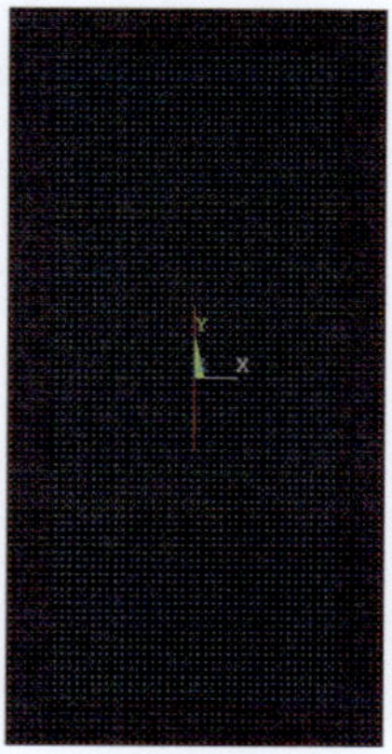

Fig. 13.93 PD-FE elements

Boundary ID=2,4 as interface is sandwiched between FE regions modeled with PLANE182 elements (shown in green).

In **Step-10,** the thermal loads in the PD region are introduced through the macro ***PD_inistrain.mac***. The procedure to use this macro and input parameters are similar to those of the preceding application.

Run the below command to enforce linear temperature variation only in X direction for $T(x) = 5x$ in the PD domain only

PD_inistrain,1,0,5,0.

Run the attached input file ***FE_inistrain. mac*** with command *FE_inistrain,1* to input the body loads due to the linear thermal load variation, $T(x) = 5x$ for the FE domain. The traction boundary conditions are naturally satisfied without any requirements. The displacement boundary conditions are directly enforced using Ansys D command by selecting the nodes on the left and right edges as shown in Fig. 13.94. The nodal DOF, UX and UY are constrained along the left edge. On the right edge of

Fig. 13.94 Boundary condition set up for the coupled PD/FE model with displacement constraints on the left edge displacement constraint on the right edge

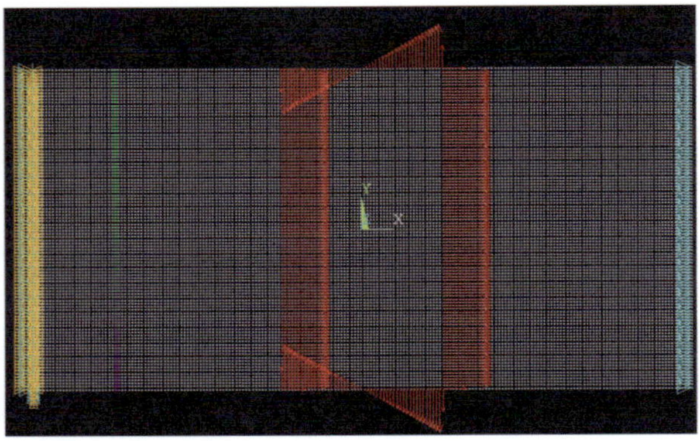

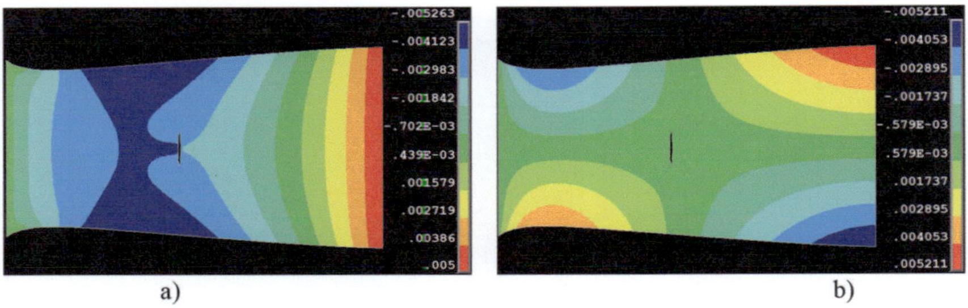

Fig. 13.95 Displacement contour plots of FE model of a plate with a central crack for ($\nu = 0.33$): (**a**) u_x and (**b**) u_y

the plate a non-zero displacement value of $u_0 = 0.005$ m is applied in X direction. The arrows shown in red color denote the traction vector due to the linear thermal load on the PD boundaries. The tractions due to the thermal load in the FEM region are treated by using the PLANE182 elements. To apply the displacement constraints and solve the model, copy the lines from 71 to 81.

Following the similar steps of post processing as in Sect. 13.7.2, Figs. 13.95 and 13.96 show the horizontal and vertical displacement variations, respectively, from FE and coupled BB-PD/FE analyses. Figures 13.97, 13.98, and 13.99 show the horizontal and vertical displacement variations, respectively, from FE, coupled OSB-PD/FE and NOSB-PD/FE analyses. Figures 13.100 and 13.101 show the comparison of coupled PD/FE displacement predictions along the center lines $y = 0$ and $x = 0$ with those of FEM. The coupled PD-FE predictions are in good agreement with FE results. Also, the predictions capture the FE results without any kinks near the PD-FE interface.

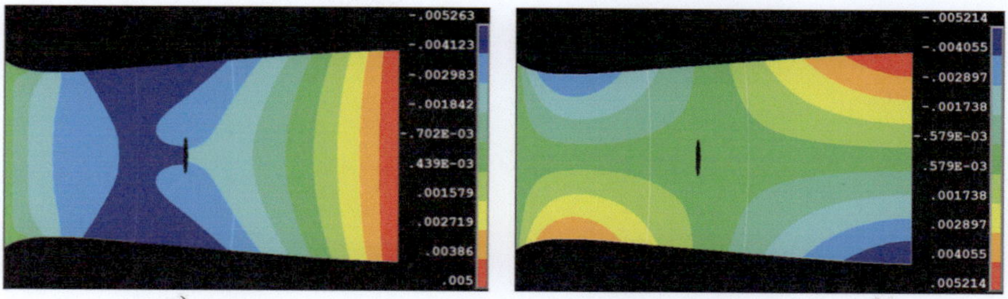

a) b)

Fig. 13.96 Displacement contour plots of coupled BB-PD/FE model of a plate with central crack: (**a**) u_x and (**b**) u_y

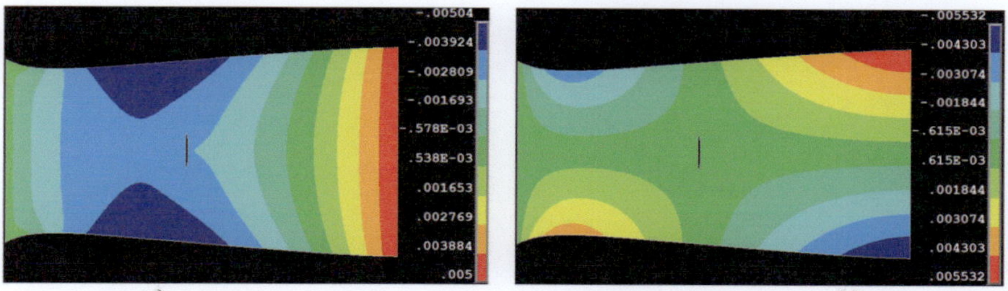

a) b)

Fig. 13.97 Displacement contour plots of FE model of a plate with a central crack ($\nu = 0.1$): (**a**) u_x and (**b**) u_y

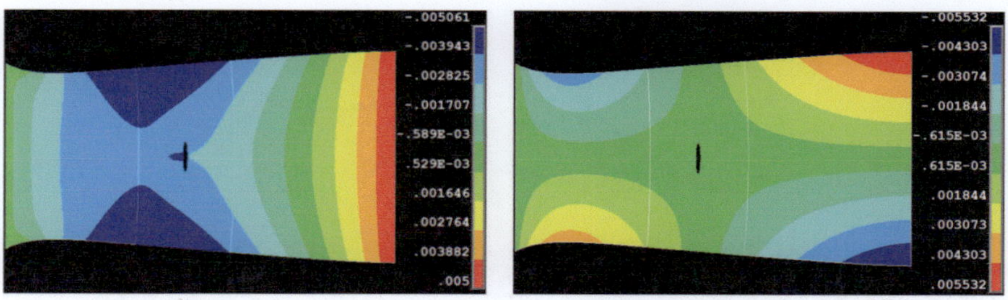

a) b)

Fig. 13.98 Displacement contour plots of coupled OSB-PD/FE model of a plate with a central crack ($\nu = 0.1$): (**a**) u_x and (**b**) u_y

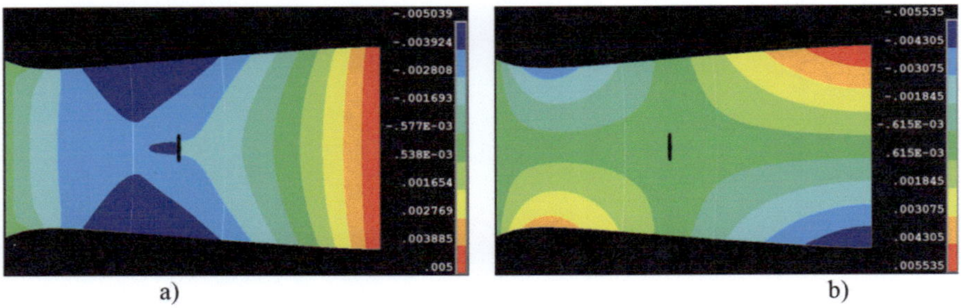

Fig. 13.99 Displacement contour plots of coupled NOSB-PD/FE model of a plate with a central crack ($\nu = 0.1$): (**a**) u_x and (**b**) u_y

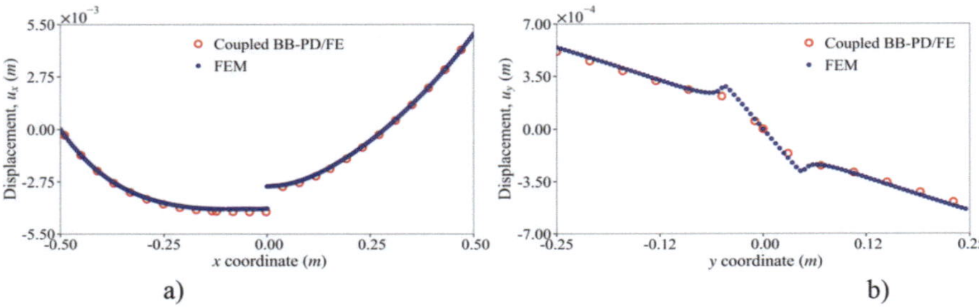

Fig. 13.100 Comparison of coupled BB-PD/FE and FEM displacement predictions along the horizontal and vertical centerlines of a plate subjected linear temperature load: (**a**) u_x and (**b**) u_y

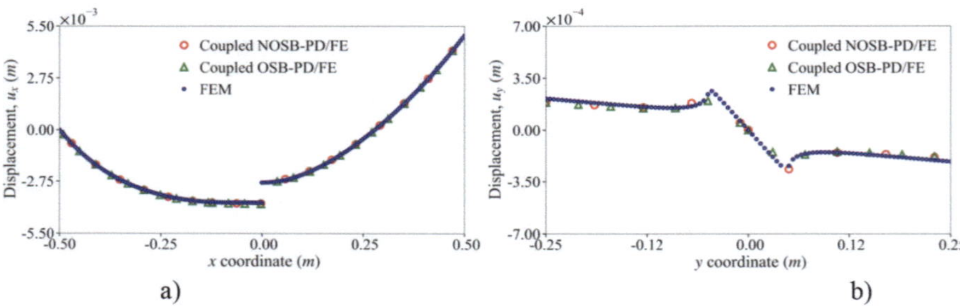

Fig. 13.101 Comparison of coupled OSB-PD/FE, NOSB-PD/FE and FEM displacement predictions along the horizontal and vertical centerlines of a plate subjected linear temperature load: (**a**) u_x and (**b**) u_y

References

Anicode, S. V. K., & Madenci, E. (2022). Seamless coupling of bond- and state-based peridynamic and finite element analyses. *Mechanics and materials., 173*, 104433.

Anicode, S. V. K., & Madenci, E. (2023). Direct coupling of dual-horizon peridynamics with finite elements for irregular discretization without an overlap zone. *Engineering with Computers, 40*, 1–31.

Chapter 14
Applications of Coupled Field Equations

14.1 Introduction

The applications in Chaps. 11–13 consider only a single field variable, such as temperature or displacement. This chapter presents applications to demonstrate the construction of solutions to the coupled field equations concerning Dual-Phase-Lag (DPL) thermal-moisture, electrodeposition, galvanic corrosion, and electromigration. The coupling among different fields is achieved through the approach explained in Chap. 7.

The DPL thermal-moisture diffusion includes the inertia and damping terms of the temperature and moisture concentration. Its specific applications concern thermal diffusion in nanoscale metal films due to ultrashort-pulse laser heating and a nonhomogeneous bar under sudden pulse of hygro-thermal loading. The modeling of electrodeposition and galvanic corrosion includes the field variables of electric potential and concentration. The governing equation of concentration includes nonlinear terms due to the gradient of the electric potential and concentration. In addition, the moving boundary due to the change of deposition or corrosion front is present. The specific applications concern deposition and corrosion of a flat metal surface. The modeling of electromigration includes the field variables of displacements, atomic concentration, electric potential, and temperature. The governing equation of concentration includes nonlinear terms due to the gradient of the bulk strain, electric potential, and concentration. It also presents moving boundary due to the void formation. Its specific application concerns deformation due to changes in atomic concentration and the prediction of void length in an unconstrained aluminum line. This chapter also provides the correspondence between the general form of coupled fields in Eq. (7.1) to those of the fields in these specific applications.

Supplementary Information The online version contains supplementary material available at https://doi.org/10.1007/978-3-031-59896-8_14.

E. Madenci et al., *Peridynamics and Its Applications Using Ansys*,
https://doi.org/10.1007/978-3-031-59896-8_14

The PD operations of the preprocessing, solution, and post-processing for each application are provided in the form of Ansys input files. The PD operations are executed through the *macro files*. The input files containing the encrypted macro files to generate the simulations are provided as supplementary material.

14.2 Coupled Dual-Phase-Lag Thermal-Moisture Fields

As derived by Xue et al. (2020), the coupled DPL thermal-moisture diffusion equation at point $\mathbf{x}$ in the domain can be cast as

$$\begin{bmatrix} \rho c_v \tau_q & -\eta \rho c_v \tau_q \\ -\lambda \tau_J & \tau_J \end{bmatrix} \begin{Bmatrix} \ddot{T}(\mathbf{x}, t) \\ \ddot{C}(\mathbf{x}, t) \end{Bmatrix} + \begin{bmatrix} \rho c_v & -\eta \rho c_v \\ -\lambda & 1 \end{bmatrix} \begin{Bmatrix} \dot{T}(\mathbf{x}, t) \\ \dot{C}(\mathbf{x}, t) \end{Bmatrix} = \begin{Bmatrix} k\nabla^2 T(\mathbf{x}, t) \\ D\nabla^2 C(\mathbf{x}, t) \end{Bmatrix} + \begin{Bmatrix} k\tau_T \nabla^2 \dot{T}(\mathbf{x}, t) \\ D\tau_C \nabla^2 \dot{C}(\mathbf{x}, t) \end{Bmatrix} + \begin{Bmatrix} S(\mathbf{x}, t) \\ G(\mathbf{x}, t) \end{Bmatrix} + \begin{Bmatrix} \tau_q \dot{S}(\mathbf{x}, t) \\ \tau_J \dot{G}(\mathbf{x}, t) \end{Bmatrix} \tag{14.1}$$

where η and λ are the coupling parameters, ρ (kg/m^3) is the mass density, c_v (J/(kg·°K)) is the specific heat capacity, T (°K) is the temperature, t (s) is the time, $S(\mathbf{x}, t)$ (J/(s · m^3)) is the volumetric thermal source, k (J/(s · m · °K)) is thermal conductivity coefficient, D (m^2/s) is the material diffusivity, τ_J (s) and τ_C (s) are the relaxation time of mass flux and phase lag of moisture gradient, respectively, τ_q (s) is the phase lag of heat flux or thermal relaxation time due to the fast transient effects of thermal inertia, and τ_T (s) represents the phase lag of temperature gradient or time delay due to the microstructural interactions. The $C(\mathbf{x}, t)$ (kg/m^3) is the moisture concentration and $G(\mathbf{x}, t)$ (kg/(s · m^3)) is the moisture source per unit time per unit mass of dry solid. The number of dots over a variable denotes the order of differentiation with respect to time. Anicode and Madenci (2023) also provides the detailed description of these equations.

This coupled DPL thermal-moisture diffusion equation involves temperature $T(\mathbf{x}, t)$ and moisture concentration $C(\mathbf{x}, t)$. It can be solved by using MATRIX27 with nonsymmetric stiffness and COMBIN14 elements native to Ansys. The stiffness matrices and damping coefficient can be constructed as presented in Chap. 7. Table 14.1 provides the correspondence between the general field variables and coefficients appearing in Eq. (7.1) to those of coupled DPL thermal-moisture diffusion, Eq. (14.1). The degree of freedom ROTX and ROTY represent the temperature and moisture concentration, respectively.

Table 14.1 The correspondence between the general form of coupled fields in Eq. (7.1) to those of DPL thermal-moisture diffusion in Eq. (14.1)

General form		DPL thermal-moisture fields
Field variables	$\mathbf{u}$	0
	ϕ	T
	φ	C
	ψ	0
Coefficients	$\mathbf{A}$	$\begin{bmatrix} \mathbf{0} & \mathbf{0} & \mathbf{0} & \mathbf{0} \\ \mathbf{0} & \rho c_v \tau_q & -\eta\rho c_v \tau_q & 0 \\ \mathbf{0} & -\lambda\tau_J & \tau_J & 0 \\ \mathbf{0} & 0 & 0 & 0 \end{bmatrix}$
	$\mathbf{B}$	$\begin{bmatrix} \mathbf{0} & \mathbf{0} & \mathbf{0} & \mathbf{0} \\ \mathbf{0} & \rho c_v & -\eta\rho c_v & 0 \\ \mathbf{0} & -\lambda & 1 & 0 \\ \mathbf{0} & 0 & 0 & 0 \end{bmatrix}$
	$\mathbf{C}$	$\mathbf{0}$
	$\mathbf{D}$	$\begin{bmatrix} 0 & 0 & 0 & 0 \\ 0 & k & 0 & 0 \\ 0 & 0 & D & 0 \\ 0 & 0 & 0 & 0 \end{bmatrix}$
	$\mathbf{P}$	$\begin{bmatrix} \mathbf{0} & \mathbf{0} & \mathbf{0} & \mathbf{0} \\ \mathbf{0} & k\tau_T & 0 & 0 \\ \mathbf{0} & 0 & D\tau_C & 0 \\ \mathbf{0} & 0 & 0 & 0 \end{bmatrix}$
	$\mathbf{Q}$	$\mathbf{0}$
	$\mathbf{H}$	$\mathbf{0}$
	s_ϕ	S
	$\dot{s}_\phi$	$\tau_q \dot{S}$
	s_φ	G
	s_ψ	$\tau_J \dot{G}$

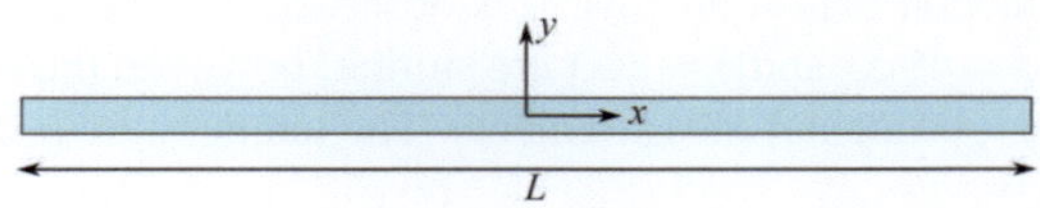

Fig. 14.1 Geometry of the nanoscale metal film

14.2.1 Ultrashort-Pulse Laser Heating of Nanoscale Metal Film

The heat conduction in a nanoscale metal film during ultrashort-pulse laser heating previously considered by Brorson et al. (1987) and Qiu and Tien (1994) and Wang et al. (2018) is modeled by specifying zero values for the coupling parameters and moisture concentration, i.e., $\eta = \lambda = C = 0$ in Eq. (14.1). It reduces to the PD DPL thermal conduction equation. The geometry and PD discretization of the film are shown in Figs. 14.1 and 14.2. The length of the film is specified as $L = 100$ nm. The

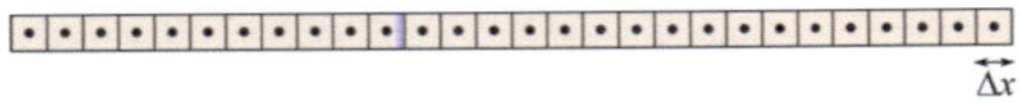

Fig. 14.2 PD discretization of the film

film insulated at both ends is subjected to a volumetric heating with a Gaussian profile as

$$S(x,t) = 0.94J_0\left(\frac{1-R}{\ell t_0}\right)e^{-\left[\frac{x}{\ell}+\frac{a_0|t-2t_0|}{t_0}\right]} \tag{14.2}$$

where $R = 0.93$, the laser influence parameter, $J_0 = 13.4$ J/m^2, the laser penetration depth, $\ell = 15.3$ nm, duration of laser pulse, $t_0 = 0.1$ ps, and the Gaussian parameter, $a_0 = 1.992$. The material properties of the film are conductivity, $k = 315$ W/(m°K), and thermal diffusivity, $\alpha = k/(\rho c_v) = 1.2 \times 10^{-4}$ m^2/s with $\rho c_v = 2625 \times 10^3$ J/(°Km^3). The phase lag parameters are specified as $\tau_q = 8.5$ ps and $\tau_T = 90$ ps.

The boundary conditions for the solution of the field equation are specified as

$$\frac{\partial T}{\partial x}(x = \pm L/2, t) = 0 \tag{14.3}$$

These boundary conditions are imposed through a boundary layer with a depth $\Delta = \Delta x$ along the boundary of the region.

The total simulation time is specified as $t = 2.5$ ps with a time step size of $\Delta t = 5 \times 10^{-5}$ ps. The discretization is achieved with 100 PD nodes with grid spacing of $\Delta x = L/100$. The horizon sizes of $\delta = 3\Delta x$ (weak nonlocality) and $\delta = 40\Delta x$ (strong nonlocality) are considered as part of the simulation.

At the center of the film ($x = 0$), the PD predictions for the normalized temperature over time for various models are shown in Fig. 14.3 along with the experimental measurements. It is observed that the temperature profiles for PD DPL with $\delta = 40\Delta x$ and $\delta = 3\Delta x$ are similar; however, the solution with $\delta = 40\Delta x$ captures the experimental measurements. The temperature profile from the PD thermal diffusion

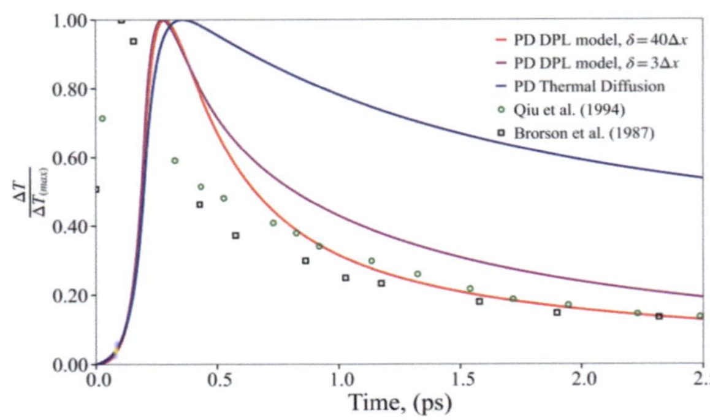

Fig. 14.3 PD predictions for normalized temperature variation at the center of the film over time

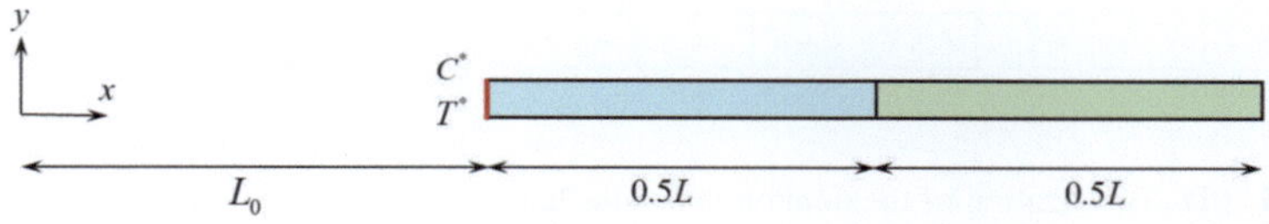

Fig. 14.4 Geometry of the nonhomogeneous bar

model with Fourier law does not capture the experimental measurements. The PD DPL model captures the nonlocal effects in heat conduction at nanoscale validating its efficacy.

14.2.2 A Bar Under Sudden Pulse of Hygro-Thermal Loading

The geometry of a nonhomogeneous bar in reference to the origin of a Cartesian coordinate system is shown in Fig. 14.4. The offset distance from the origin of the coordinate system is $L_0 = 2.7892 \times 10^{-2}$ cm. The length of the bar is specified as $L = 2L_0$. The right end of the bar is insulated, and the left end is subjected to a sudden pulse of hygro-thermal loading, i.e., $T^*(t) = 200H(t)^\circ$K and $C^*(t) = 2H(t)$ mol with $H(t)$ representing the Heaviside function. The reference temperature and moisture concentration are specified as $T_0 = 100^\circ$K and $C_0 = 1$ mol, respectively.

The thermal diffusivities of material-1 and material-2 are specified as $\alpha_1 = 1.0 \times 10^{-6}$ cm^2/h and $\alpha_2 = 7.78 \times 10^{-8}$ cm^2/h, respectively. Their moisture diffusivities are specified as $D_1 = 7.78 \times 10^{-7}$ cm^2/h and $D_2 = 7.78 \times 10^{-11}$ cm^2/h. As considered by Xue et al. (2020), the phase lag damping parameters for thermal and moisture diffusion are specified as $\tau_T = 0.02$ h and $\tau_C = 0.015$ h, respectively. The phase lag parameters for thermal and moisture inertia are $\tau_q = \tau_J = 0.2$ h. The coupling parameters are specified as $\eta = 0.25$, $\lambda = 0.0025$ respectively.

The boundary conditions for the solution of the coupled field equations are specified as

$$\frac{\partial T}{\partial x}(x = L_0 + L, t) = 0 \tag{14.4}$$

$$\frac{\partial C}{\partial x}(x = L_0 + L, t) = 0 \tag{14.5}$$

$$T(x = L_0, t) = 200H(t)^\circ\text{K} \tag{14.6}$$

and

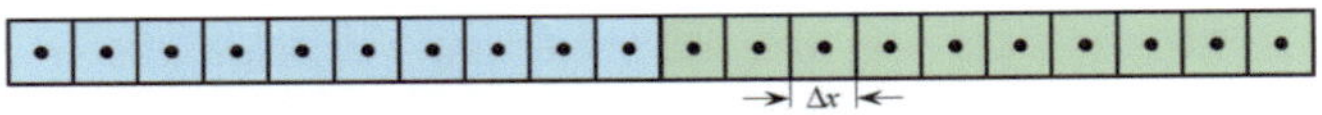

Fig. 14.5 PD discretization of the nonhomogeneous bar

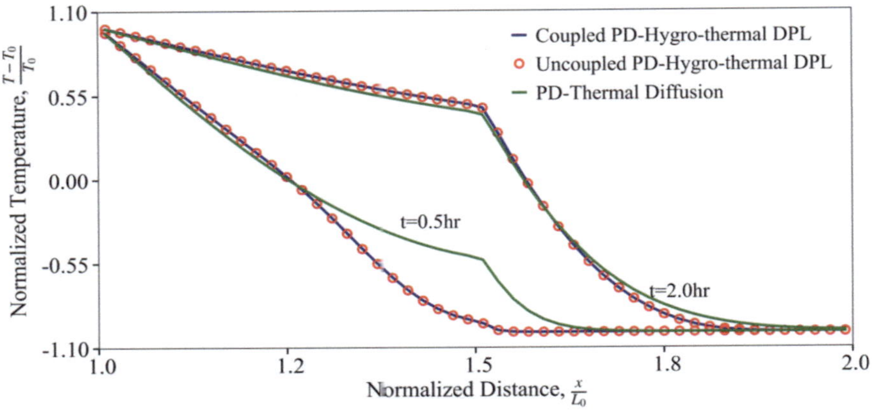

Fig. 14.6 PD DPL predictions of normalized temperature across the length of nonhomogeneous bar at varying times

$$C(x = L_0, t) = 2H(t) \text{ mol} \tag{14.7}$$

These boundary conditions are imposed through a boundary layer with a depth $\Delta = \Delta x$ along the boundary of the region. The simulation end time is $t = 2$ hr with a time step size of $\Delta t = 1 \times 10^{-3}$ hr. The discretization is achieved with 50 PD nodes with a grid spacing of $\Delta x = L_0/50$ and the horizon size of $\delta = 3\Delta x$ (Fig. 14.5).

Figures 14.6 and 14.7 show the variation of the normalized temperature and moisture profiles across the length of the bar for the coupled hygro-thermal PD DPL model. Because of the weak coupling effect of the moisture on the temperature, the temperature profiles as in Fig. 14.6 between the coupled and uncoupled PD-Hygro-thermal DPL models are almost similar. During the initial transient phase, the temperature profiles of DPL model differ significantly as compared to the PD-Thermal diffusion; however, as the time progress, the DPL and the PD-Thermal diffusion models converge.

14.3 Electrodeposition Process

The governing equations for modeling electrodeposition can be expressed as (Anicode et al., 2023)

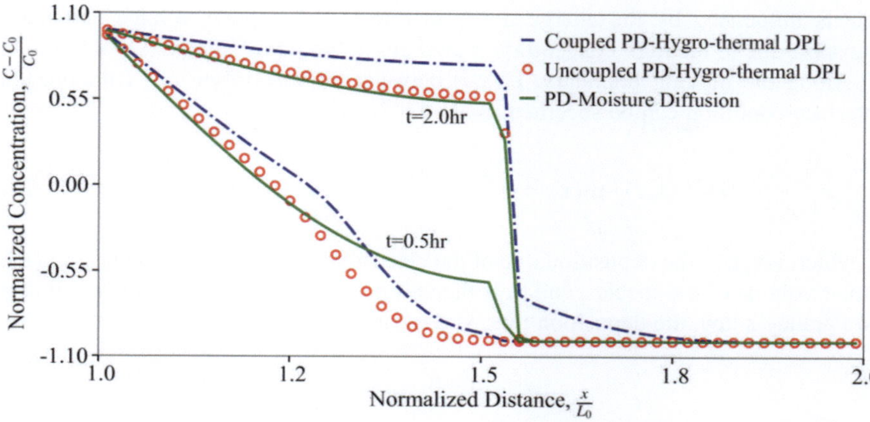

Fig. 14.7 PD DPL predictions of normalized moisture concentration across the length of nonhomogeneous bar at varying times

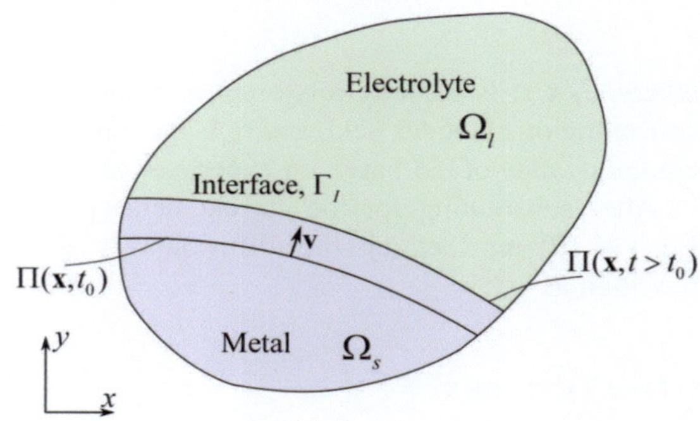

Fig. 14.8 Moving deposition front, $\Pi(\mathbf{x}, t)$ in the electrolyte domain during electrodeposition

$$\nabla^2\phi(\mathbf{x}) = 0 \qquad \forall \mathbf{x} \in \Omega_\ell \tag{14.8}$$

and

$$\frac{\partial C(\mathbf{x}, t)}{\partial t} = D\nabla^2 C(\mathbf{x}, t) + \frac{zFD}{RT}\nabla\phi(\mathbf{x}) \cdot \nabla C(\mathbf{x}, t) \qquad \forall \mathbf{x} \in \Omega_\ell \tag{14.9}$$

where $C(\mathbf{x}, t)$, D, and z represent the concentration, effective diffusion coefficient, and the charge number of the species, respectively. The electric potential of the field is $\phi(\mathbf{x}, t)$, the operating absolute temperature is T, the Faraday's constant is F, and the universal gas constant is R.

As shown in Fig. 14.8, domains Ω_ℓ and Ω_s represent the liquid and deposited metal regions, respectively. Deposition occurs on the surface of metal, which is the interface (dissolution region) between the medium (electrolyte) and metal, and its rate is influenced by the charge transport due to the applied voltage and mass transport due to the high concentration gradient in the liquid domain Ω_ℓ

Along the moving boundary, $\Gamma_I(\mathbf{x}, t)$ between the electrolyte and cathode, the interface condition can be specified as

$$-D\nabla C(\mathbf{x}, t)\cdot\mathbf{m}(\mathbf{x}) = \frac{\mathbf{v}(\mathbf{x}, t)\cdot\mathbf{m}(\mathbf{x})}{\Lambda} \qquad \forall\mathbf{x}\in\Gamma_I(\mathbf{x}, t) \tag{14.10}$$

in which $\mathbf{v}(\mathbf{x}, t)$ is the deposition rate of the electrolytic species on the metal, Λ is the molar volume of the species, and $\mathbf{m}$ is the unit normal to the cathode surface. Based on Faraday's law, the deposition rate, $\mathbf{v}(\mathbf{x}, t)$ can be expressed as

$$\mathbf{v}(\mathbf{x}, t) = \frac{\Lambda}{zF} i(\mathbf{x}, t)\mathbf{m} \tag{14.11}$$

The current density can be expressed as (Strusevich, 2013)

$$i(\mathbf{x}, t) = -i_0 \frac{C_I(\mathbf{x}, t)}{C_\infty} \exp\left[\frac{-\alpha_c\eta(\mathbf{x}, t)zF}{RT}\right] \tag{14.12}$$

where $C_I(\mathbf{x}, t)$ is the electrolyte concentration at the interface, C_∞ represents the concentration at the far field, and α_c is the transfer coefficients for the cathode with a common value of 0.5 based on experimental measurements.

After substituting for the current density from Eq. (14.12) into Eq. (14.11), Eq. (14.10) representing the diffusion flux along the interface, $\Gamma_I(\mathbf{x}, t)$, can be rewritten as

$$-D\nabla C(\mathbf{x}, t)\cdot\mathbf{m}(\mathbf{x}) = -\left(i_0 \frac{C_I(\mathbf{x}, t)}{C_\infty} \exp\left(\frac{-\alpha_c\eta(\mathbf{x}, t)zF}{RT}\right)\right) \quad \forall\mathbf{x}\in\Gamma_I(\mathbf{x}, t) \tag{14.13}$$

Based on the concentration, $C(\mathbf{x}, t)$, the volume fraction $\pi(\mathbf{x}, t)$ of the deposited material in the electrolytic domain can be determined from Eq. (14.11) along with Eq. (14.12) as

$$\pi(\mathbf{x}, t) = \frac{1}{V(\mathbf{x})}\int_{t_0}^{t} \frac{\Lambda}{zF}\left(\frac{i_0 e^{\left(\frac{-\alpha_c\eta(\mathbf{x}, t)zF}{RT}\right)}}{C_\infty}\right) C(\mathbf{x}, t)dt \tag{14.14}$$

where $V(\mathbf{x})$ is the volume of material point. As shown in Fig. 14.8, the deposition front is established by a set of points $\Pi(\mathbf{x}, t)$ defined as

$$\Pi(\mathbf{x}, t) = \{\mathbf{x} | \pi(\mathbf{x}, t) = 1\} \tag{14.15}$$

Table 14.2 The correspondence between the general form of coupled fields in Eq. (7.1) to those of electrodeposition model in Eqs. (14.8) and (14.9)

General form		Electrodeposition fields
Field variables	$\mathbf{u}$	0
	ϕ	ϕ
	φ	C
	ψ	0
Coefficients	$\mathbf{A}$	$\mathbf{0}$
	$\mathbf{B}$	$\begin{bmatrix} \mathbf{0} & \mathbf{0} & \mathbf{0} & \mathbf{0} \\ \mathbf{0} & 0 & 0 & 0 \\ \mathbf{0} & 0 & 1 & 0 \\ \mathbf{0} & 0 & 0 & 0 \end{bmatrix}$
	$\mathbf{C}$	$\mathbf{0}$
	$\mathbf{D}$	$\begin{bmatrix} 0 & 0 & 0 & 0 \\ 0 & 1 & 0 & 0 \\ 0 & 0 & D & 0 \\ 0 & 0 & 0 & 0 \end{bmatrix}$
	$\mathbf{P}$	$\mathbf{0}$
	$\mathbf{Q}$	$\mathbf{0}$
	$\mathbf{H}$	$\begin{bmatrix} \mathbf{0} & \mathbf{0} & \mathbf{0} & \mathbf{0} \\ \mathbf{0} & \mathbf{0} & \mathbf{0} & \mathbf{0} \\ \mathbf{0} & \mathbf{0} & \frac{zFD}{RT}\nabla\phi(\mathbf{x}) & \mathbf{0} \\ \mathbf{0} & \mathbf{0} & \mathbf{0} & \mathbf{0} \end{bmatrix}$
	s_ϕ	0
	$\dot{s}_\phi$	0
	s_φ	0
	s_ψ	0

The governing equations, Eqs. (14.8) and (14.9), for electrodeposition involve two field variables, i.e., the electrostatic potential $\phi(\mathbf{x})$ and electrolytic concentration $C(\mathbf{x}, t)$. Anicode and Madenci (2023) also provide the detailed description of these equations. They can be solved by using MATRIX27 with nonsymmetric stiffness and COMBIN14 elements native to Ansys. The stiffness matrices and damping coefficient can be constructed as presented in Chap. 7. Table 14.2 provides the correspondence between the general field variables and coefficients appearing in Eq. (7.1) to those of electrodeposition process, Eqs. (14.8) and (14.9). The degrees of freedom ROTX and ROTY represent the electrostatic potential and electrolytic concentration, respectively.

Electrodeposition on a Flat Metal Surface

The geometric configuration and PD discretization of the solution domain are shown in Fig. 14.9. The anode and cathode in contact with the electrolyte is separated by a distance of $H = 40\mu\text{m}$. The widths of the anode and cathode are much larger than the distance between them. Therefore, the solution domain can be modeled as infinite in

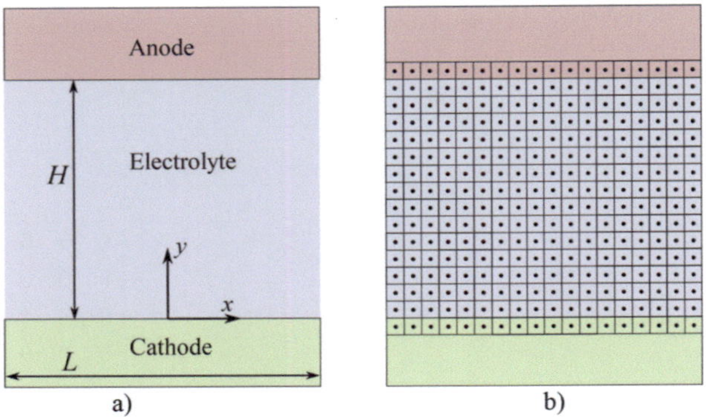

Fig. 14.9 Electrodeposition on a flat surface: (**a**) Geometric description, and (**b**) PD discretization

Fig. 14.10 PD discretization of the anode, cathode and electrolyte with additional nodes for periodic boundary conditions

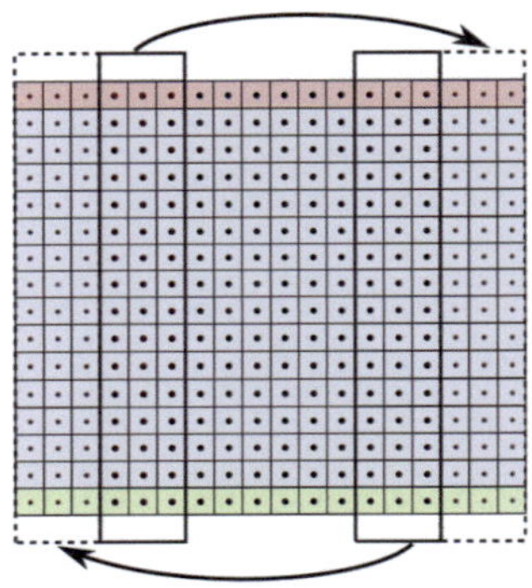

the x- direction. This condition is achieved by considering a finite width of $L = 8\mu m$ for the electrodes under periodic boundary conditions. A unique feature of the peridynamic differential operator (PDDO) is the enforcement of periodic boundary conditions. The continuity of the field variable at the boundaries defining the region is achieved by completing the family of a point, as shown in Fig. 14.10. The periodic boundary conditions are enforced without any additional constraint conditions.

The region of electrolyte between the anode and cathode is discretized with 20×100 PD nodes with a grid spacing of $\Delta x = 0.4\mu m$. In order to enforce the appropriate boundary conditions, the anode and cathode edges are modeled with a single layer of 20 PD nodes. Additional 3×100 PD nodes (shown in dashed boxes in Fig. 14.10) are introduced along the left and right boundaries of the electrolyte region to ensure the periodic boundary conditions (Madenci et al., 2017; Bekar & Madenci, 2021).

Table 14.3 Material properties for electrodeposition process

Property	Unit	Value
Cathode ion concentration, C_c	mol/m^3	1875
Copper ion concentration, C_∞	mol/m^3	1875
Anode potential, ϕ_a	V	$1.26e-4$
Cathode potential, ϕ_c	V	$-1.26e-4$
Overpotential, η	V	$-1.0e-4$
Diffusivity, D	m$^2\cdot$s^{-1}	$5.6e-10$
Temperature, T	K	303
Initial current density, i_0	A $\cdot$ m^{-2}	100

The boundary conditions for the solution of the field equations are specified as

$$\begin{aligned} &\phi(x, y=H, t)=\phi_a \quad \text{for} \quad -L/2 \leq x \leq L/2 \\ &\phi(x, y=0, t)=\phi_c \quad \text{for} \quad -L/2 \leq x \leq L/2 \\ &\phi\left(x=\frac{L}{2}, y, t\right)=\phi\left(x=-\frac{L}{2}, y, t\right) \quad \text{for} \quad 0 \leq y \leq H \end{aligned} \tag{14.16}$$

and

$$\begin{aligned} &C(x, y=H, t)=C_c \quad \text{for} \quad -L/2 \leq x \leq L/2 \\ &-D\frac{\partial C}{\partial y}(x, y(t), t)=-i_0\frac{C(x, y(t), t)}{C_\infty}\exp\left(\frac{-\alpha_c\eta(\mathbf{x}, t)zF}{RT}\right) \quad \text{for} \quad -L/2 \leq x \leq L/2 \\ &C\left(x=\frac{L}{2}, y, t\right)=C\left(x=-\frac{L}{2}, y, t\right) \quad \text{for} \quad 0 \leq y \leq H \end{aligned} \tag{14.17}$$

These boundary conditions are imposed through a boundary layer with a depth $\Delta=\Delta x=\Delta y$ along the boundary of the region. The $y(t)$ represents interface between the electrolyte and metal and changes as function of time. Its initial value is specified as $y(t=0)=0$. The simulation is performed with a time step size of $\Delta t=60s$ for a total time $t=3600s$. The material properties used for the simulation are shown in Table 14.3 (Strusevich, 2013).

Figure 14.11 shows the distribution of concentration in the electrolyte at various times. The deposition thickness variation for various times is shown in Fig. 14.12. The PD prediction of deposition rate closely recovers the previous results from COMSOL and PHYSICA (Strusevich, 2013). The velocity of the deposition front from the PD model is 0.22 μm, which compares closely to that of the values from PHYSICA and COMSOL as 0.2223 μm and 0.214 μm, respectively

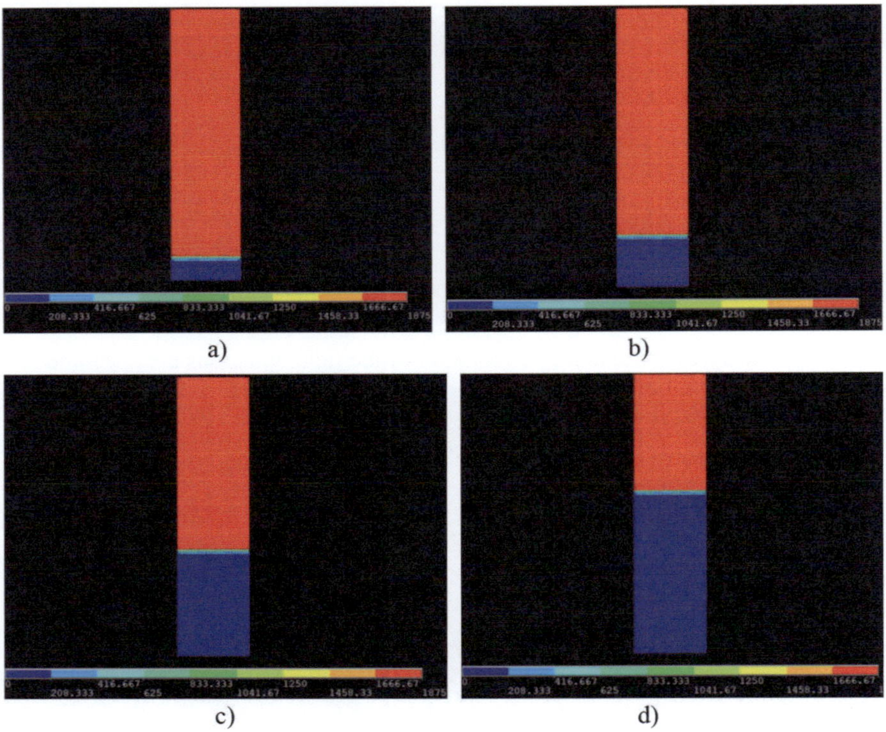

Fig. 14.11 Concentration distribution in the electrolyte at various times: (**a**) t = 10min, (**b**) t = 20min, (**c**) t = 40min, and (**d**) t = 60min

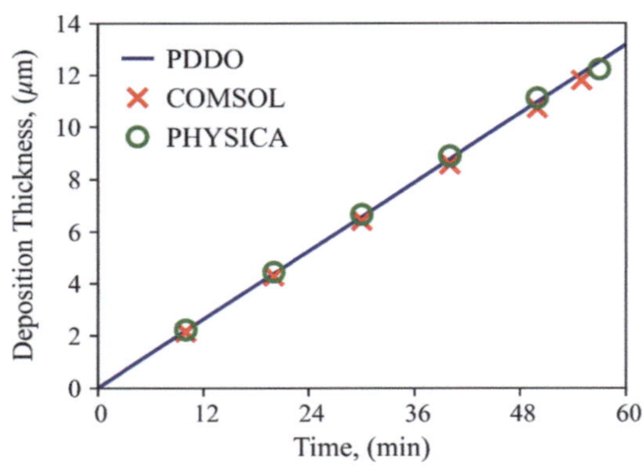

Fig. 14.12 Deposition thickness variation over time

14.4 Galvanic Corrosion Process

The governing equations for modeling galvanic corrosion can be expressed as (Anicode et al., 2023)

$$\nabla^2 \phi(\mathbf{x}, t) = 0 \qquad \forall \mathbf{x} \in \Omega_\ell \tag{14.18}$$

and

$$\frac{\partial C(\mathbf{x}, t)}{\partial t} = D\nabla^2 C(\mathbf{x}, t) + \frac{zFD}{RT} \nabla \phi(\mathbf{x}, t) \cdot \nabla C(\mathbf{x}, t) \qquad \forall \mathbf{x} \in \Omega_s \tag{14.19}$$

in which $C(\mathbf{x}, t)$ represents the concentration of aqueous solutions (electrolyte) and $\phi(\mathbf{x}, t)$ is the electric potential. In dilute aqueous solutions, the diffusion coefficient, D of most ions are similar and have values that at room temperature are in the range of 0.6×10^{-9} to 2×10^{-9} m^2/s. The parameters z, T, F, and R represent the charge number of the species, absolute temperature, Faraday's, and the universal gas constants, respectively.

As shown in Fig. 14.13, domains Ω_ℓ and Ω_s represent the liquid and metal regions, respectively. Corrosion (phase change) occurs the surface of metal, $\Gamma_f(\mathbf{x}, t)$, which is the interface (dissolution region) between the medium (electrolyte) and metal. Therefore, corrosion rate is not influenced by mass transport in the liquid domain Ω_ℓ (all liquid points always have zero metal-ion concentration). Also, mass transport in the solid Ω_s is too slow relative to the interfacial mass transport in the dissolution region. Therefore, diffusion can be disregarded in the liquid and solid regions.

For the determination of electric potential, the boundary conditions along the interfaces, $\Gamma_c(\mathbf{x}, t)$ and $\Gamma_a(\mathbf{x}, t)$ between the electrolyte and cathode and anode, respectively, can be imposed as

$$\phi(\mathbf{x}, t) = \phi_a \qquad \forall \mathbf{x} \in \Gamma_a(\mathbf{x}, t) \tag{14.20}$$

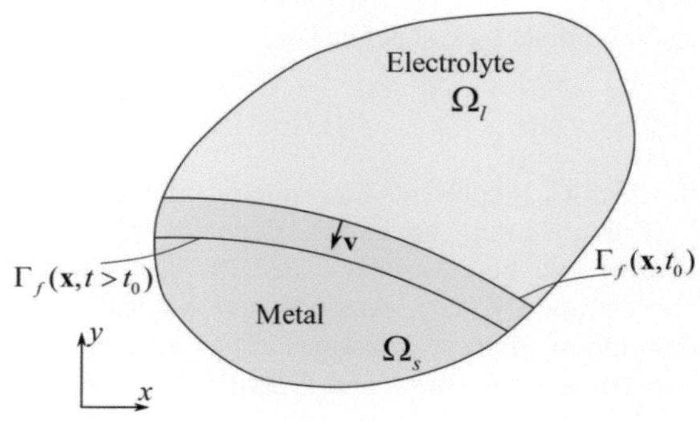

Fig. 14.13 Moving corrosion front, $\Gamma_f(\mathbf{x}, t)$ in the metallic region during galvanic corrosion

$$i(\mathbf{x},t) = -\sigma\nabla\phi(\mathbf{x},t)\cdot\mathbf{m}(\mathbf{x}) = i_0 10^{\left(\frac{\phi(\mathbf{x},t)-\phi_0}{\beta}\right)} \quad \forall\mathbf{x}\in\Gamma_c(\mathbf{x},t) \tag{14.21}$$

or in linearized form as

$$i(\mathbf{x},t) = -\sigma\nabla\phi(\mathbf{x},t)\cdot\mathbf{m}(\mathbf{x}) = i_0\left(1-\frac{\phi_0}{\beta}\log 10\right) + i_0\frac{\log 10}{\beta}\phi(\mathbf{x},t) \;\; \forall\mathbf{x}\in\Gamma_c(\mathbf{x},t) \tag{14.22}$$

In order to avoid Newton-Raphson iterations, Eq. (14.22) is used for enforcing electrode boundary condition.

Mass transport, Eq. (14.18) is considered only along the interface (dissolution region), $\Gamma_f(\mathbf{x}, t)$ as

$$\frac{\partial C(\mathbf{x},t)}{\partial t} = -D\nabla^2 C(\mathbf{x},t) - \frac{zFD}{RT}\nabla\phi(\mathbf{x},t)\cdot\nabla C(\mathbf{x},t) \quad \forall\mathbf{x}\in\Gamma_f \tag{14.23}$$

Because the migration term is negligible, this equation can be simplified as

$$\frac{\partial C(\mathbf{x},t)}{\partial t} = -D\nabla^2 C(\mathbf{x},t) \quad \forall\mathbf{x}\in\Gamma_f(\mathbf{x},t) \tag{14.24}$$

According to Faraday's law, the rate of material loss, $\dot{M}(\mathbf{x},t)$ per unit area in the progress of galvanic corrosion can be expressed as (Warkus et al., 2006)

$$\dot{M}(\mathbf{x},t) = \frac{\Lambda}{zF} i(\mathbf{x},t) \tag{14.25}$$

With this equation and Ohm's law, Eq. (14.24) can be further simplified as

$$\frac{\partial C(\mathbf{x},t)}{\partial t} = \frac{\Lambda}{zF} i(\mathbf{x},t) = -\frac{\Lambda}{zF}\sigma\nabla\phi(\mathbf{x},t)\cdot\mathbf{m}(\mathbf{x}) \quad \forall\mathbf{x}\in\Gamma_f(\mathbf{x},t) \tag{14.26}$$

It is subjected to an initial condition of $C(\mathbf{x}, t=0) = C_s$ and $\Gamma_f(\mathbf{x}, t)$ represents the moving corrosion front. As shown in Fig. 14.13, the corrosion front is established by a set of points $\Gamma_f(\mathbf{x}, t)$ defined as

$$\Gamma_f(\mathbf{x},t) = \{\mathbf{x} | C(\mathbf{x},t) = C_{sat}\} \tag{14.27}$$

in which C_{sat} represents the value of saturated concentration.

The governing equations, Eqs. (14.18) and (14.19), for galvanic corrosion involve two field variables, i.e., the electrostatic potential $\phi(\mathbf{x})$ and electrolytic concentration $C(\mathbf{x}, t)$. Anicode and Madenci (2023) also provides the detailed description of these equations. They can be solved by using MATRIX27 with nonsymmetric stiffness and COMBIN14 elements native to Ansys. The stiffness

Table 14.4 The correspondence between the general form of coupled fields in Eq. (7.1) to those of galvanic corrosion model in Eqs. (14.18) and (14.19)

General form		Galvanic corrosion fields
Field variables	$\mathbf{u}$	0
	ϕ	ϕ
	φ	C
	ψ	0
Coefficients	$\mathbf{A}$	$\mathbf{0}$
	$\mathbf{B}$	$\begin{bmatrix} \mathbf{0} & \mathbf{0} & \mathbf{0} & \mathbf{0} \\ \mathbf{0} & 0 & 0 & 0 \\ \mathbf{0} & 0 & 1 & 0 \\ \mathbf{0} & 0 & 0 & 0 \end{bmatrix}$
	$\mathbf{C}$	$\mathbf{0}$
	$\mathbf{D}$	$\begin{bmatrix} 0 & 0 & 0 & 0 \\ 0 & 1 & 0 & 0 \\ 0 & 0 & 0 & 0 \\ 0 & 0 & 0 & 0 \end{bmatrix}$
	$\mathbf{P}$	$\mathbf{0}$
	$\mathbf{Q}$	$\mathbf{0}$
	$\mathbf{H}$	$\mathbf{0}$
	s_ϕ	0
	$\dot{s}_\phi$	0
	s_φ	$\frac{\Lambda}{zF} i(\mathbf{x}, t)$ or $-\frac{\Lambda}{zF} \sigma \nabla \phi(\mathbf{x}, t) \cdot \mathbf{m}(\mathbf{x})$
	s_ψ	0

matrices and damping coefficient can be constructed as presented in Chap. 7. Table 14.4 provides the correspondence between the general field variables and coefficients appearing in Eq. (7.1) to those of electrodeposition process, Eqs. (14.18) and (14.19). The degree of freedom ROTX and ROTY represent the electrostatic potential and electrolytic concentration, respectively.

Galvanic Corrosion of a Flat Plate

The configuration of the cathode and anode with the electrolyte is shown in Fig. 14.14a, b. The solution domain is infinite in the x- direction. This condition is achieved by considering a finite width of $L = 10\text{mm}$ for the electrodes under periodic boundary conditions. The extent of anode and electrolyte domain are specified as $H_1 = H_2 = 100\text{mm}$ along the y-direction. The material properties are shown in Table 14.5.

As shown in Fig. 14.15, the region of electrolyte between the anode and cathode is discretized with 10×100 PD nodes with a grid spacing of $\Delta x = 1.0\text{mm}$. In order to enforce the appropriate boundary condition, the cathode edge is modeled with a single layer of 10 PD nodes. The anode is discretized with 10×49 nodes. Additional 3×100 nodes are introduced along the left and right boundaries (shown in dashed boxes in Fig. 14.16) of the solution domain in order to be able to enforce periodic boundary conditions (Madenci et al., 2017; Bekar & Madenci, 2021).

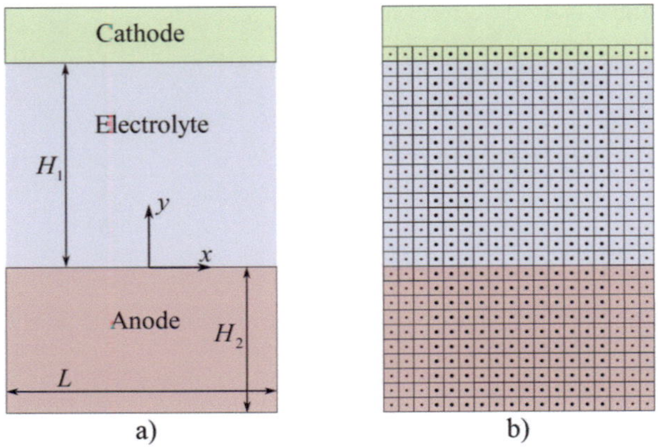

Fig. 14.14 Corrosion on a flat surface: (**a**) Geometric description, and (**b**) PD discretization

Table 14.5 Material properties for galvanic corrosion of a flat plate (Zhao et al., 2021)

Property	Unit	Value
Saturated ion concentration, C_{sat}	mol/m^3	5100
Anode ion concentration, C_s	mol/m^3	143000
Anode potential, ϕ_a	V	0.6
Cathode potential, ϕ_c	V	0.1
Electrical conductivity, σ	S. m^{-1}	2.5
Charge number, z	C	2
Faraday constant	F	96485.23

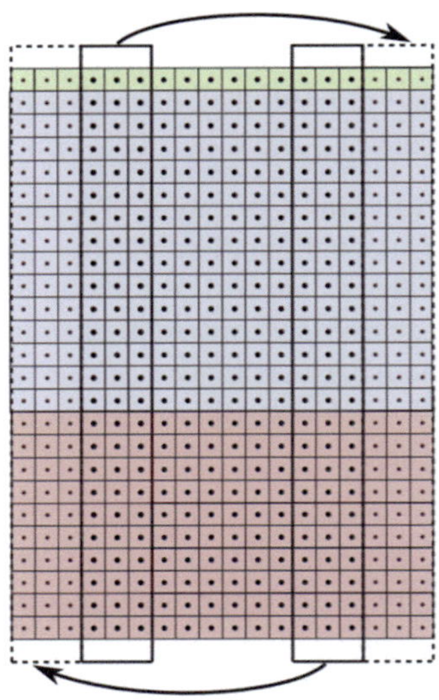

Fig. 14.15 PD discretization of electrolyte, anode and cathode boundary

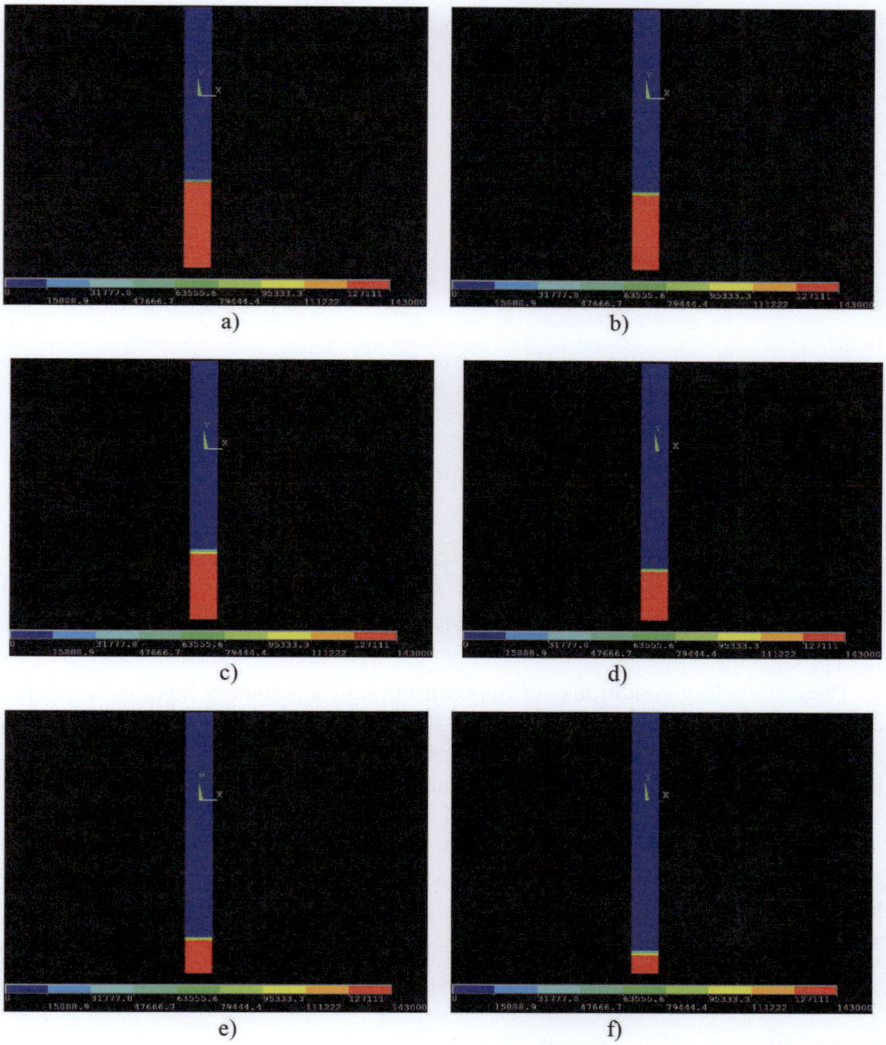

Fig. 14.16 Concentration distribution at various times during galvanic corrosion: (**a**) $t = 2.5 \times 10^4 s$, (**b**) $t = 1.25 \times 10^7 s$, (**c**) $t = 2.5 \times 10^7 s$, (**d**) $t = 5.0 \times 10^7 s$, (**e**) $t = 7.5 \times 10^7 s$, and (**f**) $t = 1.0 \times 10^8 s$

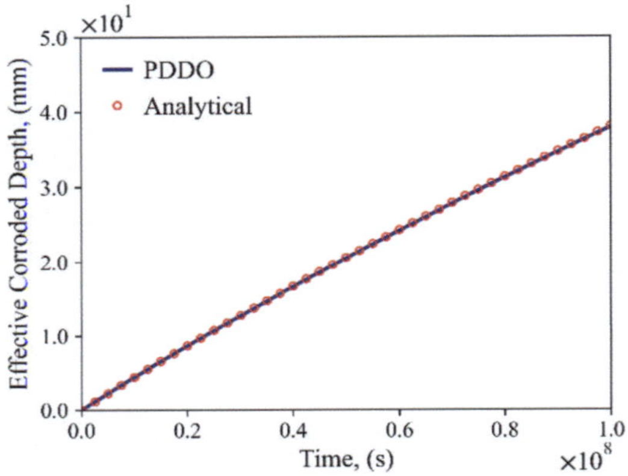

Fig. 14.17 Comparison of PD prediction with analytical solution for effective corrosion depth variation over time

The boundary conditions for the solution of the field equations are specified as

$$
\begin{aligned}
&\phi(x, y(t), t) = \phi_a \quad \text{for} \quad -L/2 \leq x \leq L/2 \\
&\phi(x, y = H_1, t) = \phi_c \quad \text{for} \quad -L/2 \leq x \leq L/2 \\
&\phi\left(x = \frac{L}{2}, y, t\right) = \phi\left(x = -\frac{L}{2}, y, t\right) \quad \text{for} \quad -H_2 \leq y \leq H_1
\end{aligned}
\tag{14.28}
$$

These boundary conditions are imposed through a boundary layer with a depth $\Delta = \Delta x = \Delta y$ along the boundary of the region. The variable, $y(t)$ represents interface between the electrolyte and metal and changes as function of time. Its initial value is specified as $y(t = 0) = 0$. The simulation is performed with a time step size of $\Delta t = 25000s$ for a total time $t = 1.0 \times 10^8 s$.

Figure 14.16 shows the distribution of the concentration in the electrolyte and anode at various times. The corrosion thickness variation for various times is shown in Fig. 14.17. The PD prediction of effective corrosion depth rate closely matches the analytical results.

14.5 Electromigration Process

Electromigration is a common phenomenon of mass transport due to the combination of interactions arising from the material and environmental characteristics, such as atomic concentration, electron wind, temperature gradient, and mechanical stress gradient. Considering self-diffusion, potential migration, mechanical stress migration, and thermal migration, the total atomic mass flux $\mathbf{J}_a(\mathbf{x}, t)$ can be expressed as (Cui et al., 2023)

$$\mathbf{J}_a(\mathbf{x},t) = D_a \begin{pmatrix} -\nabla C_a(\mathbf{x},t) - C_a(\mathbf{x},t)\dfrac{Z^* e \nabla \varphi(\mathbf{x},t)}{k_B T(\mathbf{x},t)} + C_a(\mathbf{x},t)\dfrac{\Omega \nabla \sigma^H(\mathbf{x},t)}{k_B T(\mathbf{x},t)} \\ -C_a(\mathbf{x},t)\dfrac{Q^*(\mathbf{x},t)\nabla T(\mathbf{x},t)}{k_B T^2(\mathbf{x},t)} \end{pmatrix} \tag{14.29}$$

where D_a is the atomic diffusivity, $C_a(\mathbf{x}, t)$ is the atomic concentration, Z^* is the charge number, e is the elementary charge, $\varphi(\mathbf{x}, t)$ is the electrical potential, $T(\mathbf{x}, t)$ is the temperature, Ω is atomic volume, $\sigma^H(\mathbf{x}, t)$ is the mechanical hydrostatic stress, k_B is the Boltzmann constant, and $Q^*(\mathbf{x}, t)$ is the heat of the atomic transport.

For the uniform and steady temperature and current distribution, the normalized atomic and rate of normalized atomic concentration can be written respectively as (Zhang et al., 2023)

$$\mathbf{J}(\mathbf{x},t) = D_a\left(-\left(1+\frac{\Omega B(1-f)}{k_B T}\right)\nabla C(\mathbf{x},t) - \frac{C(\mathbf{x},t) Z^* e \nabla\varphi(\mathbf{x},t)}{k_B T} + \frac{C(\mathbf{x},t)\Omega B \nabla\theta(\mathbf{x},t)}{k_B T}\right) \tag{14.30}$$

and

$$\frac{\partial C(\mathbf{x},t)}{\partial t} = \frac{C(\mathbf{x},t)}{(1-f)}\left[D_a\begin{pmatrix}\left(1+\dfrac{\Omega B(1-f)}{k_B T}\right)\nabla^2 C(\mathbf{x},t) + \dfrac{Z^* e}{k_B T}\nabla C(\mathbf{x},t)\cdot\nabla\varphi(\mathbf{x},t) \\ -\dfrac{\Omega B}{k_B T}\nabla C(\mathbf{x},t)\cdot\nabla\theta(\mathbf{x},t) - \dfrac{\Omega B C(\mathbf{x},t)}{k_B T}\nabla^2\theta(\mathbf{x},t)\end{pmatrix} - \dot{\theta}^{me}(\mathbf{x},t)\right] \tag{14.31}$$

with normalized atomic concentration $C(\mathbf{x}, t) = C_a(\mathbf{x}, t)/C_0$.

For the case of the stress-free with the uniform distribution of temperature, the total normalized atomic flux and the rate of normalized atom concentration are expressed as (Zhang et al., 2023)

$$\mathbf{J}(\mathbf{x},t) = D_a\left(-\nabla C(\mathbf{x},t) - \frac{C(\mathbf{x},t) Z^* e \nabla\varphi(\mathbf{x},t)}{k_B T}\right) \tag{14.32}$$

and

$$\frac{\partial C(\mathbf{x},t)}{\partial t} = \frac{C(\mathbf{x},t) D_a}{(1-f)}\left(\nabla^2 C(\mathbf{x},t) + \frac{Z^* e}{k_B T}\nabla C(\mathbf{x},t)\cdot\nabla\varphi(\mathbf{x},t)\right) \tag{14.33}$$

The known normalized atomic concentration, $C^*(\mathbf{x}, t)$ along the boundary (Dirichlet boundary condition), can be enforced as

$$C(\mathbf{x},t) = C^*(\mathbf{x},t) \text{ for } \mathbf{x} \in \mathscr{B}_C \tag{14.34}$$

in which $\mathscr{B}_C$ represents the boundary segment with known atomic concentration. The known normalized atomic flux, $J^*(\mathbf{x}, t)$ along the boundary (Neumann boundary condition) can be enforced as

$$\mathbf{J}(\mathbf{x}, t) \cdot \mathbf{n}(\mathbf{x}) = J^*(\mathbf{x}, t) \text{ for } \mathbf{x} \in \mathscr{B}_J \tag{14.35}$$

where $\mathbf{n}(\mathbf{x})$ is the unit normal vector to the boundary surface and $\mathscr{B}_J$ represents the boundary segment with known normalized atomic concentration.

Under the assumption of a small change in the atomic concentration, the equation of motion can be expressed as

$$\begin{aligned}\rho(\mathbf{x}, t)\ddot{\mathbf{u}}(\mathbf{x}, t) = & \mu\left(\nabla^2\mathbf{u}(\mathbf{x}, t) + 2\nabla\nabla \cdot \mathbf{u}(\mathbf{x}, t)\right) - (\mu - \lambda)\nabla\nabla \cdot \mathbf{u}(\mathbf{x}, t) \\ & - B(1 - f)\nabla C(\mathbf{x}, t) + \mathbf{b}(\mathbf{x}, t)\end{aligned} \tag{14.36}$$

It is worth noting that the term $(\mu - \lambda)$ in Eq. (14.36) disappears for 3D/2D plane strain and plane stress problems for Poisson's ratio of $\nu = 1/4$, and $\nu = 1/3$ respectively.

The deformation response of the material can be obtained by considering the applied boundary conditions

$$\boldsymbol{\sigma}(\mathbf{x}, t)\mathbf{n}(\mathbf{x}) = \mathbf{T}^*(\mathbf{x}, t) \text{ for } \mathbf{x} \in \mathscr{B}_{\boldsymbol{\sigma}} \tag{14.37}$$

and

$$\mathbf{u}(\mathbf{x}, t) = \mathbf{u}^*(\mathbf{x}, t) \text{ for } \mathbf{x} \in \mathscr{B}_{\mathbf{u}} \tag{14.38}$$

in which $\mathbf{u}^*(\mathbf{x}, t)$ is the known displacement on the boundary segment, $\mathscr{B}_{\mathbf{u}}$ and $\mathbf{T}^*(\mathbf{x}, t)$ is the known zero or non-zero force on the boundary segment $\mathscr{B}_{\boldsymbol{\sigma}}$.

Voids and hillocks develop when the concentration difference on the cathode and anode sides exceeds a critical value, which is defined as the difference between the values of $C_{hillock}(t)$ and $C_{void}(t)$. Blech (1998) derived the net drift velocity of the void front as

$$v_{void}(t) = v_{EM}(t) - v_{BF}(t) \tag{14.39}$$

in which $v_{EM}(t)$ and $v_{BF}(t)$ are the electromigration drift velocity and backflow velocity, respectively. They are expressed as

$$v_{EM}(t) = \frac{D_a}{k_B T} Z^* e\bar{\rho} j \tag{14.40}$$

and

$$v_{BF}(t) = \frac{D_b}{k_B T} \Delta F(t)/l(t) \tag{14.41}$$

in which j is the current density, $\overline{\rho}$ is the electrical resistivity, and $l(t) = L - l_{void}(t)$ is the current length with L representing the initial length of the material (strip) whose value decreases with the increase of void length $l_{void}(t)$. The free energy difference between the two ends of the strip for uniform temperature conditions is defined as $\Delta F(t) = k_B T \Delta C(t) - \Delta\sigma(t)$. Thus, the net drift velocity becomes (Belch, 1998)

$$v_{void}(t) = \frac{D_a}{k_B T}\left(-Z^* e\overline{\rho} j - \frac{k_B T \Delta C(t)}{l(t)} + \frac{\Delta\sigma(t)}{l(t)}\right) \tag{14.42}$$

where $\Delta C(t)$ and $\Delta\sigma(t)$ are the difference of atomic concentration and hydrostatic stress between two ends of the metal line, respectively.

The equation for the void length can be written as

$$l_{void}(t) = \int_0^t \frac{D_a}{k_B T}\left(-Z^* e\overline{\rho} j - \frac{k_B T(C_{hillock}(t) - C_{void}(t))}{l(t)} + \frac{(\sigma_{hillock}(t) - \sigma_{void}(t))}{l(t)}\right) dt \tag{14.43}$$

In Eqs. (14.31) and (14.33), the temperature is uniform and the current distribution is known; therefore, the governing equations for electromigration involve two field variables of displacement $\mathbf{u}(\mathbf{x}, t)$ and normalized atomic concentration $C(\mathbf{x}, t)$. Zhang et al. (2023) also provide the detailed description of these equations. They can be solved by using MATRIX27 with nonsymmetric stiffness and COMBIN14 elements native to Ansys. The stiffness matrices and damping coefficient can be constructed as presented in Chap. 7. Table 14.6 provides the correspondence between the general field variables and coefficients appearing in Eq. (7.1) to those of electromigration model, Eqs. (14.31), (14.33), and (14.36). The degree of freedom UX, UY, and ROTX represent the displacement components and normalized atomic concentration, respectively.

Unconstrained Aluminum Line

The 2D unconstrained aluminum line is shown in Fig. 14.18. The length and width of the strip are $L = 200\mu\text{m}$ and $W = 20\mu\text{m}$, respectively. The temperature in the strip is retained at $T = 500\text{K}$. The properties of the metal line are listed in Table. 14.7. There is no constraint along the boundaries; thus, allowing the strip to expand freely. Also, the total atomic flux on all boundaries is zero. The model is discretized with a uniform grid spacing $\Delta x = \Delta y = 2\mu\text{m}$. There are 1111 material points with 27112 MATRIX 27 elements and 891 dummy nodes with 891 COMBIN14 elements. The time step is specified as $\Delta t = 1\text{s}$.

Figure 14.19 shows the distributions of atomic concentration at different times. It is observed that the atomic concentration decreases on the cathode side and accumulates on the anode side. The transport direction of the atom is opposite to the

Table 14.6 The correspondence between the general form of coupled fields in Eq. (7.1) to those of electromigration model in Eqs. (14.31), (14.33), and (14.36)

General form		Electromigration: Constraint condition	Electromigration: Stress- free condition
Filed	$\mathbf{u}$	$\mathbf{u}$	
	ϕ	C	
	φ	0	
	ψ	0	
Property parameter	$\mathbf{A}$	$\mathbf{0}$	$\mathbf{0}$
	$\mathbf{B}$	$\begin{bmatrix} \mathbf{0} & \mathbf{0} & \mathbf{0} & \mathbf{0} \\ \mathbf{0} & \frac{(1-f)}{C(\mathbf{x},t)} & 0 & 0 \\ \mathbf{0} & 0 & 0 & 0 \\ \mathbf{0} & 0 & 0 & 0 \end{bmatrix}$	$\begin{bmatrix} \mathbf{0} & \mathbf{0} & \mathbf{0} & \mathbf{0} \\ \mathbf{0} & \frac{(1-f)}{C(\mathbf{x},t)} & 0 & 0 \\ \mathbf{0} & 0 & 0 & 0 \\ \mathbf{0} & 0 & 0 & 0 \end{bmatrix}$
	$\mathbf{C}$	$\begin{bmatrix} \mathbf{0} & \mathbf{0} & \mathbf{0} & \mathbf{0} \\ \mathbf{0} & \frac{\Omega B D_a \nabla^2 \theta(\mathbf{x},t)}{k_B T} & 0 & 0 \\ \mathbf{0} & 0 & 0 & 0 \\ \mathbf{0} & 0 & 0 & 0 \end{bmatrix}$	$\mathbf{0}$
	$\mathbf{D}$	$\begin{bmatrix} \mu & 0 & 0 & 0 \\ 0 & \left(1+\frac{\Omega B(1-f)}{k_B T}\right) D_a & 0 & 0 \\ 0 & 0 & 0 & 0 \\ 0 & 0 & 0 & 0 \end{bmatrix}$	$\begin{bmatrix} \mu & 0 & 0 & 0 \\ 0 & D_a & 0 & 0 \\ 0 & 0 & 0 & 0 \\ 0 & 0 & 0 & 0 \end{bmatrix}$
	$\mathbf{P}$	$\mathbf{0}$	$\mathbf{0}$
	$\mathbf{Q}$	$\mathbf{0}$	$\mathbf{0}$
	$\mathbf{H}$	$\begin{bmatrix} \mathbf{0} & \mathbf{0} & \mathbf{0} & \mathbf{0} \\ \mathbf{0} & \frac{D_a}{k_B T}\begin{pmatrix} Z^* e \nabla \varphi(\mathbf{x},t) + \\ \Omega B_a \nabla \theta(\mathbf{x},t) \end{pmatrix} & \mathbf{0} & \mathbf{0} \\ \mathbf{0} & \mathbf{0} & \mathbf{0} & \mathbf{0} \\ \mathbf{0} & \mathbf{0} & \mathbf{0} & \mathbf{0} \end{bmatrix}$	$\begin{bmatrix} \mathbf{0} & \mathbf{0} & \mathbf{0} & \mathbf{0} \\ \mathbf{0} & \frac{D_a}{k_B T} Z^* e \nabla \varphi(\mathbf{x},t) & \mathbf{0} & \mathbf{0} \\ \mathbf{0} & \mathbf{0} & \mathbf{0} & \mathbf{0} \\ \mathbf{0} & \mathbf{0} & \mathbf{0} & \mathbf{0} \end{bmatrix}$
	s_ϕ	$-\dot{\theta}^{me}$	0
	$\dot{s}_\phi$	0	0
	Note: the variables $C(\mathbf{x}, t)$ and $\theta(\mathbf{x}, t)$ in the property parameter are obtained from the previous calculation step.		

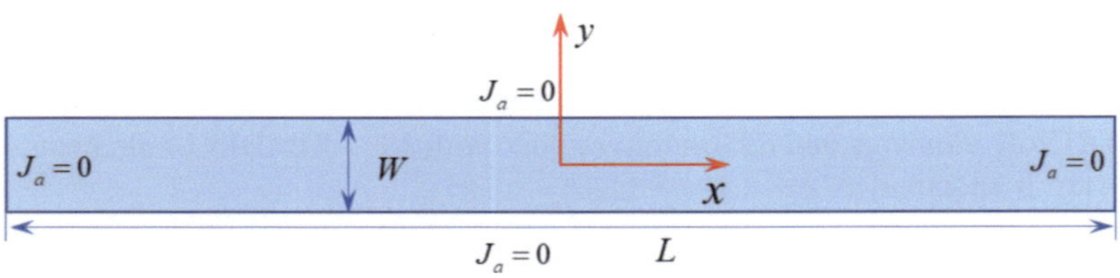

Fig. 14.18 The geometry and boundary conditions of a stress-free aluminum line

Table 14.7 Material properties for aluminum strip

Name	Value
Young's modulus (E)	70GPa
Poisson ratio (v)	0.33
Atomic diffusivity (D_a)	$1 \times 10^{-14}\,\text{m}^2/\text{s}$
Atomic volume (Ω)	$1.66 \times 10^{-29}\,\text{m}^3$
Electrical resistivity ($\overline{\rho}$)	$2.88 \times 10^{-8}\,\text{Ohm} \cdot \text{m}$
Current density (j)	$1 \times 10^{10}\,\text{A/m}^2$
Elementary charge (e)	$1.6 \times 10^{-19}\,\text{C}$
Charge number (Z^*)	1.1
Boltzmann constant (k_B)	$1.38 \times 10^{-23}\,\text{J/K}$
Vacancy relaxation factor (f)	0.7

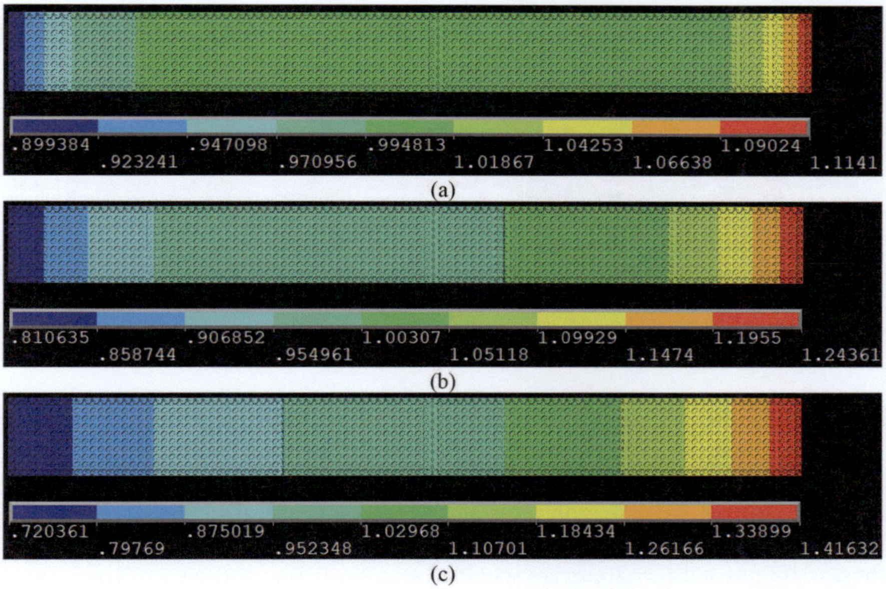

Fig. 14.19 The distribution of atomic concentration at: (**a**) 5×10^3s, (**b**) 2×10^4s, and (**c**) 5×10^4s

direction of current density. Fig. 14.20 shows the comparison of the atomic concentration along the length between the PD approach and COMSOL results at 5×10^3s, 2×10^4s, and 5×10^4s. The PD predictions of atomic concentration are in good agreement with those of COMSOL (Cui et al., 2022).

For the void nucleation and growth in the plate with a size ratio of $L/W = 10$, two cases with different lengths of $L = 100\mu\text{m}$ and $L = 20\mu\text{m}$ are simulated to compare with the experimental results. The Joule heating effect is ignored, and the temperature is retained as $T = 475\text{K}$. The Charge number is specified as $Z^* = 1.0$ and the critical concentration difference for the void formation in the strip equals to

Fig. 14.20 Comparison of the atomic concentrations between the PD and COMSOL predictions

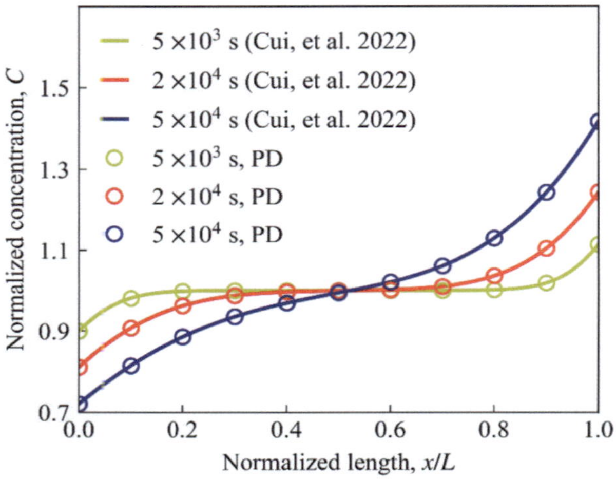

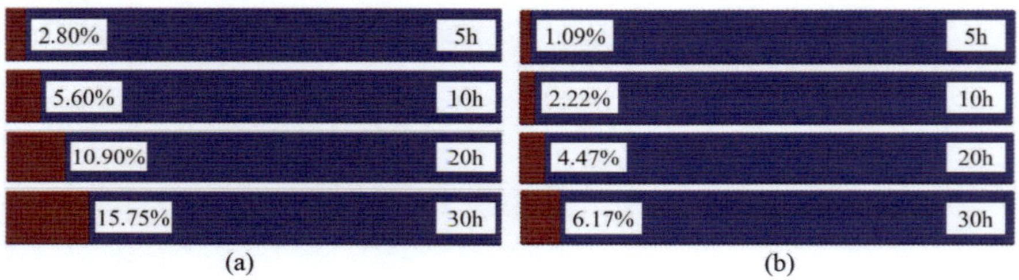

Fig. 14.21 Evolution of void at different times: (**a**) $L = 20\mu\text{m}$, and (**b**) $L = 100\mu\text{m}$

$\varDelta C = 0.075C$ (Cui, 2021). Other properties of the metal line are the same as those listed in Table. 14.7. There is no constraint on the boundary so the model can expand freely. The total atomic flux on all boundaries is specified as zero. The grid spacing for $L = 100\mu\text{m}$ and $L = 20\mu\text{m}$ are specified as $\Delta x = 0.5\mu\text{m}$ and $\Delta x = 0.1\mu\text{m}$. There are 1111 material points with 27112 MATRIX27 elements and 891 dummy nodes with 891 COMBIN14 elements. The time step is specified as $\varDelta t = 5\text{s}$. The vacancy relaxation factor f is expressed as (Cui et al., 2021)

$$f(C) = \begin{cases} 0.73, & |1 - C| \leq 0.0075 \\ 1.08\exp(-55.3|1 - C|), & |1 - C| > 0.0075 \end{cases} \tag{14.44}$$

Figure 14.21 shows the evolution of the void in the metal line at different times for the strip length $L = 20\mu\text{m}$ and$L = 100\mu\text{m}$. The void first forms on the cathode side and then propagates towards the anode side. The evolution direction of the void is similar to the transport direction of the atoms. With the same electromigration time, the degree of damage in the strip with a shorter length is more severe than that of with a longer length. After 30 hours, the proportion of the void region in the strip

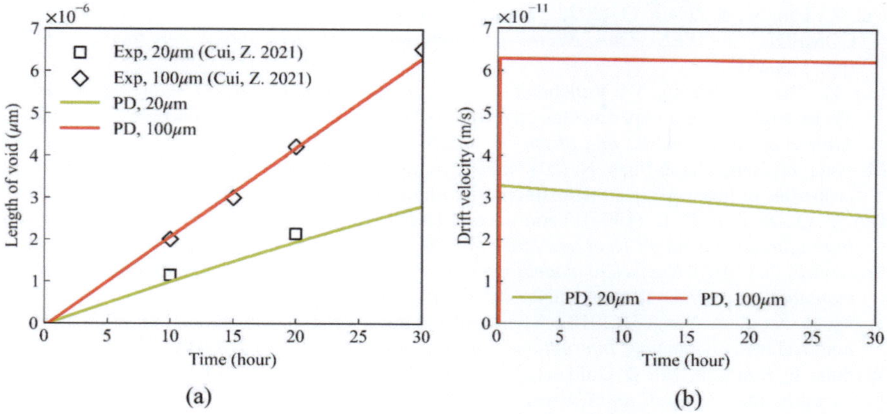

Fig. 14.22 History of: (**a**) void length, (**b**) drift velocity for different strip lengths

with length $L = 100\mu m$ is 6.17%, which is 15.75% for the strip with length $L = 20\mu m$.

Figure 14.22 shows the length of the void and drift velocity for different strip lengths. The PD predictions of the void length are in good agreement with the experimental results (Cui, 2021). With the increase in electromigration time, the void length keeps increasing almost linearly. The length of the metal line affects the drift velocity as shown in Fig. 14.22b. After the void formation, the drift velocity of the metal line with length $L = 100\mu m$ remains stable, however, the drift velocity of the metal line with length $L = 20\mu m$ decreases gradually.

References

Anicode, S. V. K., & Madenci, E. (2023). Peridynamic modeling of dual-phase-lag thermal-moisture coupling in a finite element framework. *Engineering with Computers, 39*, 911–923.

Anicode, S. V. K., Madenci, E., & Phan, N. (2023). A unified method to simulate electrodeposition and galvanic corrosion using the peridynamic differential operator. *Computer Methods in Applied Mechanics and Engineering, 408*, 115968.

Bekar, A. C., & Madenci, E. (2021). Peridynamics enabled learning partial differential equations. *Journal of Computational Physics, 434*, 110193.

Blech, I. A. (1998). Diffusional back flows during electromigration. *Acta Materialia, 46*, 3717–3723.

Brorson, S. D., Fujimoto, J. G., & Ippen, E. P. (1987). Femtosecond electronic heat-transport dynamics in thin gold films. *Physical Review Letters, 59*, 1962.

Cui, Z. (2021). *Multi-physics driven electromigration study: Multi-scale modeling and experiment* (Ph.D. dissertation). Delft University of Technology.

Cui, Z., Fan, X., & Zhang, G. (2021). Molecular dynamic study for concentration-dependent volume relaxation of vacancy. *Microelectronics Reliability, 120*, 114127.

Cui, Z., Fan, X., & Zhang, G. (2022). Implementation of fully coupled electromigration theory in COMSOL. In *IEEE 72nd electronic components and technology conference (ECTC)* (pp. 233–238).

Cui, Z., Fan, X., Zhang, Y., Vollebregt, S., Fan, J., & Zhang, G. (2023). Coupling model of electromigration and experimental verification–Part I: Effect of atomic concentration gradient. *Journal of the Mechanics and Physics of Solids, 174*, 105257.

Madenci, E., Barut, A., & Phan, N. (2017). Peridynamic unit cell homogenization for thermoelastic properties of heterogenous microstructures with defects. *Composite Structures, 188*, 104–115.

Qiu, T. Q., & Tien, C. L. (1994). Femtosecond laser heating of multi-layer metals—I. Analysis. *International Journal of Heat and Mass Transfer, 37*, 2789–2797.

Strusevich, N. (2013). *Numerical modelling of electrodeposition process for printed circuit boards manufacturing* (Ph.D. thesis). University of Greenwich, London, UK.

Wang, L., Xu, J., & Wang, J. (2018). A peridynamic framework and simulation of non-Fourier and nonlocal heat conduction. *International Journal of Heat and Mass Transfer, 118*, 1284–1292.

Warkus, J., Raupach, M., & Gulikers, J. (2006). Numerical modelling of corrosion-Theoretical backgrounds. *Materials and Corrosion, 57*, 614–617.

Xue, Z., Tian, X., & Liu, J. (2020). Hygrothermo elastic response in a hollow cylinder considering dual-phase-lag heat-moisture coupling. *Zeitschrift für angewandte Mathematik und Physik, 71*, 1–16.

Zhang, Y., Anicode, S. V. K., Fan, X., & Madenci, E. (2023). Peridynamic modeling of void nucleation and growth in metal lines due to electromigration in a finite element framework. *Computer Methods in Applied Mechanics and Engineering, 414*, 116183.

Zhao, J., Jafarzadeh, S., Rahmani, M., Chen, Z., Kim, Y. R., & Bobaru, F. (2021). A peridynamic model for galvanic corrosion and fracture. *Electrochimica Acta, 391*, 138968.

Chapter 15
Peridynamics for Failure Prediction

15.1 Introduction

Peridynamic theory enables the simulation of crack initiation and propagation by breaking the bonds between the material points. The bond breakage can be implemented based on failure criteria such as critical stretch (Silling & Askari, 2005). The critical stretch value depends on the horizon size and material property. Generally, the PD model is discretized with uniform grid spacing with a constant critical stretch value. This chapter first presents an approach to numerically determine the critical bond stretch value for bond-based (BB), ordinary state-based (OSB), and non-ordinary state-based (NOSB) peridynamic (PD) models. Subsequently, it presents an implicit bond breakage algorithm for crack growth while avoiding the potential singularity of the global stiffness matrix.

The specific applications concern double cantilever beam (DCB) and three-point bending (TPB) specimens modeled through uniform and nonuniform discretization. For computational efficiency, the models are constructed by coupling the finite element (FE) and PD regions based on the approach described in Chap. 5. The solutions to all these applications are constructed using Ansys GUI in a step-by-step fashion and compared with finite element analysis (FEA) solutions. The PD operations of the pre-processing, solution, and post-processing for each application are provided in the form of Ansys input files. The PD operations are executed through the Ansys Macro Files. The input files containing the encrypted macro files to generate the simulations are also provided as supplementary material.

Supplementary Information The online version contains supplementary material available at https://doi.org/10.1007/978-3-031-59896-8_15.

E. Madenci et al., *Peridynamics and Its Applications Using Ansys*,
https://doi.org/10.1007/978-3-031-59896-8_15

15.2 Critical Stretch Criteria for Crack Initiation and Growth

As introduced by Silling and Askari (2005), the bond breakage occurs when the stretch, s between two material points exceeds their critical values. During each load step, the interaction between the bonds is monitored through a bond status parameter given by

$$\widehat{\mu}(\mathbf{x}' - \mathbf{x}) = \begin{cases} 1, & s \leq s_c \\ 0, & s \geq s_c \end{cases} \tag{15.1}$$

where $\widehat{\mu}(\mathbf{x}' - \mathbf{x})$ represents the status of bond connecting point $\mathbf{x}$ and $\mathbf{x}'$. Therefore, damage is reflected in the equilibrium equations of motion by removing the force density vectors between the material points in an irreversible manner. As a result, the load is redistributed among the material points in the body, leading to damage nucleation and growth in an autonomous fashion. The force interactions are accounted for through the bond status parameter.

Based on the status of interaction of point $\mathbf{x}$ with all the bonds in its family, the local damage $\phi(\mathbf{x})$ is evaluated as (Silling & Askari, 2005)

$$\phi(\mathbf{x}) = 1 - \frac{\int_{H_\mathbf{x}} \widehat{\mu}(\mathbf{x}' - \mathbf{x}) dV_{\mathbf{x}'}}{\int_{H_\mathbf{x}} dV_{\mathbf{x}'}} \tag{15.2}$$

It is defined as the weighted ratio of the number of broken bonds to the total number of initial bonds of a material point with its family members. The measure of local damage is an indicator of possible crack formation within a body. Initially a material point interacts with all materials in its horizon; thus, the local damage has a value of 0. However, the creation of a crack terminates half of the interactions within its horizon resulting in a local damage value of ½.

For a linear elastic brittle material, the value of critical stretch s_c can be expressed in terms of its critical energy release rates, G_{Ic} for opening modes of fracture, respectively (Silling & Askari, 2005). It depends on the horizon size and material property. Its value can be determined analytically for BB- and OSB-PD models with uniform grid spacing (Madenci & Oterkus, 2013). However, it requires numerical evaluation for NOSB models.

15.3 Numerical Evaluation of Critical Stretch Value

As suggested by Silling and Askari (2005), the critical stretch s_c for a linear elastic brittle material can be determined by equating the energy released for a unit extension of the crack to the critical energy release rate, G_{Ic} for the opening mode of fracture. The energy released for a unit extension of the crack can be computed by summing up the energy of the total number of bonds passing through the crack of unit length/area. As shown in Fig. 15.1, for a bond $k - j$ passing through the crack opening of incremental area ΔA, $\beta_{(k)(j)}$ represents the bond angle with respect to the normal to crack surface.

The bond force component normal to the crack surface is given by

$$f^n_{(k)(j)} = \left(\mathbf{t}^\alpha_{(k)(j)} - \mathbf{t}^\alpha_{(j)(k)}\right) \cdot \mathbf{n}_{(k)(j)} \cos\left(\beta_{(k)(j)}\right) \tag{15.3}$$

where $\mathbf{t}^\alpha_{(k)(j)}$ and $\mathbf{t}^\alpha_{(j)(k)}$ are the bond forces with $\alpha = BB, OSB, NOSB$, and $\mathbf{n}_{(k)(j)}$ is the bond direction defined as

$$\mathbf{n}_{(k)(j)} = \frac{\boldsymbol{\xi}_{(k)(j)}}{\left|\boldsymbol{\xi}_{(k)(j)}\right|} \tag{15.4}$$

The displacement component normal to the crack surface is defined as

$$u^n_{(k)(j)} = s_0 \left|\boldsymbol{\xi}_{(k)(j)}\right| \cos\left(\beta_{(k)(j)}\right) \tag{15.5}$$

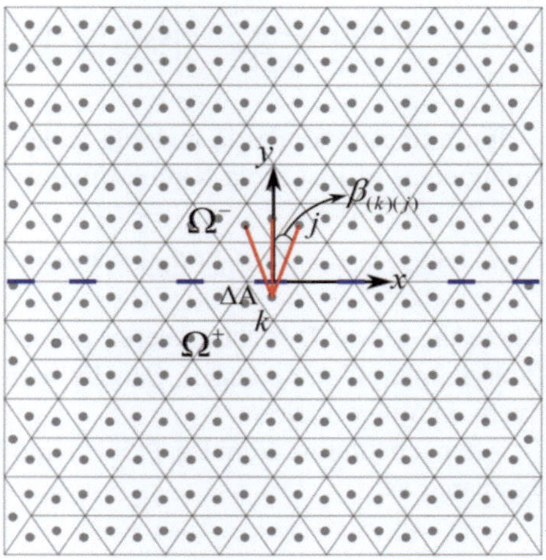

Fig. 15.1 PD bonds crossing the incremental crack surface, ΔA and their orientation angle, $\beta_{(k)(j)}$

The incremental strain energy density of the bond $k-j$ due to an isotropic expansion of s_0 is computed as

$$\Delta W_{(k)(j)}=\frac{1}{2}f^{n}_{(k)(j)}u^{n}_{(k)(j)}=\\ \frac{1}{2}\left(\mathbf{t}^{\alpha}_{(k)(j)}-\mathbf{t}^{\alpha}_{(j)(k)}\right)\cdot\mathbf{n}_{(k)(j)}\cos^{2}\left(\beta_{(k)(j)}\right)s_0\left|\boldsymbol{\xi}_{(k)(j)}\right|V_{(j)} \tag{15.6}$$

Strain energy increment per unit area due to the opening mode for all bonds passing through the crack is computed as

$$\frac{\Delta U}{\Delta A}=\frac{\sum_{k}^{\Omega^{+}}\sum_{j}^{\Omega^{-}}\Delta W_{(k)(j)}V_{(k)}}{\Delta A}=\\ \frac{1}{2\Delta A}\sum_{k}^{\Omega^{+}}\sum_{j}^{\Omega^{-}}\left(\mathbf{t}^{\alpha}_{(k)(j)}-\mathbf{t}^{\alpha}_{(j)(k)}\right)\cdot\mathbf{n}_{(k)(j)}\cos^{2}\left(\beta_{(k)(j)}\right)s_0\left|\boldsymbol{\xi}_{(k)(j)}\right|V_{(k)}V_{(j)} \tag{15.7}$$

The discrete from of force density vector in NOSB-PD can be written as

$$\mathbf{t}^{NOSB}_{(k)(j)}=\mathbf{P}_{\boldsymbol{\xi}_{(k)(j)}}\left(\mathbf{x}_{(k)}\right)\mathbf{g}\left(\mathbf{x}_{(k)},\boldsymbol{\xi}_{(k)(j)}\right) \tag{15.8}$$

with the vector, $\mathbf{g}(\mathbf{x}_{(k)},\boldsymbol{\xi}_{(k)(j)})$ defined as

$$\mathbf{g}\left(\mathbf{x}_{(k)},\boldsymbol{\xi}_{(k)(j)}\right)=w\left(|\boldsymbol{\xi}_{(k)(j)}|\right)\mathbf{K}^{-1}\left(\mathbf{x}_{(k)}\right)\boldsymbol{\xi}_{(k)(j)} \tag{15.9}$$

in which $w(|\boldsymbol{\xi}_{(k)(j)}|)$ is the weight function and the shape tensor, $\mathbf{K}(\mathbf{x}_{(k)})$ is defined as

$$\mathbf{K}\left(\mathbf{x}_{(k)}\right)=\sum_{p=1}^{N_{(k)}}w\left(\left|\boldsymbol{\xi}_{(k)(k_p)}\right|\right)\left(\boldsymbol{\xi}_{(k)(k_p)}\otimes\boldsymbol{\xi}_{(k)(k_p)}\right)\Delta V_{(k_p)} \tag{15.10}$$

For a linear and elastic material under the assumption of small deformation, the components of first Piola-Kirchhoff stress can be expressed as

$$\widehat{\mathbf{P}}_{\boldsymbol{\xi}(k)(j)}\left(\mathbf{x}_{(k)}\right)=\widehat{\boldsymbol{\sigma}}_{\boldsymbol{\xi}(k)(j)}\left(\mathbf{x}_{(k)}\right)=\mathbf{D}\widehat{\boldsymbol{\varepsilon}}_{\boldsymbol{\xi}(k)(j)}\left(\mathbf{x}_{(k)}\right) \tag{15.11}$$

where the tensor $\mathbf{D}$ is the material property matrix, and $\widehat{\mathbf{P}}_{\boldsymbol{\xi}}$, $\widehat{\boldsymbol{\sigma}}_{\boldsymbol{\xi}}$ and $\widehat{\boldsymbol{\varepsilon}}_{\boldsymbol{\xi}}$ are the first Piola-Kirchhoff stress tensor, Cauchy's stress tensor and linear strain tensor, respectively, in Voigt notation.

After substituting from Eq. (15.11) into Eq. (15.8), the force density vector can be expressed as

$$\mathbf{t}_{(k)(j)}^{NOSB} = \mathbf{G}_R\left(\mathbf{x}_{(k)}, \boldsymbol{\xi}_{(k)(j)}\right)\mathbf{D}\widehat{\boldsymbol{\varepsilon}}_{\xi(k)(j)} \tag{15.12}$$

and

$$\mathbf{t}_{(j)(k)}^{NOSB} = \mathbf{G}_R\left(\mathbf{x}_{(j)}, \boldsymbol{\xi}_{(j)(k)}\right)\mathbf{D}\widehat{\boldsymbol{\varepsilon}}_{\xi(j)(k)} \tag{15.13}$$

where $\mathbf{G}_R$ is defined as

$$\mathbf{G}_R\left(\boldsymbol{\xi}_{(k)(j)}\right) = \begin{bmatrix} g_1^{10} & 0 & g_1^{01} \\ 0 & g_1^{01} & g_1^{10} \end{bmatrix}_{2\times 3} \quad \text{for 2D} \tag{15.14}$$

and

$$\mathbf{G}_R\left(\boldsymbol{\xi}_{(k)(j)}\right) = \begin{bmatrix} g_1^{100} & 0 & 0 & g_1^{010} & g_1^{001} & 0 \\ 0 & g_1^{010} & 0 & g_1^{100} & 0 & g_1^{001} \\ 0 & 0 & g_1^{001} & 0 & g_1^{100} & g_1^{010} \end{bmatrix}_{3\times 6} \quad \text{for 3D} \tag{15.15}$$

The strain tensor associated with a bond $\boldsymbol{\xi}_{(k)(j)}$ at point $\mathbf{x}_{(k)}$ can be evaluated as

$$\begin{aligned}&\boldsymbol{\varepsilon}_{\xi(k)(j)}\left(\mathbf{x}_{(k)}\right)\\ &= \frac{1}{2}\sum_{m\in H_{(i)}\cap H_{(j)}} \begin{pmatrix} \left(\mathbf{u}_{(m)} - \mathbf{u}_{(k)}\right) \otimes \mathbf{g}_{\xi(k)(m)}\left(\mathbf{x}_{(k)}, \boldsymbol{\xi}_{(k)(m)}\right) \\ +\mathbf{g}_{\xi(k)(m)}\left(\mathbf{x}_{(k)}, \boldsymbol{\xi}_{(k)(m)}\right) \otimes \left(\mathbf{u}_{(m)} - \mathbf{u}_{(k)}\right) \end{pmatrix} \Delta V_{(m)}\end{aligned} \tag{15.16}$$

For an isotropic expansion of $\mathbf{u}_{(m)} - \mathbf{u}_{(i)} = s_0(\mathbf{x}_{(m)} - \mathbf{x}_{(i)})$, Eq. (15.16) reduces to

$$\begin{aligned}\boldsymbol{\varepsilon}_{\xi(k)(j)}\left(\mathbf{x}_{(k)}\right) &= \frac{1}{2}\sum_{m\in H_{(i)}\cap H_{(j)}} \begin{pmatrix} s_0\left(\mathbf{x}_{(m)} - \mathbf{x}_{(k)}\right) \otimes \mathbf{g}_{\xi(k)(m)}\left(\mathbf{x}_{(k)}, \boldsymbol{\xi}_{(k)(m)}\right) \\ +\mathbf{g}_{\xi(k)(m)}\left(\mathbf{x}_{(k)}, \boldsymbol{\xi}_{(i)(m)}\right) \otimes s_0\left(\mathbf{x}_{(m)} - \mathbf{x}_{(k)}\right) \end{pmatrix} \Delta V_{(m)} \\ &= \frac{s_0}{2}\sum_{m\in H_{(i)}\cap H_{(j)}} \begin{pmatrix} \left(\mathbf{x}_{(m)} - \mathbf{x}_{(k)}\right) \otimes \mathbf{g}_{\xi(k)(m)}\left(\mathbf{x}_{(k)}, \boldsymbol{\xi}_{(k)(m)}\right) \\ +\mathbf{g}_{\xi(k)(m)}\left(\mathbf{x}_{(k)}, \boldsymbol{\xi}_{(i)(m)}\right) \otimes \left(\mathbf{x}_{(m)} - \mathbf{x}_{(k)}\right) \end{pmatrix} \Delta V_{(m)} = s_0\mathbf{b}_{\xi(k)(j)}\end{aligned} \tag{15.17}$$

Substituting from Eq. (15.17) into Eqs. (15.12) and (15.13) and using the definition of $\mathbf{n}_{(k)(j)}$ yields the expression for the strain energy increment ΔU per unit area of ΔA as

$$\frac{\Delta U}{\Delta A} = \frac{\sum_{k}^{\Omega^+}\sum_{j}^{\Omega^-}\Delta W_{(k)(j)}V_{(k)}}{\Delta A}$$
$$= \frac{1}{2\Delta A}\sum_{k}^{\Omega^+}\sum_{j}^{\Omega^-}\begin{pmatrix} \mathbf{G}_R\left(\mathbf{x}_{(k)}, \boldsymbol{\xi}_{(k)(j)}\right)\mathbf{D}_{(k)}\mathbf{b}_{\xi(k)(j)} \\ -\mathbf{G}_R\left(\mathbf{x}_{(j)}, \boldsymbol{\xi}_{(j)(k)}\right)\mathbf{D}_{(j)}\mathbf{b}_{\xi(j)(k)} \end{pmatrix}\cdot \mathbf{n}_{(k)(j)}\left|\boldsymbol{\xi}_{(k)(j)}\right|\cos^2\left(\beta_{(k)(j)}\right)s_0^2 V_{(k)}V_{(j)} \tag{15.18}$$

Equating Eq. (15.18) to the critical energy release rate G_{IC} results in the critical stretch for NOSB-PD as

$$s_c = \sqrt{\frac{2G_{IC}\Delta A}{\sum_{k}^{\Omega^+}\sum_{j}^{\Omega^-}\begin{pmatrix} \mathbf{G}_R\left(\mathbf{x}_{(k)}, \boldsymbol{\xi}_{(k)(j)}\right)\mathbf{D}_{(k)}\mathbf{b}_{\xi(k)(j)} \\ -\mathbf{G}_R\left(\mathbf{x}_{(j)}, \boldsymbol{\xi}_{(j)(k)}\right)\mathbf{D}_{(j)}\mathbf{b}_{\xi(j)(k)} \end{pmatrix}\cdot \mathbf{n}_{(k)(j)}\left|\boldsymbol{\xi}_{(k)(j)}\right|\cos^2\left(\beta_{(k)(j)}\right)V_{(k)}V_{(j)}}} \tag{15.19}$$

Similarly, the resultant OSB-PD force density vector $\mathbf{t}_{(k)(j)}^{OSB} - \mathbf{t}_{(j)(k)}^{OSB}$ can be expressed based on Eq. (3.49) in Chap. 3 as

$$\mathbf{t}_{(k)(j)}^{OSB} - \mathbf{t}_{(j)(k)}^{OSB} = \begin{pmatrix} a_{(k)}\mathbf{g}_{(k)(j)}\sum_{p=1}^{N_{(k)}}\mathbf{g}_{(k)(k_p)}\cdot\left(\mathbf{u}_{(k_p)} - \mathbf{u}_{(k)}\right)V_{(k_p)} \\ -a_{(j)}\mathbf{g}_{(j)(k)}\left(\sum_{r=1}^{N_{(j)}}\mathbf{g}_{(j)(j_r)}\cdot\left(\mathbf{u}_{(j_r)} - \mathbf{u}_{(j)}\right)V_{(j_r)}\right) \\ +\frac{n}{(n+1)}\mu\frac{\delta}{\left|\boldsymbol{\xi}_{(k)(j)}\right|}\overline{\mathbf{S}}\left(\mathbf{x}_{(k)}, \mathbf{x}_{(j)}\right)\left(\mathbf{u}_{(j)} - \mathbf{u}_{(k)}\right) \end{pmatrix} \tag{15.20}$$

where

$$\overline{\mathbf{S}}\left(\mathbf{x}_{(k)}, \mathbf{x}_{(j)}\right) = \frac{\mathbf{S}\left(\mathbf{x}_{(k)}, \boldsymbol{\xi}_{(k)(j)}\right) + \mathbf{S}\left(\mathbf{x}_{(j)}, \boldsymbol{\xi}_{(j)(k)}\right)}{2} \tag{15.21}$$

For an isotropic expansion of $\mathbf{u}_{(m)} - \mathbf{u}_{(i)} = s_0(\mathbf{x}_{(m)} - \mathbf{x}_{(i)})$, Eq. (15.20) reduces to

$$\mathbf{t}_{(k)(j)}^{OSB}-\mathbf{t}_{(j)(k)}^{OSB}=\left(\begin{array}{l}\left(\begin{array}{l}a_{(k)}\mathbf{g}_{(k)(j)}\sum_{p=1}^{N_{(k)}}\mathbf{g}_{(k)(k_p)}\cdot\left(\mathbf{x}_{(k_p)}-\mathbf{x}_{(k)}\right)V_{(k_p)}\\-a_{(j)}\mathbf{g}_{(j)(k)}\sum_{r=1}^{N_{(j)}}\mathbf{g}_{(j)(j_r)}\cdot\left(\mathbf{x}_{(j_r)}-\mathbf{x}_{(j)}\right)V_{(j_r)}\end{array}\right)s_0\\+\frac{n}{(n+1)}\mu\frac{\delta}{\left|\boldsymbol{\xi}_{(k)(j)}\right|}\overline{\mathbf{S}}\left(\mathbf{x}_{(k)},\mathbf{x}_{(j)}\right)\left(\mathbf{x}_{(j)}-\mathbf{x}_{(k)}\right)s_0\end{array}\right) \tag{15.22}$$

After substituting Eq. (15.22) in Eq. (15.7), the strain energy increment ΔU per unit area of ΔA can be written as

$$\frac{\Delta U}{\Delta A}=\frac{\sum_{k}^{\Omega^+}\sum_{j}^{\Omega^-}\Delta W_{(k)(j)}V_{(k)}}{\Delta A}=\frac{1}{2\Delta A}\sum_{k}^{\Omega^+}\sum_{j}^{\Omega^-}\left(\begin{array}{l}a_{(k)}\theta_{(k)(k_p)}\mathbf{g}_{(k)(j)}\cdot\mathbf{n}_{(k)(j)}-a_{(j)}\theta_{(j)(j_r)}\mathbf{g}_{(j)(k)}\cdot\mathbf{n}_{(k)(j)}\\+\frac{n}{(n+1)}\mu\frac{\delta}{\left|\boldsymbol{\xi}_{(k)(j)}\right|}\overline{\mathbf{S}}\left(\mathbf{x}_{(k)},\mathbf{x}_{(j)}\right)\left(\mathbf{x}_{(j)}-\mathbf{x}_{(k)}\right)\cdot\mathbf{n}_{(k)(j)}\end{array}\right)\left|\boldsymbol{\xi}_{(k)(j)}\right|\cos^2(\beta_{ij})s_0^2V_{(k)}V_{(j)} \tag{15.23}$$

where

$$\theta_{(k)(k_p)}=\sum_{p=1}^{N_{(k)}}\mathbf{g}_{(k)(k_p)}\cdot\left(\mathbf{x}_{(k_p)}-\mathbf{x}_{(k)}\right)V_{(k_p)} \tag{15.24}$$

and

$$\theta_{(j)(j_r)}=\sum_{r=1}^{N_{(j)}}\mathbf{g}_{(j)(j_r)}\cdot\left(\mathbf{x}_{(j_r)}-\mathbf{x}_{(j)}\right)V_{(j_r)} \tag{15.25}$$

Equating the Eq. (15.23) to the critical energy release rate G_{IC} results in the critical stretch as

$$s_c=\sqrt{\frac{2G_{IC}\Delta A}{\sum_{k}^{\Omega^+}\sum_{j}^{\Omega^-}\left(\begin{array}{l}a_{(k)}\theta_{(k)(k_p)}\mathbf{g}_{(k)(j)}\cdot\mathbf{n}_{(k)(j)}-a_{(j)}\theta_{(j)(j_r)}\mathbf{g}_{(j)(k)}\cdot\mathbf{n}_{(k)(j)}\\+\frac{n}{(n+1)}\mu\frac{\delta}{\left|\boldsymbol{\xi}_{(k)(j)}\right|}\overline{\mathbf{S}}\left(\mathbf{x}_{(k)},\mathbf{x}_{(j)}\right)\left(\mathbf{x}_{(j)}-\mathbf{x}_{(k)}\right)\cdot\mathbf{n}_{(k)(j)}\end{array}\right)\left|\boldsymbol{\xi}_{(k)(j)}\right|\cos^2(\beta_{ij})V_{(k)}V_{(j)}}} \tag{15.26}$$

With $a_{(k)}=a_{(j)}=0$, it leads to the critical stretch for BB-PD as

$$s_c = \sqrt{\frac{2G_{IC}\Delta A}{\sum_{k}^{\Omega^+}\sum_{j}^{\Omega^-}\left(\frac{n}{(n+1)}\mu\frac{\delta}{\left|\boldsymbol{\xi}_{(k)(j)}\right|}\overline{\mathbf{S}}\left(\mathbf{x}_{(k)},\mathbf{x}_{(j)}\right)\left(\mathbf{x}_{(j)}-\mathbf{x}_{(k)}\right)\cdot\mathbf{n}_{(k)(j)}\right)\left|\boldsymbol{\xi}_{(k)(j)}\right|\cos^2\left(\beta_{ij}\right)V_{(k)}V_{(j)}}} \tag{15.27}$$

15.4 Implicit Algorithm for Crack Growth

In the absence of inertial forces, the implicit solution of the governing algebraic equations can be achieved by employing an incremental iterative solution procedure as suggested by Ni et al. (2019). The specific steps of this algorithm can be described as:

1. Specify the initial load factor, λ_1, final load factor, λ_f, incremental load factor, $\Delta\lambda$ and maximum number of internal iterations, *NINTMAX* along with the iteration number, $j = 1$ and internal iteration number $k = 1$. Note that the initial load factor, λ_1, should be small enough so that no damage occurs in any of the bonds.
2. Specify the values of critical stretch, s_c as explained in Sect. 15.3.
3. Specify the maximum allowable number of bonds, m that can break per iteration. If not specified, all bonds that exceed the critical values of s_c are broken per iteration.
4. Compute the system stiffness matrix and solve the equilibrium equations as described in Chap. 5 for λ_j from

$$\mathbf{K}_j^k\mathbf{V}_j^k = \lambda_j\mathbf{F} \tag{15.32}$$

5. Compute the stretch s_i for each bond where the subscript i denotes the bond index with $(i = 1, \cdots, N)$ and N denotes the total number of bonds in the domain. Sort the stretches s_i in descending order and store them in array a_i. Count the number of bonds, *NCOUNT* whose stretch exceeds the critical value.
6. If $NCOUNT = 0$, no bond is broken and the equilibrium is satisfied. Move to the next load step iteration by incrementing $j = j + 1$ and solve

$$\mathbf{K}_j^k\mathbf{V}_j^k = \lambda_j\mathbf{F} \tag{15.33}$$

7. If $NCOUNT > 0$, first copy all the stretches exceeding s_c from array a_i to $\widehat{a}_i$ and count the number of bonds *NBRCK* with stretch corresponding to $a_{\max} = \max(\widehat{a}_i)$. If $NBRCK < m$, break those *NBRCK* bonds by setting the status parameter to 0. Else, break only the m allowable bonds.
8. Update the stiffness matrix by modifying the real constants of the elements corresponding to the broken bonds as

$$\mathbf{K}_j^{k+1} = \mathbf{K}_j^k + \Delta\mathbf{K}_j^k \tag{15.34}$$

and solve for the internal iterations keeping the load factor constant and repeat step 7 until the equilibrium is reached and no new bonds are broken for iteration j. If the internal iteration k exceeds *NINTMAX*, exit the equilibrium loop and reset the *NINTMAX* or reduce the load factor and redo the equilibrium loop. To avoid singular shape tensor $\mathbf{K}$ of Eq. (15.10), it is not updated during crack growth.

15.5 Numerical Results

Under plane stress conditions, the quasi-static crack propagation in double cantilever beam (DCB) specimen and three point bending (TPB) specimen are simulated by using *EKILL* command in Ansys, as previously considered by Anicode and Madenci (2022, 2023). It is invoked when the stretch in the bond (modeled with MATRIX27 element) reaches its critical value. The parameter a appearing in the Gaussian type weight function, Eq. (2.1) is specified as $a = 3.0$ and $a = 3.5$ for nonuniform and uniform discretization, respectively. The first application with uniform and nonuniform discretization demonstrates the coupled PD/FE modeling for crack growth in a DCB specimen. The second application with uniform and nonuniform discretization demonstrates the coupled PD/FE modeling for crack growth in a TPB specimen.

15.5.1 *Double Cantilever Beam Specimen*

The geometry and loading of the double cantilever beam (DCB) specimen with a Cartesian coordinate system located at its center are shown in Fig. 15.2. The rectangular specimen has dimensions of $L = 0.72$mm and $W = 0.45$mm with an initial crack length of $a_0 = L/6$ and unit thickness. The right edge corresponding to $x = 0.5L$ of the specimen is fixed in both horizontal and vertical directions and the specimen is subjected to a vertical displacement constraint of u_0 at $x =- 0.5L$. As shown in Fig. 15.2, the FE region consists of areas $A_1 \ldots A_9$, and the PD region is located in the middle. The isotropic specimen has a Young's modulus and Poisson's ratio of $E = 72\,G$Pa and $\nu = 0.333$, respectively. The critical energy release rate is specified as $G_{IC} = 5.0$ N/m^2.

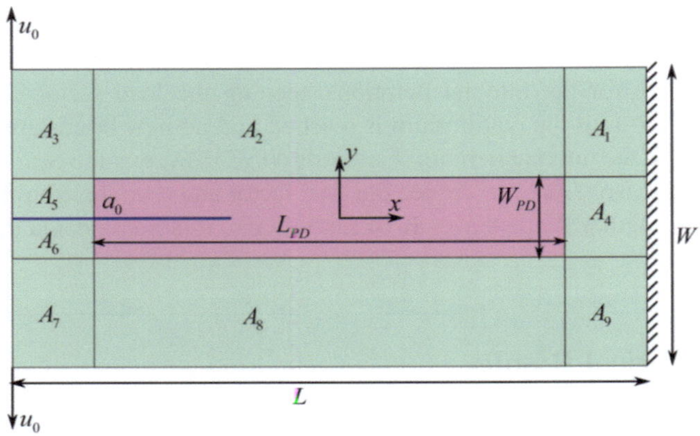

Fig. 15.2 Description of PD and FE regions of the DCB specimen, boundary, and applied loading conditions

15.5.1.1 Uniform Discretization

The FE region is meshed with varying sizes. The areas, A_1, A_3, A_7, and A_9 are meshed with an element size of 0.01395×0.01875mm resulting in 4×72 elements. The areas A_2 and A_8 are meshed with an element size of 0.005×0.005mm resulting in 2×1188 elements, and the areas A_5 and A_6 with an element size of 0.005×0.01875mm resulting in 2×66 elements. The area, A_4 is meshed with an element size 0.01875×5mm resulting in 138 elements. The PD domain around the crack has dimensions $L_{PD} = 0.5$mm and $W_{PD} = 0.12$mm. Fig. 15.3 shows the discretization of PD domain with uniform grid spacing, $\Delta x = 5 \times 10^{-3}$mm and the horizon size of $\delta = 3.015\Delta x$. The PD regions of $\mathcal{D} \cup \mathcal{R}$ and $\mathcal{B}_I$ consist of 2156 and 244 points, respectively. Based on the Eqs. (15.19) and (15.27), the critical stretch values are specified as $s_c = 2.536 \times 10^{-3}$ and $s_c = 3.6 \times 10^{-3}$ for BB and NOSB interactions, respectively.

The initial damage and its growth for NOSB and BB interactions in the PD domain are shown in Figs. 15.4, 15.5, 15.6, and 15.7 for increasing applied crack opening displacement (COD). Figure 15.8 shows the killed bonds (elements) during damage growth. The reaction forces as a function of COD from NOSB-PD/FE, BB-PD/FE, and FE models are shown in Fig. 15.9. They all have the same initial stiffness, and the peak load at the onset of damage growth. The results from FE model without damage growth serve as a reference to compare the initial stiffness of the load displacement curve. The softening behavior from the NOSB-PD/FE and BB-PD/FE are slightly different because the BB-PD is pairwise and does not have indirect interactions inherent to the NOSB-PD.

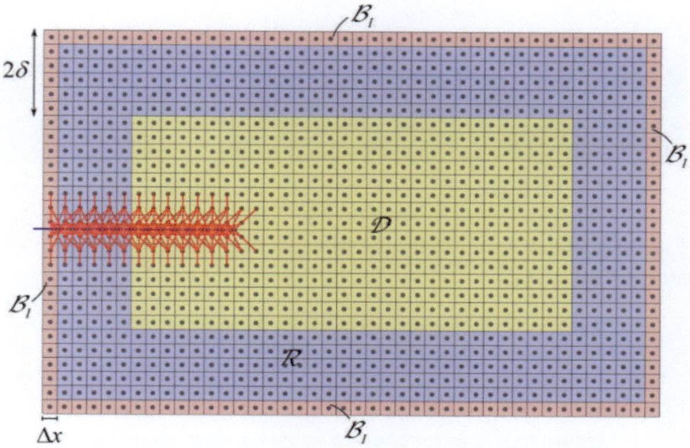

Fig. 15.3 Uniform discretization of PD domain in DCB specimen

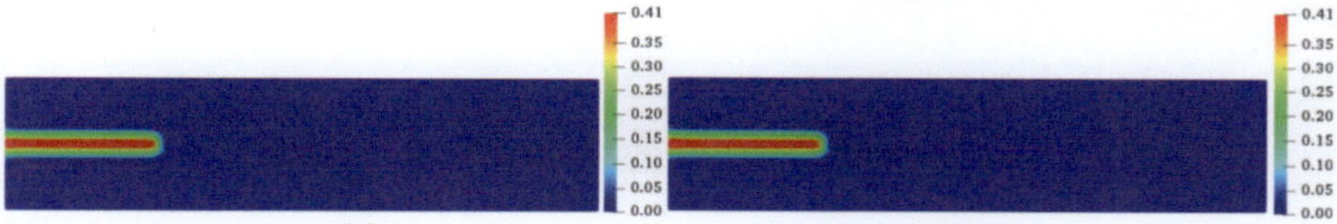

(a) (b)

Fig. 15.4 Initial damage: (**a**) NOSB-PD/FE, and (**b**) BB-PD/FE

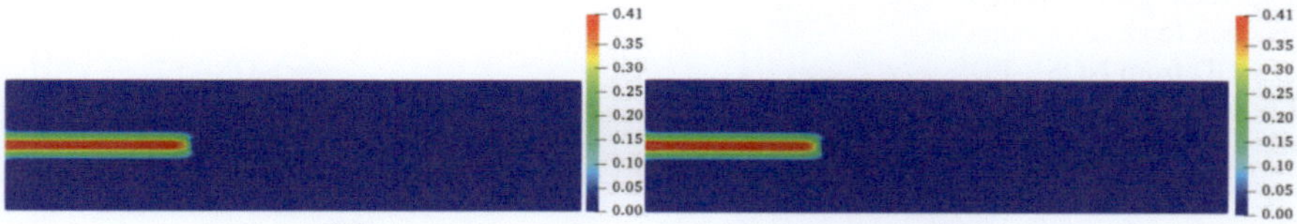

(a) (b)

Fig. 15.5 Damage growth due to crack opening displacement of $u_y = 0.5\mu\text{m}$: (**a**) NOSB-PD/FE, and (**b**) BB-PD/FE

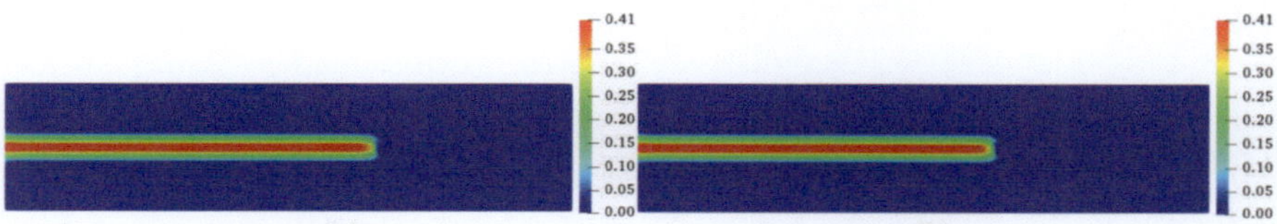

(a) (b)

Fig. 15.6 Damage growth due to crack opening displacement of $u_y = 1.0\mu\text{m}$: (**a**) NOSB-PD/FE, and (**b**) BB-PD/FE

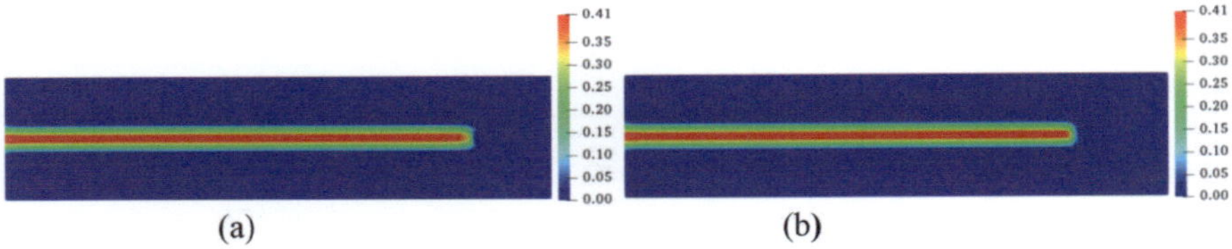

Fig. 15.7 Damage growth due to crack opening displacement of $u_y = 1.5\mu m$: (**a**) NOSB-PD/FE, and (**b**) BB-PD/FE

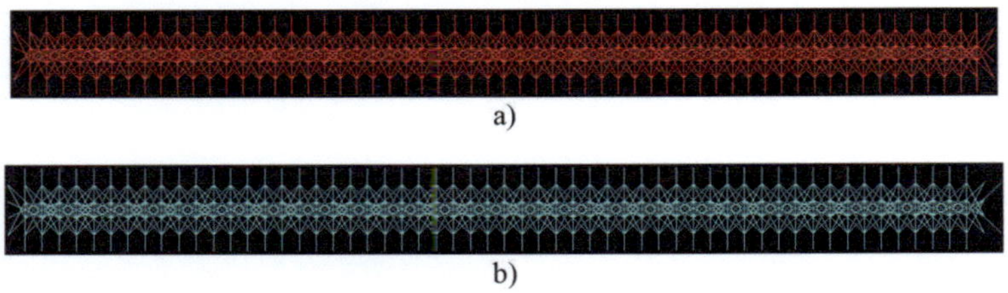

Fig. 15.8 Killed elements during crack propagation: (**a**) NOSB-PD/FE, and (**b**) BB-PD/FE

Fig. 15.9 Comparison of reaction forces as a function of COD from NOSB-PD/FE, BB-PD/FE and FE models

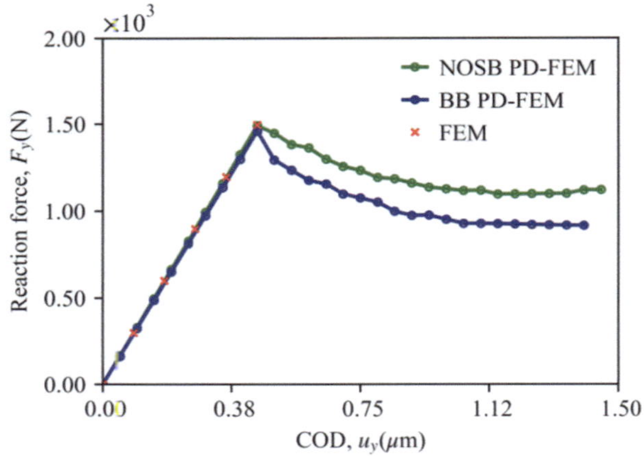

15.5.1.2 Nonuniform Discretization

Similar to the model with uniform discretization, the FE region is meshed with varying sizes. The areas, A_1, A_3, A_7, and A_9 are meshed with an element size of 0.01071mm × 0.0185mm resulting in 4 × 32 elements. The areas A_2 and A_8 are

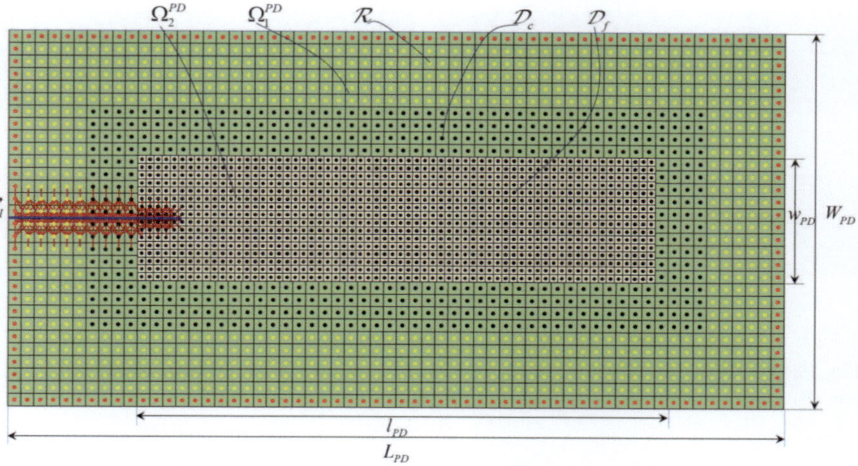

Fig. 15.10 Nonuniform discretization of PD region in DCB specimen

meshed with an element size of 0.00592mm × 0.0185mm resulting in 2 × 856 elements, and the areas A_5 and A_6 with an element size of 0.01071mm × 0.005693mm resulting in 2 × 72 elements. The area, A_4 is meshed with an element size 0.01071 × 0.005693mm resulting in 104 elements. As shown in Fig. 15.10, the PD region around the crack has the geometry corresponding to the coarse PD region, Ω_1^{PD} has a length and width of $L_{PD} = 0.6$mm and $W_{PD} = 0.16$mm while for the fine PD region, Ω_2^{PD} the length and width are $l_{PD} = 0.486$mm and $w_{PD} = 0.0342$mm. The mesh size in PD regions Ω_1^{PD} and Ω_2^{PD} are $\Delta x = 0.006$mm and 0.008mm, respectively. The fine mesh PD region consisting of $\mathcal{D}_f$ has 1344 points, the coarse mesh PD region consisting of $\mathcal{D}_c \cup \mathcal{R}$ has (476+1150), and PD boundary $\mathcal{B}_I$ has 254 points. Based on the Eq. (15.19), the critical stretch values are specified as $s_c = 4.016 \times 10^{-3}$ and $s_c = 4.165 \times 10^{-3}$ for NOSB-PD with uniform and nonuniform discretizations, respectively. In the evaluation of the critical stretch, the parameter a appearing in the Gaussian type weight function, Eq. (2.1) is specified as $a = 3.0$ for nonuniform and uniform discretizations.

Damage propagation with NOSB interactions in the PD domain are shown in Fig. 15.11 corresponding to the applied crack opening displacement (COD) of $u_y = 0.05\,\mu$m, 0.5 μm, 1.0 μm, and 1.5 μm. Figure 15.13 shows the killed bonds (elements) during damage growth. The reaction forces as a function of COD for NOSB-PD/FE with nonuniform and NOSB-PD/FE with uniform discretization are shown in Fig. 15.13 resulting the similar behavior with corresponding peak loads of 1.728kN and 1.656kN (Fig. 15.12).

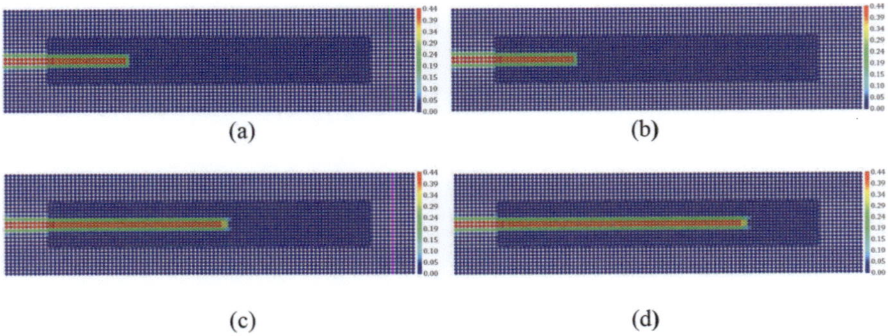

Fig. 15.11 Damage growth due to crack opening displacement using NOSB-PD/FE. (**a**) $u_y = 0.05$ µm, (**b**) $u_y = 0.5$ µm (**c**) $u_y = 1.0$ µm and (**d**) $u_y = 1.5$ µm

Fig. 15.12 Killed elements during crack propagation using NOSB-PD/FE

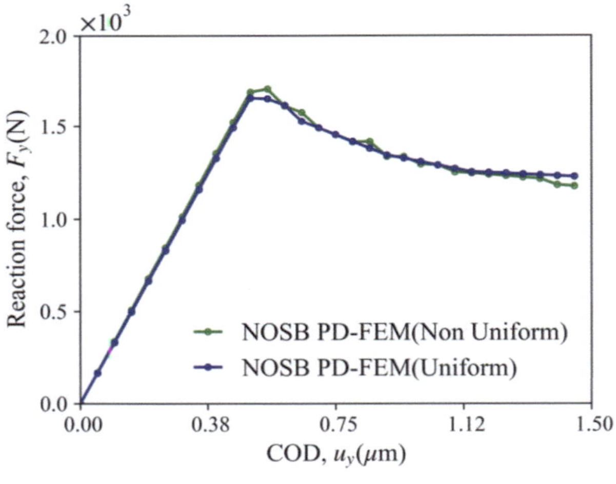

Fig. 15.13 Comparison of reaction forces as a function of COD for NOSB-PD/FE with nonuniform and NOSB-PD/FE uniform discretization

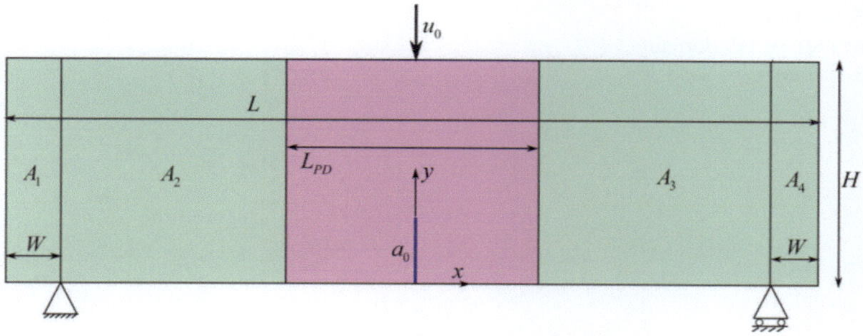

Fig. 15.14 Description of PD and FE domains of the three-point bending specimen

15.5.2 Three-Point Bending Specimen

The geometry and loading of the three-point bending (TPB) specimen is shown in Fig. 15.14. The rectangular plate has dimensions of $L = 228$mm and $H = 76$mm with an initial crack length of $a_0 = 19$mm and unit thickness. A Cartesian coordinate system is located at its center and aligned with the crack. The plate has a fixed support at a distance of $W = 12$mm from the left vertical edge and on a roller support from the right vertical edge. As shown in Fig. 15.14, the FE region consists of areas $A_1 \ldots A_4$ and the PD region is located in the middle. At the point $x = 0$, $y = H$, the displacement load of $u_0 = 0.065$mm is applied in 20 increments. The isotropic plate has a Young's modulus and Poisson's ratio of $E = 41.87\,$GPa and $\nu = 0.333$, respectively, the critical energy release rate as $G_{IC} = 31.8$ N/m^2.

15.5.2.1 Uniform Discretization

The FE region is meshed with varying sizes. The areas, A_1 and A_4 are meshed with an element size of 1mm × 3mm resulting in 2 × 304 elements. The areas, A_2 and A_3 are meshed with an element size of 5.208mm × 1mm resulting in 2 × 912 elements. The PD domain around the crack has dimensions $L_{PD} = 80$mm and $W_{PD} = 76$mm. Fig. 15.15 shows the discretization of PD domain with uniform grid spacing, $\Delta x = 1$mm and the horizon size of $\delta = 3.015\Delta x$. The PD regions of $\mathcal{D} \cup \mathcal{R}$ and $\mathcal{B}_I$ consist of 5934 and 226 points, respectively. The region $\mathcal{D}$ in PD domain is modeled using BB interactions and the critical stretch value is specified as $s_c = 9.5 \times 10^{-4}$. During crack propagation, the parameter a appearing in the Gaussian type weight function, Eq. (2.1) is specified as $a = 3.5$.

The damage growth paths are shown in Fig. 15.16. Figure 15.17 shows the killed bonds (elements) in deformed configuration during the crack propagation corresponding to the applied displacement of $u_0 = 60\mu$m. The reaction force versus the COD shown in Fig. 15.18 compares well with that of Shen et al. (2020).

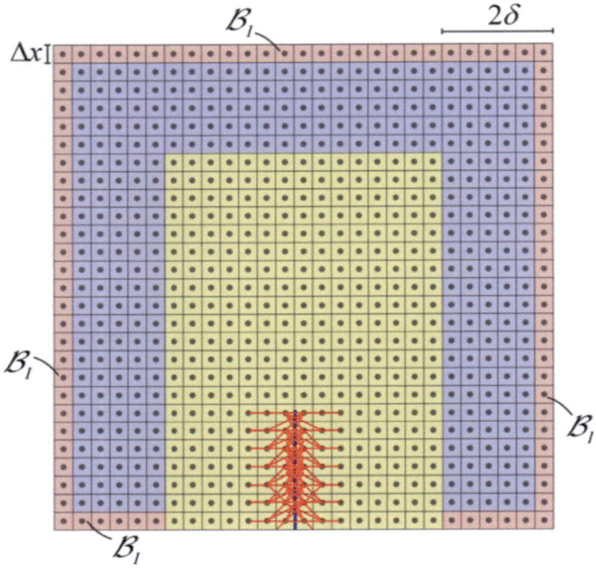

Fig. 15.15 Uniform discretization of PD domain in TPB specimen

15.5.2.2 Nonuniform Discretization

Similar to the model with uniform discretization, the FE region is meshed with varying sizes. The areas A_1, and A_4 are meshed with an element size of 1mm × 3mm resulting in 2 × 304 elements. The areas A_2 and A_3 are meshed with an element size of 5.208mm × 1mm resulting in 2 × 912 elements. As shown in Fig. 15.19, the PD region around the crack has the geometry corresponding to the coarse PD region, Ω_1^{PD}, with a length of $L_{PD} = 30$mm. The fine mesh PD region, Ω_2^{PD}, located at the $H_1 = 14$mm from the bottom edge has a length and width are $l_{PD} = 15$mm and $w_{PD} = 64$mm. The PD regions Ω_1^{PD} and Ω_2^{PD} are discretized with a mesh size of $\Delta x = 1$mm with a structured and unstructured mesh, respectively, and the horizon size of $\delta = 3.015\Delta x$. The fine mesh PD region consisting of $\mathcal{D}_f$ has 1804 points, the coarse mesh PD region consisting of $\mathcal{D}_c \cup \mathcal{R}$ has (500+950), and PD boundary $\mathcal{B}_I$ has 214 points. The region $\mathcal{D}$ in PD domain is modeled by using BB and OSB interactions respectively. The Poisson's ratio $\nu = 0.1$ is considered for the OSB-PD/FE model. The critical stretch values are computed as $s_c = 1.101\times10^{-3}$ and $s_c = 9.5\times10^{-4}$ for the OSB-PD/FE and BB-PD/FE from Eqs. (15.26) and (15.27), respectively. In the evaluation of the critical stretch, the parameter a appearing in the Gaussian type weight function, Eq. (2.1) is specified as $a = 0$. At point $x = 0$, $y = W$, the displacement load of $u_y = 80\mu$m is applied in 20 increments. During crack propagation, the parameter a appearing in the Gaussian type weight function, Eq. (2.1) is specified as $a = 3.0$.

The damage growth paths are shown in Figs. 15.20 and 15.21. Figure 15.22 shows the killed bonds (elements) in deformed configuration during the crack propagation corresponding to the applied displacement load of $u_0 = 80\mu$m. The

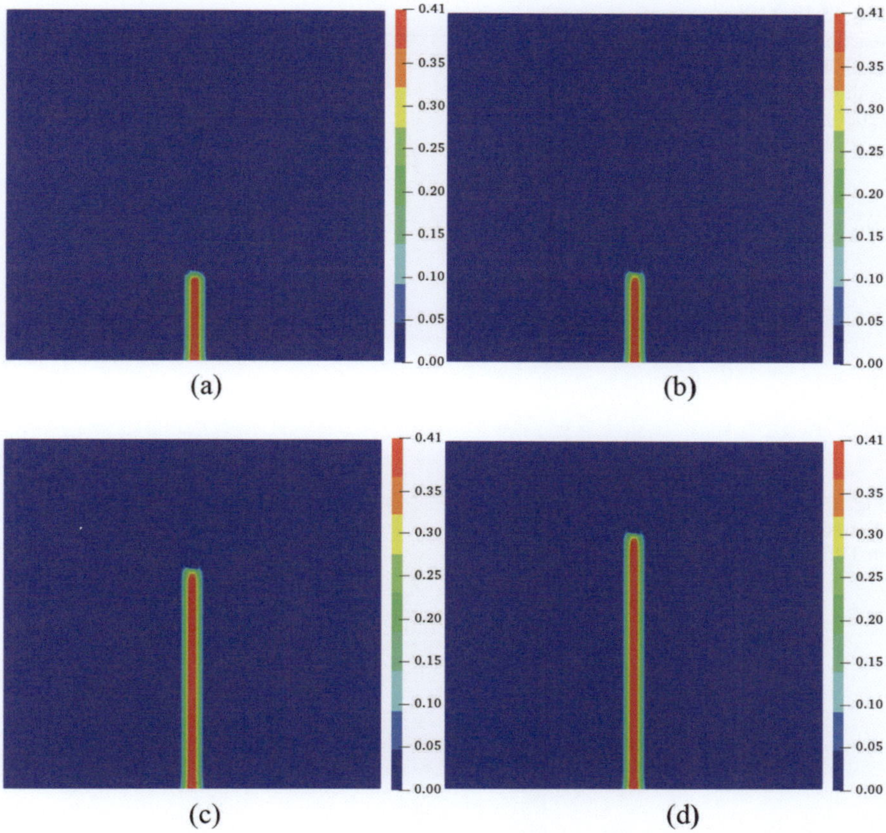

Fig. 15.16 BB-PD/FE predictions for damage growth in a three point bend test specimen due to crack opening displacement: (**a**) $u_y = 0$, $u_y = 18.768\mu m$, (**b**) $u_y = 18.768\mu m$, (**c**) $u_y = 50\mu m$, and (**d**) $u_y = 65.8\mu m$

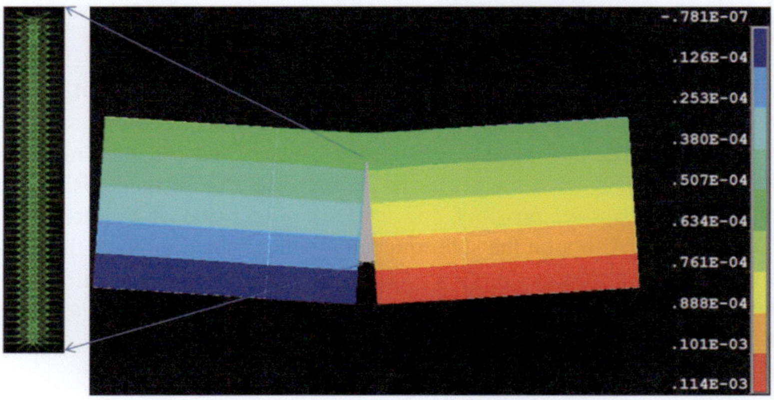

Fig. 15.17 Killed bonds (elements) in deformed configuration during the crack propagation corresponding to applied displacement of $u_0 = 60\mu m$

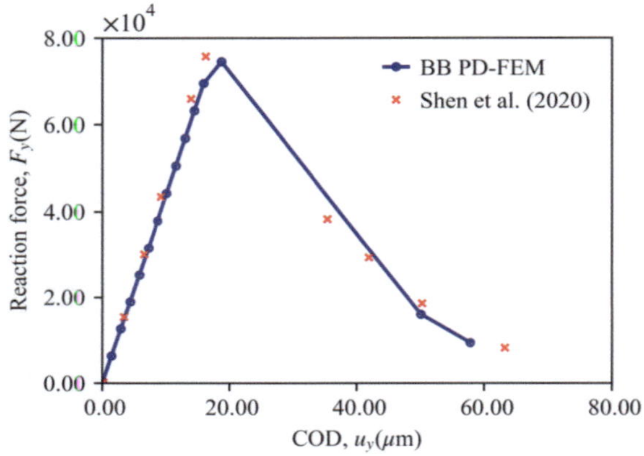

Fig. 15.18 Comparison of reaction forces as a function of COD from BB-PD/FE with that of Shen et al. (2020)

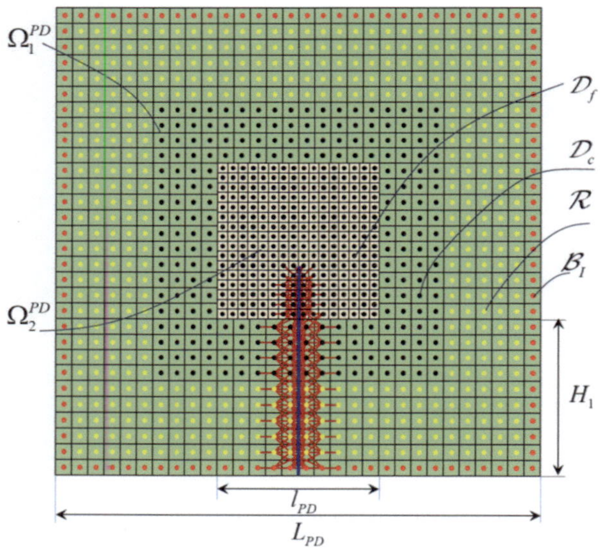

Fig. 15.19 Nonuniform discretization of PD region in TPB specimen

reaction force versus the crack mouth opening displacement (CMOD) from the BB-PD/FE and OSB-PD/FE models are shown in Fig. 15.23 along with that of Shen et al. (2020). The initial elastic stiffness before the crack initiation matches well for all three cases. However, the peak load for the OSB-PD/FE model is higher than that of BB-PD/FE model possibly due to the higher value of Poisson's ratio in BB-PD/FE model. Although the subsequent softening behaviors from the 3 models are similar, the differences in slopes could be attributed due to different load rearrangement in the OSB-PD model because of significant dilatation as compared to BB-PD model.

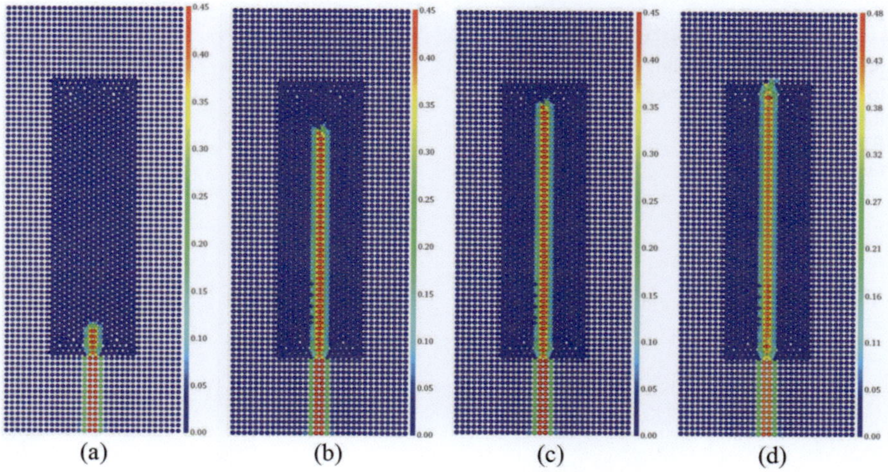

Fig. 15.20 BB-PD/FE predictions with nonuniform discretization for damage growth in a TPB specimen: (**a**) $u_y = 40\mu m$, (**b**) $u_y = 56\mu m$, (**c**) $u_y = 64\mu m$ and (**d**) $u_y = 80\mu m$

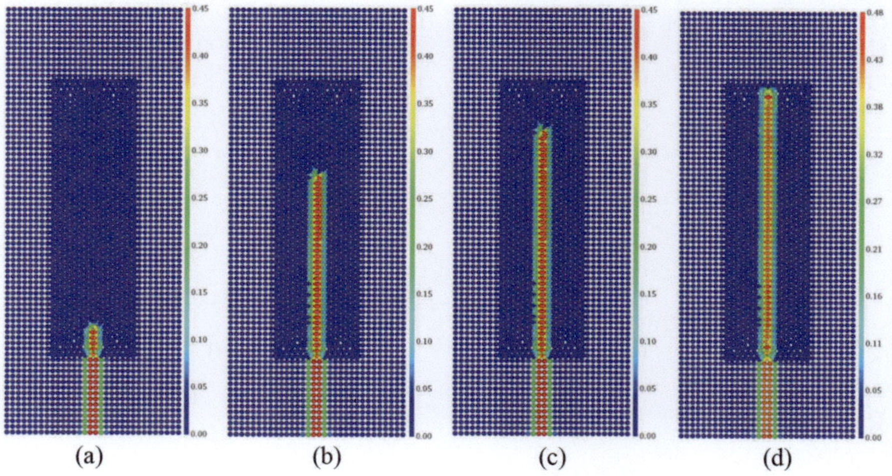

Fig. 15.21 OSB-PD/FE predictions with nonuniform discretization for damage growth in a TPB specimen: : (**a**) $u_y = 40\mu m$, (**b**) $u_y = 56\mu m$, (**c**) $u_y = 64\mu m$ and (**d**) $u_y = 80\mu m$

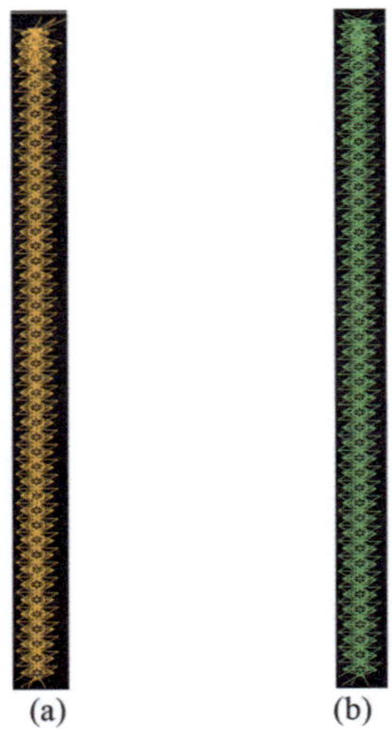

Fig. 15.22 Killed bonds (elements) during the crack propagation corresponding to the applied displacement load of $u_0 = 56\mu m$: (**a**) BB-PD/FE, and (**b**) OSB-PD/FE

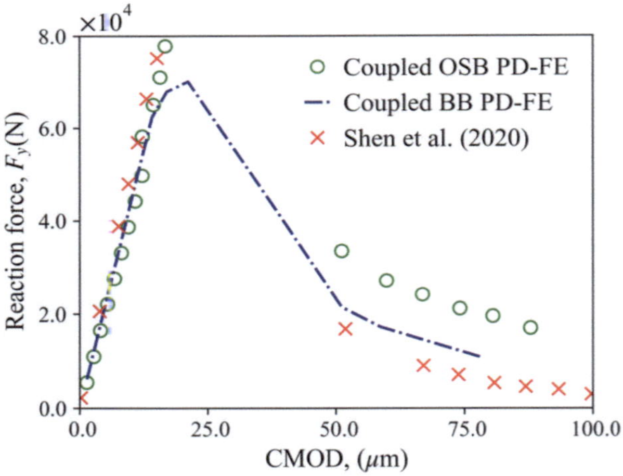

Fig. 15.23 Comparison of reaction forces as a function of CMOD for coupled dual horizon-based BB-PD/FE, OSB-PD/FE with that of Shen et al. (2020)

References

Anicode, S. V. K., & Madenci, E. (2022). Seamless coupling of bond- and state-based peridynamic and finite element analyses. *Mechanics and Materials., 173*, 104433.

Anicode, S. V. K., & Madenci, E. (2023). Direct coupling of dual-horizon peridynamics with finite elements for irregular discretization without an overlap zone. *Engineering with Computers, 40*, 1–31.

Madenci, E., & Oterkus, E. (2013). *Peridynamic theory and its applications*. Springer.

Ni, T., Zaccariotto, M., Zhu, Q. Z., & Galvanetto, U. (2019). Static solution of crack propagation problems in peridynamics. *Computer Methods in Applied Mechanics and Engineering, 346*, 126–151.

Shen, F., Yu, Y., Zhang, Q., & Gu, X. (2020). Hybrid model of peridynamics and finite element method for static elastic deformation and brittle fracture analysis. *Engineering Analysis with Boundary Elements, 113*, 17–25.

Silling, S. A., & Askari, E. (2005). A meshfree method based on the peridynamic model of solid mechanics. *Computers & Structures, 83*(17-18), 1526–1535.

Index

A
Aerodynamic, 196
Anode, 451, 452, 455, 457, 458, 460, 462, 463, 466
Ansys, 151, 196, 223–239, 241–299, 301–321, 323–326, 330, 333, 335, 342, 343, 348, 353–355, 360–362, 368, 371, 377–379, 382, 384, 385, 389, 390, 392, 397–401, 404–406, 409–417, 420–422, 425, 430, 431, 433, 437, 444, 451, 456, 463, 477
Ansys GUI, 353, 378, 384, 469
APDL commands, 226, 235, 346
Atomic concentration, 195, 197, 443, 460–463, 465, 466
Average bond property, 21

B
Blending function, 152
Body force, 10, 47, 109, 110, 119, 153, 269, 270
Body loads, 75, 109, 152, 164, 166, 217
Bond-associated deformation gradient, 60
Bond-based peridynamic (BB-PD), 2, 47, 152, 248
Bond-wise interactions, 47
Boundary conditions, 3, 9–11, 16–19, 23, 24, 47, 67, 72, 109, 199, 201, 212–216, 223, 236, 239, 250, 251, 268–287, 323, 324, 326, 328, 332–334, 340, 341, 343, 348, 350, 354, 356, 357, 361, 362, 367, 370, 371, 373, 377, 379, 381, 382, 384, 385, 390, 391, 393, 399, 401, 403, 405, 411, 412, 416, 421, 424–426, 431, 433, 435–437, 446–448, 452, 453, 455, 457, 460, 462, 464

C
Cathode, 450–452, 455, 457, 458, 462, 463, 466
Charring, 196
Chemical reactions, 195
Clamped, 11
Classical continuum mechanics (CCM), 1, 2, 59
Concrete, 195
Convection, 17, 223, 238, 239, 277–279
Corrosion, 195, 196, 455, 459, 460
Coupled field, 195, 199, 202, 219, 223, 229, 443, 445, 447, 451, 457
Coupling, 76, 80, 117, 118, 121, 122, 151, 152, 156, 161–175, 177, 196–198, 218, 285, 389, 396, 444, 469
Coupling coefficient matrices, 198, 199
Crack, 1, 2, 4, 20, 21, 151, 177, 195, 196, 257–259, 353, 369–386, 389, 408–413, 435–440, 469–472, 476–478, 481–485
Crack initiation, 1, 151, 469, 470, 486
Crack opening displacement, 379, 381, 385, 386, 478–482, 485
Crack surface, 20, 21, 257, 471
Critical stretch, 469–476, 478, 481, 483, 484

D
Damage, 2, 4, 151, 177, 195, 196, 466, 470, 476, 478, 479, 481, 483–485, 487
Damper element, 105

E. Madenci et al., *Peridynamics and Its Applications Using Ansys*,
https://doi.org/10.1007/978-3-031-59896-8

Damping effects, 224, 289, 290, 332
Damping element, 105, 106, 148
Damping terms, 209, 443
Deformation gradient, 48, 64
Degrees of freedom, 217, 223, 224, 229, 324–325, 332, 342, 354, 390, 451
Desiccation, 196
Desorption, 323, 339, 349, 351
Diffusion strain, 197
Dilatation, 2, 47, 48, 55–57, 65, 177, 186, 486
Discontinuous, 23, 238, 306
Displacement constraints, 9, 68, 110, 112, 152, 156, 190, 399, 406, 412, 417, 423, 432, 438
Displacement vector, 26, 31, 32, 38, 40, 52, 75, 110, 117, 134, 154, 162, 165, 197, 198, 224
Dissimilar, 21, 339–341
Distortion, 2, 47
Divergence, 23, 26, 31, 32, 38, 40, 44–45, 49, 68, 69, 71, 72, 110, 113, 142, 143, 153, 154, 158
Divergence theorem, 23, 110, 113, 153, 154, 158
Domain, 2, 3, 6, 10, 12, 17–19, 23–25, 27, 33, 47, 48, 67, 68, 72–74, 77, 91, 93, 97, 103, 110–116, 118–120, 152, 154–158, 160–162, 164, 165, 167, 168, 170–172, 190, 198, 201, 202, 248–250, 252, 259–261, 268, 270, 271, 274, 276, 278, 279, 281–284, 286, 289, 291–294, 296–299, 302, 325, 327, 328, 330, 331, 337, 339, 341–346, 348, 356–358, 360, 364, 368, 372, 373, 376, 377, 382, 383, 390–395, 397–401, 404, 408, 409, 411, 413–416, 420–422, 425–427, 429–431, 433, 437, 449, 455, 457, 476, 478, 479, 481, 483, 484
Double cantilever beam (DCB), 469, 477
Dual horizon, 2, 8, 9, 16, 49, 82, 86, 100, 224, 229, 233
Dual-phase-lag (DPL), 197, 228, 443–448
Dyadic product, 26, 31, 39, 42, 44, 59

E

Electrical current density, 195
Electrical potential, 195, 461
Electrodeposition, 195, 196, 443, 448, 449, 451, 453, 457
Electrolyte, 449–455, 457, 458, 460
Electromigration, 195, 196, 443, 462–464, 466, 467
Electron wind, 460
Electro-potential, 196
Element connectivity, 81, 82, 85, 87, 91, 93, 96, 99, 100, 103, 261, 263, 265, 267, 268
Energy release rates, 470
Equations of motion, 2, 3, 5–9, 25, 47–50, 68, 195–197, 224, 462, 470
Essential (forced) boundary conditions, 47
Euler-Lagrange equation, 3, 5, 7, 12, 14
External field interactions, 112, 157
External forcing function, 217

F

Failure criteria, 469
Failure prediction, 151, 469–488
Family population, 27
Fictitious, 9–11, 16–19, 47, 268, 269, 271–276, 323, 342, 343, 345, 348
Fictitious material layer, 9
Field equation, 1–3, 12–18, 47, 68–70, 72–73, 77–81, 99–106, 109, 112–115, 120–124, 151, 160, 168, 195, 197, 198, 224, 236, 238, 323, 325, 326, 333, 335, 339, 344, 351, 446, 447, 453, 460
Field flux, 16, 17, 68, 69, 71–73, 79, 80, 112, 113, 115, 122, 123, 142–148, 157, 158, 160, 161, 169, 171, 173, 199, 270, 325, 342
Field potential, 12, 13, 112, 157
Field self-stiffness matrices, 219
Field variable, 16–18, 23, 24, 47, 69, 71, 78, 101, 104, 109, 113–115, 121, 157, 160, 161, 169, 197–199, 223, 224, 233, 310, 324, 342, 443, 444, 451, 452, 456, 457, 463
Finite-difference, 10, 17, 23
Finite element, 23, 151, 177–193, 196, 223, 244, 245, 326, 343, 355, 368, 371, 389–440, 469
Finite element mesh, 244
Finite-volume, 23
First Piola-Kirchhoff stress, 49, 60, 61, 63, 66, 110, 153, 472
Flow density, 14, 69–71, 99, 101
Fluid flow, 195
Fluid transport, 196
Force density, 8, 10, 48–64, 88, 93, 153, 177–182, 184–188, 269, 270, 470, 472, 474
Freeze-thaw, 196

G

Galvanic corrosion, 196, 443, 455–460
Generalized Finite Difference Method (GFDM), 24
Ghost forces, 2
Governing equations, 1, 76, 80, 109, 197, 448, 451, 455, 456, 463
Gradient, 17, 26, 29, 31–33, 35, 38, 40, 60, 63, 66, 71, 101, 104, 271, 460
Gradient of divergence, 26, 31, 32, 38, 40
Gradient Reproducing Kernel (G-RK), 25
Granite, 195
Green's strain, 61, 63

H

Hillocks, 462
Homogeneous deformation, 64–67, 389
Horizon, 2–6, 8–12, 15, 16, 24, 26, 27, 32, 39, 40, 47–51, 53, 54, 58, 62, 64–72, 81–83, 85–87, 91–93, 99, 100, 151, 182, 185, 224, 229, 233, 256, 261, 263, 265, 266, 302, 308–310, 324, 329, 330, 341, 345, 353, 354, 358, 359, 362, 363, 366, 368–370, 374, 375, 379, 380, 385, 394, 396, 401, 406, 407, 410, 413, 428, 429, 433, 435, 438, 440, 446, 448, 469, 470, 477, 478, 483, 484, 488
Hydraulic, 196
Hydraulic fracturing, 196
Hygro-thermal loading, 443, 447–448
Hygro-thermo-mechanics, 195

I

Implicit methods, 178
Implicit transient, 224, 227, 229, 234, 332
Inertial effects, 227, 292, 293, 332
Inertial matrices, 218
Initial deformation, 178–186, 188
Initial dilatation, 179
Initial strain, 177–193, 425
Initial stretch, 178
Interaction domain, 3, 4, 6, 9, 19, 20, 24–27, 32, 40, 42, 50, 61, 63
Interface, 21, 151, 152, 156, 157, 161, 163, 165, 166, 168, 169, 171, 172, 178, 190, 196, 336, 339, 340, 398, 399, 401, 404, 406, 411, 415, 418, 421, 422, 430, 434, 437, 438, 449, 450, 453, 455, 456, 460
Internal force, 2, 47, 49–67, 74, 75, 81, 83, 86, 87, 89–92, 95, 97, 109, 111, 116, 117, 124–140, 152, 154, 156, 162, 163, 165, 178–190
Internal force density, 47, 59
Internal length, 1, 2
Internal net field flow, 15, 70, 72, 77–79, 99, 102, 120, 141–149, 168, 169
Intersection region, 60
Irregular discretization, 5, 12, 50, 69

K

Kernel functions, 24

L

Lagrangian, 5–7, 12
Lame's constants, 62
Laplacian, 26, 29, 31–33, 35, 38, 40
Least squares minimization, 24

M

Mass of moisture, 197
Mass transport, 196, 450, 455, 460
Mechanical deformation, 178, 179, 182, 183, 191
Micro-modulus, 2, 47
Mirror image, 10, 17
Moisture, 195–197, 295, 342, 343, 346–348, 351
Moisture concentration, 195, 197, 294–297, 339, 343, 344, 349, 351, 443–445, 447, 449
Moisture diffusion, 323, 339–351, 443, 447
Moisture inertia, 197, 447
Moment matrix, 25
Morphing, 152
Moving boundary, 443, 450
Multiphysics, 195–220

N

Natural boundary conditions, 47, 109, 212
Newton-Raphson iterations, 456
Non-Fickian, 195, 197
Non-Fourier, 195, 197
Nonlinear thermal conduction, 196
Non-ordinary state-based peridynamic (NOSB-PD), 2, 47
Nonuniform discretization, 3, 6, 49, 50, 178, 353, 367–368, 371–379, 381–385, 389, 403, 404, 421, 469, 477, 487

O

Ohm's law, 456
Ordinary state-based peridynamic (OSB-PD), 2, 47, 152, 248

Orthogonal functions, 24
Oxidation, 177
Oxygen diffusions, 195

P
PD mesh, 244–248, 252–260, 271, 272, 275, 279, 280, 282, 294, 295, 307, 308, 311–313, 317, 326, 343, 344, 348, 355, 356, 371, 372, 374, 376, 393, 409, 410, 425, 426, 436
PD Preprocessor, 241
Peridynamic differential operator (PDDO), 24
Peridynamic (PD) functions, 24–40, 42, 53 54, 60, 69, 201, 301–306
Peridynamic interactions, 24, 49, 50, 59, 68–70, 81, 86, 92, 93, 99, 100, 124–131, 141–142, 259–265, 330, 346, 360, 364, 376, 395, 429
Peridynamics (PD), 3, 195–220, 469–488
Periodic, 18, 19, 285, 287, 452, 457
Physical property, 157
Piola-Kirchoff stress, 9
Point-wise interactions, 47
Poisson's ratio, 2, 48, 51, 353, 354, 362, 364, 367, 369, 391, 404, 408, 412, 425, 435, 462, 477, 483, 484, 486
Polynomial basis functions, 29, 36
Porous, 195, 196
Porous medium, 195
Primary, 2, 3, 5, 6, 8, 9, 14, 16, 20, 26, 50, 51, 55, 57, 59, 69, 70, 74, 78, 81, 82, 85, 86, 88, 91, 92, 99, 100, 179, 180, 184, 185, 187, 261–264

R
Radiation, 17, 223, 236, 280–282
Relative displacement, 52
Relative position, 27, 33, 52, 302
Reproducing conditions, 24, 25
Reproducing Kernel, 24, 25
Reproducing Kernel Particle Method (RKPM), 24
Residual forces, 65–67
Roller support, 11, 353, 369, 390, 403, 483

S
Scalar valued function, 27
Secondary PD bonds, 2, 50
Second Piola-Kirchhoff stress, 61
Seepage, 196
Shape tensor, 59, 472, 477
Shrinkage, 177, 196
Singularity, 23, 311, 469
Small deformation, 52, 53, 56, 62, 64, 116, 161, 183–188, 353, 472
Smoothed particle hydrodynamics (SPH), 23, 24
Smoothness, 2, 23
Source, 12, 15–17, 47, 78, 79, 109, 112, 123, 157, 158, 171, 173, 197, 224, 234–236, 270, 283, 284, 323, 444
Spring element, 106
Stiffness matrix, 75, 76, 86, 87, 91, 93, 97, 98, 103, 104, 165, 172, 216, 219, 259, 325, 332, 469, 476
Strain energy density (SED), 6, 47, 60, 63, 472
Strain invariants, 48
Strong form, 47, 116, 249–251, 267, 325, 328, 329, 333, 334, 357–359, 373–375
Structural self-stiffness matrix, 219
Structural stiffness matrix, 220
Structural-field coupling stiffness matrices, 219, 220
Surface constraints, 151
Surface recession, 196

T
Taylor Series Expansion (TSE), 24–26
Thermal diffusion, 177, 323, 325–338, 443, 446
Thermal dilatation, 177
Thermal expansion, 177, 196, 425
Thermal inertia, 197, 444
Thermal-moisture diffusion, 444
Thermal shock, 177, 323–338
Thermal wave, 197
Thermomechanical, 177
Three point bending, 469, 477
Total deformation, 178, 179, 184, 185, 187, 188
Traction, 9–11, 49, 67, 68, 76, 97, 98, 109–112, 117, 118, 134–140, 152, 154, 156, 163, 164, 166, 190, 196, 199, 251, 268–270, 353, 361, 362, 369, 377, 379, 381, 384, 385, 389–401, 404, 411, 413, 416, 417, 421, 423, 425, 431, 437
Transient field fluctuations, 77
Transient fluctuation, 157

U
Ultrashort-pulse laser heating, 443, 445
Unified approach, 47
User Interface Design Language (UIDL), 241

V
Vapor pressure, 195, 339, 340
Virtual work, 9, 59, 74, 76, 109, 116–119, 124, 136, 140, 152, 161–166
Void formation, 197, 443, 465, 467
Voids, 462
Volume constraints, 151
Volumetric field generation, 217

W
Weight function, 24, 26, 29, 36, 41–43, 59, 307, 353, 389, 472, 477, 481, 483, 484

Y
Young's modulus, 48, 353, 369, 391, 404, 408, 412, 424, 477, 483

Z
Zero-energy mode, 60

FSC
www.fsc.org

MIX
Papier aus verantwortungsvollen Quellen
Paper from responsible sources
FSC® C105338

If you have any concerns about our products,
you can contact us on
ProductSafety@springernature.com

In case Publisher is established outside the EU,
the EU authorized representative is:
Springer Nature Customer Service Center GmbH
Europaplatz 3, 69115 Heidelberg, Germany

Printed by Libri Plureos GmbH
in Hamburg, Germany